STATISTICAL MECHANICS OF PHASES, INTERFACES, AND THIN FILMS

ADVANCES IN INTERFACIAL ENGINEERING SERIES

Microstructures constitute the building blocks of the interfacial systems upon which many vital industries depend. These systems share a fundamental knowledge base—the molecular interactions that occur at the boundary between two materials.

Where microstructures dominate, the manufacturing process becomes the product. At the Center for Interfacial Engineering, a National Science Foundation Research Center, researchers are working together to develop the control over molecular behavior needed to manufacture reproducible interfacial products.

The books in this series represent an intellectual collaboration rooted in the disciplines of modern engineering, chemistry, and physics that incorporates the expertise of industrial managers as well as engineers and scientists. They are designed to make the most recent information available to the students and professionals in the field who will be responsible for future optimization of interfacial processing technologies.

Other Titles in the Series

Edward Cohen and Edgar Gutoff (editors)
Modern Coating and Drying Technology

D. Fennell Evans and Håkan Wennerström
The Colloidal Domain: Where Physics, Chemistry, Biology, and Technology Meet

Christopher W. Macosko
Rheology: Principles, Measurements, and Applications

Neil A. Dotson, Rafael Galván, Robert L. Laurence, and Matthew Tirrell
Polymerization Process Modeling

STATISTICAL MECHANICS OF PHASES, INTERFACES, AND THIN FILMS

H. Ted Davis

WILEY-VCH

New York • Chichester • Weinheim • Brisbane • Singapore • Toronto

H. Ted Davis
Department of Chemical Engineering
and Materials Science
Institute of Technology
151 Amundson Hall
421 Washington Avenue, S.E.
University of Minnesota
Minneapolis, MN 55455-0132

This book is printed on acid-free paper.

Library of Congress Cataloging-in-Publication Data

Davis, H. Ted (Howard Ted)
Statistical mechanics of phases, interfaces, and thin films / H. Ted Davis
p.cm. – (Advances in interfacial engineering series)
Includes bibliographical references and index.
ISBN 0-471-18562-0
1. Statistical mechanics. I. Title. II. Series.
QC174.8.D38 1995
530.4'17–dc20 95-14941
CIP

ISBN 0-471-18562-0 Wiley-VCH, Inc.

10 9 8 7 6 5 4 3 2

DEDICATION

I dedicate this book to my wife, Eugenia, whose love and intellectual companionship over these many years have been my constant source of encouragement and inspiration.

PREFACE

This is a textbook on the equilibrium statistical mechanics of homogeneous and inhomogeneous systems. It can be used on a one-term course or a two-term sequence. Although simple solids and magnetic systems are treated briefly, the emphasis of the book is strongly oriented toward fluids.

Chapters 1 through 6 provide an introductory course in statistical mechanics and phase behavior. The first two chapters present the elements of kinetic theory and ensemble theory. In Chapter 3, ensemble theory is applied to numerous exactly solvable models that illustrate the power of statistical mechanics in describing the equilibrium behavior of materials. The equation of state and phase behavior of simple fluids are treated in Chapters 4 and 5. Chapter 6 is devoted to multiphase equilibria. An extensive review of the thermodynamic theory of phase stability and multiphase equilibria is given because the author has found that the typical entering graduate student from a chemical engineering, chemistry, or physics background is not very proficient in the thermodynamics of phase behavior. Chapter 6 also includes many examples of the phase diagrams of real systems and several simple statistical mechanical models that correctly capture the qualitative features of the phase behavior of a surprisingly large variety of mixtures.

Chapters 6 through 14 can be used for a one-term course of the statistical mechanics of the behavior of phases, interfaces, thin films, and colloidal systems. Either the course described in the preceding paragraph or a standard undergraduate physical chemistry course on statistical thermodynamics are adequate prerequisites. As stated above, the thermodynamics and statistical mechanics of phase equilibria are trated in Chapter 6. An introduction to interfacial thermodynamics is presented in Chapter 7. The theories of the structure and thermodynamics of homogeneous and inhomogeneous fluids are laid out in Chapters 8 and 9. Because modern statistical mechanics uses density functional theory to establish a unified theory of homogeneous and inhomogeneous materials, a primer on the calculus of functionals is presented in Chapter 9. Simple models illustrating statistical mechanical principles are also given in Chapters 8 and 9. Chapter 10 is devoted to exactly solvable one-dimensional models of inhomogeneous and confined fluids. The models afford insights into the special properties of strongly inhomogeneous fluids and are suggestive of approximate density functional theories of three-dimensional fluids. In Chapters 11 through 13, such approximate density functional theories are described and applied to the structure and properties of interfaces and thin films and to wetting and spreading transitions. Chapter 14 is dedicated to surfactant solutions. Surfactants gravitate to interfaces where they strongly affect interfacial tensions. In solution they form supramolecular microstructures or association colloidal dispersions such as micellar solutions, microemulsions, and liquid crystals.

Because the primary technique for studying the structure of materials is the scattering of elecromagnetic or matter waves, Chapter 15 has been included as a reference for those who want to better understand the meaning of the scattering results frequently mentioned or shown in connection with structure determinations.

H. Ted Davis
Minneapolis, MN
September 1995

ACKNOWLEDGMENTS

I gratefully acknowledge the graduate students who were subjected over the years to various preliminary versions of this book. Their response to the class material and exercises was very helpful in arriving at the final form of the text.

Connie Galt's assistance in the preparation of the manuscript has been invaluable. She perservered cheerfully through three different wordprocessors and several generations of computers as the book slowly and painfully took form. I would also like to thank my sister-in-law, Catherine Asimakopoulos, and Carolyn Swanson for applying their drafting and computer graphic skills to the many figures and illustrations used in the text.

CONTENTS

3 / Statistical Mechanics of Some Simple Systems *97*

4 / Statistical Thermodynamics of Simple Classical Fluids *153*

5 / Imperfect Gases *203*

6 / Phase Equilibria 247

7 / Capillarity and Interfacial Thermodynamics 333

8 / Structure and Stress in Simple Fluids and Their Mixtures: Applications to Interfaces 381

9 / Density Functional Theory of Structure and Thermodynamics 425

10 / Confined One-Dimensional Fluids 473

1

KINETIC THEORY OF DILUTE GASES IN EQUILIBRIUM

1.1 Introduction

Let us consider a pure closed system of volume V and in thermodynamic equilibrium with a temperature bath at temperature T. Thermodynamically we can characterize the system by giving its total energy U, volume V, number of particles N (or mass M), and the temperature T. The energy per unit mass $\hat{U} \equiv U/M$ and the density $\rho = M/V$ are intensive variables and for a one-phase system at equilibrium (and in the absence of nonnegligible external forces) are supposed to be constant throughout the system. Thus, if the system is divided into smaller, imaginary cells as shown below,

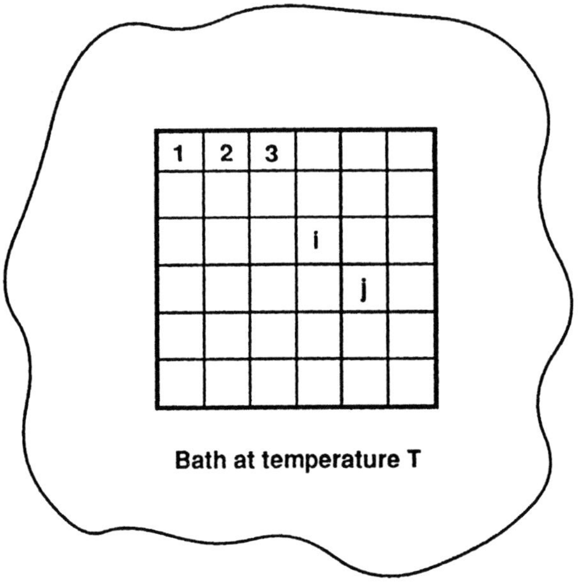

whether we measure $\hat{U}$ and ρ (as well as T, pressure P, or any other intensive variable) in the ith cell or the jth cell we should obtain the same results. Clearly, in a system composed of molecules, this thermodynamic picture will break down if the system is subdivided into too small cells. For example, if a cell is large enough to contain at most one molecule, then some of the cells will have one molecule so that $\hat{U}, \rho \neq 0$ and some will have none so that $\hat{U}, \rho = 0$. Thus, we may anticipate that a thermodynamic measurement (and characterization) would not be meaningful if the sample of material is too small. It is a role of statistical mechanics to make more precise the meaning and validity of a thermodynamic characterization of a molecular system.

Microscopically, one could completely specify the state of a system of N classical mechanical molecules by solving Newton's equation of motion describing the time evolution of the positions and internal configurations (i.e., the molecular coordinates) of the molecules. However, the problem would involve solving a set of $3N$ or more second-order differential equations for which we would need to know at some inititial time all the molecular coordinates and their rates of change. For example, for a system of monatomic molecules subject only to conservative forces, Newton's equations are of the form

$$m_i \frac{d^2\mathbf{r}_i}{\mathbf{dt}^2} = \mathbf{F}_i(\mathbf{r}_1, \ldots, \mathbf{r}_N), \quad i = 1, \ldots, N, \tag{1.1.1}$$

where m_i is the mass of a molecule whose center of mass is $\mathbf{r}_i$ and $\mathbf{F}_i$ is the force exerted on the molecule by the other molecules and by any external force fields present. To solve this set of equations we must specify at some initial time the values of the positions $\mathbf{r}_1, \ldots, \mathbf{r}_N$ and velocities $\mathbf{v}_1, \ldots, \mathbf{v}_N (\mathbf{v}_i \equiv d\mathbf{r}_i/dt, i = 1, \ldots, N)$. In practice, we can neither determine the needed $6N$ initial position and velocity components nor, even with these initial data, can we solve the set of $3N$ differential equations represented by eq. (1.1.1) because N is of the order of 10^{20} for thermodynamic systems generally encountered. One the other hand, for thermodynamic and hydrodynamic purposes we do not need the kind of detailed molecular information contained in eq. (1.1.1). Indeed, the information obtained from solution of the mechanical problem would still have to be reduced to the thermodynamic level. What we need is a description that is aimed at generating equations of state of macroscopic variables and equations of change for these variables, but that relates the macroscopic variables to the mechanics (classical or quantum) of the molecules composing the macroscopic system. We may anticipate that our interest will be in molecular quantities of a system that have been time averaged, because the corresponding quantities are obtained by studying samples containing large numbers of molecules whose molecular coordinates usually change appreciably on a time scale that is short compared to the duration of a macroscopic measurement. The science of equilibrium statistical mechanics is to provide the foundations for determining the thermodynamic laws of macroscopic systems from the mechanics of the molecules (or more generally entities such as photons, electrons, nuclei, atoms, molecules, etc.) composing the systems.

In this and the succeeding chapters we shall cover such topics as kinetic theory, ensemble theory, distribution function theory, etc. with the aim of giving the student an appreciation of the foundations and usefulness of the statistical mechanical theory of molecular systems. Often, as in the preceding paragraph, we shall discuss the mechanical behavior of a system in terms of Newton's equations require that the system behave classical mechanically. The quantum mechanical description, however, will be introduced via ensemble theory in a later chapter, and when the process in question must be treated quantum mechanically, we shall use the appropriate description. We have chosen the language of classical mechanics to develop the introductory ideas of statistical mechanics, not because a quantum mechanical development is more difficult—it is not—but because too many scientists and especially engineers interested in thermodynamics classical mechanics are more familiar with classical mechanics than quantum mechanics. A familiarity with thermodynamics is taken for granted throughout the text.

1.2 Equations of State for Pressure and Energy of an Ideal Monatomic Gas

For sufficiently dilute gases it has been found experimentally that PV/NT is equal to a universal constant k for all gases and that U/N is a function only of temperature. Here N denotes the number of molecules in the system and T the absolute thermodynamic temperature. For the particular case of a monatomic gas U/N is equal to $(3/2)kT$, so that the dilute gas equations of state for a monatomic system are

$$PV = NkT \tag{1.2.1}$$

and

$$U = \frac{3}{2}NkT. \tag{1.2.2}$$

The constant k, called Boltzmann's constant, is equal to the gas constant R,

$$R \equiv \lim_{P\to 0} \frac{PV}{N_{\text{moles}}T}$$

divided by Avogadro's number N_o. In cgs units $k = 1.38 \times 10^{-16}$ erg/K molecule.

An ideal gas is composed of point mass particles that do not interact with one another. A dilute gas is expected to behave as an ideal gas because the average separation of the particles of a dilute gas greatly exceeds the range of the interparticle interactions. The fact that eqs. (1.2.1) and (1.2.2) contain no interaction parameters partially support this expectation. More detailed analysis will also confirm this expectation and, in fact, make quantitative the density range in which a gas of interacting particles will behave as an ideal gas.

Consider a dilute monatomic gas of N identical particles confined in a cube of volume V and at a temperature T. If the kinetic energy of the ith particle in the system is $(1/2)mv_i^2$ and if the average distance of separation of the particles is very large compared to the range of the intermolecular forces of the particles, then the total energy of the system at a given instant in time is approximately

$$E = \sum_{i=1}^{N} (1/2)mv_i^2 \tag{1.2.3}$$

if there are no external forces acting on the system. In an actual dilute gas, the value of E can change with time because the particles can interact with the walls of the cube (otherwise the cube cannot confine the gas). Also, a redistribution of the energy among the kinetic energies of the particles can occur through the collisions of the particles with each other. Even though at any instant the number of particles involved in such collisions is very small compared to the number of particles in the system, over a sufficiently long period of time the collisions can profoundly affect the average value of the kinetic energy of each particle in the system. In any real measurement of E, a finite time, say τ, elapses during the measurement, so that one measures an average value of E, that is, one measures

$$\bar{E}_\tau = \frac{1}{\tau}\int_0^\tau E(t)\,dt \tag{1.2.4}$$

if a linear measuring device is used. Because a thermodynamic system can be described by a single energy function U, it must be a fact of nature that fluctuations of $E(t)$ about its mean value $\bar{E} = \lim_{\tau\to\infty} \bar{E}_\tau$ are very small and that $\bar{E}_\tau$ has already attained its limiting value in the time necessary for making a thermodynamic measurement. Otherwise, a single thermodynamic energy function will not be observed. A similar conclusion follows for other thermodynamic quantities. Thus, we postulate that a thermodynamic quantity is to be equated to the long time average of the corresponding microscopic quantity. For example, the thermodynamic energy of a dilute monatomic gas is postulated to be given by

$$U \equiv \sum_{i=1}^{N} \frac{1}{2}m\overline{v_i^2}. \tag{1.2.5}$$

Because the particles are identical, they must have the same average kinetic energy so that eq. (1.2.5) may be reduced to the form

$$U = N\overline{\frac{1}{2}mv^2}, \tag{1.2.6}$$

where $\overline{(1/2)mv^2}$ is the average kinetic energy of any particle of the gas. Comparing eq. (1.2.6) with the empirical result, eq. (1.2.1), we obtain the important result

$$\overline{\frac{1}{2}mv^2} = \frac{3}{2}kT, \tag{1.2.7}$$

that implies that in a dilute gas at thermal equilibrium the average kinetic energy of a molecule is directly proportional to the absolute temperature. Equation (1.2.7) will be shown later to be true for any classical system, that is, *the average center of mass kinetic energy per molecule is equal to* $(3/2)kT$ *for any classical mechanical gas, liquid, or solid—monatomic or polyatomic—in thermal equilibrium at temperature* T.

Equation (1.2.7) may be used to predict the observed dilute gas equation of state, eq. (1.2.1). Consider two opposing, perfectly reflecting walls oriented in the x direction. We wish to measure the pressure on one of these walls. The pressure is the force exerted on the wall per unit area, and the force is the momentum change per unit time. A particle, the x component of whose velocity is v_{ix}, will appear at the wall on which the pressure is to be measured with a frequency equal to $v_{ix}/2V^{1/3}$, or will collide with the wall $v_{ix}\delta t/2V^{1/3}$ times in the time interval δt. We have assumed a cubic system of length $V^{1/3}$ on a side. Upon each collision the change in momentum is $2mv_{ix}$ because the wall is perfectly reflecting. Thus during the time δt the total momentum change per unit area for all the molecules is

$$\sum_{i=1}^{N}(2mv_{ix})(v_{ix}\delta t/2V^{1/3})V^{-2/3}. \tag{1.2.8}$$

The momentum change per unit time per unit area, obtained by dividing eq. (1.2.8) by δt will give the pressure on the wall for a given set of velocities. Averaging this pressure over a long time interval we obtain what we identify to be the thermodynamic pressure, namely,

$$P = \frac{1}{V}\sum_{i=1}^{N}\overline{mv_{ix}^2} = \frac{2}{3V}\sum_{i=1}^{N}\overline{(1/2)mv_i^2} = \frac{NkT}{V}, \tag{1.2.9}$$

in agreement with the empirical result, eq. (1.2.1). The second equality of eq. (1.2.9) is obtained by noting that at equilibrium a gas is thermodynamically isotropic so that $\overline{v_{ix}^2} = \overline{v_{iy}^2} = \overline{v_{iz}^2} = \frac{1}{3}\overline{v_i^2}$, where $v_i^2 = v_{ix}^2 + v_{iy}^2 + v_{iz}^2$. It will be shown later that the relation $PV = NkT$ is valid for a polyatomic as well as a monatomic ideal gas even though U is not equal to $(3/2)NkT$ for a polyatomic gas.

For an ideal polyatomic gas it will be found that U/N is a function only of temperature, but the specific heat is generally greater than the monatomic value $(3/2)Nk$ and is temperature dependent.

Exercise
Assuming that $PV = NkT$ is an empirically established relationship for all sufficiently dilute gases, give an argument showing that $(1/2)mv^2 = (3/2)kT$, for all classical dilute gases (i.e., polyatomic as well as monatomic gases), where m is the molecular mass and $\mathbf{v}$ is the velocity of the center of mass of a molecule of the gas.

1.3 Replacement of Time Averages by Ensemble Averages

In the preceding section, we introduced as a basic postulate of statistical mechanics the indentification of thermodynamical quantities with the long time averages of corresponding mechanical quantities. In principle, one would calculate the time averages by following the microscopic behavior of a single system subject to whatever macroscopic constraints (fixed M, V, composition, etc.) characterize the thermodynamic situation considered. Because such a calculation is difficult, if not impossible, a device first introduced by Gibbs and Einstein, independently, is generally used in statistical mechanical calculations, namely, the device of equating the time average of a mechanical variable of a single system to the average value of the variable determined from a large collection of "thermodynamic replicas" of the given system. By thermodynamic replicas we mean systems subject to the same macroscopic constraints. Such a collection of replicas is called an ensemble or an ensemble of systems, and averages determined from such a collection are called ensemble averages.

To motivate the identification of time and ensemble averages, let us consider a system having a random variable, K, which can take on a sequence of discrete values $K_1, K_2, \ldots$ in time. Suppose the system is studied over a long time τ and the total time that K has the values $K_1, K_2, \ldots$ is $t_1, t_2, \ldots$, respectively. Then the time averaged observed value of K is

$$\bar{K} = \sum_i K_i \frac{t_i}{\tau}, \tag{1.3.1}$$

and, if τ is sufficiently large, t_i/τ represents the probability P_i that K will have the value K_i if the system is observed at an arbitrary instant in time. On the other hand, instead of starting with one system and observing it in time, we could start with a very large number M of systems identical to the first system. Then we could count, at the same instant in time, the number of systems in which K has the values K_1, $K_2, \ldots$, respectively. Denoting these numbers by $\nu_1, \nu_2, \ldots$ we obtain $\langle K \rangle$, the ensemble averaged value of K, from the relation

$$\langle K \rangle = \sum_i K_i \frac{\nu_i}{N}. \tag{1.3.2}$$

If M is sufficiently large, then ν_i/M will be the probability that K will have the value K_i in a system picked at random from the set of replicas. For a random variable $t_i/\tau \rightarrow \nu_i/M \rightarrow P_i$ as τ and M tend to infinity. Thus the time average value of K, eq. (1.3.1), will be equal to the ensemble average of K, eq. (1.3.2).

In terms of thermodynamic systems, variables K correspond to microstates (e.g., the particle positions, momenta, energy, etc.) of the system while replicas represent systems prepared or held under the same macroscopic conditions. For example, consider a system of N particles in a fixed volume V and held in a heat bath of temperature T. An ensemble of such a system (called a canonical ensemble for the given fixed thermodynamic conditions) would consist of a very large number of systems of the same N and V that are in thermal equilibrium with a heat bath at temperature T. The basic postulate of ensemble theory is that the ensemble average of a molecular variable is equal to the long time average and, therefore, the thermodynamic value of the variable. For example, the thermodynamic energy of an ideal monatomic gas is written in the form

$$U = N\langle \frac{1}{2} m v^2 \rangle, \tag{1.3.3}$$

involving the ensemble average of the kinetic energy rather than the time average given in eq. (1.2.6). The proof of the validity of this postulate, known as the ergodic hypothesis, for molecular systems obeying classical or quantum mechanics is a basic problem of ergodic theory. Under certain restrictive circumstances the ergodic hypothesis has been proved (and thereby became a theorem), but in general it remains a postulate whose validity is accepted on the basis of the vast body of data agreeing with the conclusions and implications of ensemble theory. In fact, one may take the point of view that the equivalence of thermodynamic quantities and ensemble averages of the corresponding molecular or mechanical quantities is the starting point of statistical mechanics. Thus, the question of equality of time and ensemble averages may be avoided altogether. However, such an approach ignores the fact that real experiments on thermodynamic variables involve time averaged measurements on single systems.

1.4 Some Definitions from Probability Theory

Having settled on a statistical description of the behavior of a molecular system, we are confronted with the problem of obtaining the probability or distribution functions that characterize the probable or expected value of a microscopic variable. In particular, for dilute gases in which velocity and kinetic energy averages are desired, we are interested in the probability that a typical molecule in the system has a given velocity or, because velocity is a continuous variable, the probability that the velocity lies in a given range, say with Cartesian components between v_x, v_y, v_z, and $v_x + dv_x$, $v_y + dv_y$, $v_z + dv_z$, respectively. In vector form the velocity range may be denoted as $\mathbf{v}$ to $\mathbf{v} + d\mathbf{v}$. The probability that $\mathbf{v}$ will lie between $\mathbf{v}$ and $\mathbf{v} + d\mathbf{v}$ will be seen in the next section to be governed by a Gaussian distribution. However, for those not familiar with statistical arguments involving multiple variables and continuous distributions, it is perhaps useful to digress at this point and review briefly some standard definitions of probability theory.

First, consider a system in which a discrete random variable s_i occurs with the probability p_i. Suppose we examine a large number M of replicas and find that in ν_1 of the systems the variable s has the value s_1 (i.e., ν_1 systems are in state 1), in ν_2 of the systems the variable s has the value s_2 (i.e., ν_2 systems are in state 2), etc. Then the average value of the variable s is

$$\langle s \rangle = \lim_{M \to \infty} \sum_i s_i \frac{\nu_i}{M} = \sum_i s_i p_i. \qquad (1.4.1)$$

If $h(s)$ is some function of s, for example, s^2 or $\sin(s)$, then when the system is in the ith state, that is, when $s = s_i$ the corresponding value of h is $h(s_i)$. The average value of this function will be

$$\langle h \rangle = \lim_{M \to \infty} \sum_i h(s_i) \frac{\nu_i}{M} = \sum_i h(s_i) p_i. \qquad (1.4.2)$$

Suppose next that the state or random variable of the system is characterized by more than one quality or index, say by the three indices ijk. Then in studying a large number of replicas of such a system we may record the number ν_{ijk} of systems in which the random variable $\mathbf{s}$ has the value s_{ijk}. The average value of any function $h(\mathbf{s})$ will then be

$$\langle h \rangle = \lim_{M \to \infty} \sum_{i,j,k} h(s_{ijk}) \frac{\nu_{ijk}}{M} = \sum_{i,j,k} h(s_{ijk}) p_{ijk} \qquad (1.4.3)$$

where p_{ijk} is the probability that a system picked at random will be in the state ijk. The function h might depend only on one of the three indices ijk, say i, in which case we need only the probability

Exercise

If two outcomes are possible with p the probability of 1 and q that of 2 and $p + q = 1$, then after N trials the probability that 1 would have occurred n_1 times is

$$P_N(n_1) = \frac{N!}{n_1!(N-n_1)!} p^{n_1} q^{N-n_1}.$$

This is the *binomial distribution*. Show that the mean value of n_1 is

$$< n_1 > = pN$$

and that the mean square of n_1 is

$$< n_1^2 > = (Np)^2 + Npq.$$

Thus, the variance is

$$\sigma_N^2 \equiv < n_1^2 > - < n_1 >^2$$
$$= Npq$$

and the fractional mean square deviation is

$$\frac{\sigma_N}{< n_1 >} = \sqrt{\frac{q}{Np}}.$$

Plot $\sigma_N / < n_1 >$ versus N for the occurrence of heads ($p = q = 1/2$) in flipping a coin.
Show that for very large N the distribution is well approximated by

$$P = \frac{1}{(2\pi\sigma_N^2)^{1/2}} e^{-\frac{(N-pN)^2}{2\sigma_N^2}}$$

This result is an example of the *central limit theorem*.

that the first index of the state of the system is i. This probability is given by the expression

$$p_i = \sum_{j,k} p_{ijk}. \tag{1.4.4}$$

To appreciate this property, note that in the ensemble M, the total number of systems ν_i for which the first index determining the state is i, regardless of what the second two indices are, is equal to the sum $\sum_{j,k} \nu_{ijk}$ so that the fraction of such systems is $\sum_{j,k} \nu_{ijk}/M$. Thus, because by definition $p_i \equiv \lim_{M\to\infty} \nu_i N$, eq. (1.4.4) follows. If h is a function only of the first index, then

$$\langle h \rangle = \sum_{i,j,k} h_i p_{ijk} = \sum_i h_i p_{ijk} = \sum_i h_i p_i. \tag{1.4.5}$$

Summation of probability functions over "degrees of freedom" or "state indices" not appearing in the quantity h being averaged will turn out to be very useful in our studies of the distributions occurring in statistical mechanics.

Exercise

Show that if the probability of outcome 1 is small, i.e., if $p << 1$, the binomial distribution is well approximated by the *Poisson distribution*

$$P_N(n_1) = \frac{(Np)^{n_1} e^{-Np}}{n_1!}.$$

This distribution describes some nuclear decay processes and some polymerization processes.

Example

Suppose we have four squares, one black with green dots, two black with red dots, and one green with red dots. If we were to pick one of the squares at random, the probability that it is black with green dots is $\rho_{11} = 1/4$, or black with red dots is $\rho_{12} = 1/2$ or green with red dots is $\rho_{22} = 1/4$. One the other hand, the probability that the chosen square is green regardless of dots is $\rho_2 = \rho_{21} + \rho_{22} = 1/4$. Note that $\rho_{21} = 0$ because green with green dots is not a state of the system.

Consider next a system in which a continuous random variable s is an n-dimensional vector with independent components $s_1, s_2, \ldots, s_n$ ranging over the values $a_1 < s_1 < b_1, \ldots, a_n < s_n < b_n$, respectively. One can no longer label the states of the system by a sequence of integers because a continuous variable can take on a nondenumerable infinity of values. What one does instead is to divide the n-dimensional volume spanned by s into a very large but finite number (or denumerably infinite number if the a_i or b_i are infinite small cells labelled 1,2,3, ... and having volume $\Delta^n s^{(1)} = \Delta s_1^{(1)} \Delta s_2^{(1)} \ldots \Delta s_n^{(1)}$, $\Delta^n s^{(2)} = \Delta s_1^{(2)} \ldots \Delta s_n^{(2)}, \ldots$, respectively, where $\Delta s_j^{(k)}$ is the length of the jth side of the kth cell. The system is said to be in the kth state if the vector **s** lies somewhere in the volume $\Delta^n s^{(k)}$, that is, if s lies between $s_1^{(k)}, \ldots, s_n^{(k)}$ and $s_1^{(k)} + \Delta s_1^{(k)}, \ldots, s_n^{(k)} + \Delta s_n^{(k)}$ or, in abbreviated notation, between $\mathbf{s}^{(k)}$ and $\mathbf{s}^{(k)} + \Delta \mathbf{s}^{(k)}$. If in an ensemble of N replicas, there are ν_1 systems with **s** lying in $\Delta^n s^{(1)}$, ν_2 with **s** in $\Delta^n s^{(2)}$, etc., then the probability that s will be between $\mathbf{s}^{(k)}$ and $\mathbf{s}^{(k)} + \Delta\mathbf{s}^{(k)}$ will be $p(\mathbf{s}; \Delta^n s^{(k)}) \equiv \lim_{N\to\infty} \nu_k/N$ and the average value of some continuous function h of s may be estimated to be

$$\langle h\rangle = \sum_k h(\mathbf{s}\ in\ \Delta^n s^{(k)}) p(\mathbf{s}; \Delta^n s^{(k)}). \qquad (1.4.6)$$

The reason eq. (1.4.6) is only an estimate is that for a given state k the value of h will not be known exactly because the value of s can range over all possible vectors in the cell $\Delta^n s^{(k)}$.

If the total volume of the system is denoted by Ω and if s is distributed with equal probability throughout Ω, then the probability $p(\mathbf{s}; \Delta^n s^{(k)})$ is equal simply to the ratio $\Delta^n s^{(k)}/\Omega$, that is, p is directly proportional to the volume $\Delta^n s^{(k)}$ of the kth cell. In the general case in which s is not distributed with equal probability, we expect $p(\mathbf{s}; \Delta^n s^{(k)})$ to still be proportional to $\Delta^n s^{(k)}$, but the proportionality factor will not be a constant, that is, $p(\mathbf{s}; \Delta^n s^{(k)})$ will be of the form

$$p(\mathbf{s}; \Delta^n s^{(k)}) = \omega(\mathbf{s}; \Delta^n s^{(k)}) \Delta^n s^{(k)} \qquad (1.4.7)$$

where $\omega(\mathbf{s}; \Delta^n s^{(k)})$ represents a probability density or a probability per unit volume of s space. For sufficiently smooth distributions, the limit

$$\lim_{\Delta s_1^{(k)},\dots,\Delta s_n^{(k)}\to 0} \frac{p(\mathbf{s}; \Delta^n s^{(k)})}{\Delta^n s^{(k)}} \to \omega(\mathbf{s}) \qquad (1.4.8)$$

Exercise
The probability of an archer hitting the target is one in 5. If she shoots ten times, what is the probability of hitting the target four times? At least four times?

exists and will be a continuous function of s. Substituting eq. (1.4.7) into eq. (1.4.6) and considering a smaller and smaller subdivision of s subspace, that is, considering a larger and larger number of smaller and smaller cells $\Delta^n s_k$, we obtain for distributions obeying eq. (1.4.8) the result

$$\begin{aligned}\langle h\rangle &= \lim_{\Delta^n s\to 0} \sum_k h(\mathbf{s} \in \Delta^n s^{(k)}) \omega(\mathbf{s}; \Delta^n s^{(k)}) \Delta^n s^{(k)} \\ &= \int_{a_1}^{b_1} \cdots \int_{a_n}^{b_n} h(\mathbf{s})\omega(\mathbf{s}) d^n s,\end{aligned} \qquad (1.4.9)$$

where the second equality follows because the first equality defines a Riemann integral if h and ω are piecewise continuous functions.

If $d^n s$ is an infinitesimal volume fixed on the vector s, then $\omega(\mathbf{s})d^n s$ represents the probability that the variable s will be between s and $\mathbf{s} + d\mathbf{s}$. If we want the probability $\omega_i(s_i)ds_i$ that s_i lie between s_i and $s_i + ds_i$, regardless of what the other components of s are, we "sum," that is, integrate over all components of s except s_i. Thus,

$$\omega_i(s_i)ds_i \equiv ds_i \int \cdots \int \omega(\mathbf{s}) ds_1 \cdots ds_{i-1} ds_{i+1} \cdots ds_n. \qquad (1.4.10)$$

Similarly, the probability $\omega_{ij}(s_i, s_j)ds_i ds_j$ that s_i and s_j lie in the indicated ranges is

$$\omega_{ij}(s_i, s_j)ds_i ds_j \equiv ds_i ds_j \int \cdots \int \omega(\mathbf{s}) \prod_{\alpha \neq i,j}^{n} ds_\alpha, \qquad (1.4.11)$$

where the notation $\prod_{\alpha \neq i,j}^{n}$ means the product is over all α except i and j.

In the case of discrete variables the sum of the probability function over all states equals one. Similarly, in the case of continuous variables, the condition

$$\int_{a_1}^{b_1} \cdots \int_{a_n}^{b_n} \omega(\mathbf{s}) d^n s = 1 \qquad (1.4.12)$$

is required if $\omega(\mathbf{s})$ is a distribution function or probability density.

Some variables may be distributed independently, that is, the probability that one variable has a value s_i is independent of the fact that another variable has a value s_j. We say s_i and s_j are statistically independent if

$$\omega_{ij} = \omega_i \omega_j. \qquad (1.4.13)$$

The Cartesian components of the velocity distribution of classical particles provide an important example of statistical independence of physical variables. We examine the velocity distribution in considerable detail in this chapter.

Let us end this section with examples drawn from the statistical mechanics of a rotating electric dipole as an example.

Example

If an object undergoes an isotropic random walk in g-dimensional space, after a large number N of steps, the probability distribution will be

$$P(R) = \left(\frac{1}{(2\pi N \overline{r^2}/3)^{g/2}} \right) e^{-\left[\frac{3R^2}{2N\overline{r^2}}\right]},$$

where $P(R)d^g R$ is the probability that R lies in the volume interval $d^g R$ after N steps and $\overline{r^2}$ is the mean square displacement in a single step. This distribution is obtained as the *central limit theorem* for rather general probability distributions of each step (in particular $\overline{r^2}$ must be finite).

If the mean frequency of a step is ν, then N can be transformed to time by $t = N/\nu$ and $D = \nu \overline{r^2}/6$ so that the random walk distribution becomes

$$P(R) = \left(\frac{1}{(4\pi Dt)^{g/2}} \right) e^{-\left[\frac{R^2}{4Dt}\right]},$$

the probability that a diffusing particle has reached the distance R in time t. The mean square displacement obeys the relation

$$< R^2 >= 2gDt$$

which can be used to measure the diffusivity D of the particle.

Example

a) Consider a dipole with a moment μ free to rotate about its center of mass. The value of the dipole moment projected along the z axis is $\mu_z = \mu \cos\theta$, expressed in polar coordinates.
If the dipole is free from external influence, it can orient itself equally well in any direction. Therefore, the probability that the dipole is oriented in the solid angle $d^2 A = \sin\theta d\theta d\eta$ is simply $d^2 A/4\pi$, or

$$p = \frac{\sin\theta d\theta d\eta}{4\pi} = \omega \sin\theta d\theta d\eta,$$

so that $\omega = 1/4\pi$.
Calculate the average of μ_z and μ_z^2, that is, the mean and mean square values of μ_z. For $w = 1/4\pi$, we find

$$\langle \mu_z \rangle = \frac{1}{4\pi} \int_0^\pi \int_0^{2\pi} \mu \cos\theta \sin\theta d\theta d\eta = 0$$

$$\langle \mu_z^2 \rangle = \frac{\mu^2}{4\pi} \int_0^\pi \int_0^{2\pi} \cos\theta \sin\theta \sin\theta d\theta d\eta = \frac{1}{3}\mu^2.$$

Thus, the moment of a dipole free to rotate without external influence will have an average value $\langle \mu \rangle = \mathbf{0}$, because $\langle \mu_x \rangle = \langle \mu_y \rangle = \langle \mu_z \rangle$, and a mean square value $\langle \mu^2 \rangle = \mu^2$ that follows from the fact that $\langle \mu_x^2 \rangle = \langle \mu_y^2 \rangle = \langle \mu_z^2 \rangle$ and so $\langle \mu^2 \rangle = 3\langle \mu_z^2 \rangle$.

b) Find the integral expressions for the mean and mean square values of μ_z in the case for which the dipole of (a) is at thermal equilibrium in the presence of an electric field E oriented along the z axis.

As can be shown later in the text, the distribution function ω for this case is

$$\omega = \frac{\mu E \exp(-\mu E \cos\theta / kT)}{2\pi kT[\exp(\mu E/kT) - \exp(-\mu E/kT)]}.$$

Thus,

$$\langle \mu_z \rangle = \frac{\mu E \int_0^\pi \exp(-\mu E \cos\theta/kT)\mu \cos\theta \sin\theta d\theta}{kT[\exp(\mu E/kT) - \exp(-\mu E/kT)]}$$

and

$$\langle \mu_z^2 \rangle = \frac{\mu E}{kT} \frac{\int_0^\pi \exp(-\mu E \cos\theta/kT)(\mu \cos\theta)^2 \sin\theta d\theta}{[\exp(\mu E/kT) - \exp(-\mu E/kT)]}.$$

Several statistics can be relatively easily established for the rotating dipole described above as is illustrated by the following exercise.

Exercise

For case (b) of the preceding example, show that $\langle \mu_y \rangle = \langle \mu_x \rangle = 0$, $\langle \mu_x^2 \rangle = \langle \mu_y^2 \rangle = \mu^2/3$, and

$$\langle \mu_z \rangle = \frac{kT}{E}[1 - \frac{\mu E}{kT} \coth \frac{\mu E}{kT}]$$
$$\langle \mu_z^2 \rangle = (\frac{kT}{E})^2[2 + (\frac{\mu E}{kT})^2 - 2\frac{\mu E}{kT} \coth \frac{\mu E}{kT}].$$

1.5 Maxwell Velocity Distribution

Consider a dilute gas system of N particles in a volume V and at temperature T. What we wish to calculate is the probability $p(\mathbf{v}; \Delta^3 v)$ that a typical particle of such a system chosen at random from an ensemble of systems will have velocity components lying between v_x and $v_x + \Delta v_x$, v_y and $v_y + \Delta v_y$ and v_z and $v_z + \Delta v_z$. The quantity $\Delta^3 v = \Delta v_x \Delta v_y \Delta v_z$ represents the volume in velocity space in which the velocity vector lies. If the probability distribution of velocities is continuous and differentiable, then the limit

$$\phi(\mathbf{v}) \equiv \lim_{\Delta v_x, \Delta v_y, \Delta v_z \to 0} \frac{p(\mathbf{v}; \Delta^3 v)}{\Delta v_x \Delta v_y \Delta v_z} \tag{1.5.1}$$

exists, and the quantity $\phi(\mathbf{v})$ defined here is called the velocity distribution function or the velocity probability density. Thus, if $d\mathbf{v}$ is an infinitesimal volume element, the probability that $\mathbf{v}$ will lie between $\mathbf{v}$ and $\mathbf{v} + d\mathbf{v}$ is

$$\begin{aligned} p(\mathbf{v}; d^3 v) &= \phi(\mathbf{v}) dv_x dv_y dv_z \\ &= \phi(\mathbf{v}) d^3 v. \end{aligned} \tag{1.5.2}$$

The normalization condition

$$\int_{-\infty}^{\infty}\int_{-\infty}^{\infty}\int_{-\infty}^{\infty} \phi(\mathbf{v}) dv_x dv_y dv_z = 1 \tag{1.5.3}$$

must be imposed because the probability is unity that the velocity has some value in its allowed range.

If $h(\mathbf{v})$ is some function of the velocity of a particle, then the ensemble average value, and by hypothesis the thermodynamic value, of this quantity may be written in the form 1

$$\langle h(\mathbf{v}) \rangle = \int\int\int h(\mathbf{v}) \phi(\mathbf{v}) d^3 v. \tag{1.5.4}$$

In a systematic development of ensemble theory, one introduces one more hypothesis (postulate of equal *a priori probabilities*) and then, using the laws of large numbers, one determines explicit expressions for the distribution functions for various kinds of ensembles (the closed, diathermal, and open systems of thermodynamics have their corresponding ensembles: the microcanonical, canonical, and grand canonical ensembles, respectively). At this point, we take a much less ambitious approach to the problem, giving the original derivation devised by Maxwell to deduce the form of ϕ for an isothermal system. This derivation, although simple and perhaps intuitively acceptable, lacks the rigor of kinetic theoretical and ensemble theoretical derivations. That Maxwell's result itself was correct has been verified by numerous experiments and by rigorous theories.

Maxwell's assumptions were that:

1 ϕ depends on velocity only through the kinetic energy of the particle, that is,

$$\phi = \phi(mv^2/2),$$

2 Cartesian components of the kinetic energy distribute independently, that is,

$$\phi(mv^2/2) dv_x dv_y dv_z = [\gamma(mv_x^2/2) dv_x][\gamma(mv_y^2/2) dv_y][\gamma(mv_z^2/2) dv_z], \tag{1.5.5}$$

in which $mv^2/2 = mv_x^2/2 + mv_y^2/2 + mv_z^2/2$, and

3 γ is differentiable.

Assumption 1 is quite acceptable, because it actually requires only that the velocity distribution be a function of the magnitude and not the direction of the particle velocity. This simply expresses the fact that an equilibrium gas is isotropic. That the Cartesian components of the kinetic energy distribute independently is difficult to justify *a priori*, and, therefore, assumption 2 represents the weak point in Maxwell's derivation. The final, and not very restrictive, assumption is that the distribution functions ϕ and γ be differentiable functions. These three assumptions suffice to totally determine the form of ϕ.

Consider the differentiable functions f and g having the property

$$f(x+y) = g(x)g(y)$$

for all values of x and y. We differentiate both sides of this expression with respect to x to obtain

$$f'(x+y) = g'(x)g(y) \tag{1.5.6}$$

and with respect to y to obtain

$$f'(x+y) = g(x)g'(y). \tag{1.5.7}$$

Combining eqs. (1.5.6) and (1.5.7), we find the equality $g'(x)g(y) = g(x)g'(y)$, which when rearranged gives

$$\frac{g'(x)}{g(x)} = \frac{g'(y)}{g(y)}. \tag{1.5.8}$$

Because the left-hand side of this equation depends only on x and the right hand-side (rhs) only on y, which can be varied independently of x, then both sides must be equal to a constant $-\beta$ (the minus sign is chosen for later convenience). Thus, g satisfies the simple equation

$$\frac{g'(x)}{g(x)} = -\beta \tag{1.5.9}$$

that has the solution

$$g(x) = \alpha e^{-\beta x}, \tag{1.5.10}$$

where α is a constant of integration. A similar argument easily shows that the condition

$$f(x+y+z) = g(x)g(y)g(z)$$

on differentiable functions f and g implies again that g is of the form given by eq. (1.5.10). Thus, substituting ϕ for f, γ for g, and mv_x^2, mv_y^2, and $mv_z^2/2$ for x, y, and z, respectively, we obtain Maxwell's result:

$$\phi(mv^2/2) = \alpha^3 e^{-\beta mv^2/2}, \tag{1.5.11}$$

that states that the velocity of a particle in an isothermal system is distributed according to a Gaussian distribution about a mean velocity of zero.

Because of the normalization condition on ϕ, only one of the parameters, α or β, is independent. The independent parameter, say β, must be determined from the thermodynamic conditions under which the system is prepared. Consider first the normalization condition:

$$1 = \int \phi d^3v = \alpha^3 \int_{-\infty}^{\infty} dv_x \int_{-\infty}^{\infty} dv_y \int_{-\infty}^{\infty} dv_z e^{-(\beta/2)(mv_x^2+mv_y^2+mv_z^2)} \quad (1.5.12)$$

where $mv^2/2$ and d^3v have been written in terms of their Cartesian components. To evaluate the rhs of eq. (1.5.12), the following result, obtained from a standard table of integrals, is useful:

$$\int_{-\infty}^{\infty} e^{-ct^2} dt = \sqrt{\pi/c} \,. \quad (1.5.13)$$

From this and eq. (1.5.12) it follows that

$$\alpha = \sqrt{\beta m/2\pi} \,. \quad (1.5.14)$$

Another condition on ϕ is the equation,

$$\langle \frac{1}{2} mv^2 \rangle = \frac{3}{2} kT \,, \quad (1.5.15)$$

that we obtained earlier in comparing kinetic theoretical and thermodynamic results for dilute monatomic gases. Because $\langle mv_x^2 \rangle = \langle mv_y^2 \rangle = \langle mv_z^2 \rangle = 1/3 \langle mv^2 \rangle$, eq. (1.5.15) implies $\langle mv_x^2 \rangle = kT$. Using this expression and the form of ϕ given by eq. (1.5.11), with α determined by eq. (1.5.14), we obtain

$$kT = \frac{1}{2} \left(\frac{\beta m}{2\pi} \right)^{3/2} \int_{-\infty}^{\infty} dv_x e^{-\beta m v_x^2/2} mv_x^2 \int_{-\infty}^{\infty} dv_y e^{-\beta m v_y^2/2} \int_{-\infty}^{\infty} dv_z e^{-\beta m v_z^2} . \quad (1.5.16)$$

The integrals over dv_y and dv_z may be evaluated by comparison with eq. (1.5.13). To evaluate the integral over dv_x, we make use of a trick. Define $I(c)$ to be

$$I(c) = \int_{-\infty}^{\infty} e^{-ct^2} dt. \quad (1.5.17)$$

Thus, the first derivative of I with respect to c is

$$\frac{dI}{dc} = - \int_{-\infty}^{\infty} e^{-ct^2} t^2 dt, \quad (1.5.18)$$

which is of the form of the integration over dv_x in eq. (1.5.1). On the other hand, from eq. (1.5.13) it follows that $dI/dc = -(1/2c)(\pi/c)^{1/2}$, and so

$$\int_{-\infty}^{\infty} e^{-ct^2} t^2 dt = (1/2c)\sqrt{\pi/c}. \tag{1.5.19}$$

By manipulations similar to those leading to eq. (1.5.18) it is easy to show that

$$\int_{-\infty}^{\infty} e^{-ct^2} t^{2n} dt = (-1)^n \frac{d^n I}{dc^n} \tag{1.5.20}$$

$$\int_{0}^{\infty} e^{-ct^2} t^{2n+1} dt = (-1)^n \frac{d^n J}{dc^n}, \tag{1.5.21}$$

where $I = (\pi/c)^{1/2}$ and $J = 1/2c$.

Exercise

Prove eq. (1.5.13) by squaring $\int_0^\infty e^{-ct^2} dt$ and using polar coordinates to evaluate the resulting two-dimensional integral.

Returning to eq. (1.5.16) and comparing the integrals in this equation with eqs. (1.5.13) and (1.5.19), we obtain the result

$$kT = \frac{1}{\beta}, \tag{1.5.22}$$

that leads to the form

$$\phi(v) = \left(\frac{m}{2\pi kT}\right)^{3/2} e^{-(mv^2/2kT)} \tag{1.5.23}$$

for the Maxwell distribution. This very important result implies that the distribution of the velocity of particles in an isothermal dilute gas of monatomic particles is a Gaussian distribution whose mean dispersion is determined solely by the absolute temperature T and the mass m of the particles. Even more surprising is that this distribution can be shown by ensemble theory to hold for the center of mass motion of any isothermal system of classical particles independently of the state of matter of the system even if the particles are polyatomic.

1.6 Distribution and Mean Values Derived from Maxwell Distribution

In certain instances, for example in calculating the pressure on a flat wall with a given orientation or the rate of leakage of molecules through a pinhole in the wall, we need the distribution of a component of the velocity in a given direction. Suppose the desired component is the x component. The probability that the x component of velocity will lie between v_x and $v_x + dv_x$, regardless of the values of the velocity components v_y and v_z, is

$$\phi_x(v_x)dv_x = dv_x \int_{-\infty}^{\infty}\int_{-\infty}^{\infty} \phi(\mathbf{v})dv_y dv_z = \left(\frac{m}{2\pi kT}\right)^{1/2} e^{-(mv_x^2/2kT)}dv_x, \tag{1.6.1}$$

where, as discussed in Section 1.4, we have integrated over those indices or degrees of freedom (v_y and v_z) not specified in the definition of $\phi_x(v_x)$. Thus, in a system of N particles, the probable number of particles having a velocity between $\mathbf{v}$ and $\mathbf{v} + d\mathbf{v}$ is $N\phi(\mathbf{v})dv_x dv_y dv_z$, and the probable number having an x component of velocity between v_x and $v_x + dv_x$ but having any value of the y and z components of velocity is $N\phi(v_x)dv_x$.

For the Maxwell distribution the average of the x component of the particle velocity is of the form

$$\langle v_x \rangle = \int_{-\infty}^{\infty} v_x \phi_x(v_x)dv_x = \left(\frac{m}{2\pi kT}\right)^{1/2} \int_{-\infty}^{\infty} e^{-(mv_x^2/2kT)} v_x dv_x. \tag{1.6.2}$$

If we introduce the change of variable $u \equiv -v_x$ into eq. (1.6.2), we obtain the result

$$\langle v_x \rangle = -\left(\frac{m}{2\pi kT}\right)^{1/2} \int_{-\infty}^{\infty} e^{-u^2/2kT} u du. \tag{1.6.3}$$

Aside from having a different symbol, u, for the dummy variable, v_x, the integral in eq. (1.6.3) is identical to that in eq. (1.6.2). Consequently, we conclude that $\langle v_x \rangle$ is equal to $-\langle v_x \rangle$, and so

$$\langle v_x \rangle = 0. \tag{1.6.4}$$

Similarly, we can show that $\langle v_y \rangle = \langle v_z \rangle = 0$, or, in fact, $\langle \mathbf{v} \rangle = 0$. Thus, in a gas obeying the Maxwell distribution of velocities, the average value of a Cartesian component of the particle velocity is zero, and the distribution function obeyed by a Cartesian component is a Gaussian distribution symmetric about zero. In Figure 1.1 is shown a plot of $\phi_x(v_x)$ versus v_x for various temperatures.

Figure 1.1
Distribution of a Cartesian component of the velocity of a particle obeying the Maxwell distribution law $T_1 < T_2 < T_3$.

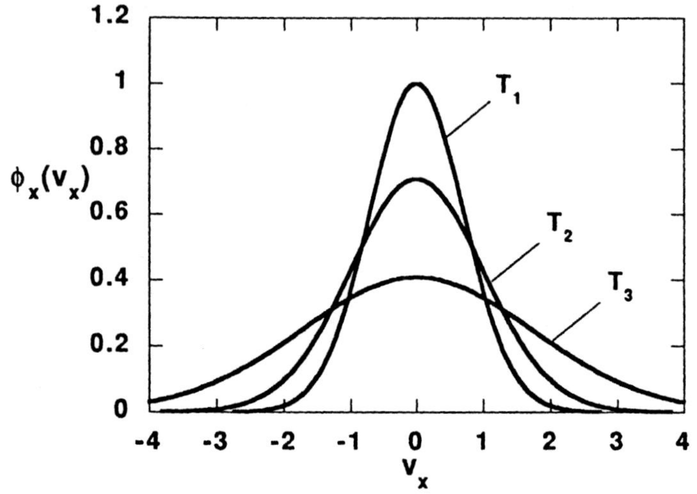

The root mean square (rms) width (or root mean square deviation) or variance of the Gaussian ϕ_x is defined by

$$v_x^{\rm rms} = \sqrt{\langle (v_x - \langle v_x \rangle)^2 \rangle} \tag{1.6.5}$$

and for the Maxwell distribution has the value $v_x^{\rm rms} = \sqrt{kT/m}$. Thus, the lower the temperature, the more sharply the distribution of v_x will be peaked about a mean value of zero.

Exercise
Prove that $\langle v_x^k v_y^l v_z^m \rangle = 0$, if k, l and m are positive integers whose sum is odd.

In using Maxwell's distribution in the form given above, we have actually restricted ourselves to systems that are rigid vessels fixed in the inertial frame of the observer. If, however, a rigid vessel containing N particles at temperature T is moving with a constant velocity $\mathbf{u}$ relative to the observer, then the Maxwell distribution measured by the observer would be of the form

$$\phi = \left(\frac{m}{2\pi kT}\right)^{3/2} \exp\left[\frac{-m(\mathbf{v} - \mathbf{u})^2}{2kT}\right], \tag{1.6.6}$$

where $\mathbf{v}$ is the velocity of a particle measured relative to a frame at rest relative to theobserver.

Although the average velocity of a Maxwellian molecule is zero (returning to situations in which the system and observer are fixed in the same frame of reference), the average speed of a molecule is not zero and is an important parameter in understanding the details of the constant thermal movements of molecular systems. The speed of a molecule of velocity $\mathbf{v}$ is by definition the magnitude v of the velocity. To compute the average value of the speed, it is convenient to express the velocity in terms of the spherical coordinates v, θ, and η shown in Figure 1.2.

In terms of the spherical coordinates, $v_x = v \sin\theta \cos\eta$, $v_y = v \sin\theta \sin\eta$, $v_z = v \cos\theta$,

$$\mathbf{v} = v(\sin\theta \cos\eta \hat{\imath} + \sin\theta \sin\eta \hat{\jmath} + \cos\theta \hat{k}), \qquad (1.6.7)$$

and

$$v \equiv \sqrt{\mathbf{v} \cdot \mathbf{v}} = \sqrt{v_x^2 + v_y^2 + v_z^2}. \qquad (1.6.8)$$

And finally, by the theory of coordinate transformation, it can be shown that

$$d^3v = dv_x dv_y dv_z = J d\eta d\theta dv = v^2 \sin\theta d\eta d\theta dv, \qquad (1.6.9)$$

Figure 1.2
Spherical coordinates of a vector, $\mathbf{v}$.

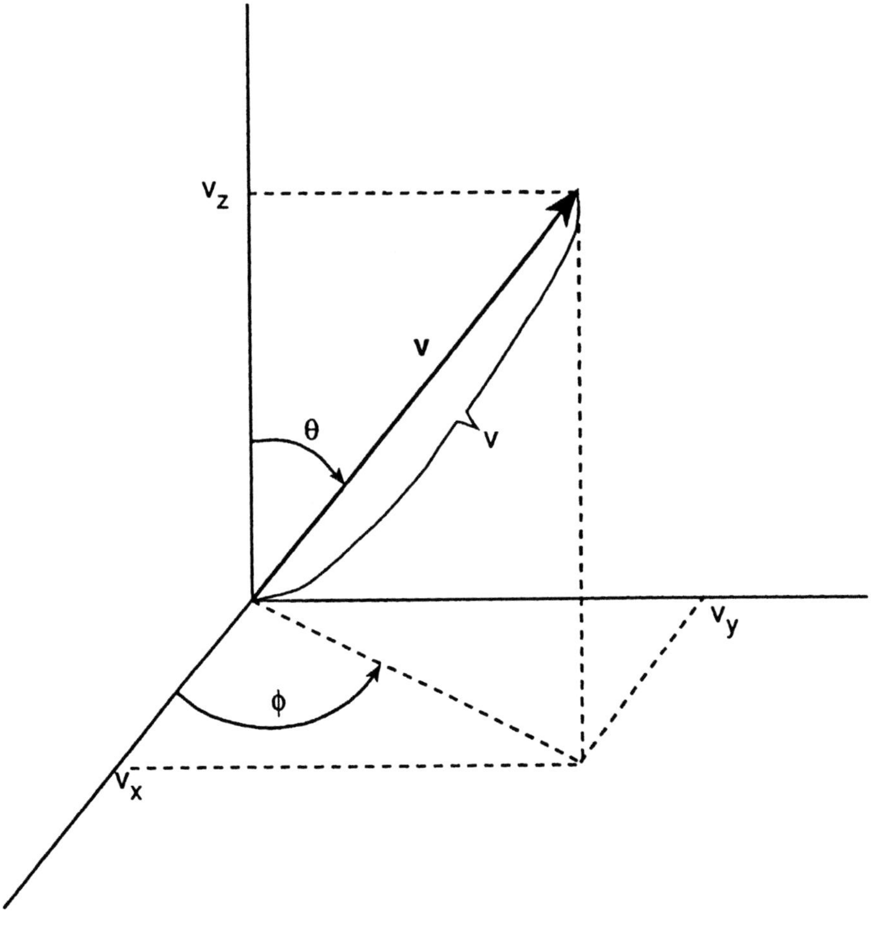

where $J = (v^2 \sin\theta)$ is the Jacobian of transformation from Cartesian to spherical coordinates. The ranges of η, θ, and v are

$$0 \le \eta \le 2\pi; \quad 0 \le \theta \le \pi; \quad 0 \le v < \infty. \tag{1.6.10}$$

We are now ready to compute the average speed $\langle v \rangle$, which is defined by the relation

$$\langle v \rangle = \left(\frac{m}{2\pi kT}\right)^{3/2} \int_0^{2} \pi \int_0^{\pi} \int_0^{\infty} e^{-mv^2/2kT} v^3 \sin\theta d\eta d\theta dv. \tag{1.6.11}$$

The exponential factor of the integrand in eq. (1.6.11) depends only on the magnitude v, and so the angular parts of the integral can be done immediately to obtain a factor of 4π. The remaining integral over dv can be evaluated by integrating by parts or from the general formula given in the exercise in Section 1.5. The final result is

$$\langle v \rangle = \sqrt{8kT/\pi m}. \tag{1.6.12}$$

For a nitrogen (N_2) gas at room temperature (300K), eq. (1.6.12) predicts that $\langle v \rangle \simeq 4.7 \times 10^4$cm/s. Thus, in a dilute nitrogen gas contained in a liter vessel (cubic for simplicity) a molecule on average will cross the vessel 4700 times each second. That molecules were zooming about their vessels at such high speeds must have been a fascinating discovery for the original kinetic theorists. The speed of sound (i.e., the speed of transmission of a density or pressure disturbance in air, about $\sim 10^4$cm/s, could then be understood in terms of molecular motions as well as could many other phenomena such as the Tyndall effect and thermal effusion.

Another quantity of physical interest is the distribution of speeds $\Phi(v)$, where

$\Phi(v)dv =$ the probability that a molecule picked at random from the system will have a speed lying between v and $v + dv$.

An alternative statement of the definition of $\Phi(v)dv$ is that this quantity represents the probability that the magnitude (or radial component) of $\mathbf{v}$ will lie between v and $v + dv$ regardless of the orientation (i.e., regardless of the values of η and θ) of $\mathbf{v}$. Thus, we can obtain $\Phi(v)dv$ by integrating $\phi(\mathbf{v})d^3v$ over all values of η and θ, that is,

$$\Phi(v)dv = \int_0^{2\pi} \int_0^{\pi} \phi(\mathbf{v}) v^2 \sin\theta d\eta d\theta dv, \tag{1.6.13}$$

from which it follows that

$$\Phi(v) = 4\pi v^2 \phi(\mathbf{v}), \tag{1.6.14}$$

because ϕ depends only on the magnitude of $\mathbf{v}$.

Still another way of thinking of Φ is that $N\Phi(v)dv$ represents the probable number of molecules of the system whose velocities lie in the spherical shell contained between two concentric spheres of radii v and $v + dv$, respectively, in velocity space. In Figure 1.3, $\Phi(v)$ in units of $(m/2kT)^{1/2}$ is plotted versus v in units of $(2kT/m)^{1/2}$. The distribution goes through a maximum value corresponding to the most probable speed, $\tilde{v}$, defined by the condition

$$\frac{d\Phi}{dv}(v) = 0 \quad \text{at} \quad v = \tilde{v}. \tag{1.6.15}$$

Using the form of Φ given by eq. (1.6.14) and differentiating with respect to v, we obtain the expression

$$\frac{d\Phi}{dv} = 4\pi\left(\frac{m}{2\pi kT}\right)^{3/2}\left[2v - \frac{mv^3}{kT}\right]e^{(-mv^2/2kT)} \tag{1.6.16}$$

that, when set equal to zero, gives

$$\tilde{v} = \left(\frac{2kT}{m}\right)^{1/2}. \tag{1.6.17}$$

Thus, the average molecular speed is somewhat larger than the most probable molecular speed. In fact,

$$\langle v\rangle/\tilde{v} = \sqrt{4/\pi} = 1.128. \tag{1.6.18}$$

The root mean square velocity v_{rms}, defined by the expression

$$\begin{aligned} v_{rms} &\equiv [\langle(\mathbf{v} - <\mathbf{v}\rangle)\cdot(\mathbf{v} - \langle\mathbf{v}\rangle)\rangle]^{1/2} \\ &= \sqrt{\langle v^2\rangle} = \sqrt{3kT/m}, \end{aligned} \tag{1.6.19}$$

is larger than the average and most probable speeds, although the difference is not great as witnessed by the ratios

$$v_{rms}/\tilde{v} = 1.224; \quad v_{rms}/\langle v\rangle = 1.085. \tag{1.6.20}$$

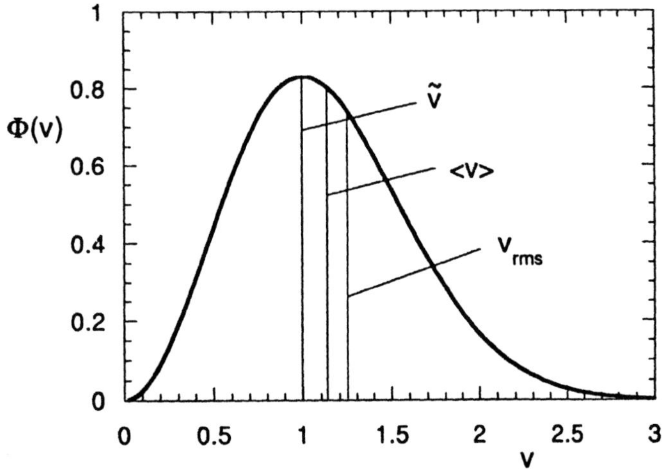

Figure 1.3
Distribution of speeds Φ in units of $(m/2kT)^{1/2}$ versus the speed v in units of $(2kT/m)^{1/2}$.

Example

Application of the Maxwell distribution. To attain very high temperatures one sometimes creates (by detonation for example) a shock front in a duct. Then, assuming the gases in the shock front are at local thermal equilibrium, one measures the shock front temperature by "spectral line broadening" due to the thermal motions of the molecule.

To illustrate this technique, let us assume that the gas contains atomic sodium that, when excited by high temperature collisions, emits radiation at the wavelength $\Lambda_o = 5889.97A$. If a very sensitive spectrometer were used to measure the intensity of the sodium emission in the shock front, one would observe emission not just at the wavelength, $\Lambda = \Lambda_o$, but in continuous range of Λ near Λ_o. In fact, measuring intensity I of emitted radiation versus wavelength, one would observe the distribution shown qualitatively in the diagram below.

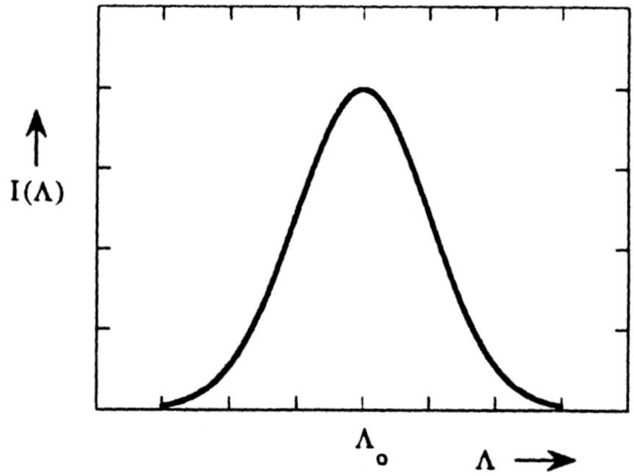

If local equilibrium holds, the intensity will be a Gaussian distribution centered on $\Lambda_o = 5889.97A$ with a half-width that is temperature dependent.

If the sodium atoms were motionless, the emission would be observed only at $\Lambda = \Lambda_o$. However, the thermal motion of an emitting atom will give rise to a Doppler shift in the wavelength observed by the spectrometer. Because there is a distribution of atomic velocities, there will be a distribution of Doppler shifts.

Suppose the spectrometer is located in the x direction relative to the shock front at the instant of the measurement of the radiation. If an atom in the shock front is moving with a velocity component v_x relative to the observer, then a photon emitted by the atom at a wavelength Λ_o will be observed to have the wavelength Λ, where

$$\Lambda = \Lambda_o\left(1 + \frac{v_x}{c}\right),$$

giving a Doppler shift

$$\Delta\Lambda = \Lambda - \Lambda_o = \Lambda_o \frac{v_x}{c}.$$

The quantity $c(=3 \times 10^{10} cm/s)$ is the speed of light. If the atom is in local thermal equilbrium at temperature T, the probable number of atoms whose x components of velocity will lie between v_x and $v_x + dv_x$ is proportional to $exp(-mv_x^2/2kT)$. But since $v_x = c(\Lambda - \Lambda_o)/\Lambda_o$, the probable number of photons emitted and, therefore, the intensity of radiation that will be observed in the wavelength range Λ and $\Lambda + d\Lambda$ obeys the proportionality

$$I(\Lambda) \propto [-\frac{mc^2}{2\Lambda_o^2 kT}(\Lambda - \Lambda_o)^2],$$

in agreement with the Gaussian form of $I(\Lambda)$ illustrated above. The half width $(\Delta\Lambda)_{1/2}$ that is, the root mean square average predicted by this distribution is $(\Delta\Lambda)_{1/2} = \Lambda_o(kT/m)^{1/2}/c = (\Lambda_o/(3)^{1/2})(v_{rms}/c)$. From this result we predict that if the observed half-width is 0.12A for the sodium spectral line $\Lambda_o = 5889.97A$, the temperature of the shock front is $10^5 K$.

1.7 Mean Free Path and Mean Collision Frequency for Rigid Sphere Molecules

In a dilute gas, collisions occur between molecules even though the mean distance of separation of the molecules is large compared to the range of the intermolecular interactions. An important parameter for characterizing dilute gas collisions is the mean free path λ, defined as the mean distance a molecule moves between collisions. If we assume the molecules to be perfectly elastic (rigid) spheres (like billiard balls) of diameter d, it is not too difficult to obtain an expression for λ as a function of the molecular diameter d and the number density n. In reality the forces among the molecules are more complicated than rigid sphere forces. In particular, there are attractive as well as repulsive interactions among the molecules. However, the main effect of including realistic interactions in the computation of λ is that the molecular diameter becomes a weak function of temperature that depends on the details of the intermolecular potential of interaction. Consequently, we adopt the simple rigid sphere model for discussion of the mean free path.

To avoid obscuring the physical principles with the mathematical manipulation let us present first a heuristic derivation of the formula for λ. Consider a one component dilute gas of rigid sphere molecules at a temperature T and a density n. Consider a small cylinder of length $\delta t v_{av}$, radius d, and fixed such that molecule 1 is moving with the average speed v_{av} along its axis. δt is a small interval of time and d is equal to the diameter of the gas molecules. Any molecules having their centers in the cylinder will be struck by molecule 1 in the time δt. The average number of such molecules is $n(\delta t v_{av} \pi d^2)$ so that the number of collisions per unit time that molecule 1 experiences is $\nu = n(v_{av} \pi d^2)$. This result is based on the tacit assumption that the other molecules of the system are motionless. If the relative motions of the molecules are accounted for properly, the collision frequency is given by

$$\nu = \sqrt{2} n v_{av} d^2 = \sqrt{2} n \left[\frac{8kT}{\pi m}\right]^{1/2} \pi d^2, \qquad (1.7.1)$$

the factor $\sqrt{2}$ arising from the relative motions of colliding molecules. The average distance λ a molecule travels between collisions, that is, the *mean free path*, is defined by the expression $\lambda = v_{av} \nu^{-1}$ and according to eq. (1.7.1) is

$$\lambda = \frac{1}{\sqrt{2} \pi d^2 n}. \qquad (1.7.2)$$

Equations (1.7.1) and (1.7.2) allow us to estimate mean free paths and collision frequencies from the thermodynamic state (T and n) and the masses and diameters of the molecules of the gas. The molecular diameters can be estimated from solid- state densities, equation of state data, transport coefficients, etc.

Before going on with the discussion of λ, we present a rigorous derivation of eq. (1.7.2) for rigid spheres, thus accounting for the factor $\sqrt{2}$. Suppose molecules 1 and 2 are approaching each other with a relative velocity $\mathbf{v}_{21} = \mathbf{v}_2 - \mathbf{v}_1$. For them to actually collide during the short time interval, their line of centers must not be separated by a distance greater than $|\mathbf{v}_{21} \cdot \hat{\mathbf{r}}|\delta t$, where $\hat{\mathbf{r}}$ is a unit vector lying along the line of centers of the molecules at contact. Thus, if a coordinate system is centered on molecule 1 and the orientation of the z axis is aligned with $\mathbf{v}_{21}$, molecule 2 will collide with molecule 1 if it is in the volume element shown in Figure 1.4 and if $\mathbf{v}_{21} \cdot \hat{\mathbf{r}} < 0$. The volume of the element is $d^2 d\Omega |\mathbf{v}_{21} \cdot \hat{\mathbf{r}}|\delta t$, where $d\Omega (= \sin\theta d\theta d\eta)$ is an element of solid angle.

Calculated first is the probable number of collisions that molecule 1 with a velocity in the range $\mathbf{v}_1$ to $\mathbf{v}_1 + d\mathbf{v}_1$ will have with molecules whose velocities are in the range $\mathbf{v}_2$ to $\mathbf{v}_2 + d\mathbf{v}_2$ during the time interval δt. This probable number is equal to the probability that the velocity of molecule 1 is in the range $\mathbf{v}_1$ to $\mathbf{v}_1 + d\mathbf{v}_1$ times the probable number of molecules in the volume element $d^2 |\mathbf{v}_{21} \cdot \hat{\mathbf{r}}\ |r d\Omega \delta t$ times the probability that the velocity of one of these molecules is in the range $\mathbf{v}_2$ to $\mathbf{v}_2 + d\mathbf{v}_2$, that is, it equals

$$(\phi(\mathbf{v}_1)d^3v_1)(nd^2|\mathbf{v}_{21}\cdot\hat{\mathbf{r}}|d\Omega\delta t)(\phi(\mathbf{v}_2)d^3v_2). \tag{1.7.3}$$

The average number δn_{coll} of collisions of molecule 1 during the time δt is obtained by integrating over the variables $\mathbf{v}_1$, $\mathbf{v}_2$ and Ω in eq. (1.7.3), keeping in mind the constraint placed on $\mathbf{v}_{21}\cdot\hat{\mathbf{r}}$ to assure a collisional event. The average collision frequency $\nu(\equiv \delta n_{\text{coll}}/\delta t)$ is then

$$\nu = d^2 n \underset{\mathbf{v}_{21}\cdot\hat{\mathbf{r}}<0}{\int\int\int} \phi(\mathbf{v}_1)\phi(\mathbf{v}_2)|\mathbf{v}_{21}\cdot\hat{\mathbf{r}}|d\Omega d^3v_1 d^3v_2, \tag{1.7.4}$$

where the requirement that molecules 1 and 2 be approaching each other is satisfied by the condition $\mathbf{v}_{21}\cdot\hat{\mathbf{r}} < 0$.

Setting $d\Omega = \sin\theta d\theta d\eta$, where θ and η are shown in Figure 1.4, we see that $\mathbf{v}_{21}\cdot\hat{\mathbf{r}} = |\mathbf{v}_{21}|\cos\theta$. Thus, the constraint $\mathbf{v}_{21}\cdot\hat{\mathbf{r}} < 0$ implies that $\cos\theta < 0$ or the range of θ is from $\pi/2$ to π. Equation (1.7.4) can be rewritten in the form

$$\begin{aligned}\nu &= nd^2 \int_o^{2\pi} d\eta \int_{\pi/2}^{\pi} d\theta \sin\theta\cos\theta \int\int \phi(\mathbf{v}_1)\phi(\mathbf{v}_2)|\mathbf{v}_{21}|d^3v_1 d^3v_2 \\ &= \pi n d^2 \int\int \phi(\mathbf{v}_1)\phi(\mathbf{v}_2)|\mathbf{v}_{21}|d^3v_1 d^3v_2.\end{aligned} \tag{1.7.5}$$

To finish the evaluation of eq. (1.7.5) it is convenient to transform the variables $\mathbf{v}_1$ and $\mathbf{v}_2$ to the new variables

$$\mathbf{G} \equiv \frac{1}{2}(\mathbf{v}_1 + \mathbf{v}_2), \quad \mathbf{g} \equiv \mathbf{v}_2 - \mathbf{v}_1 \tag{1.7.6}$$

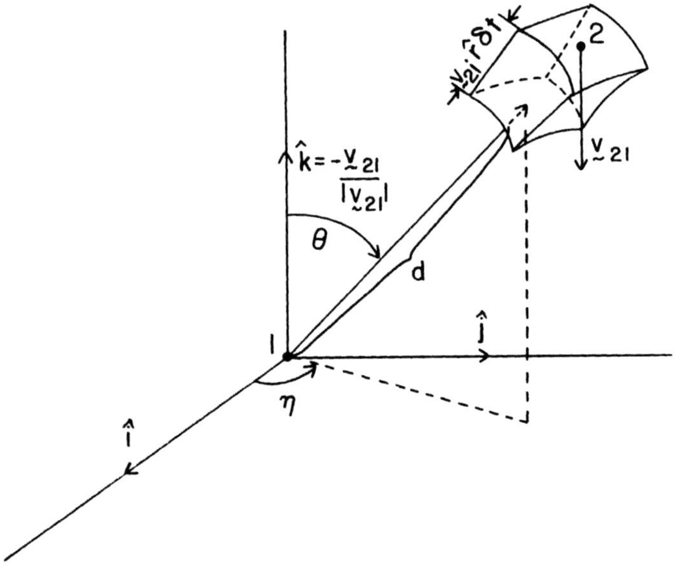

Figure 1.4
Volume element in which molecule 2 must lie if it is to collide with molecule 1 in the time δt

The quantity $\mathbf{g}$ is equal to $\mathbf{v}_{21}$ and has been introduced only to simplify notation somewhat. One can easily verify that

$$v_1^2 + v_2^2 = g^2/2 + 2G^2 \tag{1.7.7}$$

and, from the theory of coordinate transformations, that

$$d^3v_1 d^3v_2 = d^3g d^3G. \tag{1.7.8}$$

Exercise

Verify eq. (1.7.7) from the definitions (1.7.6). Using a Cartesian coordinate system, verify eq. (1.7.8), that is, show that the Jacobian of the transformation $\mathbf{v}_1, \mathbf{v}_2 \to \mathbf{G}, \mathbf{g}$ is unity.

At thermal equilibrium the distribution functions $\phi(\mathbf{v}_1)$ and $\phi(\mathbf{v}_2)$ are Maxwell distributions so that, with the aid of eqs. (1.7.7) and (1.7.8), we may reduce eq. (1.7.5) to the form

$$\nu = \pi n d^2 \left(\frac{m}{2\pi kT}\right)^3 \left[\int e^{-(mG^2/kT)} d^3G\right]\left[\int e^{-(mg^2/4kT)} g d^3g\right]. \tag{1.7.9}$$

If d^3G and d^3g are expressed in polar coordinates, the angular integrals can be done immediately yielding

$$\nu = \pi n d^2 \left(\frac{m}{2\pi kT}\right)^3 \left[4\pi \int_0^\infty e^{-(mG^2/kT)} G^2 dG\right]\left[4\pi \int_0^\infty e^{-(mg^2/4kT)} g^3 dg\right]. \tag{1.7.10}$$

The remaining integrals are of the form evaluated in the exercise in Section 1.5. Using the results given in the exercise, we recover eqs. (1.7.1) and (1.7.2), and thus confirm the $\sqrt{2}$ factor.

Consider next a two component gas of rigid spheres. In this case it can be shown that the frequency of collision of a molecule of type A with molecules of type B is given by

$$\nu_{AB} = n_B \left(\frac{d_A + d_B}{2}\right)^2 \iiint_{\mathbf{v}_{AB}\cdot\hat{\mathbf{r}}<0} \phi(\mathbf{v}_A)\phi(\mathbf{v}_B)|\mathbf{v}_{AB}\cdot\hat{\mathbf{r}}| d\Omega d^3v_A d^3v_B, \tag{1.7.11}$$

which may be integrated to yield

$$\nu_{AB} = \pi n_B \left(\frac{d_A + d_B}{2}\right)^2 \left(\frac{8kT}{\pi\mu_{AB}}\right)^{1/2}, \tag{1.7.12}$$

where $\mu_{AB} = m_A m_B/(m_A + m_B)$.

By introducing an intuitively attractive assumption, we can show that the probability that a particle in a dilute gas will move a distance x without undergoing a collision is

$$f = e^{-x/\lambda}, \tag{1.7.13}$$

where λ is the mean free path. Similarly, the probability that a particle will move without collision for a time t is

$$f = e^{-\nu t} = e^{-v_{av} t/\lambda}. \tag{1.7.14}$$

The assumption used in deriving eq. (1.7.13) is that the probability of a collision of a molecule in moving the distance dx is $-\alpha dx$, where α is a proportionality constant. Thus, if $f(x)$ is the probability that no collision has occurred after the molecule has moved a distance x, the probability that it has not had a collision after moving a distance $x + dx$ is

$$f(x + dx) = (1 - \alpha dx) f(x).$$

Rearranging this expression and passing to the limit $dx \to 0$, we find the differential equation

$$f(x) = -\frac{1}{\alpha} \lim_{dx \to 0} \frac{f(x + dx) - f(x)}{dx} = -\frac{1}{\alpha}\frac{df}{dx},$$

whose solution, with the condition $f(0) = 1$, is

$$f(x) = e^{-\alpha x}.$$

The probability that a collision occurs when a particle moves a distance x is $p(x) = 1 - f(x)$, so that the probability of a collision occurring between x and $x + dx$ is $dp = +\alpha f(x) dx$. The mean free path λ of particle is given by

$$\lambda = \int_0^\infty x dp = \alpha \int_0^\infty x e^{-\alpha x} dx = \alpha^{-1},$$

from which eq. (1.7.13) follows.

It is interesting to compare for dilute gases the mean free path λ with the average intermolecular separation $l \equiv (V/N)^{1/3} = n^{-1/3}$ and the range r_f typical of intermolecular forces. In nonpolar gases, the intermolecular forces are negligible for intermolecular separations greater than about $2d$. In Table 1.1, λ and l are given for nitrogen gas at 300°C and at pressures of 10^{-3} atm and 1 atm, respectively.

Several points can be made concerning the entries in Table 1.1. Because the value of l is much greater than $2d$ for the pressures considered, the interaction will be negligible on the average and

TABLE 1.1 Collison Frequencies, Mean Free Paths, and Mean Molecular Spacings for N_2 at 300K

P (atm)	ν (collisions/s)	l (cm)	λ (cm)
10^{-6}	2.03×10^3	3.46×10^{-5}	$2.34 \times 10^{+1}$
10^{-3}	2.03×10^6	3.46×10^{-6}	2.34×10^{-2}
1	2.03×10^9	3.46×10^{-7}	2.34×10^{-5}

Molecular diameter is assumed to be $d = 2 \times 10^{-8}$ cm.

the gas will behave as an ideal gas. The mean free paths in the table are much longer than the mean molecular spacing. In fact, for gases behaving ideally, the ordering

$$\lambda > l > r_f$$

is generally observed. Satisfaction of the condition $l > r_f$, being necessary for ideal behavior, usually guarantees the ordering $\lambda \gg l$. We note that a molecule, even in a gas behaving ideally, undergoes a large number of collisions in a short time. Indeed, in moving 10 cm, an N_2 molecule in a 1 atm gas will collide $10\text{cm}/\lambda = 10\text{cm}/2.34 \times 10^{-5} cm = 4.38 \times 10^5$ times. This result, incidentally, implies that our heuristic derivation of eq. (1.2.9) for the pressure of a gas is only valid at extremely low pressures, because we assumed in the derivation that a molecule could move from one side of the box to another without undergoing a collision with another molecule. In the next paragraph we present a rigorous derivation that fortunately gives the same result as eq. (1.2.9).

Example

What are the probabilities that in nitrogen gas at 300K and 10^{-5} atm and 10^{-7} atm, respectively, a molecule would travel all the way across a 10 cm vessel without collision?

Under these conditions the mean free paths are 2.34 and 234 cm, respectively. Thus, from eq. (1.7.13), we estimate that the desired probabilitites are

$$f = e^{-10/2.34} = 0.0139$$

and

$$f = e^{-10/234} = 0.956.$$

Suppose the gas is in a cube with perfectly reflecting walls and consider a face of the cube oriented in the $x$direction. In a very small duration of time δt, a molecule with the $x$component of velocity equal to v_x (or in the range v_x and $v_x + dv_x$) will collide with the wall of the cube only if it is within a distance $v_x \delta t$ of the wall, that is, only if it is in a volume increment equal to $V^{2/3} v_x \delta t$. The probability that one of these molecules has a velocity such that v_x is between v_x and $v_x + dv_x$ is $\phi_x(v_x)dv_x$. Thus, the probable number of molecules in the volume increment and with v_x in the given velocity range is

$$n(V^{2/3} v_x \delta t)\phi_x(v_x)dv_x. \tag{1.7.15}$$

If δt is sufficiently small, $v_x \delta t$ will be shorter than a mean free path and a molecule can collide with the wall without colliding with another molecule. In this case the momentum change at the wall will be $2mv_x$ for a molecule in the volume increment with the x component of velocity equal to v_x if v_x is positive. Thus, the total average momentum change at the wall during δt will be the product of $2mv_x$ and eq. (1.7.15) integrated over dv_x subject to the condition $v_x > 0$. The pressure (momentum change per area per unit time) is obtained by dividing the total average momentum change by $V^{2/3}\delta t$. The final result is

$$P = n \int_0^\infty \phi_x(v_x) 2mv_x^2 dv_x = nkT, \tag{1.7.16}$$

the same expression as eq. (1.2.9).

In closing this section let us mention that free path is very important in distinguishing between fluid flow that may be characterized by hydrodynamics (in which case the mean free path λ is short compared to the linear dimensions of the vessel, $V^{1/3}$) and the fluid flow that has no hydrodynamical description (in which case $\lambda > V^{1/3}$), known as Knudsen flow. Also, transport properties of dilute gases are intimately related to the mean free path.

1.8 Effusion

Consider an element of area dA whose normal is $\hat{\varepsilon}$. If a molecule with velocity $\mathbf{v}$ is moving toward dA, that is, if $\hat{\varepsilon} \cdot \mathbf{v} < 0$, then it will reach dA in the time δt if it is in the cylinder shown in Figure 1.5. We wish to determine $H_{\hat{\varepsilon}}(\mathbf{v})$, where

$H_{\hat{\varepsilon}}(\mathbf{v})d^3v =$ the number of molecules with the velocity between $\mathbf{v}$ and $\mathbf{v} + d\mathbf{v}$, striking a surface with an orientation $\hat{\varepsilon}$ per unit area per unit time. (1.8.1)

The volume of the cylinder shown in Figure 1.5 is $|\hat{\varepsilon} \cdot \mathbf{v}|\delta t A$. Thus, the probable number of molecules, in the specified velocity range, striking dA (we are considering the back side of dA to be another surface) in the time interval δt is the product of $n|\hat{\varepsilon} \cdot \mathbf{v}|\delta t dA$, the probable number of molecules in the cylinder, and $\phi(\mathbf{v})d^3v$, the

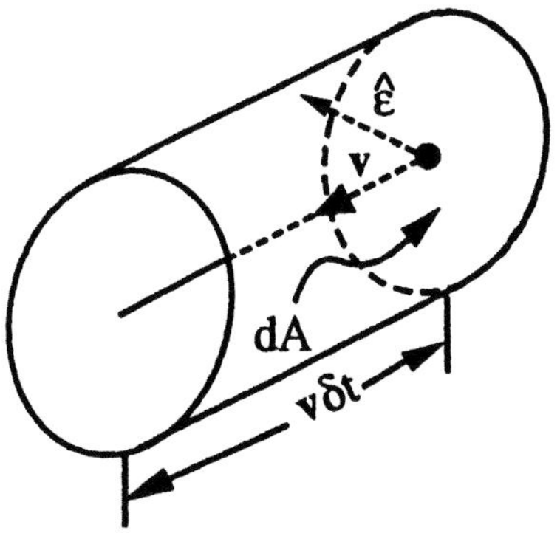

Figure 1.5
Cylinder containing molecules having velocity $\mathbf{v}$ and that will strike the surface dA in the time δt.

probability that the molecules have a velocity in the specified range. Dividing this product by $\delta t dA$, we obtain the desired result

Figure 1.6
Effusion.

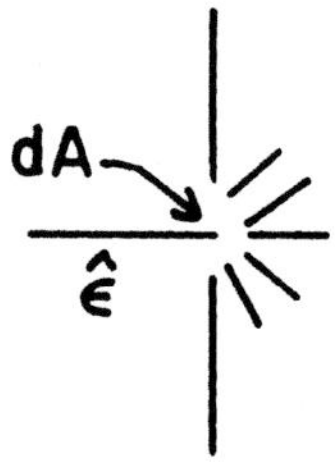

$$H_{\hat{\varepsilon}}(\mathbf{v})d^3v = n|\mathbf{v}\cdot\hat{\varepsilon}|\phi(\mathbf{v})d^3v, \quad \mathbf{v}\cdot\hat{\varepsilon} < 0, \tag{1.8.2}$$

where the condition $\mathbf{v}\cdot\hat{\varepsilon} < 0$ assures us that the molecules counted are those striking the side of dA indicated by the direction of $\hat{\varepsilon}$.

Suppose the surface dA is a small hole in a thin plate to the left of which is a dilute gas at temperature T and density n and to the right of which is a vacuum. The gas particles will escape from the hole in directions and at a rate determined by the Maxwell distribution of velocities. Such a process is called effusion and is illustrated schematically in Figure 1.6.

The total flux or total rate per unit area of effusive escape of the molecules through dA can be obtained by integrating eq. (1.8.2) over all values of $\mathbf{v}$ with the condition that $\mathbf{v}\cdot\hat{\varepsilon} < 0$, that is, that the molecules be moving from the gas to the vacuum side of the wall. Expressing d^3v in spherical coordinates in a Cartesian frame whose polar axis lies along $\hat{\varepsilon}$, we obtain the following expression for the escape flux F:

$$F = \int_{v\cdot\hat{\varepsilon}<0} H_{\hat{\varepsilon}}(\mathbf{v})d^3v = n\int_{\pi/2}^{\pi}\int_0^{2\pi}\int_0^{\infty} |\cos\theta| v\phi(\mathbf{v})v^2\sin\theta\, d\theta\, d\eta\, dv$$

$$= \frac{n}{2}\left(\frac{2kT}{\pi m}\right)^{1/2} = \frac{n}{4}\langle v\rangle. \tag{1.8.3}$$

Example

Let us determine the distribution of velocities of the mole cules effusing from a small hole as shown in Figure 1.6. We want to find $\phi_E d^3v$, the fraction of effusing particles having velocities between $\mathbf{v}$ and $\mathbf{v} + d\mathbf{v}$. The total number of particles effusing in a short time δt is

$$F dA\delta t = \frac{n}{4}\langle v\rangle dA\delta t \tag{1.8.3a}$$

whereas the number escaping ($\mathbf{v}\cdot\varepsilon < 0$) with velocities between $\mathbf{v}$ and $\mathbf{v} + d\mathbf{v}$ is

$$H_{\hat{\varepsilon}}(\mathbf{v})d^3v\, dA\delta t. \tag{1.8.3b}$$

The desired fraction is equal to the ratio of eq. (1.8.3b) to eq. (1.8.3a), that is,

$$\phi_E d^3v = 4\frac{v}{\langle v\rangle}|\cos\theta|\phi(v)d^3v, \tag{1.8.3c}$$

where θ is the angle between $\mathbf{v}$ and $\hat{\varepsilon}$ and is restricted to the range $\frac{\pi}{2} < \theta < \pi$ to insure that the molecules move out through the small hole. ϕ_E is seen to obey a "cosine law."

It is interesting to compare the average speed v_E of the effusing beam with $\langle v \rangle$, the average speed of the molecules in the vessel. The quantity v_E is given by

$$v_E = \int\limits_{\frac{\pi}{2}<\theta<\pi} 4\frac{v^2}{\langle v \rangle}|cos\theta|\phi(v)d^3v$$

$$= \frac{8\pi}{\langle v \rangle}\int_0^1\int_0^\infty \mu v^4 \phi(v) d\mu dv$$

$$= v_{rms}^2/\langle v \rangle = (3\pi/8)\langle v \rangle.$$

The speed v_E is thus about 19% higher than $\langle v \rangle$, the reason being that proportionally more high speed molecules escape through the effusion hole than low speed ones during the same period of time.

One of the implications of eq. (1.8.3) was known over a century ago as *Graham's law of diffusion, that stated that the rates of effusion of different gases at the same temperature and pressure vary inversely as the square roots of the molecular weights or mass densities.* This is the basis of the effusiometer devised by Bunsen to measure relative densities of two gases subjected to the same temperature and pressure. One lets a fixed volume of gas 1 effuse through a small hole in the time t_1. The same amount of the second gas is then let effuse through the same size hole to displace the same volume. The time for the second gas to flow is t_2. For equal volume displacements $t_1 F_1 = t_2 F_2$, or

$$\frac{t_1}{t_2} = \left(\frac{m_1}{m_2}\right)^{1/2} = \left(\frac{nm_1}{nm_2}\right)^{1/2} = \left(\frac{\rho_1}{\rho_2}\right)^{1/2}. \qquad (1.8.4)$$

Because the temperature and pressure of each gas are the same, the number densities are equal too, and so the ratio of measured displacement times t_1 and t_2 is the ratio of the mass densities ρ_1 and ρ_2 of the gases.

Perhaps the most important use of the thermal effusion effect is isotope separation. Consider a two component dilute gas system at temperature T and pressure P. If a small hole of area δA in the containing vessel allows effusive flow, the fluxes of components 1 and 2 are, respectively, $F_1 = n_1(kT/2\pi m_1)^{1/2}$ and $F_2 = n_2(kT/2\pi m_2)^{1/2}$. If the effusing molecules are pumped off into a collection vessel as they appear, in a short time δt the number of molecules of component 1 effusing will be $F_1\delta t\delta A = N_1'$ and of component 2 will be $F_2\delta t\delta A = N_2'$. Thus, the ratio of the densities of Components 1 and 2 in the collection vessel will be

$$\frac{n_1'}{n_2'} = \frac{N_1'}{N_2'} = \frac{n_1}{n_2}\left(\frac{m_2}{m_1}\right)^{1/2}, \qquad (1.8.5)$$

with the result that the effused gas is richer in the component having the smaller molecular weight (this enrichment arises from the fact that the lighter molecules have a larger average speed than the heavier ones). The enrichment implied by eq. (1.8.5) provides the basis of isotopic separation processes through gaseous effusion. Of course, if an appreciable quantity of gas effuses from the original vessel the overall ratio n_1'/n_2' will not be as large as predicted by eq. (1.8.5) because the densities of components 1 and 2 will decrease and their ratio will change with time. For example, if half of the gas initially in the container effuses, then the overall ratio n_1'/n_2' is given by

$$\frac{n_1'}{n_2'} = \frac{n_1}{n_2}\left\{0.69\left[\left(\frac{m_2}{m_1}\right)^{1/2} - 1\right] + 1\right\}, \tag{1.8.6}$$

where we have chosen $m_1 < m_2$.

During World War II, it was desired to enrich the U^{235} content of naturally occurring uranium (in natural uranium the ratio $n_{U^{235}}/n_{U^{238}} = 1/140$) until the mixture of U^{235} and U^{238} had 90%U^{235}. By converting the uranium into the uranium fluoride UF_6, which has a vapor pressure of 1 atm at 56°C, a gaseous effusion separation process could be used. For $U^{235}F_6$ and $U^{238}F_6$, the ratio $(m_2/m_1)^{1/2} \simeq 1.0043$, and so in a one stage effusion process in which half of the initial gas effuses, the separation factor is, according to eq. (1.8.6),

$$\frac{n_1'/n_2'}{n_1/n_2} = 1.003. \tag{1.8.7}$$

Thus, if $n_1/n_2 = 1/140$ initially, then $n_1'/n_2' = 1.003/140$. This represents a rather meager enrichment of the higher component. However, the actual process used to obtain the desired 90%U^{235} product made use of a multistage or cascade system of the type illustrated in Figure 1.7. A feed gas of initial composition n_1/n_2 enters stage 1 and effuses through a porous barrier (an inert material with tiny holes). In the uranium separation process used during the war, half of the feed gas at each stage was allowed to effuse through the porous barrier and become the feed gas of the next higher stage, while the other (impoverished) half was returned to the feed of the next lower stage. The theoretical single stage separation factor of uranium hexafluoride is 1.003 as given by eq. (1.8.7). However, due to nonidealities (back diffusion, imperfect mixing, and imperfections in the barrier), the experimental factor is about 1.0014. The number of stages used to obtain the desired enrichment was about 4000.

Another interesting effusion effect is the so-called transpiration effect. Suppose Vessels 1 and 2 containing the same kind of dilute gas and being held at temperature T_1 and T_2, respectively, are connected by a tube of constant cross section A. If the flow through the tube is effusive, then at steady state the rate, $n_1\langle v_1\rangle A/4$, at which molecules enter the tube from vessel 1 equals the rate,

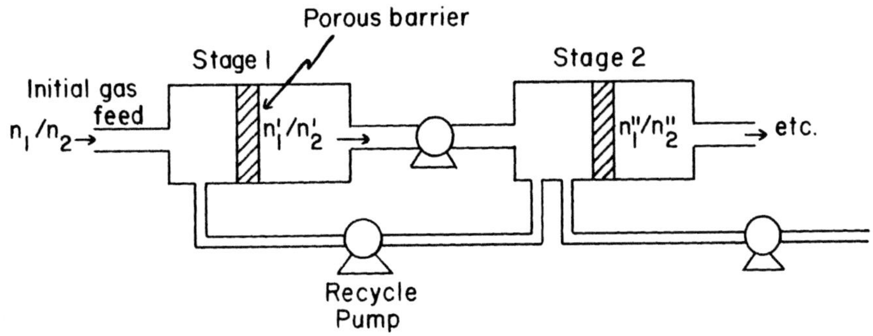

Figure 1.7
Schematic of a multistage effusion process for separating isotopes in the gas phase.

$n_2\langle v_2\rangle A/4$, at which molecules enter the tube from vessel 2. Or, because $\langle v\rangle = (8kT/\pi m)^{1/2}$ and because $n = P/kT$ for a dilute gas, the flux balance can be written in the form

$$\frac{P_1}{\sqrt{T_1}} = \frac{P_2}{\sqrt{T_2}}. \tag{1.8.8}$$

Thus, the steady-state pressures of the two vessels will be different if the temperatures are different. This effect is often important in manometers designed to measure, by displacement of a liquid at room temperature, the vapor pressure of a low temperature liquid or solid. For example, if T_1 is 100K and T_2 is 300K, then in the effusion range of pressures, $P_2/P_1 = \sqrt{3} = 1.73$, so that P_2, the pressure observed by a room temperature (300K) manometer, will actually be 1.73 times as large as the vapor pressure P_1 of a system held at 100K.

Let us close this section with a brief discussion of sources of nonidealities in effusive processes. As is clear from the derivation of eq. (1.8.3), we have made the assumption that a gas molecule entering the small effusion hole will not be reflected back (back diffusion) into the system. For this to be true, it is necessary for the mean free path to be long compared to the thickness of the wall in which the hole is located, that is, the wall must be thin compared to a mean free path, as was more than likely the case in the uranium separation described above, thermal separation will nevertheless occur. It will simply be a much more complicated process (involving the mean free path, pore dimensions, molecule–pore wall collisional characteristics, etc.) than the effusion process described below.

The diameter (or whatever dimension characterizes the size of the hole) of the hole must also be small compared to the mean free path for purely effusive flow to occur. When the mean free path is long compared to the diameter of the hole, the loss of a molecule through the hole will disturb the molecules in the vicinity of the hole because the distance over which the molecules affect each other, namely, the mean free path, is long compared to the diameter of the hole. Thus, the loss of a molecule through the hole will not enhance the probability that another molecule will escape through the hole.

On the other hand, suppose the mean free path is short compared to the diameter of the hole. The molecules are thermalized over a few mean free paths and, therefore, in local cells larger than a few cubic mean free paths the average velocity of a particle will be approximately zero due to the random changes in the direction of the velocity arising from collisions with background molecules. If a molecule near the hole is lost through the hole rather than rebounding off the wall, then the molecule it should have collided with after the rebound will have nothing to cancel out its outward bound motion. Thus, the second molecule will escape out of the hole as a result of the escape of the first molecule. The process is repeated until finally a large number of molecules are moving simultaneously through the hole. This is a hydrodynamical flow rather than an effusive flow (to visualize the extreme hydrodynamic limit imagine removal of an entire wall of the vessel).

Knudsen studied effusion through platinum sheets from 0.0025 to 0.005 mm thick, the 0.0025 mm sheet having a hole $5 \times 10^{-6}\text{cm}^2$ in area and the 0.005 mm sheet having a hole $6.6 \times 10^{-5}\text{cm}^2$ in area. He concluded that the measured effusion flux agreed very well with the theory *when the mean free path was not less than 10 times the diameter of the hole.* For somewhat smaller mean free paths the flux was somewhat greater than that predicted for effusion, and finally at sufficiently high pressures the flow rate became that predicted from hydrodynamics.

1.9 Evaporation and Chemical Reaction Rates

Consider a solid or a liquid in equilibrium with its vapor. The average flux F of vapor molecules impinging on the solid or liquid surface is given by

$$F = n_v \left(\frac{kT}{2\pi m}\right)^{1/2} = P_v \left(\frac{1}{2\pi mkT}\right)^{1/2}, \tag{1.9.1}$$

where P_v is the vapor pressure of the solid or liquid at the temperature T. At equilibrium the rate of condensation is equal to the rate of evaporation, so that eq. (1.9.1) would be equal to the rate of evaporation if every impinging vapor molecule were condensed. In reality only a fraction α of impinging vapor molecules condense so that the evaporation flux F_e is

$$F_e = \alpha P_v (1/2\pi mkT)^{1/2}. \tag{1.9.2}$$

The factor α is known as the coefficient of evaporation. Because $\alpha \leq 1$, substitution of unity for α in eq. (1.9.2) provides a simple calculation of the upper limit of the evaporation rate of a solid or liquid at a given temperature. Where α is known evaporation rate measurements can be used to determine vapor pressures.

Theories of the quantity α are not well developed so that one must either approximate it by unity or determine it experimentally. In one such experiment, Brönsted and Hevesy (1920, 1922) found that α is approximately 1 for liquid mercury and for solid mercury below $-140°C$, whereas for solid mercury above $-100°C$, α ranged from about 0.90 to 0.92. The rate of evaporation was obtained by measuring the rate at which mercury is deposited on the walls of a high vacuum glass vessel held at liquid air temperature and containing a thermostatted sample of mercury. In such a system, any molecules escaping the surface of the condensed mercury would travel to and attach to the wall without collision with other molecules (low temperature walls provide a cryogenic pump that would keep the pressure of vapor in the vessel very low).

Example

Transmission electron microscopy can be performed on sufficiently thin slices of biological substances. It is desirable that a sample of muscle be studied in the hydrated form. To prevent dehydration the sample must be processed at low temperature.

Suppose a sample of muscle 2000 Å thick and 1 mm length and width is going to be observed under the microscope. If the total time of sampling and observation is to be 45 min, estimate the maximum temperature at which the sample can be handled (under vacuum or in inert, dry atmosphere) in order that less than 25% of the water in the sample will be lost by evaporation. Muscle is about 75% water by weight.

A number of approximations will be made to obtain the desired estimate. First, we assume that the water occupies a volume fraction equal to its mass fraction and that the area fraction of water on an exposed surface is equal to the volume fraction of water. Second, we assume that the edge evaporation can be neglected (the area of the edges is only $8 \times 10^{-6}cm^2$ compared to the top and bottom areas of $10^{-2}cm^2$ each). Third, we assume that the exposed surface area remains constant as the evaporation front moves through the sample and that the exposed protein and fat pose no barrier to evaporative losses. Fourth, we assume that the water in the meat behaves ideally so that the vapor pressure of ice in meat is the same as that of pure ice at the same temperature. Fifth, we assume that evaporation occurs isothermally. The rate of change of thickness $\dot{L}$ of a planar ice surface is related to the evaporation flux by the expression

$$\dot{L} = \alpha F m / \rho_s = (\alpha P_v / \rho_s)(m / 2\pi k T)^{1/2}, \qquad (1.9.3)$$

where m is the mass of a molecule of water and ρ_s is the mass density of ice. (This expression follows from the mass balance $\dot{L}\delta A\delta t\rho_s = mF_e\delta t\delta A$, where δA is an element of surface area and δt is an increment of time.) The International Critical Tables give the following expression for the vapor pressure of low temperature ice:

$$\log_{10} P_v = -\frac{2445.5646}{T} + 8.2312\log_{10} T - 0.01677006T + 1.20514 \times 10^{-5}T^2 - 6.757169 \tag{1.9.4}$$

where P_v is in mm Hg and T is in K. With the aid of this expression the maximum value of $\dot{L}$ (obtained by setting $\alpha = 1$) has been computed and listed in Table 1.2 for various temperatures.

With the assumptions outlined above, the maximum temperature of operation of the microscope may be estimated from

$$0.25(200\text{Å}) = (45\ \text{min})(2\dot{L}_{max}(T)).$$

We find $T = 150.7°K$.

Exercise

Plot the maximum time available for experimentation versus temperature of the microscope sample chamber for the experiment described in the above example.

Turning now to another application of kinetic theory, let us consider a gaseous species A which undergoes a reaction on a solid surface with a species B present on the solid surface (B may be the surface molecules of a homogeneous solid or the molecules of

TABLE 1.2 Thickness Loss Rate of Ice Estimated from Eq. (1.9.3) With $\alpha = 1$

T (K)	P_v (dyne/cm^2)	$\dot{L}_{max}$(Å/s)
160	7.07×10^{-4}	1.05
155	2.03×10^{-4}	0.305
150	5.37×10^{-5}	8.20×10^{-2}
145	1.30×10^{-5}	2.01×10^{-2}
140	2.83×10^{-6}	4.47×10^{-3}
135	5.52×10^{-7}	8.89×10^{-4}
130	9.51×10^{-8}	1.56×10^{-4}
120	1.83×10^{-9}	3.12×10^{-6}
110	1.73×10^{-11}	3.09×10^{-8}
100	6.56×10^{-14}	1.23×10^{-10}
90	7.37×10^{-17}	1.45×10^{-13}

an absorbed species). We wish to compute the reaction rate for a heterogeneous reaction under the following conditions.

a) Species B is distributed uniformly on the solid surface with a surface fraction w_B.

b) A gaseous molecule of type A will react with one of type B upon collision with the solid surface only if the magnitude of its velocity component normal to the surface is greater than v^*, that is, the kinetic energy of the colliding molecule relative to the surface must be greater than $E^* = (1/2)m_A(v*)^2$ for the reaction to occur.

From eq. (1.8.2), we find that the flux (number per unit area per unit time) of gaseous molecules colliding with a surface with impinging velocities great enough to react with B (that is, with velocities such that $|\mathbf{v}\cdot\hat{\varepsilon}| > v^*$, where $\hat{\varepsilon}$ is the normal to the surface) is given by

$$
\begin{aligned}
F_A^* &= \int_{\mathbf{v}\cdot\hat{\varepsilon}<-\mathbf{v}*} H_{\hat{\varepsilon}} d^3v \\
&= n_A\left(\frac{m_A}{2\pi kT}\right)^{3/2} \int_{-\infty}^{\infty}\int_{-\infty}^{\infty}\int_{-\infty}^{-v^*} |v_z| e^{-(m_A/2kT)[v_x^2+v_y^2+v_z^2]} dv_x dv_y dv_z \qquad (1.9.5) \\
&= n_A\left(\frac{kT}{2\pi m_A}\right)^{1/2} e^{-E^*/kT}.
\end{aligned}
$$

To perform the integrals of F_A^*, we introduced a Cartesian coordinate system with the zaxis along the surface normal $\hat{\varepsilon}$. If the area of the reaction surface is A, then the rate R_A, at which molecules of type A react with species B is the product of $F_A^* A$, the number of surface collisions capable of reaction, and w_B, the probability that a colliding gaseous molecule will hit a molecule of species B. Thus,

$$
R_A = k_{AB} n_A A, \qquad (1.9.6)
$$

where k_{AB} is a rate constant of the form

$$
k_{AB} = w_B \left(\frac{kT}{2\pi m_A}\right)^{/2} e^{-E^*/kT}. \qquad (1.9.7)
$$

For a fixed concentration w_B of B, k_{AB} has the so-called Arrhenius temperature dependence (ignoring the pre-exponential factor of $T^{1/2}$) with an *activation energy* E^*. In some cases, species B is present in the vapor phase and w_B is determined by a "Langmuir isotherm" resulting from the condition of vapor–solid equilibrium for species B, that is,

$$
w_B = \frac{k_1 P_B}{k_2 + k_1 P_B} \qquad (1.9.8)
$$

where $k_1(1 - w_B)P_B$, the rate of adsorption of B on the solid surface, is equated to $k_2 w_B$, the rate of evaporation from the solid surface. k_1 and k_2 are temperature dependent constants and $P_B(= n_B kT$ for a dilute gas) is the partial pressure of species B in the vapor phase.

Exercise

An experimentalist studied reaction rates of species A with species B in a gaseous mixture at low pressures in a certain metallic vessel. He found that reaction rate was proportional to the product of concentrations $n_A n_B$ and concluded that the reaction was a homogeneous bimolecular reaction. However, when a colleague carried out the same experiment in a glass vessel he found that species A and B did not react appreciably.

Can you explain the results of the two experimentalists? How could the first experimentalist have avoided his erroneous conclusion?

Consider next a bimolecular homogeneous reaction between species A and B of a dilute gas. We shall calculate the reaction rate for the following special case.

a) The molecules of both species A and B are hard spheres with diameters d_A and d_B, respectively.

b) A reaction occurs only if the relative velocity along the line of centers of a colliding pair is greater than v_r^*, that is, only if $|\mathbf{v}_{AB} \cdot \hat{\mathbf{r}}| \geq v_r^*$, where $\mathbf{v}_{AB} = \mathbf{v}_A - \mathbf{v}_B$ and $\hat{\mathbf{r}}$ is a unit vector lying along the line centers of A and B at the instant of collision.

Analogously to the computation of the average collision frequence ν_{AB} in eq. (1.7.13), we find the frequency ν_{AB}^* of "reacting collisions" of a molecule of type A with those of type B to be given by the expression

$$\nu_{AB}^* = n_B d_{AB}^2 \int\int\int_{\mathbf{v}_{AB}\,\hat{\mathbf{r}}<-v_r^*} \phi(v_A)\phi(v_B)|\mathbf{v}_{AB} \cdot \hat{\mathbf{r}}| d\Omega d^3 v_A d^3 v_B, \tag{1.9.9}$$

where

$$d_{AB} \equiv \frac{d_A + d_B}{2}. \tag{1.9.10}$$

To cast eq. (1.9.7) into a form convenient for integration let us define the variables

$$\mathbf{g} = \mathbf{v}_A - \mathbf{v}_B, \tag{1.9.11}$$

and

$$\mathbf{G} = \frac{m_A \mathbf{v}_A + m_B \mathbf{v}_B}{m_A + m_B}. \tag{1.9.12}$$

In terms of these variables, for which $d^3v_A d^3v_B = d^3g d^3G$ and $m_A v_A^2 + m_B v_B^2 = \mu_{AB} g^2 + M_{AB} G^2$, eq. (1.9.9) becomes

$$\nu_{AB}^* = n_B d_{AB}^2 \left(\frac{m_A m_B}{(2\pi kT)^2}\right)^{3/2} \iiint_{\mathbf{g}\cdot\hat{\mathbf{r}}<-v_r^*} e^{-(\mu_{AB}g^2/2kT)-(M_{AB}G^2/2kT)} |\mathbf{g}\cdot\hat{\mathbf{r}}| d\Omega d^3g d^3G. \quad (1.9.13)$$

We have used the notation

$$\mu_{AB} = m_A m_B/(m_A + m_B) \quad (1.9.14)$$

and

$$M_{AB} = m_A + m_B. \quad (1.9.15)$$

The integral over d^3G may be performed to obtain

$$\int e^{-M_{AB}G^2/2kT} d^3G = \left(\frac{2\pi kT}{M_{AB}}\right)^{3/2}. \quad (1.9.16)$$

It is convenient to express $\mathbf{g}$ in a Cartesian coordinate system in which $\hat{\mathbf{r}}$ is the z axis, that is,

$$\mathbf{g} = g_x \hat{\imath} + g_y \hat{\jmath} + g_z \hat{\mathbf{r}}. \quad (1.9.17)$$

The $d^3g = dg_x dg_y dg_z$ and the constraint $\mathbf{g}\cdot\hat{\mathbf{r}} < -v_r^*$ becomes $g_z < -v_r^*$. Thus, eq. (1.9.13) reduces to the form

$$\begin{aligned}\nu_{AB}^* &= n_B d_{AB}^2 \left(\frac{\mu_{AB}}{2\pi kT}\right)^{3/2} \int d\Omega \int_{-\infty}^{-v_r^*} dg_z |g_z| e^{-(\mu_{AB}g_z^2/2kT)} \iint_{-\infty}^{\infty} dg_x dg_y e^{-\mu_{AB}/2kT(g_x^2+g_y^2)} \\ &= n_B \pi d_{AB}^2 \left(\frac{8kT}{\pi\mu_{AB}}\right)^{1/2} e^{-E_{AB}^*/kT},\end{aligned} \quad (1.9.18)$$

where

$$E_{AB}^* = (1/2)\mu_{AB} v_r^{*2}. \quad (1.9.19)$$

The integral over $d\Omega$ is over the entire unit sphere and therefore gives 4π in eq. (1.9.16) because the constraint on $\mathbf{g}\cdot\hat{\mathbf{r}}$ is accounted for in the integration over dg_z.

If the volume of the vessel in which the reaction is being carried out is V, the rate R_A of reaction of species A in the vessel is the product of $n_A V$, the number of molecules of A in the vessel, and ν_{AB}^*, the average frequency of reactive collisions of A with species B. Thus,

$$R_A = K_{AB} n_A n_B V, \quad (1.9.20)$$

where the homogeneous biomolecular rate constant K_{AB} is defined by the relation

$$K_{AB} = \pi d_{AB}^2 \left(\frac{8kT}{\pi\mu_{AB}}\right)^{1/2} e^{-E_{AB}^* kT}. \quad (1.9.21)$$

Again, aside from the pre-exponential factor of $T^{1/2}$, the rate constant is of the Arrhenius form, $E^*_{AB} = (1/2)\mu_{AB}(v^*_r)^2$ being the activation energy.

A more rigorous treatment of heterogeneous and homogeneous biomolecular reactions than that given here is necessary for a true quantitative understanding of these processes (e.g., what are the roles of internal energy states and nonrigid sphere interactions in the reaction process and what is the origin of the activation energy E^* or E^*_{AB}). However, the simple considerations outlined above do provide us with a qualitative appreciation of the molecular theory of reaction rates.

1.10 Experimental Tests of Maxwell's Distribution Law

Many experimental tests of the Maxwell distribution have been developed since 1889. Spectral line broadening due to the Doppler shift, an effect discussed in Section 1.7, was perhaps the earliest evidence introduced to support the validity of the Maxwell distribution. Lord Rayleigh (1889) gave the first discussion of the importance of the Maxwell distribution in line broadening or frequency (wavelength) dispersion. Other indirect methods and several direct methods of testing the Maxwell distribution were devised over a period of some 45 years following Rayleigh's article. These experiments leave no doubt of the validity of Maxwell's distribution law.

The first direct measurement of the distribution of velocities was made by Stern (1920). A silver coated platinum wire W (Fig. 1.8) was aligned along the axis of a cylinder D, with a narrow slit S. Cylinder D was concentric to a larger outer cylinder E. The wire and two cylinders rotated as a rigid body about their common axis. By heating the platinum wire, a low pressure gas of silver atoms could be evaporated from the wire. If the system were not rotating, then the silver atoms passing through the slit would be deposited on the outer drum over the solid angle $d\Omega$ centered on the point P. However, if the system were rotating, a constant angular velocity ω, then a particle passing from the wire through the center of the slit S with a velocity $\mathbf{v}$ in the instantaneous direction of $d\omega$ would arrive at a point P' on the outer drum, where P' lies a distance $r_E\omega t$ away from P, and where t is the time it takes a particle with a speed v to travel the distance $r_E - r_D$ between the two cylinders. This time is given by the relation $t = (r_E - r_D)/v$. Thus, the distance away from P that an atom is deposited is directly related to the velocity of the particle. Using this relationship, Stern, by measurement of the intensity of silver atom deposition versus position on the outer drum at a known rotational speed, obtained a direct measurement of the velocity distribution of the silver atoms.

Figure 1.8
Stern's experiment (1920).

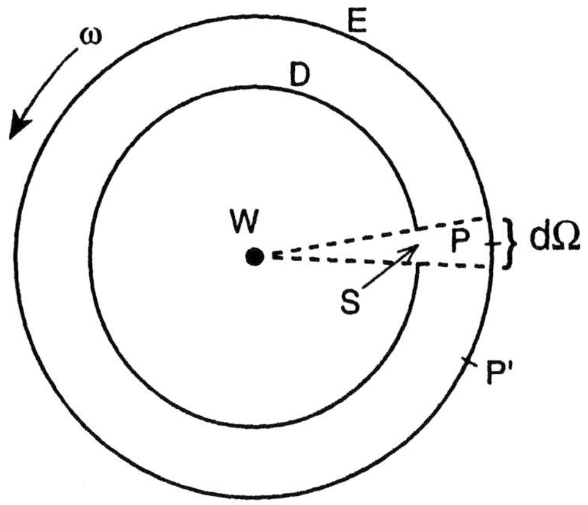

Perhaps the most accurate and certainly the most useful device for determining the velocity distribution make use of rotating velocity selector disks, which, by varying rotational speeds, can select only the molecules in a desired velocity range. A velocity selection apparatus for studying the velocity distribution of a gas at temperature T is illustrated in Figure 1.9. A dilute gas is held at temperature T in the oven shown in the figure. A molecule leaving the effusion hole in the oven will arrive at the collector if it passes through the small collimating slit, through a similar slit in Disk 1 of the rotating disk assembly shown in Figure 1.9 and through an identical slit in Disk 2. If the slits in the disks were aligned with the collimating slit and were not rotating, the number of molecules per unit time reaching the collector with speeds in the range $\mathbf{v}$ to $\mathbf{v} + d\mathbf{v}$ would be (from eq. [1.8.2)]

$$\frac{dN}{dt} = \int\int_{\Delta\Omega} A H_{\hat{\varepsilon}}(\mathbf{v}) d^3v = \int\int_{\Delta\Omega} A \cos\phi v \phi(v) \sin\theta d\theta d\eta v^2 dv \approx A \Delta\Omega v^3 \phi(v) dv, \qquad (1.10.1)$$

where A is the area of the effusion hole and $\Delta\Omega$ is the small solid angle (in which $\theta \sim 0°$ and $\cos\theta \sim 1$) determined by the cone in which a molecule must remain if it is to move from the effusion hole through the three narrow slits and into the detector.

Suppose now that the disks are rotating with an angular speed ω (in revolutions/unit time) and assume for the following argument that the slots are of zero width. A molecule getting through the slot in disk 1 will get through disk 2 if the time it takes the molecule to travel L, the axial distance between the disks, is equal to the time it takes the slots in the disks to make one revolution. Consequently, a molecule of speed v in the axial direction, getting through the slot in disk 1, will make it through the slot in disk 2 only if $1/\omega = L/v$. Although slower molecules whose velocities are equal to v/j, where $j = 2, 3, \ldots$, would get through the second slot in principle, in practice, for a number of reasons beyond the scope of the present discussion, these "higher harmonics" contribute very little to the molecular current passed at a given angular speed. In

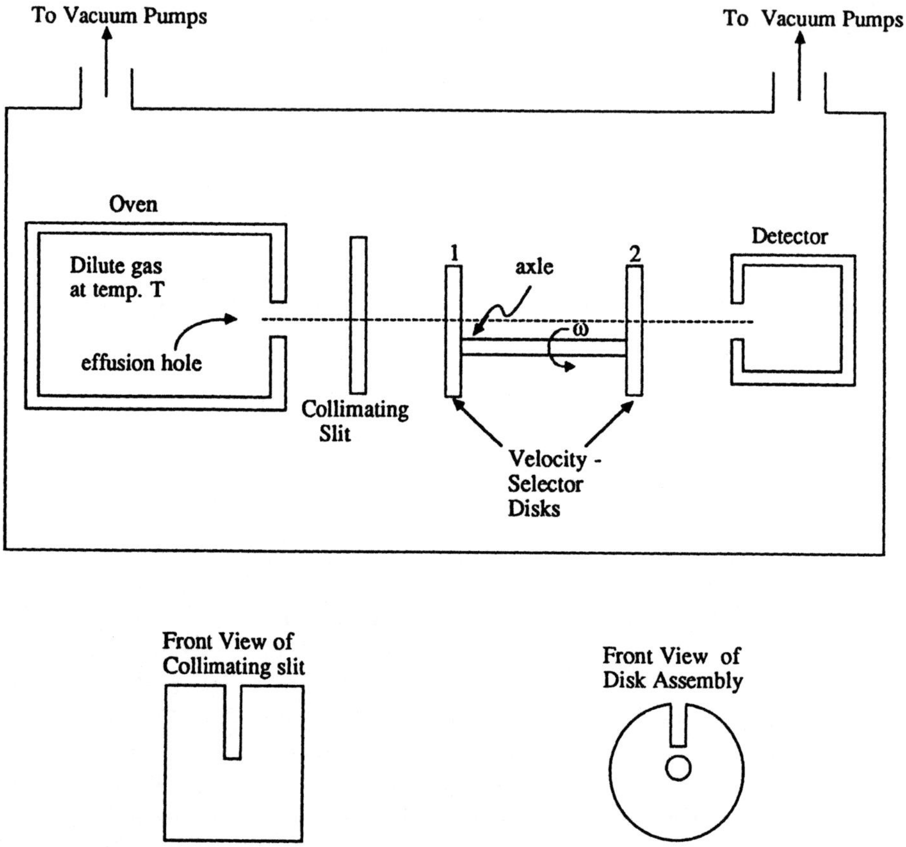

Figure 1.9
Velocity-selector apparatus for studying velocity distribution of a dilute gas at temperature T. To prevent scattering of the beam of molecules effusing from the oven, the entire apparatus is kept pumped down to a very low pressure (Marcus and McFee, 1959).

an actual experiment the slots have a finite width so that for a given ω molecules will reach the collector with speeds lying in the range $\omega L - \delta v/2$ to $\omega L + \delta v/2$. The interval δv will be a function of the slot dimensions and geometry and of ω. Thus, the rate at which particles accumulate in the collector for a given rotational velocity ω is given by the expression

$$\frac{dN}{dt} \sim A\Delta\Omega(L\omega)^3\phi(v = L\omega)\delta v. \tag{1.10.2}$$

With proper attention to the role of higher harmonics, eq. (1.10.2) has been used to verify the Maxwellian form of $\phi(v)$ from data generated by the rotating-disk velocity-selector apparatus.

Perhaps more important than its role in verifying the Maxwell velocity distribution has been the use of the velocity selector for producing a "molecular beam," that is, a beam of molecules all of whose velocities are almost the same. Such beams have been invaluable to the physicist in measurements of magnetic moments of atoms and molecules, of dipole moments (and other electric moments) of molecules, the wave-like behavior of particles predicted by quantum mechanics, scattering cross sections of molecules, etc. Chemists have also profited from beam experiments, obtaining such things as the energy dependence of reaction rate constants and heats of dissociation of molecules and radicals. The chemical information makes its way into engineering analysis of such processes as combustion, explosion, photochemical reactions occurring between flue gases and the atmosphere, etc.

1.11 The Boltzmann Factor and Barometric Formula

In the previous sections, we have assumed that no external forces (such as gravitational, electric and magnetic fields) were acting on the system, and, therefore, the density n is constant throughout the system. If there is an external field, however, the density need no longer be constant. For example, in the earth's gravitational field, the density at the bottom of the vessel will be somewhat larger than at the top of the vessel. The density at the bottom of the ocean is higher than at the surface. In this section we derive an expression for the variation of the density of a dilute *isothermal gas* in the presence of a conservative external force field $\mathbf{F}$. F is conservative if it can be written in the form $\mathbf{F} = -\nabla u(\mathbf{r})$, where $u(\mathbf{r})$ is a (scalar) potential energy function depending only on the position $\mathbf{r}$ of the particle on which the force is acting. If the field is gravity, the force on a particle is $\mathbf{F} = -mg\hat{\mathbf{k}}$, where m is the mass of the particle, g the acceleration of the gravity and $\hat{k}$ the unit vector pointing vertically from the earth's surface. For this case, $u(\mathbf{r}) = mg\hat{\mathbf{k}} \cdot \mathbf{r} = mgz$.

The relation of the local hydrostatic pressure $P(\mathbf{r})$ of a fluid at equilibrium to the external force $\mathbf{F}(\mathbf{r})$ can be obtained by performing a local force balance on a small volume element fixed at some arbitrary point in space.

As is illustrated in Figure 1.10, the pressure of the fluid surrounding the volume element exerts on each face of the element a force normal to the face and directed into the element. The net surface force exerted by the surrounding fluid is

$$\begin{aligned}
&-\hat{k}\int_{x}^{x+\Delta x}\int_{y}^{y+\Delta y}[P(x',y',z+\Delta z)-P(x',y',z)]dx'dy' \\
&-\hat{j}\int_{x}^{x+\Delta x}\int_{z}^{z+\Delta z}[P(x',y,\Delta y+z')-P(x',y,z')dx'dz' \qquad (1.11.1)\\
&-\hat{i}\int_{y}^{y+\Delta y}\int_{z}^{z+\Delta z}[P(x'+\Delta x,y',z')-P(x,y',z')dy'dz'.
\end{aligned}$$

Figure 1.10
Illustration of body and surface forces on a small rectangular parallelepiped in an equilibrium fluid. Surface forces shown only for faces of parallelepiped whose normals are parallel to the z axis.

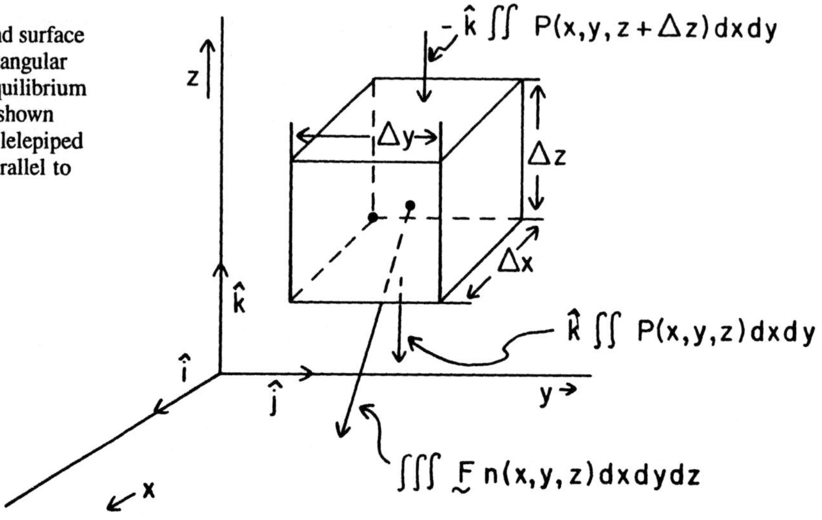

Expressed in Cartesian coordinates, the net body force on the subject volume is

$$\int_{x}^{x+\Delta x}\int_{y}^{y+\Delta y}\int_{z}^{z+\Delta z}\Big[\hat{\imath} n F_x(x',y',z') + \hat{\jmath} n F_y(x',y',z') + \hat{k} n F_z(x',y',z')\Big]\, dx'dy'dz'. \tag{1.11.2}$$

Because the fluid is a hydrostatic equilibrium, the sums of the respective components of the surface and body force are zero. For the x components, the equilibrium condition yields

$$-\frac{1}{\Delta y \Delta z}\int_{y}^{y+\Delta y}\int_{z}^{z+\Delta z}\left[\frac{P(x+\Delta x, y'z') - P(x,y',z')}{\Delta x}\right] dy'dz' - \frac{1}{\Delta x \Delta y \Delta z}\int_{x}^{x+\Delta x}\int_{y}^{y+\Delta y}\int_{z}^{z+\Delta z} n F_x(x',y',z')dx'dy'dz' = 0 \tag{1.11.3}$$

where we have divided the force-balance equation by $\Delta x \Delta y \Delta z$. Taking the limit of eq. (1.11.3) as Δx goes to zero converts the pressure term in the brackets to the partial derivative $\partial d P(x, y', z')/\partial x$ and with the use of the mean value theorem gives

$$\lim_{\Delta x \to 0}\int_{x}^{x+\Delta x} n F_x(x',y',z')dx' = n F_x(x,y',z'). \tag{1.11.4}$$

Thus, eq. (1.11.3) becomes in this limit

$$\frac{1}{\Delta y \Delta z}\int_{y}^{y+\Delta y}\int_{z}^{z+\Delta z}[-\frac{\partial P}{\partial x}(x, y', z') + nF_x(x, y', z')]dy'dz' = 0. \qquad (1.11.5)$$

Taking the limits Δy and $\Delta z \to 0$ successively, and using the mean value theorem with each limiting process, we obtain from eq. (1.11.5) the result

$$-\frac{\partial P}{\partial x}(x, y, z) + nF_x(x, y, z) = 0. \qquad (1.11.6)$$

A similar result is obtained for the y and z components of the force-balance equation. Thus, the condition of hydrostatic equilibrium can be summarized by the vector equation

$$-\nabla P + n\mathbf{F} = 0, \qquad (1.11.7)$$

where ∇ is the gradient operator, which in Cartesian coordinates has the form

$$\nabla = \hat{\imath}\frac{\partial}{\partial x} + \hat{\jmath}\frac{\partial}{\partial y} + \hat{k}\frac{\partial}{\partial z} \qquad (1.11.8)$$

Using the ideal gas relation $n = P/kT$ and the relation $\mathbf{F} = -\nabla u$ for a conservative force, we can rewrite eq. (1.11.6) in the form

$$0 = -\frac{1}{kT}\nabla u - \nabla \ln P. \qquad (1.11.9)$$

The solution to eq. (1.11.9) for an isothermal system is

$$P(\mathbf{r}) = P(\mathbf{r}_o)e^{-(u(\mathbf{r})-u(\mathbf{r}_o)/kT)}, \qquad (1.11.10)$$

where $P(\mathbf{r}_o)$ and $u(\mathbf{r}_o)$ are the pressure and potential energy at some point $\mathbf{r}_o$.

In the case of a gravitational field, eq. (1.11.9) becomes

$$P(z) = P(z_o)e^{-(mg(z-z_o)/kT)}, \qquad (1.11.11)$$

which is the so-called barometric formula and $z - z_o$ is the vertical distance between two points above the earth's surface. The idea that an atmosphere pressure would decrease with increasing vertical distance was first presented by Pascal, who had his brother-in-law test the idea by carrying a barometer to the top of a mountain. The experiment was outlined on November 15, 1647 in a letter to F. Perier. Perier reported the results in a letter to Pascal on September 22, 1648. Pascal published the work in 1663 in his book, *Great Experiment Concerning the Equilibrium of Fluids.*

Using $P = nkT$, we can transform eq. (1.11.10) into Boltzmann's equation:

$$n(\mathbf{r}) = n(\mathbf{r}_o)e^{-[u(\mathbf{r})-u(\mathbf{r}_o)/kT]}. \qquad (1.11.12)$$

Thus, the ratio $n(\mathbf{r})/n(\mathbf{r}_o)$, the relative number densities of particles at $\mathbf{r}$ and $\mathbf{r}_0$, is given by the so-called Boltzmann factor $\exp\{-[u(\mathbf{r}) - u(\mathbf{r}_o)]/kT\}$. This turns out to be a special case of a very general law governing the relative distribution of particles among energy states in an isothermal system.

We can interpret eq. (1.11.12) in statistical terms. If d^3r is a volume element centered on the position $\mathbf{r}$, the quantity $n(\mathbf{r})d^3r$ represents the probable number of particles in the volume, that is, the probable number of particles whose positions lie between $\mathbf{r}$ and $\mathbf{r} + d\mathbf{r}$. Assuming there are N molecules in the volume V, we have the condition

$$N = \int_V n(\mathbf{r})d^3r = n(\mathbf{r}_o)e^{[u(\mathbf{r}_o)/kT]} \int_V e^{-u(\mathbf{r})/kT} d^3r, \qquad (1.11.13)$$

that allows us to eliminate $n(\mathbf{r}_o)e^{u(\mathbf{r}_o)/kT}$ in eq. (1.11.12) to obtain

$$n(r) = Ne^{-u(\mathbf{r})/kT} / \int_V e^{-u(\mathbf{r})/kT} d^3r. \qquad (1.11.14)$$

Because $n(\mathbf{r})d^3r$ is the probable number of particles in the volume d^3r, the quantity $p(\mathbf{r})d^3r \equiv n(\mathbf{r})d^3r/N$ represents the probability that a particle in the N particle system will be in the volume d^3r. Thus, $p(\mathbf{r})$ is a probability density or distribution function in position space.

The probable number of particles having a position between $\mathbf{r}$ and $\mathbf{r} + d\mathbf{r}$ and velocities between $\mathbf{v}$ and $\mathbf{v} + d\mathbf{v}$ will be $n(\mathbf{r})d^3r\phi(\mathbf{v})d^3v$, so that the probability of finding such a particle in d^3r is

$$p(\mathbf{r})\phi(\mathbf{v})d^3rd^3v = \text{constant}\, e^{-[(1/2)mv^2 + u(\mathbf{r})]/kT} d^3rd^3v. \qquad (1.11.15)$$

This result is interesting because it is suggestive of a very general result. Suppose we want the probability

$$P_N(\mathbf{r}_1, \mathbf{r}_2, \ldots, \mathbf{r}_N, \mathbf{v}_1, \mathbf{v}_2, \ldots, \mathbf{v}_N)d^3r_1d^3r_2, \ldots, d^3r_Nd^3v_1d^3v_2, \ldots, d^3v_N \qquad (1.11.16)$$

that in an N-particle system particle 1 is between $\mathbf{r}_1$ and $\mathbf{r}_1 + d\mathbf{r}_1$ with velocity between $\mathbf{v}_1$ and $\mathbf{v}_1 + d\mathbf{v}$, particle 2 is between $\mathbf{r}_2$ and $\mathbf{r}_2 + d\mathbf{r}_2$ with velocity between $\mathbf{v}_2$ and $\mathbf{v}_2 + d\mathbf{v}_2$, etc. If it is assumed that the ith particle is distributed as in eq. (1.11.15), but that the potential on it is that due to all the other particles and any external potentials, then we would obtain for P_N

$$P_N = \text{constant}\, e^{-(\Sigma_i 1/2m_i v_i^2 + u^N)/kT} \qquad (1.11.17)$$

where $u^N = \Sigma_{i=1}^N[u_{ei}(\mathbf{r}_i) + u(\mathbf{r}_i; \{\mathbf{r}_{j\neq i}\})]$. Here u_{e_i} denotes the external potential on particle i and $u(\mathbf{r}_i; \{\mathbf{r}_{j\neq i}\})$ denotes the interaction potential of particle i with all the other particles in the fluid. If the latter are pairwise additive, then $u^N = \Sigma_{i=1}^N u_{e_i}(\mathbf{r}_i) + \Sigma_{i>j=1}^N u(\mathbf{r}_i, \mathbf{r}_j)$. In spite of the heuristic way we have just derived the N-body distribution function P_N, the result is exact for a classical fluid. We shall see this in a later chapter.

1.12 Waterston's Contribution to Kinetic Theory

For historical perspective on the origins of kinetic theory, the reader is urged to read the introductory chapter of J. H. Jeans' *The Dynamical Theory of Gases*, (1954). In addition to the ideas of the ancients, based on Democritus' concept of atoms, Jeans quotes a sequence of fundamental works by Gasendi (published in 1658), D. Bernoulli (1738), Herapath (1821), Waterston (1845), Joule (1848), Kronig (1856), and Clausius (1857) leading up to Maxwell's discovery of the law of distribution of velocities. Among the contributors listed by Jeans, Waterston is exceptional in that his contribution was not widely recognized until his original manuscript was published in the *Philosophical Transactions* by Lord Rayleigh in 1892 about 50 years after Waterston completed the manuscript.

Waterston presented his paper to the Royal Society in 1845. The Society refused publication of the paper. According to Jeans, it was not published because it "contained certain inaccuracies." This claim, however, does not square with comments of the original referees quoted in Lord Rayleigh's introduction to Waterston's paper. One referee expressed opinions such as "the paper is nothing but nonsense, unfit even for reading before the Society." Another referee asserted

> that the whole investigation is confessedly founded on a principle entirely hypothetical, from which it is the object to deduce a mathematical representation of the phenomena of elastic media. It exhibits much skill and many remarkable accordances with the general facts, as well as numerical values furnished by observation.... The original principle itself involves an assumption which seems to be very difficult to admit, and by no means a satisfactory basis for a mathematical theory, viz., that the elasticity of a medium is to be measured by supposing its molecules in vertical motion, and making a succession of impacts against an elastic gravitating plane.

In view of these referees' remarks, it seems more likely that instead of the reason given by Jeans, the paper was not published because the Society was reluctant to accept new and somewhat speculative ideas from a young author who had not yet received the blessings of the scientific establishment. In proposing this latter reason, Lord Rayleigh even suggested that "Had he (Waterston) put forward his investigation as a development of the theory of D. Bernoulli, a referee might have hesitated to call it nonsense." Lord Rayleigh speculates that failure to publish Waterston's work on kinetic theory "probably retarded the development of the subject by ten or fifteen years." We can take two lessons from this episode in the history of science: first, that excessive scientific conservatism may impede the development of a new theory and second, that an idea whose time has come cannot be stopped—delayed maybe—but not stopped. What then did Waterston's paper contain? To answer this question we can do nothing better than to quote from Lord Rayleigh's introduction of the 1892 publication of Waterston's paper:

> At the present time the interest of Waterston's paper can, of course, be little more than historical. What strikes one most is the marvellous courage with which he attacked questions, some of which even now present serious difficulties. To say that he was not always successful is only to deny his claim to rank among the very foremost theorists of all ages. The character of the advance to be dated from this paper will be at once understood when it is realized that Waterston was the first to introduce into the theory the conception that heat and temperature are to be measured by *vis vivam.*
>
> This enabled him at a stroke to complete Bernoulli's explanation of pressure by showing the accordance of the hypothetical medium with the law of Dalton and Gay-Lussac. In the second section the great feature is the statement (VII.), that in mixed media the mean square molecular velocity is inversely proportional to the specific weight of the molecules. The proof which Waterston gave is doubtless not satisfactory; but the same may be said of that advanced by Maxwell fifteen years later. The law of Avogadro follows at once, as well as that of Graham relative to diffusion. Since the law of equal energies was actually published in 1851, there can be no hesitation, I think, in attaching Waterston's name to it. The attainment of correct results in the third section, dealing with adiabatic expansion, was only prevented by a slip of calculation.
>
> In a few important respects Waterston stopped short. There is no indication, so far as I can see, that he recognized any other form of motion, or energy, than the translatory motion, though this is sometimes spoken of as vibratory. In this matter the priority in a wider view rests with Clausius. According to Waterston the ratio of specific heats should be (as for mercury vapor) 1.67 in all cases. Again, although he was well aware that the molecular velocity cannot be constant, there is no anticipation of the law of distribution of velocities established by Maxwell.
>
> A large part of the paper deals with chemistry, and shows that his views upon that subject were much in advance of those generally held at the time. . . .

In a 1901 postscript to his 1892 introduction, Lord Rayleigh added:

> It may be added that Waterston's memoir contains the first calculation of the molecular velocity, and further that it points out the relation of this velocity to the velocity of sound. The earliest actual publication of such a calculation is that of Joule, who gives for the velocity of hydrogen molecules at 0°C 6055 feet per second (Manchester Memoirs, Vol. IX. p. 107, Oct. 1848; Phil. Mag. Ser. 4, Vol. XIV. P. 2311; Joule's Scientific Papers, Vol. I. p. 295), thus anticipating by eight or nine years the first paper of Clausius (Pogg. Ann. 1857), to whom priority is often erroneously ascribed.

To remind the reader, the physical laws referred to above are:

1 *Dalton's Law:* The pressure exerted by a mixture of gases is equal to the sum of the pressures exerted separately by the several components of the mixture.

2 *Avogadro's Law:* Two different gases or mixtures of gases, when at the same temperature and pressure, contain equal numbers of molecules in equal volume.

3 *Graham's Law:* The rates of effusion of different gases at the same temperature and pressure vary inversely as the square roots of the molecular weights or mass densities.

Supplementary Reading

Brush, S. G. 1983. *Statistical Physics and the Atomic Theory of Matter, from Boyle and Newton to Landau and Onsager*, Princeton University Press, Princeton.

Lay, J. E. 1990. *Statistical Mechanics and Theromodynamics of Matter*, Harper & Row, New York.

Partington, J. R. 1962. *Advanced Treatise on Physical Chemistry*, Longmans and Green, New York.

Present, R. D. 1958. *Introduction to the Kinetic Theory of Gases*, McGraw-Hill, New York.

Reif, F. 1965. *Statistical and Thermal Physics*, McGraw Hill, New York.

Exercises

1 Consider the continuous random variable s. Suppose the probability that s lies in between s and $s + ds$ is $w(s)ds$ where

$$w(s) = \frac{1}{\pi}\frac{b}{s^2 + b^2}$$

b being a positive constant and s ranging from $-\infty$ to $+\infty$. Compute the mean of s, $|s|$, and s^4. A distribution of the form $w(s)$ is called a Lorentzian.

2 A gas of molecules, each of mass m, is in thermal equilibrium at the absolute temperature T. Denote the velocity of a molecule by $\mathbf{v}$, its three Cartesian components by v_x, v_v, and v_z, and its speed by v. What are the following mean values:

a) $\overline{v_x}$ b) $\overline{v_x^2}$

c) $\overline{v^2 v_x}$ d) $\overline{v_x^3 v_y}$

e) $\overline{(v_x + bv_y)^2}$, b is a constant f) $\overline{v_x^2 v_y^2}$

This problem can be done without evaluating many integrals explicitly.

3 Show that

$$I_n \equiv \int_{-\infty}^{\infty} e^{-ct^2} t^{2n} dt = (-1)^n \frac{d^n I}{dc^n},$$

where $I = \int_{-\infty}^{\infty} e^{-ct^2} dt$, and

$$J_n \equiv \int_{0}^{\infty} e^{-ct^2} t^{2n+1} dt = (-1)^n \frac{d^n J}{dc^n},$$

where $J = \int_0^{\infty} e^{-ct^2} dt$. Use this result to evaluate I_n and J_n.

4 Prove that

$$\langle v_x^k v_y^l v_z^m \rangle = 0$$

if k, l, and m are positive integers and any of k, l, and m are odd.

5 Derive an expression for the distribution of the energy of a dilute gas of particles obeying the Maxwell velocity distribution, that is, determine $g(\varepsilon)$, where $g(\varepsilon)d\varepsilon$ represents the probability that a particle in the gas has an energy between ε and $\varepsilon + d\varepsilon$. Compute the most probable and the average energies at 100K.

6 Starting with the definitions

$$G \equiv \frac{1}{2}(\mathbf{v}_1 + \mathbf{v}_2)$$

and

$$\mathbf{g} = \mathbf{v}_2 - \mathbf{v}_2$$

that were used in determining the expression for mean free path, verify

$$v_1^2 + v_2^2 = \frac{1}{2}g^2 + 2G^2.$$

Also, using a Cartesian coordinate system, verify that

$$d^3v_1 d^3v_2 = d^3g d^3G,$$

that is, show that the Jacobian of the transformation

$$\mathbf{v}_1, \mathbf{v}_2 \rightarrow \mathbf{G}, \mathbf{g}$$

is unity.

7 Given the probability that a particle in a dilute gas can move a distance x without undergoing a collision is

$$f(c) = e^{-x/\lambda}$$

where λ is the mean free path, compute the probabilities that in Ar at 3 atm and in He at 3 atm an atom will go 100 Å and 1 μ m, respectively. Assume the gas is at room temperature. Use the 6–12 Lennard-Jones parameters (see Chapter 5) for atomic diameters.

8 By introducing the concept of a mean free path, the thermal conductivity of a dilute monatomic gas can be expressed as

$$K = \frac{1}{2} nk\langle v \rangle \lambda$$

where K is the thermal conductivity, n the number of molecules per unit volume, k the Boltzmann constant, v the molecular speed, and λ the mean free path. Using experimental data for the thermal conductivity of argon, evaluate its mean free path and estimate its atomic diameter. Compare this diameter with the length parameter of the 6–12 Lennard–Jones potential (see Chapter 5).

9 A thin walled vacuum jar sprang a pinhole leak (pinhole of diameter 5×10^{-4}cm) to a low pressure space-pressure held at 10^{-3} atm ($= 10^3$ dyne/cm^2). At room temperature ($T = 300$ K) how long will it take for the pressure of the vacuum jar to reach 10^{-6} atm? The gas in the low pressure space is nitrogen with a molecular weight of 28.

10 A space craft is preparing to make a 10 day trip to the moon and back. It will carry 10 pounds more than enough oxygen to sustain the astronauts for the 10 days. Assuming that the space craft operates at 1 atm and 72°F, estimate the biggest pinhole leak that the space craft could spring during launching and still successfully complete the mission. After determining the size of the pinhole, check the data in Table 1.1 to see if effusion or diffusion would occur under the circumstances you have calculated.

11 In a molecular beam experiment, the source is a tube containing hydrogen at a pressure $P_s = 0.15$ mm of mercury and at a temperature $T = 300$ K. In the tube wall is a slit 20×0.025 mm, opening into a highly evacuated region. Opposite the source slit and 1 m away from it is a second detector slit parallel to the first and of the same size. This slit is in the wall of a small enclosure in which the pressure $\overline{P}$ can be measured. How many H_2 molecules leave the source slit per second? How many H_2 molecules arrive at the detector slit per second? Consider the source slit as a point source and compute the number of molecules with speed in the range v and $v + dv$ that emerge into a solid angle $d\Omega$ after striking a unit area of the detector slit wall. What is the pressure P_d in the detector chamber when a steady state has been reached so that P_d is independent of time, that is, the number of molecules entering the detection chamber is equal to the number being emitted?

12 An ideal monatomic gas (total mass M temperature T) is in a cylindrical vessel (radius R, length L), rotating with an angular velocity ω about its axis. Find the density of the gas as a function of r, the distance from the axis of the vessel. What is the total energy of the gas? What is the difference between the entropy in its rotating state and in a nonrotating state at the same temperature?

Hint: Assume the thermodynamic relation between entropy and T and P, or T and $\tilde{V}$, that is,

$$d\tilde{S} = \tilde{C}_p dT + \left(\frac{\partial \tilde{S}}{\partial P}\right)_T dP \quad = \quad \tilde{C}_v dT + \left(\frac{\partial \tilde{S}}{\partial \tilde{V}}\right)_T d\tilde{V},$$

holds locally, $\tilde{S}, \tilde{C}$, and $\tilde{V}$ are the entropy, specific heat and volume per molecule, respectively.

13 At 25°C what fraction of the molecules in hydrogen gas have a kinetic energy within $kT \pm 10\%$? What fraction at 500°C? What fraction of molecules in mercury vapor?

14 Show that the number of collisions per second per unit volume between unlike molecules, A and B, in a dilute gas is

$$z_{AB} = \pi n_A n_B \left(\frac{d_A + d_B}{2}\right)^2 \sqrt{\frac{8kT}{\pi \mu}}$$

where the reduced mass, $\mu = (m_A m_B)/(m_A + m_B)$. In an equimolar mixture of H_2 and I_2 at 500°K and 1 atm calculate the number of collisions per second per milliliter between H_2 and H_2, H_2 and I_2, I_2 and I_2. For H_2 take $d = 2.18$Å, for I_2, $d = 3.76$ Å.
Hint: Use the relation $z_{AB} = n_A \nu_{AB}$ and the coordinates $\mathbf{G} = (m_A \mathbf{v}_A + m_B \mathbf{v}_B)/(m_A + m_B)$ and $\mathbf{g} = \mathbf{v}_B - \mathbf{v}_A$.

15 The permeability constant at 20°C of pyrex glass to helium is given as 6.4×10^{-12} cm^3 s^{-1}/cm^2 area/mm thickness/cm Hg pressure difference. The helium content of the atmosphere at sea level is about 5×10^{-4} mol %. Suppose a 100 ml round pyrex flask with walls 2 mm thick was evacuated to 10^{-10} mm and sealed. What would be the pressure at the end of 1 year due to inward diffusion of helium? Estimate the size of a single pinhole in the flask that would give the same rate of helium leakage as the reported rate (that is, as that given by the permeability constant of pyrex).

16 Imagine a furnace containing sodium vapor at 3000K and suppose a spectrometer is used to observe the spectral intensity $I(\Lambda)$ versus wavelength of the spectral line at $\Lambda_o = 5889.97$Å. Plot the expected normalized spectral intensity versus wavelength.

17 Molecules of species A and B are constrained to move on a flat solid surface. Aside from being constrained to stay on the solid surface, the molecules behave as an ideal gas. If a molecule of type A collides with one of type B, they react immediately to form a product C that is released from the surface and collected elsewhere. Assuming steady-state surface concentrations of $n_A = 10^{13}$ molecules/cm^2 and $n_B = 1.5 \times 10^{13}$ molecules/cm^2 and that A and B behave as hard spheres, estimate the number of molecules of C produced on 1 cm^2 of solid surface in 1 h. The following data may be used for computations: $d_A = 3 \times 10^{-8}$ cm, $d_B = 2 \times 10^{-8}$ cm, $T = 450$K, and $m_A = m_B = 5 \times 10^{-23}$ gm/molecule.

References

Brönsted, J. N. and Hevesy, G. V. 1920. *Nature*, **106**, 144; 1922. Phil. Mag., **43**, 31.

Jeans, J. H. 1953. *The Dynamical Theory of Gases*, Dover Pub., Inc. New York.

Lord Rayleigh. 1889. *Phil. Mag.*, **27**, 298.

Lord Rayleigh. 1892. *Phil. Trans.*, **183A**, 1.

Marcus, P. M. and McFee, J. H., in I. Estermann, 1959. *Recemt Research in Molecular Beam*, Academic Press, New York, p. 43.

Pascal, B. 1648. Récit de la Grande Expërience de l' ëquilibre des liquers, Chez Charles Savreus, Paris, see also Pascal: Oeuvres Complete, 1954 ed. J. Chevalier, Editions Gallimard, Bruges, Belgique.

Stern, O. 1920. Z. *Physik*, **2**, 49.

2

THE ELEMENTS OF ENSEMBLE THEORY

2.1 Introduction

Because this is a textbook on the statistical mechanics of materials at equilibrium aimed at advanced undergraduates and beginning graduate students, it is appropriate to review the fundamentals upon which the molecular theory of matter is based. Intermolecular forces, the quantum mechanics of energy levels, and the ensemble theory of thermodynamic and stationary processes will be treated in this chapter. It is anticipated that the reader's familiarity with thermodynamics corresponds to that gained in an undergraduate course.

2.2 Intermolecular Forces

The theory of intermolecular forces is a relatively complicated application of the principles of quantum mechanics to the assembly of nuclei and electrons composing atoms and molecules. A full treatment of this subject lies outside the scope of this book. Instead, we describe in this section several model representations of the potential of intermolecular forces that capture the most important features of these forces.

The fundamental interaction between atomic particles is the Coulomb force between charged particles. For a pair of ions separated by the distance r the potential of this force is

$$u(r) = z_i z_j \frac{e^2}{r}, \tag{2.2.1}$$

where e is the unit electronic charge (4.80×10^{-10} esu and z_i is the valence of particle i). Two unit charges of opposite sign, $z_i = -z_j = 1$, separated by a distance of $r = 3.4$Å will have a Coulumbic potential energy of -6.78×10^{-12} erg. For $r = 1.06$Å, the Coulombic potential energy is -2.18×10^{-11} erg or -13.6

eV, which is equal in magnitude to the ionization potential of a hydrogen atom. As a point of reference we note that $kT = 4.11 \times 10^{-14}$ erg at $T = 298$K.

A neutral molecule is a charged nucleus surrounded by a cloud of electrons. If a charged particle interacts with this electron cloud it will deform (polarize) the cloud. The interaction between the charge of the particle and the polarized molecule is represented by the polarization potential

$$u(r) = -z_i^2 \frac{e^2 \alpha}{r^4}, \tag{2.2.2}$$

where z_i is the valence of the charged particle and α is the polarizability of the molecule. α is roughly of the order of magnitude of the molecular volume of the molecule. If the molecule is nonspherical, then α represents the mean of the polarizability α_1, α_2, and α_3 of the principal directions of polarizability of the molecules. For example, for molecules of cylindrical symmetry $\alpha = (2\alpha_T + \alpha_L)/3$, where α_L is the polarizability parallel to the major axis of symmetry and α_T is the polarizability along a direction orthogonal or transverse to the major axis. If the angular dependence of the polarization potential is included for a cylindrically symmetric molecule, then

$$u(r) = -z_i^2 \frac{e^2 \alpha}{r^4} - z_i^2 \frac{e^2 \alpha_a}{r^4} P_2(\cos\theta), \tag{2.2.3}$$

where $\alpha = (2\alpha_T + \alpha_L)/3$, $\alpha_a = \alpha_L - \alpha_T$, and $P_2(\cos\theta)$ is a Legendre polynomial defined by

$$P_2(\cos\theta) = \frac{3}{2}\cos^2\theta - \frac{1}{2}. \tag{2.2.4}$$

θ is the angle between the vector connecting the centers of the charged particle and the molecule and a vector lying on the major axis of the molecule.

Examples of α are $\alpha = 1.7 \times 10^{-24}$ cm^3 for Ar; $\alpha_L = 23.8 \times 10^{-25}$cm^3, $\alpha_T = 14.5 \times 10^{-25}$cm^3 for N_2; $\alpha = 26 \times 10^{-25}$cm^3 for CH_4; $\alpha_L = 54.8 \times 10^{-25}$cm^3 and $\alpha_T = 39.7 \times 10^{-25}$cm^3 for C_2H_6; and $\alpha_L = 63.5 \times 10^{-25}$cm^3 and $\alpha_T = 123.1 \times 10^{-25}$cm^3 for C_6H_6. Argon and a unit charge, $z_i = \pm 1$, held a distance 3.4 Å away will have a polarization potential of -2.93×10^{-13} erg.

If in a neutral molecule the nuclear and electronic charge distribution is such that there is a separation of the center of positive charge, then the molecule possesses a permanent dipole moment D, which has dimensions of charge times distance. In a point charge model a dipole is represented as a pair of equal and opposite charges. The value of D is then the product of the magnitude of one of the charges and the distance of separation of the charges.

The potential of interaction between a dipole and charged particle is

$$u(\mathbf{r}, \mathbf{D}) = -z_i e \frac{\mathbf{D} \cdot \mathbf{r}}{r^3} \tag{2.2.5}$$

for distances that are large compared to D/e, the effective separation of the charges of the dipole. $\mathbf{D}$ is a vector of magnitude D lying in the direction of the line between the centers of the charges defining the dipole. $\mathbf{r}$ is the vector distance between the charge $z_i e$ and the center of the dipole. For most modeling one imagines $\mathbf{D}$ to be a point dipole, that is, a dipole of finite value and definite orientation, but of negligible length.

For a pair of interacting point dipoles, the potential of interaction is of the form

$$u(\mathbf{r}, \mathbf{D}_i, \mathbf{D}_j) = -\frac{[\mathbf{D}_i \cdot \mathbf{D}_j r^2 - 3(\mathbf{D}_i \cdot \mathbf{r})(\mathbf{D}_j \cdot \mathbf{r})]}{r^5}, \tag{2.2.6}$$

r being the separation of the dipoles and $\mathbf{D}_i$ and $\mathbf{D}_j$ their dipole moments.

Fluids of molecules possessing dipole moments are referred to as polar fluids. Common examples are water, alcohols, alkyl halides, etc. Some values of dipole moments are $D = 1.85$, 1.69, and 1.91 Debyes for water, ethanol, and iodoethane, respectively. A Debye equals 10^{-18} esu-cm. If $D_i = D_j = 1.85$ Debyes, $\mathbf{D}_i \cdot \mathbf{r} = D_i r$, $z_i = 1$, $\mathbf{D}_i \cdot \mathbf{D}_j = -D_i D_j$, and $r = 3.4$Å, the ion–dipole potential equals -7.68×10^{-13} erg and the dipole–dipole potential is -1.74×10^{-13} erg.

Other asymmetries of the charge distribution of neutral molecules lead to quadrupole moments, octapole moments, etc. The potential of interaction between a dipole and quadrupole falls off as the fourth power of separation r and depends on the relative orientations of the dipolar and quadrupolar entities. The quadrupole–quadrupole potential falls off as the fifth power of r and depends on the relative orientation of the quadrupolar species. It is sometimes assumed that the dipolar and quadrupolar interactions can be angle averaged over a thermal distribution weighted according to the Boltzmann factor $\exp(-u/kT)$ approximated by $1 - u/kT$. Such angle averaged dipole–dipole, dipole–quadrupole, and quadrupole–quadrupole interaction potentials are

$$u(r) = -\frac{2}{3kT}\frac{D_i^2 D_j^2}{r^6} \tag{2.2.7}$$

$$u(r) = -\frac{1}{kT}\frac{D_i^2 Q_j^2}{r^8} \tag{2.2.8}$$

$$u(r) = -\frac{7}{40kT}\frac{Q_i^2 Q_j^2}{r^{10}} \tag{2.2.9}$$

Q_i is an angle averaged quadrupole moment and is of the order of 10^{-26} esu-cm^2 for molecules such as water, benzene, and ammonia. For $T = 298K$, $D_i = D_j = 1.85$ Debye, $Q_i = 10^{-26}$ esu-cm^2, and

$r = 3.4Å$, eq. (2.2.7)—(2.2.9) yield pair potentials of -1.23×10^{-13}, -4.66×10^{-15}, and -2.06×10^{-17} erg, respectively.

An important class of interactions are the so-called dispersion forces. The polarization interaction is an example of these forces. The distinction between these and the permanent multipolar forces is that the former are dynamical, induced by molecular proximity, whereas the latter are static, arising from asymmetries of the unperturbed charge distribution of interacting molecules.

As two spherically symmetric, neutral molecules are brought together, the charge distribution about one of the molecules induces a dipole in the other and vice versa. The induced dipoles give rise to an attractive potential of which the dominant term gives the so-called London (or van der Waals) dispersion interaction

$$u(r) = -\frac{A}{r^6}, \qquad (2.2.10)$$

where A is a constant related to the polarizabilities α_i and ionization potentials I_i of the interacting molecules. London estimated $A = 3I_iI_j\alpha_i\alpha_j/2(I_i + I_j)$. This approximation for the dispersion forces is generally quoted as the theoretical justification for the r^{-6} term in the Lennard–Jones potential. There are further terms contributing to the dispersion interactions that vary as r^{-8}, r^{-10}, etc. Also for nonspherical molecules the dispersion interactions depend on molecular orientations.

For argon $A = 1.029 \times 10^{-58}$ erg-cm^6. Thus, for a separation $r = 3.4Å$, the London dispersion potential for a pair of argon molecules is -6.66×10^{-14} erg.

At close distances of approach the interactions between molecules become strongly repulsive (from Coulomb interactions, but especially the Pauli exclusion forces). This repulsion is sometimes modeled as a hard sphere potential,

$$\begin{aligned} u(r) &= \infty, \qquad r < d, \\ &= 0, \qquad r > d, \end{aligned} \qquad (2.2.11)$$

as an exponential potential

$$u(r) = Ae^{-Br}, \qquad A, B > 0, \qquad (2.2.12)$$

or as an inverse power potential

$$u(r) = Ar^{-\nu}, \qquad \nu > 0, \qquad (2.2.13)$$

or where ν is a positive integer of the order of 10 (e.g., 10, 11, 12, 13 have been used).

Certain models incorporating, at least qualitatively, the repulsive and attractive interactions of molecules have received a lot of attention. The square-well model,

$$u(r) = \infty, \qquad r < d_1$$

$$= -\varepsilon \qquad (2.2.14)$$
$$= 0, \qquad r > d_2,$$

is perhaps the simplest of these.

The model used most often for nonpolar fluids is the 6–12 Lennard–Jones potential:

$$u(r) = 4\varepsilon\left[\left(\frac{\sigma}{r}\right)^{12} - \left(\frac{\sigma}{r}\right)^{6}\right]. \qquad (2.2.15)$$

$u(r)$ is a minimum at $r_m = 2^{1/6}\sigma$. For this separation $u(r_m) = -\varepsilon$. For argon $\varepsilon = 1.65 \times 10^{-14}$ erg, $\sigma = 3.405$Å.

For dipolar fluids a popular model is the Stockmayer potential

$$u = 4\varepsilon\left[\left(\frac{\sigma}{r}\right)^{12} - \left(\frac{\sigma}{r}\right)^{6}\right] - \frac{1}{r^5}(\mathbf{D}_i \cdot \mathbf{r})(\mathbf{D}_j \cdot \mathbf{r}). \qquad (2.2.16)$$

Although much of the emphasis in the statistical physics of fluids proceeds under the assumption that intermolecular forces are pairwise additive, this is not strictly the case. In particular, the three-body dispersion potential, whose dominant contribution is the Axilrod–Teller (1943) term

$$u_{123} = \varepsilon_{123}(1 + \hat{r}_{12} \cdot \hat{r}_{13} + \hat{r}_{13} \cdot \hat{r}_{23} + \hat{r}_{12} \cdot \hat{r}_{23})/r_{12}^3 r_{13}^3 r_{23}^3, \qquad (2.2.17)$$

has been shown to be important in connection with the crystal structure of solids. For argon $\varepsilon_{123} = 7.32 \times 10^{-83}$ erg-cm^9. If the centers of three molecules are configured on an equilateral triangle of length $3.4 \times 2^{1/6}$Å on each side, then $u_{123} = 1.05 \times 10^{-15}$ erg, compared to the total pair energy -4.95×10^{-14} erg of the three atoms in this configuration. (The pair energy is computed from the Lennard–Jones potential.)

For highly nonspherical molecules, it is necessary to introduce noncentral potentials. This is an area of active research currently. However, it lies outside the scope of this book and will not be pursued further herein. In fact, in trying to implement a statistical mechanical theory of phase behavior and of interfaces and thin films, we will frequently retreat to the simplest of models, such as hard rod, hard sphere, or Lennard–Jones potential, in order to minimize the complexity and severity of approximation in deducing tractable equations from theory.

2.3 Quantum Mechanics of Simple Systems

The ensemble theory of statistical mechanics is most logically developed within the context of quantum mechanics. However, only a relatively elementary knowledge of quantum mechanics is necessary to feel comfortable with the concepts of statistical mechanics. Thus in this section we outline a few especially useful, but simple, models, of sufficient generality to meet the needs of the present text.

If the classical energy of a system of particles is of the form

$$H_{Cl} = \sum_{i=1}^{N} \frac{1}{2} m_i v_i^2 + u^N(\mathbf{r}_1, \ldots, \mathbf{r}_N). \tag{2.3.1}$$

then the quantum mechanical energy state functions ψ and energies E are determined from the Schrödinger equation

$$H\psi = E\psi \tag{2.3.2}$$

where H is the Hamiltonian operator of the form

$$H = [-\sum_{i}^{N} \frac{\hbar^2}{2m_i} \nabla_{r_i}^2 + u^N]. \tag{2.3.3}$$

ψ is known as the wave function of the system. For a given E, the wave function ψ_E has the interpretation of a probability amplitude, that is, $|\psi_E|^2 (d^3r)^N$ denotes the probability that a system of particles of energy E is in the configuration $\mathbf{r}_1$ to $\mathbf{r}_1 + d\mathbf{r}_1$, $\mathbf{r}_2$ to $\mathbf{r}_2 + d\mathbf{r}_2, \ldots, \mathbf{r}_N$ to $\mathbf{r}_N + d\mathbf{r}_N$.

The particles $i = 1, \ldots, N$ may be the collection of electrons and nuclei composing the atoms and molecules of the system, the collection of atoms composing the system, or any collection of particles whose classical energy can be expressed satisfactorily in the form given by eq. (2.3.1). If H_{Cl} is different from eq. (2.3.1) Schrödinger's equation will be valid but the form of H will be different from (2.3.3).

For finite systems eq. (2.3.2) generally has solutions only for a discrete set of energies E. This "quantization" of energies sets quantum mechanical systems apart from classical systems, in which energy can have a continuous set of values. In illustrating this point, we present the solutions of the Schrödinger equation for several model systems. The models are chosen to provide some perspective for actual molecular systems. The solutions are given without derivation. That they are solutions can be verified by inserting them into the Schrödinger equation. Numerous books on differential equations or quantum mechanics contain the mathematical methods used to find the solution.

2.3.1 Particle in a Box

Consider a particle free to move within a rectangular box of lengths L_x, L_y, and L_z along its three sides. The Schrödinger equation for this situation is

$$-\frac{\hbar^2}{2m} \nabla_r^2 \psi = E\psi \tag{2.3.4a}$$

with the boundary conditions

$$\psi(x, y, z) = 0 \tag{2.3.4b}$$

For the boundary conditions given, eq. (2.3.4) has the solutions

$$\psi_{\mathbf{n}} = A_{\mathbf{n}} \sin \frac{n_x \pi x}{L_x} \sin \frac{n_y \pi y}{L_y} \sin \frac{n_z \pi z}{L_z} \tag{2.3.5}$$

corresponding to the energy values

$$E_{\mathbf{n}} = \left(\frac{n_x^2}{L_x^2} + \frac{n_y^2}{L_y^2} + \frac{n_z^2}{L_z^2} \right) \frac{\pi^2 \hbar^2}{2m}, \tag{2.3.6a}$$

or, if the box is cubic (that is, $L_x = L_y = L_z = V^{1/3}$)

$$E_{\mathbf{n}} = \frac{(n_x^2 + n_y^2 + n_z^2)\pi^2\hbar^2}{2mV^{2/3}}, \tag{2.3.6b}$$

where $n_x = 1, 2, \ldots$; $n_y = 1, 2 \ldots$; and $n_z = 1, 2, \ldots$; and $A_{\mathbf{n}}$ is an arbitrary constant. Thus, the particle can have only discrete energy values. The energy spacing between successive levels for a cubic system is

$$\frac{(2n+1)}{V^{2/3}} \frac{\pi^2\hbar^2}{2m} \tag{2.3.7}$$

where n is the minimum of the set (n_x, n_y, n_z).

If $V = 1\ \text{cm}^3$ and m is the mass of a nitrogen molecule, then

$$\frac{1}{V^{2/3}} \frac{\pi^2\hbar^2}{2m} \approx 1.06 \times 10^{-31}\ \text{erg}. \tag{2.3.8}$$

At 300K, $kT = 4.14 \times 10^{-14}$ ergs. Thus, the energy spacing for a free molecule in a macroscopic box is very small except for exceedingly large values of n_x, n_y, and n_z. This fact will be seen later to imply the validity of the classical approximation in treating the translational motion of the centers of mass of fluid molecules in macroscopic systems at all but extremely low temperatures.

Exercise

Show that if by separation of variables one seeks a solution to eq. (2.3.4) of the form $\psi = XYZ$, where $X = a_x \sin \alpha_x x + b_x \cos \beta x$, $Y = a_y \sin \alpha_y y + b_y \cos \beta y$ and $Z = a_z \sin \alpha_z z + b_z \cos \beta z$, the boundary conditions require the solution to be of the form of eq. (2.3.5) and yield the energy E given by eq. (2.3.6a).

2.3.2 Noninteracting Particles in a Box

The Schrödinger equation for this case is of the form

$$\sum_{i=1}^{N} -\frac{\hbar^2}{2m_i}\nabla_{r_i}^2 \psi = E\psi, \qquad (2.3.9)$$

with the boundary condition

$$\psi(x_1, y_1, z_1, x_2, \ldots, z_N),$$

for $x_i = 0$, or L_x, for $y_i = 0$ or L_y, or $z_i = 0$ for L_z for any $i = 1, \ldots, N$. x_i, y_i and z_i are Cartesian coordinates of the position vector $\mathbf{r}_i$ of the ith particle.

As can be shown by substitution, eq. (2.3.9) has the solutions

$$\psi_{\{\mathbf{n}\}} = \pi_{i=1}^{N} \psi_{\mathbf{n}_i}, \qquad (2.3.10)$$

where

$$\psi_{\mathbf{n}_i} = A_{\mathbf{n}_i} \sin\frac{n_{x_i}\pi x_i}{L_x} \sin\frac{n_{y_i}\pi y_i}{L_y} \sin\frac{n_{z_i}\pi z_i}{L_z}. \qquad (2.3.11)$$

$A_{\mathbf{n}_i}$ is an arbitrary constant and each of n_{x_i}, n_{y_i}, and n_{z_i} can have the values 1, 2, 3,.... The energy corresponding to (2.3.11) is

$$E_{\{\mathbf{n}\}} = \sum_{i=1}^{N}\left(\frac{n_{x_i}^2}{L_x^2} + \frac{n_{y_i}^2}{L_y^2} + \frac{n_{z_i}^2}{L_z^2}\right)\frac{\pi^2\hbar^2}{2m_i}, \qquad (2.3.12)$$

which is just the sum of the individual particle energies.

2.3.3 Rigid Rotator

Consider a mass point m constrained to move about an origin 0 at a constant radius r_o. The position $\mathbf{r}$ of the particle can be expressed in the polar coordinate system shown in Figure 2.1. The form of ∇_r^2 subject to the constraint $r = r_o$ is

$$\nabla_r^2 = \frac{1}{r_o^2}\left[\frac{1}{\sin\theta}\frac{\partial}{\partial\theta}\left(\sin\theta\frac{\partial}{\partial\theta}\right) + \frac{1}{\sin^2\theta}\frac{\partial^2}{\partial\phi^2}\right]. \qquad (2.3.13)$$

The corresponding form of Schrödinger's equation is

$$-\frac{\hbar^2}{2I}\left[\frac{1}{\sin\phi}\frac{\partial}{\partial\theta}\left(\sin\theta\frac{\partial\psi}{\partial\theta}\right) + \frac{1}{\sin^2\theta}\frac{\partial^2\psi}{\partial\phi^2}\right] = E\psi, \qquad (2.3.14)$$

where

$$I \equiv mr_o^2. \qquad (2.3.15)$$

Equation (2.3.14) is well known in the study of spherical harmonics and has solutions of the form

$$\psi_{lm} = A_{lm}Y_{lm}(\theta, \phi), \qquad (2.3.16)$$

Figure 2.1
Polar coordinate representation of mass point constrained to move on a sphere of radius r_o. $\mathbf{r} = r_o[\sin\theta \cos\phi \hat{\imath} + \sin\theta \sin\phi \hat{\jmath} + \cos\theta \hat{k}]$.

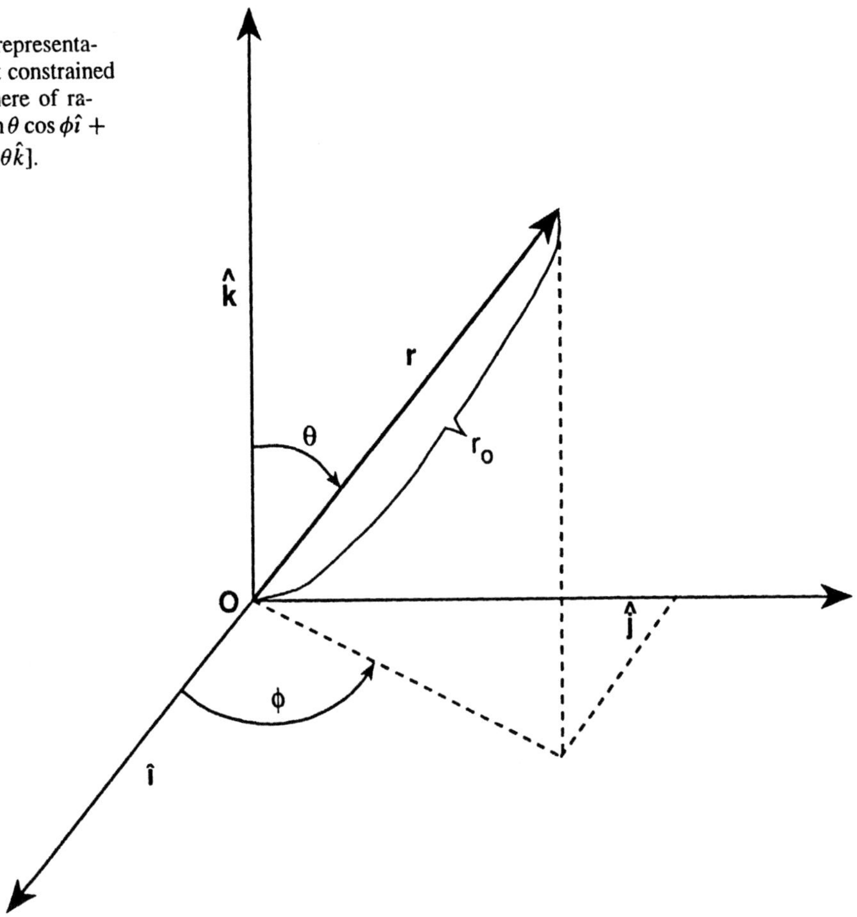

for

$$l = 0, 1, 2, \ldots$$
$$m = -l, -l+1, \ldots, 0, 1, \ldots, l,$$

where

$$Y_{lm} = P_l^m(\cos\theta)e^{im\phi} \qquad (2.3.17)$$

$$P_l^m(z) = (1 - z^2)^{|m|/2} \frac{d^{|m|}}{dz^{|m|}} P_l(z) \qquad (2.3.18)$$

$$P_l(z) = \frac{1}{2^l l!} \frac{d^l}{dz^l}(z^2 - 1)^l. \qquad (2.3.19)$$

P_l is a Legendre polynominal, P_l^m is an associated Legendre polynomial, and A_{lm} is an arbitrary constant. The energy corresponding to ψ_{lm} is

$$E_{lm} = \frac{\hbar^2}{2I} l(l+1). \qquad (2.3.20)$$

Note that E_{lm} is independent of m. Thus, there are $2l+1$ states represented by $\psi_{l,m=-l}, \psi_{l,m=-l+1}, \ldots, \psi_{l,m=l}$, corresponding to the same energy. For this reason, we say the energy state ψ_{lm} of the rotator considered here is $(2l+1)$–fold degenerate. In general, if more than one state function ψ corresponds to the same energy E, we say such states are degenerate.

2.3.4 One-Dimensional Harmonic Oscillator

The classical energy of a one-dimensional harmonic oscillator is of the form

$$H_{Cl} = \frac{1}{2}mv^2 + \frac{1}{2}\lambda x^2, \qquad (2.3.21)$$

where x is the displacement of a mass m from its equilibrium position and v is its velocity. The potential energy term, $u = \lambda x^2/2$, yields the well-known result that the restoring force of a harmonically bound particle is proportional to the displacement, that is, $F = -\partial u/\partial x = -\lambda x$. The Schrödinger equation corresponding to eq. (2.3.21) is

$$-\frac{\hbar^2}{2m}\frac{d^2}{dx^2}\psi + \frac{\lambda}{2}x^2\psi = E\psi, \qquad (2.3.22)$$

and has the solutions

$$\psi_\nu = A_\nu H_\nu\left(\frac{x}{x_o}\right) e^{-(x^2/2x_o^2)}, \qquad (2.3.23)$$

where A_ν is an arbitrary constant,

$$x_o \equiv (\hbar^2/m\lambda)^{1/4}, \qquad (2.3.24)$$

and H_ν are Hermite polynomials defined by

$$H_\nu(z) = (-1)^\nu e^{z^2}\frac{d^\nu}{dz^\nu}(e^{-z^2}). \qquad (2.3.25)$$

The allowed values of ν are $\nu = 0, 1, 2, \ldots$ and the energy corresponding to the state ψ_ν is

$$E_\nu = (\nu + 1/2)\hbar\omega \qquad (2.3.26)$$

with

$$\omega = (\lambda/m)^{1/2}. \qquad (2.3.27)$$

An interesting aspect of the classical harmonic oscillator is that the minimum vibrational energy, $E_o = \hbar\omega/2$, is nonzero. This is a purely quantum mechanical result, the minimum possible vibrational energy in classical mechanics being 0 (corresponding to $v = x = 0$).

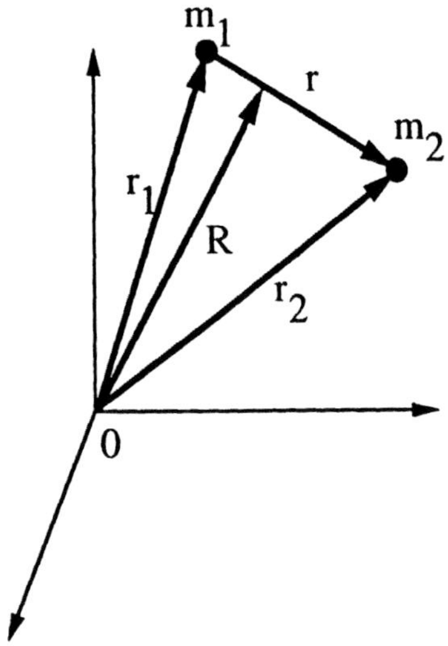

Figure 2.2
Position vectors of atoms of a diatomic molecule.

2.3.5 A Model Diatomic Molecule in a Box

Let us consider a diatomic molecule to be two mass points m_1 and m_2 held together by the potential energy $u(\mathbf{r})$, where $\mathbf{r}$ is a vector with its origin in particle 1 and extending to particle 2. The situation is illustrated in Figures 2.2 and 2.3.

The vector **R** denotes the center of mass of the pair of particles 1 and 2 and **r** the position of 2 relative to 1. The classical energy function

$$H_{Cl} = \sum_{i=1}^{2} \frac{1}{2} m_i v_i^2 + u(r) \qquad (2.3.28)$$

can be expressed in terms of the coordinates **R**, r, θ and ϕ, where θ and ϕ are polar coordinates denoting the orientation of m_1 and m_2

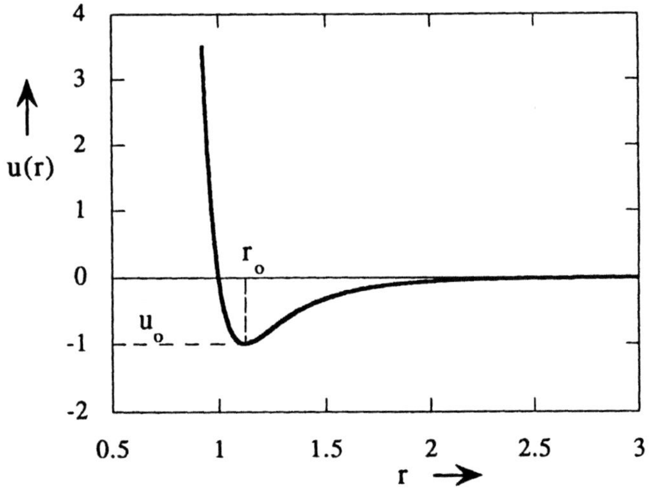

Figure 2.3
Illustration of the bond interaction potential of a diatomic molecule.

relative to a coordinate frame with its origin at the center of mass of the particles. These angles are shown in Figure 2.4.

The center of mass is defined as

$$\mathbf{R} = \frac{m_1\mathbf{r}_1 + m_2\mathbf{r}_2}{M} \tag{2.3.29}$$

and the center of mass velocity is

$$\mathbf{v}_{\mathrm{CM}} = \frac{m_1\mathbf{v}_1 + m_2\mathbf{v}_2}{M} \tag{2.3.30}$$

where

$$M = m_1 + m_2. \tag{2.3.31}$$

The vector $\mathbf{r}$ may be expressed as

$$\mathbf{r} = \mathbf{r}_2 - \mathbf{r}_1 = r(\sin\theta\cos\phi\hat{\imath} + \sin\theta\sin\phi\hat{\jmath} + \cos\theta\hat{k}) \tag{2.3.32}$$

yielding for its time rate of change

$$\begin{aligned} \mathbf{v}_r &\equiv \frac{d\mathbf{r}}{dt} = \mathbf{v}_2 - \mathbf{v}_1 \\ &= \dot{r}\hat{r} + r\omega, \end{aligned} \tag{2.3.33}$$

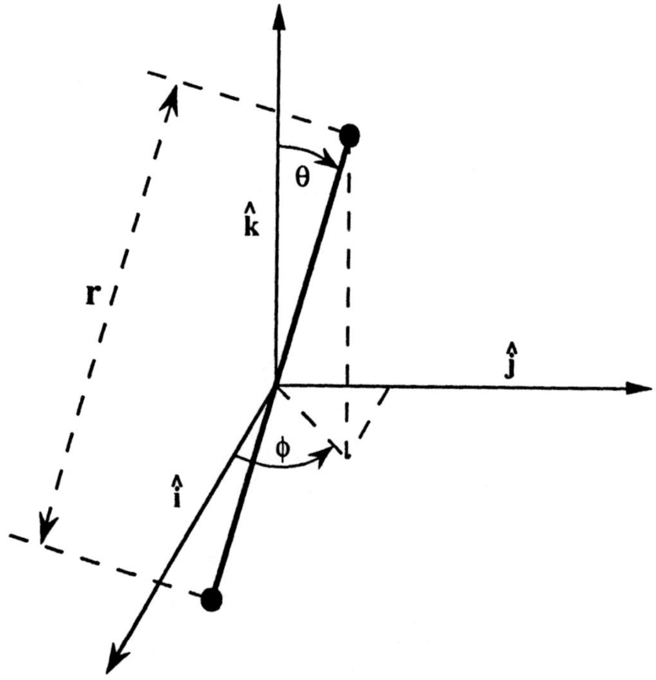

Figure 2.4
A polar coordinate system representing the orientation of the atoms of a diatomic molecule relative to an origin fixed on the center of mass of the molecule.

where ω is the angular velocity of the rotation of particles 1 and 2 relative to their center of mass. In the polar coordinates given in eq. (2.3.32) the angular velocity ω has the Cartesian components

$$\begin{aligned}\omega_x &= \dot{\theta}\cos\theta\cos\phi - \dot{\phi}\sin\theta\sin\phi \\ \omega_y &= \dot{\theta}\cos\theta\sin\phi + \dot{\phi}\sin\theta\cos\phi \\ \omega_z &= -\dot{\theta}\sin\theta.\end{aligned} \tag{2.3.34}$$

In eqs. (2.3.33) and (2.3.34) we have used the abbreviations

$$\dot{r} = \frac{dr}{dt}, \quad \dot{\theta} = \frac{d\theta}{dt}, \quad \text{and } \dot{\phi} = \frac{d\phi}{dt}. \tag{2.3.35}$$

Using eqs. (2.3.30) and (2.3.33), we can rewrite the classical energy of the model diatomic molecule in the form

$$H_{Cl} = H_{Cl}^T + H_{Cl}^R + H_{Cl}^V, \tag{2.3.36}$$

where the energy of translation of the center of mass of the diatomic molecule is

$$H_{Cl}^T \equiv \frac{1}{2}Mv_{\text{CM}}^2, \tag{2.3.37}$$

the energy of rotation of the molecule relative to the center of mass is

$$H_{Cl}^R = \frac{1}{2}I\omega^2 = \frac{1}{2}I(\dot{\theta}^2 + \sin^2\theta\dot{\phi}^2), \tag{2.3.38}$$

with

$$I \equiv m_r r^2 \tag{2.3.39}$$

and

$$m_r \equiv m_1 m_2/(m_1 + m_2), \tag{2.3.40}$$

and the energy of vibration of m_2 relative to m_1 is

$$H_{Cl}^V \equiv \frac{1}{2}m_r\dot{r}^2 + u(r). \tag{2.3.41}$$

If it is assumed that the relative vibrations are of a fairly small amplitude, H_{Cl}^V can be simplified by expanding $u(r)$ about the point r_o of its minimum and truncating after terms of order $(r - r_o)^2$,that is,

$$u(r) \approx u_o + \frac{(r - r_o)^2}{2}\left(\frac{\partial^2 u}{\partial r^2}\right)_{r_o}, \tag{2.3.42}$$

where we have used the fact that $\partial u/\partial r = 0$ at the minimum of u. With the definitions

$$x = r - r_o \tag{2.3.43}$$

and

$$\lambda \equiv \left(\frac{\partial^2 u}{\partial r^2}\right)_{r_o}, \tag{2.3.44}$$

approximation eq. (2.3.42) allows us to write eq. (2.3.41) as

$$H_{Cl}^V \approx \frac{1}{2}m_r\dot{x}^2 + \frac{1}{2}\lambda x^2 + u_o, \tag{2.3.45}$$

the same form as the harmonic oscillatory energy. If we make the further assumption that the relative vibrations are small enough to justify the approximation

$$m_r r^2 \approx m_r r_o^2 \equiv I_o, \tag{2.3.46}$$

then rotational energy H_{Cl}^R is independent of the vibrational energy H_{Cl}^V. Thus, the translational, rotational, and vibrational motions contribute H_{Cl}^T, H_{Cl}^R, and H_{Cl}^V, respectively, and are independent of one another.

A corresponding separation of the Hamiltonian operator in Schrödinger's equation is possible. In terms of $\mathbf{r}_1$ and $\mathbf{r}_2$, Schrödinger's equation has the form

$$-\sum_{i=1}^{2}\frac{\hbar^2}{2m_i}\nabla_{r_i}^2\psi + u(|\,\mathbf{r}_1 - \mathbf{r}_2\,|)\psi = E\psi. \tag{2.3.47}$$

In terms of the coordinates $\mathbf{R}$, r, θ, and ϕ defined above, eq. (2.3.47) can be transformed to the form

$$[H^T + H^R + H^V]\psi = E\psi, \tag{2.3.48}$$

where

$$H_T^\psi = -\frac{\hbar^2}{2M}\nabla_R^2\psi \tag{2.3.49}$$

$$H^R\psi = -\frac{\hbar^2}{2I}\left[\frac{1}{\sin\theta}\frac{\partial}{\partial\theta}\left(\sin\theta\frac{\partial\psi}{\partial\theta}\right) + \frac{1}{\sin^2\theta}\frac{\partial^2\psi}{\partial\phi^2}\right] \tag{2.3.50}$$

and

$$H^V\psi = -\frac{\hbar^2}{2m_r r^2}\frac{\partial}{\partial r}\left(r^2\frac{\partial\psi}{\partial r}\right) + u(r)\psi. \tag{2.3.51}$$

For small amplitude vibrations,

$$H_V^\psi \approx -\frac{\hbar^2}{2I}\frac{\partial}{\partial r}\left(r^2\frac{\partial\psi}{\partial r}\right) + \left[\frac{\lambda}{2}(r - r_o)^2 + u_o\right]\psi \tag{2.3.52}$$

and

$$I \approx m_r r_o^2 \equiv I_o. \tag{2.3.53}$$

so that H^T, H^R, and H^V are independent operators. For this case the solutions to the Schrödinger equation for a diatomic molecule in a box are

$$\psi_{\mathbf{n},lm,\nu} = \psi_{\mathbf{n}}^T \psi_{lm}^R \psi_\nu^V, \tag{2.3.54}$$

where $\psi_{\mathbf{n}}^T$ is of the form of eq. (2.3.11), ψ_{lm}^R is of the form of eq. (2.3.16), and

$$\psi_\nu^V = \frac{A_\nu}{r} H_\nu \left(\frac{r - r_o}{x_o} \right) e^{-(r-r_o)^2/2x_o}, \tag{2.3.55}$$

with $x_o = (\hbar^2/m_r\lambda)^{1/4}$. To see that ψ_ν^V is of the form given here, we let $\psi_\nu^V = \chi/r$ and find that the equation

$$H_\nu^V \psi_\nu^V = E_\nu^V \psi_\nu^V$$

transforms to the harmonic oscillator form

$$\left[-\frac{\hbar^2}{2m_r} \frac{d^2}{dr^2}\chi + \frac{\lambda}{2}(r - r_o)^2 \chi + u_o \chi \right] = E_\nu^v \chi.$$

The energy corresponding to the solution $\psi_{\mathbf{n},\mathbf{l}\mathbf{m},\nu}$ is

$$E_{\mathbf{n},lm,\nu} = E_{\mathbf{n}}^T + E_{lm}^R + E_\nu^V + u_o, \tag{2.3.56}$$

where $E_{\mathbf{n}}^T$ is of the form given by eq. (2.3.6) with atomic mass m replaced by molecular mass M, $E_{lm}^R = l(l+1)\hbar^2/2I_o$, and $E_\nu^V = (\nu + 1/2)\hbar\omega$, with $\omega = (\lambda/m_r)^{1/2}$. The allowed values of $\mathbf{n}$, l, m and ν are the same as noted previously in cases 1, 3, and 4.

2.3.6 *N* Noninteracting Diatomic Molecules of the Type Considered in Case 5

For this case the Schrödinger equation can be written in the form

$$\sum_{i=1}^{N} H_i \psi = E\psi, \tag{2.3.57}$$

where H_i is hamiltonian of the ith molecule and of the form given by eqs. (2.3.48)–(2.3.52). Because the molecules are independent (noninteracting), the solutions to eq. (2.3.57) are of the form

$$\psi_{\{\mathbf{n},lm,\nu\}} = \prod_{i=1}^{N} \psi_{\mathbf{n}_i,l_i m_i,\nu_i}, \tag{2.3.58}$$

and the corresponding energies are

$$E_{\{\mathbf{n},lm,\nu\}} = \sum_{i=1}^{N} E_{\mathbf{n}_i,l_i m_i,\nu_i}, \tag{2.3.59}$$

where the form of $E_{\mathbf{n}_i,l_i m_i,\nu_i}$ for molecule i is that given by eq. (2.3.56).

If the vibrations are not sufficiently small for approximations in eqs. (2.3.42) and (2.3.46), the rotational and vibrational motions are coupled, and the energy of a molecule is of the form

$$E_{\mathbf{n},\alpha_i} = E^T_{\mathbf{n}_i} + E^{R,V}_{\alpha_i} \tag{2.3.60}$$

where $E^R_{\alpha_i}, V$ is the energy corresponding to a rotational–vibrational state $\psi^{R,V}_{\alpha_i}$. This case is not as simple as the situation we have considered above and usually requires numerical analysis. α_i is an abbreviation for whatever collection of integers or "quantum numbers" characterize the state function $\psi^{R,V}_{\alpha_i}$. The case of vibrational–rotational coupling is treated in many standard texts on quantum mechanics and spectroscopy.

2.3.7 Energy of Hydrogen Atom in a Box

A hydrogen atom is a special case of the diatomic model discussed above. However, the Coulombic interaction between the electron and the proton cannot be well approximated by the harmonic oscillator potential. In terms of the center of mass position $\mathbf{R}$ and the position $\mathbf{r}$ of the electron relative to the proton, the Schr'odinger equation has the form

$$[H^T + H^e]\psi = E\psi, \tag{2.3.61}$$

where

$$H^T = -\frac{\hbar^2}{2M}\nabla^2_{\mathbf{R}}$$

and

$$H^e = -\frac{\hbar^2}{2m_r}\nabla^2_r - \frac{e^2}{r}.$$

The operator H^T arises from the translational motion of the hydrogen atom as a unit ($M = m_p + m_e \approx m_p$) and H^e is the Hamiltonian of a particle of mass m_r moving in a Coulumbic field centered on the proton. [Note that $m_r = m_p m_e/(m_p + m_e) \approx m_e$ and $\mathbf{R} = (m_p\mathbf{r}_p + m_e\mathbf{r}_e)/(m_e + m^p) \approx \mathbf{r}_p$].

With the boundary conditions that $\psi = 0$ for $\mathbf{R}$ on the boundary of the box and that $\int |\psi|^2 d^3r < +\infty$, the solutions to eq. (2.3.61) are

$$\psi_{\mathbf{n},n^e} = \psi^T_{\mathbf{n}}(\mathbf{R})\psi_{n^e,l^e,m^e}(r,\theta,\phi), \tag{2.3.62}$$

where $\psi^T_{\mathbf{n}}$ is of the form of eq. (2.3.5) and the electronic part of the wave function is of the form

$$\psi_{n^e,l^e,m^e} = A_{n^e,l^e,m^e}(2\kappa r)^{l^e}\exp(-\kappa r)L^{2l^e+1}_{n^e+l^e}(2\kappa r)Y_{l^e m^e}(\theta,\phi) \tag{2.3.63}$$

with the allowed quantum numbers

$$n^e = 1, 2, \ldots$$

$$l^e = 0, 1, 2, \ldots, n^e - 1 \tag{2.3.64}$$
$$m^e = -l^e, -l^e + 1, \ldots, l^e$$

The quantity κ is defined by

$$\kappa^2 = -\frac{2m_r}{\hbar^2} E^e_{n^e} \tag{2.3.65}$$

where

$$E^e_{n_e} = -\frac{m_r e^4}{2(n^e)^2 \hbar^2}. \tag{2.3.66}$$

$Y_{l^e m^e}$ is the spherical harmonic defined for the rigid rotator case and the quantity $L^{2l^e+1}_{n^e+l^e}$ is generated by the formula

$$L^s_r(z) = \frac{d^s}{dz^s} L_r(z), \qquad r > s \tag{2.3.67}$$

$$L_r(z) = e^z \frac{d^r}{dz^r}(z^r e^{-z}). \tag{2.3.68}$$

The energy corresponding to $\psi_{\mathbf{n},n^e,l^e m^e}$ is

$$E_{\mathbf{n},n^e} = E^T_{\mathbf{n}} + E^e_{n^e}, \tag{2.3.69}$$

where $E^T_{\mathbf{n}}$ is of the form given by eq. (2.3.6) with the mass m replaced by $M = m_p + m_e$.

Because the energy $E^e_{n^e}$ is independent of l^e and m^e, there are $n^e(2l^e + 1)$ different electronic states ψ_{n^e,l^e,m^e} corresponding to the same energy. Thus, the electronic energy level is $n^e(2l^e + 1)$–fold degenerate.

For electronic energies greater than zero, the electron is ionized, that is, it is not bound to the proton. In this case, the electron and proton behave as separate particles rather than as an atom.

2.3.8 *N* Noninteracting Hydrogen Atoms in a Box

As in examples above

$$H = \sum_{i=1}^{N} H_i, \quad \psi_{\{\mathbf{n},\mathbf{n}^e\}} = \prod_{i=1}^{N} \psi_{\mathbf{n}_i,n^e_i}, \quad \text{and } E_{\{\mathbf{n},\mathbf{n}^e\}} = \sum_{i=1}^{N} E_{\mathbf{n}_i,n^e_i},$$

where $\psi_{\mathbf{n}_i,n^e_i}$ and $E_{\mathbf{n}_i,n^e_i}$ are given by eqs. (2.3.62) and (2.3.69) for the ith atom.

2.3.9 A Polyatomic Model that Includes Electronic States

Our treatment of the diatomic model is appropriate for the translational, rotational, and vibrational motions of the nuclei of diatomic molecules as long as no electronic excitations occur. However, the bond potential energy $u(r)$ depends on the particular quantum state the electrons of the diatomic system are in. Thus, the energy state of a diatomic molecule undergoing small vibrations should be written in the form (in the remainder of this chapter we denote the collection of quantum numbers necessary to specify the total energy of a molecule by α and a system of molecules by $\{\alpha\}$)

$$E_\alpha = E^T_{\mathbf{n}} + E^R_{l,\mathbf{n}^e} + E^V_{\nu,\mathbf{n}^e} + E^e_{\mathbf{n}^e} + u_{o,\mathbf{n}^e} \tag{2.3.70}$$

where $E^T_{\mathbf{n}}$ is given by eq. (2.3.6) with m replaced by the total mass of the diatomic molecule;

$$E^R_{L,\mathbf{n}^e} = \frac{\hbar^2}{2I_{\mathbf{n}_e}} l(l+1) \tag{2.3.71}$$

with allowed $l = 0, 1, 2, \ldots$;

$$E^V_{\nu,\mathbf{n}^e} = (\nu + \frac{1}{2})\hbar\omega_{\mathbf{n}^e} \tag{2.3.72}$$

with allowed $\nu = 0, 1, 2, \ldots$; and $E^e_{\mathbf{n}^e}$ is the electron energy of the quantum state described by the collection of quantum numbers denoted by $\mathbf{n}^e$. Equation (2.3.71) is appropriate for the small vibration approximation as discussed earlier. However, because the bond potential depends on the electronic state so also do the equilibrium moment of inertia $I_{\mathbf{n}_e}$ and the characteristic vibration frequency $\omega_{\mathbf{n}^e}$. The electronic wave functions and energies of polyatomic molecules must be computed numerically, so we will not offer any examples here. $u_{o,\mathbf{n}^e}$ is the minimum value of the bond potential energy for the electronic energy state $\mathbf{n}^e$.

A polyatomic molecule in the small vibration approximation will undergo c independent vibrational motions (normal modes), so that its energy for a state α will be of the form

$$E_\alpha = E^T_{\mathbf{n}} + E^R_{l,\mathbf{n}^e} + \sum_{j=1}^{c}(E^V_{\nu_j,\mathbf{n}^e} + u^{(j)}_{o,\mathbf{n}^e}) + E^e_{\mathbf{n}^e}. \tag{2.3.73}$$

The rotational energy is given by eq. (2.3.71) for a linear molecule. For molecules with one axis of symmetry (for which $I_{1,\mathbf{n}^e}$ and $I_{2,\mathbf{n}^e}$ denote the two principal moments of inertia of the molecule):

$$\begin{aligned} E^R_{l,\mathbf{n}^e} &= \frac{\hbar^2}{2}\left[\frac{J(J+1)}{I_{1,\mathbf{n}^e}} + K^2\left(\frac{1}{I_{2,\mathbf{n}^e}} - \frac{1}{I_{1,\mathbf{n}^e}}\right)\right], \\ J &= 0, 1, 2, \ldots \\ K &= 0, +1, \ldots, J \end{aligned} \tag{2.3.74}$$

For unsymmetrical molecules $E^R_{l,\mathbf{n}^e}$ must be determined numerically and the rotational state l is specified by three integers.

The vibrational energy $E^v_{\nu j,\mathbf{n}^e}$ of the jth mode is of the form given by eq. (2.3.72), characterized by the fundamental frequency $\omega_{j,\mathbf{n}^e}$ and with $\nu_j = 0, 1, 2, \ldots$. The quantity $u^{(j)}_{o,\mathbf{n}^e}$ is the minimum value of the potential energy of the jth bond of the molecule in the electron state $\mathbf{n}^e$. The number of independent vibrational modes is

$$\begin{aligned} c &= n_a - 5 \quad \text{for linear molecules} \\ &= n_a - 6 \quad \text{for nonlinear molecules.} \end{aligned} \tag{2.3.75}$$

where n_a is the number of atoms in the molecule.

For a system of N noninteracting polyatomic molecules, the total energy is $E_{\{\alpha\}} = \sum_{i=1}^{N} E_{\alpha_i}$ where E_{α_i} is given by eq. (2.3.73) for the ith molecule in the small vibration limit. For large vibrations, the rotational and vibrational motions are not independent so that a decomposition of the form given by eq. (2.3.73) is not possible. However, this complication as well as the complication arising from the interactions among the molecules of an N-molecule system are only practical difficulties and do not alter the fact that the energy of finite quantum mechanical systems takes on a denumerable or discrete set of values. This is the only fact on which the general development of ensemble theory of statistical mechanics depends on.

2.4 Postulates of Ensemble Theory

In the quantum mechanical description of a molecular system, the microstate or quantum state of the system is determined by the wavefunction ψ, that is, the solution to the Schrödinger equation. Different quantum states i correspond to different wavefunctions ψ_i. The energy corresponding to a given ψ_i will be denoted by E_i. Although some distinct states (different ψ_i) have the same energy, we shall label the energy E_i by the state index so that we can refer to quantum states as the "ith quantum state" or as the "energy state E_i" without ambiguity.

After Gibbs it is convenient to define an ensemble: an ensemble is a very large number M of systems, all identical on a thermodynamic level, that is, all subject to the same thermodynamic constraints (e.g., same V, N, and T). An ensemble is classified according to its thermodynamic constraints. Common examples are: the *microcanonical ensemble*, a large collection of isolated systems ($\mathbf{N}$, V, and E same for all systems); (2) *canonical ensemble*, a large collection of closed, isothermal systems ($\mathbf{N}$, V, T same for all systems); and the *grand canonical* ensemble, a large collection of open, isothermal systems ($\boldsymbol{\mu}$, V, T same for all systems). The quantity $\mathbf{N}$ denotes $N_1, N_2, \ldots,$ the numbers of particles of the various species and $\boldsymbol{\mu}$ denotes $\mu_1, \mu_2, \ldots,$ the chemical potentials of the various species.

Subject to a given set of external (thermodynamics) constraints, say fixed V, $\mathbf{N}$ and T, a system, via intra- and interparticle interactions can generally pass in time through a great number of quantum states consistent with these constraints. The thermodynamic value of a given mechanical variable is the long time average of the values this variable takes on as the system evolves through the allowed microstates. The *first postulate* of ensemble theory of statistical mechanics *is that the thermodynamic or long time average value of a mechanical variable is equal to the ensemble average value of the variable.*

The second postulate of ensemble theory is *in an ensemble representative of an isolated thermodynamic system, the systems of the ensemble are distributed uniformly, that is, with equal probability or frequency over the possible quantum states of the system.* This is known as the postulate of equal *a priori* probability. It states that over a long period the fraction of time an isolated system spends in any one quantum state, consistent with the constraint of fixed **N**, V, and E, is the same as the fraction of time it spends in any other distinct quantum state.

The use of ensemble averages was first introduced by Gibbs (1902) and, independently, Einstein (1902, 1903). The work of Tolman (1938) was especially important in interpreting the hypotheses and implications of ensemble theory. Hill's 1960 textbook, contains a beautifully simple exposition of the theory. The validity of the postulates of ensemble theory of thermodynamics may be argued for on intuitive grounds. However, the real support of the theory derives from comparison between experiment and prediction. Since quantum mechanics was discovered, there have been no contradictions between the predictions of ensemble theory and experiment. Let us turn next to a systematic development of the implications of the two basic postulates of ensemble theory. The development draws heavily from that given by Hill.

2.5 Canonical Ensemble

Consider a large collection M of systems of volume V and composition **N** that have been equilibrated at temperature T. Suppose these systems are able to communicate through rigid diathermal walls (that is, walls impenetrable to matter but which allow heat to pass so that the systems are in thermal equilibrium) and the entire collection of systems is insulated from the rest of the universe. The result, illustrated in Figure 2.5, is a large isolated supersystem of which the original systems form subsystems.

According to quantum mechanics, a system of particles can have quantum states with discrete energies E_i, $i = 1, 2, \ldots$, which depend on the volume V and composition **N** of the system. Thus the supersystem shown in Figure 2.5 can be characterized by the set

Figure 2.5
Isolated system composed of M subsystems, each of volume V, composition **N** and at temperature T.

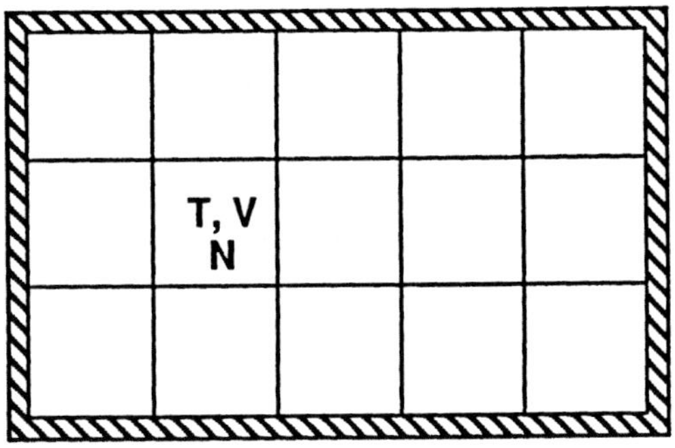

of numbers $\nu_1, \nu_2, \ldots$, where ν_i denotes the number of subsystems in the ith quantum state. The set of numbers obeys the condition

$$M = \sum_i \nu_{i.} \tag{2.5.1}$$

Because the supersystem is isolated its total energy, E_T,

$$E_T = \sum_i \nu_i E_i, \tag{2.5.2}$$

is constant.

For a given set of numbers, $\nu_1, \nu_2, \ldots$, which are called a distribution, the supersystem of Figure 2.5 can be in any of

$$\Omega(\{\nu\}) = \frac{M!}{\prod_i \nu_i!} \tag{2.5.3}$$

distinct states. To see this, consider first the special case that every subsystem is in a different quantum state. For this case there are $M!$ different states of the supersystem that can be generated by permuting the different quantum states E_i among the M subsystems. If, however, ν_1 of the subsystems are in the same quantum state, then $M!$ overcounts the number of different states of the supersystem by the factor $\nu_1!$ corresponding to the number of permutations of the ν_1 subsystems in the same state. Reasoning in this manner, we conclude that eq. (2.5.3) gives the number of distinct states of the supersystem corresponding to the distribution $\nu_1, \nu_2, \ldots$.

The total number of distinct states possible for the supersystem is

$$\Omega_T = \sum_{\nu_1, \nu_2, \ldots} \Omega(\{\nu\}), \tag{2.5.4}$$

where the sum is taken over all distributions $\nu_1, \nu_2, \ldots$ consistent with the constraints eqs. (2.5.1) and (2.5.2) of a fixed number of subsystems and constant total energy.

We can now take advantage of the postulate of equal *a priori* probabilities to compute the probability p_i that a system of volume V, of composition **N**, and at temperature T will be in quantum state E_i. For a state of the supersystem corresponding to a given distribution $\nu_1, \nu_2, \ldots$, the fraction of subsystems in state E_i is

$$\frac{\nu_i}{M}. \tag{2.5.5}$$

There are $\Omega(\{\nu\})$ states of the supersystem corresponding to the distribution $\nu_1, \nu_2, \ldots$ and there are Ω_T different states of the supersystem. According to the postulate that any possible state of an isolated system is equally likely, the probability that the supersystem, picked at random from a microcanonical ensemble of such supersystems, is in a state corresponding to the distribution $\nu_1, \nu_2, \ldots$ is

$$\frac{\Omega(\{\nu\})}{\Omega_T}. \tag{2.5.6}$$

Thus, in a microcanonical ensemble of supersystems, the average fraction of subsystems in state E_i is

$$\langle\frac{\nu_i}{M}\rangle = \sum_{\nu_1,\nu_2,\ldots} \frac{\nu_i \Omega(\{\nu\})}{\Omega_T}. \tag{2.5.7}$$

Because the supersystem itself is composed of a canonical ensemble of subsystems of volume V, of composition $\mathbf{N}$, and at temperature T, in the limit $M \to \infty$, the quantity $< \nu_i/M >$ becomes the probability that such a subsystem picked at random from a canonical ensemble will be in the quantum state i, that is,

$$p_i = \lim_{M\to\infty} \sum_{\nu_1,\nu_2\ldots} \frac{\nu_i \Omega(\{\nu\})}{\Omega_T}. \tag{2.5.8}$$

The form of p_i simplifies greatly because of the property that for large $\Omega(\{\nu\})$ is dominated by its largest value. To appreciate this fact, consider the special case that there are only two quantum states possible for the subsystem. Then

$$\Omega(\nu) = \frac{M!}{\nu_1!(M - \nu_1)!}. \tag{2.5.9}$$

Ω is a maximum at $\nu_1 = M/2$. With the aid of Stirling's approximation, $\nu! \approx (\nu/e)^\nu$, can be expressed in the form

$$\Omega(\{\nu\}) = \frac{M^M}{\nu_1^{\nu_1}(M - \nu_1)^{M-\nu_1}}. \tag{2.5.10}$$

Let m denote the deviation of ν_1 from the value $M/2$, for which Ω is maximum, that is, $m = \nu_1 - M/2$. Then the ratio of Ω to its maximum value becomes

$$\frac{\Omega}{\Omega_{\max}} = \frac{(1 - \frac{2m}{M})^{m-\frac{M}{2}}}{(1 + \frac{2m}{M})^{m+\frac{M}{2}}}. \tag{2.5.11}$$

From the defining expression for the base e of natural logarithms,

$$e^x = \lim_{n\to\infty}\left(1 + \frac{x}{n}\right)^n, \tag{2.5.12}$$

we can reduce eq. (2.5.11) to

$$\frac{\Omega}{\Omega_{max}} = \frac{(1 - \frac{2m}{M})^m}{(1 + \frac{2m}{M})^m} \tag{2.5.13}$$

for sufficiently large M. Setting $m = \xi M/2$, where ξ is a positive fraction, we obtain

$$\lim_{M\to\infty} \frac{\Omega}{\Omega_{max}} = \lim_{M\to\infty}\left(\frac{1-\xi}{1+\xi}\right)^{\xi M/2} = 0, \quad \text{if } \xi > 0 \text{ and } = 1 \text{ if } \xi = 0. \tag{2.5.14}$$

Therefore, for sufficiently large M, sums such as those occurring in eq. (2.5.8) will be totally dominated by the terms corresponding to the largest value of $\Omega(\{\nu\})$, and so

$$p_i = \lim_{M\to\infty} \frac{\nu_i^* \Omega_{max}}{M\Omega_{max}} = \lim_{M\to\infty} \frac{\nu_i^*}{M}, \tag{2.5.15}$$

where $\nu_1^*, \ \nu_2^*, \ldots$ is the distribution for which $\Omega(\{\nu\})$ is maximum. To determine this distribution for the present case, we must seek the maximum of $\Omega(\{\nu\})$ subject to the constraints, eqs. (2.5.1) and (2.5.2), of fixed M and E_T.

The distribution that maximizes Ω is the same as the one that maximizes $\ln \Omega$. If $\ln \Omega$ is a maximum for $\nu_1^*, \ \nu_2^*, \ldots$ then

$$\sum_i \left(\frac{\partial \ln \Omega}{\partial \nu_i}\right)_{\{\nu^*\}} \delta\nu_i = 0, \tag{2.5.16}$$

for allowed variations of $\delta\nu_i$. Because of the constraints of fixed M and E_T, not all ν_i's are independent, and so the $\delta\nu_i$ must satisfy the equations

$$\sum_i \delta\nu_i = 0 \tag{2.5.17}$$

and

$$\sum_i E_i \delta\nu_i = 0. \tag{2.5.18}$$

The method of Lagrange multipliers can be used to handle these constraints. Two of the variations, say $\delta\nu_1$ and $\delta\nu_2$, are fixed by the others. Multiplying eqs. (2.5.17) and (2.5.18) by the numbers α and β and subtracting these from eq. (2.5.16), we obtain

$$\sum_i \left[\left(\frac{\partial \ln \Omega}{\partial \nu_i}\right)_{\{\nu^*\}} - \alpha - \beta E_i\right] \delta\nu_i = 0. \tag{2.5.19}$$

We now choose α and β such that

$$\left(\frac{\partial \ln \Omega}{\partial \nu_i}\right)_{\{\nu^*\}} - \alpha - \beta E_i = 0, \quad i = 1 \text{ and } 2. \tag{2.5.20}$$

As a consequence, eq. (2.5.19) becomes

$$\sum_{i\neq 1,2}\left[\left(\frac{\partial \ln \Omega}{\partial \nu_i}\right)_{\{\nu^*\}} - \alpha - \beta E_i\right]\delta\nu_i = 0. \qquad (2.5.21)$$

Because eq. (2.5.20) holds for arbitrary $\delta\nu_3, \delta\nu_4, \ldots$, we conclude that

$$\left(\frac{\partial \ln \Omega}{\partial \nu_i}\right)_{\{\nu^*\}} - \alpha - \beta E_i = 0, \quad i = 3, 4, \ldots, \qquad (2.5.22)$$

the same form of the extremum condition as eq. (2.5.20) for $i = 1$ and 2.

Next we need to evaluate the derivative of $\ln \Omega$. Introducing Stirling's approximation and recalling that $M = \sum_i \nu_i$, we find the expression

$$\ln \Omega = M \ln M - \sum_i \nu_i \ln \nu_i \qquad (2.5.23)$$

whose partial derivative with respect to ν_i is

$$\frac{\partial \ln \Omega}{\partial \nu_i} = \ln M - \ln \nu_i. \qquad (2.5.24)$$

Combined with eqs. (2.5.20) and (2.5.22) this result yields

$$\ln \frac{M}{\nu_i^*} - \alpha - \beta E_i = 0, \quad i = 1, 2, 3, \ldots$$

or

$$p_i = \lim_{M\to\infty} \frac{\nu_i^*}{M} = e^{-\alpha-\beta E_i}. \qquad (2.5.25)$$

One of the parameters α and β can be related to the other by imposing the normalization condition on p_i, that is,

$$\sum_i p_i = 1 = e^{-\alpha} \sum_i e_i^{-\beta E}$$

or

$$e^{-\alpha} = \frac{1}{Q_N},$$

so that

$$p_i = \frac{e^{-\beta E_i}}{Q_N}, \qquad (2.5.26)$$

where the normalization constant Q_N is defined by

$$Q_N = \sum_i e^{-\beta E_i}. \qquad (2.5.27)$$

The quantity Q_N will be seen later to contain all thermodynamic functions, and is called the partition function indicating partitioning among quantum states.

The internal energy and pressure computed as the canonical ensemble average are

$$U = \sum_i E_i p_i \qquad \text{and} \qquad P = \sum_i P_i p_i, \tag{2.5.28}$$

where

$$P_i = -\left(\frac{\partial E_i}{\partial V}\right)_{\mathrm{N}}. \tag{2.5.29}$$

One obtains eq. (2.5.29) for the pressure corresponding to the quantum state i by noting that the reversible work that has to be done on a closed system in state i to increase its volume by dV is $dE_i = (\partial E_i/\partial V)_{\mathrm{N}} dV \equiv -P_i dV$. From the forms of Q_{N}, U, and P, it follows that

$$U = -\left(\frac{\partial \ln Q_{\mathrm{N}}}{\partial \beta}\right)_{\mathrm{N},V} \tag{2.5.30}$$

and

$$P = \frac{1}{\beta}\left(\frac{\partial \ln Q_{\mathrm{N}}}{\partial V}\right)_{\mathrm{N},\beta}. \tag{2.5.31}$$

At this point the Lagrange multiplyer is undetermined. To find what it is, it is convenient to consider an ideal monatomic gas of distinguishable particles, that is, each particle has some mechanically distinguishing characteristic. The particles are also assumed to have no internal energy states. Whether such a gas exists in reality is not important, the implications of a legitimate thought experiment being as valid as those of an actual experiment. The energy of a given quantum state of this gas, in a cubic container, is

$$E = \sum_{j=1}^{N} \frac{\hbar^2\pi^2}{2m_j V^{2/3}}(n_{x_j}^2 + n_{y_j}^2 + n_{z_j}^2), \tag{2.5.32}$$

where $n_{x_1}, n_{y_1}, \ldots, n_{z_N}$ is some set of positive integers. The sum over quantum states demanded in the definition of Q_N is then a sum of $n_{x_1}, \ldots, n_{z_N}$ over every possible set of integer $3N$-tuples. Thus,

$$Q_{\mathrm{N}} = \sum_{n_{x_1}=1}^{\infty} \cdots \sum_{n_{z_N}=1}^{\infty} \exp\left[-\beta \sum_{j=1}^{N} \frac{\hbar^2\pi^2}{2m_j V^{2/3}}(n_{x_j}^2 + n_{y_j}^2 + n_{z_j}^2)\right] = \prod_{j=1}^{N} q_j, \tag{2.5.33}$$

where

$$q_j = \sum_{n_x=1}^{\infty}\sum_{n_y=1}^{\infty}\sum_{n_z=1}^{\infty} \exp\left[-\beta \frac{\hbar^2\pi^2}{2m_j V^{2/3}}(n_x^2 + n_y^2 + n_z^2)\right]. \tag{2.5.34}$$

q_j can be rewritten in the form

$$q_j = \sum_{n_x=1}^{\infty}\sum_{n_y=1}^{\infty}\sum_{n_z=1}^{\infty} \Delta n_x \Delta n_y \Delta n_z \exp\left[-\frac{\beta\hbar^2\pi^2}{2m_j V^{2/3}}(n_x^2 + n_y^2 + n_z^2)\right], \tag{2.5.35}$$

where $\Delta n_x = \Delta n_y = \Delta n_z = 1$. Define the variables

$$\xi = \frac{\beta^{1/2}}{V^{1/3}} n_x, \qquad \zeta = \frac{\beta^{1/2}}{V^{1/3}} n_y, \qquad \eta = \frac{\beta^{1/2}}{V^{1/3}} n_z, \tag{2.5.36}$$

so that

$$q_j = \frac{V}{\beta^{3/2}} \sum_{\xi=\frac{\beta^{1/2}}{V^{1/3}}}^{\infty} \sum_{\zeta=\frac{\beta^{1/2}}{V^{1/3}}}^{\infty} \sum_{\eta=\frac{\beta^{1/2}}{V^{1/3}}}^{\infty} \Delta\xi \Delta\zeta \Delta\eta \, e^{-\frac{\hbar^2\pi^2}{2m_j}(\xi^2 + \zeta^2 + \eta^2)} \tag{2.5.37}$$

ξ, ζ, and η range from $\beta^{1/2}/V^{1/3}$ to ∞ in multiples of $\beta^{1/2}/V^{1/3}$. Imagine a system with V sufficiently large that the sum in eq. (2.5.37) becomes an integral well approximated by

$$\int_0^\infty \int_0^\infty \int_0^\infty d\xi d\zeta d\eta \, e^{-\frac{\hbar^2\pi^2}{2m_j}(\xi^2 + \zeta^2 + \eta^2)} = \frac{1}{8}\left(\frac{2\pi m_j}{\hbar^2\pi^2}\right)^{3/2}, \tag{2.5.38}$$

a result yielding for Q_N

$$Q_N = \frac{V^N}{\beta^{3N/2}} \prod_{j=1}^{N} \left(\frac{\pi m_j}{2\hbar^2\pi^2}\right)^{3/2}. \tag{2.5.39}$$

Evaluating P from eq. (2.5.31), we obtain

$$PV = \frac{N}{\beta}, \tag{2.5.40}$$

that, when compared to the thermodynamic ideal gas equation of state, $PV = NkT$, yields

$$\beta = 1/kT. \tag{2.5.41}$$

Thus, the Lagrange multiplier of a canonical ensemble is a measure of the absolute temperature T. From eq. (2.5.30), we compute for this gas the expression

$$U = \frac{3}{2} NkT. \tag{2.5.42}$$

It turns out that $\beta = 1/kT$ for an arbitrary thermodynamic system. The systems of the canonical ensemble considered in arriving at eqs. (2.5.26) and (2.5.27) were not assumed to have any particular substructure. Assume now that the systems were actually of the form shown in Figure 2.6, that is, each of the systems of the canonical ensemble is a composite of subsystems A and B, subsystem A being some arbitrary thermodynamic system and subsystem B the special ideal gas considered above. For such a composite system, the quantum energy of the composite system can be decomposed as

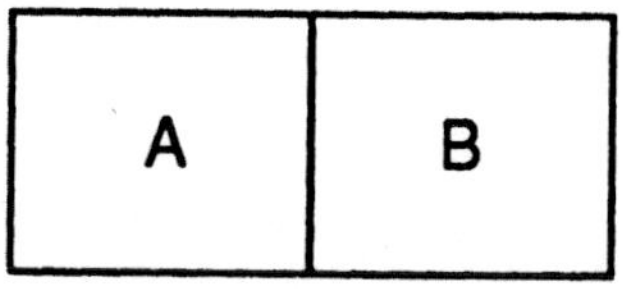

Figure 2.6
A is an arbitrary system and *B* is the special ideal gas for which eq. (2.5.42) holds. *A* and *B* are separated by a rigid, impermeable, diathermal wall.

$$E_i = E_i^A + E_i^B, \tag{2.5.43}$$

where E_i^A and E_i^B may take on all possible values independently of one another. In this case,

$$Q_N = Q_N^A Q_N^B, \qquad (2.5.44)$$

so that eq. (2.5.30) yields

$$U = U^A + U^B, \qquad (2.5.45)$$

$$U^A = -\left(\frac{\partial \ln Q_N^A}{\partial \beta}\right)_{N,V} \qquad (2.5.46)$$

$$U^B = -\left(\frac{\partial \ln Q_N^B}{\partial \beta}\right)_{N,V} = \frac{3N}{2\beta}. \qquad (2.5.47)$$

But, comparing eqs. (2.5.47) and (2.5.42), both valid for system B, we conclude that $\beta = 1/kT$ for system A and, therefore, for any system at temperature T.

Summarizing the results so far in this section, we have found for an arbitrary isothermal system

$$p_i = \frac{e^{-E_i/kT}}{Q_N} \qquad (2.5.48)$$

$$Q_N = \sum_i e^{-E_i/kT} \qquad (2.5.49)$$

$$U = \sum_i E_i p_i = kT^2 \left(\frac{\partial \ln Q_N}{\partial T}\right)_{N,V} \qquad (2.5.50)$$

$$P = -\sum_i \left(\frac{\partial E_i}{\partial V}\right)_N p_i = kT \left(\frac{\partial \ln Q_N}{\partial V}\right)_{N,T}. \qquad (2.5.51)$$

Next, to identify the formula for entropy, consider an isothermal expansion process for a closed system. From thermodynamics,

$$dU = T\,dS - P\,dV \qquad (2.5.52)$$

for the process. From eq. (2.5.50), it follows that

$$dU = \sum_i E_i\, dp_i + \sum_i p_i\, dE_i$$
$$= -kT\,d(\sum_i p_i \ln p_i) - P\,dV. \qquad (2.5.53)$$

Comparison of eqs. (2.5.53) and (2.5.54) yields

$$dS = -k\,d\left(\sum_i p_i \ln p_i\right). \qquad (2.5.54)$$

Example

From eq. (2.5.50) it follows that $U = \overline{E}$ and

$$C_V = \left(\frac{\partial U}{\partial T}\right)_{N,T}$$

$$= \frac{\overline{E^2} - \overline{E}^2}{kT^2}.$$

Thus, if $\tilde{C}_V$ is the heat capacity per molecule it follows that the mean square deviation of energy $\sigma_E^2 = \overline{E^2} - \overline{E}^2$ obeys

$$\sigma_E^2 = N\tilde{C}_V kT^2.$$

Since $\overline{E}$ goes as N and σ_E goes as $N^{1/2}$, it follows from the *central limit theorem* that the probability distribution of the energy in a closed isothermal system is well approximated by

$$P(E) =$$

$$\left[\frac{1}{(2\pi N\tilde{C}_V kT^2)^{1/2}}\right] e^{-\frac{(E-U)^2}{2N\tilde{C}_V kT^2}}.$$

Thus, the probability $P(E)\delta E$ that the energy deviates from U by $10^{-4} NkT$ is of the order of $N^{1/2} \exp(10^{-10^{-9}N})$, since $\tilde{C}_V/k$ is of the order of 5. Plainly, in macroscopic systems the deviation of the energy from its mean is unimportant.

According to thermodynamics dS is an exact differential, and so it follows from eq. (2.5.54) that $d\sum_i p_i \ln p_i$ is also an exact differential. Consequently, integration of eq. (2.5.54) yields

$$\begin{aligned} S &= -k\sum_i p_i \ln p_i + C \\ &= \frac{U}{T} - k\ln Q_{\mathrm{N}} + C, \end{aligned} \tag{2.5.55}$$

where C is a constant of integration that is independent of the quantities on which S depends, namely, **N** V, T. Without loss of generality (because only changes in S can be measured experimentally), one can set $C = 0$ to get the result

$$S = -k\sum_i p_i \ln p_i = \frac{U}{T} - \ln Q_{\mathrm{N}}, \tag{2.5.56}$$

which, when compared with the thermodynamic formula

$$S = \frac{U}{T} - \frac{F}{T}, \tag{2.5.57}$$

yields

$$F = -kT\ln Q_{\mathrm{N}}, \tag{2.5.58}$$

a fundamental connection between Q_N and the thermodynamic functions.

The relations

$$P = -\left(\frac{\partial F}{\partial V}\right)_{\mathrm{N},T}, \quad S = -\left(\frac{\partial F}{\partial T}\right)_{\mathrm{N},V}, \quad \text{and} \quad \mu_i = \left(\frac{\partial F}{\partial N_i}\right)_{N_{j\neq i},V,T} \tag{2.5.59}$$

translate through eq. (2.5.58) to

$$P = kT\left(\frac{\partial \ln Q_{\mathrm{N}}}{\partial V}\right)_{\mathrm{N},T}, \quad S = \left(\frac{\partial (kT\ln Q_{\mathrm{N}})}{\partial T}\right)_{\mathrm{N},V}, \quad \text{and} \quad \mu_i = -kT\left(\frac{\partial \ln Q_{\mathrm{N}}}{\partial N_i}\right)_{N_{j\neq i},V,T} \tag{2.5.60}$$

where μ_i is the chemical potential of component i. Thus, the calculation of thermodynamic functions is reduced to the evaluation of the canonical partition function Q_{N}.

2.6 Grand Canonical Ensemble

In this section we investigate a grand canonical ensemble, that is, a large collection M of identical systems in a permeable container of volume V, thermally equilibrated with a bath at temperature T, and chemically equilibrated (we allow mass transfer between the system and the bath) with a bath of chemical potentials $\boldsymbol{\mu}$, where $\boldsymbol{\mu}$ denotes the set of chemical potentials $\mu_1, \mu_2, \ldots$ of the species composing the system. A supersystem of the collection M is assembled and isolated as shown in Figure 2.7.

Because the systems of a grand canonical ensemble are open (permeable), $\mathbf{N}$ is not the same in each, but fluctuates about a mean value of $\overline{\mathbf{N}}(=\overline{N}_1, \overline{N}_2, \ldots)$. Consequently, a given state of the supersystem shown in Figure 2.7 corresponds to a distribution $\nu_i(\mathbf{N})$, $i = 1, 2, \ldots,\ j = 1, 2, \ldots,$ where $\nu_i(\mathbf{N})$ denotes the number of systems having composition $\mathbf{N}$ and being in the quantum state with energy $E_i\ (\mathbf{N}, \mathrm{V})$.

Let us specialize temporarily to a one-component system. For a given distribution $\{\nu\}$, that is, for a set of numbers,

$$\begin{aligned}
&\nu_1(0)\\
&\nu_1(1), \nu_2(1), \nu_3(1), \ldots\\
&\nu_1(2), \nu_2(2), \nu_3(2), \ldots,
\end{aligned}$$

the number of possible states of the supersystem is

$$\Omega(\{\nu\}) = \frac{M!}{\prod_{i,N} \nu_i(N)!}, \tag{2.6.1}$$

with

$$M = \sum_i \sum_N \nu_i(N). \tag{2.6.2}$$

In addition to eq. (2.6.2), a distribution must obey the additional constraints

$$E_T = \sum_{i,N} \nu_i(N) E_i(N, V) \tag{2.6.3}$$

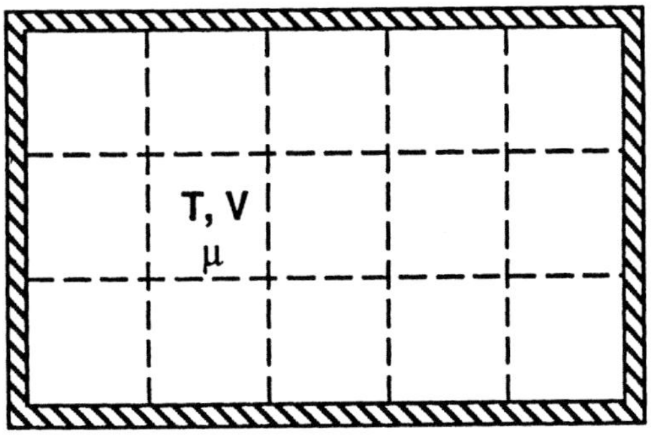

Figure 2.7
Isolated system composed of M subsystems, each with volume V, temperature T, and chemical potentials $\boldsymbol{\mu}$.

and

$$N_T = \sum_{i,N} N\nu_i(N) \tag{2.6.4}$$

because the supersystem is isolated.

Similar to the argument given in Section 2.5, we can show that the distribution $\{\nu^*\}$ for which $\Omega(\{\nu\})$ is maximum is obtained from

$$\left[\frac{\partial \ln \Omega(\{\nu\})}{\partial \nu_i(N)}\right]_{\{\nu^*\}} - \alpha - \beta E_i - \gamma N = 0, \tag{2.6.5}$$

where α, β, and γ are Lagrange multipliers arising from the constraints of fixed M, E_T, and N_T. From eq. (2.6.5), we get

$$\nu_i^*(N) = Me^{-\alpha-\beta E_i-\gamma N}. \tag{2.6.6}$$

$e^{-\alpha}$ is found from the condition $M = \sum_{i,N} \nu_i(N)$ to be

$$e^{-\alpha} = 1/\sum_{i,N} e^{-\beta E_i(N,V)-\gamma N}. \tag{2.6.7}$$

The probability $p_i(N)$ that an open isothermal system have N particles and be in energy state $E_i(N, V)$ is then

$$\begin{aligned} p_i(N) &= \lim_{M\to\infty} \frac{\sum_{\{\nu\}} \nu_i(N)\Omega(\{\nu\})}{\sum_{\{\nu\}} \Omega(\{\nu\})} \\ &= \lim_{M\to\infty} \frac{\nu_i^*(N)}{M} \\ &= \frac{e^{-\beta E_i(N,V)-\gamma N}}{\Xi}, \end{aligned} \tag{2.6.8}$$

where Ξ, the grand canonical partition function, is

$$\Xi = \sum_{i,N} e^{-\beta E_i(N,V) - \gamma N}. \tag{2.6.9}$$

The thermodynamic energy, pressure, and average number of particles in V may be computed as follows:

$$U = \sum_{i,N} E_i(N, V)p_i(N) \tag{2.6.10}$$

$$P = -\sum_{i,N} \left(\frac{\partial E_i}{\partial V}\right)_N p_i(N) \tag{2.6.11}$$

and

$$\overline{N} = \sum_{i,N} Np_i(N). \tag{2.6.12}$$

Following arguments similar to those presented in Section 2.5, we can again demonstrate that $\beta = 1/kT$. Then, taking the derivative of the expression for U given by eq. (2.6.10), we find for an isothermal process

$$dU = -kTd\left(\sum_{i,N} p_i(N) \ln p_i(N)\right) - PdV - kT\gamma d\overline{N}, \tag{2.6.13}$$

a result that, when compared to the thermodynamic expression

$$dU = TdS - PdV + \mu d\overline{N}, \tag{2.6.14}$$

yields

$$\gamma = -\mu/kT \tag{2.6.15}$$

and

$$S = -k\sum_{i,N} p_i(N) \ln p_i(N). \tag{2.6.16}$$

Again a constant of integration has been set equal to zero in writing eq. (2.6.16) for S.

The grand canonical partition function can be expressed in the following interesting form

$$\begin{aligned}\Xi &= \sum_{i,N} e^{-E_i/kT} e^{N\mu/kT} \\ &= \sum_N e^{N\mu/kT} \sum_i e^{-E_i(N,V)/kT} \\ &= \sum_N e^{N\mu/kT} Q_N(V,T),\end{aligned} \tag{2.6.17}$$

where Q_N is the canonical partition function for a system of N particles in volume V at temperature T.

The probability $p(N)$ that a system of volume V, temperature T, and chemical potential have N particles regardless of the energy state is given by

$$p(N) = \sum_i p_i(N) = \frac{Q_N}{\Xi} e^{N\mu/kT}. \tag{2.6.18}$$

The average number of particles $\overline{N}$ in an open system with V, T, can be expressed in terms of a $p(N)$,

$$\begin{aligned}\overline{N} &= \sum_{i,N} Np_i(N) = \sum_N N \sum_i p_i(N) \\ &= \sum_N Np(N) = \sum_N N\frac{Q_N}{\Xi} e^{N\mu/kT}.\end{aligned} \tag{2.6.19}$$

From the final form given for $\overline{N}$, we see that

$$\overline{N} = kT\left(\frac{\partial \ln \Xi}{\partial \mu}\right)_{V,T}. \tag{2.6.20}$$

Similarly, P and S can be computed from the grand canonical partition function as follows:

$$P = kT\left(\frac{\partial \ln \Xi}{\partial V}\right)_{\mu,T} \quad \text{and} \quad S = \left[\frac{\partial}{\partial T}(kT \ln \Xi)\right]_{V,\mu}. \tag{2.6.21}$$

Combining eqs. (2.6.8) and (2.6.16), we find

$$S = \frac{U}{T} - \frac{\overline{N}\mu}{T} + k \ln \Xi. \tag{2.6.22}$$

If the system is large enough for surface effects to be negligible, then the thermodynamic eq. (2.6.14) can be integrated to obtain

$$S = \frac{U}{T} - \frac{\overline{N}\mu}{T} + \frac{PV}{T}. \tag{2.6.23}$$

In this case, comparison of eqs. (2.6.22) and (2.6.23) yields

$$PV = kT \ln \Xi. \tag{2.6.24}$$

It is important to keep in mind that eq. (2.6.24) is only valid asymptotically, that is,

$$P = \frac{kT \ln \Xi}{V}\left[1 + O\left(\frac{1}{V}\right)\right], \tag{2.6.25}$$

whereas the pressure can be computed from the expression at eq. (2.6.21) for a system of any size.

Let us close this system with a summary of the grand canonical ensemble results for a multicomponent system.

$$p_i(\mathbf{N}) = \frac{e^{-(E_i(\mathbf{N},V) - \mathbf{N}\cdot\boldsymbol{\mu})/kT}}{\Xi} \tag{2.6.26}$$

$$p(\mathbf{N}) = \frac{Q_{\mathbf{N}} e^{\mathbf{N}\cdot\boldsymbol{\mu}/kT}}{\Xi} \tag{2.6.27}$$

$$\overline{N}_\alpha = kT\left(\frac{\partial \ln \Xi}{\partial \mu_\alpha}\right)_{V,T,\mu_{\beta\neq\alpha}} \tag{2.6.28}$$

$$S = \left[\frac{\partial}{\partial T}(kT \ln \Xi)\right]_{V,\boldsymbol{\mu}} \tag{2.6.29}$$

$$P = kT\left(\frac{\partial \ln \Xi}{\partial V}\right)_{\boldsymbol{\mu},T} \tag{2.6.30}$$

$$\Xi = \sum_{\mathbf{N}} Q_{\mathbf{N}} e^{\mathbf{N}\cdot\boldsymbol{\mu}/kT} \tag{2.6.31}$$

$$P = \frac{kT}{V} \ln \Xi[1 + O(\frac{1}{V})], \tag{2.6.32}$$

where $\mathbf{N}$ denotes the set of compositions $N_1, N_2 \ldots, \boldsymbol{\mu}$ the set of chemical potentials $\mu_1, \mu_2, \ldots$ and

$$\mathbf{N}\cdot\boldsymbol{\mu} \equiv \sum_\alpha N_\alpha \mu_\alpha. \tag{2.6.33}$$

Example

From eq. (2.6.20) it follows that $kT\partial\overline{N}/\partial\mu = \overline{N^2} - \overline{N}^2 = \overline{N}nkT\kappa_T$, where $n = \overline{N}/V$ and κ_T is the isothermal compressibility, $\kappa_T = -V^{-1}(\partial V/\partial P)_{\overline{N},T}$. In an ideal gas $nkT\kappa_T = 1$ and in condensed phase it is typically much smaller than 1. Thus, $\sigma_N \sim \overline{N}^{1/2} << \overline{N}$ and so according to the central limit theorem the probability distribution of occupancy N is

$$P(N) = \frac{1}{(2\pi\overline{N}\chi)^{1/2}} e^{-\frac{(N-\overline{N})^2}{2\overline{N}\chi}},$$

where $\chi = nkTk_T$. In a macroscopic system, say $\overline{N} = O(10^{20})$, the probability that N deviate from $\overline{N}$ is small indeed.

The details of the derivation of eqs. (2.6.26)–(2.6.32) closely follow those given above for the one-component case. Similar to the canonical ensemble situation, the calculation of thermodynamic functions is reduced to evaluation of the grand canonical partition function Ξ.

2.7 Microcanonical Ensemble

The isolated system is an important idealization in thermodynamics. Its statistical mechanical analogue is the microcanonical ensemble, that is, the ensemble systems at fixed energy E, volume V, and **N**.

Following the technique of Hill (1960) we can obtain the thermodynamic functions of a microcanonical ensemble from the results of the canonical ensemble. Denote by $W(\mathbf{N}, V, E)$ the number of quantum states of a system corresponding to the same energy E. This number is generally very large for large **N** because of the high degeneracy of the quantum states (owing to the uncertainty principle it is preferable to think of $W(\mathbf{N}, V, E)$ as the number of quantum states in a narrow energy range centered on E (for discussion of this point an advanced book on statistical mechanics should be consulted).

From a canonical ensemble, let us pick out only those systems with energy E and thermally insulate each of them. This new collection of systems can be thought of as a canonical ensemble (because all the systems are in thermal equilibrium with one another) of systems whose allowed quantum states have the same energy. The entropy of this ensemble is

$$S(\mathbf{N}, V, E) = -k \sum_j p_j \ln p_j \,. \tag{2.7.1}$$

But $e^{-E_j/kT}$ is the same for all the systems because $E_j = E$. Therefore, $1 = \sum_j p_j = p \sum_j = pW$, or $p = 1/W$, and so

$$S = -k \sum_j \frac{1}{W} \ln \frac{1}{W} = k \ln W. \tag{2.7.2}$$

This formula, connecting the entropy of an isolated system to the number of quantum states available to the isolated system, is introduced by many authors as the basic postulate of ensemble theory.

The temperature can be computed from the microcanonical ensemble from the thermodynamic relationship

$$T^{-1} = \left(\frac{\partial S}{\partial E(\mathbf{N}, V, E)} \right)_{\mathbf{N}, V} . \tag{2.7.3}$$

Example
The number of quantum states $W(E, N, V)$ of a classical monatomic gas in its lowest electronic and nuclear state is

$$W = \omega_o \left(\frac{4\pi m E}{3h^2 N} \right)^{\frac{3N}{2}} \left(\frac{V}{N} \right)^N e^{\frac{5}{2}},$$

where ω_o is the degeneracy of the ground state. The entropy of the gas in the microcanonical ensemble (closed system) is $S = k \ln W$. The temperature is obtained from eq. (2.7.3) and the pressure from eq. (2.7.4). Thus, it follows that $T^{-1} = (3Nk/2E)$ and $P = 2E/3V$, or $E = 3NkT/2$ and $P = NkT/V$.

And the pressure follows from

$$P = T\left(\frac{\partial S}{\partial V(\mathbf{N}, \mathbf{V}, E)}\right)_{\mathbf{N},E}. \qquad (2.7.4)$$

For systems macroscopic (large compared to molecular dimensions) in all dimensions, the mean values of thermodynamic quantities obey the same equations of state, with negligible differences, in all the ensembles. For example, $S(\mathbf{N}, V, E)$ in the microcanonical ensemble is for all practical purposes the same function as $S(\mathbf{N}, V, \bar{E})$ in the canonical ensemble and as $S(\bar{\mathbf{N}}, V, \bar{E})$ in the grand canonical ensemble. For systems that are small (of order of molecular dimensions) in at least one dimension, mean values can differ significantly among ensembles. This is an issue of importance with regard to thin films or fluids in micropores.

2.8 Isobaric Ensemble

In the isobaric ensemble we imagine systems kept at fixed temperature T, composition $\mathbf{N}$, and pressure P. The volume is allowed to fluctuate among the systems of this ensemble. To derive the properties of the isobaric ensemble, one must discretize the volume, that is, assume that the volumes accessible to the system are $V_j = j v_0$, where v_0 is some small element of volume.

A given state of a system then is given by its volume V_j and its energy $E_i(\mathbf{N}, V_j)$. The probability that in an isolated supersystem of fixed total energy E_T and fixed total volume V_T a system will be found in the state $E_i(\mathbf{N}, V_j)$ and V_j, can be shown by arguments similar to those of the grand canonical ensemble to be

$$P_{ij} = \frac{e^{-\beta E_i(\mathbf{N}, V_j) - \gamma V_j}}{\Delta}, \qquad (2.8.1)$$

where

$$\Delta \equiv \sum_{i,j} e^{-\beta E_i(\mathbf{N}, V_j) - \gamma V_j} \qquad (2.8.2)$$

and β and γ are Lagrange multipliers arising from conservation of total energy E_T and volume V_T of the supersystem.

Using arguments similar to those advanced in Sections 2.5 and 2.6, one can show that $\beta = 1/kT$ and $\gamma = P/kT$, and so the isobaric partition function is

$$\Delta = \sum_{i,j} e^{-\beta E_i(\mathbf{N}, V_j) - \beta P V_j}. \qquad (2.8.3)$$

The average volume in this ensemble can be computed from

$$\bar{V} = -\left(\frac{\partial \ln \Delta}{\partial \beta P}\right)_{\beta,\mathbf{N}}. \qquad (2.8.4)$$

Exercise
The monomeric units of a certain linear polymer have two configurations 1 and 2. In state 1 the monomeric energy is ϵ_1 and its length is ℓ_1. In state 2, the monomeric energy is ϵ_2 and its length is ℓ_2. Suppose the polymer has N monomers, is at temperature T and is under constant tension P. Evaluate the isobaric partition function Δ for this one-dimensional system and show that the equation of state for the mean length $\bar{L}$ of the polymer is

$$\bar{L} = \frac{N[\ell_1 e^{(P\ell_1-\epsilon_1)/kT} + \ell_2 e^{(P\ell_2-\epsilon_2)/kT}]}{[e^{(P\ell_1-\epsilon_1)/kT} + e^{(P\ell_2-\epsilon_2)/kT}]}.$$

Plot $\bar{L}/N\ell_1$ versus $P\ell_1/kT$ for $\ell_2/\ell_1 = 1.5$ and $(\epsilon_2 - \epsilon_1)/kT = 5$. Wool fibers behave somewhat as predicted by this model.

The Helmholtz free energy $F(\mathbf{N}, V_j, T)$ of a closed isothermal system of volume V_j is given by

$$F(\mathbf{N}, V_j, T) = -kT \ln \sum_i e^{-\beta E_i(\mathbf{N}, V_j)}, \tag{2.8.5}$$

and the chemical potential of species α in the system

$$\mu_\alpha(\mathbf{N}, T, V_j) = \left(\frac{\partial F}{\partial N_\alpha}(\mathbf{N}, V_j, T) \right)_{N_{\beta \neq \alpha}}. \tag{2.8.6}$$

Thus, the chemical potential in the isobaric ensemble is

$$\mu_\alpha(\mathbf{N}, T, P) = \sum_{i,j} \mu_\alpha(\mathbf{N}, T, V_j) P_{ij}, \tag{2.8.7}$$

or

$$\mu_\alpha(\mathbf{N}, T, P) = -kT \left(\frac{\partial \ln \Delta}{\partial N_\alpha} \right)_{T, P, N_{\beta \neq \alpha}}. \tag{2.8.8}$$

Because volume is really a continuous variable, it is better to replace the sum over discrete values $v_0, 2v_0, 3v_0, \ldots$ by an integral. We assume v_0 is small enough that $v_0 \Delta j = dV$, where $\Delta j = 1$, and transform eq. (2.8.3) to

$$\Delta = \sum_i \frac{1}{v_0} \int_0^\infty e^{-\beta E_i(\mathbf{N}, V) - \beta P V} dV$$

or

$$\Delta = \frac{1}{v_0} \int_0^\infty Q_\mathbf{N}(V, T) e^{-\beta P V} dV, \tag{2.8.9}$$

where $Q_\mathbf{N}$ is the canonical ensemble partition function. In this form, we see that the isobaric ensemble partition function is just the Laplace transform of the canonical ensemble partition function, with the Laplace variable βP replacing the original variable V.

2.9 Classical Mechanical Limit

Frequently some or all of the degrees of freedom of a system can be treated classically. This occurs when the spacing between successive energy levels is very small compared to kT. Particular examples will be illustrated in the next chapter. Here we simply assume that the classical limit exists and present a heuristic derivation of the conversion factor from quantum levels to phase space.

First let us introduce the probability $P_{E,\delta E}$ that an isothermal system have energy between E and $E + \delta E$, where δE is an infinitesimal quantity. Denote by i_l and i_u the value of i such that $E < E_i < E + \delta E$. Then

$$P_{E,\delta E} = \sum_{i=i_l}^{i_u} e^{-E_i/kT}/Q_N$$
$$\equiv W(E;\delta E)e^{-E/kT}/Q_N. \tag{2.9.1}$$

In classical mechanics the molecular state of a system is identified by giving the positions $\mathbf{r}_i, \ldots, \mathbf{r}_N$ and momenta $\mathbf{p}_1, \ldots, \mathbf{p}_N$ of all the particles. The energy or classical Hamiltonian H of the system is a continuous function of the positions and momenta. The general correspondence principle of quantum mechanics is that the number of quantum states in the element $dqdp$ of classical phase space is

$$\frac{dqdp}{h}, \tag{2.9.2}$$

where q is a position coordinate and p is its canonical momentum. q and p could be, for example, the x, y or z coordinates of $\mathbf{r}_i$ and $\mathbf{p}_i$. The general derivation of the correspondence principle lies outside the scope of this text, although we will give a particular example yielding eq. (2.9.2) at the end of this section.

The correspondence principle given by eq. (2.9.2) implies that $W(H;\delta H)$, the number of quantum states with energy lying between H and $H + \delta H$, is proportional to

$$\frac{1}{h^{3N}} d^3r_1 \cdots d^3r_N d^3p_1 \cdots d^3p_N \tag{2.9.3}$$

for a classical mechanical system of N atoms. If the system is a fluid, the number of quantum states $W(H, \delta H)$ is not, however, equal to this quantity. The reason is that quantum mechanical states differing only in the permutation of mechanically identical particles are not distinct states. If there are N_i particles of species N_i, then eq. (2.9.3) would over count by a factor of $N_i!$ the number of quantum states in the energy range H to $H + \delta H$. Thus, the translation of eq. (2.9.1) to the classical mechanical limit for a fluid is

$$W(E;\delta E)\frac{e^{-E/kT}}{Q_N} \to \frac{e^{-H/kT}}{Q_N}\frac{d^3r_1 \cdots d^3p_N}{\prod_i N_i! h^{3N_i}} \equiv P_N(\mathbf{r}_1, \ldots, \mathbf{p}_N)d^3r_1 \cdots d^3p_N, \tag{2.9.4}$$

where $P_N(\mathbf{r}_1, \ldots, \mathbf{p}_N)d^3r_1 \cdots d^3p_N$ denotes the classical mechanical probability that the positions and momenta of the particles in a closed isothermal system lie between $\mathbf{r}_i$ and $\mathbf{r}_i + d\mathbf{r}_i$ and $\mathbf{p}_i$ and $\mathbf{p}_i + d\mathbf{p}_i$. When P_N is of the form given by eq. (2.9.4), we say the system obeys classical mechanics and Boltzmann's statistics. This is the result derived heuristically in Chapter 1 [eq. (1.11.17)].

The positions and momenta are continuous variables and so P_N is normalized by integrating over all $\mathbf{r}_i$'s and $\mathbf{p}_i$'s, that is,

$$\int \cdots \int P_N(\mathbf{r}_i, \ldots, \mathbf{p}_N) d^3 r_1 \cdots d^3 p_N = 1. \tag{2.9.5}$$

From this it follows that the classical mechanical partition function for a system of N atoms is of the form

$$Q_\mathrm{N} = \frac{1}{\prod_i N_i! h^{3N_i}} \int \cdots \int e^{-H/kT} d^3 r_1 \cdots d^3 p_N. \tag{2.9.6}$$

For a Hamiltonian of the form

$$H = \sum_j \frac{p_j^2}{2m_j} + u^N(\mathbf{r}_1, \ldots, \mathbf{r}_N),$$

where u^N is the potential energy of the system, eq. (2.9.6) reduces to

$$Q_\mathrm{N} = \prod_i \frac{1}{N_i!} \left(\frac{2\pi m_i kT}{h^2} \right)^{3N_i/2} Z_\mathrm{N}, \tag{2.9.7}$$

where Z_N is the "configurational partition function,"

$$Z_N \equiv \int \cdots \int e^{-u^N/kT} d^3 r_1 \cdots d^3 r_N. \tag{2.9.8}$$

If the system is a classical solid, the particles are distinguishable because they are identifiable in terms of the lattice sites they occupy. Thus, the partition function of a classical solid is the same as eq. (2.9.7) except that the factor $N_i!$ is not present.

In many circumstances some of the degrees of freedom of every molecule (e.g., center of mass, translational motion and rotational motion) behave classically whereas other degrees of freedom (e.g., nuclear, electronic, and vibrational motions) behave quantum mechanically. If the part $H^{(1)}$, the Hamiltonian representing the classical mechanical degrees of freedom, is independent of the part $H^{(2)}$, representing the quantum mechanical degrees of freedom, the probability distribution function and the partition function become

$$P_{i,N_{(1)}}(q_1, \ldots, p_{N^{(1)}}) = \frac{e^{-[H^{(1)} + E_i^{(2)}]/kT}}{Q_N} \frac{dq_1 \cdots dp_{N^{(1)}}}{h^{N^{(1)}}}, \tag{2.9.9}$$

and

$$Q_N = \frac{1}{N! h^{N^{(1)}}} \left(\sum_i e^{-E_i^{(2)}/kT} \right) \int \cdots \int e^{-H^{(1)}/kT} dq_1 \cdots dp_{N^{(1)}}. \tag{2.9.10}$$

Here q_j and p_j denote canonical position and momentum coordinate of the jth classical degree of freedom. Throughout most of this textbook we shall assume that the center of mass translational motion in fluids is classical, that is, all fluid particles have degrees of freedom behaving classically (only in liquid helium at a few degrees Kelvin do atomic or molecular center of mass motions cease to behave classically to any appreciable extent).

As an example, for pure fluid nitrogen, $q_1, \ldots, p_{N^{(1)}}$ could represent the $3N$ center of mass and $2N$ rotational coordinates and momenta of nitrogen approximated as classical degrees of freedom, whereas $E_i^{(2)}$ could denote the nuclear, electronic and vibrational energy of nitrogen. The internal energy partition function $\sum_i e^{-E_i^{(2)}/kT}$ is independent of the volume of the system, and so the pressure of a system described by eq. (2.9.9) is affected only by the classical degrees of freedom. The fact that an empirical equation of state is often equally accurate for fluids having very different internal energy states is an indication that the approximation proposed by eq. (2.9.9) is a reasonable one.

Before leaving this section let us present a specific example in which the correspondence principle given at eq. (2.9.2) results. We want to find the proportionality constant C_N such that

$$\sum_i e^{-E_i/kT} \to C_N \int e^{-H/kT} d^3r_1 \cdots d^3r_N d^3p_1 \cdots d^3p_N, \qquad (2.9.11)$$

that is, the classical mechanical partition function is required to be the classical limit of the quantum mechanical one. The example is Einstein's model of a solid.

Einstein's model of a solid is that it is composed of N identical but distinguishable three-dimensional harmonic oscillators ($H = \sum_{i=1}^N p_i^2 2m + \sum_{i=1}^N (\lambda/2)(\mathbf{r}_i - \mathbf{R}_i)^2$) where $\mathbf{R}_i$ is the ith lattice site. In this case the sums in the quantum mechanical partition function can be done explicitly. In particular,

$$\begin{aligned}\sum_i e^{-E_i/kT} &= \sum_{\nu_1=0}^{\infty} \cdots \sum_{\nu_{3N}=0}^{\infty} e^{-\hbar\omega[\nu_1+(1/2)]/kT} \cdots e^{-\hbar\omega[\nu_{3N}+(1/2)]/kT} \\ &= \frac{e^{-(3/2)N\hbar\omega/kT}}{(1-e^{-\hbar\omega/kT})^{3N}},\end{aligned} \qquad (2.9.12)$$

where $\omega = (\lambda/m)^{1/2}$. By direct integration it can also be shown that

$$\int e^{-H/kT} d^3r_1 \cdots d^3p_N = [(2\pi kT)^2 m/\lambda]^{\frac{3N}{2}}, \qquad (2.9.13)$$

and so the correspondence at eq. (2.9.10) becomes

$$\left(\frac{e^{-\hbar\omega/2kT}}{1-e^{-\hbar\omega/kT}}\right)^{3N} \to C_N \left(\frac{2\pi kT}{(\lambda/m)^{1/2}}\right)^{3N} \quad \text{as } T \to \infty. \qquad (2.9.14)$$

The precise meaning of the correspondence arrow in eq. (2.9.14) is that the left-hand side goes to the right-hand side as T

goes to infinity, or $\hbar\omega/kT$ becomes sufficiently small. For small $\hbar\omega/kT$, eq. (2.9.14) becomes

$$\left[\frac{kT}{\hbar}\omega\left(1-\frac{1}{24}\left(\hbar\frac{\omega}{kT}\right)^2+\cdots\right)\right]^{3N} \to C_N\left(\frac{2\pi kT}{(\lambda/m)^{1/2}}\right)^{3N}. \tag{2.9.15}$$

Dividing each side of this expression by T^{3N}, taking for each side the limit $T \to \infty$, and equating the results, we obtain

$$C_N = \frac{1}{h^{3N}}, \tag{2.9.16}$$

in agreement with the general correspondence principle asserted at eq. (2.9.2).

Supplementary Reading

Bates, D. R., Ed. 1962 *Quantum Theory*, Academic Press, New York.

Eisberg, R. and Resnick, R. 1985. *Quantum Physics of Atoms, Molecules, Solids, Nuclei and Particles* 2nd ed., Wiley, New York.

Jeans, J. H. 1954. *The Dynamical Theory of Gases*, Dover Pub., Inc., New York.

Lay, J. E. 1990. *Statistical Mechanics and Theromodynamics of Matter*, Harper & Row, New York.

Partington, J. R. 1962. *Advanced Treatise on Physical Chemistry*, Longmans and Green, New York.

Present, R. D. 1958. *Introduction to the Kinetic Theory of Gases*, McGraw–Hill, New York.

Reif, F. 1965. *Statistical and Thermal Physics* , McGraw–Hill, New York.

Exercises

1 The Gibbs free energy is connected to the constant **N**, T, P ensemble partition function Δ by the relation

$$G(T, P, \mathbf{N}) = -kT ln \Delta.$$

For an ideal gas evaluate Δ and find expressions for G, μ_i, F, and P.

2 Consider a classical monotomic solid composed of N particles interacting with the potential energy

$$u^N = u_o^N + \sum_{j=1}^{N} \frac{\lambda}{2}(\mathbf{r}_j - \mathbf{R}_j)^4,$$

where $u_o^N (\equiv u_o^N(\{\mathbf{r}_i = \mathbf{R}_i\}))$ is the value of the potential energy of the system when the particles are at the lattice sites $\mathbf{R}_1, \ldots, \mathbf{R}_N$. λ is a constant. For this solid, derive expressions for:

a) the constant volume specific heat,

b) the change in entropy in a process in which the temperature goes from T_1 to T_2 at constant volume.

The value of the following integral may be useful to you:

$$\int_{-\infty}^{\infty} e^{-t^4} dt = 0.906.$$

3 A box of volume V contains a classical ideal gas of N_1 particles of type 1 and N_2 of type 2:

a) find the total number of states $W(\mathbf{N}, V, E)$ available to a system whose energy lies between E and $E + \sigma E, \sigma \ll E$;

b) find from the pressure P of the system as function V and T.

4 Show that the energy fluctuation in a canonical ensemble is related to the constant volume beat capacity by

$$\langle (E - \langle E \rangle)^2 \rangle = kT^2 C_V,$$

where $\langle \cdots \rangle$ denotes a canonical ensemble average. Note that $\sqrt{\langle (E - \langle E \rangle)^2 \rangle} \propto N^{1/2}$ for an ideal gas.

5 Show that the fluctuation in the number of particles in a one-component system in a grand canonical ensemble obeys the relation:

$$\langle (N - \langle N \rangle)^2 \rangle = \langle N \rangle kT \left(\frac{\partial n}{\partial P} \right)$$

where $\langle \cdots \rangle$ denotes a grand canonical ensemble average. Note that for an ideal gas $\sqrt{\langle (N - \langle N \rangle)^2 \rangle} = \sqrt{\langle N \rangle}$.

6 A solid may be represented as a system of harmonic oscillators having the Hamiltonian (energy)

$$H = \sum_{i=1}^{N} \frac{1}{2} m v^2 + \sum_{i=1}^{N} \frac{\lambda_i}{2} (\mathbf{r}_i - \mathbf{R}_i)^2 + u_o^N,$$

where $\mathbf{R}_i$ denotes the lattice point (fixed in space) about which the ith atom vibrates.

To the oscillator approximation, the intermolecular potential energy has been expanded in a Tayler's series about $\mathbf{r}_i = \mathbf{R}_i$ and truncated after the quadratic term. The force constants λ_i are related to the curvature of the potential energy function. If only nearest neighbor interactions are included, we find for u_o^N the expression

$$u_o^N = \frac{1}{2} N \nu u(l),$$

where $u(l)$ is the pair potential, ν is the number of nearest neighbors, and l is the distance of separation of nearest neighbors. For a simple cubic lattice $N/V = 1/l^3$ or $l^3 = V/N$ and $\nu = 6$.

For the harmonic oscillator approximation for a simple cubic lattice of atoms interacting via the Lennard–Jones potential,

$$u(r) = 4\varepsilon \left[\left(\frac{\sigma}{r} \right)^{12} - \left(\frac{\sigma}{r} \right)^{6} \right],$$

compute the following: (a) Z_N, (b) F, (c) U, (d) C_V, (e) P.

7 In organic molecules having conjugated double bonds, some electrons move freely from one end of the molecule to the other (or from one end of the conjugation sequence to the other). Thus, the electrons behave approximately like particles in a one-dimensional box. When white visible light shines on such molecules, photons

with the frequency $\nu = (\varepsilon_{2_e} - \varepsilon_{1_e})/h$ will be absorbed, where ε_{i_e} is the ith electronic energy state. Thus, molecules with conjugated bonds can be used as dyes. Using the quantum theory of a particle in a one-dimensional box, estimate the length L that an organic molecule with conjugated bonds must have to absorb photons with wavelengths $\lambda = 6500$ and 4000Å. These correspond to red and violet, respectively, in the spectrum of white light.

8 With reference to Problem 2.7, at what temperature would 5% of the electrons populate the first excited state (ε_{2_e}) for each molecular length L? Ignore the fact that the molecules will decompose if they are raised to these temperatures.

References

THE ELEMENTS OF ENSEMBLE THEORY

Axilrod, B. M. and Teller, E. 1943. *J. Chem. Phys.*, **11**, 2991.

Einstein, A. *Ann. Physik*, 1902. **9**, 417; 1903 **11**, 170.

Gibbs, J. W., 1902. *Elementary Principles in Statistical Mechanics*, Yale Unversity Press, New Haven, CT.

Hill, T. L. 1960. *Introduction to Statistical Thermodynamics*, Addison–Wesley, Reading, MA.

Tolman, R. C., 1938. *The Principles of Statistical Mechanics*, Oxford University Press, London.

3

STATISTICAL MECHANICS OF SOME SIMPLE SYSTEMS

3.1 Boltzmann Statistics

Without discussing certain subtleties, in Section 2.9 we gave the distribution and partition functions of a classical system obeying Boltzmann's statistics. A better appreciation of the meaning of Boltzmann's statistics can be gained by examining in detail an ideal gas.

Consider an ideal gas of identical particles. The energy of such a system in a particular quantum state is

$$E_j = \varepsilon_{1j_1} + \cdots + \varepsilon_{Nj_N} = \sum_{i=1}^{N} \varepsilon_{ij_i}, \tag{3.1.1}$$

where ε_{ij_i} is the energy of the ith particle in the jth quantum state of a single particle. The partition function is

$$Q_N = \sum{}' e^{-\sum_{i=1}^{N} \varepsilon_{ij_i}/kT}, \tag{3.1.2}$$

where $\sum'$ denotes the sum over all distinct quantum states.

Because the particles are noninteracting, one might expect to sum eq. (3.1.2) independently over each of the single particle quantum states, so that

$$Q_N = \left(\sum_{j_1} e^{-\varepsilon_{1j_1}/kT}\right)\left(\sum_{j_2} e^{-\varepsilon_{2j_2}/kT}\right)\cdots\left(\sum_{j_N} e^{-\varepsilon_{Nj_N}/kT}\right) \tag{3.1.3}$$

$$= \pi_{i=1}^{N} q_i$$

where

$$q_i = \sum_{j_i} e^{-\varepsilon_{ij_i}/kT} \tag{3.1.4}$$

is the single particle partition function. If the particles were not identical, they would be distinguishable and eq. (3.1.3) would be correct. However, if the particles are identical, they are indistinguishable according to the basic principles of quantum mechanics. The effect of the indistinguishability principle is that the quantum state represented by $\mathbf{j}' = \{j_1, \ldots, j_N\}$ is identical to that represented by $\mathbf{j}' = \{j_1', \ldots, j_N\}$ if $\mathbf{j}'$ and $\mathbf{j}$ differ only by the interchange of the identical particles among the same set of single particle states.

As a consequence of the principle of indistinguishability of identical particles, the quantity q^N grossly overestimates the partition function Q_N because nondistinct quantum states have been included in q^N. For example, consider a two particle system in which there only are the two single particle quantum states a and b. For this case, $\varepsilon_{ia} = \varepsilon_a$ and $\varepsilon_{ib} = \varepsilon_b$, and

$$Q_N = e^{-(\varepsilon_{1a}+\varepsilon_{2a})/kT} + e^{-(\varepsilon_{1a}+\varepsilon_{2b})/kT} + e^{-(\varepsilon_{1b}+\varepsilon_{2b})/kT}. \tag{3.1.5}$$

In the sum, the term $\exp[-(\varepsilon_{1b} + \varepsilon_{2a})/kT]$ does not appear, because for identical particles the quantum state corresponding to $\varepsilon_{1b} + \varepsilon_{2a}$ is the same as the one represented by $\varepsilon_{1a} + \varepsilon_{2b}$. The quantity q^2, on the other hand, is

$$\begin{aligned} q^2 = q_1 q_2 &= \left[e^{-\varepsilon_{1a}/kT} + e^{-\varepsilon_{1b}/kT}\right]\left[e^{-\varepsilon_{2a}/kT} + e^{-\varepsilon_{2b}/kT}\right] \\ &= e^{-(\varepsilon_{1a}+\varepsilon_{2a})/kT} + e^{-(\varepsilon_{1a}\varepsilon_{2b})/kT} \\ &+ e^{-(\varepsilon_{1b}+\varepsilon_{2a})/kT} + e^{-(\varepsilon_{1b}+\varepsilon_{2b})/kT} > Q_N. \end{aligned} \tag{3.1.6}$$

As illustrated in the above example, $Q_N < q^N$. In general evaluation of Q_N will be a difficult process of assessing each state for indistinguishability and evaluating the contributions to Q_N term by term. However, for gases or liquids of identical atoms or molecules at temperatures greater than a few degrees Kelvin, Q_N can be adequately estimated by

$$Q_N = \frac{q^N}{N!}. \tag{3.1.7}$$

To understand the origin of eq. (3.1.7), consider in eq. (3.1.2) the particular term $E_j = \varepsilon_{1j_1} + \varepsilon_{2j_2} + \cdots + \varepsilon_{Nj_N}$ for which all the single particle quantum states are different. There are $N!$ ways the particles $1, \ldots, N$ can be permuted among the states represented by $j_1, \ldots, j_N$. In the sum giving Q_N such states must be counted only once. In the product q^N they are counted N! times. Continuing the argument, one can see that states E_j in which $s_\alpha (\sum_\alpha s_\alpha = N)$ particles are in the same single states are overcounted in q^N by a factor differing from $N!$. Thus, in general Q_N and q^N are not related in any simple way. However, if the overwhelming contribution to Q_N comes from states E_j for which the individual particles are all in different single particle states, then eq. (3.1.7) is a valid approximation. We will derive the conditions under which eq. (3.1.7) is an acceptable approximation.

When eq. (3.1.7) is valid we say the system obeys *Boltzmann* or *classical statistics*. For an ideal gas mixture containing N_α molecules of species α, $\alpha = 1, \ldots, \nu$, the generalization of the one component result is

$$Q_N = \prod_{\alpha=1}^{\nu} \frac{q_\alpha^{N_\alpha}}{N_\alpha!} \tag{3.1.8}$$

where q_α is the single particle function for species α.

Equation (3.1.8) yields a very general result for ideal gas mixtures. The translational motion of the center of mass of the molecules is always independent of the internal degrees of freedom. Thus, it follows that $q_\alpha = q_{\alpha,I} q_{\alpha,T}$, where $q_{\alpha,I}$ is the partition function of the internal degrees of freedom (nuclear, electronic, vibrational, and rotational) of a molecule of species α and $q_{\alpha,T}$ is the partition function of the center of mass translation motion. $q_{\alpha,T}$ depends on temperature but is independent of the volume V of the system. $q_{\alpha,T}$ is usually well approximated by its classical limit, namely,

$$q_{\alpha,T} = \left(\frac{2\pi M_\alpha kT}{h^2}\right)^{3/2} V. \tag{3.1.9}$$

Because $F = -kT \ln Q_N$, it follows from eqs. (3.1.8) and (3.1.9) that

$$P = -\left(\frac{\partial F}{\partial V}\right)_{T,\mathrm{N}} = \frac{NkT}{V} \tag{3.1.10}$$

and

$$\mu_\alpha = -\left(\frac{\partial F}{\partial N_\alpha}\right)_{T,N_{\beta\alpha},V} = \mu_\alpha^+(T) + kT \ln \frac{N_\alpha}{V}, \tag{3.1.11}$$

where

$$\mu_\alpha^+ \equiv -kT \ln \left[q_{\alpha,I}(T) \left(\frac{2\pi M_\alpha kT}{h^2}\right)^{3/2} \right]. \tag{3.1.12}$$

Thus, the equation of state and the concentration, which depend on the chemical potential of a component of an ideal gas, are universal properties of all systems obeying Boltzmann statistics and the classical mechanical center of mass motion.

In Chapter 2 we introduced Boltzmann statistics in the classical limit of the partition function without really justifying it. We now investigate the conditions under which Boltzmann statistics hold. Because the probability that a particle is in the state ε is proportional to $e^{-\varepsilon/kT}$, we conclude that the "easily accessible" states are those for which ε lies below or near kT. Thus, if $\Phi(\varepsilon)$ denotes the number of quantum states with energy less than or equal to ε, it is sufficient to require

$$N \ll \Phi(\epsilon = kT) \tag{3.1.13}$$

to assure the validity of eq. (3.1.7). This is because under condition (3.1.13) there are many more easily accessible single particle quantum states than there are particles in the system. Thus, most occupations involve the situation in which all N particles are in different quantum states, and so eqs. (3.1.7) and (3.1.8) are good approximations.

We can obtain a lower bound to $\Phi(\varepsilon)$ by considering only the translational energy states. If eq. (3.1.13) is satisfied for these states, then the existence of any internal energy states only strengthens the inequality. The translational energy for a particle in a cubic box is

$$\varepsilon = \frac{h^2}{8mV^{2/3}}(n_x^2 + n_y^2 + n_z^2), \tag{3.1.14}$$

with n_x, n_y, n_z restricted to the positive integers. The states are represented in two dimensions in Figure 3.1. Each point represents a quantum state. There is one quantum state per unit volume in the quantum number space $\mathbf{n}$, as illustrated in Figure 3.1. Thus, if we define

$$R_\varepsilon^2 \equiv n_x^2 + n_y^2 + n_z^2 = \frac{8mV^{2/3}}{h^2}\varepsilon, \tag{3.1.15}$$

we can compute the number of quantum states with energy between 0 and ε as the volume of the segment of a sphere of radius R_ϵ in the positive octant of $\mathbf{n}$-space, that is,

$$\Phi(\varepsilon) = \frac{1}{8}\left[\frac{4}{3}\pi R_\epsilon^3\right] = \frac{\pi}{6}\left(\frac{8m\epsilon}{h^2}\right)^{3/2} V. \tag{3.1.16}$$

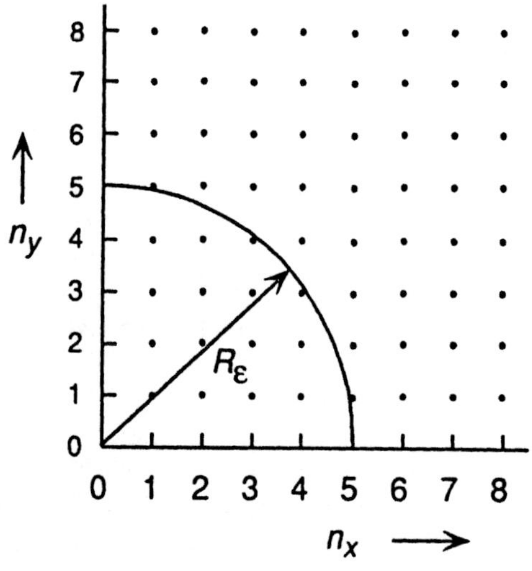

Figure 3.1
Translational energy quantum number space for a particle in a box.

With the aid of eq. (3.1.16) the condition for validity of Boltzmann statistics, eq. (3.1.13) can be expressed (aside from a factor of order of unity) as

$$\Lambda^3 \frac{N}{V} \ll 1, \quad \Lambda \equiv h/(2\pi mkT)^{1/2} \tag{3.1.17}$$

where Λ is known as the thermal de Broglie wavelength. Condition (3.1.17) means that Boltzmann statistics would maintain if the de Broglie wavelength is small compared to the average distance between the particles. This is the condition that matter would behave as discrete particles instead of overlapping waves.

In Table 3.1, values of $\Lambda^3 N/V$ are given for several substances at 1 atm in the dilute gas state and in the liquid state. The temperatures chosen are the normal boiling points of the substances. It follows from the entries in the table that classical statistics will be a good approximation for dilute gases down to extremely low temperatures, say below about 10K. Even at liquid densities, the condition that $\Lambda^3 N/V$ be very small compared to unity is well satisfied for all but light elements (molecular weights less than about 20) at temperatures lower than about 30K. Thus, in most situations involving fluids, Boltzmann statistics can be applied. In the next section we will see that when the translational degrees of freedom behave classically, Boltzmann statistics are always operative, a fact legitimizing our use of the factor $1/\pi N_i!$ in the classical mechanical limit of the partition function given in Chapter 3.

At very low temperatures or in the case of small particles such as electrons, alternatives to eqs. (3.1.7) and (3.1.8) must be considered. These alternatives involve Fermi–Dirac or Bose–Einstein statistics, the properties of which will not be pursued in this text.

TABLE 3.1 Values of $\Lambda^3 N/V$ for Substances in Vapor and Liquid State at Normal Boiling Points

	Temperature	$\Lambda^3 N/V$	
Substance	*(K)*	*Vapor*	*Liquid*
He	4.2	0.12	1.5
H_2	20.4	7.5×10^{-3}	0.44
Ne	27.2	1.11×10^{-4}	0.015
N_2	77.4	4.91×10^{-6}	0.00091
A	87.4	2.12×10^{-6}	0.00054
Hexane	342.1	2.30×10^{-8}	4.90×10^{-6}

Pressure = 1 atm at normal boiling points.

3.2 Monatomic Gases

The energy of an ideal monatomic particle can be expressed as

$$\varepsilon = \varepsilon^n + \varepsilon^e + \varepsilon^T, \tag{3.2.1}$$

where ε^n, ε^e, and ε^T are the nuclear, electronic, and translational energy of the particle. As the spacing between nuclear energy levels is generally equivalent to tens or hundreds of trillions of degrees Kelvin, molecules remain in their lowest energy state in fluids at conditions of interest in this book. At such conditions, the nuclear states do not contribute to the heat capacity or equation of state behavior of the fluid and will usually be ignored in the theory of these quantities. If the particles remain mostly in their lowest energy state (ground state), then to a good approximation it can be assumed that the electronic and nuclear states are independent. Thus, the single particle partition function becomes

$$q = q_n q_e q_T, \tag{3.2.2}$$

where

$$q_n = \sum_{\substack{\text{nuclear}\\ \text{states}}} e^{-\varepsilon_i^n/kT} \tag{3.2.3}$$

$$q_e = \sum_{\substack{\text{electronic}\\ \text{states}}} e^{-\varepsilon_i^e/kT}, \tag{3.2.4}$$

and

$$q_T = \sum_{n_x=1}^{\infty} \sum_{n_y=1}^{\infty} \sum_{n_z=1}^{\infty} e^{-\frac{h^2}{8mV^{2/3}}(n_x^2+n_y^2+n_z^2)}. \tag{3.2.5}$$

Because the thermodynamic functions of a macroscopic system are not affected by the shape of the containing volume, it is assumed that the gas is contained in a cube of volume V.

As noted in Chapter 2, quantum states are often degenerate, that is, more than one state can have the same energy. If we denote by w_i the degeneracy or the number of states having the same energy ε_i, then q_n and q_e can be expressed in the more useful forms

$$q_n = \sum_{\substack{\text{energy}\\ \text{levels}}} w_{ni} e^{-\varepsilon_i^n/kT} = e^{-\varepsilon_1^n/kT} \sum_{\substack{\text{energy}\\ \text{levels}}} w_{ni} e^{-\Delta\varepsilon_i^n/kT}, \tag{3.2.6}$$

$$q_e = \sum_{\substack{\text{energy}\\ \text{levels}}} w_{ei} e^{-\varepsilon_i^e/kT} = e^{-\varepsilon_1^n/kT} \sum_{\substack{\text{energy}\\ \text{levels}}} w_{ei} e^{-\Delta\varepsilon_i^n/kT}, \tag{3.2.7}$$

and $\Delta\varepsilon_i \equiv \varepsilon_i - \varepsilon_1$.

By convention we assign $\varepsilon_1^n = 0$. Because $\varepsilon_i^n/k =$ is of the order of 10^9 to 10^{10}K for values of $i > 1$, it follows that $e^{-\varepsilon_i^n/kT} \approx 0$ for $i > 1$, and so

$$q_n = w_{n1} \tag{3.2.8}$$

to an excellent approximation. Thus, only the degeneracy of the nuclear ground state is needed for computing thermodynamic properties.

Similarly, for $i > 1$, the ratio $\Delta\varepsilon_i^e/kT$, is often much greater than unity, in which case

$$q_e = w_{e1} e^{-\varepsilon_1^e/kT} \tag{3.2.9}$$

is a good approximation. For example, for the hydrogen atom $(\varepsilon_i^e - \varepsilon_1^e)/k = 1.56 \times 10^5[1 - i^{-2}]$K. Clearly, for temperatures smaller than 10^4K, eq. (3.2.9) is an excellent representation of q_e. Similarly, for $i > 0$, ε_i^e/k is of order 10^5 to 10^6K for the inert gas atoms, is of order 10^4K for alkali metal atoms, and is of order 5×10^3 to 10^4K for halogen atoms.

For temperatures on the order of 1000K, the first two terms in q_e must be kept for some of the halogens if high accuracy is required, that is,

$$q_e = e^{-\varepsilon_1^e/kT}[w_{e1} + w_{e2} e^{-\Delta\varepsilon_2^e/kT}]. \tag{3.2.10}$$

Pertinent parameters for these atoms are given in Table 3.2. The data were obtained from spectroscopy. At $T = 1000$K, $q_e/w_{e1}e^{-\varepsilon_1^e/kT} = 1.28, 1.14, 1.003$, and 1.00001 for F, Cl, Br, and I, respectively.

The probability that a particle is in the electronic energy level ε_i^e is

$$p_i^e = \frac{w_{ei} e^{-\varepsilon_i^e/kT}}{q_e}. \tag{3.2.11}$$

When eq. (3.2.9) is a good approximation, we have $p_1^e = 1$, $p_i^e = 0$, $i > 1$, that is, all the particles in the system are in their ground electronic state. For the halogens at 1000K,

$$\begin{aligned} p_1^e &= 0.78,\ 0.88,\ 0.997,\ 0.999999 \quad \text{and} \\ p_2^e &= 0.22,\ 0.12,\ 0.003,\ 10^{-15}, \end{aligned} \tag{3.2.12}$$

respectively.

Thus, almost all bromine and iodine atoms are in their ground state, whereas 22% of the fluorine and 12% of the chlorine atoms are in the first excited state ($i = 2$).

TABLE 3.2 Spectroscopic Data for Halogen Atoms

Atom	w_{e1}	w_{e2}	$\Delta\epsilon_2^e/k$ (K)
F	4	2	571.7
Cl	4	2	1,257.8
Br	4	2	5,259.6
I	4	2	10,747.8

Let us consider now the evaluation of q_T. Transforming the sum similarly to what we did in Section 2.5 of Chapter 2 we obtain

$$q_T = \left(\frac{8mkT}{h^2}\right)^{3/2} V \sum_{\xi} \sum_{\zeta} \sum_{\eta} \Delta\xi \Delta\zeta \Delta\eta e^{-(\xi^2+\zeta^2+\eta^2)} \quad (3.2.13)$$

where ξ, ζ, and η range from $(h^2/8mkT)^{1/2}/V^{1/3}$ to infinity in multiples of $(h^2/8mkT)^{1/2}/V^{1/3}$. Because the major contribution to the summation in eq. (3.2.13) comes from the values of ξ, ζ, and η in which $e^{-\xi^2}$, $e^{-\zeta^2}$, $e^{-\eta^2}$ are of the order of unit, the summation can be replaced by integration if

$$\left(\frac{h^2}{8mkT}\right)^{1/2} / V^{1/3} \ll 1. \quad (3.2.14)$$

With the definition $\Lambda = h/(2\pi mkT)^{1/2}$, the condition in eq. (3.2.14) becomes

$$\frac{\pi}{2} \frac{\Lambda}{V^{1/3}} \ll 1. \quad (3.2.15)$$

This expresses the condition that the translational energies are continuous, that is, that the translational motion of the particles can be treated classically. We saw in the previous section that for atoms and molecules above about 20K, the condition $N\Lambda^3/V < 0$ holds, or

$$\frac{\Lambda}{V^{1/3}} < N^{-1/3}. \quad (3.2.16)$$

Because $N^{-1/3} \ll 1$, for macroscopic systems eq. (3.2.15) is strongly obeyed. Thus, we can pass the integral in eq. (3.2.13) and carry out the integrations to obtain

$$q_T = \left(\frac{2\pi mkT}{h^2}\right)^{3/2} V. \quad (3.2.17)$$

To summarize then, the partition function for a one component ideal gas obeying Boltzmann's statistics can be accurately estimated by

$$Q_N = \frac{q^N}{N!} = \frac{(q_n q_e q_T)^N}{N!} = \frac{(w_{n1} q_e)^N}{N!} \left[\frac{2\pi mkT}{h^2}\right]^{3N/2} V^N, \quad (3.2.18)$$

where q_e is given by either eq. (3.2.9) or (3.2.10). The corresponding formulas for the Helmholtz free energy, entropy, energy, heat capacity, and pressure q_N in the system are [with $N! = (N/e)^N$, q_e from eq. (3.2.10)]

$$\begin{aligned} F &= F_T + F_e + F_n \\ &= -NkT \ln\left[\left(\frac{2\pi mkT}{h^2}\right)^{3/2} e\frac{V}{N}\right] - NkT \ln\left\{e^{-\varepsilon_1^e/kT}\left[w_{e1} + w_{e2} e^{-\Delta\varepsilon_2^e/kT}\right]\right\} \quad (3.2.19) \\ &\quad -NkT \ln w_{n1}; \end{aligned}$$

$$S = S_T + S_e + S_n$$

$$= Nk \ln\left[\left(\frac{2\pi mkT}{h^2}\right)^{3/2} e^{5/2}\frac{V}{N}\right] + Nk \ln\left\{e^{-\varepsilon_1^e/kT}\left[w_{e1} + w_{e2}e^{-\Delta\varepsilon_2^e/kT}\right]\right\} \tag{3.2.20}$$

$$+\frac{N}{T}\frac{\varepsilon_1^e w_{e1} + \varepsilon_2^e w_{e_2} e^{-\Delta\varepsilon_2^e e/kT}}{w_{e1} + w_{e2}e^{-\Delta\varepsilon_2^e/kT}} + Nk \ln w_{n1};$$

$$U = U_T + U_e + U_n$$

$$= \frac{3}{2}NkT + N\frac{\varepsilon_1^e w_{e1} + \varepsilon_2^e w_{e2}e^{-\Delta\varepsilon_2^e/kT}}{w_{e1} + w_{e2}e^{-\Delta\varepsilon_2^e/kT}} + 0; \tag{3.2.21}$$

Exercise
The energy of an ideal gas particle in the presence of gravity is $mv^2/2 + mgz$, where z is the verticle position above the earth's surface. If the gas behaves classically and is contained in an infinitely tall cylinder at temperature T, then the canonical ensemble partition function is

$$Q_N = \frac{1}{N!}\left(\frac{AkT}{mg}\right)^N \left(\frac{2\pi mkT}{h^2}\right)^{\frac{3N}{2}},$$

where A is the cross-section of the cylinder. The heat capacity of the gas is then

$$C_V = \frac{5}{2}Nk,$$

Nk greater than an ideal gas in the absence of gravity. Consider the gas confined to a cylinder of height $z = L$, where $L << kT/mg$. Evaluate the partition function for this case and show why we do not observe the effect of gravity on heat capacity in laboratory experiments.

$$C_V = C_{V,T} + C_{V,e} + C_{V,n}$$

$$= \frac{3}{2}Nk + \frac{N(w_{e1}w_{e2})(\Delta\varepsilon_2^e)^2}{kT^2}\frac{e^{-\Delta\varepsilon_2^e/kT}}{[w_{e1} + w_{e2}e^{-\Delta\varepsilon_2^e/kT}]^2} + 0; \tag{3.2.22}$$

$$PV = NkT. \tag{3.2.23}$$

The ideal gas equation of state is independent of internal states of the gas particles.

The only contribution of the nuclear states is to the free energy, namely, $F_n = -NkT \ln w_{n1}$ and entropy $S_n = Nk \ln w_{nl}$. These contributions arise from the degeneracy of the ground state. In the limit $\Delta\varepsilon_2^e/kT = \infty$, $F_e = -NkT \ln w_{e1}$, $S_e = Nk \ln w_{e1}$, $U_e = N\varepsilon_{e1}$, and $C_{V,e} = 0$. Approximation by this limit is usually adequate. At $T = 1000$K, for example, the ratio $C_{V,e}/C_{V,T} = 0.04$ and 0.09 for F and Cl and is extremely small for Br and I. The ratio is even smaller for inert gases. It is no surprise then that, to a very good approximation, $C_V = 3Nk/2$ for monatomic gases

For a multicomponent ideal gas obeying Boltzmann's statistics

$$Q_N = \prod_{\alpha=1}^{\nu} \frac{q_\alpha^{N_\alpha}}{N_\alpha!}, \tag{3.2.24}$$

where q_α is the single particle partition function for species α, the equation of state of the system is still $PV = NkT$, but the Helmholtz free energy becomes

$$F = \sum_{\alpha=1} F_\alpha(N_\alpha, P, T) + F_M \tag{3.2.25}$$

where $F_\alpha(N_\alpha, P, T)$ denotes the free energy of pure species α at temperature T and pressure P; and F_M, the free energy of mixing, is

$$F_M = \sum_{\alpha=1} N_\alpha kT \ln(N_\alpha/N). \tag{3.2.26}$$

The formula for $F_\alpha(N_\alpha, P, T)$ can be determined from those of the one component case by setting $V/N = kT/P$ in them.

The thermodynamic energy and entropy of the ideal gas mixture are

$$U = \sum_{\alpha=1} U_\alpha(N_\alpha, T) \tag{3.2.27}$$

and

$$S = \sum_{\alpha=1} S_\alpha(N_\alpha, P, T) + S_M \tag{3.2.28}$$

with

$$S_M = \sum_{\alpha=1} N_\alpha k \ln(N_\alpha/N). \tag{3.2.29}$$

These imply what is already known from thermodynamics: the energy of mixing of an ideal gas is zero and the entropy of mixing is positive and of the form given by eq. (3.2.29).

3.3 Polyatomic Gases

We consider gases in which only the ground state of the nuclear and electronic energies contribute to the partition function and in which the center of mass translational, the rotational, and the vibrational motions are independent of one another. In this case the single molecule partition function is of the form

$$q = q_n q_e q_T q_R q_V, \tag{3.3.1}$$

where the subscripts n, e, T, R, and V denote nuclear, electronic, translational, rotational, and vibrational partition functions. In the case considered here $q_n = w_{n1}$ and $q_e = w_{e1}\exp(-\varepsilon_1^e/kT)$. For the translational partition function the classical mechanical approximation (which as we saw in the previous section is almost always a good approximation for molecular systems) is

$$q_T = \left(\frac{2\pi MkT}{h^2}\right)^{3/2} V. \tag{3.3.2}$$

Often the rotational partition function and sometimes the vibrational partition function behave classically, and so we examine their classical mechanic limits now. If for a diatomic molecule we choose the polar and azimuthal angle θ and ϕ, then the corresponding canonical momenta are $p_\theta \equiv \partial H^R/\partial\dot{\theta} = I\dot{\theta}$ and $p_\phi \equiv \partial H^R/\partial\dot{\phi} = I\dot{\phi}\sin^2\theta$. Thus the classical mechanical Hamiltonian rotational energy is

$$H^R = \frac{p_\theta^2}{2I} + \frac{p_\phi^2}{2I\sin^2\theta}, \tag{3.3.3}$$

and the rotational partition function is

$$q_R = \int_{-\infty}^{\infty}\int_{-\infty}^{\infty}\int_{0}^{2\pi}\int_{0}^{\pi} e^{-H^R/kT}\,\frac{d\theta d\phi dp_\theta dp_\phi}{\sigma h^2} = \frac{8\pi^2 IkT}{\sigma h^2}. \qquad (3.3.4)$$

The factor σ is 1 if the molecule is heteronuclear (e.g., HCl) and is 2 if it is homonuclear (e.g., N_2). The reason for the factor is that classical phase space counts configurations involving interchange of identical atoms in the homonuclear case.

Equation (3.3.4) is the classical mechanical rotational partition function of any linear polyatomic molecule (e.g., N_2, HCl, CO_2, C_2H_2 , etc.) if the effect of vibrational motion on the moment of inertia can be neglected. In the case of a nonlinear polyatomic molecule, Euler angles must be used to evaluate the classical mechanical rotational partition function. The details are rather tedious, but the final result is simply

$$q_R = \frac{(\pi I_A I_B I_C)^{1/2}}{\sigma}\left(\frac{8\pi^2 kT}{h^2}\right)^{3/2} \qquad (3.3.5)$$

where I_A, I_B, and I_C are the principal moments of inertia, remembering that identical atoms are indistinguishable and σ is the number of identical configurations attained by rotation of the molecule about its center of mass ($\sigma = 12$ for methane and benzene).

The classical mechanical vibrational partition function is also quite simple. The classical mechanical vibrational energy of a diatomic molecule is

$$H^V = \frac{p^2}{2m_r} + \frac{\lambda x^2}{2} + u_o \qquad (3.3.6)$$

and so the classical mechanical vibrational partition function is

$$q_V = \int_{-\infty}^{\infty}\int e^{-H^V/kT}\frac{dxdp}{h} = \frac{2\pi kT}{h}\left(\frac{m_r}{\lambda}\right)^{1/2} e^{-u_o/kT}. \qquad (3.3.7)$$

The lower limit of x is actually $-r_o$, the average bond length of the molecule. However, it is consistent with the harmonic oscillator approximation that $\lambda r_o^2/2$ is large compared to kT and so letting x range from $-\infty$ to ∞ in eq. (3.3.7) is consistent with eq. (3.3.6).

Mechanically, the motion of any polyatomic molecule can be described in terms of translation of its center of mass, rigid body rotation about the center of mass of the molecule in its lowest energy configuration, and "vibrations" or motions of the atoms that deform the molecule from its lowest energy configuration. In many substances such as water, methane, benzene, and acetone, the "vibrational" motions can be well approximated as small displacements from the lowest energy configuration. In these cases, the classical vibrational energy is well approximated as a system of coupled harmonic oscillators. By appropriate coordinate transformation the classical mechanical Hamiltonian of n_V coupled harmonic oscillators can be expressed in the form

$$H^V = \sum_{i=1}^{n_v} \left[\frac{p_i^2}{2m_i} + \frac{\lambda_i}{2} \xi_i^2 \right] + u_o \tag{3.3.8}$$

where m_i has units of mass and λ_i has units of force per unit length. The vibrational partition function corresponding to eq. (3.3.8) can be evaluated easily,

$$\begin{aligned} q_V &= \frac{e^{-u_o/kT}}{h^{n_v}} \int_{-\infty}^{\infty} \cdots \int \prod_{i=1}^{n_v} e^{-\left(\frac{p_i^2}{2m_i} + \frac{\lambda_i}{2}\xi_i^2\right)/kT} d\xi_i dp_i \\ &= \left(\frac{2\pi kT}{h} \right)^{n_v} \prod_{i=1}^{n_v} (m_i \lambda_i)^{1/2} e^{-u_o/kT}. \end{aligned} \tag{3.3.9}$$

n_v is the number of vibrational degrees of freedom. Recall from Chapter 2, Section 2.3, that $n_v = 3n_a - 5$ for linear molecules and $3n_a - 6$ for nonlinear molecules, where n_a is the number of atoms composing the molecule.

The partition function Q_N of a one component ideal polyatomic gas obeying Boltzmann's statistics and having a partition function of the form of eq. (3.3.1) is

$$Q_N = \frac{(q_n q_e q_T q_R q_V)^N}{N!}. \tag{3.3.10}$$

The Helmholtz free energy is of course given by $F = -kT \ln Q_N$ and the pressure and entropy can be computed from $P = -(\partial F/\partial V)_{T,N}$ and $S = -(\partial F/\partial T)_{V,N}$.

Because only q_T depends on volume, the equation of state is

$$PV = NkT. \tag{3.3.11}$$

As long as the translational partition function can be approximated by its classical mechanical formula, the ideal gas equation of state is of the form given by eq. (3.3.11), whether or not the nuclear electronic vibrational and rotational motions are independent or behave classically or quantum mechanically.

The free energy, entropy, and energy corresponding to eq. (3.3.10) are

$$\begin{aligned} F &= F_n + F_e + F_T + F_R + F_V \\ &= -NkT \ln w_{n1} - NkT \ln w_{e1} + N\varepsilon_1^e \\ &\quad -NkT \ln \left[\left(\frac{2\pi MkT}{h^2} \right)^{3/2} e\frac{V}{N} \right] - NkT \ln \left[\frac{8\pi IkT}{\sigma h^2} \quad \text{or} \quad \frac{(\pi I_A I_B I_C)^{1/2}}{\sigma} \left(\frac{8\pi^2 kT}{h^2} \right)^{3/2} \right] \\ &\quad -NkT \ln \left[\left(\frac{2\pi kT}{h} \right)^{n_v} \prod_{i=1}^{n_v} (m_i/\lambda_i)^{1/2} \right] + Nu_o, \end{aligned} \tag{3.3.12}$$

$$
\begin{aligned}
S &= S_n + S_e + S_T + S_R + S_V \\
&= Nk \ln w_{n1} + Nk \ln w_{e1} + Nk \ln \left[\left(\frac{2\pi MkT}{h^2} \right)^{5/2} e^{3/2} \frac{V}{N} \right] \\
&+ kT \left\{ \ln \left(\frac{8\pi^2 eIkT}{\sigma h^2} \right) \quad \text{or} \quad \ln \left[\frac{e^{3/2} (\pi I_A I_B I_C)^{1/2}}{\sigma} \left(\frac{8\pi^2 kT}{h^2} \right)^{3/2} \right] \right\} \\
&+ Nk \ln \left[\left(\frac{2\pi ekT}{h} \right)^{n_v} \prod_{i=1}^{n_v} (m_i/\lambda_i)^{1/2} \right],
\end{aligned}
\tag{3.3.13}
$$

and (from $U = F + TS$)

$$
\begin{aligned}
U &= U_n + U_e + U_T + U_R + U_V \\
&= O + N\varepsilon_1^e + \frac{3}{2} NkT + \left\{ NkT \quad \text{or} \quad \frac{3}{2} NkT \right\} + n_v NkT + u_0.
\end{aligned}
\tag{3.3.14}
$$

Equation (3.3.14) illustrates the so-called equipartition theorem of classical statistical mechanics, namely, each translational and rotational degree of freedom contributes $kT/2$ to the thermodynamic energy and each vibration degree of freedom contributes kT. There are always three translational degrees of freedom, the x, y and z components of the kinetic energy. A linear polyatomic has two rotational degrees of freedom (θ and ϕ) and a nonlinear polyatomic has three rotational degrees of freedom (three Euler angles). The number of vibrational degrees for vibrations $n_v = 3n_a - 5$ for linear and $3n_a - 6$ for nonlinear molecules. In terms of the constant volume λ heat capacity, the equipartition theorem reads

$$
\begin{aligned}
C_V^T &= \frac{3}{2} Nk \\
C_V^R &= Nk \quad \text{or} \quad \frac{3}{2} Nk \\
C_V^V &= N(3n_a - 5)k \quad \text{or} \quad N(3n_a - 6)k.
\end{aligned}
\tag{3.3.15}
$$

For the monatomic gases Ar, Kr and Xe at room temperature, experiments yielded the expected $C_v = 3Nk/2$. However, in the case of N_2 and O_2 at room temperature, it was found that $C_V \approx 5Nk/2$, not the expected value $C_V = 7Nk/2$. Moreover, the observed heat capacities of dilute polyatomic gases tended to be temperature dependent. This failure of classical statistical mechanics turns out to be a failure of classical mechanics. It is not a failure of statistical mechanics as we shall learn next by considering the quantum mechanical versions of q_R and q_V.

Let us turn again to the quantum mechanical forms of the rotational and vibrational partition functions of a heteronuclear diatomic molecule. In the case of rotations

$$
\begin{aligned}
q_R &= \sum_{l=0}^{\infty} \sum_{m=-l}^{l} e^{-l(l+1)h^2/8\pi^2 IkT} \\
&= \sum_{l=0}^{\infty} (2l+1) e^{-l(l+1)\Theta_R/T}
\end{aligned}
\tag{3.3.16}
$$

where Θ_R is a characteristic rotational energy expressed in units of temperature, namely

$$\Theta_R \equiv \frac{h^2}{8\pi^2 Ik}. \tag{3.3.17}$$

If Θ_R is sufficiently small compared to temperature, the sum over l can be approximated by the integral

$$\begin{aligned} q_R &\approx \int_0^\infty e^{-l(l+1)\Theta_R/T} d[l(l+1)] \\ &= \frac{T}{\Theta_R} = \frac{8\pi^2 IkT}{h^2}, \end{aligned} \tag{3.3.18a}$$

which is the classical mechanical approximation. Equation (3.3.18a) is the large T/Θ_R limit of the sum given at eq.(3.3.16). For homonuclear diatomics, the classical mechanical partition function is

$$q_R = \frac{T}{2\Theta_R}, \tag{3.3.18b}$$

where the factor of 2 arises from the degeneracy of the homonuclear diatomic molecule to the interchange of indistinguishable atoms.

To order $(\Theta_R/T)^4$ the quantum mechanical partition function for $\Theta_R/T < 1$ is (Mulholland's formula)

$$q_R \approx \frac{T}{\Theta_R}\left[1 + \frac{1}{3}\frac{\Theta_R}{T} + \frac{1}{15}\left(\frac{\Theta_R}{T}\right)^2 + \frac{4}{315}\left(\frac{\Theta_R}{T}\right)^3\right]. \tag{3.3.19}$$

Equation (3.3.19) estimates the quantum mechanical partition function to better than 0.05% for $\Theta_R/T < 0.5$ and to better than 0.5% for $\Theta_R/T < 1$.

Referring to Table 3.3, whose entries are computed from spectroscopic measurement, we see that the classical mechanical approximation to the rotational partition function is accurate for most molecules except at very low temperatures. Hydrogen, with a characteristic rotational temperature $\Theta_R = 85.4$K is an exception, but even in this case at room temperature (say 295K) quantum effects represent about a 10% correction to the classical mechanical partition function. For homonuclear diatomic molecules the nuclear spin states must be taken into account in evaluating q_R. This is a complication that will not be addressed here because except for hydrogen the classical mechanical approximation is accurate at usually encountered temperatures.

The rotational temperatures of methane (CH_4) and deuterated methane (CD_4) are $\Theta_R = 7.5$ and 2.75K, respectively. Larger molecular weight polyatomics have even smaller Θ_R's.

Whereas in most circumstances, the rotational motions can be treated classically, the opposite is true of vibrational motions. Fortunately, this does not represent a problem in evaluating the

TABLE 3.3 Parameters for Selected Diatomic Molecules.

	Θ_V (K)	Θ_R (K)	r_o(A)	u_o(eV)
H_2	6210	85.4	0.740	4.454
N_2	3340	2.86	1.095	9.76
O_2	2230	2.07	1.204	5.08
CO	3070	2.77	1.128	9.14
NO	2690	2.42	1.150	5.29
HCl	4140	51.2	1.275	4.43
HBr	3700	12.1	1.414	3.60
HI	3200	9.0	1.604	2.75
Cl_2	810	0.346	1.989	2.48
Br_2	470	0.116	2.284	1.97
I_2	310	0.054	2.667	1.54

Determined from spectroscopic measurement.

quantum mechanical vibrational partition function, which for a diatomic molecule

$$q_v = e^{-u_0/kT} \sum_{v=0} e^{-\hbar\omega(v+\frac{1}{2})/kT}$$
$$= \frac{e^{-(u_0+\frac{1}{2}\hbar\omega)/kT}}{1-e^{-\hbar\omega/kT}} = e^{-u_0/kT} \frac{e^{-\Theta_V/2T}}{1-e^{-\Theta_V/T}} \qquad (3.3.20)$$

where Θ_V is the characteristic vibrational energy in units of temperature:

$$\theta_V \equiv \hbar\omega/k = h\omega/2\pi k.$$

The known property of the geometric series, $\sum_{v=0}^{\infty} x^v = 1/(1-x), 0 < x < 1$, is used in evaluating q_v. When $\Theta_V/T \ll 1$, eq. (3.3.20) reduces to the limit

$$q_V \approx e^{-u_o/kT} e^{-\Theta_V/2T} \frac{T}{\Theta_V}. \qquad (3.3.21)$$

This differs from the classical mechanical result in that the energy $u_0 + \hbar\omega/2$ has replaced u_0 in the classical mechanical partition function. This results from the fact that the lowest quantum mechanical energy of a harmonic oscillator is $\hbar\omega/2$, even in the limit of zero temperature.

As illustrated in Table 3.3, the classical mechanical approximation is in many cases not valid for the vibrations of most diatomics at temperatures normally encountered. This is true for the majority of the vibrational degrees of freedom of all polyatomic molecules. There are, however, lower energy motions, for example the bindered rotation of one CH_3 group relative to the other CH_3 group in ethane, which cannot be well approximated by a harmonic oscillation or a free rotation. Such degrees of freedom cannot be analyzed in terms of either a rigid rotator or a harmonic oscillator and so their quantum statistical mechanics becomes a complex numerical problem.

The contribution of vibration to the free energy, entropy, energy, and constant volume specific heat in an ideal diatomic gas can be computed from the partition function given at eq. (3.3.20). The results are

$$F_V = N[u_0 + \frac{1}{2}k\Theta_V + kT\ln(1 - e^{-\Theta_V/T})], \qquad (3.3.22)$$

$$S_V = Nk\left[-\ln(1 - e^{-\Theta_V/kT}) + \frac{\Theta_V/T}{e^{\Theta_V/T} - 1}\right], \qquad (3.3.23)$$

$$U_V = N\left[u_0 + \frac{1}{2}k\Theta_V + \frac{k\Theta_V}{e^{\Theta_V/T} - 1}\right], \qquad (3.3.24)$$

and

$$\frac{C_V^V}{Nk} = \frac{(\Theta_V/T)^2 e^{\Theta_V/T}}{\left(e^{\Theta_V/T} - 1\right)^2}. \qquad (3.3.25)$$

As $T/\Theta_V \to 0$, the vibrational heat capacity goes to zero as $(\Theta_V/T)^2\exp(-\Theta_V/T)$ and as $T/\Theta_V \to \infty$, C_V^V/Nk approaches the classical value of unity. A plot of C_V/Nk versus T/Θ_V is presented in Figure 3.2. At 300K, the contribution of vibrational motion to the specific heats of N_2 and Cl_2 are $1.6 \times 10^{-4} Nk$, which amount to $6 \times 10^{-3}\%$ and 7%, respectively, of the total specific heats at that temperature.

In Figure 3.2 the heat capacity of a dilute gas of HCl is plotted using the values of Θ_R and Θ_V given in Table 3.3. The contribution of the translational motion to the heat capacity is $3Nk/2$ in the temperature range reported in Figure 3.2. Above about 5K the rotational heat capacity begins increasing and goes through at maximum at about 12K. Between about 20K and 1000K the heat

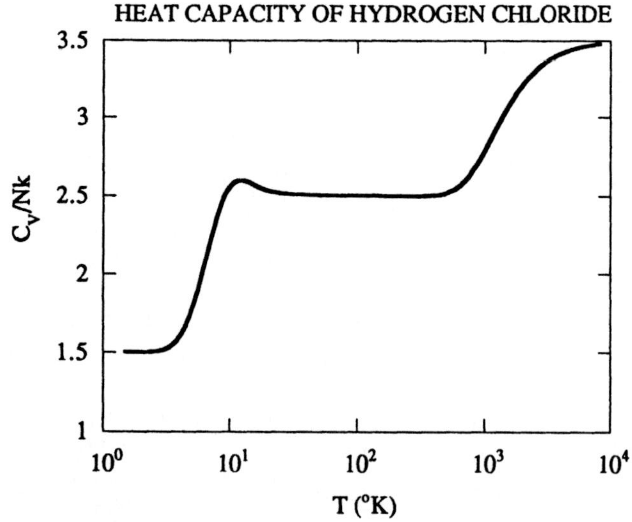

Figure 3.2
Heat capacity of a dilute gas of HCl predicted form spectroscopically measured values of Θ_R and Θ_V.

capacity is about constant at $5Nk/2$, the translational and rotational degrees of freedom being fully thermally activate. Above 8000K the vibrational degrees of freedom are fully thermally activated and the heat capacity is about $7Nk/2$, obeying in this range the classical mechanical equipartition rule. The vibrational partition function of a polyatomic molecule whose internal motions relative to the center of mass can be approximated as coupled harmonic oscillations is simply

$$q_V = e^{-u_0/kT} \prod_{i=1}^{n_v} \frac{e^{-\Theta_{V,i}/2kT}}{1 - e^{-\Theta_{V,i}/kT}}. \tag{3.3.26}$$

It follows from this that the vibrational contributions to free energy, entropy, energy, and specific heat can be determined by substituting $\Theta_{V,i}$ for Θ_V in eqs. (3.3.22)–(3.3.25) and summing over the n_v vibrational degrees of freedom.

3.4 Ideal Solids

We define an ideal solid as a substance in which N atoms are assigned to the vicinity of sites fixed in space at positions $\mathbf{R}_1, \ldots, \mathbf{R}_N$. The atoms can be identical or different. In any case they are distinguishable because they are assigned to particular lattice sites. The potential energy $u^N(\mathbf{r}_1, \ldots, \mathbf{r}_N)$ is a minimum when the atoms are at the site positions, that is, $u^N(\mathbf{r}_1, \ldots, \mathbf{r}_N) > u^N(\mathbf{R}_1, \ldots, \mathbf{R}_N) \equiv u_o^N$. It is further assumed that the atoms are sufficiently tightly bound that we can expand the potential energy in a Taylor's series about the minimum energy configuration and truncate the series after second order in the displacement $\mathbf{r}_i - \mathbf{R}_i$ of atoms from their lattice sites, that is,

$$u^N(\mathbf{r}_1, \ldots, \mathbf{r}_N) \approx u_o^N + \frac{1}{2} \sum_{i,j=1}^{N} \left[\nabla_{\mathbf{R}_i} \nabla_{\mathbf{R}_j} u_o^N\right] : (\mathbf{r}_i - \mathbf{R}_i)(\mathbf{r}_j - \mathbf{R}_j). \tag{3.4.1}$$

Terms linear in $\mathbf{r}_i - \mathbf{R}_i$ are missing because u_o^N is a minimum of u^N. Equation (3.4.1) defines the ideal solid to be investigated here. According to this potential energy an ideal solid is a set of $3N$ coupled harmonic oscillators. It has not been necessary to assume that the lattice sites are periodically arrayed or that forces are central or pairwise additive.

Let $\xi_1, \ldots, \xi_{3N}$ be the x, y, and z coordinates of the positions of the atoms relative to the lattice sites (that is, $\xi_1 = r_{1x} - R_{1x}$, $\xi_2 = r_{1y} - R_{1y}, \xi_3 = r_{1z} - R_{1z}, \xi_4 = r_{2x} - R_{2x}, \ldots, \xi_{3N} = r_{Nz} - R_{Nz}$). The classical mechanical Hamiltonian of a system of atoms obeying eq. (3.4.1) can be expressed in the form

$$H = \sum_{i=1}^{3N} \frac{1}{2} m_i \dot{\xi}_i^2 + \frac{1}{2} \sum_{i,j=1}^{3N} a_{ij} \xi_i \xi_j + u_o^N. \tag{3.4.2}$$

$\dot{\xi}_i \equiv d\xi_i/dt$ is the velocity of coordinate ξ_i, m_i is the mass of the atom associated with coordinate ξ_i, and

$$a_{ij} \equiv \left(\frac{\partial^2 u^N}{\partial \xi_i \partial \xi_j}\right)_{\mathbf{r}^N=\mathbf{R}^N}. \tag{3.4.3}$$

Because $a_{ij} = a_{ji}$, it follows from a basic theorem in linear algebra that there exists an orthogonal transformation of the coordinates $\xi_1, \ldots, \xi_{3N}$ into a new set

$$\zeta_i = \sum_{j=1}^{3N} \alpha_{ij}\xi_j, \quad i = 1, \ldots, 3N \tag{3.4.4}$$

in terms of which eq. (3.4.2) becomes

$$H = \sum_{i=1}^{3N} \frac{1}{2}\left[m\dot{\zeta}_i^2 + \lambda_i \zeta_i^2\right] + u_o^N, \tag{3.4.5}$$

where m is some reference atomic mass.

Exercise
The observed tension on a cylinder of rubber of length L obeys the expression

$$f = \frac{NkT}{L_o}\left(\frac{L}{L_o} - \frac{L_o^2}{L^2}\right),$$

where L_o is the length of the unstressed system. Using the relation $f = -\partial F/\partial L$, find the Helmholtz free energy, the entropy and the energy of the rubber relative to the unstressed state. Why is f called an entropic force?

All one needs to know to develop the statistical mechanics of an ideal solid is that eq.(3.4.2) can be recast in the form of eq.(3.4.5). For the sake of some who may be curious, the steps in the transformation will be given in a bit more detail. Others can skip past eq. (3.4.12) to resume statistical mechanics. First define $\eta_i \equiv (m_i/m)^{1/2}\xi_i$ and $\tilde{a}_{ij} = (m^2/m_i m_j)^{1/2} a_{ij}$ so that

$$H = \sum_{i=1}^{3N} \frac{1}{2} m\dot{\eta}_i^2 + \frac{1}{2}\sum_{i,j=1}^{3N} \tilde{a}_{ij}\eta_i\eta_j + u_o^N. \tag{3.4.6}$$

The matrix $\mathbf{A} = [\tilde{a}_{ij}]$ is symmetric, $\tilde{a}_{ij} = \tilde{a}_{ji}$, and real and is therefore self-adjoint and so it has $3N$ orthonormal eigenvectors $\mathbf{s}_i$, that is,

$$\mathbf{A}\mathbf{s}_i = \lambda_i \mathbf{s}_i \tag{3.4.7}$$

where λ_i is an eigenvalue of $\mathbf{A}$ (real because $\mathbf{A}$ is self-adjoint) and $\mathbf{s}_i$ is a $3N$ dimensional column vector with components $s_{i1}, s_{i2}, \ldots, s_{i,3N}$ that obey the orthonormality condition

$$\mathbf{s}_i^T \mathbf{s}_j \equiv \sum_{k=1}^{3N} s_{ik}s_{jk} = \delta_{ij}. \tag{3.4.8}$$

$\delta_{ij} = 0$ if $i = j$ and $= 0$ if $i \neq j$. In eq. (3.4.8), $\mathbf{s}_i^T$ is the row vector formed by taking the transpose of the column vector $\mathbf{s}_i$. If $\mathbf{S}$ is a matrix whose column vectors are $\mathbf{s}_1, \ldots, \mathbf{s}_{3N}$, then

$$\mathbf{S}^T\mathbf{S} = \mathbf{I} = [\delta_{ij}], \tag{3.4.9}$$

$$\mathbf{S}^T\mathbf{A}\mathbf{S} = \mathbf{\Lambda} = [\lambda_i \delta_{ij}]. \tag{3.4.10}$$

In matrix notation,

$$H = \frac{1}{2}m\dot{\eta}^T\dot{\eta} + \frac{1}{2}\eta^T\mathbf{A}\eta + u_o^N,$$

which, with the definition $\eta \equiv \mathbf{S}\zeta$, becomes

$$\begin{aligned} H &= \frac{1}{2}m\dot{\zeta}^T\mathbf{S}^T\mathbf{S}\dot{\zeta} + \frac{1}{2}\zeta^T\mathbf{S}^T\mathbf{A}\mathbf{S}\zeta + u_o^N \\ &= \frac{1}{2}\sum_{i=1}^{3N}[m\dot{\zeta}_i^2 + \lambda_i\zeta_i^2] + u_o^N, \end{aligned} \tag{3.4.11}$$

where the final form is obtained by using eqs. (3.4.9) and (3.4.10) and the rules of matrix multiplication. This completes the proof of eq. (3.4.5).

From the definition of $\zeta(\eta \equiv \mathbf{S}\zeta)$ and eq. (3.4.9) it follows that $\zeta = \mathbf{S}^T\eta$. Thus,

$$\zeta_i = \sum_{j=1}^{3N} s_{ji}\eta_j = \sum_{j=1}^{3N}\left(\frac{m_j}{m}\right)^{1/2} s_{ji}\xi_j, \tag{3.4.12}$$

and so the coefficients α_{ij} claimed in eq. (3.4.4) are the product of $(m/m_j)^{1/2}$ and the components of the eigenvectors of the matrix $\mathbf{A} = [\tilde{a}_{ij}]$.

With the form of H given by eq. (3.4.11) and the fact that the momentum conjugate to ζ_i is $p_i = m\dot{\zeta}_i$, the classical mechanical partition function can be easily evaluated:

$$\begin{aligned} Q_N &= e^{-u_o^N/kT}\int_{-\infty}^{\infty}\cdots\int_{-\infty}^{\infty} e^{-\sum_{i=1}^{3N}\left(\frac{p_i^2}{2m}+\frac{1}{2}\lambda_i\zeta_i^2\right)/kT}\frac{d\zeta_1\cdots d\zeta_{3N}dp_1\cdots dp_{3N}}{h^{3N}} \\ &= \left(\frac{2\pi kT}{h}\right)^{3N}\prod_{i=1}^{3N}(m/\lambda_i)^{1/2}e^{-u_o^N/kT}. \end{aligned} \tag{3.4.13}$$

This yields for the Helmholtz free energy formula

$$F = -3NkT\ln\left(\frac{2\pi kT}{h}\right) - \frac{1}{2}kT\sum_{i=1}^{3N}\ln(m/\lambda_i) + u_o^N, \tag{3.4.14}$$

and for the constant volume specific heat the relation

$$C_V = 3Nk. \tag{3.4.15}$$

Equation (3.4.15) is again an example of the equipartition theorem of classical statistical mechanics. It was known historically as the law of Dulong and Petit and as was the case with dilute gases, it was only valid above certain temperatures for certain substances. For example, for a 1 mol sample $3Nk = 25$ J/mol K in reasonable agreement with the experimental value of 25.5 for Cu and 24.4 for Al but in strong disagreement with the experimental value of 6.1 for diamond at 298K.

In fact many solids at room temperature obey approximately the law of Du.Long and Petit, but many do not, and all solids exhibit a decreasing heat capacity with decreasing temperature. This temperature dependence results from the failure of classical mechanics for vibrational motion. For the ideal solid considered here, the Schrödinger equation is

$$\sum_{i=1}^{3N} H_i\psi + u_o^N\psi = E\psi \tag{3.4.16}$$

where

$$H_i\psi = -\frac{\hbar^2}{2m_i}\frac{\partial^2\psi}{\partial\zeta_i^2} + \frac{1}{2}\lambda_i\zeta_i^2\psi. \tag{3.4.17}$$

The quantum energy of each of the $3N$ independent harmonic oscillators is

$$E_{\iota\nu} = \hbar\omega_i(\nu + \frac{1}{2}), \quad \nu = 0, 1, 2, \ldots, \tag{3.4.18}$$

where $\omega_i = (\lambda_\iota/m)^{1/2}$. Because the atoms of the ideal solid are assigned to distinguishable sites, the partition function of the solid is

$$\begin{aligned} Q_N &= e^{-u_o^N/kT}\prod_{i=1}^{3N} q_i \\ q_\iota &= \sum_{\nu=0}^{\infty} e^{-\hbar\omega_\iota(\nu+\frac{1}{2})/kT} = \frac{e^{-\hbar\omega_\iota/2kT}}{1-e^{-\hbar\omega_\iota/kT}}. \end{aligned} \tag{3.4.19}$$

The specific heat resulting from eq. (3.4.19) is

$$C_V = \sum_{i=1}^{3N} \frac{[(\hbar\omega_i)^2/kT^2]e^{-\hbar w_\iota/kT}}{(1-e^{-\hbar\omega_\iota/T})^2}. \tag{3.4.20}$$

To proceed further, we need to find the eigenvalues of the matrix **A** given at eq. (3.4.7). The eigenvalues will depend on the ideal solid (e.g., simple cubic, face centered cubic, diamond, etc.), m_i, and the force constants a_{ij}. In general, because of its size and complexity one cannot solve the eigenvalue problem. Einstein circumvented the difficulty by assuming that all the eigenvalues were the same, that is, that all the independent modes of oscillation had the same fundamental frequency ω_E. In this case, eq. (3.4.20) becomes

$$\mathcal{C}_V = 3Nk\left(\frac{\Theta_E}{T}\right)^2 \frac{e^{-\Theta_E/T}}{(1-e^{-\Theta_E/T})^2}, \tag{3.4.21}$$

where $\Theta_E \equiv \hbar\omega_E/k$. At high temperatures, that is, $T \ll \Theta_E$, this expression for C_V approaches of course the classical value $3Nk$.

A fit of the Einstein equation to the data for diamond is shown in Figure 3.3. At the time, (1907), Einstein's model compared favorably with the heat capacity data and, in fact, did much to enhance the acceptance of quantum mechanics. However, careful experimentation later revealed that Einstein's predicted heat capacities approach zero more strongly with temperature ($C_V \approx 3Nk(\Theta_E/T)^2e^{-\Theta_E/T}$) than do the experimental data. The reason is that the eigenvalues

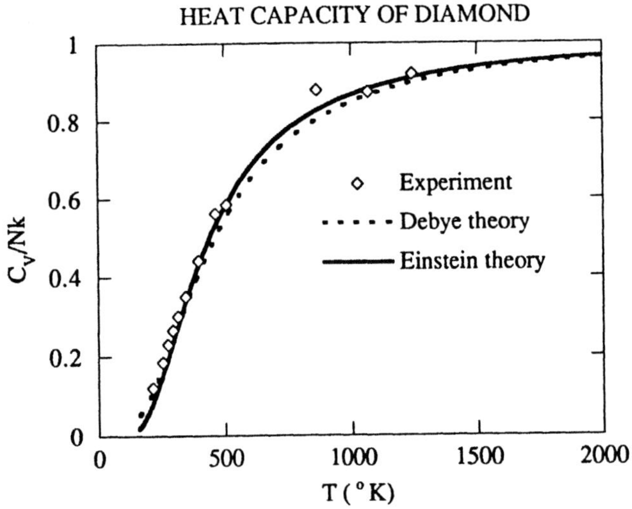

Figure 3.3
Temperature dependence of constant volume heat capacity of diamond. Experimental results (points) are compared to Einstein's theory with $\Theta_E = 1320$K (after Einstein, 1907) and to Debye's theory (1912) with $\Theta_D = 1860$K.

representing the various vibrational modes of a solid are not the same, but instead range from nearly zero to some maximum value.

To get a qualitative idea of the range of the eigenvalues of a system of coupled harmonic oscillators, consider a one-dimensional system of N identical mass points obeying the following equations of motion

$$m\frac{d^2x_i}{dt^2} = a(x_{i+1} - 2x_i + x_{i-1}), \quad i = 1, \ldots, N, \tag{3.4.22}$$

with $x_o = x_{N+1} = 0$. The quantity a is the force constant of the oscillators. These equations describe a ring of particles in which the force on particle i is zero when the particle lies midway between its nearest neighbors and experiences a restoring force linear in its displacement from the midpoint position. In matrix notation eq. (3.4.22) is

$$m\frac{d^2\mathbf{x}}{dt^2} = -\mathbf{A}\mathbf{x}, \tag{3.4.23}$$

where the ith component of $\mathbf{x}$ is x_i and $\mathbf{A}$ is the tridiagonal matrix

$$\mathbf{A} = -a\begin{bmatrix} -2 & 1 & 0 & & 0 \\ 1 & -2 & 1 & 0 & \\ 0 & 1 & -2 & 1 & 0 \\ & 0 & 1 & -2 & \\ & & 0 & & 1 \\ 0 & & & 1 & -2 \end{bmatrix}. \tag{3.4.24}$$

The eigenvectors and eigenvalues of this matrix are known. The jth element of the ith eigenvector $\mathbf{s}_i$ is

$$s_{ij} = (-1)^{j-1}\frac{\sin[i\pi(N-j)/(N+1)]}{\sin[\pi i/(N+1)]} \tag{3.4.25}$$

and the eigenvalue corresponding to $\mathbf{s}_i$ is

$$\lambda_j = 2a\left[1 + \cos\left(\frac{\pi j}{N+1}\right)\right]. \tag{3.4.26}$$

Thus, the minimum eigenvalue is $\lambda_N = 2a\{1 + \cos[N\pi/(N+1)]\}$ and the maximum eigenvalue is $\lambda_1 = 2a\{1 + \cos[\pi/(N+1)]\}$. For sufficiently large N these are approximately zero and $4a$.

If the column vectors $\mathbf{s}_i$ of $\mathbf{S}$ are normalized eigenvectors of $\mathbf{A}$, that is, $\mathbf{s}_i^{\mathrm{T}}\mathbf{s}_j = \delta_{ij}$, then $\mathbf{S}^T\mathbf{S} = \mathbf{I}$ and

$$\mathbf{AS} = \mathbf{S\Lambda} \quad \text{or} \quad \mathbf{S}^T\mathbf{AS} = \mathbf{\Lambda}. \tag{3.4.27}$$

$\mathbf{\Lambda}$ is a matrix whose only nonzero elements are the eigenvalues λ_i lying along the main diagonal. With the coordinate transformation

$$\mathbf{x} = \mathbf{S}\boldsymbol{\zeta}, \tag{3.4.28}$$

Newton's equations eq. (3.4.23) becomes

$$m\frac{d^2\boldsymbol{\zeta}}{dt^2} = -\mathbf{\Lambda}\boldsymbol{\zeta} \quad \text{or} \quad m\frac{d^2\zeta_i}{dt^2} = -\lambda_i\zeta_i, \quad i = 1, \ldots, N. \tag{3.4.29}$$

Similarly, the classical mechanical energy of the system is

$$H = \sum_{i=1}^{N}\left[\frac{1}{2}m\left(\frac{d\zeta_i}{dt}\right)^2 + \frac{1}{2}\lambda_i\zeta_i^2\right], \tag{3.4.30}$$

and, correspondingly, the quantum mechanical partition function and specific heat are

$$Q_N = \prod_{i=1}^{N}\frac{e^{-\hbar\omega_i/2kT}}{1 - e^{-\hbar\omega_i/kT}} \tag{3.4.31}$$

and

$$C_V = \sum_{i=1}^{N}\frac{(\hbar\omega_i)^2}{kT^2}\frac{e^{-\hbar\omega_i/kT}}{[1 - e^{-\hbar\omega_i/kT}]^2}, \tag{3.4.32}$$

with

$$\omega_i = (\lambda_i/m)^{1/2} = \left(\frac{2a}{m}\right)^{1/2}\left\{1 + \cos\left[\frac{\pi i}{N+1}\right]\right\}^{1/2}. \tag{3.4.33}$$

For sufficiently large N (certainly 10^{23} is sufficiently large) the sum in eq. (3.4.32) can be well approximated by an integral. For the purpose, consider the transformation (with $\Delta i = 1$)

$$\sum_{i=1}^{N} f(\omega_i)\Delta i \approx \int_{i=1}^{N} f(\omega_i)di = \int_{\omega_{\min}}^{\omega_{\max}} f(\omega_i)\frac{1}{|d\omega_i/di|}d\omega_i$$

$$= \int_{\omega_{\min}}^{\omega_{\max}} f(\omega)\phi(\omega)d\omega. \tag{3.4.34}$$

Here $\omega_{min} = (\lambda_N/m)^{1/2} \approx 0, \omega_{max} = (\lambda_1/m)^{1/2} \approx 2(a/m)^{1/2}$ and $\phi(\omega)d\omega$ denotes the number of modes of oscillation lying between ω and $\omega + d\omega$. It is computed from $|d\omega_i/di|$ by eliminating the index i in favor of ω_i. We note that

$$\frac{d\omega_i}{di} = -\frac{1}{2}\left(\frac{\pi}{N+1}\right)(2a/m)^{1/2}\frac{\sin[\pi i/(N+1)]}{\{1+\cos[\pi i/(N+1)]\}^{1/2}}. \tag{3.4.35}$$

But

$$\cos\left(\frac{\pi i}{N+1}\right) = \frac{m}{2a}\omega_i^2 - 1, \tag{3.4.36}$$

the result of which allows transformation of eq. (3.4.35) to

$$\frac{d\omega_i}{di} = -\left(\frac{\pi}{N+1}\right)\left(\frac{a}{m}\right)\frac{[1-(1-2\omega_i^2/\omega_o^2)^2]^{1/2}}{\omega_i} \tag{3.4.37}$$

where $\omega_o \equiv (4a/m)^{1/2}$.

Hence, we obtain

$$\phi(\omega) = \frac{2(N+1)}{\pi\omega_o}\frac{1}{(1-\omega^2/\omega_o^2)^{1/2}}. \tag{3.4.38}$$

For large N, $\omega_{max} \approx \omega_o$ and $\phi(\omega)$ obeys the normalization condition

$$\int_0^{\omega_o} \phi(\omega)d\omega = N \tag{3.4.39}$$

representing the fact that there are N particles and therefore N independent modes of oscillation (known as normal modes in classical mechanics).

The distribution of normal modes for this system is plotted in Figure 3.4. Approximating the sum in eq. (3.4.32) by the integral over frequency ω and introducing the characteristic temperature $\Theta_o = \hbar\omega_o/kT$, we obtain

$$C_V = \frac{2(N+1)kT}{\pi\Theta_o}\int_0^{\Theta_o/T}\frac{x^2}{(1-T^2x^2/\Theta_o^2)^{1/2}}\frac{e^{-x}}{(1-e^{-x})^2}dx. \tag{3.4.40}$$

At high temperatures $T \gg \Theta_o$, this expression yields the classical mechanical result, $C_V \simeq Nk$. At low temperatures, $T \ll \Theta_o$,

$$C_V = (2\pi/3)(N+1)T/\Theta_o, \tag{3.4.41}$$

where the coefficient in the expression comes from

$$\frac{2}{\pi}\int_0^{\infty} x^2\frac{e^{-x}}{(1-e^{-x})^2}dx = \frac{2\pi}{3}. \tag{3.4.42}$$

Figure 3.4
Frequency distribution for one-dimensional coupled harmonic oscillators (or a 1-D crystal with Hookian nearest neighbor interactions).

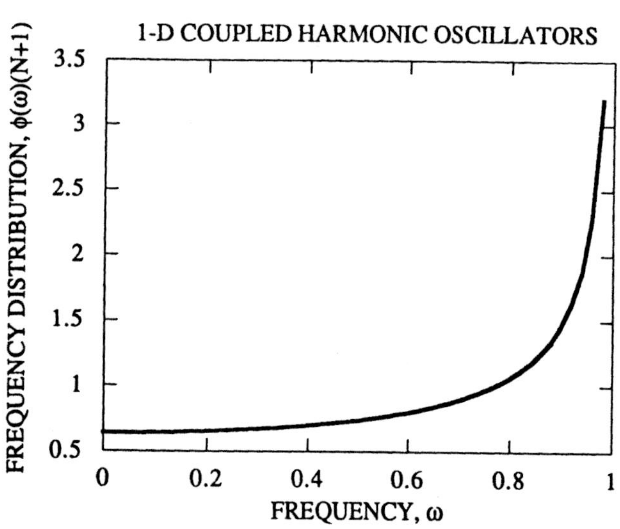

The flaw in Einstein's model is apparent in Figure 3.4. He assumed that all the modes of oscillation were the same and occurred at a finite frequency ω_E. Thus, when the temperature drops below $h\omega_E/k$, the probability of excitation from the ground state to the next quantum state goes to zero as $e^{-\Theta_E/T}$. The result is that as the temperature decreases in the range $T \ll \Theta_E$, the probability that the system can absorb energy and, therefore, the heat capacity goes to zero as $e^{-\Theta_E/T}$. In the case of the rigorous distribution of frequencies $\phi(\omega)$ for the one-dimensional model studied here, there is a distribution of normal mode frequencies extending to zero frequency. Thus, at any given T there is a finite population of normal modes with frequencies ω such that $\hbar\omega/k < T$. These modes can absorb energy easily and are the ones controlling the heat capacity. The population, equal to $\int_0^{kT/\hbar} \phi(\omega)d\omega$, decreases with decreasing temperature. The net result is a heat capacity that decreases to zero linearly with temperature, much more slowly than is predicted by the Einstein approximation.

Although the qualitative implications of the one-dimensional model are correct, the quantitative behavior of a real, three-dimensional solid cannot be captured by the model. In particular, the specific heat of nonmetallic solids is observed to approach zero as T^3 as the temperature approaches absolute zero. Recognizing that it is the low frequency and therefore long wavelength and small deformation modes of vibration that contribute primarily to the very low temperature vibrational heat capacity, Debye argued that the frequency distribution can be estimated from the linear elastic continuum theory of a solid. According to this theory the distribution of vibrational modes is

$$\phi(\omega)d\omega = 3\frac{V}{2\pi^2 c_s^3}\omega^2 d\omega. \tag{3.4.43}$$

V is volume and c_s is the effective sound velocity, namely,

$$\frac{3}{c_s^3} \equiv \frac{1}{c_l^3} + \frac{2}{c_t^3}, \tag{3.4.44}$$

where c_l and c_t are the velocities of propagation of a longitudinal and a transverse deformation, respectively. As in the case of the one-dimensional system discussed above, the total number of vibrational modes has to be equal to the total number of degrees of freedom of the solid. For a three-dimensional solid with N atoms this is $3N$, and so the maximum vibrational frequency ω_D is determined by the condition

$$\int_0^{\omega_D} \phi(\omega)d\omega = 3N, \tag{3.4.45}$$

which yields $\omega_D = c_s(6\pi^2 N/V)^{1/3}$. This "Debye frequency," ω_D, is determined by the velocities of sound and the density of the solid.

Passing the sum to an integral in eq. (3.4.20) and using the Debye frequency distribution, we obtain the Debye theory of the specific heat:

$$C_V = 3Nkf_D(\Theta_D/T), \tag{3.4.46}$$

where Θ_D is the Debye temperature, $\Theta_D \equiv h\omega_D/k$, and

$$f_D(y) = \frac{3}{y^3}\int_0^y \frac{e^{-x}}{(1-e^{-x})^2}x^4 dx. \tag{3.4.47}$$

As $T/\Theta_D \to \infty$, eq. (3.4.46) yields the classical mechanical limit, $C_V = 3Nk$, whereas as $T/\Theta_D \to 0$,

$$\begin{aligned} f_D\left(\frac{\Theta_D}{T}\right) &= 3\left(\frac{T}{\Theta_D}\right)\left[\int_0^\infty \frac{e^{-x}}{(1-e^{-x})^2}x^4 dx + O(T/\Theta_D)\right] \\ &= 3\left(\frac{T}{\Theta_D}\right)^3\left[\frac{4\pi^4}{15} + O(T/\Theta_D)\right]. \end{aligned} \tag{3.4.48}$$

Thus, in the low temperature range, the Debye model predicts

$$C_V = \frac{12\pi^4}{5}Nk\left(\frac{T}{\Theta_D}\right)^3 \tag{3.4.49}$$

plus terms of higher order in T/Θ_D.

Debye's theory is compared in Figure 3.5 with Einstein's model and with the experimental data available to Einstein for diamond. Debye's theory appears a bit better than Einstein's, but with the data in Figure 3.5 no strong conclusion could be drawn. However, since Einstein's early work, accurate experiments have left no doubt that at sufficiently low temperatures Debye's theory is superior to Einstein's theory. The T^3 dependence of C_V predicted

Figure 3.5
Plot of C_V/T versus T^2 for solid KCl at low temperatures. Data demonstrate the validity of Debye's theory at low temperatures. (Data from Keeson and Pearlman, 1953).

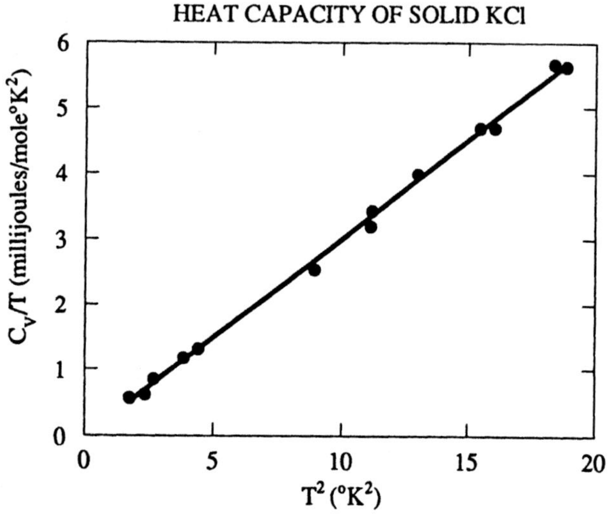

at low temperatures has been verified experimentally for numerous solids, including metals, alkali halides, and diamond. A typical example of the behavior of C_V at very low temperature is shown in Figure 3.5 for solid KCl.

The accuracy of Debye's theory rises from the fact that at low temperatures the distribution of active vibrational modes (active ones being those of low energy and long wavelength) can be well-approximated by the continuum theory of elasticity. At high frequencies Debye's distribution is not quantitatively correct, but its failure is somewhat muted because the specific heat results from an average over the entire frequency range and C_V is bound by $3Nk$ as $T/\Theta_D \to \infty$. Finally, at ultralow temperatures, say below 1K, there is an electronic contribution to C_V that is linear in T. Thus, at a very low temperature the Debye model fails for a metal.

3.5 Paramagnetism

Unlike ferromagnetism, paramagnetism is not a collective phenomena, that is, the magnetic moments in paramagnetism are independent of one another. To study such magnetism, consider a system of N identical atoms whose magnetic spins are independent. In the presence of an external magnetic field **H**, the magnetic energy of an atom is

$$\varepsilon = -\boldsymbol{\mu} \cdot \mathbf{H}, \tag{3.5.1}$$

where $\boldsymbol{\mu}$ is the magnetic moment of the atom. The magnetic moment is proportional to the angular momentum $k\mathbf{J}$ of the atom, namely,

$$\boldsymbol{\mu} = g\mu_0 \mathbf{J} \tag{3.5.2}$$

where g is a dimensionless number, of the order of 1 for electron magnetic moments and 10^{-3} for nuclear magnetic moments, and μ_0 is the Bohr magneton ($\mu_0 = e\hbar/2mc$, with e and m the electronic charge and mass and c the speed of light). Because $\hbar$ has units of angular momentum, the vector **J** is dimensionless.

If the field **H** is oriented in the z direction, then eqs. (3.5.1) and (3.5.2) yield

$$\varepsilon = -g\mu_0 J_z H. \tag{3.5.3}$$

According to quantum mechanics, the z component of the angular momentum vector **J** can have only the values

$$J_z = -J,\ -J+1,\ \ldots,\ J-1,\ J, \tag{3.5.4}$$

where the characteristic spin J is a whole or half integer. For example, for an atom with a single unpaired electron $J = 1/2$ and $J_z = -1/2$ and $1/2$ are the only allowed states. For an atom (e.g., Cr^{+3} with three unpaired electrons with $J = 3/2$, the four states $J = -3/2, -1/2$, 1/2, and 3/2 are allowed. In general there are $2J+1$ allowed quantum states for J_z.

The canonical ensemble partition function for an independent magnetic moment is

$$q_m = \sum_{m=-J}^{J} e^{g\mu_0 mH/kT}. \tag{3.5.5}$$

Using the property $\sum_{i=0}^{2J+1} x^i = x^{-J}(1-x^{2J+1})/(1-x)$ of a geometric series, we obtain

$$q_m = \frac{e^{-J\zeta} - e^{(J+1)\zeta}}{1-e^{\zeta}}, \tag{3.5.6}$$

where

$$\zeta \equiv g\mu_0 H/kT. \tag{3.5.7}$$

The average magnetic moment for an atom can be computed from

$$\langle\mu_z\rangle = \sum_{m=-J}^{J} g\mu_0 m e^{g\mu_0 mH/kT}/q_m, \tag{3.5.8}$$

or equivalently, from

$$\langle\mu_z\rangle = kT\frac{\partial \ln q_m}{\partial H}. \tag{3.5.9}$$

The magnetization $\overline{M}$ of a sample with N atoms in the volume V is

$$\overline{M} = \frac{N}{V}\langle\mu_z\rangle,$$

and so, with the aid of eq. (3.5.6), eq. (3.5.9) yields for the magnetization

$$\overline{M} = \frac{N}{V} g\mu_0 J B_J(\zeta), \tag{3.5.10}$$

where the "Brillouin function" B_J is defined by

$$B_J(\zeta) = \frac{1}{J}\left\{\left(J + \frac{1}{2}\right)\coth\left[\left(J + \frac{1}{2}\right)\zeta\right] - \frac{1}{2}\coth\frac{1}{2}\zeta\right\}. \tag{3.5.11}$$

The magnetic susceptibility, χ, is defined by

$$\overline{M} \equiv \chi H, \tag{3.5.12}$$

or

$$\chi = \frac{Ng\mu_0 J}{VH} B_J(\zeta). \tag{3.5.13}$$

$B_J(\zeta)$ is a monotonically increasing function, being zero at $\zeta = 0$ and increasing to unity as $\zeta \to \infty$. At small H (or large T), ζ is small and $B_J(\zeta) \approx (J+1)\zeta/3$ and so the susceptibility obeys *Curie's law*,

$$\chi = \frac{N}{V}\frac{g^2\mu_0^2 J(J+1)}{3kT}. \tag{3.5.14}$$

At sufficiently large H (or low T), the magnetism becomes independent of H, that is, $\overline{M} = (N/V)g\mu_0 J$, corresponding to the alignment of all the magnetic dipoles with the external magnetic field.

Theoretical and experimental average magnetic moments are plotted versus H/T in Figure 3.6 for three difference substances.

The magnetic internal energy of the system, $U_m = -NH\langle\mu_z\rangle = -V\overline{M}H$, is given by

$$U_m = -Ng\mu_0 JHB_J(\zeta), \tag{3.5.15}$$

and so the magnetic constant volume specific heat is

$$C_{V,m} = \frac{N(g\mu_0 H)^2 J}{kT^2}\frac{dB_J}{d\zeta} = -NkJ\zeta^2\frac{dB_J}{d\zeta} \tag{3.5.16}$$

where

$$\frac{dB_J}{d\zeta} = -\frac{1}{J}\left\{\frac{(J+\frac{1}{2})^2}{\sinh^2[(J+\frac{1}{2})\zeta]} - \frac{(\frac{1}{2})^2}{\sinh^2\frac{1}{2}\zeta}\right\}. \tag{3.5.17}$$

At low field H (or high T), where ζ is small, the specific heat becomes

$$C_{V,m} \simeq \frac{N(g\mu_0 H)^2 J(J+1)}{3kT^2}, \tag{3.5.18}$$

and at high H (or low T) the specific heat approaches zero as $\zeta^2\exp[-\zeta]$. the specific heat is plotted versus ζ in Figure 3.7 for the case $J = 1/2$. The specific heat goes through a maximum of

Figure 3.6
Plot of the mean magnetic moment $\overline{\mu}_s$ of an ion (in units of the Bohr magneton μ_0) as a function of H/T. The solid curves are Brillouin functions. The experimental points are those for (I) potassium chromium alum, (II) iron ammonium alum, and (III) gadolinium sulfate octahydrate. In all cases, $J = S$, the total electron spin of the ion, and $g = 2$. Note that at 1.3K a field of 50,000 Gauss is sufficient to produce more than 99.5% magnetic saturation. Redrawn from Reif, (1965).

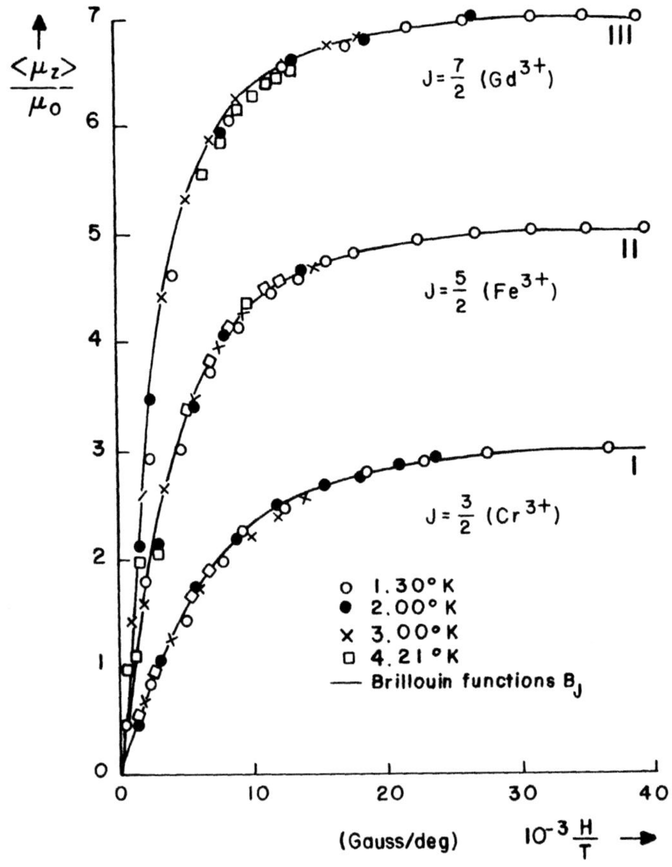

about $0.44Nk$ at $\zeta \simeq 2.4$. At small ζ little energy is stored by magnetization and at large ζ transitions between quantum states are rare; in either case, then, the specific heat will be small.

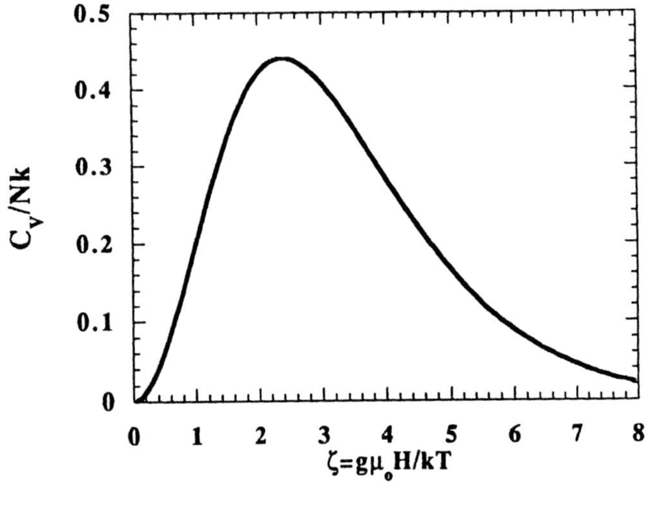

Figure 3.7
Specific heat of a paramagnetic system versus H/T.

3.6 Langmuir Adsorption

When a fluid is in contact with a solid surface, it can be adsorbed on by the solid. Langmuir adsorption is perhaps the simplest example of this. According to the Langmuir model, a solid surface consists of M sites, each of which can adsorb exactly one molecule. The adsorbed molecules are assumed to be independent of one another. The energy of an adsorbed molecule is $-\varepsilon_0$ relative to its energy in the fluid phase.

Consider a fluid at chemical potential μ in contact with a Langmuir surface. The canonical ensemble partition function of N adsorbed molecules is

$$Q_N = \frac{M![q_a^I(T)]^N}{N!(M-N)!} e^{N\varepsilon_0/kT}, \quad (3.6.1)$$

where $q_a^I(T)$ is the partition function of the internal degrees of freedom of an adsorbed molecule and $M!/N!(M-N)!$ gives the number of ways the N molecules can be distributed among the M identical sites. The grand canonical ensemble partition function can be evaluated explicitly, namely,

$$\begin{aligned} \Xi &= \sum_{N=0}^{M} \frac{M![q_a^I(T)]^N}{N!(M-N)!} e^{N(\varepsilon_0+\mu)/kT} \\ &= [1 + q_a^I(T) e^{(\varepsilon_0+\mu)/kT}]^M. \end{aligned} \quad (3.6.2)$$

The average number of adsorbed molecules can be computed from

$$\overline{N} = kT \left(\frac{\partial \ln \Xi}{\partial \mu} \right)_T \quad (3.6.3)$$

to obtain

$$\overline{N} = \frac{M q_a^I(T) e^{(\varepsilon_0+\mu)/kT}}{1 + q_a^I(T) e^{(\varepsilon_0+\mu)/kT}}, \quad (3.6.4)$$

known as the Langmuir adsorption isotherm.

A qualitative plot of the fraction of adsorption sites occupied, $\overline{N}/M$, versus activity, $e^{\mu/kT}$, is given in Figure 3.8. At high activity (chemical potential) $\overline{N}/M$ approaches unity, corresponding to the saturation of the adsorption sites. At sufficiently small e^{μ}/kT the denominator in eq. (3.6.4) is approximately 1, and so the equation can be rearranged to the ideal solution result

$$\mu = -\varepsilon_0 + kT \ln q_a^I(T) + kT \ln(\overline{N}/M). \quad (3.6.5)$$

There are some known systems (a dilute gas of ammonia on charcoal and dilute aqueous solutions of fatty acids on silica gel) that obey the Langmuir isotherm.

Figure 3.8
Langmuir adsorption isotherm.

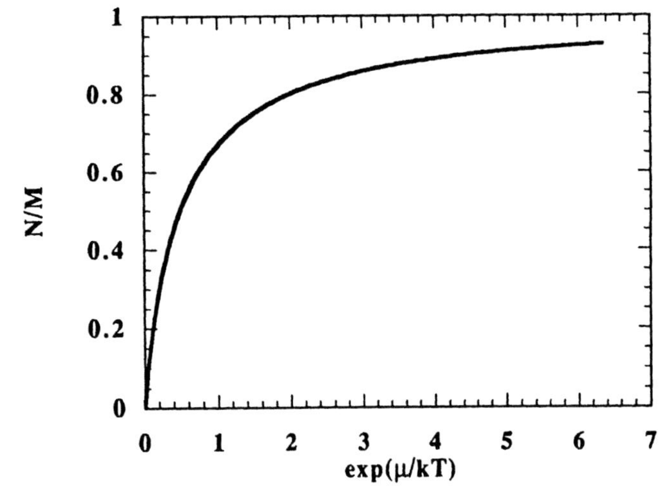

When the fluid phase is a dilute gas, behaving ideally, its chemical potential obeys the equation

$$\mu = kT \ln\left[\frac{\Lambda^3}{kTq^I(T)}\right] + kT \ln P, \tag{3.6.6}$$

where $\Lambda = h/(2\pi mkT)^{1/2}$ and P is the pressure of the gas. Combining eqs. (3.6.4) and (3.6.6) we obtain

$$\frac{\overline{N}}{M} = \frac{K(T)P}{1 + K(T)P}, \tag{3.6.7}$$

the form of the Langmuir isotherm commonly introduced to describe adsorption of dilute gases on a solid substrate. We derived this expression in Chapter 1, but did not give a theoretical expression for the adsorption constant $K(T)$. We find here that it is equal to $[q_a^I \Lambda^3/kTq^I]\exp(\varepsilon_o/kT)$.

If $K(T)$ is measured in controlled experiments, then the amount of gas adsorbed can be used with eq. (3.6.7) to determine the total surface area of particle dispersion or porous media. This is frequently done to characterized dispersed catalysts. However, it is more common to use gases obeying the BET (Brunauer–Emmett–Teller) equation than those obeying the Langmuir equation. The BET equation is obtained by assuming that an adsorption site can adsorb a pile of molecules, the first one having one energy state, $-\varepsilon_0$, and all successive ones having a different energy states, $-\varepsilon_1$. For adsorption of a dilute gas the BET isotherm is

$$\frac{\overline{N}}{M} = \frac{C(T)(P/P_s)}{(1 - P/P_s)[1 + (C(T) - 1)(P/P_s)]}, \tag{3.6.8}$$

where $P_s(T)$ is the vapor pressure of saturated vapor at temperature T and $C(T)$ is a constant similar to $K(T)P_s$. In the limit of small P/P_s and large $C(T)$ (that is, low temperature), the BET isotherm

Figure 3.9
Comparison of BET and Langmuir isotherms

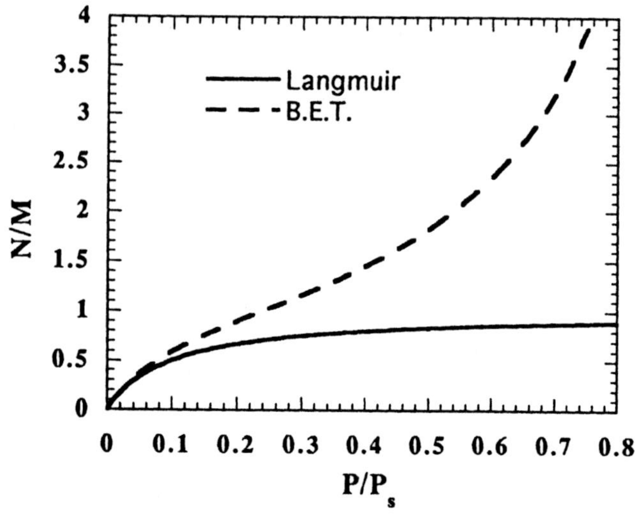

reducts to the Langmuir isotherm. The Langmuir and BET isotherm are shown in Figure 3.9 for $K(T)P_s = C(T) = 10$.

3.7 One-Dimensional Ising Model

A one-dimensional Ising model consists of a chain of N spins that can have values $s_i = +1$ or $s_i = -1$ (spin up or spin down, we say). The spins interact with nearest neighbors only and obey the energy function

$$E_s = -J \sum_{i=1}^{N-1} s_i s_{i+1}, \tag{3.7.1}$$

where J is a parameter measuring the strength of interaction of a pair of spins. If J is *positive*, then neighboring spin alignment, $s_i = s_{i+1}$, is favored and the model can be thought of as a one-dimensional analog of a ferromagnetic. If J is *negative*, antialignment ($s_i = -s_{i+1}$) of neighboring spins is preferred and so the model is analogous to an antiferromagnetic.

The partition function of the system is

$$Z_N = \sum_{s_1=-1}^{1} \cdots \sum_{s_N=-1}^{1} e^{\beta J \sum_{i=1}^{N-1} s_i s_{i+1}}. \tag{3.7.2}$$

The summations can be readily carried term by term. For example,

$$\begin{aligned} Z_2 &= \sum_{s_1=-1}^{1} \sum_{s_2=-1}^{1} e^{\beta J s_1 s_2} \\ &= 2(e^{\beta J} + e^{-\beta J}) = 4\cosh \beta J. \end{aligned} \tag{3.7.3}$$

For arbitrary N the result is

$$Z_N = 2(2\cosh\beta J)^{N-1}. \tag{3.7.4}$$

Thus, the Helmholtz free energy, internal energy, and specific heat of the system is

$$F = -kT\ln 2 - (N-1)kT\ln(2\cosh\beta J) \tag{3.7.5}$$

$$U = -(N-1)J\tanh\beta J \tag{3.7.6}$$

$$C_V = (N-1)k(\beta J)^2(1-\tanh^2\beta J). \tag{3.7.7}$$

At high temperatures, $|\beta J| \ll 1$, the specific heat obeys the asymptotic formula

$$C_V \simeq (N-1)\frac{J^2}{kT^2}, \tag{3.7.8}$$

whereas at low temperature, $|\beta J| \gg 1$, the asymptotic formula is

$$C_V \simeq 2(N-1)\frac{J^2}{kT^2}e^{-2|J|/kT}. \tag{3.7.9}$$

The specific heat thus goes through a maximum and vanishes as the temperature approaches zero or infinity (see Fig. 3.10).

As $T \to 0$ the specific heat vanishes because the spins enter the lowest energy state and cannot be excited to the first excited state if kT is too small compared to the coupling parameter $|J|$. The ground state when $J > 0$ is one with all spins aligned in the same sense (that is, all $s_i = +1$ or all $s_i = -1$). When $J < 0$, the ground state is such that all neighboring spins are oppositely aligned ($s_i = -s_{i+1}$). As $T \to \infty$, the spins become randomly ordered and so their contributions to the internal energy cancel out with the result that the specific heat also vanishes.

Although the specific heat goes through a maximum in the range $0 < T < \infty$, it is infinitely differentiable with respect to temperature and so no magnetic transitions (such as the order–disorder transition of a ferromagnetic) are predicted by the model.

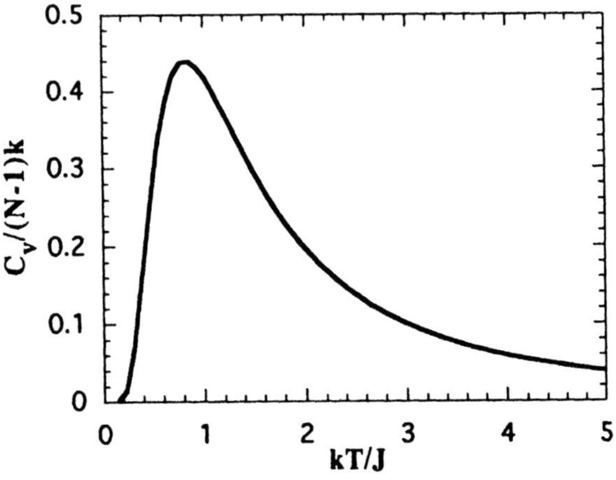

Figure 3.10
Specific heat of the one-dimensional Ising system versus temperature.

It is interesting to see what effect an external magnetic field H has on the one-dimensional Ising model. The energy in this case is

$$E_s = -J\sum_{i=1}^{N-1} s_i s_{i+1} - h\sum_{i=1}^{N} s_i, \tag{3.7.10}$$

where $h = \mu H$ and μ denotes the magnitude of the magnetic spins. The evaluation of the partition function for this case is somewhat more difficult. Kramers and Wannier (1941) used a transfer matrix technique to solve the problem. Their result for the Helmholtz free energy per spin is

$$\tilde{F} = \lim_{N\to\infty} F_N/N = -kT\ln\{e^{\beta J}\cosh\beta h + [e^{2\beta J}\sinh^2\beta h + e^{-2\beta J}]^{1/2}\}. \tag{3.7.11}$$

Exercise
The canonical partition function corresponding to eq. (3.7.10) can be expressed as

$$Z_N = \sum_{s_1}\cdots\sum_{s_N} a_{s_1 s_2}\cdots a_{s_N s_1}$$
$$= TrA^N = \lambda_+^N + \lambda_-^N,$$

where A is the 2×2 matrix

$$A = \begin{bmatrix} e^{\beta(J+h)} & e^{-J} \\ e^{-J} & e^{\beta(J+h)} \end{bmatrix}$$

and λ_+ and λ_- are the largest and smallest eigenvalues of A. Show that the property

$$\lim_{N\to\infty} \frac{\ell n Z_N}{N} = \ell n\lambda_+$$

yields the free energy at eq. (3.7.11).

The magnetization M of the system can be computed from the relation $M = n\langle\mu s_1\rangle = -n\mu(\partial\tilde{F}/\partial h)_\beta$, where n is the number density of spins. The result is

$$M = \frac{n\mu\sinh\beta h}{[\sinh^2\beta h + e^{-4\beta J}]^{1/2}}. \tag{3.7.12}$$

The magnetization, $M/n\mu$, is plotted in Figure 3.11 versus magnetic field h for several values of βJ. As βJ ranges from 0 to ∞ the function $M/n\mu$ ranges from a hyperbolic tangent to a step function of $\beta\mu H$.

The magnetization is a smooth function of H for all finite temperatures and is zero when $H = 0$. Thus, the system exhibits no order–disorder transitions and does not possess spontaneous magnetization.

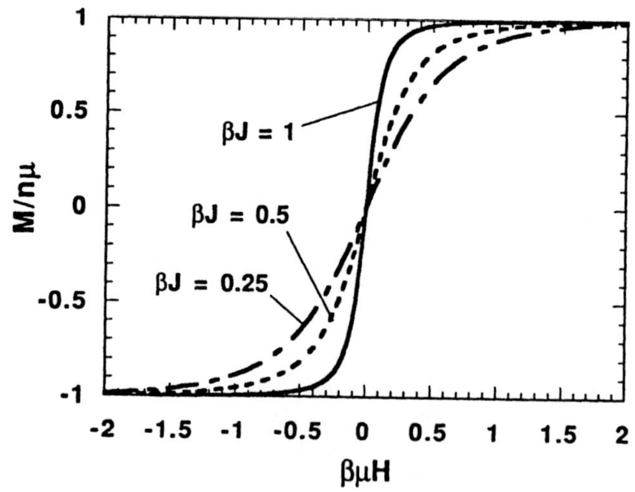

Figure 3.11
Magnetization isotherms of a one-dimensional Ising system as a function of the external magnetic field.

3.8 Two-Dimensional Ising Model

None of the model systems treated thus far in this chapter is capable of a phase transition. Models capable of capturing the cooperative phenomena necessary for a phase transition are usually not analytically solvable. However, the two-dimensional Ising model exhibits cooperative phenomena and has the attractive feature that it can be solved exactly. It has a direct bearing on order–disorder transitions occurring in alloys and ferromagnetic materials.

Imagine a square lattice (Fig. 3.12) whose sites can be occupied by atom A or atom B or by magnetic spins that can be oriented up, ↑, or down, ↓. We assume that the interactions between atoms or spins is the nearest neighbor and can be either $-J$ or J. We assume that the energy is $-J$ for like pairs (AA and BB or ↑↑ and ↓↓) and is J for unlike pairs (AB or ↑↓). Thus, the energy of the system can be expressed as

$$E_s = -J \sum_{(ij)} s_i s_j, \tag{3.8.1}$$

where the sum is over neighboring pairs (ij) and where $s_i = +1$ if the atom is A or spins are ↑ and $s_i = -1$ if the atom is B or spins are ↓. Equation (3.7.1) defines the configurational energy of the Ising model. The canonical ensemble configuration function is

$$Z_N = \sum_{\{s\}} e^{\frac{J}{kT} \sum_{(i,j)} s_i s_j}, \tag{3.8.2}$$

where $\{s\}$ denotes a sum over all possible configurations $s_1, \ldots, s_N$.

The energy parameter J can be chosen to be positive or negative. If J is positive the ordered state of the alloy favors the surrounding of like atoms by like atoms and in the magnetic system the ordered state is ferromagnetic. If J is negative the ordered state of the alloy favors unlike neighboring pairs and the ordered phase of the magnetic system is antiferromagnetic.

Although the two-dimensional Ising model can be solved exactly (first done by Onsager using a transfer matrix technique), the mathematical details of the solution are beyond the scope of this

A	B	B	A		↓	↑	↑	↓
A	A	A	B		↓	↓	↓	↑
B	A	B	A		↑	↓	↑	↓
A	B	A	B		↓	↑	↓	↑
BINARY ALLOY					MAGNET			

Figure 3.12
Ising model of a two-dimensional binary alloy and a two-dimensional magnet.

text. We simply give some results here and refer the reader to the now enormous literature on the Ising model. The results we quote are for $J > 0$, in which case the configuration partition function is

$$\frac{1}{N} \ln Z_N = \ln[2\cosh(2\beta J)] + \frac{1}{2\pi} \int_0^{\pi} d\eta \ln\left\{\frac{1}{2}[1 + (1 - \kappa^2 \sin^2 \eta))^{1/2}]\right\} \tag{3.8.3}$$

where $\beta = 1/kT$ and

$$\kappa \equiv \frac{2\sinh 2\beta J}{\cosh^2 2\beta J}. \tag{3.8.4}$$

The configurational energy of the system, $U_c = \partial \ \ln Z_N/\partial\beta$, is found to be

$$U_c = -NJ\coth\left\{2\beta J\left[1 + \kappa' K_1(\kappa)\right]\right\} \tag{3.8.5}$$

where

$$\kappa' \equiv 2\tanh^2 2\beta J - 1, \tag{3.8.6}$$

and where $K_1(\kappa)$ is an elliptic integral of the first kind,

$$K_1(\kappa) \equiv \int_0^{\pi/2} \frac{d\varphi}{\sqrt{1 - \kappa^2 \sin^2 \varphi}}. \tag{3.8.7}$$

Note that $\kappa'^2 + \kappa^2 = 1$. The specific heat, $C_V = \partial U_c/\partial T$, is

$$C_V = \frac{2Nk}{\pi}(\beta J \coth 2\beta J)^2\{2K_1(\kappa) - 2E_1(\kappa) - (1 - \kappa')[\frac{\pi}{2} + \kappa' K_1(\kappa)]\}, \tag{3.8.8}$$

where $E_1(\kappa)$ is an elliptic integral of the second kind,

$$E_1(\kappa) = \int_0^{\pi/2} \sqrt{1 - \kappa^2 \sin^2 \phi}\, d\phi. \tag{3.8.9}$$

The critical temperature at which the order–disorder transition occurs, that is, at which the specific heat diverges, is given by the condition

$$\sinh \frac{J}{kT_c} = 1, \tag{3.8.10}$$

or $T_c = 2.269 J/k$.

The configuration energy and the corresponding specific heat are plotted qualitatively in Figure 3.13. Because the elliptic integral $K_1(\kappa)$ is logarithmically singular near $\kappa = 1$ (where $T = T_c$), the energy goes as $-|T - T_c| \ln |T - T_c|$ for T near T_c. Thus, the specific heat has a logarithmic singularity ($C_V \sim \ln |T - T_c|$) at the order–disorder critical point T_c. In fact, near the critical point T_c,

$$C_V \approx N(8k/\pi)(\beta J)^2 \ln |1/(T - T_c)|. \tag{3.8.11}$$

Figure 3.13
Configuration energy and specific heat of two-dimensional Ising ferromagnet (or antiferromagnet) or binary alloy versus temperature.

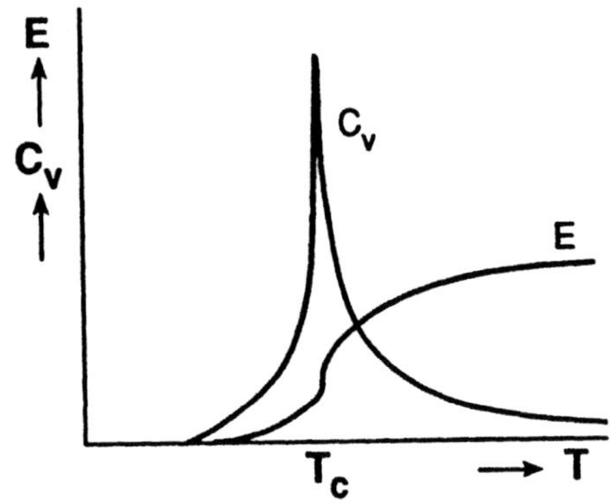

This is an enormously important result in the the theory of phase behavior because it is exact and is in disagreement with the classical results of mean field theory.

The magnetization ($M = \sum_i \langle s_i \rangle$) of the two-dimensional Ising model was determined by Yang (1952). For zero external field his result for a ferromagnetic ($J > 0$) is

$$M/N = \frac{(1 + e^{-4\beta J})^{1/4}(1 - 6e^{-4\beta J} + e^{-8\beta J})^{1/8}}{(1 - e^{-4\beta J})^{1/2}}, \quad T < T_c$$
$$= 0, \quad T > T_c. \tag{3.8.12}$$

Thus, below the critical point T_c (known as the Curie point in the literature of magnetism), the two-dimensional Ising model possesses residual magnetism even in the absence of an external field. The residual magnetism spin goes monotonically from unity to zero as the temperature ranges from zero to the critical point T_c (see Fig. 3.14). Near the critical point, the magnetization obeys the scaling law

$$M/N \propto |T_c - T|^{1/8}, \quad T < T_c. \tag{3.8.13}$$

Let us close this section with a qualitative discussion of the cooperative nature of the order–disorder transition. Consider the ferromagnetic case. At sufficiently low temperature virtually all of the spins are parallel and there is therefore long-range order in

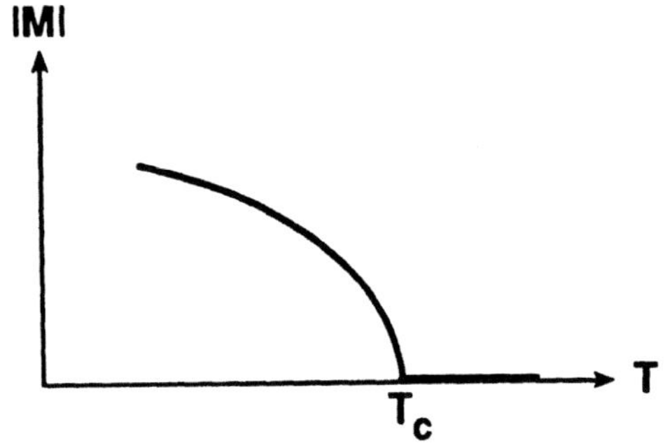

Figure 3.14
Magnetization of a two-dimensional Ising ferromagnetic versus temperature in the absence of an external magnetic field.

spin alignment. At somewhat higher temperature a few spins will have an alignment opposite to the others. This reduces somewhat the long-range order in the system. Below the critical temperature, the majority of the spins remain aligned in the same direction and there is substantial long-range order. However, any misaligned spin present makes it easier for its neighbor to flip its spin. Thus, the higher the degree of spin disorder (misalignment) the easier it is for increased temperature to create further disorder (with significant consumption of energy as shown in Fig. 3.13). The result is a cooperative or cascade process in which as the temperature increases toward the critical point, the long-range order sharply decreases and, at the critical point, gives way totally to disorder on the long-range scale.

3.9 Tonks–Takahashi Fluid

A Tonks–Takahashi fluid is a one-dimensional fluid in which the molecular interactions are pair additive and of the form

$$
\begin{aligned}
u(x_{ij}) &= \infty, \quad |x_{ij}| < a \\
&= \psi(x_{ij}), \quad a < |x_{ij}| < \nu a \\
&= 0, \quad |x_{ij}| > \nu a.
\end{aligned}
\tag{3.9.1}
$$

x_i is the position of the center of molecule i, ψ is an arbitrary function, and α a number lying between 1 and 2. In such a fluid, a given particle interacts only with its nearest neighbors because the next nearest neighbor must be further away than $2d$, which is greater than the range of the pair potential.

If the internal degrees of freedom of the molecules do not affect $u(x_{ij})$ and if the center of mass motion of molecules obeys classical mechanics, then the canonical ensemble partition function is of the form

$$
Q_N = \frac{1}{N!}\left(\frac{2\pi mkT}{h^2}\right)^{N/2} q_I(T)^N Z_N, \tag{3.9.2}
$$

where $q_I(T)$ is the partition function of all the internal energies of a molecule except for the kinetic energy and Z_N is the configuration partition function, namely,

$$
Z_N = \int_0^L \cdots \int_0^L e^{-\beta \sum_{i>j=1}^{N} u(x_{ij})} dx_1 \cdots dx_N. \tag{3.9.3}
$$

L is the length of the box containing the fluid.

Because the potential energy depends only on the absolute value of x_{ij}, the independent integrations in eq. (3.9.3) can be transformed to the ordered integral

$$Z_N = N! \int_0^L dx_N \int_0^{x_N} dx_{N-1} \cdots \int_0^{x_2} dx_1 e^{-\beta[u(x_{12})+u(x_{23})+\cdots+u(x_{N-1,N})]}. \tag{3.9.4}$$

The proof of this transformation will be given in Chapter 5.

The form of eq. (3.9.4) allows us to evaluate analytically the partition function Δ of the isobaric ensemble. First, recall from the theory of Laplace transforms that, if $\tilde{g}(s)$ is the Laplace transform of $g(x)$, it is defined by

$$\tilde{g}(s) = \int_0^\infty e^{-sx} g(x) dx, \tag{3.9.5}$$

then the Laplace transform of

$$f(L) = \int_0^L dx g(L-x) k(x) \tag{3.9.6}$$

is

$$\tilde{f}(s) = \tilde{g}(s)\tilde{k}(s). \tag{3.9.7}$$

Repeated use of property eq. (3.9.7) yields for

$$f(L) = \int_0^L dx_N \int_0^{x_N} dx_{N-1} \cdots \int_0^{x_2} dx_1 g(L-x_N) \prod_{i=1}^{N-1} k(x_{i+1}-x_i)\, g(x_1) \tag{3.9.8}$$

the Laplace transform

$$\tilde{f}(s) = [\tilde{g}(s)]^2 [\tilde{k}(s)]^{N-1}. \tag{3.9.9}$$

The isobaric ensemble partition function for a one-dimensional fluid is

$$\Delta = \frac{1}{v_0} \int_0^\infty e^{-\beta P L} Q_N dL. \tag{3.9.10}$$

Moreover, aside from a factor depending on T and N, Q_N is of the form of eq. (3.9.8) with

$$g(x) = 1 \quad \text{and} \quad k(x) = e^{-\beta u(x)}. \tag{3.9.11}$$

Thus, it follows that

$$\Delta = \frac{1}{v_0} \left(\frac{2\pi m k T}{h^2} \right)^{N/2} q_{_I}(T)^N \frac{1}{\beta P} [\tilde{k}(\beta P)]^{N-1}, \tag{3.9.12}$$

where

$$\tilde{g}(P) = \int_0^\infty e^{-\beta P x} dx = \frac{1}{\beta P} \quad \text{and} \quad \tilde{k}(\beta P) = \int_0^\infty e^{-\beta[Px + u(x)]} dx. \tag{3.9.13}$$

The chemical potential of the Tonks–Takahashi fluid can be computed from the formula,

$$\mu = -kT \left(\frac{\partial \ln \Delta}{\partial N} \right)_{T,P}, \tag{3.9.14}$$

to obtain

$$\mu = \mu^+(T) - kT \ln \tilde{k}(\beta P), \tag{3.9.15}$$

where

$$\mu^+(T) = -kT \ln \left[\left(\frac{2\pi mkT}{h^2} \right)^{1/2} q_I(T) \right]. \tag{3.9.16}$$

The density n $(=N/\bar{L})$ of the fluid can be computed from the Gibbs–Duhem equation, $n^{-1} = (\partial \mu / \partial P)_T$ to obtain

$$n^{-1} = -\tilde{k}'(\beta P)/\tilde{k}(\beta P). \tag{3.9.17}$$

$\tilde{k}'$ denotes the derivative of $\tilde{h}(\beta P)$ with respect to βP.

The simplest example of a fluid obeying eq. (3.9.1) is the hard-rod fluid for which $\psi(x_{ij}) \equiv 0$. In this case,

$$\tilde{k}(\beta P) = \int_a^\infty e^{-\beta P x} dx = \frac{e^{-\beta P a}}{\beta P}. \tag{3.9.18}$$

Inserting this result into eq. (3.9.15), we find

$$\mu = \mu^+(T) + kT \ln \beta P + Pa, \tag{3.9.19}$$

and from eq. (3.9.17) we obtain $n^{-1} = d + (\beta P)^{-1}$, or

$$P = \frac{nkT}{1 - na} = \frac{NkT}{\bar{L} - Na}. \tag{3.9.20}$$

Thus, the hard-rod equation of state differs from the equation of state of an ideal gas, $P = NkT/\bar{L}$, only in that the volume $\bar{L}$ of the system has been replaced by the volume $\bar{L} - Na$ which is the volume in which the rods are free to translate. The volume Na excluded by the hard-rod centers has been removed.

Two other simple models that can be solved analytically are the square-well model, $\psi(x) = -\varepsilon$, and the triangle-well model, $\psi(x) = -\varepsilon(\nu a - x)/a(\nu - 1)$, shown in Figure 3.15. In the case of the square-well fluid

$$\tilde{k}(\beta P) = \frac{e^{-\beta P a + \beta \varepsilon} + (1 - e^{\beta \varepsilon}) e^{-\beta P \nu a}}{\beta P}; \tag{3.9.21}$$

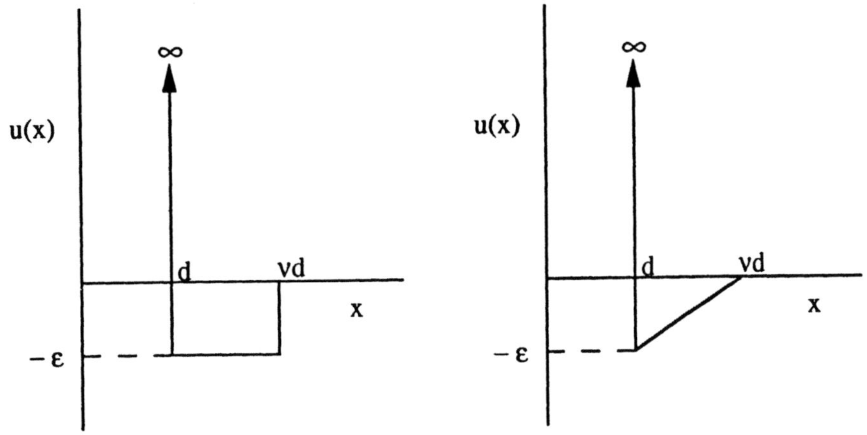

Figure 3.15
Pair potential versus separation for two simple Tonks-Takahashi fluids.

in the case of triangle-well fluid

$$\tilde{k}(\beta P) = \frac{e^{-\beta Pa+\beta\varepsilon} - e^{-\beta P\nu a}}{\beta P + \beta\varepsilon/(\alpha - 1)a} + \frac{e^{-\beta P\nu a}}{\beta P}. \qquad (3.9.22)$$

From these results, we find for the *square-well fluid*

$$\mu = \mu^+(T) + kT \ln \beta P + Pd - \ln\{e^{\beta\varepsilon} + (1 - e^{\beta\varepsilon})e^{-\beta P(\nu-1)a}\} \qquad (3.9.23)$$

and

$$n^{-1} = a + \frac{1}{\beta P} + \frac{(\nu - 1)a(1 - e^{\beta\varepsilon})e^{-\beta P(\nu-1)a}}{e^{\beta\varepsilon} + (1 - e^{\beta\varepsilon})e^{-\beta P(\nu-1)a}} \qquad (3.9.24)$$

and for the *triangle-well fluid*

$$\begin{aligned}\mu &= \mu^+(T) + kT \ln\{\beta P[\beta P + \beta\varepsilon/(\nu - 1)a]\} \\ &\quad + Pa - kT \ln\{\beta P e^{\beta\varepsilon} + [\beta\varepsilon/(\nu - 1)a]e^{-\beta P(\nu-1)a}\}\end{aligned} \qquad (3.9.25)$$

and

$$\begin{aligned}n^{-1} &= a + \frac{2\beta P + \beta\varepsilon/(\nu - 1)a}{\beta P[\beta P + \beta\varepsilon/(\nu - 1)a]} \\ &\quad - \frac{e^{\beta\varepsilon} - \beta\varepsilon e^{-\beta P(\nu-1)a}}{\beta P e^{\beta\varepsilon} + [\beta\varepsilon/(\nu - 1)a]e^{-\beta P(\nu-1)a}}.\end{aligned} \qquad (3.9.26)$$

In the dimensionless units βPa and na, the hard sphere equation of state is a universal function, that is, it is independent of T or any molecular parameters. In the case of the square-well and triangle-well fluids the equation of state in the dimensionless units depends only on the dimensionless parameters $\beta\varepsilon$ and $(\nu - 1)$.

Plots of $\beta\mu$ versus βPa and na versus βPa are given in Figure 3.16 for the special case $\beta\varepsilon = 2$ and $\nu = 2$. Note that the density-pressure plot is monotonic for all three fluids. This means that no liquid-vapor phase split can occur in a hard-rod fluid (not surprising because there are no attractive forces in the hard-rod

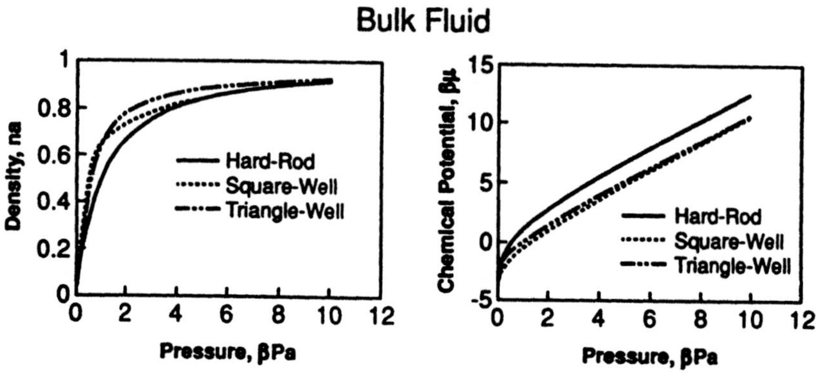

Figure 3.16
Density versus pressure and chemical potential versus pressure for homogeneous fluids. Units of density, chemical potential, and pressure are a^{-1}, β^{-1}, and $(\beta a)^{-1}$. For the square-well and triangle-well fluids $\beta\varepsilon = 2$ and $\nu = 2$.

system) nor can the phase split occur in the square-well or triangle-well fluid at the conditions $\beta\varepsilon = 2$ and $\nu = 2$. It turns out that further analysis shows that the density–pressure curve is monotonic for all values of $\beta\varepsilon$ and α and, therefore, these fluids cannot have a phase transition. In fact, this is a special case of a general theorem proved by van Hove: a one-dimensional fluid cannot undergo a phase transition if the interparticle forces have a finite range.

3.10 Ideal Photon Gas (Black Body Radiation)

If instead of atoms or molecules, a container of volume V contains photons, which can be adsorbed or emitted by the walls of the container, there is no indistinguishability principle and no conservation of the total number of photons. The energy states $\varepsilon_1, \varepsilon_2, \ldots$ can then be occupied by $n_1, n_2, \ldots$ particles, respectively, where $n_i = 0, 1, 2, \ldots$ The partition function for such a system is then

$$Q = \sum_{n_1, n_2 \cdots} e^{-\beta n_1 \varepsilon_1 - \beta n_2 \varepsilon_2 \cdots}$$
$$= \prod_m \sum_{n=0}^{\infty} e^{-\beta n \varepsilon_m} = \prod_m \frac{1}{(1 - e^{-\beta \varepsilon_m})} \tag{3.10.1}$$

The free energy of the system is thus

$$F = kT \sum_m \ln(1 - e^{-\beta \varepsilon_m}). \tag{3.10.2}$$

Because

$$\frac{\partial \ln Q}{\partial \varepsilon_s} = -\beta \sum_{n_1, n_2 \cdots} \frac{n_s e^{-\beta n_1 \varepsilon_1 - \beta n_2 \varepsilon_2 \cdots}}{Q} \tag{3.10.3}$$

it follows that the average number $\bar{n}_s$ of photons in state s is given by

$$\bar{n}_s = \frac{e^{-\beta\varepsilon_s}}{1 - e^{-\beta\varepsilon_s}} = \frac{1}{e^{\beta\varepsilon_s} - 1}. \tag{3.10.4}$$

This is called the Planck distribution. According to the theory of electromagnetic radiation the photon energy and momentum obey the equations

$$E\varepsilon = \hbar\omega, \quad \mathbf{p} = \hbar\mathbf{k}, \quad |\mathbf{p}| = \frac{\hbar\omega}{c}. \tag{3.10.5}$$

where ω is the frequency, and $\mathbf{k}$ the wave vector (quantized) of the photon. Because the number of quantum states in the momentum d^3p and volume V is

$$\frac{d^3pV}{h^3}, \tag{3.10.6}$$

it follows that the probable number of photons per unit volume in the wave vector range d^3k is

$$\begin{aligned} f(\mathbf{k})d^3k &= \bar{n}_S\frac{d^3pV}{h^3V} \\ &= \frac{1}{e^{\beta\hbar\omega} - 1}\frac{d^3k}{(2\pi)^3}. \end{aligned} \tag{3.10.7}$$

Averaging eq. (3.10.7) over all directions for $\mathbf{k}$ and multiplying by $2\hbar\omega$ to account for the energy in the two directions of electric polarization of the photons, and using the relation $\omega = ck$, we obtain the energy distribution function $e(\omega)$, where

$$e(\omega)dw = \frac{\hbar}{\pi^2c^3}\frac{\omega^3 d\omega}{e^{\beta\hbar\omega} - 1}. \tag{3.10.8}$$

Here, $e(\omega)d\omega$ denotes the porbable energy per unit volume arising from photons having an energy between ω and $\omega + d\omega$. $e(\omega)$ is plotted versus $\hbar\omega/kT$ in Figure 3.17. With energy density in the units $(kT)^4/\pi^2c^3\hbar^3$, the $e(\omega)$ curve is a universal plot.

The maximum in $e(\omega)$ occurs where $\partial e(\omega)/\partial\omega = 0$. From this equation we find that the frequency of maximum radiation is given by

$$\hbar\omega_{\text{max}}/kT = 2.821439. \tag{3.10.9}$$

This is known as Wein's displacement law. To appreciate the implications of Wein's displacement law for black body radiation, let us convert eq. (3.10.9) to an equation for the wavelength λ_{max} of the most intense radiation at the temperature T. Recall $\nu = c/\lambda$ or $\omega = 2\pi c/\lambda$, so that

$$\lambda_{\text{max}} = hc/(2.821439kT). \tag{3.10.10}$$

Figure 3.17
Radiation energy distribution versus radiation frequency ω.

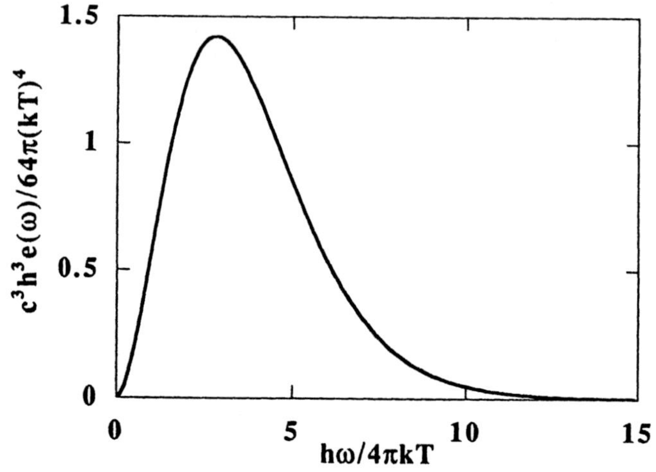

Thus a black body source will radiate ultraviolet light (say 10,000 Å) at $T = 14,400$K, visible light (say $\lambda_{max} = 3000$ Å) at $T = 4800$K, infrared radiation (say $\lambda_{max} = 300,000$ Å$= 30\mu$m) at $T = 480$K, microwaves (say $\lambda_{max} = 3$ cm) at $T = 0.48$K, and radio waves (say $\lambda_{max} = 30$ m) at $T = 4.8 \times 10^{-5}$K.

The total energy density at a given temperature is

$$\bar{e}(T) \equiv \int_0^\infty e(\omega)d\omega = \frac{\hbar}{\pi^2 c^3}\left(\frac{kT}{\hbar}\right)^4 \int_0^\infty \frac{\eta^3 d\eta}{e^\eta - 1}$$
$$= \frac{\pi^2}{15}\frac{(kT)^4}{(c\hbar)^3}. \tag{3.10.11}$$

The result $\bar{e}(T) \propto T^4$ is known as the Stefan–Boltzmann law. It was through trying to understand this law and the low frequency behavior of $e(\omega)$ that Planck discovered the fundamental constant h of quantum physics.

We can compute the thermodynamic properties of black body radiation in a large cavity or reservoir. Assuming that the electric field vector of black body radiation obeys Maxwell's equation

$$\nabla^3 \mathbf{E} = \frac{1}{c^2}\frac{\partial^2 E}{\partial t^2} \tag{3.10.12}$$

and solving for $\mathbf{E}$ is a large cubic cavity of length L on the side with the boundary conditions

$$\mathbf{E}(x+L, y, z) = \mathbf{E}(x, y+L, z) = \mathbf{E}(x, y, z+L) = \mathbf{E}(x, y, z), \tag{3.10.13}$$

we find by separation of variables solution

$$\mathbf{E} = \sum_{n_x, n_y, n_z} e^{i(\nu_n t - \mathbf{k}_n \cdot \mathbf{r})}\mathbf{E}_o, \tag{3.10.14}$$

where

$$\nu_{\mathbf{n}}^2 = c^2\mathbf{k}_{\mathbf{n}}^2 = c^2/\lambda_{\mathbf{n}}^2 \tag{3.10.15}$$

where

$$k_{\mathbf{n}}^2 = \frac{1}{L^2}(n_x^2 + n_y^2 + n_z^2); \quad n_\alpha = 0, \pm 1, \pm 2, \ldots, \quad \alpha = x, y, z. \tag{3.10.16}$$

$\mathbf{k}_n$ is the wave vector and $\lambda_{\mathbf{n}} = 1/|\mathbf{k}_{\mathbf{n}}|$ is the wavelength of the **n**th photon vibrational mode.

For a large system $\nu_{n_x,n_y,n_z}/c\,[= (n_x^2 + n_y^2 + n_z^2)^{1/2}/L]$ varies very little with increase or decrease of n_x, n_y, or n_z by ± 1. Thus, we can approximate the number of states with frequency between 0 and ν by

$$N(\nu) \approx \int\limits_{R_n^2 \le L^2\nu^2/c^2} dn_x dn_y dn_z, \tag{3.10.17}$$

where

$$R_n^2 = n_x^2 + n_y^2 + n_z^2 = L^2\nu^2/c^2. \tag{3.10.18}$$

Using spherical coordinates, $n_x = R_n \cos\theta$, $ny = R_n \sin\theta \sin\varphi$, and $n_y = R_n \sin\theta \cos\varphi$, $R_n = L\nu/c$, we obtain

$$\begin{aligned} N(\nu) &= \int_0^{L\nu/c}\int_0^{2\pi}\int_0^{\pi} \frac{L^2\nu^2}{c^2} \sin\theta d\theta d\varphi d\left(\frac{L\nu}{c}\right) \\ &= \frac{4\pi}{3}\left(\frac{L\nu}{c}\right)^3. \end{aligned} \tag{3.10.19}$$

But

$$dN(\nu) = \Phi(\nu)d\nu = 4\pi\left(\frac{L}{c}\right)^3 \nu^2 d\nu = 4\pi\frac{V}{c^3}\nu^2 d\nu \tag{3.10.20}$$

where $\Phi(\nu)d\nu$ represents the number of photons lying between ν and $\nu + d\nu$.

With this frequency distribution, the free energy, eq. (3.10.2), can be evaluated as

$$F = -2kT\int_0^\infty \ln(1 - e^{-h\nu/ckT})\frac{4\pi V}{c^3}\nu d\nu. \tag{3.10.21}$$

The factor of 2 in front of kT is necessary because photons possess two independent polarizations. Equation (3.10.21) can be reduced analytically. The result is

$$F = -\frac{4\sigma}{3c}VT^4, \tag{3.10.22}$$

where $\sigma = 2\pi^5k^4/15c^2h^3 = 5.67 \times 10^{-5}$ erg/s cm^2 K^4 is the Stefan–Boltzmann constant. Of course, $Q = \exp(-F/kT)$. From eq. (3.10.22) all the other thermodynamic properties of black body radiation can be obtained. These are

$$S = \frac{16\sigma}{3c} V T^3 \tag{3.10.23}$$

$$U = \frac{4\sigma}{c} V T^4 \tag{3.10.24}$$

$$P = \frac{4\sigma}{3c} T^4 = \frac{1}{3}\frac{U}{V} \tag{3.10.25}$$

$$H = U + PV = \frac{16\sigma}{3c} V T^4 \tag{3.10.26}$$

$$C_V = \frac{16\sigma}{c} V T^3 \tag{3.10.27}$$

$$C_P = \frac{64\sigma}{3c} V T^3 = \frac{4}{3} C_V \tag{3.10.28}$$

$$G = F + PV = 0, \quad \mu = 0. \tag{3.10.29}$$

The temperature of the sun is about 10^4K and so its radiation pressure is about 0.63 dyn/cm^2, small but nonzero.

3.11 Ideal Gases Obeying Fermi–Dirac and Bose–Einstein Statistics

According to the principles of quantum mechanics, no two identical particles having a half-integer spin angular momentum (e.g., electrons, protons, He^3 atoms, etc.) can occupy the same single-particle state. This is Pauli's exclusion principle. It provides the theoretical basis for Fermi–Dirac statistics.

If the single particle energy states are $\varepsilon_1, \varepsilon_2, \ldots,$ and n_m is the number of particles in state m, then the energy corresponding to a given quantum state is

$$E = n_1\varepsilon_1 + n_2\varepsilon_2 + \cdots = \sum_m n_m \varepsilon_m \tag{3.11.1}$$

and the total number of particles in the system is given by

$$N = \sum_m n_m. \tag{3.11.2}$$

For Fermi–Dirac statistics in any state, n can only be 0 or 1. Thus, the canonical ensemble partition function is

$$Q_N = \sum_{\{n_m\}}' e^{-\beta[n_1\varepsilon_1 + n_2\varepsilon_2 + \cdots]} \tag{3.11.3}$$

where the prime on the summation means that any allowed set of $\{n_1, n_2, \ldots\}$ has to obey eq. (3.11.2). The grand canonical ensemble partition function is given by

$$\begin{aligned}\Xi &= \sum_{N=0}^{\infty} e^{\beta N \mu} Q_N \\ &= \sum_{N=0}^{\infty} \sum_{\{n_m\}}{}' e^{\beta[n_1(\mu-\varepsilon_1)+n_2(\mu-\varepsilon_2)\cdots]}.\end{aligned} \tag{3.11.4}$$

With sum over N any set of occupation numbers $\{n_1, n_2, \ldots\}$ will be allowed. Thus, summing over N is equivalent to dropping the constraint on the summation over $\{n_m\}$. The result is

$$\begin{aligned}\Xi &= \sum_{n_1=0}^{1} \sum_{n_2=0}^{1} \cdots e^{\beta[n_1(\mu-\varepsilon_1)+n_2(\mu-\varepsilon_2)+\cdots]} \\ &= \prod_m \left(1 + e^{\beta(\mu-\varepsilon_m)}\right)\end{aligned} \tag{3.11.5}$$

The probable number of particles $\bar{n}_s$ in the energy state ε_s is given by $\bar{n}_s = \partial \ln \Xi / \partial \beta \varepsilon_s$, or

$$\bar{n}_s = \frac{1}{e^{\beta(\varepsilon_s-\mu)} + 1}. \tag{3.11.6}$$

The chemical potential is related to the average number of particles in the system by the condition

$$\bar{N} = \sum_s \bar{n}_s = \sum_s \frac{1}{e^{\beta(\varepsilon_s-\mu)} + 1}. \tag{3.11.7}$$

The Fermi–Dirac occupancy distribution, eq. (3.11.6), is illustrated in Figure 3.18 for finite and zero temperature. At absolute zero temperature, all energy levels are filled with probability 1 up to $\varepsilon = \mu$ and there are no particles excited to an energy state larger

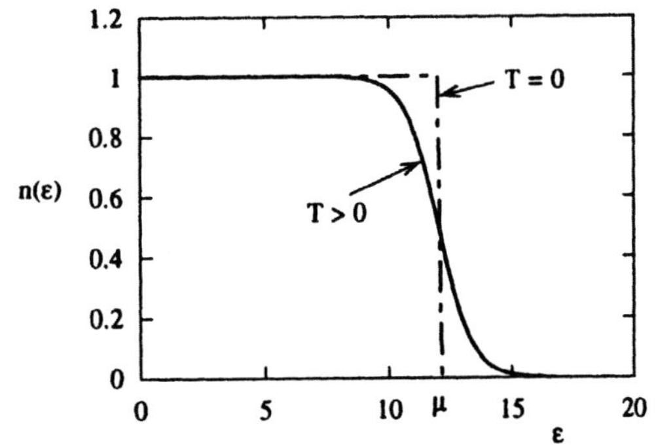

Figure 3.18
Fermi–Dirac occupancy distribution.

than the chemical potential μ. At a finite temperature, there is a distribution of particles partially occupying the states in the energy range of about $\pm kT$ of $\varepsilon = \mu$.

The chemical potential of conduction electrons in metals can be roughly approximated by the formula

$$\mu \simeq \frac{\hbar^2}{2m}\left(3\pi^2 \frac{N}{V}\right)^{2/3}, \tag{3.11.8}$$

where m is the electron mass and N/V is the member density of electrons. Assuming one conduction electron per sodium atom and using the density of sodium metal, we estimate for the conduction electrons in sodium

$$\mu \simeq 5.12 \times 10^{-12}\text{erg}. \tag{3.11.9}$$

The Fermi energy ε_F is by definition equal to μ at 0K, and so the characteristic Fermi temperature, $T_F = \varepsilon_F/k = \mu/k$, for sodium metal is $T_F = 3.7 \times 10^4$K. Thus, at room temperature the energy distribution of conduction electrons in a metal resembles closely the $T = 0K$ distribution, or the step function.

Electricity is conducted in metals by the small population of electrons thermally excited to energies above μ. It follows from the Fermi–Dirac distribution that this population is approximately $N_{ex} \simeq NkT/\mu \simeq NT/T_F$. These excited electrons are also the ones that contribute to the heat capacity, because they can store energy through thermal excitation. The electronic heat capacity is approximately $C_V^e \simeq \frac{3}{2}N_{ex}k$, or

$$C_V^e \simeq \frac{3}{2}Nk\frac{T}{T_F}. \tag{3.11.10}$$

It should be noted that because $T/T_F \ll 1$, the electronic heat capacity of a metal is tiny compared to atomic contributions at room temperature. However, as T approaches zero, the atomic contribution goes to zero as T^3 (see Section 3.4), and so at extremely low temperatures the heat capacity of a metal should become a linear function of temperature. Or, stated differently, a plot of C_V^e/T versus T should have a nonzero intercept at $T = 0$K. This has indeed been observed to be the case for metals.

The other case of quantum mechanical statistics is that for particles having an integral total spin angular momentum (e.g., neutrons, photons, He^4, etc.). In this case there is no restriction on how many particles occupy the same single particle quantum states, and so the grand canonical partition function becomes

$$\begin{aligned}\Xi &= \sum_{n_1=0}^{\infty}\sum_{n_2=0}^{\infty}\cdots e^{-\beta[n_1(\varepsilon_1-\mu)+n_2(\varepsilon_2-\mu)+\cdots]} \\ &= \prod_m \left(1 - e^{-\beta(\varepsilon_m-\mu)}\right)\end{aligned} \tag{3.11.11}$$

where the property $\sum_{n=0}^{\infty} x^n = 1/(1-x)$, $x = e^{-\beta(\varepsilon-\mu)}$, has been used. Again computing $\bar{n}_s$ from $-\partial \ln \Xi/\partial\beta\varepsilon_s$, we find

$$\bar{n}_s = \frac{1}{e^{\beta(\varepsilon_s-\mu)} - 1}. \tag{3.11.12}$$

This result is known as the Bose–Einstein distribution.

The most common example of a system obeying the Bose–Einstein distribution is the photon gas discussed in the previous section. In that case $\mu = 0$, reflecting the fact that there is no restriction on the number of particles occupying a given volume. Otherwise μ has to obey the relationship

$$\bar{N} = \sum_s \frac{1}{e^{\beta(\varepsilon_s-\mu)} - 1}. \tag{3.11.13}$$

Supplementary Reading

Chandler, D. 1987. *Introduction to Modern Statistical Mechanics*, Oxford University Press, New York.

Hill, T. L. 1960. *An Introduction to Statistical Thermodynamics*, Addison–Wesley, New York.

Kestin, J. and Dorfman, J. R. 1971. *A Course in Statistical Thermodynamics*, Academic Press, New York.

Reif, F. 1965. *Statistical and Thermal Physics*, McGraw–Hill, New York.

Robertson, H. S. 1993. *Statistical Thermophysics*, Prentice–Hall, New York.

Ter Haar, D. 1954. *Elements of Statistical Mechanics*, Holt, Rinehart and Winston, New York.

Exercises

1 Consider an ideal gas of N particles obeying classical mechanics and Boltzmann statistics. Suppose the energy of a gas particle obeys the equation.

$$\varepsilon = \alpha|\mathbf{p}|^\nu$$

where α and ν are positive numbers.

a) Find the equation of state and the heat capacity of the ideal gas for arbitrary α and ν.

b) When $\nu = 1$ the gas is relativistic and when $\nu = 2$ it is Newtonian. Give expressions for the internal energy and the entropy for these two cases.

c) Plot the specific heat per particle, C_V/Nk, versus ν.

2 In the presence of an electric field $\mathbf{E}$, the energy of a dipole μ is $-E\mu\cos\theta$, where θ is the orientation of the dipole relative to the

direction of the field. The probability that a dipolar molecule in an ideal gas has the orientation between θ and $\theta + d\theta$ is given by

$$p(\theta)\sin\theta d\theta = \frac{\mu E e^{E\mu\cos\theta/kT}}{e^{E\mu/kT} - e^{-E\mu/kT}} \sin\theta d\theta.$$

The electric polarization P per unit volume of N molecules in volume V is

$$P = \frac{N\mu}{V}\overline{\cos\theta}.$$

a) Show that the polarization of the system obeys the equation

$$P = \frac{N\mu}{V}\left[\coth\left(\frac{\mu E}{kT}\right) - \frac{kT}{\mu E}\right].$$

b) Plot $PV/N\mu$ versus $\mu E/kT$.

c) The electric displacement is defined by

$$D \equiv E + 4\pi P.$$

The dielectric constant is defined by

$$\varepsilon = \lim_{E\to 0} D/E.$$

Show that

$$\varepsilon = 1 + \frac{4\pi N\mu^2}{3kTV}.$$

Thus, in an ideal gas of permanent dipoles, the dielectric constant deviates from its vacuum value of unity by a term that is proportional to number density and the square of the dipole moment and is inversely proportional to temperature.

d) The dipole moment of a water molecule is $\mu = 1.25$ Debye. Estimate the dielectric constant of water vapor at 10^{20} molecules/cm^3 and $T = 500$K.

3 A simple harmonic one-dimensional oscillator has energy levels given by $E_m = (n + \frac{1}{2})\hbar\omega$, where ω is the characteristic (angular) frequency of the oscillator and where the quantum number N can assume the possible integral values $n = 0, 1, 2, \ldots$. Suppose that such an oscillator is in thermal contact with a heat reservoir at temperature T low enough so that $kT/(\hbar\omega) \ll 1$.

a) Find the ratio of the probability of the oscillator being in the first excited state to the probability of its being in the ground state.

b) Assuming that only the ground state and first excited state are appreciably occupied, find the mean energy of the oscillator as a function of the temperature T.

4 A sample of mineral oil is placed in an external magnetic field H. Each proton has spin $\frac{1}{2}$ and a magnetic moment μ; it can, therefore, have two possible energies $\varepsilon = \pm\mu H$, corresponding to the two possible orientations of its spin. An applied radio-frequency field can induce transitions between these two energy levels if its frequency ν satisfies the Bohr condition $h\nu = 2\mu H$. The power absorbed from this radiation field is then proportional to the *difference* in the number of nuclei in these two energy levels. Assume that the protons in the mineral oil are in thermal equilibrium at a temperature T that is so high that $\mu H \ll kT$. How does the absorbed power depend on the temperature T of the sample?

5 A system consists of N weakly interacting particles, each of which can be in either of two electronic states with respective energies ε_1 and ε_2, where $\varepsilon_1 < \varepsilon_2$.

a) Without explicit calculation, make a qualitative plot of the mean energy $\bar{E}$ of the system as a function of its temperature T. What is $\bar{E}$ in the limit of very low and very high temperatures? Roughly near what temperature does $\bar{E}$ change from its low to its high temperature limiting values?

b) Using the result of (a), make a qualitative plot of the heat capacity C_v (at constant volume) as a function of the temperature T.

c) Calculate explicitly the mean energy $\bar{E}(T)$ and heat capacity $C_v(T)$ of this system. Verify that your expressions exhibit the qualtative features discussed in (a) and (b).

6 The nuclei of atoms in a certain crystalline solid have spin one. According to quantum theory, each nucleus can therefore be in any one of three quantum states labeled by the quantum number m, where $m = 1, 0$, or -1. This quantum number measures the projection of the nuclear spin along a crystal axis of the solid. Because the electric charge distribution in the nucleus is not spherically symmetrical but ellipsoidal, the energy of a nucleus depends on its spin orientation with respect to the internal electric field existing at its location. Thus a nucleus has the same energy $E = \varepsilon$ in the state $m = 1$ and the state $m = 1$ compared with an energy $= 0$ in the state $m = 0$.

a) Find an expression, as a function of absolute temperature T, of the nuclear contribution to the molar internal energy of the solid.

b) By directly counting the total number of accessible states, calculate the nuclear contribution to the molar entropy of the solid at very low temperatures. Calculate it also at very high temperatures. Show that the expression in part (b) reduces propertly to these values as $T \to 0$ and $T \to \infty$.

c) Make a qualitative graph showing the temperature dependence of the nuclear contribution to the molar heat capacity of the solid. Calculate its temperature dependence explicity. What is its temperature dependence for large values of T?

7 Two atoms of mass M interact with each other by a force derivable from a mutual potential energy of the form

$$U = U_o\left[\left(\frac{a}{x}\right)^{12} - 2\left(\frac{a}{x}\right)^{6}\right]$$

where x is the separation between the two particles. The particles are in contact with a heat reservoir at a temperature T low enough so that $kT \ll U_o$, but high enough so that classical statistical mechanics is applicable. Derive an approximate formula to the mean separation $\bar{x}(T)$ of the particles and show how it can be used to compute the quantity

$$\alpha = \frac{1}{\bar{x}}\frac{\partial \bar{x}}{\partial T}.$$

(This illustrates the fundamental procedure for calculating the coefficient of linear expansion of a solid.) Your calculation should make approximations based on the fact that the temperature is fairly low; thus retain only the lowest order terms that yield a value of $\alpha \neq 0$.

8 Calculate the fraction of Cl_2 molecules in the first three vibrational states at 300, 1000 and 3000K.

9 Consider a Tonks–Takahashi fluid for which

$$\psi(x) = kT \ln\left(\frac{x}{\nu a}\right).$$

Calculate and plot the density and chemical potential versus pressure for $\nu = 2$. Use a convenient set of units.

10 a) Estimate the equilibrium constant for the dissociation reaction of iodine,

$$I_2 \rightleftharpoons 2I,$$

at 1000K.

b) Suppose 1 mol of I_2 is heated to 1200K and allowed to equilibrate at 1 atm. Estimate the amount of I at equilibrium.

11 Adsorption in a pore may be modeled as a 1-dimensional gas. The adsorbates obey the following potential in a pore of length L

$$\phi(x) = -|a|x, \quad 0 \leq x \leq L$$

Consider that the adsorbate particles are noninteracting with each other.

a) Calculate the partition function, Ξ, of the adsorbed gas. (Remember to include kinetic energy.) *Note:*

$$\Xi = \sum_{N=0}^{\infty} \frac{e^{\beta N \mu} Q_N}{N! \Lambda^N}.$$

b) Determine the adsorption isotherm (that is, obtain $\overline{N}$) by taking on appropriate derivative of Ξ.

c) How does $\overline{N}$ vary with pressure? [Assume $\mu = kT \ln\left(P/P_o\right)$]. What effect would an increase in $|a|$ have?

d) Is the isotherm realistic at high pressures? What model might be an improvement?

12 Suppose the particles of a Tonks–Takahaski fluid have the pair potential

$$\begin{aligned} u(x) &= \infty, \quad x < d \\ &= -\varepsilon \ln\left[\frac{(\alpha + 1)d - x}{d}\right], \quad d < x < \alpha d \\ &\quad 0, \quad x > \alpha d. \end{aligned}$$

Find μ as a function of P and n as a function of P for this fluid. Let $\alpha = 2$ and plot nd versus $\beta P d$ for $\beta\varepsilon = 0.1, 1, 2,$ and 10.

13 The characteristic Debye temperature for solid aluminum is 390K. Calculate the molar entropy and heat capacity at 200K using the Debye model. Experimentally the molar entropy is 19.13 J/K mol.

14 The characteristic Debye temperature for diamonds is 1860K. Use the Debye theory to predict the molar entropy at 298.15K. Experimentally the molar entropy is 1.439 J/K mol.

15 The characteristic Einstein temperature for solid aluminum is 260K. Use the Einstein model to estimate the molar entropy and heat capacity of aluminum at 200K. Experimentally the molar entropy is 19.13 J/K mol.

16 The entropy of vaporization is by definition $\Delta S_{vap} \equiv S_v - S_l$, where S_v and S_l are the entropy of vapor and liquid phase at the coexisting temperature. At not too high vapor pressure, where $PV \approx NkT$, $S_v \gg S_l$ and the translational part of S_v dominates. Thus, the entropy of vaporization of a monatomic fluid is well approximated by the ideal gas result

$$\Delta S_{vap} \approx S_v^T = \frac{5}{2}Nk + Nk \ln\left[\frac{(2\pi mkT)^{3/2}}{h^3}\frac{V}{N}\right].$$

This is known as the Sackur–Tetrode equation. Because the enthalpy of vaporization ΔH_v is related to the entropy by $\Delta H_{vap} = T\Delta S_{vap}$, the Sackur–Tetrode equation is a useful estimate of the heat of vaporization at low vapor pressure.

Estimate the enthalpy of vaporization of argon, nitrogen, and butane at their boiling points. Compare the results with the experiment and explain the trends.

17 The characteristic rotational temperature of the linear N_2O molecule is $\theta_r = 0.6$K.

a) Find the moment of inertia of N_2O.

b) Compute the sum of the translational and rotational entropies of N_2O at 298.15K and 1 atm. The measured value is 110 J/K mol. Explain the difference between the theory and experiment.

18 Define the equilibrium constant for the reaction

$$\sum_{\text{reactants}} \nu_i A_i \rightleftharpoons \sum_{\text{products}} \nu_i B_i$$

conducted at ideal gas conditions as

$$K_p = \frac{\prod_{\text{prod.}} P_i^{\nu_i}}{\prod_{\text{react.}} P_i^{\nu_i}}$$

where P_i is the partial pressure (mole fraction times pressure) of component i.
Estimate K_p for the reaction (H, hydrogen, D, deuterium, O, oxygen)

$$H_2O + D_2O \rightleftharpoons HOD$$

at 298 and 800K. Neglect vibrational excitations and use the data given below.

Molecular	Molecule Weight	I_A	I_B (kgm^2 × 10^{47})	I_c	Zero-Point Energy (kJ/mol)
H_2O	18.01	1.02	1.92	2.94	231.91
D_20	20.03	1.84	3.83	5.67	169.58
HOD	19.02	1.21	3.06	4.27	201.79

The rotational symmetry factors of these molecules are $\sigma_{H_2O} = \sigma_{D_2O} = 2$ and $\sigma_{HOD} = 1$.

19 The surface temperature of the sun is $T = 5800K$. Estimate the wavelength at which the emitted power is a maximum. What color does this wavelength correspond to? What is the radiation pressure of a black body at 5800K?

20 Consider the diatomic molecule AB, whose atoms have atomic weights 20 and 40, respectively. Suppose the bond potential energy is of the form

$$u(r) = 4\varepsilon\left[\left(\frac{\sigma}{r}\right)^{12} - \left(\frac{\sigma}{r}\right)^{6}\right],$$

with $\varepsilon/k = 2000K$ and $\sigma = 1.5 \times 10^{-8}$ cm. Ignoring electronic excitations, esimate for the dilute gas state

a) The characteristic rotational temperature $\theta_R \equiv \hbar^2/2Ik$ and the rotational specific heat at 300K.

b) The characteristic vibrational tempeature $\theta_V \equiv \hbar\omega/k$ and the vibrational specific heat at 300K.

c) The total specific heat at 300K and at 10K.

d) The enthalpy of dissociation AB $\rightarrow$A+B at 300K and 1 atm.

e) The temperature at which 1 mol of AB will be in chemical equilibrium with 0.5 mol each of A and B in a volume of 45 L. The ground electronic states of AB, A and B are not degenerate.

21 Estimate the mean free path of N_2 in air at the earth's surface and 15 miles above the earth's surface.

22 Consider a two-dimensional, classical monatomic solid composed of particles interacting with the potential energy

$$u^N = u_o^N + \sum_{j=1}^{N} \frac{\lambda}{2}(\mathbf{r}_j - \mathbf{R}_i)^2,$$

where u_o^N and λ are constants, $\mathbf{R}_i$ is lattice site positions, and $\mathbf{r}_j$, $j = 1, \ldots, N$ denote the particle positions.
Calculate the pressure, specific heat and entropy expressions for the solid. What is the qualitative explanation for the result for the pressure?

23 a) Consider an ideal gas of H_2 molecules. Estimate the heat capacity C_V at $T = 5K$, 50K, and 500K.

$$H_2 + T_2 \rightleftharpoons 2HT,$$

where T is the tritium isotope (atomic weight of 3) of hydrogen.

b) Assuming electronic structures and bond potentials are identical for H_2, T_2, and HT, derive an expression for the equilibrium constant $K(T)$ for the above reaction. You may assume $(\varepsilon_{2e} - \varepsilon_{1e})/kT \gg 1$.

Hint: At equilibrium $\sum_i \nu_i\mu_i = 0$, where ν_i is the stoichiometric coefficient of component i and μ_i is its chemical potential. Thus, for the reaction considered $-\mu_{H_2} - \mu_{T_2} + 2\mu_{HT} = 0$. By definition $K = \prod_i e^{\nu_i\mu_i^+/kT}$, where

$$\mu_i = \mu_i^+ + kT\ln x_i P$$

for an ideal gas.

c) If one mixed H_2 and T_2 and allowed the reaction to reach equilibrium at 1000K and 1 atm, what would be the partial pressure of HT at equilibrium?

d) Estimate the entropy of vaporization of a monatomic liquid at 300K if the vapor pressure is 1 atm at 300K. Assume the molecular weight to be 40. Data for H_2 : $\theta_V(\equiv \frac{\hbar w}{k}) =$ 6210K; $\theta_R(\equiv \frac{\hbar^2}{2Ik}) = 85.4$K.

24 The following problem illustrates the magnetic cooling process. Consider a monatomic, one component solid whose potential energy is of the form given by the Einstein model, that is,

$$u^N = u_o^N + \sum_{i=1}^{N} \frac{\lambda}{2}(\mathbf{r}_i - \mathbf{R}_i)^2,$$

where $u_o^N = u^N(\mathbf{r}_1, \ldots, \mathbf{r}_N) \mid_{\{\mathbf{r}_i\}=\{\mathbf{R}_i\}}$ and λ is a force constant (independent of T). Assume $(\varepsilon_{2e} - \varepsilon_{1e})/kT \gg 1$ throughout this problem.

Suppose each atom of the solid has a spin of $\frac{1}{2}$ with a magnetic moment μ. The magnetic moments of the atoms do not interact with one another and interact only weakly with the atomic motions (by weakly interacting we mean that the energy states of the lattice and the magnetic spins may be treated as independent although there is an exceedingly weak interaction between lattice and spins that allows thermal equilbrium between lattice and spin states, that is, the lattice and spins have the same temperature at equilibrium). In the presence of a magnetic field of strength H there are two spin states of energies, H and μH, respectively, available to each of atom of the lattices.

a) Derive an expression for the specific heat C_V of the solid. Give a qualitative plot of C_V versus T for nonzero H. What is the high temperature limit of C_V?

b) Suppose the solid, originally at 1.3K and subject to a field $H = 50,000$ Gauss, is put into a rigid, thermally insulated container. If the magnetic field is then turned off and the solid is allowed to equilibrate in isolation, its temperature will fall. This process is called magnetic cooling. Give equations determining the final temperature of the solid. Compute the final temperature of the solid.

c) The magnetization M of a substance is defined by the expression

$$M = -\left(\frac{\partial F}{\partial H}\right)_{T,N,V}$$

and the magnetic susceptibility X is defined by the equation

$$M = XH.$$

At sufficiently large temperatures

$$X = \alpha/T,$$

where α is a constant. Find α for the present problem. What is the limiting value of M as $T \to 0$K or $H \to +\infty$? Data for this problem: $\theta_E(\equiv \hbar\omega_E/k) = 80$K. $H = 5 \times 10^{-16}$erg at $H =$ 50, 000 Gauss.

25 The one-dimensional Langmuir adsorption model can be modified by adding nearest neighbor interactions. The energy of the model is

$$E = -J \sum_{i=1}^{M} s_i s_{i+1} - \varepsilon_o \sum_{i=1}^{M} s_i$$

where $s_i = 0$ or 1 and $s_{N+1} = s_1$ and M is the number of adsorption sites. The molecule-substrate interaction energy is $-\varepsilon_o$ and the pair interaction energy of nearest neighbor molecules is $-J$. The grand partition of the system is

$$\Xi = \sum_{s_i=0}^{1} \cdots \sum_{s_N=0}^{1} \prod_{i=1}^{M} a_{s_i s_{i+1}} = Tr\mathbf{A}^N,$$

where $\mathbf{A}$ is the 2×2 matrix

$$\mathbf{A} = \begin{bmatrix} 1 & 1 \\ e^{\beta(\mu+\varepsilon_o)} & e^{\beta(\mu+\varepsilon_o+J)} \end{bmatrix}.$$

μ is the chemical potential. Use the method of the marginalia exercise on p. 103 to show that

$$\lim_{M\to\infty} \frac{ln\Xi}{M} = ln\lambda_+,$$

where the maximum eigenvalue of $\mathbf{A}$ is

$$\lambda_+ = \frac{1}{2}\left[1 + e^{\beta(\mu+\varepsilon_o+J)} + W\right];$$

$$W = \left\{\left[1 - e^{\beta(\mu+\varepsilon_o+J)}\right]^2 + 4e^{\beta(\mu+\varepsilon_o)}\right\}^{1/2}.$$

The pressure of this one-dimensional system is given by

$$\frac{Pl}{kT} = \lim_{M\to\infty} \frac{ln\Xi}{M} = ln\lambda_+$$

where l is the distance between nearest neighbor sites. The average number of adsorbed particles per site is given by

$$\theta = kT\frac{\partial}{\partial\mu} \lim_{M\to\infty} \frac{ln\Xi}{M} = kT\frac{\partial ln\lambda_+}{\partial\mu}$$

$$= \frac{e^{\beta(\mu+\varepsilon_o+J)}W - e^{\beta(\mu+\varepsilon_o+J)}[1 - e^{\beta(\mu+\varepsilon_o+J)}] + 2e^{\beta(\mu+\varepsilon_o)}}{[1 + e^{\beta(\mu+\varepsilon_o+J)}]W + W^2}.$$

Plot Pl/kT and θ versus $e^{\beta(\mu+\varepsilon_o)}$ for $\beta J = 0$, 1 and 10.

References

Debye, P. 1912. *Ann. Physik,* **39,** 789.

Einstein, A. 1907. *Ann. Physik,* **22,** 186.

Keeson, P. H. and Pearlman, N. 1953. *Phys. Rev.,* **91,** 1354.

Kramers, H. A. and Wannier, G. H. 1941. *Phys. Rev.,* **60,** 252.

Reif, F. 1965. *Fundamentals of Statistical and Thermal Physics,* McGraw–Hill, New York, p. 262.

Yang, C. N. 1952. *Phys. Rev.,* **85,** 809.

4

STATISTICAL THERMODYNAMICS OF SIMPLE CLASSICAL FLUIDS

4.1 Distribution Functions and Thermodynamic Quantities

In this chapter we focus our attention on the statistical thermodynamics of classical particles with no internal structure. We assume they are identical structureless particles interacting via conservative, pairwise additive, centrally symmetric intermolecular forces.

By centrally symmetric forces, we mean that the force $\mathbf{F}_{12}$ exerted on molecule 1 by molecule 2 is directed from the center of 2 to the center of 1, and the magnitude of the force depends only on the distance r_{12} apart of the centers, that is,

$$\mathbf{F}_{12} = \frac{\mathbf{r}_{12}}{r_{12}} F(r_{12}). \tag{4.1.1}$$

The vector $\mathbf{r}_{12}/r_{12}$ is a unit vector lying in the direction of the vector $\mathbf{r}_{12}$ shown in Figure 4.1. If the separation of the pair of molecules is held fixed, then the magnitude of their force of interaction will not change if the pair is translated or rotated as a rigid body. That the forces are conservative means there exists a potential energy function u whose derivative generates the force, that is,

$$\mathbf{F}_{12} = -\nabla_1 u. \tag{4.1.2}$$

Examples of contributions to the intermolecular potential energy were given in Chapter 2. To satisfy the form given in eq. (4.1.1), u must be a function only of r_{12}. Thus,

$$\mathbf{F}_{12} = -\nabla_1 u(r_{12}) = -\nabla_1 r_{12} \frac{du(r_{12})}{dr_{12}} = -\frac{\mathbf{r}_{12}}{r_{12}} \frac{du(r_{12})}{dr_{12}} \tag{4.1.3}$$

Figure 4.1
The positions $\mathbf{r}_1$ and $\mathbf{r}_2$ of the centers of molecules 1 and 2 relative to an arbitrary reference point 0, and the position $\mathbf{r}_{12}$ of the center of molecule 1 relative to the center of molecule 2.

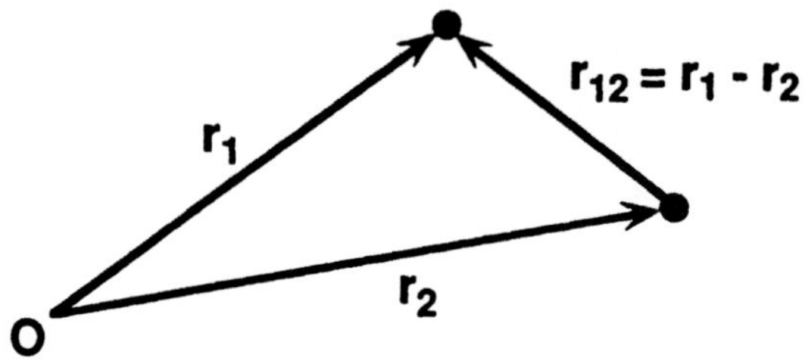

where the property

$$\nabla_1 r_{12} = \frac{\mathbf{r}_{12}}{r_{12}} \tag{4.1.4}$$

has been used. Comparing eqs. (4.1.1) and (4.1.3), we obtain the expression

$$F(r_{12}) = -\frac{du(r_{12})}{dr_{12}}. \tag{4.1.5}$$

Exercise

Using a Cartesian coordinate representation of ∇_1 and r_{12}, that is,

$$\nabla_1 = \hat{\imath}\,\frac{\partial}{\partial x_1} + \hat{\jmath}\,\frac{\partial}{\partial y_1} + \hat{k}\,\frac{\partial}{\partial z_1}$$

and $r_{12} = \sqrt{(x_1 - x_2)^2 + (y_1 - y_2)^2 + (z_1 - z_2)^2}$, derive eq. (4.1.4).

Experimentally and theoretically (from quantum mechanics), it is known that neutral molecules strongly repel each other at short distances (say $r_{12} \leq \sigma$, where σ is a length parameter of the order of magnitude of the spatial extent of the electronic charge distribution around a molecule), attract each other at an intermediate range ($\sigma \leq r \leq$ a few multiples of σ), and have negligible influence on each other at large separations ($r \gg \sigma$). A plot of the $F(r)$ and $u(r)$ typical of monatomic molecules and $u(r)$ typical of monatomic molecules is shown in Figure 4.2. The force can be computed as $-du(r)/dr$ if the potential $u(r)$ is known, or if the force is given then $u(r)$ can be computed by integrating eq. (4.1.5) from $r = +\infty$ (where u may be equated to zero by convention) to an arbitrary value of r, that is,

$$u(r) = \int_r^\infty F(\xi)d\xi. \tag{4.1.6}$$

The Coulombic interaction was first established as a force. For molecules, however, it is $u(r)$ that must be computed from quantum

Figure 4.2
Illustration of the pair force $F(r)$ and potential energy $u(r)$ of spherically symmetrical particles. 6–12 Lennard–Jones model. r is in units of σ, u of ε and F of ε/σ.

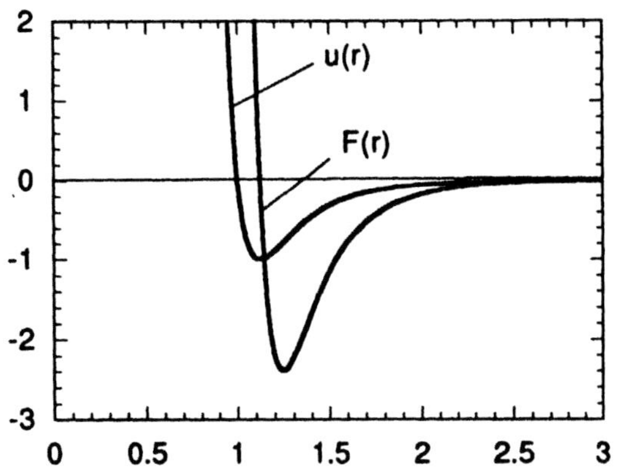

theory and is also the quantity one normally assumes known for the purpose of modeling molecular systems.

As we pointed out in Chapter 1 the 6–12 Lennard–Jones potential, which has some theoretical basis, is often used to describe the behavior of simple molecules such as Ar, Kr, and Xe. The model is a two parameter function (called the 6–12 Lennard–Jones model),

$$u(r) = 4\varepsilon\left[(\sigma/r)^{12} - (\sigma/r)^{6}\right] \tag{4.1.7}$$

in which the parameter ε gives the minimum value of the potential energy and σ is the point at which the potential energy becomes zero as a result of the balance between repulsive and attractive interactions. For argon, the values $\sigma = 3.405 \times 10^{-8}$ cm and $\varepsilon = 1.64 \times 10^{-14}$ erg/molecule have been estimated from fits of the theoretical equation of state to experimental data. Because the potential energy given by eq. (4.1.7) becomes very large (and the corresponding force becomes strongly repulsive) for $r < \sigma$, the parameter σ can be thought of as the molecular "size," similar to the rigid sphere diameter d introduced to approximate the mean free path behavior of molecules in Chapter 1.

It should be noted that Newton's law of equal action and reaction is automatically satisfied if u is a function only of the distance of separation r_{12} of a pair of particles. The force $\mathbf{F}_{21}$ exerted on molecule 2 by molecule 1 is

$$\begin{aligned}\mathbf{F}_{21} &= -\nabla_2 u(r_{12}) = -\nabla_2 r_{12} \cdot \frac{du(r_{12})}{dr_{12}} \\ &= -\frac{\mathbf{r}_{21}}{r_{21}}\frac{du(r_{12})}{dr_{12}} = +\frac{\mathbf{r}_{12}}{r_{12}}\frac{du(r_{12})}{dr_{12}},\end{aligned} \tag{4.1.8}$$

where the properties $r_{12} = r_{21}$, $\mathbf{r}_{12} = -\mathbf{r}_{21}$, and $\nabla_i r_{ij} = \mathbf{r}_{ij}/r_{ij}$ have been used. Comparing eqs. (4.1.8) and (4.1.3), we see that $\mathbf{F}_{21} = -\mathbf{F}_{12}$, reflecting Newton's law of equal action and reaction.

The pairwise additive assumption placed on the intermolecular forces requires that the force on a molecule be the sum of the individual pair forces between the molecule and each of the other molecules of the system. Thus, molecule i interacts with the other $N-1$ particles of an N-particle system with a total force

$$\mathbf{F}_i = \sum_{j=1, i\neq j}^{N} \mathbf{F}_{ij}(\mathbf{r}_{ij}), \tag{4.1.9}$$

where $\mathbf{F}_{ij}$ is the force that j would exert on i if the pair of molecules were isolated. The total potential energy u^N of the system will also be pairwise additive, every pair ij of molecules contributing an amount $u(r_{ij})$ to the potential energy,

$$u^N = \sum_{i>j=1} u(r_{ij}). \tag{4.1.10}$$

Let us now discuss distribution functions for the system. It follows from the ensemble theory presented in Chapter 3 that the probability of finding a system with energy between E and $E + dE$ is proportional to $e^{-E/kT}\Phi(E, dE)dE$, where $\Phi(E, dE)dE$ is the number of quantum states with energy lying between E and $E + dE$. In the classical mechanical limit E is replaced by the Hamiltonian H and ΦdE by $h^{-3N} d^3r_1 \cdots d^3r_N d^3p_1 \cdots d^3p_N$. In the case of identical and structureless particles, the particle momentum and velocity are simply related, namely, $\mathbf{p}_i = m\mathbf{v}_i$ and so velocity can be used instead of momentum to describe the state of a particle. With central pair forces the Hamiltonian is

$$H = \sum_{i=1}^{N} \frac{1}{2} m v_i^2 + \sum_{i>j=1}^{N} u(r_{ij}). \tag{4.1.11}$$

$u(r_{ij})$ is the potential of interaction between the pair of particles i and j. Thus, in the classical mechanical limit, the probability density $P_N(\mathbf{r}_i, \cdots, \mathbf{r}_N, \mathbf{v}_1, \cdots, \mathbf{v}_N)$ that particle 1 lies between $\mathbf{r}_1$ and $\mathbf{r}_1 + d\mathbf{r}_1$ with velocity between $\mathbf{v}_1$ and $\mathbf{v}_1 + d\mathbf{v}_1$, and that particle 2 lies between $\mathbf{r}_2$ and $\mathbf{r}_2 + d\mathbf{r}_2$ with velocity between $\mathbf{v}_2$ and $\mathbf{v}_2 + d\mathbf{v}_2$, etc. is of the form $P_N d^3v_1 \cdots d^3r_N d^3v_1 = A_N e^{-H/kT} d^3r_1 \cdots d^3v_N$, where A_N is the normalization constant obtained from the condition

$$\int \cdots \int P_N d^3r_1 \cdots d^3v_N = 1. \tag{4.1.12}$$

Thus, it follows that

$$P_N = \frac{e^{-H/kT}}{\int \cdots \int e^{-H/kT}(d^3r d^3v)^N}, \tag{4.1.13}$$

where we have introduced the abbreviated notation $(d^3r d^3v)^N \equiv d^3r_1 \cdots d^3r_N d^3v_1 \cdots d^3v_N$.

From the N-body distribution function lower order distributions can be derived. The configuration distribution function

$p_N(\mathbf{r}_1, \ldots, \mathbf{r}_N)$ is obtained by integrating P_N over all particle velocities, that is, $p_N = \int P_N (d^3v)^N$. With the aid of the relation

$$\int e^{-mv_i^2/2kT} d^3v_i = \left(\frac{2\pi kT}{m}\right)^{3/2}, \tag{4.1.14}$$

it is easy to show that

$$p_N = \frac{e^{-u^N/kT}}{Z_N}, \tag{4.1.15}$$

where Z_N is the configuration partition function,

$$Z_N \equiv \int e^{-u^N/kT} (d^3r)^N. \tag{4.1.16}$$

Similarly, the N-body velocity distribution function, defined by $\phi_N(\mathbf{v}_1, \ldots, \mathbf{v}_N)$
$= \int P_N (d^3r)^N$, can be shown to obey

$$\phi_N(\mathbf{v}_1, \ldots, \mathbf{v}_N) = \prod_{i=1}^{N} \phi(\mathbf{v}_i), \tag{4.1.17}$$

where

$$\phi(\mathbf{v}_i) = \left(\frac{m}{2\pi kT}\right)^{3/2} e^{-mv_i^2/2kT}. \tag{4.1.18}$$

This result could have been anticipated from Chapter 1, where we proved heuristically that a particle obeys the Maxwell distribution of velocities: because there is no restraint on the position of the particles their velocities are not correlated and so the N-body velocity distribution function is simply the product of one-body or singlet distributions.

The s-body configuration distribution function, $p_s(\mathbf{r}_1, \ldots, \mathbf{r}_s)$, is given by

$$p_s(\mathbf{r}_1, \ldots, \mathbf{r}_s) = \int P_N (d^3r)^{N-s} (d^3v)^N = \int p_N (d^3r)^{N-s}$$
$$= \frac{\int e^{-u^N/kT} d^3r_{s+1} \cdots d^3r_N}{Z_N}. \tag{4.1.19}$$

For a system with pair, central forces the singlet and doublet density functions suffice to determine the thermodynamic energy and pressure.

As a homogeneous fluid is isotropic, the local density $n(\mathbf{r}_1) = Np(\mathbf{r}_1)$ is the same throughout the system and is given by $n = N/V$. Thus, it must be true that $p(\mathbf{r}_1) = 1/V$ just as was the case for an ideal gas. As is intuitively clear, the existence of a homogeneous system is possible only if the boundary induced structure (caused by interaction between the fluid and its container) does not persist far from the boundary.

The quantity $p_2(\mathbf{r}_1, \mathbf{r}_2)d^3r_1d^3r_2$ represents the probability a particle will be located between $\mathbf{r}_1$ and $\mathbf{r} + d\mathbf{r}_1$ and another will be between $\mathbf{r}_2$ and $\mathbf{r}_2 + d\mathbf{r}_2$ regardless of where the other particles of the system are. If the particles of the system are independent, that is, if they are noninteracting, the $p_2(\mathbf{r}_1, \mathbf{r}_2)$ is the product of the singlet probabilities $p(\mathbf{r}_1)$ and $p(\mathbf{r}_2)$. The deviation of p_2 from this product, called the pair correlation function $g^{(2)}$, is defined by the equation

$$p_2(\mathbf{r}_1, \mathbf{r}_2)d^3r_1d^3r_2 = [p(\mathbf{r}_1)d^3r_1][p(\mathbf{r}_2)d^3r_2]g^{(2)}(\mathbf{r}_1, \mathbf{r}_2). \qquad (4.1.20)$$

For a homogeneous fluid, the condition of isotropy on p_2 is that $p_2(\mathbf{r}_1, \mathbf{r}_2)$ depend only on the distance of separation r_{12} of the points $\mathbf{r}_1$ and $\mathbf{r}_2$. Thus, because $p(\mathbf{r}_i) = V^{-1}$, the pair correlation function for a homogeneous fluid is defined by the expression

$$g^{(2)}(r_{12}) = V^2 \int e^{-\sum_{i>j=1}^{N} u(r_{ij})/kT} (d^3r)^{N-2}/Z_N. \qquad (4.1.21)$$

To gain some appreciation of the general properties of $g^{(2)}$, let us consider a two particle system (actually the pair correlation function for this case is the dilute gas limit). For this system

$$g^{(2)}(r_{12}) = \frac{V^2 e^{-u(r_{12})/kT}}{\int e^{-u(r_{12})/kT} d^3r_1 d^3r_2}. \qquad (4.1.22)$$

The integration over d^3r_2 in the denominator of eq. (4.1.22) may be performed in a coordinate system with an origin fixed on particle 1. This is equivalent to a change of variable from $\mathbf{r}_1, \mathbf{r}_{12}$ for which $d^2r_1d^3r_2 = d^3r_1d^3r_{12}$ so that

$$\begin{aligned}\int e^{-u(r_{12})/kT} d^3r_1 d^3r_2 &= \int d^3r_1 \left[\int e^{-u(r_{12})/kT} d^3r_{12}\right] \\ &= V\left\{V + \int \left[e^{-u(r_{12})/kT} - 1\right] d^3r_{12}\right\}\end{aligned} \qquad (4.1.23)$$

where the integral over d^3r_1 has been performed explicitly and the integral over d^3r_{12} has been rearranged into the sum of V and the term

$$\int [e^{-u(r_{12})/kT} - 1] d^3r_{12}.$$

This term is of the order of $\frac{4}{3}\pi r_c^3$, where r_c is the separation at which u is negligible compared to kT. Characteristically r_c is of the order of 10^{-7} cm, so that $\frac{4}{3}\pi r_c^3 \simeq 4 \times 10^{-21}$ cm^3, a volume truly negligible compared to the macroscopic volume V. Thus, the right-hand side (rhs) of eq. (4.1.23) may be equated to V^2 with negligible error, giving for $g^{(2)}(r_{12})$ the dilute gas limit

$$g^{(2)}(r_{12}) = e^{-u(r_{12})/kT}. \qquad (4.1.24)$$

Recalling from Figure 4.2 the typical form of $u(r_{12})$, we obtain for $g^{(2)}(r_{12})$ the qualitative form shown in Figure 4.3. When the particle separation is less than σ, the distance at which the strongly repulsive forces begin to manifest themselves, $g^{(2)}(r_{12})$ is essentially zero, meaning the probability p_2 of two particles penetrating each other is essentially zero. When r_{12} is large compared to the range of u ($\sim$ 2 or 3 times σ for neutral molecules), $g^{(2)}$ is unity, implying that at such separations the particles of the system behave independently, so that p_{12} is just the product of the singlet probabilities. At a distance of separation corresponding to the minimum of the potential energy, $g^{(2)}$ is a maximum, meaning that this separation is favorable because of the attraction the particles have for each other at this point.

In a dense fluid (dense gas or liquid) the properties,

$$g^{(2)}(r_{12}) = 0\,, \quad r_{12} < \sigma \quad \text{and} \quad g^{(2)}(r_{12}) \longrightarrow 1, \quad r_{12} \longrightarrow \infty, \tag{4.1.25}$$

are still observed and occur for the same reasons as in the dilute gas. However, there is considerably more structure in $g^{(2)}$ between the regions of small and large separations. A typical plot of the pair correlation function of a liquid is shown in Figure 4.4. In particular, there are several peaks of $g^{(2)}$ corresponding to the average nearest neighbor distance (first and strongest peak arising from the local ordering of the nearest neighbor molecules about the central molecule), the average next nearest neighbor distance (the second peak) and so on. The peaks become weaker and weaker as the order of successive neighboring shell becomes less and less well defined. In a perfect solid, the pair correlation function would simply consist of evenly spaced sharp spikes representing the lattice sites and would be zero elsewhere. A liquid, however, only has the short range order of the type represented by the sequence of decreasing peaks in Figure 4.4.

In a system composed of monatomic particles acting with pairwise additive forces, all the thermodynamic functions can be

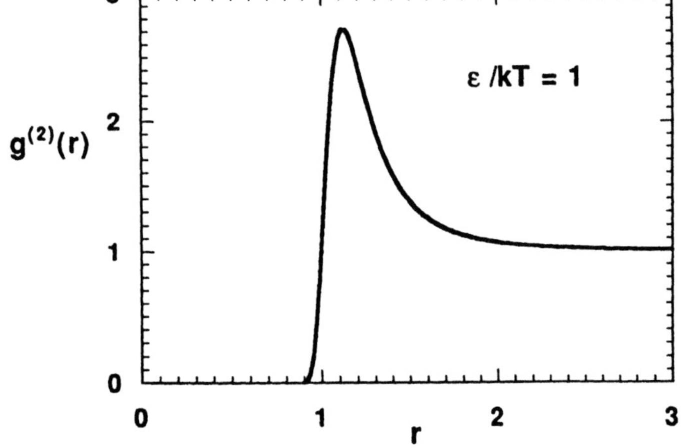

Figure 4.3
Dilute gas pair correlation function as a function of distance of separation. For 6–12 Lennard-Jones potential, distance in units of σ.

Figure 4.4
Illustration of the behavior of the pair correlation function of a liquid as a function of distance of separation. The plot is a smooth curve drawn from neutron scattering data for argon at 133 K and 1.11 g/cm^3 (from Khan, 1964).

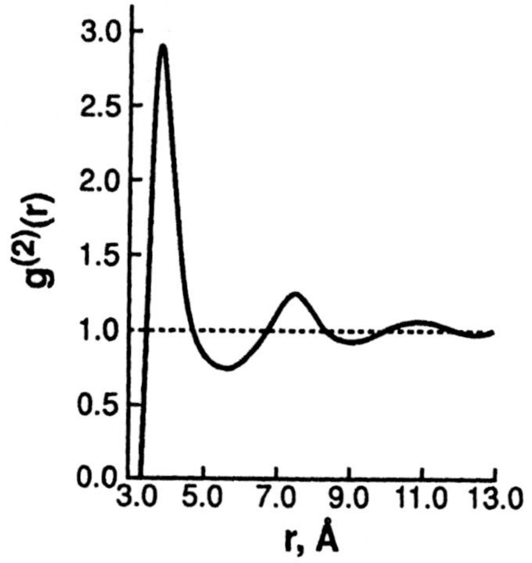

computed given only the Maxwell velocity distribution function and the pair correlation function. The latter function can be obtained, in principle, from the definition, eq. (4.1.21). However, for any realistic potential model the $3N-6$ dimensional integrals in eq. (4.1.21) cannot be done exactly. Thus, one must develop approximate techniques for computing $g^{(2)}$, a problem that has gotten a lot of attention during the last three decades, or one must measure $g^{(2)}$ experimentally by X-ray or neutron scattering techniques. The theories and measurement of $g^{(2)}$ will be discussed in a later chapter.

In closing this section, let us demonstrate the use of the pair correlation function to calculate a thermodynamic quantity. The total energy of an N-particle system of the type considered here is the sum of the kinetic energy and the pair potential energies. The average kinetic energy of one particle is $(3/2)kT$, so that the total average kinetic energy is $(3/2)NkT$. The average potential energy of an interacting pair, say molecules 1 and 2, is

$$\int P_2(\mathbf{r}_1, \mathbf{r}_2)u(r_{12})d^3r_1d^3r_2 = \frac{1}{V^2}\int g^{(2)}(r_{12})u(r_{12})d^3r_1d^3r_2$$

$$= \frac{1}{V^2}\int g^{(2)}(r)u(r)d^3r_1d^3r = \frac{1}{V}\int g^{(2)}(r)u(r)d^3r, \tag{4.1.26}$$

where the coordinate transformation $\mathbf{r}_1, \mathbf{r}_2 \rightarrow \mathbf{r}_1, \mathbf{r}\,(\equiv \mathbf{r}_1 - \mathbf{r}_2)$, for which $d^3r_1d^3r_2 = d^3r_1d^3r$, has been used in arriving at the final form of the rhs of eq. (4.1.26). Using this coordinate transformation is equivalent to evaluating [as we did in eq. (4.1.23)] the integration over d^3r_2 in a coordinate frame fixed on the particle at $\mathbf{r}_1$. Because there are $N(N-1)/2$ pairs of interacting particles, the total average potential energy is $N(N-1)/2$ times eq. (4.1.26). Thus, adding the average kinetic and potential energies we obtain

$$U = \frac{3}{2}NkT + \frac{N(N-1)}{2V}\int g^{(2)}(r)u(r)d^3r. \tag{4.1.27}$$

We can also obtain eq. (4.1.27) by averaging the total energy H with respect to P_N. By definition

$$U = \langle H \rangle = \int P_N(r_1, \ldots, v_N) \left[\sum_{i=1}^{N} m v_i^2/2 + \sum_{i>j=1}^{N} u(r_{ij}) \right] (d^3r)^N (d^3v)^N$$

$$= \sum_{i=1}^{N} \frac{\int e^{-H/kT} (m v_i^2/2)(d^3r)^N (d^3v)^N}{\int e^{-H/kT} (d^3r)^N (d^3v)^N} \tag{4.1.28}$$

$$+ \sum_{i>j=1}^{N} \frac{\int e^{-H/kT} u(r_{ij})(d^3r)^N (d^3v)^N}{\int e^{-H/kT} (d^3r)^N (d^3v)^N}.$$

In the term in eq. (4.1.28) involving the average kinetic energy, the integrals over $(d^3r)^N$ and over all $(d^3v)^N$ except d^3v_i are identical in the numerator and denominator and, consequently, cancel. Similarly the integrals over $(d^3v)^N$ cancel in the term involving the average of the potential energy. Thus, eq. (4.1.28) can be reduced as follows:

$$U = \sum_{i=1}^{N} (m/2\pi kT)^{3/2} \int e^{-mv_i^2/2kT} (1/2) m v_i^2 d^3v_i$$

$$+ \sum_{i>j=1}^{N} \frac{1}{V^2} \int d^3r_i d^3r_j u(r_{ij}) \frac{V^2 \int e^{-u^N/kT} (d^3r)^{N-2}}{Z_N}$$

$$= \sum_{i=1}^{N} \frac{3}{2} kT + \sum_{i>j=1}^{N} \frac{1}{V^2} \int d^3r_i d^3r_j u(r_{ij}) g^{(2)}(r_{ij}) \tag{4.1.29}$$

$$= \frac{3}{2} NkT + \frac{N(N-1)}{2V} \int g^{(2)}(r) u(r) d^3r.$$

The steps in arriving at this final result involve (1) identifying the quantity in the square brackets in the first equality with the pair correlation function $g^{(2)}(r_{ij})$ defined by eq. (4.1.21) and (2) noting that the average of the potential energy $u(r_{ij})$ will be the same for any pair of particles, and so the summation $\sum_{i>j=1}$ gives $N(N-1)/2$ terms identical to the pair average treated in eq. (4.1.26).

Next consider the pressure. Because the shape of the container does not affect the properties of a macroscopic isotropic fluid, we suppose that the container is a cube of length $V^{1/3}$ on each side. The classical mechanical partition function for the system is

$$Q_N = \frac{m^{3N}}{N! h^{3N}} \int \cdots \int e^{-(\sum_i m v_i^2/2 + u^N)/kT} d^3r_1 \cdots d^3v_N$$

$$= \frac{1}{N!} \left(\frac{2\pi m kT}{h^2} \right)^{3N/2} Z_N. \tag{4.1.30}$$

From the thermodynamic relation

$$P = -\left(\frac{\partial F}{\partial V}\right)_{N,T} \tag{4.1.31}$$

and the canonical ensemble result $F = -kT \ln Q_N$, it follows that

$$P = kT\left(\frac{\partial \ln Z_N}{\partial V}\right)_{N,T}. \tag{4.1.32}$$

If the volume integrals in Z_N are expressed in Cartesian coordinates fixed at one corner of the cube and aligned with the cubic axes, then

$$Z_N = \int_0^{V^{1/3}} \cdots \int_0^{V^{1/3}} e^{-\sum\limits_{i>j=1} u(r_{ij})/kT} dx_1 dy_1 dz_1 dx_2 \cdots dz_N \tag{4.1.33}$$

and $r_{ij} = \{(x_i - x_j)^2 + (y_i - y_j)^2 + (z_i - z_j)^2\}^{1/2}$. Introduce next the scaled coordinates

$$\xi_i = x_i / V^{1/3}, \quad \zeta_i = y_i / V^{1/3}, \quad \eta_i = z_i / V^{1/3},$$
$$\tilde{r}_{ij} = r_{ij} / V^{1/3} = \{(\xi_i - \xi_j)^2 + (\zeta_i - \zeta_j)^2 + (\eta_i - \eta_j)^2\}^{1/2}.$$

With this transformation, we obtain

$$Z_N = V^N \int_0^1 \cdots \int_0^1 e^{-\sum\limits_{i>j=1}^{N} u(V^{1/3}\tilde{r}_{ij})/kT} d\xi_1 d\zeta_1 d\eta_1 d\xi_2 \cdots d\eta_N, \tag{4.1.34}$$

that when inserted into eq. (4.1.32) yields

$$P = \frac{NkT}{V} - \frac{V^N}{3V^{2/3}Z_N} \sum_{i>j=1}^{N} \int_0^1\int_0^1 e^{-\sum\limits_{i>j=1}^{N} u(V^{1/3}r_{ij})/kT} \tilde{r}_{ij} u'(V^{1/3}\tilde{r}_{ij}) d\xi_1 \cdots d\eta_N. \tag{4.1.35}$$

To get the second term on the rhs of this equation, the chain rule of differentiation, $\partial u(V^{1/3}\tilde{r}_{ij})/\partial V = u'(V^{1/3}\tilde{r}_{ij})\partial(V^{1/3}\tilde{r}_{ij})/\partial V = u'(V^{1/3}\tilde{r}_{ij})(V^{-2/3}\tilde{r}_{ij}/3)$, was used. u' denotes the derivative of u with respect to its argument. Transforming coordinates back to the original ones and carrying out steps similar to those used above in evaluating the thermodynamic energy, we find

$$\begin{aligned} P &= \frac{NkT}{V} - \frac{1}{3V} \sum_{i>j=1}^{N} \frac{\int \cdots \int e^{-u^N/kT} r_{ij} u'(r_{ij}) (d^3r)^N}{Z_N} \\ &= \frac{NkT}{V} - \frac{N(N-1)}{6V} \int g_2(r_{12}) r_{12} u'(r_{12}) d^3 r_{12}. \end{aligned} \tag{4.1.36}$$

This is the desired result. It is known as the virial equation of state and enables one to compute, for a classical mechanical fluid of particles interacting via central pair forces, the pressure from the pair potential and the pair correlation function.

4.2 One-Dimensional Fluid of Rigid Rods

The one-dimensional fluid of rigid rods is the one-dimensional analog of the three-dimensional fluid of rigid spheres. The potential energy of interaction u of a pair of rigid rods and the Boltzmann factor $e^{-u/kT}$ are shown in Figure 4.5 as functions of the distance x_{12} of the centers of the rods. The value of studying the one-dimensional system is that the partition function can be evaluated explicitly for this case and the result can be used as a guide in approximating the effects of the strongly repulsive interactions in the more complicated three-dimensional fluid systems, which have thus far escaped exact treatment.

In Chapter 3 we studied in the isobaric ensemble the hard-rod fluid as a special case of a Tonks–Takahashi fluid. Because in this chapter we are using the canonical ensemble, we will consider the hard-rod fluid in this ensemble. The starting point is

$$F = -kT \ln Q_N, \tag{4.2.1}$$

where the partition function is

$$Q_N = \frac{1}{N!}\left(\frac{2\pi mkT}{h^2}\right)^{N/2} Z_N; \quad Z_N = \int_0^L \cdots \int_0^L e^{-\sum_{i<j=1}^{N} u(x_{ij})/kT} dx_1 \cdots dx_N. \tag{4.2.2}$$

It will turn out to be convenient in what follows to rewrite eq. (4.2.2) in the form

$$Z_N = N! \int_0^L dx_N \int_0^{x_N} dx_{N-1} \cdots \int_0^{x_2} dx_1 e^{-\beta u^N}. \tag{4.2.3}$$

This form results from a standard coordinate transformation (details are given at the end of this section).

In one-dimensional systems, the pressure P is defined from the reversible work of one-dimensional expansion, that is,

$$dW_r = (dU)_{dS=0} = -PdL. \tag{4.2.4}$$

Figure 4.5
The rigid-rod pair potential u and the Boltzmann factor $e^{-u/kT}$ versus the distance of separation x_{12}.

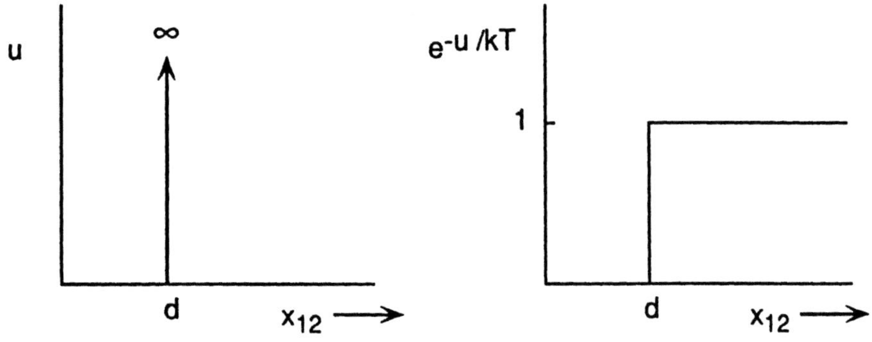

Combining the reversible work and the reversible heat transfer, $dq_r = TdS$, we obtain from the first law of thermodynamics

$$dU = TdS - PdL, \tag{4.2.5}$$

that, along with the definition

$$F = U - TS,$$

implies

$$S = -\left(\frac{\partial F}{\partial T}\right)_{N,L} = k\left(\frac{\partial}{\partial T}(T \ln Q_N)\right)_{N,L} \tag{4.2.6}$$

$$P = -\left(\frac{\partial F}{\partial L}\right)_{N,T} = kT\left(\frac{\partial \ln Q_N}{\partial L}\right)_{N,T} = kT\left(\frac{\partial \ln Z_N}{\partial L}\right)_{N,T} \tag{4.2.7}$$

$$U = kT^2\left(\frac{\partial \ln Q_N}{\partial T}\right)_{N,L}, \tag{4.2.8}$$

the one-dimensional analogues of the three-dimensional results introduced in previous chapters.

Before evaluating the partition function explicitly, let us derive, by heuristic arguments, expressions for the thermodynamic functions of the hard-rod system. Consider first the equation of state. For an ideal gas the equation of state is $PL = NkT$. If the particles interact with the hard rod potential shown in Figure 4.5, then the effect of the interaction simply reduces the free volume (length) available to the particles. Instead of being able to move an overall distance L, a particle can move only the distance $L - Nd$, the distance Nd being taken up by the neighboring particles. We may expect, therefore, that the ideal gas equation of state still holds if the free length $L - Nd$ is used rather than total length L, that is,

$$P(L - Nd) = NkT. \tag{4.2.9}$$

Also, because the potential energy of hard rods only serves to keep the particles from interpenetrating, it will behave as a complicated boundary but will not contribute to the average energy of the particles. The average energy is, therefore, purely kinetic, giving the one-dimensional result

$$U = (1/2)NkT. \tag{4.2.10}$$

A general thermodynamic result for a process at constant N is

$$dS = C_V\frac{dT}{T} + \left(\frac{\partial P}{\partial T}\right)_{N,L} dL, \tag{4.2.11}$$

where

$$C_V = \left(\frac{\partial U}{\partial T}\right)_{N,L}. \tag{4.2.12}$$

Evaluating C_V and $(\partial P/\partial T)_{N,L}$ in terms of eqs. (4.2.10) and (4.2.9), we obtain a form for dS that can be integrated to yield

$$S = Nk \ln\left[T^{1/2}(L - Nd)\right] + S^o(N), \tag{4.2.13}$$

where $S^o(N)$ is a constant of integration. Other quantities, such as enthalpy and the free energies can be obtained from the given forms of P, U, and S.

Entropy is directly related to the randomness of a fluid system: the more random the system the greater the entropy. Equation (4.2.13) bears out this property as the volume Nd excluded to the particles increases, the volume $L - Nd$ accessible to the particles decreases, the system becomes more and more ordered, and the entropy from eq. (4.2.13) decreases.

Let us now turn to the rigorous evaluation of Z_N for the hard-rod system. The effect of the Boltzmann factor

$$e^{-\sum_{i<j=1} u(x_{ij})/kT}$$

in Z_N is to restrict the domains over which the variables $x_1, \ldots, x_N$ may range, being zero when any $x_{ij} < d$ and being unity otherwise. Expressing explicitly the fact that the rods cannot interpenetrate and that the Boltzmann factor, $\exp(-\beta u(x_{ij}))$, is unity when particle i and j are not in contact, we obtain

$$Z_N = N! \int_{(N-1)d+\frac{d}{2}}^{L-\frac{d}{2}} dx_N \int_{(N-2)d+\frac{d}{2}}^{x_N-d} dx_{N-1} \cdots \int_{\frac{3}{2}d}^{x_3-d} dx_2 \int_{\frac{d}{2}}^{x_2-d} dx_1. \tag{4.2.14}$$

By inspection of Figure 4.6, we see that for a given position x_2 of the center of particle 2, the range of the center of particle 1 is $d/2 < x_1 < x_2 - d$, for a given x_3 of the center of particle 3, the range of the center of particle 2 is $3/2d < x_2 < x_3 - d$, and so on until we deduce the ranges of $x_1, \ldots, x_N$ given in eq. (4.2.14). Introducing the coordinate transformation

$$\xi_j = x_j - \left[(j-1)d + \frac{d}{2}\right], \tag{4.2.15}$$

we can rewrite Z_N in the form

$$Z_N = N! \int_0^{L-Nd} d\xi_N \int_0^{\xi_N} d\xi_{N-1} \cdots \int_0^{\xi_3} d\xi_2 \int_0^{\xi_2} d\xi_1, \tag{4.2.16}$$

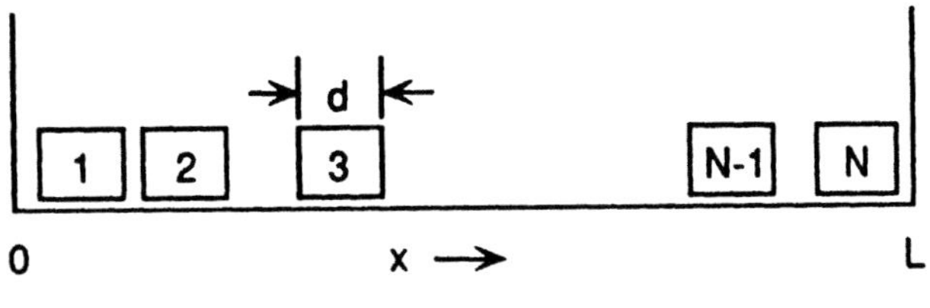

Figure 4.6
N rigid rods in a one-dimensional box of length *L*.

that can easily be evaluated to yield

$$Z_N = (L - Nd)^N. \tag{4.2.17}$$

Substitution of eq. (4.2.17) into eq. (4.2.7) verifies the equation of state given by eq. (4.2.9). Use of the same result in eq. (4.2.8) verifies the form of the energy given by eq. (4.2.10) and in eq. (4.2.6) leads to

$$S = Nk \ln\left(\frac{T^{1/2}(L - Nd)}{N}\right) + \frac{1}{2} Nk \ln\left(\frac{2\pi mke^3}{h^2}\right), \tag{4.2.18}$$

verifying the form of eq. (4.2.12) and identifying $S^o(N)$.

For imperfect gases, it has been found experimentally that the pressure can be expressed in a power series in number density. This series is called the virial expansion and for a one-dimensional fluid is written in the form

$$P = \frac{N}{L} kT + \sum_{i=2}^{\infty} kT\left(\frac{N}{L}\right)^i B_i(T). \tag{4.2.19}$$

The quantity B_i is called the ith virial coefficient. The exact expression for P, eq. (4.2.8), can be expanded in the following power series

$$P = \frac{NkT}{L} \sum_{i=0}^{\infty}\left(\frac{N}{L}\right)^i d^i, \tag{4.2.20}$$

that, when compared to eq. (4.2.19), yields for the virial coefficients the result

$$B_i = d^{i-1}, \quad i = 2, 3, \ldots. \tag{4.2.21}$$

For noninteracting rods $d = 0$ and $B_i = 0$, so that B_i is seen to arise directly from intermolecular interactions. B_i is proportional to particle volume (which is d in one dimension) to the $i - 1$st power.

In the case of a one-dimensional rigid rod system, $Nd < L$ or $Nd/L < 1$ except at the close packed limit, so that the virial expansion converges for all densities except the close packed limit. For real, three-dimensional fluids the virial expansion of a gas diverges for a given isotherm when the density is reached at which phase separation occurs. The lack of divergence of eq. (4.2.20) at any but the close packed density implies the absence of a phase transition for the rigid rod system.

Returning to eq. (4.2.17), we note that the effect of the rigid-rod interactions on the partition function is simply to reduce the volume, L, available to each molecule by the excluded volume Nd. We expect a similar effect from the volume excluded by the molecules in a three-dimensional system of rigid spheres. Thus, in analogy with eq. (4.2.17), we may guess that the three-dimensional configuration partition function for rigid spheres is

$$Z_N = (V - Nb)^N, \tag{4.2.22}$$

where b denotes the excluded volume per molecule. The pressure can be computed from the relation $P = kT(\partial \ln Z_N/\partial V)_{N,T}$. The result is

$$P = \frac{NkT}{V - Nb}. \tag{4.2.23}$$

This expression, which we will call the "Clausius equation," although not an exact result, gives a good qualitative picture of the effect on strongly repulsive interactions on the pressure of a fluid. [In 1857 Clausius suggested a correction of the form of eq. (4.2.22) for repulsive forces and in 1863 Hirn wrote down the equation. See Clausius, 1857, and Hirn, 1863.] At temperatures much higher than the critical temperature the effect of the attractive interactions between molecules is not important, the average kinetic energies of the molecules being large compared to the attractive parts of the intermolecular potentials. In this temperature region, the isotherm of real fluids are similar to those predicted by the Clausius equation.

Expressing the Clausius equation of state in the form of the virial expansion

$$P = \frac{nkT}{V} + \sum_{i=2}^{\infty} kT\left(\frac{N}{V}\right)^i B_i, \tag{4.2.24}$$

we find $B_i = b^{i-1}$, that is, the ith virial coefficient is proportional to the $i-1$st power of the molecular excluded volume.

Let us close this section with a proof of eq. (4.2.3). We shall proceed by induction. Consider the integral expression

$$\begin{aligned} I_2 \equiv \int_0^L dx_2 \int_0^L dx_1 f(x_1, \ldots, x_N) &= \int_0^L dx_2 \int_0^{x_2} dx_1 f + \int_0^L dx_2 \int_{x_2}^L dx_1 f \\ &= \int_0^L dx_2 \int_0^{x_2} dx_1 f + \int_0^L dx_1 \int_0^{x_1} dx_2 f, \end{aligned} \tag{4.2.25}$$

where the interchange of order of integration of the term $\int_0^L dx_2 \int_{x_2}^L dx_1 f$ must be accompanied with the change in the range of the variables x_1 and x_2 as shown. But, because x_1 and x_2 are equivalent dummy indices, their subscripts may be interchanged in the second term on the rhs of eq. (4.2.25) to obtain

$$I_2 = 2\int_0^L dx_2 \int_0^{x_2} dx_1 f. \tag{4.2.26}$$

Assume now that

$$I_{N-1} \equiv \int_0^L dx_{N-1} \int_0^L dx_{N-1} \cdots \int_0^L dx_1 f = (N-1)! \int_0^L dx_{N-1} \int_0^{x_{N-1}} dx_{N-2} \cdots \int_0^{x_2} dx_1 f. \tag{4.2.27}$$

Then

$$I_N \equiv \int_0^L dx_N I_{N-1} = (N-1)! \int_0^L dx_N \int_0^{x_N} dx_{N-1} \cdots \int_0^{x_2} dx_1 f$$

$$+ (N-1)! \int_0^L dx_N \int_{x_N}^L dx_{N-1} \int_0^{x_{N-1}} dx_{N-2} \cdots \int_0^{x_2} dx_1 f \tag{4.2.28}$$

$$= (N-1)! \int_0^L dx_N \int_0^{x_N} dx_{N-1} \cdots \int_0^{x_2} dx_1 f + (N-1)! \int_0^L dx_{N-1} \int_0^{x_{N-1}} dx_N \cdots \int_0^{x_2} dx_1 f.$$

The second term on the rhs of eq. (4.2.28),

$$(N-1)! \int_0^L dx_{N-1} \int_0^{x_{N-1}} dx_N \int_0^{x_{N-1}} dx_{N-2} \int_0^{x_{N-2}} dx_{N-3} \cdots \int_0^{x_2} dx_1 f,$$

under the transformations leading to I_2, becomes

$$(N-1)! \int_0^L dx_{N-1} \int_0^{x_{N-1}} dx_N \int_0^{x_N} dx_{N-2} \int_0^{x_{N-2}} dx_{N-3} \int_0^{x_{N-3}} dx_{N-4} \cdots \int_0^{x_2} dx_1 f$$

$$+(N-1)! \int_0^L dx_{N-1} \int_0^{x_{N-1}} dx_{N-2} \int_0^{x_{N-2}} dx_N \int_0^{x_{N-2}} dx_{N-3} \int_0^{x_{N-3}} dx_{N-4} \cdots \int_0^{x_2} dx_1 f.$$

The process continues until there are no identical upper limits of integration at which point I_N will have been decomposed into a sum of N identical terms. Thus, we obtain

$$I_N = N! \int_0^L dx_N \int_0^{x_N} dx_{N-1} \cdots \int_0^{x_2} dx_1 f. \tag{4.2.29}$$

That the result hypothesized for I_{N-1} in eq. (4.2.27) implies the corresponding result for I_N and that the case for $N = 2$ was proved explicitly completes the induction proof that Z_N can be expressed in the form given by eq. (4.2.3).

4.3 van der Waals Model

The model equation of state introduced about a century ago by van der Waals today still provides the simplest picture of the relationship between intermolecular interactions and the qualitative aspects of the behavior of gases and liquids. The pair potential for the van der Waals (VDW) model is assumed to be of the form shown in Figure 4.7. The molecules are assumed to have rigid cores ($u(r) = \infty,\ r < d$) surrounded by a spherically symmetric, slowly varying, attractive potential field $-w(r)$. In this section we

Figure 4.7
Illustration of the pair potential of the molecules of a VD fluid.

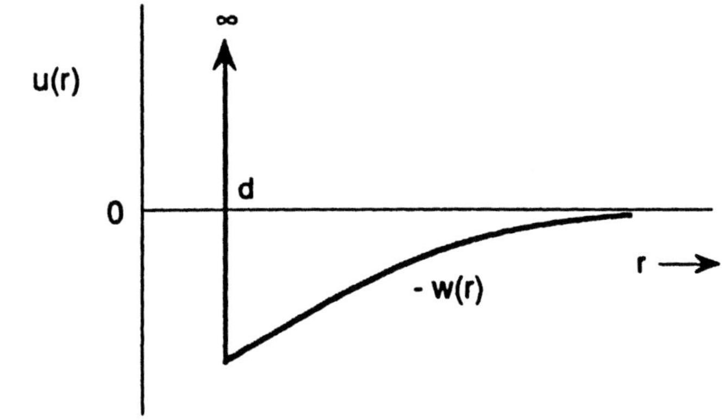

restrict our attention to homogeneous one-phase fluids. The two-phase coexistence regions will be determined from the one-phase results. A detailed treatment of the interaction regions of two-phase (inhomogeneous) systems will be treated in a later section.

The VDW equation of state can be obtained by a simple heuristic argument. The first assumption of the model is that the only effect of the hard sphere interactions is to reduce the free volume available to the molecules. Thus, in the absence of attractive interactions, we assume that the equation of state is of the form of eq. (4.2.23), namely,

$$P(V - Nb) = NkT, \tag{4.3.1}$$

where Nb is the average volume excluded to the center of a molecule due to interaction with the hard cores of the other molecules. In what follows, we relate b to the rigid sphere volume $4\pi d^3/3$.

Another assumption of the VD Waals model is that eq. (4.3.1) still holds when attractive interactions are present, if the pressure appearing there is the "ideal" pressure P_I obtained by adding to the actual pressure P the correction $\delta P = aN^2/V^2$, representing the reduction, due to attractive interactions, in the ideal pressure that would exist if there were no attractive interactions. Thus, we obtain the VDW famous equation of state

$$\left(P + \frac{N^2 a}{V^2}\right)(V - Nb) = NkT. \tag{4.3.2}$$

The reason that the correction δP is proportional to n^2 is that the interactions are two body interactions and the probability of a pair being close enough together to interact is proportional to n^2.

Let us next obtain eq. (4.3.2) from approximations to partition function. This derivation has the advantages that the nature of the approximate treatment of the attractive interactions is made clearer, giving an explicit form for a, and that the partition function provides a rigorous starting point from which a systematic improvement on the VDW approximation could be attempted.

Writing the VDW pair potential in the form

$$u(r) = u_R(r) - w(r) \tag{4.3.3}$$

where

$$\begin{aligned} u_R(r), &= \infty, \quad r \le d \\ &= 0, \quad r > d \end{aligned} \tag{4.3.4}$$

and $-w(r)$ is a negative, slowly varying potential energy accounts for the attractive interactions between molecules of the fluid. The corresponding form for the configuration partition functions is

$$Z_N = \int_V^{\cdots}\int e^{-u_R^N/kT} e^{\sum_{i>j=1}^N w(r_{ij})/kT} (d^3r)^N. \tag{4.3.5}$$

The first approximation of the VD Waals model is the so-called "mean field" approximation. One argues that because the potential function $w(r_{ij})$ is long-range and slowly varying, the sum $\sum_{i,j} w(r_{ij})$ in the integrand of eq. (4.3.5) can be replaced by its average value $\langle \sum_{ij} w(r_{ij}) \rangle$, which in turn becomes $N(N-1)\langle w(r_{12})\rangle/2$ since the particles are identical. Thus, Z_N is approximated as

$$Z_N \approx e^{N(N-1)\langle w(r_{12})\rangle/2kT} \int_V^{\cdots}\int e^{-u_R^N/kT} (d^3r)^N, \tag{4.3.6}$$

where

$$\begin{aligned} \langle w(r_{12})\rangle &= V^{-2}\int w(r_{12}) g^{(2)}(r_{12}) d^3r_1 d^3r_2 \\ &= V^{-1}\int w(r_{12}) g^{(2)}(r_{12}) d^3r_{12}. \end{aligned} \tag{4.3.7}$$

Defining

$$a = (1/2)\int w(r_{12}) g^{(2)}(r_{12}) d^3r_{12}, \tag{4.3.8}$$

we obtain

$$\begin{aligned} Z_N &\approx e^{\frac{N^2 a}{VkT}} Z_N^R \\ Z_N^R &= \int_V^{\cdots}\int e^{-u_R^N/kT} (d^3r)^N. \end{aligned} \tag{4.3.9}$$

The second approximation of the VDW model is the assumption that the hard core interactions in the integrand of eq. (4.3.9) simply reduce the free volume available to the variables of integration to $V - Nb$, that is, that Z_N^R is of the form given by eq. (4.2.23). Under this assumption, we find

$$Z_N \approx e^{\frac{N^2 a}{VkT}} (V - Nb)^N. \tag{4.3.10}$$

The approximations of the VDW model can be improved upon, but with a great loss of simplicity. Thus, we reserve the improvement

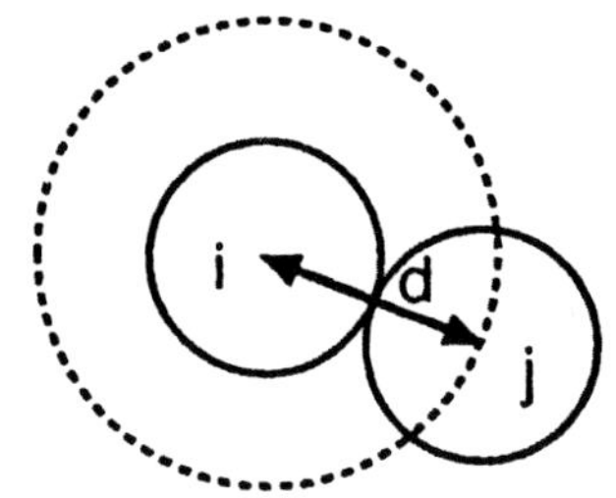

Figure 4.8
Illustration of volume (dashed circle) excluded to the center of molecule j due to the presence of molecule i.

until we return in a later chapter to a more advanced study of fluid properties. The equation of state obtained from the formula,

$$P = -kT\left(\frac{\partial \ln Z_N}{\partial V}\right)_{N,T}, \tag{4.3.11}$$

agrees with eq. (4.3.2) if a and b are assumed to be independent of V.

We have not yet specified the relationship of b to the hard core diameter d. Because of the rigid sphere interaction between a pair of molecules, a volume of magnitude $4/3\pi d^3$ around the center of the ith molecule is excluded to the center of the jth molecule (see the dashed circle in Figure 4.8). Because there are $N(N-1)/2$ pairs of molecules in the system, the total volume excluded to the molecules is $[N(N-1)/2](4\pi d^3/3)$. On the one hand, by definition Nb is the volume excluded to each molecule, so the total excluded volume is also equal to $N^2 b$. Thus, we find that b is four times the rigid sphere volume, that is,

$$b = \frac{2\pi}{3} d^3 = 4\left[\frac{4\pi}{3}\left(\frac{d}{2}\right)\right]. \tag{4.3.12}$$

Determination of b from the equation of state, eq. (4.3.2) and employment of eq. (4.3.12) afforded an early means of estimating molecular sizes from equation of state data. Equation (4.3.12) actually overestimates the excluded volume, the reason being that N particles can arrange themselves in configurations in which they share the pair volumes assumed additive in arriving at eq. (4.3.12). However, as b enters the model as an adjustable parameter, we do not try to improve upon the molecular estimate of b.

For the Hamiltonian view at eq. (4.1.10), the classical mechanical partition function factors are given by eq. (4.1.29). Assuming that a and b are constant, combining eqs. (4.3.10) and (4.1.29), and using the connection $F = -kT \ln Q_N$, with the thermodynamic relations $S = -(\partial F/\partial T)_{N,V}$ and $U = F + TS$, we obtain for a VDW fluid the results

$$F = -NkT \ln\left[\left(\frac{2\pi mkT}{h^2}\right)^{3/2} e\left(\frac{V-Nb}{N}\right)\right] - \frac{N^2 a}{V}, \tag{4.3.13}$$

$$S = Nk \ln\left[\left(\frac{2\pi mkT}{h^2}\right)^{3/2} e^{5/2}\left(\frac{V-Nb}{N}\right)\right], \tag{4.3.14}$$

and

$$U = \frac{3}{2} NkT - \frac{N^2 a}{V}. \tag{4.3.15}$$

It is interesting to note that the VDW free energy is of the form $F = F^R + U^A$, where F^R is the Helmholtz free energy of a hard sphere system and $U^A = -N^2 a/V$ is the contribution to thermodynamic energy from the attractive interactions. We could imagine starting with a system of N hard spheres of volume V at temperature T with free energy F^R. If the attractive interactions between molecules are then turned on reversibly, the contribution to the Helmholtz free energy will be $U^A - TS^A$, and the total free energy will be $F^R + U^A - TS^A$. Thus, a basic assumption of the VDW model is that the entropy of the system is determined by the repulsive interactions, that is, that S^A is negligible compared to S^H. In other words, in a structural (order versus disorder) interpretation of entropy, the VDW model is based on the assumption that the repulsive forces are the primary factor determining fluid structure.

Equations (4.3.2) and (4.3.13)–(4.3.15) are often useful for estimating thermodynamic properties of real fluids if a and b are known. We see in the next section that a and b can be obtained from the critical pressure and temperature. An example of the application of VDW theory is given below.

Example

If gas flows through a nozzle from a pressure P_1 to a pressure P_2 at constant enthalpy $H(\equiv U + PV)$, it will undergo a change of temperature whose sign will depend on the Joule–Thomson coefficient.

$$\mu_{JT} \equiv \left(\frac{\partial T}{\partial P}\right)_H. \tag{4.3.16}$$

Such a constant enthalpy flow process, called a throttling or Joule–Thomson process, is useful in cooling gases for liquefaction if μ_{JT} is greater than zero. If this is the case, the compressed gas at P_1 can flow through a throttle valve to a low pressure side at $P_2 < P_1$, with an accompanying decrease in the temperature of the gas $\delta T \propto \mu_{JT}\delta P$, so that $\delta P < 0$ implies $\delta T < 0$ if $\mu_{JT} > 0$. Thus, it is useful to know at what pressures $\mu_{JT} > 0$ for a given temperature. We shall use the VDW equation of state to estimate the density–temperature curve across which μ_{JT} goes from positive to negative, that is, the curve along which $\mu_{JT} = 0$.

Thermodynamic identities can be used to rewrite μ_{JT} as follows:

$$\begin{aligned}\mu_{JT} &= \frac{-\left(\frac{\partial H}{\partial P}\right)_T}{\left(\frac{\partial H}{\partial T}\right)_P} = -\frac{T\left(\frac{\partial S}{\partial P}\right)_T + V}{C_P} \\ &= -\frac{-T\left(\frac{\partial V}{\partial T}\right)_P + V}{C_P} \\ &= \frac{V}{C_P}(T\alpha - 1)\end{aligned} \tag{4.3.17}$$

where α is the isobaric thermal expansivity.

Computing α from the VDW equation of state and inserting the result into eq. (4.3.17), we obtain

$$\mu_{JT} = \frac{N[-kTb + 2a(2 - nb)^2]}{C_P[kT - 2an(1 - nb)^2]} \qquad (4.3.18)$$

which is zero if

$$-kTb + 2a(1 - nb)^2 = 0$$

or

$$n = (1 - \sqrt{kTb/2a})/b, \qquad (4.3.19)$$

where n is the fluid density N/V. Equation (4.3.19) gives density as a function of temperature along the curve for which $\mu_{JT} = 0$. For a given temperature T, $\mu_{JT} > 0$ for densities smaller than that of eq. (4.3.19) and $\mu_{JT} < 0$ for densities larger than that of eq. (4.3.20). The van der Waals pressure–temperature curve for $\mu_{JT} = 0$ obeys the equation

$$P = \frac{a}{b^2}\left(1 - \sqrt{\frac{kTb}{2a}}\right)\left(3\sqrt{\frac{kTb}{2a}} - 1\right) \qquad (4.3.20)$$

If there are no attractive interactions, that is, $a = 0$, then from eq. (4.3.18) we conclude that μ_{JT} is always negative. The experimental observation of positive values of μ_{JT} was used in the early development of the molecular theory of fluids as evidence of the existence of attractive forces between neutral molecules.

Equations (4.3.13)–(4.3.15) are valid only for monatomic systems because rotational and vibrational motion have been neglected in evaluating Q_N. On the other hand, even for polyatomic molecules a centrally symmetric pair potential is in some cases (e.g., fluids of small, compact nonpolar molecules such as N_2, O_2, H_2, and CH_4) a reasonable approximation to the actual intermolecular interaction potential. The rapid rotational and vibrational motions average or smooth out to some extent the noncentrally symmetric part of the actual pair potential energy function. In view of this argument and the fact that the pressure depends only on the interactions (through Z_N), we may expect VDW equation of state, eq. (4.3.2), to give a correct qualitative description of many polyatomic fluids, even though the energies and entropy predicted by eqs. (4.3.13)–(4.3.15) would have to be modified to account for the internal degrees of freedom (e.g., rotations and vibrations) of polyatomic molecules.

Exercise
Given that a system obeys VDW equations of state, prove that the constant volume specific heat C_V is a function only of temperature. Show that this result leads to the following forms for the internal energy and entropy, respectively,

$$U = \int C_V(T)dT - \frac{N^2 a}{V} + constant \qquad (4.3.21)$$

$$S = \int \frac{C_V(T)}{T} dT + \int \frac{Nk}{V - Nb} dV + constant \qquad (4.3.22)$$

Note that in the case of a monatomic system, $C_V = \frac{3}{2}Nk$ and, consequently, eq. (4.3.14) is recovered from eq. (4.3.22).

4.4 Phase Behavior of Pure Fluids

The VDW model provides a remarkably simple and qualitatively accurate description of the phase behavior of fluids. For this reason the qualitative conclusions presented in this section will lean heavily upon VDW model.

For convenience of discussion, let us define the molar volume, $\tilde{V}$, and the molar quantities, $\tilde{a}$ and $\tilde{b}$ as follows:

$$\tilde{V} = V/(N/N_o) \qquad (4.4.1)$$

$$\tilde{b} = N_o b \qquad (4.4.2)$$

$$\tilde{a} = N_o^2 a, \qquad (4.4.3)$$

where N_o $(= 6.02 \times 10^{23})$ is Avogadro's number. In terms of these quantities the VDW equation of state is of the form

$$P = \frac{RT}{\tilde{V} - \tilde{b}} - \frac{\tilde{a}}{\tilde{V}^2}, \qquad (4.4.4)$$

in which R $(= N_o k)$ is the gas constant.

The VDW equation of state is a cubic equation in molar volume. The qualitative dependence of pressure on volume for such cubic equations at different temperatures is illustrated in Figure 4.9. At temperatures above a characteristic or critical temperature T_c, the pressure isotherms are monotonically decreasing functions of the molar volume (the T_2 isotherm is an example of this case). Above T_c only one homogeneous fluid phase exists at a given pressure. Below T_c, there exists for every isotherm a range of pressures (e.g., pressures lying between P' and P''' along the T_1 isotherm) for which there are three molar volumes corresponding to each value of pressure. Thus, for each isotherm below T_c, VDW theory predicts the existence of a pressure range in which there are three homogeneous fluid phases corresponding to each pressure in that

Exercise
If the density of liquid argon is 1.4 g/cm^3 at 86.5K and 1 atm, use corresponding states to estimate the densities of liquid krypton and xenon at 1.13 atm and 120.25K and 1.2 atm and 166.05K, respectively. Compare the results with experiment. The critical points of Ar, Kr and Xe are T_c, P_c, $\tilde{V}_c$ = 151, 48, 75.2; 209.4, 54.3, 92.2; and 289.75, 58.0, 118.8 K, atm, cm^3/gmol, respectively.

Figure 4.9
Qualitative pressure versus molar volume isotherms (constant T) predicted by cubic equations of state. The P–V states between a and b on the T_1 isotherm are thermodynamically unstable.

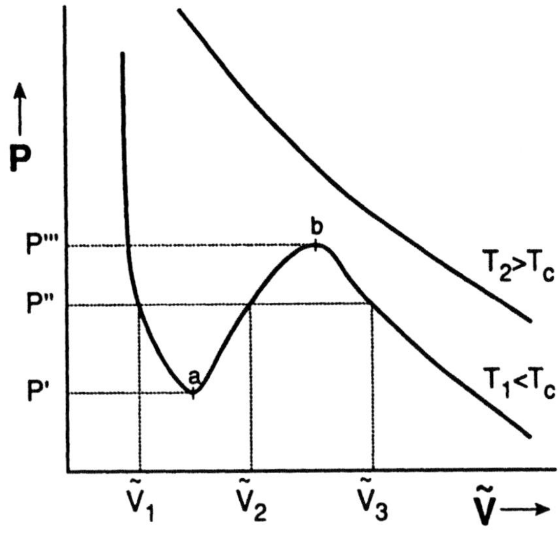

range. For example, at T_1 and $P_{p''}$ phases of molar volumes $\tilde{V}_1$, $\tilde{V}_2$, and $\tilde{V}_3$, respectively, are predicted. The phase of the intermediate density, $\tilde{V}_2^{-1}$, is, however, not thermodynamically stable, because the stability condition

$$\left(\frac{\partial P}{\partial \tilde{V}}\right)_T < 0 \tag{4.4.5}$$

is violated at $\tilde{V}_2$. All of the states between a and b along the T_1 isotherm are thermodynamically unstable.

Furthermore, for a given $T < T_c$, there is only one pressure $P(T)$ for which the two stable phases are in thermodynamic equilibrium with each other. This is because the two phases at the same pressure are not in thermodynamic equilibrium unless their chemical potentials are the same.

Because the variation of the chemical potential along an isotherm is given by the thermodynamic expression

$$d\mu = \tilde{V} dP, \tag{4.4.6}$$

the chemical potential difference for any two states $\tilde{V}^\alpha$, $\tilde{V}^\beta$ along an isotherm is

$$\mu^\beta - \mu^\alpha = \int_{\tilde{V}^\alpha}^{\tilde{V}^\beta} \tilde{V} dP. \tag{4.4.7}$$

For a given isotherm T, α and β will correspond to two phases at equilibrium with each other if $P^\alpha = P^\beta$ and $\mu^\alpha = \mu^\beta$, the latter condition being equivalent to

$$0 = \int_{\tilde{V}^\alpha}^{\tilde{V}^\beta} \tilde{V}\,dP. \tag{4.4.8}$$

Equation (4.4.8) implies a simple geometric construction for finding the two-phase equilibrium states (that is, the vapor–liquid coexistence curve, the low and high density states being referred to as the vapor and the liquid phases, respectively). Consider a line (called a vapor–liquid tie line) drawn parallel to the $\tilde{V}$ axis and cutting a particular isotherm in such a way that the enclosed area below the line and between the line and the isotherm is equal to the area above the line and between the line and the isotherm. Such a construction is shown in Figure 4.10 for the isotherm T_1. The parallel pressure line for which $A_1 = A_2$ in the figure satisfies eq. (4.4.8) and the condition $P^V(T_1) = P^L(T_1) = P_b(T_1)$ and, therefore, determines the molar volumes $\tilde{V}^L(T_1)$ and $\tilde{V}^V(T_1)$ of the liquid and vapor that can coexist at temperature T_1 and pressure $P(T_1)$. By such a geometric construction, called the Maxwell equal area tie-line construction, one can determine the vapor–liquid coexistence curve or dome shown in Figure 4.10. It is plain from the geometric construction that there is only one coexistence pressure per isotherm.

The point P_c, $\tilde{V}_c$ at which the isotherm T_c is tangent to the top of the two-phase coexistence dome shown in Figure 4.10 is at a point of inflection, satisfying the "critical point" conditions

$$\left(\frac{\partial P}{\partial \tilde{V}}\right)_T = \left(\frac{\partial^2 P}{\partial \tilde{V}^2}\right)_T = 0 \quad \text{at} \quad T_c. \tag{4.4.9}$$

For temperatures above this so-called critical isotherm, only a single fluid phase exists. The quantities P_c and $\tilde{V}_c$, fixed by the conditions of eq. (4.4.9), are called the critical pressure and critical molar volumes, respectively.

Exercise

Using the VDW equation of state and the critical point conditions, show that

$$\tilde{V}_c = 3\tilde{b}, \quad T_c = \frac{8\tilde{a}}{27R\tilde{b}}, \quad P_c = \frac{\tilde{a}}{27\tilde{b}^2} \tag{4.4.10}$$

and therefore

$$\tilde{a} = \frac{27R^2T_c^2}{64P_c}, \quad \tilde{b} = \frac{RT_c}{8P_c}. \tag{4.4.11}$$

Figure 4.10
Schematic phase diagram of a VDW fluid, illustrating the liquid–vapor coexistence dome, the metastable one-phase region, the unstable one-phase region, and the Maxwell equal area construction (shaded area A_1 equals shaded area A_2) of the liquid–vapor tie-lines.

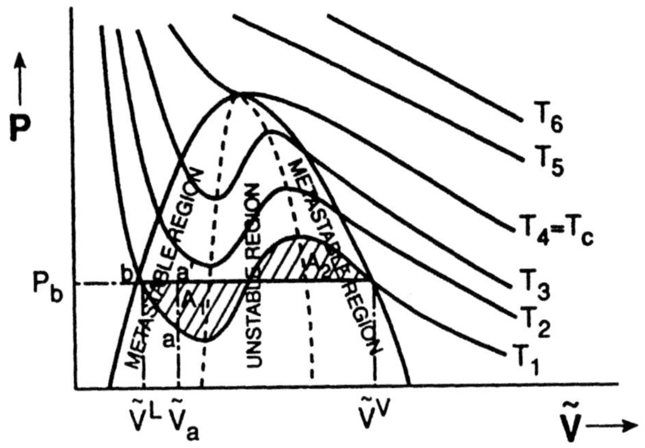

Calculation of $\tilde{b}$ from the formula in eq. (4.4.11) and comparison with the molecular expression $\tilde{b} = N_0 \frac{2\pi}{3} d^3$ provided one of the early estimates of molecular sizes from equation of state data.

The region labeled "metastable region" in Figure 4.10 (lying between the two-phase dome and the unstable dome) contains one-phase supercooled vapor states or one-phase superheated liquid states. These one-phase states are stable according to eq. (4.4.5). However, a system in such a one-phase state may break up into a two-phase state that is at a lower Helmholtz free energy and is therefore the more stable of the two states. To see this property, consider N particles in a rigid, diathermal container of volume V at equilibrium with a heat bath at temperature T_1. Let us compare the Helmholtz free energy F_a of a one-phase system in the state denoted by the point a in Figure 4.10. The state of the two-phase system at the same volume and temperature is denoted by the point a' in Figure 4.10. Neglecting a small interfacial free energy (computed in the next section), we can write the Helmholtz free energy of the two-phase system as

$$\begin{aligned} F_{a'} &= \left[\mu^L(T_1) - P_b \widetilde{V}^L(T_1)\right]N^L + \left[\mu^V(T_1) - P_b \widetilde{V}^V(T_1)\right]N^V \\ &= N\mu^L(T_1) - P_b\left[N^L \widetilde{V}^L(T_1) + N^V \widetilde{V}^V(T_1)\right] \\ &= F^L - N^V P_b\left(\widetilde{V}^V - \widetilde{V}^L\right), \end{aligned} \tag{4.4.12}$$

where F^L is the Helmholtz free energy of N molecules of the liquid phase at T_1 and P_b. N^V and N^L denote the number of molecules in the vapor and liquid phases, respectively, corresponding to a system in state a'.

Using the thermodynamic relation

$$(d\widetilde{F})_{T,N} = -P d\widetilde{V}, \tag{4.4.13}$$

we find

$$\tilde{F}_a - \tilde{F}^L = -\int_{\tilde{V}^L}^{\tilde{V}_a} P d\tilde{V}$$

or, because $F_a = N\tilde{F}_a$ and $F^L = N\tilde{F}^L$,

$$F_a - F^L = -N \int_{\tilde{V}^L}^{\tilde{V}_a} P d\tilde{V}. \tag{4.4.14}$$

Combining eqs. (4.4.12) and (4.4.14), we obtain

$$\begin{aligned} F_a - F_{a'} &= -N \int_{\tilde{V}^L}^{\tilde{V}_a} P d\tilde{V} + N^V P_b(\tilde{V}^V - \tilde{V}^L)] \\ &= N \int_{\tilde{V}^L}^{\tilde{V}_a} (P_b - P) d\tilde{V} - \{P_b[N\tilde{V}_a - N^V\tilde{V}^V - N^L\tilde{V}^L]\} \end{aligned} \tag{4.4.15}$$

The term in the braces in eq. (4.4.15) vanishes because by definition $V = N\tilde{V}_a$ and $V = N^V\tilde{V}^V + N^L\tilde{V}^L$. Thus, because $P_b \geq P$ along the T_1 isotherm from b to a, we find

$$F_a - F_{a'} = N \int_{V^L}^{\tilde{V}_a} [P_b - P] d\tilde{V} \geq 0, \tag{4.4.16}$$

or $F_{a'} \leq F_a$, implying that the two-phase state a' is more stable than the one-phase state. Similar considerations show that the two-phase state is more stable than a corresponding one-phase state on the vapor side of the unstable region.

The reason for the lasting importance of the VDW equation is that the qualitative features of the phase diagram shown in Figure 4.10 are universal characteristics of all real fluids. Moreover, phenomena such as the Joule–Thomson coefficient inversion (change of sign), liquid–vapor phase transitions, and liquid mixture–liquid mixture phase transitions (predicted from the multicomponent version of the VDW equation) are very clearly demonstrated by the VDW equation to depend on the balance between the repulsive and attractive forces among the particles.

The metastable regions predicted in Figure 4.10 have long been known to exist. By very carefully heating liquids in clean vessels with polished surfaces, so as not to have bubble nucleation sites on the vessel surfaces, investigators were able to heat liquids somewhat past the vaporization point without vaporization occurring. Similarly, vapors could be cooled past the condensation point without condensing. Until a few years ago, however, as far as the author has determined, no one had experimentally followed the metastable one-phase isotherms all the way to the unstable region (at least in controlled laboratory experiments, no one had gone to the unstable region). Experiments approaching the unstable region from the liquid side can be performed with the aid of the simple apparatus shown in Figure 4.11.

A droplet of the liquid of interest is injected by a hypodermic needle into a heated column of an immiscible liquid having a slightly higher density than the subject liquid. The column is heated differentially so that the temperature of the carrier liquid increases with height above the injection point. Because the injected droplet is surrounded by the carrier liquid, there are no nucleation sites for bubble formation. Thus, the droplet can easily be superheated. As the droplet rises up the column of carrier liquid, it heats up until it eventually reaches the temperature at which the sign of $(\partial P/\partial V)_T$ goes from minus to plus. At this point, the droplet is no longer thermodynamically stable and, consequently, suddenly vaporizes explosively. Thus by measuring the maximum temperature at which the droplet explodes in the column as a function of pressure, one obtains the liquid side envelope of the unstable region. This envelope is determined from the equation of state by the condition

$$\left(\frac{\partial P}{\partial \tilde{V}}\right)_T = 0. \tag{4.4.17}$$

Exercise

Show that the envelope of the unstable homogeneous fluid region is predicted by the VDW equation to be of the form

$$P = \frac{\tilde{a}}{\tilde{V}^2}\left(1 - \frac{2\tilde{b}}{\tilde{V}}\right). \tag{4.4.18}$$

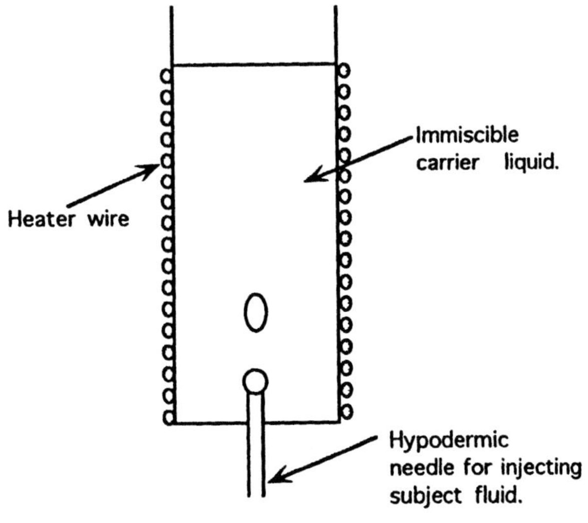

Figure 4.11
Apparatus for measuring the liquid side of the envelope of the unstable PVT region.

It is important to note that the metastable states of fluid are not simply esoteric novelties of the physicist. As a matter of fact, the popping noise (and splattering) heard when food containing water is cooked in hot grease or oil can be from explosive vaporization of superheated water droplets reaching their unstable temperature. Thus, an engineer concerned with large scale deep-frying processes might want to consider this effect in the design criteria. Understanding the thermodynamically unstable region is also important in the liquid phase commercial transport of low temperature fluids such as N_2, the freons, and light hydrocarbons. Also, as we see later, the states in the unstable region play a role in determining the interfacial properties of fluids.

Let us close this section with a brief discussion of the behavior of fluids near the critical point. One of the interesting phenomena occurring near the critical point is "critical opalescence." At the critical point the compressibility,

$$\kappa_T = -\frac{1}{\tilde{V}}\left(\frac{\partial \tilde{V}}{\partial P}\right)_T \tag{4.4.19}$$

is infinite and sufficiently near the critical point it is very large. This means that near the critical point, small fluctuations in the pressure will cause large fluctuations in the density. Compressive fluctuations ($\delta P > 0$) cause tiny droplet (or mist) formation, and these droplets scatter impinging light, giving the fluid an opalescent appearance near the critical point. This phenomenon is known as critical opalescence.

The behavior of the isothermal compressibility κ_T and the heat capacities C_V and C_P near the critical point have been studied in many carefully conceived experiments and have been the subject of extensive theoretical investigation. All of these quantities become infinite as the critical point is approached from either the liquid or the gas phase. For example, for several fluids the constant volume specific heat C_V has been found experimentally to depend on T as shown in Figure 4.12. The C_V data have been fitted to the expression

$$C_V = B^{\pm} \ln\left|1 - T/T_c\right| + C^{\pm}, \tag{4.4.20}$$

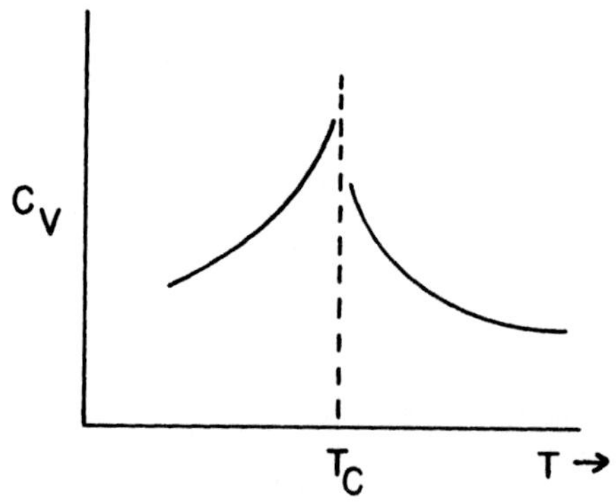

Figure 4.12
Typical behavior of the specific heat of a fluid near the critical point.

where the plus and minus superscripts refer, respectively, to the temperature regions above and below the critical point. In Chapter 3, we saw that the logarithmic singularity was predicted exactly for the specific heat of a 2-dimensional Ising model. Near the critical point the isothermal compressibility has been observed to be of the form

$$\kappa_T \propto |T_c - T|^{\gamma^{\pm}}, \tag{4.4.21}$$

where $\gamma^+ \simeq 1.1$ and $\gamma^- \simeq 1.2$.

There are two interesting features of eqs. (4.4.20) and (4.4.21) from a theoretical standpoint. The first is that the quantities C_V and κ_T are nonanalytic functions of temperature near the critical point, and the second is that these quantities approach infinity with respect to temperatures at different rates, depending on whether $T \to T_c$ from above or below. The singular behavior of the thermodynamic quantities κ_T, C_V and C_P and the phenomenon of critical opalescence are related to the large fluctuations and long-ranged intermolecular correlations occurring near the critical point. We have not yet included fluctuations in the treatment of the VDW model, so we cannot at this point discuss the molecular aspects of critical phenomena. However, one can show that van der Waals mean field theory, if fluctuations of the molecular potential energy about the mean field value are included, predicts temperature divergent heat capacities and compressibility as T approaches T_c, although the critical exponents of the theory differ from the experiment and from more rigorous statistical mechanics based on renormalization theory.

Exercise

The vapor pressure of xenon as a function of its melting point is well estimated by

$$P = -3089.9 + 12.726T + 0.040141T^2$$

in the temperature range 170 to 255K. The units of P and T are atm and K. To estimate the vapor pressure at which argon melts at 106K we use corresponding states. The xenon temperature corresponding to 106K for argon is $T = 106(289.75/151) = 203.4K$. The vapor pressure of xenon at this temperature is 1159.3 atm. The corresponding vapor pressure of argon is $P = 1159.3(48.0/58.0) = 959.4$ atm.

Use the law of corresponding states to convert the above equation for xenon to a vapor pressure versus melting point equation for argon, krypton and methane.

4.5 Semiempirical and Empirical Equations of State

In connection with the VDW equation of state, we should mention the law of corresponding states that it spawned. If we define the reduced variables,

$$\begin{aligned} V_r &= \tilde{V}/\tilde{V}_c \\ T_r &= T/T_c \\ P_r &= P/P_c \end{aligned} \tag{4.5.1}$$

and express eq. (4.4.4) in terms of these variables, we find the "universal" equation of state,

$$P_r = \frac{8}{3V_r - 1} - \frac{3}{V_r^2}. \tag{4.5.2}$$

This equation expresses the law of corresponding states: All fluids obeying van der Waals equation of state will have the same reduced pressure P_r is they are compared at the same reduced volume

V_r and temperature T_r. Although VDW equation of state is only qualitatively correct, the law of corresponding states, that P_r is *some* universal function of V_r and T_r, is obeyed quite well for a variety of fluids. Systematic application of the law will be discussed fully in a later chapter. Pressure–volume isotherms and Maxwell tie-lines are shown in the next section for a VDW fluid. The reduced variables used for the figure differ by numerical factors from those defined by eq. (4.5.1).

From purely dimensional considerations one can show that a law of corresponding states will be obeyed by any class of fluids described by a two parameter equation of state. Three such equations of state that have been of practical importance are the equation of Berthelot,

$$\left(P + \frac{\tilde{a}}{T\widetilde{V}^2}\right)(\widetilde{V} - \tilde{b}) = RT, \tag{4.5.3}$$

that of Dieterici,

$$Pe^{\tilde{a}/\widetilde{V}RT}(\widetilde{V} - \tilde{b}) = RT, \tag{4.5.4}$$

and that of Redlich and Kwong (RK),

$$\left[P + \frac{\tilde{a}}{T^{1/2}\widetilde{V}(\widetilde{V} + \tilde{b})}\right](\widetilde{V} - b) = RT. \tag{4.5.5}$$

Exercise

Derive for eqs. (4.5.3)–(4.5.5) reduced forms corresponding to eq. (4.5.2).

Exercise

The equation of state of a certain lattice gas model is

$$P = -\frac{kT}{b}ln(1-nb) - n^2a.$$

Find the critical pressure, temperature, and density as a function of the parameters a and b. Show that this equation of state obeys a law of corresponding states. Predict and plot the temperature–density binodal for liquid–vapor coexistence. Use the reduced variables of the corresponding states for the plot.

The Berthelot and RK equations are empirical modifications of the VDW equation of state. Equations (4.5.3)–(4.5.5) yield universal equations for P_r as a function of V_r and T_r. The parameters $\tilde{a}$ and $\tilde{b}$ in these equations can be determined from the critical conditions, eq. (4.4.9), to be functions of P_c, T_c, and V_c. The relations between the two parameters of the equations of state and critical quantities are given in Table 4.1.

The critical compressibility factor, P_cV_c/RT_c, predicted by the Dieterici equation agrees quite well with the average value, 0.273, observed (Beattie and Stockmayer, 1940) for 25 nonpolar gases. This is the best performance of the four two-parameter equations of state considered here. In overall performance, however, the RK equation appears to be the best all-around two-parameter equation of state. With appropriate mixing rules relating mixture parameters to pure fluid parameters, the RK equation is one of the most popular equations of state (another is the Benedict–Webb–Rubin, BWR, equation) in industrial use.

TABLE 4.1 Relation of Thermodynamic Critical Quantities and Parameters of Three Two-Parameter Equations of State

	van der Waals	*Berthelot*	*Dieterici*	*Redlich–Kwong*
P_c	$\tilde{a}/27\tilde{b}^2$	$(\tilde{a}R/216\tilde{b}^3)^{1/2}$	$\tilde{a}/29.56\tilde{b}^2$	$0.0299\left[R^{1/2}\tilde{a}/\tilde{b}^{5/2}\right]^{2/3}$
$\tilde{V}_c$	$3\tilde{b}$	$3\tilde{b}$	$2\tilde{b}$	$3.841\tilde{b}$
T_c	$8\tilde{a}/27\tilde{b}R$	$(8\tilde{a}/27\tilde{b}R)^{1/2}$	$\tilde{a}/4\tilde{b}R$	$\left[\tilde{a}/4.934R\tilde{b}\right]^{2/3}$
$P_c\tilde{V}_c/RT_c$	3/8	3/8	0.2706	1/3
$\tilde{a}$	$27R^2T_c^2/64P_c$	$27R^2T_c^3/64P_c$	$0.5412R^2T_c^2/P_c$	$0.4278R^2T_c^{2.5}/P_c$
$\tilde{b}$	$RT_c/8P_c$	$RT_c/8P_c$	$0.1353RT_c/P_c$	$0.0867RT_c/P_c$

Let us compare the performance of the above equations by applying them to the prediction of the internal energy of a gas over a wide range of densities. The molar internal energy of a fluid can be obtained from the equation of state through the thermodynamic expression

$$\tilde{U} = \tilde{U}^I(T) + \int_{\infty}^{\tilde{V}} \left[T\left(\frac{\partial P}{\partial T}\right)_{\tilde{V}} - P\right] d\tilde{V}, \qquad (4.5.6)$$

where $\tilde{U}^I(T)$ is the molar energy of the fluid in the ideal gas state ($\tilde{V} = \infty$). The results for the above two-parameter equations of state are

1 VDW

$$\tilde{U} = \tilde{U}^I - \frac{\tilde{a}}{\tilde{V}} \qquad (4.5.7)$$

2 Berthelot

$$\tilde{U} = \tilde{U}^I - \frac{2\tilde{a}}{T\tilde{V}} \qquad (4.5.8)$$

3 Dieterici

$$\tilde{U} = \tilde{U}^I - \frac{\tilde{a}}{\tilde{b}} e^{\tilde{a}/\tilde{b}RT}\left\{Ei\left(\frac{\tilde{a}}{\tilde{b}RT}\right) - Ei\left[\frac{\tilde{a}}{\tilde{b}RT}\left(1 - \frac{\tilde{b}}{\tilde{V}}\right)\right]\right\}, \qquad (4.5.9)$$

where

$$Ei(x) \equiv \int_{-\infty}^{x} \frac{e^t}{t}\, dt \qquad (4.5.10)$$

and

4 RK

$$\tilde{U} = \tilde{U}^I - \frac{3\tilde{a}}{2\tilde{b}T^{1/2}} \ln\left(\frac{\tilde{V} + \tilde{b}}{\tilde{V}}\right). \qquad (4.5.11)$$

In Figure 4.13 predictions of these equations are compared with an experiment for CO_2 at 150°C. The parameters $\tilde{a}$ and $\tilde{b}$ were determined by T_c and P_c from the formulas given in the last

Figure 4.13
A comparison of the experimental internal energy of CO_2 at 150° with values calculated using five empirical equations of state.

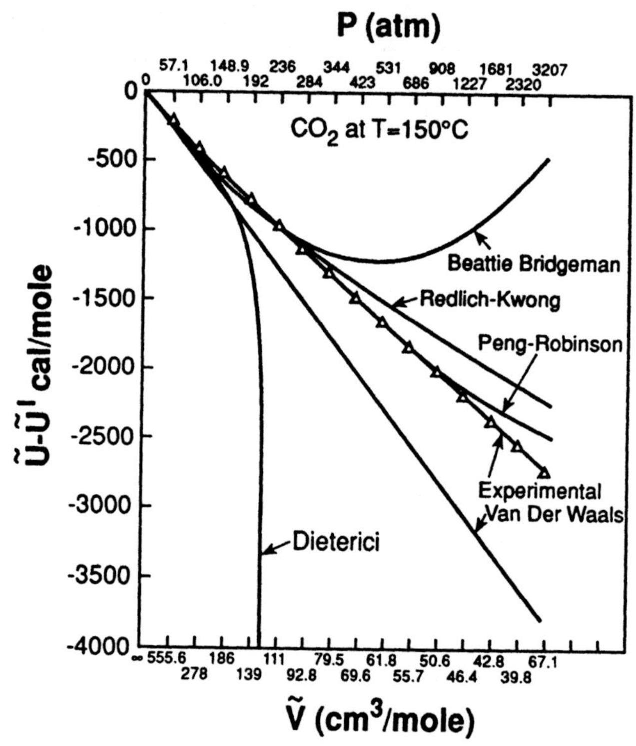

two rows of Table 4.2. Critical parameters for several compounds are given in Table 4.2. As seen in Figure 4.14, the RK predictions are better than the predictions of the other three over the whole pressure range considered, although the Dieterici predictions are quite accurate at moderate pressures (say less than 150 atm). At sufficiently high densities the Dieterici equation breaks down completely because the exponential factor $\exp(\tilde{a}/\tilde{V}RT)$ corrects too strongly for the effect of attractive interactions on pressure. The comparisons made in Figure 4.14 are typical for nonpolar fluids. All four of the two-parameter equations of state perform poorly for polar and hydrogen bonding fluids.

All of the two-parameter equations of state are more accurate for gas phase than for liquid phase predictions, although here again the RK equation is superior (the Dieterici equation being totally unacceptable at high densities). A test of the RK equation for vapor pressure consistency has been carried out by Horvath (1972). He took experimental vapor pressures, calculated $\tilde{a}$ and $\tilde{b}$ from critical data, and predicted the liquid–vapor coexistence densities for a variety of substances. If P, $\tilde{a}$, and $\tilde{b}$ are known, one can rearrange eq. (4.5.5) into the cubic equation

$$\tilde{V}^3 - \frac{RT}{P}\tilde{V}^2 + \left(\frac{\tilde{a}}{PT^{1/2}} - \tilde{b}^2 - \frac{RT\tilde{b}}{P}\right)\tilde{V} - \frac{\tilde{a}\tilde{b}}{PT^{1/2}} = 0. \qquad (4.5.12)$$

TABLE 4.2 Critical Constants and Acentric Factor for Several Substances

Compound	$\tilde{V}_c$ *(cm³/mol)*	$T_c(K)$	P_c *(atm)*	$P_c\tilde{V}_c/RT_c$	ω_a
Methane	99.3	191	45.8	0.290	0.013
Ethane	147.0	306	48.2	0.284	0.105
Propane	199.5	370	42.0	0.276	0.152
n-Butane	254.9	425	37.5	0.274	0.201
Isobutane	262.3	408	36.0	0.282	0.192
n-Pentane	310.4	470	33.3	0.268	0.252
Isopentane	308.2	461	32.9	0.268	0.206
Neopentane	302.1	434	31.6	0.268	0.195
n-Hexane	369.3	508	29.8	0.264	0.290
n-Heptane	426.8	540	27.0	0.260	0.352
n-Octane	489.8	569	24.6	0.258	0.408
Ethylene	124.1	282	50.0	0.268	0.073
Propylene	181.3	365	45.6	0.276	0.143
1-Butene	239.6	420	39.7	0.276	0.203
Argon	75.5	151.2	48	0.292	−0.0027
Bromine	144.3	584	102	0.307	0.132
Carbon dioxide	94.5	304	72.9	0.276	0.225
Carbon monoxide	81.2	133	39.5	0.294	0.049
Chlorine	124.1	417	76.1	0.276	0.074
Helium	57.7	5.3	2.26	0.300	0
Hydrogen	64.9	33.3	12.8	0.304	0
Hydrogen chloride	87.1	325	81.5	0.266	0.133
Krypton	92.1	209.4	54.1	0.290	−0.002
Nitrous oxide	96.2	310	71.7	0.271	0.160
Nitrogen	89.8	126	33.5	0.291	0.040
Oxygen	73.6	155	50.1	0.29	0.021
Sulfur dioxide	121.8	431	77.8	0.268	0.273
Sulfur trioxide	126.0	491	83.8	0.262	0.510
Xenon	118.9	289.8	57.6	0.288	0.002
Ammonia	72.6	406	111	0.242	0.250
Nitric oxide	57.7	180	64	0.25	0.600
Water	56.0	647	218	0.23	0.344

This equation has one real root when $q < 0$ and three real roots when $q > 0$,

$$q = \frac{\tilde{a}}{PT^{1/2}} - \tilde{b}^2 - \frac{RT\tilde{b}}{P} \tag{4.5.13}$$

If the RK equation were exact, then whenever a coexistence vapor pressure were used in eq. (4.5.12), it would have three roots, the largest of which would be the vapor molar volume, the smallest the liquid molar volume, and the intermediate one a mechanically unstable state. The results of Horvath's calculations are shown in Figure 4.14. In the case of NH_3 and HCl hydrogen bonded fluids, there are appreciable regions of vapor pressure for which eq. (4.5.12) has only one root. Thus, the RK model is worse for hydrogen bonded fluids than for others. Of course, if one constructs from

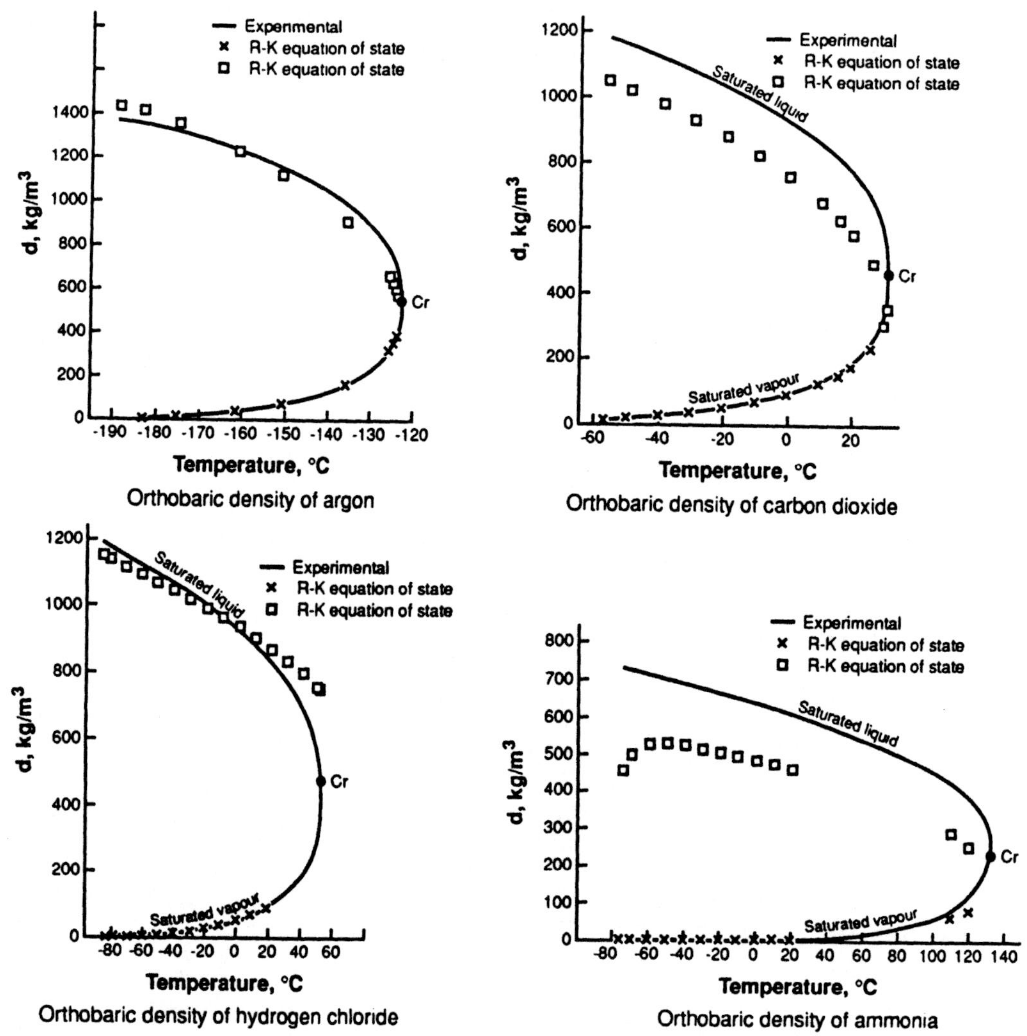

Figure 4.14
Liquid–vapor coexistence curves. Adapted from Horvath (1972).

the RK theory the liquid–vapor density–temperature coexistence curve, there will not be regions under the critical point for which eq. (4.5.12) has only one solution. The RK equation and its variations are used broadly in practical estimations of pressure-volume-temperature (PVT) relations, enthalpy and vapor-liquid equilibrium data (especially for multicomponent systems).

The basic assumption of the VDW model and its variations, the Berthelot and RK models, is that the pressure can be expressed as the sum

$$P = P_R + P_A, \tag{4.5.14}$$

where P_R is the pressure of a fluid of strongly repulsive molecules and P_A is a correction arising from the attractive interaction. These models all use the Clausius approximation $P_R = RT/(\tilde{V} - \tilde{b})$ for the repulsive force contribution. For one-dimensional hard rods the Clausius form is exact. However, it is not exact for three-dimensional hard spheres. Theoretical studies and computer simulations of recent years have generated a very accurate hard sphere (HS) equation of state, namely, the following one formulated by Carnahan and Starling (1969):

$$P_{HS} = \frac{RT}{\tilde{V}} \frac{1 + y + y^2 - y^3}{(1 - y)^3}, \tag{4.5.15}$$

where

$$y = \frac{Nb}{4V} = \frac{\tilde{b}}{4\tilde{V}}. \tag{4.5.16}$$

Carnahan and Starling (1972) suggested that the two parameter equations will be improved by substituting P_{HS} for P_R. For example, the suggested modified forms of VDW and RK equations are the hard sphere (HS) VDW equation

$$P = P_{HS} - \frac{\tilde{a}}{\tilde{V}^2} \tag{4.5.17}$$

and the hard sphere RK (HSRK) equation

$$P = P_{HS} - \frac{\tilde{a}}{T^{1/2}\tilde{V}(\tilde{V} + \tilde{b})}. \tag{4.5.18}$$

The molar energies $\tilde{U}$ corresponding to these equations are given by eqs. (4.5.7) and (4.5.11), the substitution of P_{HS} for $RT/(\tilde{V} - b)$ not changing the molar energies. Carnahan and Starling present calculations of enthalpies and densities of a variety of substances and the vapor pressure of methyl chloride that tend to indicate that HSVDW and HSRK are improvements over VDW and RK. For example, in a comparison of density predictions for N_2, CO_2, H_2S, C_2H_4, i-C_5H_{12}, $CH_4C_3H_4$, C_4H_{10}, and C_5H_{12}, the average deviation of predicted density from the experiment was 4.495, 2.898, 1.496, and 1.133% for VDW, HSVDW, RK, and HSRD, respectively. Their vapor pressure comparisons are shown in Table 4.3.

Several three-parameter equations of state have been proposed that are substantial improvements over the two-parameter equations. One of these, introduced by Soave (1972), is a modification of the RK equation, namely,

$$P = \frac{RT}{\tilde{V} - \tilde{b}} - \frac{\tilde{a}(T)}{\tilde{V}(\tilde{V} + \tilde{b})}, \tag{4.5.19}$$

where

$$\tilde{a}(T) \equiv \tilde{a}(T_c)\{1 + m(1 - T_r^{1/2})\}^2, \tag{4.5.20}$$

with $T \equiv T/T_c$ and

$$m \equiv 0.480 + 1.574\omega_a - 0.176\omega_a^2. \tag{4.5.21}$$

TABLE 4.3 Prediction of Methyl Chloride Pressures from Experimental Temperatures and Vapor Specific Volumes

T °C	*P* (atm)	*VDW* (atm)	*HSVDW* (atm)	*RK* (atm)	*HSRK* (atm)
125.0	6.975	7.063	7.051	7.02	7.006
	20.049	20.629	20.523	20.555	20.132
	30.644	32.017	31.749	31.109	30.814
	39.817	42.215	1.728	40.630	40.124
	40.015	52.848	52.039	50.407	49.649
143.0	9.567	9.632	9.616	9.571	9.549
Critical	30.398	31.288	31.123	30.670	30.449
Isotherm	51.077	53.818	53.421	52.291	51.731
	69.954	251.92	145.77	103.16	83.291
	101.510	936.55	328.59	200.86	125.46
	206.85	4265.0	666.61	398.17	196.88

Adapted from Hsu and McKetta (1972).

ω_a is the acentric factor, a parameter introduced by Pitzer into the law of corresponding states to account for nonsphericity of molecules. The definition of the acentric factor is

$$\omega_a = -1 + \log_{10} \frac{P_c}{P^{\text{sat}}} \quad \text{at} \quad T_r = 0.7. \tag{4.5.22}$$

Values of ω_a are given in Table 4.2 for a variety of compounds. An enormous tabulation of ω_a is available (Reid 1966).

As in the two-parameter equations of state, $\tilde{a}(T_c)$ and $\tilde{b}$ of the Soave equation are evaluated from critical point data. The critical point conditions yield

$$\tilde{b} = 0.08664RT_c/P_c, \quad \tilde{a}(T_c) = 0.42747R^2T_c^2/P_c, \quad P_c\tilde{V}_c/RT_c = 1/3. \tag{4.5.23}$$

Following the lead of Soave, Peng and Robinson (PR: 1976) introduced the equation of state

$$P = \frac{RT}{\tilde{V} - \tilde{b}} - \frac{\tilde{a}(T)}{\tilde{V}^2 + \tilde{b}(2\tilde{V} - \tilde{b})}, \tag{4.5.24}$$

where

$$\tilde{a}(T) \equiv \tilde{a}(T_c)\{1 + m(1 - T_r^{1/2})\}^2, \tag{4.5.25}$$

with

$$m \equiv 0.37464 + 1.54226\omega_a - 0.26992\omega_a^2. \tag{4.5.26}$$

The critical point conditions yield for the PR equation

$$\tilde{b} = 0.07780RT_c/P_c, \quad \tilde{a}(T_c) = 0.4574R^2T_c^2/P_c, \quad P_c\tilde{V}_c/RT_c = 0.307. \tag{4.5.27}$$

The molar energy derived from the Soave equation is

$$\tilde{U} = \tilde{U}^I - \frac{3}{2\tilde{b}}\left[\tilde{a}(T) - T\frac{d\tilde{a}(T)}{dT}\right]\ln\left(\frac{\tilde{V} + \tilde{b}}{\tilde{V}}\right) \tag{4.5.28}$$

and from the PR equation is

$$\widetilde{U} = \widetilde{U}^I - \frac{1}{2^{3/2}\tilde{b}}\left[\tilde{a}(T) - T\frac{d\tilde{a}(T)}{dT}\right]\ln\left[\frac{\widetilde{V} + (1+2^{1/2})\tilde{b}}{\widetilde{V} + (1-2^{1/2})\tilde{b}}\right]. \tag{4.5.29}$$

The results of the PR equation is compared with the experiment and with the two-parameter equations in Figure 4.15. The PR equation is superior to the two-parameter models. So is the Soave equation, although it is not featured in Figure 4.15.

Sanchez and Lacombe (1976) have produced a three-parameter equation of state by a different approach. Instead of empirically modifying the VDW equation, they found an approximate solution to a fluid in which molecules occupy lattice points. The equation of state, which they call a lattice fluid (LF) equation, is

$$P = -\frac{RT}{v^*}\left[\ln\left(1 - \frac{rv^*}{\widetilde{V}}\right) + \left(1 - \frac{1}{r}\right)\frac{rv^*}{\widetilde{V}}\right] - \frac{\varepsilon r^2 v^*}{\widetilde{V}^2}, \tag{4.5.30}$$

where v^* and ε are volume and energy parameters and r a coordination parameter, related to the number of lattice sites assigned to a molecule. Instead of fitting their parameters to the critical point, Sanchez and Lacombe fitted them by least squares fit of pressure to vapor pressure data. These parameters are given in Table 4.4 for several alkanes. The molar energy derived from the LF model is

$$\widetilde{U} = \widetilde{U}^I - \frac{\varepsilon r^2 v^*}{\widetilde{V}}. \tag{4.5.31}$$

Mohanty et al. (1980) computed the vapor pressure, the coexistence liquid densities, and the heats of vaporization ($\equiv \widetilde{U}^V + P\widetilde{V}^V - \widetilde{U}^L - P\widetilde{V}^L$) with the PR and LF equations and compared these to experiment for several alkanes. The results are

TABLE 4.4 Lattice Fluid Parameters

Alkane	ε *(kcal/mol)*	v^* *(cm³/mol)*	*r*
CH_4	0.446	7.52	4.26
C_2H_6	0.627	8.00	5.87
C_3H_8	0.737	9.84	6.50
n-C_4H_{10}	0.802	10.40	7.59
n-C_5H_{12}	0.876	11.82	8.09
n-C_6H_{12}	0.947	13.28	8.37
n-C_8H_{18}	0.998	13.55	10.34
n-$C_{10}H_{22}$	1.053	14.47	11.75
n-$C_{12}H_{26}$	1.098	15.28	13.06
n-$C_{14}H_{30}$	1.132	15.99	14.36
n-$C_{17}H_{36}$	1.184	17.26	15.83

Figure 4.15
Experimental vapor pressure versus predictions of lattice fluid and equations.

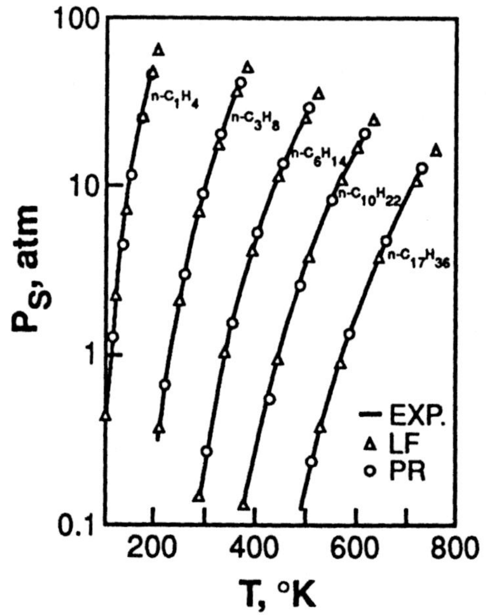

shown in Figures 4.16–4.18. The equations are of comparable accuracy (as also is the Soave equation) for the systems examined. The agreement between model and experiment is typical of alkane predictions.

Another well-studied empirical equation of state for gases is the following five-parameter Beattie–Bridgeman equation (1927), (1928):

$$P\tilde{V}^2 = RT\left(1 - \frac{c}{\tilde{V}T^3}\right)\left(\tilde{V} + B_o - \frac{bB_o}{\tilde{V}}\right) - A_o\left(1 - \frac{a}{\tilde{V}}\right). \tag{4.5.32}$$

The internal energy equation for this model is

$$\tilde{U} = \tilde{U}^I - \left(A_o + \frac{3Rc}{T^2}\right)\frac{1}{\tilde{V}} - \frac{1}{2\tilde{V}^2}\left(-aA_o + 3\frac{RcB_o}{T^2}\right) + \frac{RcbB_o}{T^2\tilde{V}^3}. \tag{4.5.33}$$

The model is accurate for gases up to pressures of the order of 250 atm. At higher pressures the equation performs more poorly than does the VDW equation. For example, the internal energy predicted by the Beattie–Bridgeman equation begins to deviate upward from the experiment on CO_2 at 150° at about 225 atm and errs a factor of 2.5 by about 1200 atm. In fact, at high pressures the Beattie–Bridgeman internal energy dependence on P is even qualitatively wrong, giving an increasing internal energy where a decreasing one is observed. Values for the constants A_o, B_o, a, b, and c have been determined for several gases. These are given in Table 4.5.

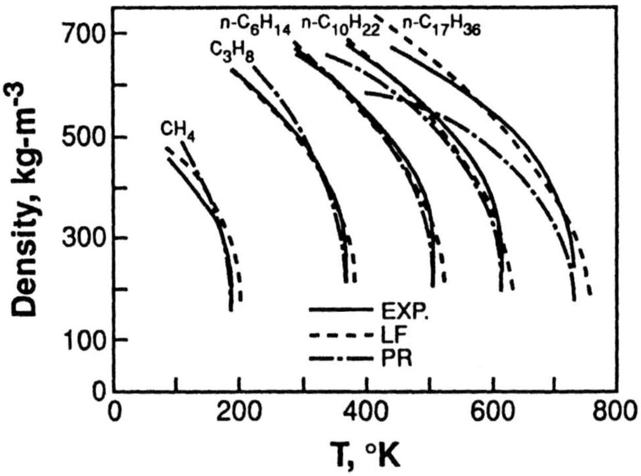

Figure 4.16
Experimental saturated liquid density versus predictions of lattice fluid and equations.

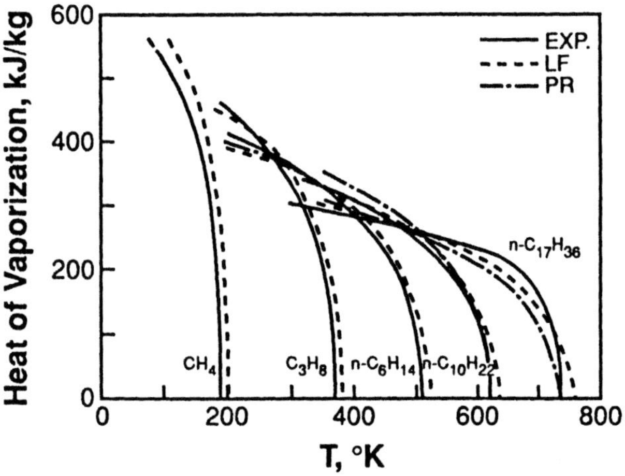

Figure 4.17
Experimental heats of vaporization versus predictions of lattice fluid and equations.

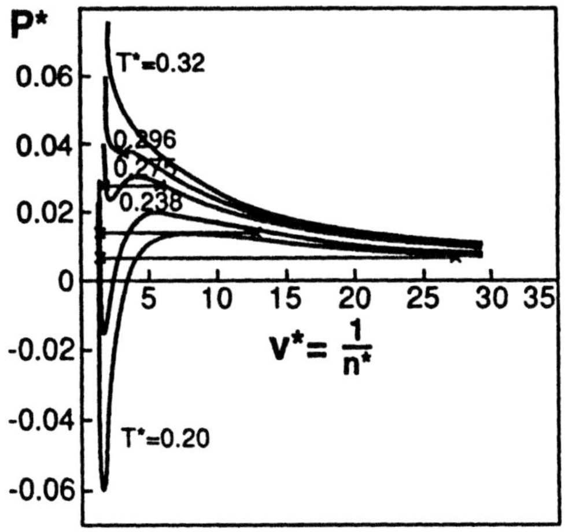

Figure 4.18
Reduced pressure $P^* = Pb^2/a$ versus reduced volume $v^* = 1/n^* = 1/nb$ of a one-component VDW fluid [eq. (4.6.22)] for several values of reduced temperature $T^* = bkT/a$. The symbols x represent the bulk liquid and vapor phase reduced volumes and are connected by Maxwell tie lines. $T^* = 0.296$ at the critical point.

TABLE 4.5 Values of Constants of Beattie–Bridgeman Equation of State for Several Gases

$$P = \left[RT(1-\delta)/\tilde{V}^2\right]\left[\tilde{V}+B\right] - A/\tilde{V}^2$$

$$A = A_o(1 - a/\tilde{V}) \quad B = B_o(1 - b/\tilde{V}), \quad \delta = c/\tilde{V}T^3$$

Gas	A_o	a	B_o	b	$c \times 10^{-4}$
He	0.0216	0.05984	0.01400	0.0	0.0040
Ne	0.2125	0.02196	0.02060	0.0	0.101
A	1.2907	0.02328	0.03931	0.0	5.99
Kr[b]	2.4230	0.02865	0.05261	0.0	14.89
Xe[c]	4.6715	0.03311	0.07503	0.0	30.02
H_2	0.1975	−0.00506	0.02096	−0.04359	0.504
N_2	1.3445	0.02617	0.05046	−0.00691	4.20
O_2	1.4911	0.02562	0.04624	0.004208	4.80
Air	1.3012	0.01931	0.04611	−0.01101	4.34
I_2	17.0	0.0	0.325	0.0	4000.
CO_2	5.0065	0.07132	0.10476	0.07235	66.00
NH_3	2.3930	0.17031	0.03415	0.019112	476.87
CH_4	2.2769	0.01855	0.05587	−0.15870	12.83
C_2H_4	6.1520	0.04964	0.12156	0.03597	22.68
C_2H_6	5.8800	0.05861	0.09400	0.01915	90.00
C_3H_8	11.9200	0.07321	0.18100	0.04293	120.00
1-C_4H_8	16.6979	0.11988	0.24046	0.10690	300.00
iso-C_4H_8	16.9600	0.10860	0.24200	0.08750	250.00
n-C_4H_{10}	17.7940	0.12161	0.24620	0.09423	350.00
iso-C_4H_{10}	16.6037	0.11171	0.23540	0.07697	300.00
n-C_5H_{12}	28.2600	0.15099	0.39400	0.13960	400.00
neo-C_5H_{12}	23.3300	0.15174	0.33560	0.13358	400.00
n-C_7H_{16}	54.520	0.20066	0.70816	0.19179	400.00
CH_3OH	33.309	0.09246	0.60362	0.09929	32.03
$(C_2H_5)_2O$	31.278	0.12426	0.45446	0.11954	33.33

Adapted from Hirschfelder et al. (1954) Units: normal atmospheres, liters/mole, K.

Benedict, et al. (1940, 1942) generalized the Beattie–Bridgeman equation to an eight-parameter equation that was able to fit hydrocarbon gaseous data up to densities twice critical density. Their equation is of the form

$$P = \frac{RT}{\tilde{V}} + \frac{1}{\tilde{V}^2}\left(RT\left(B_o + \frac{b}{\tilde{V}}\right) - \left(A_o + \frac{a}{\tilde{V}} - \frac{a\alpha}{\tilde{V}^4}\right)\right.$$

$$\left. - \frac{1}{T^2}\left[C_o - \frac{c}{\tilde{V}}\left(1 + \frac{\gamma}{\tilde{V}^2}\right)e^{-\gamma/\tilde{V}^2}\right]\right). \tag{4.5.34}$$

Parameters for the equation are given in Table 4.6 for several gases. The internal energy formula is

$$\widetilde{U} = \widetilde{U}^I + RT - P\widetilde{V} + \frac{1}{\widetilde{V}}\left(B_oRT - 2A_o - \frac{4C_o}{T^2}\right) + \frac{1}{2\widetilde{V}^2}(2bRT - 3a) + \frac{6a\alpha}{5\widetilde{V}^5}$$
$$+ \frac{c}{\widetilde{V}^2T^2}\left[\frac{3\widetilde{V}^2}{\gamma}\left(1 - e^{-\gamma/\widetilde{V}^2}\right) - \frac{1}{2}e^{-\gamma/\widetilde{V}^2} + \frac{\gamma}{\widetilde{V}^2}e^{-\gamma/\widetilde{V}^2}\right]. \qquad (4.5.35)$$

Modifications of the BWR equations involving as many as 14 parameters are used in thermodynamic modeling in industry.

The Tait equation (Gibson and Loeffler, (1939, 1949) and a variation introduced by Kirkwood (Richardson et al. (1947) are especially useful formulas for liquid state correlations. The Tait equation is of the form

$$\ln\left[\frac{P + L(T)}{P_o + L(T)}\right] = \frac{\widetilde{V}_o - \widetilde{V}}{K}, \qquad (4.5.36)$$

where K is a constant and $L(T)$ is a function of the temperature, P_o is some standard pressure and $\widetilde{V}_o$ the corresponding liquid volume. The equation has been successfully applied to quite a variety of liquids (water, benzene and its derivatives, salt solutions, etc.) over a wide range of temperatures and pressures (up to 1000 atm). Kirkwood's variation on the Tait equation, in which entropy is taken instead of temperature as an independent variable, has been shown to give excellent correlations of the equation of state behavior of water up to 25,000 atm. In spite of its success, the Tait equation has little theoretical justification.

For further discussion of empirical and theoretical equations of state (including equations applicable at very high pressures, say in the 10^3–10^6 atm range) the reader is referred to Hirschfelder et al. (1954).

TABLE 4.6 Empirical Constants for Benedict–Webb–Rubin Equation

Gas	A_o	B_o	$C_o \times 10^{-6}$	a	b	$c \times 10^{-6}$	$\alpha \times 10^3$	$\gamma \times 10^2$
Nitrogen	1.19250	0.0458000	0.00588907	0.0149000	0.00198154	0.000548064	0.291545	0.75000
Methane	1.85500	0.0426000	0.0225700	0.494000	0.00338004	0.00254500	0.124359	0.60000
Ethylene	3.33958	0.0556833	0.131140	0.259000	0.0086000	0.021120	0.178000	0.92300
Ethane	4.15556	0.0627724	0.179592	0.345160	0.0111220	0.0327670	0.243389	1.18000
Propylene	6.11220	0.0850647	0.439182	0.774056	0.0187059	0.102611	0.455696	1.82900
Propane	6.87225	0.0973130	0.508256	0.947700	0.0225000	0.129000	0.607175	2.20000
i-Butane	10.23264	0.137544	0.849943	1.93763	0.0424352	0.286010	1.07408	3.40000
i-Butylene	8.95325	0.116025	0.927280	1.69270	0.0348156	0.274920	0.910889	2.95945
n-Butane	10.0847	0.124361	0.992830	1.88231	0.0399983	0.316400	1.10132	3.40000
i-Pentane	12.7959	0.160053	1.74632	3.75620	0.0668120	0.695000	1.70000	4.63000
n-Pentane	12.1794	0.156751	2.12121	4.07480	0.0668120	0.824170	1.81000	4.75000
n-Hexane	14.4373	0.177813	3.31935	7.11671	0.109131	1.51276	2.81086	6.66849
n-Heptane	17.5206	0.199005	4.74574	10.36475	0.151954	2.47000	4.35611	9.00000

Adapted from Benedict et al. (1951). Units: atmospheres, liters, moles, K. Gas constants· $R = 0.08207$; $T = 273.13 + t$ (°C).

4.6 Computation of Liquid–Vapor Phase Diagrams from Equations of State

For studying phase behavior one needs the chemical potential as well as the equation of state. The Helmholtz free energy can be derived from the equation of state using the relationship

$$d(\widetilde{F} - \widetilde{F}^I)_T = -\left[P - \frac{RT}{\widetilde{V}}\right] d\widetilde{V} \tag{4.6.1}$$

and the chemical potential can be computed from the expression

$$\mu = \left(\frac{\partial F}{\partial N}\right)_{T,V}. \tag{4.6.2}$$

The ideal gas Helmholtz free energy F^I is given by

$$F^I = N\mu^+(T) + NkT \ln(N/V), \tag{4.6.3}$$

where $\mu^+(T)$ is the volume independent contribution to the free energy arising from the kinetic and internal energies (rotational, vibrational, electronic, and nuclear) of the molecules. μ^+ does not affect phase behavior. In molar form eq. (4.6.3) becomes

$$\widetilde{F}^I = \tilde{\mu}^+(T) - RT \ln \widetilde{V}. \tag{4.6.4}$$

For a Clausius fluid, $P_{\text{Cl}} = NkT/(V - Nb)$, and so eq. (4.6.1) yields for the free energy

$$F_{\text{Cl}} = N\mu^+ + NkT \ln[n/(1 - nb)] \tag{4.6.5}$$

and the chemical potential

$$\mu_{\text{Cl}} = \mu^+ + kT/(1 - nb) + kT \ln[n/(1 - nb)]. \tag{4.6.6}$$

For a hard sphere fluid, the Carnahan–Starling equation of state [eq. (4.5.15)] leads to the free energy

$$F_{HS} = N\mu^+ + NkT\,[\ln n - 1] + NkTy(4 - 3y)/(1 - y)^2 \tag{4.6.7}$$

and the chemical potential

$$\mu_{HS} = \mu^+ + kT \ln n + kTy(8 - 9y + 3y^2)/(1 - y)^3. \tag{4.6.8}$$

For the VDW, RK, Soave, and PR equations of state, the Helmholtz free energy and chemical potential are of the form

$$F = F_{\text{Cl}} + F_A \quad \text{and} \quad \mu = \mu_{\text{Cl}} + \mu_A, \tag{4.6.9}$$

where the attractive force contributions, F_A and μ_A, are given in Table 4.7. The hard sphere modification of these models is obtained by substituting F_{HS} and μ_{HS} for F_{Cl} and μ_{Cl} in eqs. (4.6.7) and (4.6.8).

TABLE 4.7 Contribution of Attractive Interactions to Helmholtz Free Energy and Chemical Potential for Several Equations of State

Model	F_A	μ_A
van der Waals	$-Nna$	$-2na$
Redlich–Kwong	$-\left(Na/bT^{1/2}\right)\ln(1+nb)$	$-\left(a/bT^{1/2}\right)\ln(1+nb)-na/\left[T^{1/2}(1+nb)\right]$
Soave	$-(Na/b)\ln(1+nb)$	$-(a/b)\ln(1+nb)-na/(1+nb)$
Peng–Robinson	$-\frac{Na}{2^{3/2}b}\ln\left[\frac{1+(1+2^{1/2})nb}{1+(1-2^{1/2})nb}\right]$	$-\frac{a}{2^{3/2}b}\ln\left[\frac{1+(1+2^{1/2})nb}{1+(1-2^{1/2})nb}\right]-\frac{na}{1+2nb-(nb)^2}$

The formulas for the Helmholtz free energy and the chemical potential of the lattice-fluid model of Sanchez and Lacombe are

$$F = N\mu^+(T) + NkT \ln n + \frac{NkT}{nv^*}(1 - nrv^*)\ln(1 - nrv^*) - Nn\varepsilon r^2 v^* \qquad (4.6.10)$$

and

$$\mu = \mu^+(T) + kT(1 - r + \ln n) - rkT \ln(1 - nrv^*) - 2n\varepsilon r^2 v^*. \qquad (4.6.11)$$

The parameters ε and v^* in these formulas are on a per molecule basis. They are smaller by a factor of Avogadro's number than the entries given in Table 4.4. The liquid–vapor phase diagram, is determined by finding the liquid and vapor densities, n_l and n_g, satisfying the equilibrium equations.

$$\mu(n_l) = \mu(n_g) \quad \text{and} \quad P(n_l) = P(n_g) \qquad (4.6.12)$$

at temperatures below the critical temperature. Equation (4.6.10) denotes a nonlinear system of equations of the form

$$\mathbf{f}(\mathbf{x}) = 0, \qquad (4.6.13)$$

where $\mathbf{f}$ is a vector function with components $f_1 = \mu(n_l) - \mu(n_g)$ and $f_2 = P(n_l) - P(n_g)$ and $\mathbf{x}$ is a vector with components $x_1 = n_l$ and $x_2 = n_g$. The Newton–Raphson method is efficient for solving a nonlinear system if a good first guess $\mathbf{x}^0$ of the solution can be found. According to this method $\mathbf{f}$ is expanded in a Taylor series about $\mathbf{x} = \mathbf{x}^0$ and truncated after the linear terms. Thus, the next estimate of $\mathbf{x}$ is obtained by solving the equation

$$\mathbf{J}(\mathbf{x} - \mathbf{x}^0) = -\mathbf{f}(\mathbf{x}^0), \qquad (4.6.14)$$

where the Jacobian $\mathbf{J}$ is a matrix with components

$$J_{ij} = \frac{\partial f_i}{\partial x_j}. \qquad (4.6.15)$$

The derivatives of f_i are evaluated at $\mathbf{x}^0$. The solution is obtained by iteration of eq. (2.7.14), that is, the kth estimate is obtained from the $k - 1$st estimate by solving

$$\mathbf{J}\left(\mathbf{x}^{k-1}\right)\left(\mathbf{x}^k - \mathbf{x}^{k-1}\right) = -\mathbf{f}\left(\mathbf{x}^{k-1}\right). \qquad (4.6.16)$$

An acceptable solution is obtained when $\|\mathbf{x}^k - \mathbf{x}^{k-1}\|$ and $\|\mathbf{f}(\mathbf{x}^{k-1})\|$ are smaller than some small, preset values. For convenience one can define the vector norm as $\|\mathbf{x}\| \equiv \left(\sum_i x_i^2\right)^{1/2}$.

For the one-component liquid–vapor phase diagram problem, the Jacobian is

$$\mathbf{J} = \begin{bmatrix} \frac{1}{n_l}\frac{\partial P(n_l)}{\partial n_l} & -\frac{1}{n_g}\frac{\partial P(n_g)}{\partial n_g} \\ \frac{\partial P(n_l)}{\partial n_l} & -\frac{\partial P(n_g)}{\partial n_g} \end{bmatrix} \tag{4.6.17}$$

In the first row of $\mathbf{J}$ we have used the Gibbs–Duhem equation, $nd\mu = dP$, to eliminate derivatives of the chemical potential.

When $\mathbf{x}^0$ lies within the radius of convergence, the Newton–Raphson equation is very efficient. Otherwise, it fails. Thus, a good first guess is essential. An effective strategy is to plot P versus $\widetilde{V}$ at the temperature $T = 0.8T_c$, use the Maxwell equal-area tie-line construction to get a first estimate of $\widetilde{V}_g$ and $\widetilde{V}_l$) (and thus of n_g and n_l), and use this estimate in the Newton–Raphson equations to obtain the coexistence densities at $0.8T_c$. Then continue the Newton–Raphson computations by advancing temperature in small increments toward higher and lower values. If the incremental change in temperature is small enough, the solution at the previous temperature provides an adequate first estimate of the solution at the next temperature.

In the computation of phase diagrams from equations of state it is always advantageous to put the variables in dimensionless form, either using the critical point quantities or parameters of the model for reducing to dimensionless form. For example, with the dimensionless variables

$$\mu_r = n_c\mu/P_c, \quad P_r = P/P_c, \quad T_r = T/T_c, \quad \text{and} \quad n_r = n/n_c, \tag{4.6.18}$$

the VDW equations become

$$P_r = \frac{8n_rT_r}{3 - n_r} - 3n_r^2 \tag{4.6.19}$$

and

$$\mu_r = \mu_r^+ + \frac{8T_r}{3 - nr} + \frac{8T_r}{3}\ln\left[\frac{n_r}{3 - n_r}\right] - 6n_r. \tag{4.6.20}$$

Or, with the dimensionless variables

$$\mu^* = \mu b/a, \quad P^* = Pb^2/a, \quad T^* = bkT/a, \quad \text{and} \quad n^* = nb, \tag{4.6.21}$$

the VDW equations become

$$P^* = \frac{n^*T^*}{1 - n^*} - n^{*2} \tag{4.6.22}$$

and

$$\mu^* = \mu^{+*} + \frac{T^*}{1 - n^*} + T^*\ln\left[\frac{n^*}{1 - n^*}\right] - 2n^*. \tag{4.6.23}$$

The phase diagram of the VDW fluid is shown in Figure 4.18 in the dimensionless units defined at eq. (4.6.21). The liquid-vapor binodal was determined by the Newton–Raphson method described above. Expressed in the dimensionless variables, the phase diagram is of course a universal diagram because eqs. (4.6.22) and (4.6.23) are free of parameters of a given fluid. This property is the law of corresponding states mentioned in the preceding section.

4.7 Microscopic Derivation of Law of Corresponding States

For a one-component system of particles interacting with central, pairwise additive forces that are independent of the internal degrees of freedom, the pressure is given by

$$P = kT\left(\frac{\partial \ln Z_N}{\partial V}\right)_{T,N} \tag{4.7.1}$$

where

$$Z_N = \int_V \cdots \int e^{-\beta u_N} d^3r_1 \cdots d^3r_N \tag{4.7.2}$$

and

$$u_N = \sum_{i>j=1}^{N} u(r_{ij}). \tag{4.7.3}$$

We assume that for the class of fluids of interest the pair potential is of the form

$$u(r) = \varepsilon u^*(r/\sigma), \tag{4.7.4}$$

where $u^*(x)$ is the same function as the argument x for all fluids in the class.

Because pressure is independent of shape for a large system, the shape can be chosen to be a cube of length $L = V^{1/3}$ on a side. In eq. (4.7.3) let us introduce Cartesian coordinates so that $d^3r_i = dx_i dy_i dz_i$ and

$$Z_N = \int_0^{V^{1/3}} \cdots \int_0^{V^{1/3}} e^{-\beta \sum_{i>j=1}^{N} u(r_{ij})} dx_1 dy_1 dz_1 \cdots dx_n dy_N dz_N. \tag{4.7.5}$$

Introducing the dimensionless variables

$$\rho_i = r_i/\sigma, \quad \xi_i = x_i/\sigma, \quad \zeta_i = y_i/\sigma, \quad \eta_i = z_i/\sigma,$$
$$T^* = kT/\varepsilon, \quad \text{and} \quad V^* = V/\sigma^3, \tag{4.7.6}$$

we obtain

$$Z_N = \sigma^{3N} \int_0^{(V^*)^{1/3}} \cdots \int_0^{(V^*)^{1/3}} e^{-\sum_{i>j=1}^{N} u^*(\rho_{ij})/T^*} d\xi_1 \cdots d\eta_N$$
$$= \sigma^{3N} Z_N^*. \tag{4.7.7}$$

Z_N^* depends only on the dimensionless variables V^* and T^* at a given N. Thus, eq. (4.7.1) becomes

$$P = kT\left(\frac{\partial \ln Z_N^*}{\partial V}\right)_{T^*,N} = \frac{kT}{\sigma^3}\left(\frac{\partial \ln Z_N^*}{\partial V^*}\right)_{T^*,N} \tag{4.7.8}$$

Introducing the dimensionless pressure,

$$P^* = \frac{P\sigma^3}{\varepsilon}, \tag{4.7.9}$$

we obtain

$$P^* = T^*\left(\frac{\partial \ln Z_N^*}{\partial V^*}\right)_{T^*,N} \tag{4.7.10}$$

This, plus the fact that pressure is an intensive variable, implies that P^* is a universal function of $\tilde{V}^* = V^*/N$ and T^* for the class of fluids having pair potentials of the form of eq. (4.7.4). The 6–12 Lennard–Jones potential provides an example of such a class.

The implications of this result are that (1) P^* will be the same for all fluids compared at the same values of $\tilde{V}^*$ and T^* and (2) that a macroscopic law of corresponding states holds. Because P^* is a universal function of $\tilde{V}^*$ and T^* and because the critical point is a point in a universal phase diagram, it follows that

$$P_c^* = \frac{P_c\sigma^3}{\varepsilon}, \quad \tilde{V}_c^* = \tilde{V}_c/\sigma^3, \quad \text{and} \quad T_c^* = kT_c/\varepsilon \tag{4.7.11}$$

are universal constants. Thus, because $\sigma^3 = \tilde{V}_c/\tilde{V}_c^*$, $\varepsilon = T_c^*/kT_c$, and $\varepsilon/\sigma^3 = P_c/P_c^*$, if we define the macroscopic dimensionless variables, $Pr = P/P_c$, $\tilde{V}_r = \tilde{V}/\tilde{V}_c$, and $T_r = T/T_c$, it follows that

$$P_r = \frac{P}{P_c} = \frac{P^*}{P_c^*} = \frac{P^*\left(\tilde{V}_c^*\tilde{V}_r, T_c^*T_r\right)}{P_c^*}. \tag{4.7.12}$$

Thus, because P_c^*, $\tilde{V}_c^*$, and T_c^* are universal constants, it follows that the rhs of eq. (4.7.12) is a universal function of $\tilde{V}_r$ and T_r. This establishes the macroscopic law of corresponding states for all fluids obeying a pair potential of the form of eq. (4.7.4).

Supplementary Reading

Egelstaff, P. A. 1967. *An Introduction to the Liquid State,* Academic Press, New York.

Guggenheim, E. A. 1957. *Thermodynamics: An Advanced Treatment for Chemists and Physicists,* Interscience, New York.

Kestin, J. and Dorfman, J. R. 1971. *A Course in Statistical Thermodynamics,* Academic Press, New York.

Rice, S. A. and Gray, P. 1965. *The Statistical Mechanics of Simple Liquids,* Interscience, New York.

Sandler, S. I. 1989. *Chemical and Engineering Thermodynamics,* Wiley, New York.

Exercises

1 Consider a system consisting of N particles of magnitude spin 1/4 fixed on lattice sites. Assume the spins interact only with an external magnetic field H. In the presence of the field, each spin has two energy states, μH or $-\mu H$, where μ is the magnetic moment of the spin.
Find the canonical ensemble partition function Z_N, the internal energy U, the entropy S, and the heat capacity.

2 Evaluate Q_N for an ideal gas in the absence of external fields. Derive from the expression

$$p_N(\mathbf{r}_1, \ldots, \mathbf{v}_N)(d^3r)^N(d^3v)^N = \frac{e^{-H/kT}(d^3r)^N(d^3v)^N}{Q_N}$$

an expression for $p_N(\mathbf{r}, \ldots, \mathbf{r}_N)$ where $p_N(\mathbf{r}_1, \ldots, \mathbf{r}_N)d^3r_1, \ldots, d^3r_N$ is the probability that one particle is located between $\mathbf{r}_1$ and $\mathbf{r}_1 + d\mathbf{r}_1$, another at $\mathbf{r}_2$ and $\mathbf{r}_2 + d\mathbf{r}_2$, etc. Evaluate the function p_N, determined above, for the case of an ideal gas in the absence of external fields. Does the result make sense intuitively? Explain how.

3 Using a Cartesian coordinate representation of ∇_1, that is,

$$\nabla_1 = \hat{\imath}\frac{\partial}{\partial x_1} + \hat{\jmath}\frac{\partial}{\partial y_1} + \hat{k}\frac{\partial}{\partial z_1}$$

show that

$$\nabla_1 r_{12} = \frac{\mathbf{r}_{12}}{r_{12}}.$$

Prove that $g^{(2)}(\mathbf{r}_1, \mathbf{r}_2)$, the pair correlation function, is a function only of $r_{12} = |\mathbf{r}_1 - \mathbf{r}_2|$ in systems interacting via pairwise additive centrally symmetric forces.
Hint: Proceed by evaluating the integral through judicious selection of coordinate systems.

4 The configuration partition function for a VDW fluid can be approximated by

$$Z_N \simeq \exp\left\{\frac{N^2a}{VkT}\right\}(V - Nb)^N$$

where a and b are parameters specific to a particular gas. Using the partition function given above, derive an expression for the equation of state, the Helmholtz free energy, entropy and internal energy of a VDW fluid.

5 The critical temperature and critical pressure of propane are 370K and 42 atm, respectively. Assuming the VDW equation of state to be valid, predict the following quantities:

a) the critical molar volume;

b) the critical compressibility factor, Z_c;

c) the entropy, energy, and heat (enthalpy) of vaporization at 1 atm.

Make a table comparing the predicted values with experimental values. (Smith and van Ness, 19XX, and Perry, 19XX, are likely sources for the experimental data.)

6 If the gas is propane at 30 atm, what is the temperature at which the Joule–Thompson coefficient, μ_{JT}, changes sign?
Suppose propane gas at 30 atm and 450K passes through a throttle valve to a pressure of 1 atm. Assuming that propane obeys the VDW equation of state, does the gas cool down? Assuming that propane obeys the PR equation of state, does the gas cool down? Estimate the temperature of the throttled gas. The low pressure constant pressure heat capacity of propane is:

$$C_P^* = \left[2.41 + 57.195 \times 10^{-3}T - 17.533 \times 10^{-6}T^2\right] \frac{\text{cal}}{\text{g} - \text{molK}}.$$

7 From the appropriate (thermodynamic) Maxwell relations, demonstrate the validity of the following equations:

$$dS = \frac{C_v}{T}dT + \left(\frac{\partial P}{\partial T}\right)_{V,N} dV$$

$$= \frac{C_P}{T}dT - \left(\frac{\partial V}{\partial T}\right)_{P,N} dP$$

$$dU = C_v dT + \left[T\left(\frac{\partial P}{\partial T}\right)_{V,N} - P\right] dV$$

$$dH = C_P dT + \left[V - T\left(\frac{\partial V}{\partial T}\right)_{P,N}\right] dP.$$

Given that a system obeys the VDW equation of state, prove that the constant volume specific heat C_v, is a function only of temperature and that this result leads to the following form for the thermodynamic energy.

$$u = \int C_v(T)dT - \frac{N^2 a}{V} + \text{constant}.$$

Using the VDW equation of state and the critical point conditions, show that

$$\tilde{V}_c = 3\tilde{b}, \quad T_c = \frac{8\tilde{a}}{27R\tilde{b}}, \quad P_c = \frac{\tilde{a}}{27\tilde{b}^2},$$

and, therefore

$$\tilde{a} = \frac{27R^2T_c^2}{64P_c}, \tilde{b} = \frac{RT_c}{8P_c}.$$

Show that the envelope of the unstable region predicted by the VDW equation of state is of the form

$$P = \frac{\tilde{a}}{\tilde{V}^2}\left(1 - \frac{2\tilde{b}}{\tilde{V}}\right).$$

8 Estimate the atomic radius of argon from the critical point parameters and compare this with the handbook value.

9 Assume that the pair potential obeyed by Ar and N_2 is of the form

$$u(r) = \varepsilon\left[\left(\frac{\sigma}{r}\right)^{12} - \left(\frac{\sigma}{r}\right)^{6}\right]$$

where ε and σ are constants and ε/k, σ = 119.8K, 3.405 Å, and 95.05K, 3.698 Åfor Ar and N_2, respectively. If the second virial coefficient B_2 is −46.93 Å^3/molecule for nitrogen at 223K, what is the value of B_2 for Ar at 281K?

10 Liquid mercury at 1 atm pressure and 0°C has a molar volume of 14.72 cm^2/mol and a specific heat of constant pressure of $C_p = 28$ J/molK. Its coefficient of expansion is $\alpha = 1.81 \times 10^{-4}$ K^{-1} and its isothermal compressiblity is $\kappa = 3.88 \times 10^{-12}$ cm^2/dyn. Estimate the constant volume specific heat C_v of liquid mercury.

11 Compute the density versus temperature curves along which the Joule–Thomson coefficient equals zero using the Dieterici, RK, and Soave equations of state.

12 Find the second virial coefficient for the VDW, Berthelot, Dieterici, RK, PR, Soave, Sanchez–Lacombe, Beattie–Bridgeman, and HSRK equations of state.

13 The Boyle temperature is the temperature at which the second virial coefficient equals zero. For pentane, estimate the Boyle temperature from the equations of state listed in Problem 11.

References

Beattie, J. A. and Bridgeman, O. C. 1965. *J. Am. Chem. Soc.*, **49,** 1665; 1928. *Proc. Am. Acad. Arts Sci.*, **63,** 229.

Beattie, J. A. and Stockmayer, W. H. 1940. *Reports Prog. Phys.* **7,** 195.

Benedict, M., Webb, G. B., and Rubin, L. C. 1940. *J. Chem. Phys.*, **8,** 334; 1942. **10,** 747.

Benedict, M., Webb, G. B., and Rubin, L. C. 1951. *Chem. Eng. Prog.*, **47,** 419.

Carnahan, N. F. and Starling, K. E. 1969. *J. Chem. Phys.*, **51,** 635.

Carnahan, N. F. and Starling, K. E. 1972. *AIChE J.*, **18,** 1184.

Clausius, R. J. E. 1857. *Ann. Phys.*, **100,** 353.

Gibson, R. E. and Loeffler, O. H. 1949. *Ann. N.Y. Acad. Sci.*, **51,** 727; 1939. *J. Am. Chem. Soc.*, **61,** 2515.

Hirn, J. 1863. *Cosmos,* **22,** 283.

Hirschfelder, J. O., Curtiss, C. F., and Bird, R. B. 1954. *Molecular Theory of Gases and Liquids,* Wiley, New York.

Horvath, A. L. 1972. *Chem. Eng. Sci.*, **27,** 1185.

Hsu, C. C. and McKetta, J.J. 1972. *J. Chem. Eng. Sci.*, **27,** 1197.

Khan, A. A. 1964. *Phys. Rev.*, **134,** 367.

Mohanty, K. K., Dombrowski, M., and Davis, H. T. 1980. *Chem. Eng. Commun.,* **5,** 85.

Peng, D.-Y. and Robinson, D. B. 1976. *Ind. Eng. Chem. Fund.,* **15,** 59.

Reid, R. C. and Sherwood, T. K. 1966. *The Properties of Gases and Liquids,* McGraw–Hill, New York.

Richardson, J. M., Arons, A. B., and Halverson, R. R. 1947. *J. Chem. Phys.,* **15,** 785.

Sanchez, I. C. and Lacombe, R. H. 1976. *J. Chem. Phys.,* **80,** 2352.

Soave, G. 1972. *Chem. Eng. Sci.,* **27,** 1197.

5

IMPERFECT GASES

5.1 Virial Expansion of Thermodynamic Functions

In the dilute gas limit, the equation of state is $P = nkT$, with $n = N/V$ and $N = \sum_{\alpha=1}^{c} N_\alpha$, where N_α is the number of particles of species α. In this chapter, we shall denote the set of mole fractions by either $x_1, \ldots, x_c$ or $\{\mathbf{x}\}$. At finite densities it is sometimes useful to express P in the following power series, called the virial expansion,

$$\frac{P}{nkT} = 1 + \sum_{i=2}^{\infty} n^{i-1} B_i, \tag{5.1.1}$$

where B_i is the so-called ith virial coefficient. In general the B_i depend on T and $\{\mathbf{x}\}$, but not on n.

In principle, a power series such as eq. (5.1.1) may be employed at all densities for which it converges. At a gas–liquid or gas–solid phase boundary, the virial expansion must fail because the pressure becomes a double valued function of density in a coexistence region.

The virial expansion is most useful when eq. (5.1.1) can be well approximated by neglecting terms of order n^3 and higher, that is, when

$$\frac{P}{nkT} \simeq 1 + B_2 n + B_3 n^2. \tag{5.1.2}$$

Higher order virial coefficients become quite difficult to determine experimentally and are expensive to compute theoretically for any molecular model but the hard sphere model.

Instead of expanding P/nkT as a power series in density, we may equally well expand it in a power series of P/kT, that is,

$$\frac{P}{nkT} = 1 + \sum_{i=2}^{\infty} \left(\frac{P}{kT}\right)^{i-1} B_i'. \tag{5.1.3}$$

The "pressure" virial coefficient B_i' can be related to the "density" virial coefficients B_i. B_i' is a function of B_j, $j = 1, \ldots, i$. The relationship is easily established by noting that eq. (5.1.3) can be rearranged into the form

$$n = f(y) \tag{5.1.4}$$

with the definitions

$$y = P/kT \tag{5.1.5}$$

and

$$f(y) = \frac{y}{1 + \sum_{i=2}^{\infty} y^{i-1} B_i'}. \tag{5.1.6}$$

Expanding eq. (5.1.4) in a Taylor's series about $n = 0$, we obtain

$$n = \sum_{k=0}^{\infty} \frac{n^k}{k!} \left[\frac{\partial^k f(y)}{\partial n^k} \right]_{n=0}. \tag{5.1.7}$$

Setting the coefficient of each power of n equal to zero [because eq. (5.1.7) is valid for a continuous range of n], we find

$$\begin{aligned} [f(y)]_{n=0} &= 0 \\ \left[\frac{\partial f(y)}{\partial n} \right]_{n=0} &= 1 \\ \left[\frac{\partial^k f(y)}{\partial n^k} \right]_{n=0} &= 0, \quad k > 1. \end{aligned} \tag{5.1.8}$$

A term by term evaluation of these equations, using eq. (5.1.1) to determine the value of $[\partial^k y/\partial n^k]_n = 0$, will lead to the desired relationship between B_i and B_i'. For example, because $[y]_{n=0} = 0$, $[\partial y/\partial n]_{n=0} = 1$ and $[\partial^k y/\partial n^k]_{n=0} = k!B_k$, $k > 1$, it is straightforward to show that eq. (5.1.8) yields the results

$$\begin{aligned} B_2' &= B_2 \\ B_3' &= B_3 - B_2^2 \\ B_4' &= B_4 - 3B_2B_3 + 2B_2^3. \end{aligned} \tag{5.1.9}$$

As the choice between eqs. (5.1.1) and (5.1.3) will be determined by thermodynamic convenience, it is useful to have formulas relating the B_i's and B_i''s.

With the aid of eq. (5.1.1) and the thermodynamic relation

$$(dU)_{T,N} = \left[T\left(\frac{\partial P}{\partial T}\right)_{V,N} - P \right] dV, \tag{5.1.10a}$$

$$U = U^I + NkT^2 \sum_{i=2}^{\infty} \left(\frac{\partial B_i}{\partial T}\right) \frac{n^{i-1}}{i-1}. \tag{5.1.10b}$$

Similarly, eq. (5.1.3) and the thermodynamic relation

$$(dH)_{T,N} = \left[V - T\left(\frac{\partial P}{\partial V}\right)_{T,N} \right] dP \tag{5.1.11a}$$

combine to yield

$$H = H^I - NkT^2 \sum_{i=2}^{\infty} \left[\frac{\partial}{\partial T} \left(\frac{B'_i}{(kT)^{i-1}} \right) \right] \frac{P^{i-1}}{i-1}, \qquad (5.1.11b)$$

where H is the enthalpy ($= U + PV$). Similarly, the entropy can be put in the form

$$S = S^I(\mathbf{N}, T, P) - Nk \sum_{i-2}^{\infty} \left\{ \frac{B'_i}{(kT)^{i-1}} + T \left[\frac{\partial}{\partial T} \left(\frac{B'_i}{kT} \right) \right] \right\} \frac{P^{i-1}}{i-1}, \qquad (5.1.12)$$

or

$$S = S^I(\mathbf{N}, T, V) - Nk \sum_{i=2}^{\infty} \left[B_i + T \left(\frac{\partial B_i}{\partial T} \right) \right] \frac{n^{i-1}}{i-1}. \qquad (5.1.13)$$

U^I, H^I, and S^I are the energy, enthalpy and entropy of the fluid in the ideal gas state. Because $F = U - TS$ and $G = H - TS$, eqs. (5.1.10)–(5.1.13) can be used to get the Helmholtz and Gibbs free energy as functions of the virial coefficients.

Another frequently used thermodynamic quantity is the fugacity f_α that is related to the chemical potential by the defining expression

$$\mu_\alpha = \mu_\alpha^+ + kT \ln f_\alpha \qquad (5.1.14)$$

where μ_α^+ is the chemical potential of pure α in the ideal gas state at temperature T and at a pressure of 1 atm. Using the thermodynamic relation

$$(d\mu_\alpha)_{T,\mathbf{N}} = \left(\frac{\partial V}{\partial N_\alpha} \right)_{T,P,N_{S \neq \alpha}} dP, \qquad (5.1.15)$$

one can derive for f_α the expression

$$\begin{aligned} \ln(f_\alpha / x_\alpha P) &= \frac{1}{kT} \int_0^P \left[\left(\frac{\partial V}{\partial N_\alpha} \right)_{T,P,N_{\beta \neq \alpha}} - \frac{kT}{P} \right] dP \\ &= \int_0^P \left[N \left(\frac{\partial Z}{\partial N_\alpha} \right)_{T,P,N_{\beta \neq \alpha}} + Z - 1 \right] \frac{dP}{P}, \end{aligned} \qquad (5.1.16)$$

where the compressibility factor Z is defined by $Z = P/nkT$. Inserting into this equation the pressure virial expansion for Z and carrying out the integrations over P, we get

$$\ln(f_\alpha / x_\alpha P) = \sum_{i-2}^{\infty} \frac{1}{i-1} \left(\frac{P}{i-1} \right)^{i-1} \left[N \left(\frac{\partial B'_i}{\partial N_\alpha} \right)_{T,N_{\beta \neq \alpha}} + B'_i \right]. \qquad (5.1.17)$$

For a one-component fluid, B_i' depends only on temperature and eq. (5.1.17) reduces to the simple form

$$\ln(f_\alpha/P) = \sum_{i=1}^{\infty} \frac{B_i'}{i-1} \left(\frac{P}{kT}\right)^{i-1}. \tag{5.1.18}$$

In the remainder of this section we concentrate on the second and third virial coefficients. Plots of the second virial coefficients (plotted as $\tilde{B}_2 \equiv N_o B_2$, N_o = Avogadro's number) versus temperature are shown in Figures 5.1 and 5.2 for nitrogen and methyl chloride, respectively. The negative and positive regions of $\tilde{B}_2$ arise from a balance of the effects of the attractive and repulsive interactions, a balance that shifts in favor of the repulsive interactions as the kinetic energy of the molecules becomes large compared to the attractive potential energy. The following example demonstrates this balance in terms of the van der Waals model.

Example

Determine B_i for van der Waals equation of state. Writing the equation in the form

$$\frac{P}{nkT} = \frac{1}{1-nb} - \frac{na}{kT}, \tag{5.1.19}$$

inserting for $(1-nb)^{-1}$ the series expansion

$$\frac{1}{1-nb} = \sum_{j=0}^{\infty} (nb)^j, \tag{5.1.20}$$

and comparing with eq. (5.1.1), we obtain

$$B_2 = b - \frac{a}{kT} \tag{5.1.21}$$

and

$$B_j = b^{j-1}, \quad j > 2. \tag{5.1.22}$$

Note that $B_2 < 0$ for $T < a/kb$ and $B_2 > 0$ for $T > a/kb$, in qualitative agreement with Figures 5.1 and 5.2. In the density region where $P/nkT \approx 1 + B_2 n$, the ideal gas law holds for the temperature $T_B = a/kb$. The temperature at which $P/nkT = 1$ is known as the Boyle temperature.

For nitrogen, the van der Waals formulas, $N_o b = RT_c/8P_c$ and $N_o^2 a = 27 P_c N_o^2 b^2$, yield the values $N_o b =$ 38.5 cm^2/mol and $N_o^2 a = 1.34 \times 10^6 cm^6$ atm/(mol)^2. With these values $\tilde{B}_2$ ($\equiv N_o B_2$) is compared to experiment in Figure 5.1. The agreement is better than one expects given the crudeness of the van der Waals model.

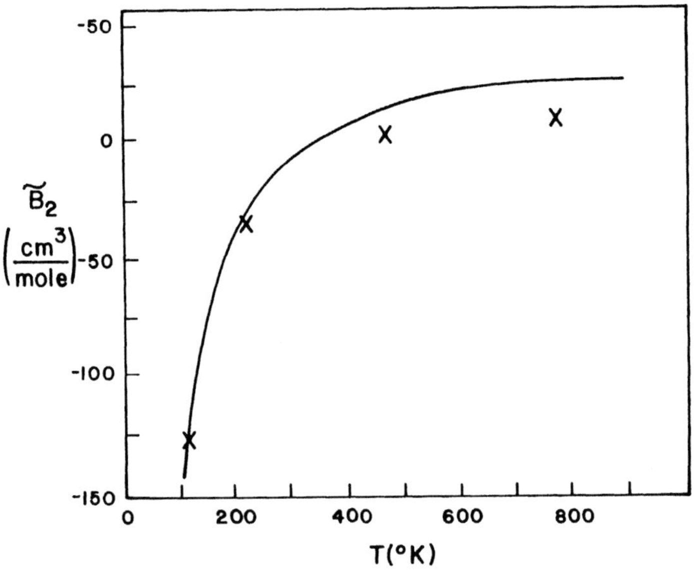

Figure 5.1
Second virial coefficient versus temperature for nitrogen. Curve represents a smooth curve drawn through the data of Holborn and Otto, (1925). Crosses are from van der Waals theory.

The third virial coefficient is plotted in Fig. 5.3. The conversion from the variables $B_3/(r^*)^6$ and kT/ε to the macroscopic corresponding state variables $N_o^2 B_3/(v_c)^2$ and T/T_c can be done using the parameters in Table 5.1. The qualitative features of the plot will be unchanged. In the units used in Figure 5.3, the van der Waals prediction is that $B_3/(r^*)^6 = 1/2$. Experimentally this quantity varies sharply with temperature. The 6–12 Lennard–Jones (LJ) model prediction of B_3 is also shown in Figure 5.3. The theory will be discussed in a later section.

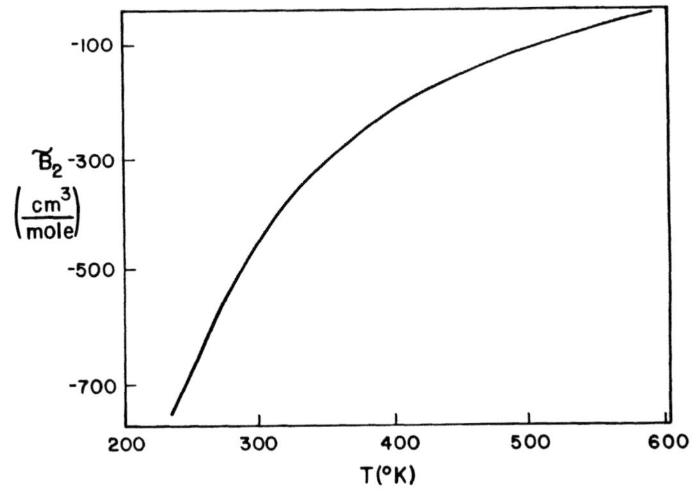

Figure 5.2
Second virial coefficient versus temperature for methyl chloride. Plot represents a smooth curve drawn through the data of Hirschfelder et al. (1942).

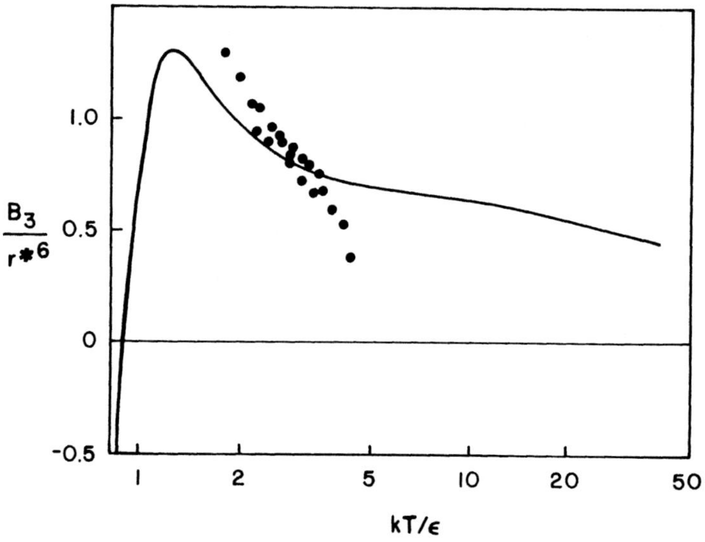

Figure 5.3
B_3/r^{*6} as a function of kT/ε, calculated from theory for the 6–12 LJ potential. The experimental points are a mixture of points for A, N_2, and CH_4.

If a class of fluids obeys a law of corresponding states, so also will the virial coefficients. McGlashan and Potter (1962) found that the empirical expression ($T_r \equiv T/T_c$)

$$\frac{B_2}{V_c} = 0.430 - 0.886T_r^{-1} - 0.694T_r^{-2}, \tag{5.1.23}$$

where V_c is the critical volume per molecule, correlates the second virial coefficients' data quite well for methane, argon, krypton, and xenon. For the normal paraffins and α-olefins, McGlashan et al. found that the expression

$$\frac{B_2}{V_c} = 0.430 - 0.886T_r^{-1} - 0.694T_r^{-2} - 0.0375(\nu - 1)T_r^{-4.5}, \tag{5.1.24}$$

correlates data well. ν represents the number of carbon atoms in the molecules considered. Figure 5.4 illustrates the agreement between eq. (5.1.24) and thee experiment.

TABLE 5.1 Values of l

Molecule	l
Argon	0.0
Hydrogen	0.0
Xenon	0.6
Nitrogen	1.0
Ethane	1.0
Neopentane	1.8
Benzene	2.5
n-Octane	4.25

Figure 5.4
Corresponding states correlation of McGlashan *et al.* for second virial coefficients of normal paraffins α-olefins. Adapted from Prausnitz 1969.

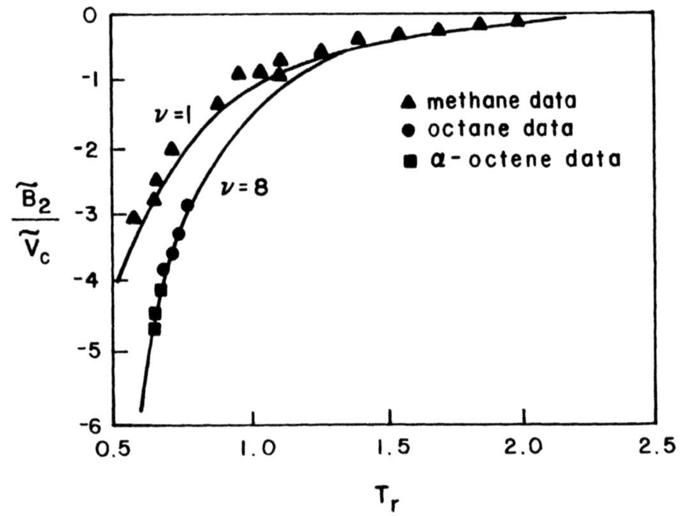

Chueh and Prausnitz (1967) found that the third virial coefficient data for many nonpolar fluids can be correlated well with the equation

$$\frac{B_3}{\tilde{V}_c^2} = (0.232T_r^{-0.25} + 0.464T_r^{-5})(1 - e^{(1-1.89T_r^2)}) + le^{-(2.49-2.30T_r+2.70T_r^2)}, \qquad (5.1.25)$$

where the quantity depends on the fluid. Some values of l are found in Table 5.2. At temperatures such that $T_r \geq 1.5$, eq. (5.1.25) is essentially independent of l, so that the parameter need not be obtained for high temperature correlations. In range $T < 1.5$, one should try to guess the value of l for molecules not listed. The parameter l seems to vary with the polarizability α of the molecules. Thus, one might plot l versus α for the molecules given and estimate l for an unknown molecule from its polarizability. The polarizabilities of ethane and n-octane are 4.4×10^{-24} cm^3 and 15.2×10^{-24} cm^3, respectively, being in the ratio 1:3.46, whereas the ratio of l's is 1:4.5 for the same molecules.

TABLE 5.2 Reduced Critical Constants

Fluid	$\frac{\tilde{V}_c}{r^{*3}}$	$\frac{kT_c}{\varepsilon}$	$\frac{P_c r^{*3}}{\varepsilon}$	$\frac{P_c \tilde{V}_c}{kT_c}$
Ne	2.35	1.25	0.157	0.295
A	2.23	1.26	0.164	0.290
Xe	2.05	1.31	0.187	0.293
N_2	2.09	1.33	0.185	0.291
O_2	1.90	1.33	0.201	0.292
CH_4	2.09	1.29	0.178	0.288
Average	2.12	1.29	0.179	0.292

In a c-component mixture, the second and third virial coefficients are of the forms

$$B_2 = \sum_{\alpha,\beta} x_\alpha x_\beta B_2^{\alpha\beta} \tag{5.1.26}$$

and

$$B_3 = \sum_{\alpha,\beta,\gamma} x_\alpha x_\beta x_\gamma B_3^{\alpha\beta\gamma}, \tag{5.1.27}$$

where $B_2^{\alpha\beta}$ and $B_3^{\alpha\beta\gamma}$ depend only on the temperature T.

Attempts have been made to relate the mixed virial coefficients to those measured in pure systems. One such scheme, based on insight gained from the theoretical expressions for the virial coefficients, is to estimate $B_2^{\alpha\beta}$ from the pure fluid correlation, eq. (5.1.29), by assuming that $B_2^{\alpha\beta}$ is equal to the second virial coefficient of a pseudo-one-component fluid whose critical parameters are given by

$$\tilde{V}_c^{\alpha\beta} = \frac{1}{8}\left[\left(\tilde{V}_c^{\alpha}\right)^{1/3} + \left(\tilde{V}_c^{\beta}\right)^{1/3}\right]^3 \tag{5.1.28}$$

and

$$T_c^{\alpha\beta} = \left(T_c^{\alpha} T_c^{\beta}\right)^{1/2}, \tag{5.1.29}$$

where $\tilde{V}_c^{\alpha}$ is the molecular critical volume (volume/molecule) of species α. In Table 5.3 experimental values of B_2 are compared to values predicted by using eqs. (5.1.28) and (5.1.29) to estimate $B_2^{\alpha\beta}$ from eq. (5.1.23).

For the more complicated molecules obeying eq. (5.1.24), Prausnitz suggests that in addition to eqs. (5.1.28) and (5.1.29) the mixing rule

$$\nu^{\alpha\beta} = \left(\nu^{\alpha} + \nu^{\beta}\right)/2 \tag{5.1.30}$$

be used to define a pseudo-one-component fluid obeying eq. (5.1.24), where ν^{α}, ν^{β} are the pure fluid values of ν. This scheme has been shown to give good agreement for mixtures of propene and α-heptane, and propane and octane.

TABLE 5.3 Equimolar Argon-Nitrogen Mixture

T (°C)	N_0B_2 (Calcd) (cm³/mol)	N_0B_2 (Exp.) (cm³/mol)
0	−16.5	−16.3
− 70	−41.3	−40.4
−130	−89.4	−88.3

N_0 = Avogadro's number. Adapted from Prausnitz (1969).

Prausnitz states that the following approximation is the best available method for estimating $B_3^{\alpha\beta\gamma}$ from pure fluid data:

$$B_3^{\alpha\beta\gamma} \approx \left(B_3^{\alpha\beta} B_3^{\alpha\gamma} B_3^{\beta\gamma}\right)^{1/3}, \tag{5.1.31}$$

where $B_3^{\nu\mu}$ is obtained from eq. (5.1.25) for the pseudo-one-component fluid defined by eqs. (5.1.28) and (5.1.29) with

$$l^{\nu\mu} \equiv (l^{\nu} + l^{\mu})/2. \tag{5.1.32}$$

Predictions from this scheme are shown in Table 5.4.

Finally let us discuss the practical problem of reduction of data for determination of the virial coefficients. The virial expansion of $\mathcal{Z}$ can be rearranged into the form

$$\frac{1}{n}(\mathcal{Z} - 1) = B_2 + nB_3 + \cdots. \tag{5.1.33}$$

Thus, if $n^{-1}(\mathcal{Z} - 1)$ is plotted versus n, then the intercept and slope at $n = 0$ yield B_2 and B_3, respectively. Having determined B_2 and B_3, we can then plot $n^{-1}[n^{-1}(\mathcal{Z} - 1) - B_2]$ versus n to obtain a curve whose intercept is B_3 (a check on the previous step) and slope is B_4. The process may be continued as far as desired, subject of course to the uncertainties or limitations of available data.

In Figures 5.5 and 5.6, the observed values of $\mathcal{Z}$ versus n or P are compared to the approximations

$$\mathcal{Z} = 1 + B_2 n + B_3 n^2 \tag{5.1.34}$$

and

$$\mathcal{Z} = 1 + B_2'(P/kT) + B_3'(P/kT)^2, \tag{5.1.35}$$

respectively. It is interesting that although eqs. (5.1.34) and (5.1.35) are both of second order in the expansion parameter, the density

TABLE 5.4 Experimental and Calculated Third Virial Cross-Coefficients for Some Binary Mixtures

Components 1	2	*Temp.* (K)	$N_A^2 B_3^{122}$ (cm^3/mole)^2 *Experimental*	*Calculated*
Ar	N_2	273	1349	510
		203	1706	1770
		163	2295	2420
N_2	Ar	273	1399	340
		203	1780	1750
CF_4	CH_4	273	4900	5250
		373	3400	3360
		473	2600	2700
		573	2400	2400
N_2	C_2H_4	323	2300	2300

Adapted from Orentlicher and Prausnitz (1967).

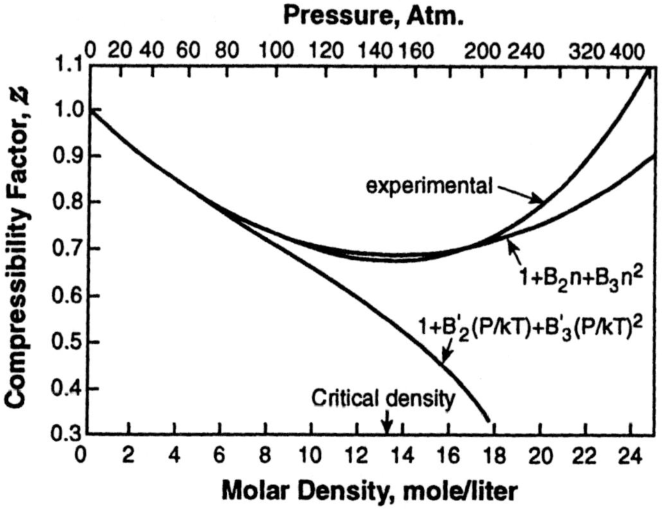

Figure 5.5
Compressibility factor for argon at 203.15K. Adapted from Prausnitz (1969).

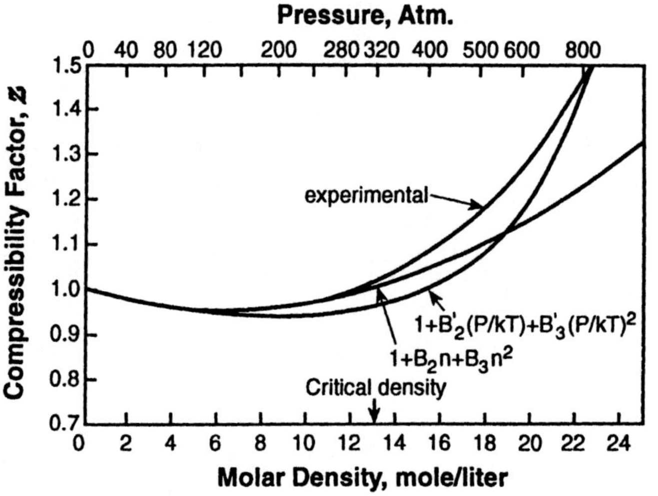

Figure 5.6
Compressibility factor for argon at 298.15K. Adapted from Prausnitz (1969).

expansion is accurate over a broader range of thermodynamic states than is the pressure expansion. This result implies the desirability of using the density expansions instead of pressure expansion to estimate thermodynamic functions from virial coefficient data.

5.2 Theory of Virial Coefficients

Let us consider a multicomponent fluid of particles whose canonical ensemble partition function is of the form

$$Q_N = \prod_{\alpha=1}^{c} \frac{(q_\alpha^T q_\alpha^I / V)^{N_\alpha}}{N_\alpha!} Z_N, \tag{5.2.1}$$

where q_α^I denotes the partition function of the internal energy states (including the rotational states) of a particle of species α and q_α^T is the translational energy partition function. Assume further that the potential energy function depends on the positions of the centers of mass $\mathbf{r}_i$ and the orientations $\mathbf{e}_i$ of the particles so that

$$Z_N = \frac{1}{\prod_{\alpha=1}^{c} [4\pi(2\pi)^{\chi_\alpha}]^{N_\alpha}} \int \cdots \int e^{-\beta u^N} d^3r_1 \cdots d^3r_N d^{2+\chi}e_1 \cdots d^{2+\chi}e_N, \tag{5.2.2}$$

where $\chi = 1$ if the molecular species are nonlinear and $\chi = 0$ if they are linear. If the intermolecular potential is independent of orientation, then the orientation variables can be integrated to obtain $\prod_\alpha^c [4\pi(2\pi)^{\chi\alpha}]^{N\alpha}$ that cancels with the prefactor in eq. (5.2.2). For nonlinear molecules e_i can be expressed in polar coordinates, and so $d^2e_i = \sin\theta_i d\theta_i d\phi_i$. For nonlinear molecules, one can use Eulerian angles in which case $d^3e_i = \sin\theta_i d\theta_i d\phi_i d\eta_i$, where $0 < \theta_i < \pi$, $0 < \phi_i < 2\pi$ and $0 < \eta_i < 2\pi$. θ_i and ϕ_i correspond to the usual polar angles of some axis fixed in molecule i and ψ_i corresponds to rotation of the molecule about the fixed axis.

The grand canonical partition function for a fluid obeying eq. (5.2.1) is

$$\Xi = \sum_{N_1=0}^{\infty} \cdots \sum_{N_c=0}^{\infty} \prod_{\alpha=1}^{c} \frac{\zeta_\alpha^{N_\alpha}}{N_\alpha!} Z_N, \tag{5.2.3}$$

where

$$\zeta_\alpha \equiv e^{\mu_\alpha/kT} q^T q^I / V. \tag{5.2.4}$$

In Chapter 2 it was shown that $PV = kT \ln \Xi$, and so

$$\begin{aligned}\frac{PV}{kT} = \ln \Xi = \ln\Bigg[1 &+ \sum_{\alpha=1}^{c} \zeta_\alpha Z_1^\alpha \\ &+ \sum_{\alpha=1}^{c}\sum_{\beta=1}^{c} \frac{\zeta_\alpha \zeta_\beta}{2} Z_2^{\alpha\beta} \\ &+ \sum_{\alpha=1}^{c}\sum_{\beta=1}^{c}\sum_{\gamma=1}^{c} \frac{\zeta_\alpha\zeta_\beta\zeta_\gamma}{6} Z_3^{\alpha\beta\gamma} + \cdots\Bigg].\end{aligned} \tag{5.2.5}$$

With the aid of the Taylor's series

$$\ln(1+x) = x - \frac{x^2}{2} + \frac{x^3}{3} + \cdots, \tag{5.2.6}$$

we can expand $\ln \Xi$ or PV/kT in a series in $\{\zeta_\alpha\}$. The first few terms of the series are

$$\begin{aligned}\frac{PV}{kT} = \ln \Xi = &\sum_{\alpha=1}^{c} \zeta_\alpha Z_1^\alpha + \sum_{\alpha=1}^{c}\sum_{\beta=1}^{c} \frac{1}{2}\zeta_\alpha\zeta_\beta(Z_2^{\alpha\beta} - Z_1^\alpha Z_1^\beta) \\ &+ \sum_{\alpha=1}^{c}\sum_{\beta=1}^{c}\sum_{\gamma=1}^{c} \zeta_\alpha\zeta_\beta\zeta_\gamma \left(\frac{Z_3^{\alpha\beta\gamma}}{6} - \frac{Z_1^\alpha Z_2^{\beta\gamma}}{2} + \frac{Z_1^\alpha Z_1^\beta Z_1^\gamma}{3}\right) + \cdots.\end{aligned} \tag{5.2.7}$$

The average number $\overline{N}_\lambda$ of particles of species λ in the system at temperature T and in volume V is given by

$$\overline{N}_\lambda = \left(\frac{\partial \ln \Xi}{\partial \ln \zeta_\lambda} \right)_{T,V} \tag{5.2.8}$$

so that the density $n_\lambda \equiv \overline{N}_\lambda / V$ has the series expansion

$$\begin{aligned} n_\lambda &= \zeta_\lambda \frac{Z_1^\lambda}{V} + \sum_\alpha \frac{\zeta_\alpha \zeta_\lambda}{V} (Z_2^{\alpha\lambda} - Z_1^\alpha Z_1^\lambda) \\ &+ \sum_{\alpha,\beta} \frac{\zeta_\lambda \zeta_\alpha \zeta_\eta}{V} \left(\frac{Z_3^{\lambda\alpha\beta}}{2} - \frac{Z_1^\lambda Z_2^{\alpha\beta}}{2} - Z_1^\alpha Z_2^{\lambda\beta} + Z_1^\lambda Z_1^\alpha Z_1^\beta \right) + \cdots \\ &\equiv \zeta_\lambda + \sum_\alpha \zeta_\lambda \zeta_\alpha a_{\lambda\alpha} + \sum_{\alpha,\beta} \zeta_\lambda \zeta_\alpha \zeta_\beta a_{\gamma\alpha\beta} + \cdots . \end{aligned} \tag{5.2.9}$$

The form of $\ln \Xi$ given by the right-hand side (rhs) of eq. (5.2.7) was used along with the fact that $Z_1^\lambda = V$ to get the final result in eq. (5.2.9). The quantities

$$a_{\lambda\alpha} = (Z_2^{\alpha\lambda} - Z_1^\alpha Z_1^\lambda) / V = (Z_2^{\alpha\lambda} - V^2) / V \tag{5.2.10}$$

and

$$a_{\lambda\alpha\beta} = \left[\frac{Z_3^{\lambda\alpha\beta}}{2} - \frac{V Z_2^{\alpha\beta}}{2} - V Z_2^{\lambda\beta} + V^3 \right] / V \tag{5.2.11}$$

will be shown to be independent of volume V for a macroscopic system.

The series expansions of n_λ in $\zeta_1 \cdots \zeta_c$, $\lambda = 1, \ldots c$, can be inverted to a series expansions of ζ_λ in $n_1, \ldots n_c$. Formally,

$$\zeta_\lambda = A_\lambda + \sum_\alpha A_{\lambda\alpha} n_\alpha + \sum_{\alpha,\beta} A_{\lambda\alpha\beta} n_\alpha n_\beta + \cdots . \tag{5.2.12}$$

Substituting for n_α the series given by eq. (5.2.9), we obtain

$$\begin{aligned} \zeta_\lambda = A_\lambda &+ \sum_\alpha A_{\lambda\alpha} \zeta_\alpha \left[1 + \sum_\gamma \zeta_\gamma a_{\alpha\gamma} + \sum_{\gamma,\delta} \zeta_\gamma \zeta_\delta a_{\alpha\gamma\delta} + \cdots \right] \\ &+ \sum_{\alpha,\beta} A_{\lambda\alpha\beta} \zeta_\alpha \zeta_\beta \left[1 + \sum_\gamma \zeta_\gamma a_{\alpha\gamma} + \sum_{\gamma,\delta} \zeta_\gamma \zeta_\delta a_{\alpha\gamma\delta} + \cdots \right] \\ &\times \left[1 + \sum_\kappa \zeta_\kappa a_{\beta\kappa} + \sum_{\kappa,\varepsilon} \zeta_\kappa \zeta_\varepsilon a_{\beta\kappa\varepsilon} + \cdots \right] \\ &+ \sum_{\alpha\beta\gamma} A_{\lambda\alpha\beta\gamma} \zeta_\alpha \zeta_\beta \zeta_\gamma + O(\zeta^4). \end{aligned} \tag{5.2.13}$$

Equating coefficients of like power of $\{\zeta_1, \ldots, \zeta_c\}$ on each side of eq. (5.2.13), we find

$$\begin{aligned}
A_\lambda &= 0 \\
A_{\lambda\alpha} &= \delta_{\lambda,\alpha} \\
A_{\lambda\alpha\beta} &= -A_{\lambda\alpha}a_{\alpha\beta} = -\delta_{\lambda,\alpha}a_{\alpha\beta} \\
A_{\lambda\alpha\beta\gamma} &= -A_{\lambda\alpha\beta}\left(a_{\alpha\gamma}+a_{\beta\gamma}\right) - A_{\lambda\alpha}a_{\alpha\beta\gamma} \\
&= \delta_{\lambda,\delta}a_{\alpha\beta}\left(a_{\alpha\gamma}+a_{\beta\gamma}\right) - \delta_{\gamma,\alpha}a_{\alpha\beta\gamma},
\end{aligned} \tag{5.2.14}$$

where $\delta_{\lambda,\alpha}$ is a Kronecker delta, and so

$$\zeta_\lambda = n_\lambda - \sum_\beta a_{\lambda\beta}n_\lambda n_\beta + \sum_{\beta,\gamma}\left[a_{\lambda\beta}\left(a_{\lambda\gamma}+a_{\beta\gamma}\right) - a_{\lambda\beta\gamma}\right]n_\lambda n_\beta n_\gamma + \cdots. \tag{5.2.15}$$

Finally, because

$$\frac{P}{kT} = \sum_{\alpha=1}\zeta_\alpha + \sum_{\alpha,\beta}\frac{1}{2}\zeta_\alpha\zeta_\beta a_{\alpha\beta} + \sum_{\alpha,\beta,\gamma}\frac{\zeta_\alpha\zeta_\beta\zeta_\gamma}{3}a_{\alpha\beta\gamma} + \cdots, \tag{5.2.16}$$

it follows that

$$\begin{aligned}
\frac{P}{kT} = {} & \sum_\alpha n_\alpha - \sum_{\beta,\alpha} a_{\alpha\beta}n_\alpha n_\beta + \sum_{\alpha,\beta,\gamma}\left[a_{\alpha\beta}\left(a_{\alpha\gamma}+a_{\beta\gamma}\right) - a_{\alpha\beta\gamma}\right]n_\alpha n_\beta n_\gamma \\
& + \sum_{\alpha,\beta}\frac{1}{2}n_\alpha n_\beta a_{\alpha\beta} - \sum_{\alpha,\beta,\gamma} n_\alpha n_\beta n_\gamma a_{\alpha\beta}a_{\alpha\gamma} + \sum_{\alpha,\beta,\gamma}\frac{n_\alpha n_\beta n_\gamma}{3}a_{\alpha\beta\gamma} + \cdots.
\end{aligned} \tag{5.2.17}$$

Comparing this result with the virial expansion, eq. (5.1.1), we find, finally

$$B_2 = -\frac{1}{2}\sum_{\alpha,\beta} x_\alpha x_\beta\left(Z_2^{\alpha\beta} - V^2\right)/V \tag{5.2.18}$$

and

$$B_3 = \sum_{\alpha,\beta,\gamma} x_\alpha x_\beta x_\gamma\left[Z_2^{\alpha\beta}\left(Z_2^{\alpha\gamma} - V^2\right)/V^2 + \left(V^3 - Z_3^{\alpha\beta\gamma}\right)/3V\right]. \tag{5.2.19}$$

Up to this point we have not restricted ourselves to pairwise additive forces. In general, u^N can be decomposed as follows

$$u^N = u_{(2)}^N + u_{(3)}^N + \cdots = \sum_{i=2}^N u_{(i)}^N, \tag{5.2.20}$$

where $u_{(i)}^N$ denotes the contribution of i-body forces to u^N and is of the form

$$u_{(i)}^N = \sum_{j_i > j_{i-1} > \cdots > j_1 \geq 1} \cdots \sum u_{j_1, j_2, \ldots, j_i}. \tag{5.2.21}$$

$u_{j_1,\ldots,j_i}$ denotes the part of the interaction potential among the i particles that cannot be accounted for by the $2, 3, \ldots, i-1$ body interaction potentials.

With the decomposition of u^N given by eq. (5.2.20), B_2 and B_3 can be expressed in the forms

$$B_2 = \sum_{\alpha,\beta} x_\alpha x_\beta B_2^{\alpha\beta} \tag{5.2.22}$$

$$B_3 = \sum_{\alpha,\beta,\gamma} x_\alpha x_\beta x_\gamma B_3^{\alpha\beta\gamma}, \tag{5.2.23}$$

where we have defined the composition independent quantities

$$B_2^{\alpha\beta} \equiv -\frac{1}{2V}\int \gamma_{12}^{\alpha\beta} d^3r_1 d^3r_2 \frac{d^{2+\chi_\alpha}e_1 d^{2+\chi_\beta}e_3}{(4\pi)^2(2\pi)^{\chi_\alpha+\chi_\beta}} \tag{5.2.24}$$

and

$$B_3^{\alpha\beta\gamma} \equiv -\frac{1}{3V}\int \left[\gamma_{12}^{\alpha\beta}\gamma_{23}^{\beta\gamma}\gamma_{13}^{\alpha\gamma} + e^{-\beta\left(u_{12}^{\alpha\beta}+u_{13}^{\alpha\gamma}+u_{23}^{\beta\gamma}\right)}\left(e^{-\beta u_{123}^{\alpha\beta\gamma}}-1\right)\right]\frac{d^3r_1\cdots d^{2+\chi_\gamma}e_3}{(4\pi)^3(2\pi)^{\chi_\alpha+\chi_\beta+\chi_\gamma}} \tag{5.2.25}$$

with

$$\gamma_{12}^{\alpha\beta} = e^{-\beta u_{12}^{\alpha\beta}} - 1. \tag{5.2.26}$$

Because in general $u_{12}^{\alpha\beta}$ depends on $\mathbf{r}_1$ and $\mathbf{r}_2$ only through $\mathbf{r}_{12}$ and $u_{123}^{\alpha\beta\delta}$ on $\mathbf{r}_1, \mathbf{r}_2$ only through $\mathbf{r}_{12}, \mathbf{r}_{13}$ and $\mathbf{r}_{23} = \mathbf{r}_{13} - \mathbf{r}_{12}$, we can transform the variables $\mathbf{r}_1, \mathbf{r}_2$ into $\mathbf{r}_1$ and $\mathbf{r}_{12}$ and eliminate the integral over d^3r_1 in eq. (5.2.24) to obtain

$$B_2^{\alpha\beta} = \frac{1}{2}\int \gamma_{12}^{\alpha\beta} d^3r_{12} \frac{d^{2+\xi_\alpha}e_1 d^{2+\chi_\beta}e_2}{(4\pi)^2(2\pi)^{\chi_\alpha+\chi_\beta}} \tag{5.2.27}$$

and we can transform $\mathbf{r}_1, \mathbf{r}_2$, and $\mathbf{r}_3$ to $\mathbf{r}_1, \mathbf{r}_{12}$, and $\mathbf{r}_3$ and obtain

$$B_2^{\alpha\beta\gamma} = -\frac{1}{3}\int \left[\gamma_{12}^{\alpha\beta}\gamma_{23}^{\beta\gamma}\gamma_{13}^{\alpha\gamma} + e^{-\beta\left(u_{12}^{\alpha\beta}+u_{13}^{\alpha\gamma}+u_{23}^{\beta\gamma}\right)}\left(e^{-\beta u_{123}^{\alpha\beta\gamma}}-1\right)\right]\frac{d^3r_{12}d^3r_3\cdots d^{2+\chi_\gamma}e_3}{(4\pi)^3(2\pi)^{\chi_\alpha+\chi_\beta+\chi_\gamma}}. \tag{5.2.28}$$

For monatomic fluids eqs. (5.2.27) and (5.2.28) become

$$B_2^{\alpha\beta} = \frac{1}{2}\int \gamma_{12}^{\alpha\beta} d^3r_{12} \tag{5.2.29}$$

$$B_3^{\alpha\beta\gamma} = -\frac{1}{3}\int\int \left[\gamma_{12}^{\alpha\beta}\gamma_{23}^{\beta\gamma}\gamma_{13}^{\alpha\gamma} + e^{-\beta\left(u_{12}^{\alpha\beta}+u_{13}^{\alpha\gamma}+u_{23}^{\beta\gamma}\right)}\left(e^{-\beta u_{123}^{\alpha\beta\gamma}}-1\right)\right] d^3r_{12}d^3r_3. \tag{5.2.30}$$

Defining the range r as the value of r_{ij} beyond which a two- or three-body interaction is negligible, we see by inspection that $B_2 = O(V_r)$ and $B_3 = O(V_r^2)$, where $V_r \equiv \frac{4}{3}\pi r^3$. Of course, different species will be characterized by different values of r. More generally we could show that $B_k = O\left(V_r^{k-1}\right)$, a result implying that the virial expansion is an expansion in the ratio of the molecular interaction volume V_r (of the order of 100 Å^3 for small nonionic molecules to the volume per molecule $\tilde{V} \equiv V/N$. At dilute gas densities $V_r/\tilde{V} = O(10^{-2})$, implying rapid convergence of the virial expansion, and at dense fluid or liquid densities $V_r/\tilde{V} = O(1)$ indicating slow convergence of the virial expansion.

5.3 Potential Models and Virial Coefficients

Owing to their simplicity, pairwise additive potential models have received much more attention than higher order potentials. In dense fluids, the general theory of structure and thermodynamic functions is not sufficiently well developed to justify the expenditure of a great deal of effort in determining the higher order interactions. For such systems, researchers have generally been content with focusing on pairwise potentials whose parameters may be adjusted to absorb somewhat the many-body effects. However, in moderately low density gases and in low temperature solids many-body effects are demonstrably present and can contribute appreciably to certain properties. For example, three-body interactions contribute appreciably (as much as half) to the third virial coefficient and have been shown to contribute 7–10% of the cohesive energy of solid argon, krypton, and xenon. For nonpolar molecules the Axibrod–Teller three-body potential,

$$u_{123} = -\varepsilon_{123}(1 + \hat{r}_{12} \cdot \hat{r}_{13} + \hat{r}_{13} \cdot \hat{r}_{23} + \hat{r}_{12} \cdot \hat{r}_{23})/r_{12}^3 r_{13}^3 r_{23}^3, \tag{5.3.1}$$

gives the dominant contribution. Similarly to the r^{-6} two-body potential, u_{123} arises from interactions associated with induced dipolar configurations of the electronic distributions about interacting molecules. We give examples of ε_{123} values later in this section where we discuss attempts to accurately determine the potential energy for the inert gas atoms. For the next several paragraphs, however, we consider many of the simple two-body models that have been used to describe fluid behavior.

5.3.1 Hard Sphere Fluid

Perhaps the simplest potential model is the hard sphere model ($u_{12} = \infty,\ r_{12} < d;\ u_{12} = 0,\ r_{12} > d$), which is useful for understanding (a) the behavior of high temperature fluids in which attractive interactions are unimportant compared to repulsive interactions and (b) the structure of very dense fluids in which the repulsive interactions may play a dominant role. The second virial coefficient for hard spheres of diameter d is

$$B_2 = -\int \gamma_{12} d^3 r_{12} = \frac{1}{2} \int_{r_{12}<d} d^3 r_{12} = \frac{2\pi}{3} d^3. \tag{5.3.2}$$

To evaluate B_3 we introduce the bipolar coordinate transformation shown in Figure 5.7 to express $d^3 r_3$ in the form

$$d^3 r_3 = \frac{2\pi}{r_{12}} r_{13} dr_{13} r_{23} dr_{23}, \tag{5.3.3}$$

with

$$0 < r_{23} < \infty, \quad |r_{12} - r_{23}| < r_{13} < r_{12} + r_{23}.$$

Figure 5.7
Bipolar coordinates. The volume $d^3r_3 = 2\pi y dx dy$ can be shown to be of the form given by eq. (5.3.3) by noting that $x^2 \times y^2 = r_{13}^2$ and $y^2 + (r_{12} - x)^2 = r_{23}^2$ and seeking the Jacobian J for the transformation $dxdy = J dr_{13} dr_{23}$. The volume element considered is a ring formed by rotating the area $dxdy$ around the x axis.

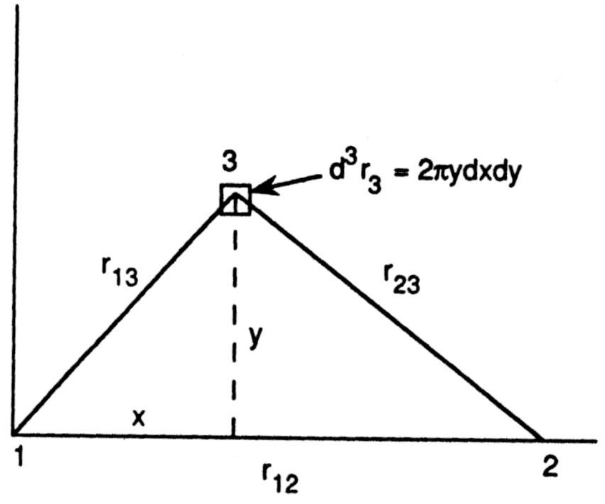

With this transformation we find

$$B_3 = -\frac{1}{3} 4\pi \int_0^d dr_{12} r_{12}^2 \frac{2\pi}{r_{12}} \int_0^d dr_{23} r_{23} \int_{|r_{12}-r_{23}|}^{r_{12}+r_{23}} dr_{13} r_{13} \gamma_{13}$$

$$= \frac{8\pi}{3} \int_0^d dr_{12} r_{12} \int_0^{d-r_{12}} dr_{23} r_{23} \int_{|r_{12}-r_{23}|}^{d} dr_{13} r_{13}, \tag{5.3.4}$$

where the second from of the rhs of eq. (5.3.4) is obtained by noting that $\gamma_{13} = 0$ for $r_{23} \geq d - r_{12}$ and $\gamma_{13} = -1$ for $d < r_{13} < |r_{12} - r_{23}|$. The integrations in eq. (5.3.4) can be performed easily to yield

$$B_3 = \frac{5}{8} \left(\frac{2\pi d^3}{3} \right)^2. \tag{5.3.5}$$

B_4 can be evaluated analytically, B_5 and B_6 have been evaluated numerically, and B_7 has been estimated (see for example, Curtiss, 1967. A summary of these results is presented in Table 5.5. Also given in the table are the van der Waals values corresponding to the identification $b \equiv \frac{2\pi}{3} d^3$ and for $a = 0$.

From the entries in Table 5.5, we see that the van der Waals model yields hard sphere virial coefficients that are, aside from B_2, increasingly too large, the error increasing with the order of the virial coefficient. The error of the van der Waals model arises from the assumption that the excluded volume of k interacting hard spheres is equal to $k(k-1)/2$ times the excluded volume of a pair of hard spheres. Such an assumption yields, for example, the erroneous conclusion that at hexagonal close packing the density would be $(3/2)\pi d^3$ instead of the known value $(2/\sqrt{2})d^3$. Actually, the $(k+1)$th hard spheres virial coefficient may be interpreted as the kth power of the average excluded volume per sphere in a

TABLE 5.5 Hard Sphere Virial Coefficients ($b_o \equiv \frac{2\pi}{3d^3}$)

	Exact	*van der Waals* $b \equiv b_o$, $a \equiv 0$
B_2	b_0	b_0
B_3	$\frac{5}{8}b_0^2$	b_0^2
B_4	$0.2869b_0^3$	b_0^2
B_5	$(0.1103 \pm 0.0003)b_0^4$	b_0^4
B_6	$(0.0386 \pm 0.0004)b_0^5$	b_0^5
B_7	$\approx 0.0127b_0^6$	b_0^6

system of $k+1$ hard spheres. For a seven particle system the average excluded volume is about half of that of the two particle system.

Because of the decrease in the average excluded volume with increase in the order of virial coefficient, the hard sphere virial expansion converges faster than indicated by the reduced density nb_o. If one were to demand an accuracy of less than 0.5% for $\mathcal{Z}$, then the seven virial coefficients in Table 5.5 would be adequate up to about $nb_o = (0.005/0.0127)^{1/3} = 0.735$, an estimate obtained by assuming that the neglected higher order terms contribute about the same as the last term kept ($n^6 B_7$) in the expansion. To appreciate the density range over which the seventh order virial expansion is valid for hard spheres, note that at the critical point nb_o is about $3/4$ for nonpolar fluids of small molecules.

5.3.2 Square-Well Fluid

Whereas in hard spheres (or purely repulsive particles) the virial coefficients are determined solely by the excluded volume effect, in a real fluid, composed of molecules having attractive as well as repulsive interactions, the virial coefficients arise from a balance between the excluded volume effect and a cohesive effect arising from attractive interactions. The square-well pair potential, defined by the expression

$$\begin{aligned} u(r) &= \infty, \quad r < d_1 \\ &= -\varepsilon, \quad d_1 < r < d_2 \\ &= 0, \quad r > d_2, \end{aligned} \tag{5.3.6}$$

and shown in Figure 5.8, is perhaps the simplest model to examine in trying to understand the relative roles of the repulsive and attractive interactions equation of state behavior.

Figure 5.8
The square-well potential.

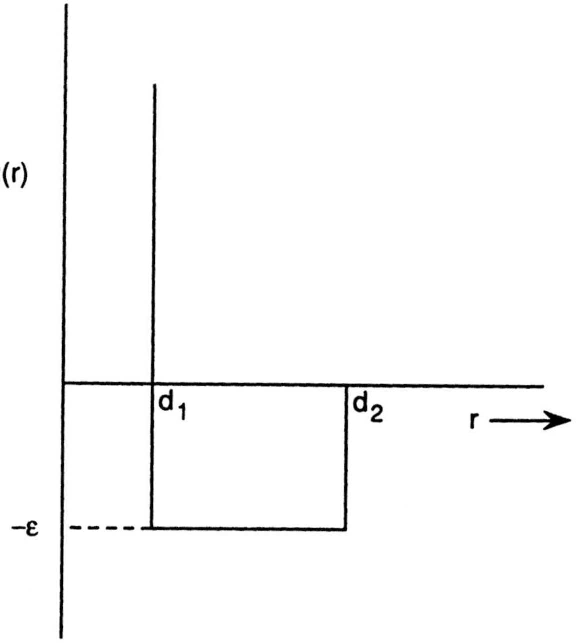

The second virial coefficient is easily evaluated for this model.

$$B_2 = -\frac{1}{2}\int (e^{-\beta u(r)} - 1)d^3r = -\int_{r_{12}<d_1} d^3r - (e^{\beta\varepsilon} - 1)\int_{d_1<r_{12}<d_2} d^3r$$

$$= \frac{2\pi}{3}d_1^3\left[1 - \left(\frac{d_2^3}{d_1^3} - 1\right)\left(e^{\beta\varepsilon} - 1\right)\right]. \tag{5.3.7}$$

In the limit of high temperature, where the attractive energy ε is small compared to the kinetic energy $\frac{3}{2}kT$, the excluded volume effect dominates and $B_2 \simeq \frac{2\pi}{3}d - 1^3$. At the other extreme, where $kT \ll \varepsilon$, $B_2 \simeq -\frac{2\pi}{3}d_2^3e^{\beta\varepsilon}$. In the low temperature limit there is a tendency of the particles to associate and, thereby, reduce the effective number of particles exerting pressure to $\sim N(1 - \frac{2\pi}{3}d_2^3ne^{\beta\varepsilon})$. Incidentally, only in the limit that $e^{\beta\varepsilon} - 1 \simeq \beta\varepsilon$ will eq. (5.3.7) have the temperature dependence predicted by the van der Waals model.

Exercise

For the square-well model, the third virial coefficient may be expressed in the form

$$B_3 = -\frac{8\pi}{3}\int_0^{d_2} dr_{12}r_{12}\gamma_{12}\int_0^{d_2} dr_{23}r_{23}\gamma_{23}\int_{|r_{12}-r_{23}|}^{r_{12}+r_{23}} dr_{13}r_{13}\gamma_{13}, \tag{5.3.8}$$

where $\gamma_{ij} = e^{-\beta u_{ij}} - 1$.

Evaluate eq. (5.3.8) and plot B_3/d_1^3 versus kT/ε for $d_2 = 1.5d_1$.

5.3.3 LJ Fluid

The model pair potential most often used for nonpolar fluids is the 6–12 LJ model. The attractive part $-4\varepsilon(\sigma/r)^6$ is of the form of the dominant attractive term predicted from quantum mechanics and the repulsive part is chosen to be steeply repulsive compared to r^{-6}. There is no theoretical basis to the form r^{-12} and only tradition distinguishes it from many other choices.

In the LJ model σ is a good measure of the distance of separation at which the repulsive interactions begin to become extremely strong, so that σ may be thought of as the "size" or "diameter" of the molecule. The quantity ε is the minimum of the pair potential and can be thought of as the strength of the attractive potential. The $\varepsilon - \sigma$ parameters of the 6–12 LJ model have been determined from second virial coefficient data for a large collection of fluids. Some of these are given in Table 5.6.

Values of B_2 determined by the parameters given in Table 5.6 are compared with experimental values in Table 5.7 for a number of fluids. The comparison is, of course, only a test of the ability of the LJ model to fit B_2 to experimental data. For this purpose the

TABLE 5.6 Parameters ε and σ

Substance	ε/k (K)	σ(Å)
Ar	119.8	3.405
Kr	171	3.60
Xe	221	4.10
N_2	95.05	3.698
O_2	117.5	3.58
CO	100.2	3.763
CO_2	192.25	4.416
CH_4	142.87	4.01
AsF_3	164.02	15.35
SiF_4	147.48	5.647
C_2H_6	169.02	5.607
C_3H_8	187.22	6.648
C_4H_{10}	223.74	7.152
$C(CH_3)_4$	233.66	7.420
C_6H_6	187.82	10.17
NH_3	190.1	5.77
H_2O	809.1	2.641

Parameters are of the Lennard–Jones potential energy model, $u(r) = 4\varepsilon[(\sigma/r)^{12} - (\sigma/r)^6]$, determined from second virial coefficient data. k is Boltzmann's constant. Parameters taken from Hirschfelder et al. and Kunz (1968).

TABLE 5.7 Second Virial Coefficients Predicted by Lennard–Jones Model Compared With Experiment

		$\tilde{B}_2$ (cm^3/gmol)		
Substance	*T* (K)	*Calculated*	*Observed*	*% Error*
Xe	389.2	−130.2	−128.4	1.40
	348.2	−94.5	−94.5	0.00
	373.2	−81.2	−81.5	−0.37
	498.2	−39.1	−38.9	0.51
	648.3	−28.0	−28.0	0.00
	573.3	−23.5	−23.4	0.43
AsF_3	353.79	−2327.6	−2320.0	3.82
	373.63	−1973.6	−1880.0	4.98
	404.84	−1497.0	−1410.0	6.17
	433.15	−1132.3	−1220.0	−7.19
SiF_4	293.16	−144.8	−145.6	−0.52
	323.16	−110.7	−109.4	1.22
	353	−83.2	−83.7	−0.54
CH_4	303.16	−43.5	−41.6	4.59
	353.16	−26.6	−26.6	0.00
	403.16	−14.3	−15.4	−6.83
C_2H_6	273.16	−229.0	−223.0	2.70
	333.16	−144.9	−139.9	3.56
	410.94	−77.2	−90.4	−14.62
	510.94	−24.0	−47.4	−49.36
C_2H_8	295.40	−401.4	−399.0	0.60
	357.90	−260.8	−265.0	−1.60
	412.90	−177.2	−182.0	−2.62
C_6H_6	308.36	−1321.7	−1394.0	−5.19
	393.20	−737.0	−729.0	1.09
	457.66	−456.9	−521.0	−12.31

Parameters taken from Table 5.6.

LJ model appears to be quite adequate. However, the values of ε and σ determined from virial coefficient data disagree somewhat with those determined from other quantities, for example, from transport coefficient data.

As was the case for the general equation of state, the virial coefficients of LJ fluids obey a law of corresponding states. To see this we define $B_2^* = B_2/\sigma^3$ and B_3^*/σ^6 and

$$B_2^* = -\frac{8\pi}{3}\int_0^\infty \gamma(x, T^*)x^2 dx \tag{5.3.9}$$

and

$$B_3^* = -\frac{8\pi}{3}\int_0^\infty dx x\gamma(x, T^*)\int_0^\infty dy y\gamma(y, T^*)\int_{|x-y|}^{x+y} dz z\gamma(z, T^*), \tag{5.3.10}$$

where $T^* = kT/\varepsilon$ and

$$\gamma(x, T^*) = e^{-4[x^{-12}-x^{-6}]/T^*} - 1. \tag{5.3.11}$$

One can in fact show that 6–12 LJ fluids obey the law of corresponding states at any density. The reduced pressure $p^* = p\sigma^3/\varepsilon$ is a universal function of T^* and $V^* = v/\sigma^3$ for these fluids.

5.3.4 Kihara Fluids

A potential model designed to describe interactions among nonspherical molecules was introduced by Kihara 1953, 1958, 1963. He assumed that molecules have an inner hard core of some geometrical shape and that pairs of molecules interact via a potential depending only on the distance ρ, defined as the shortest distance between two cores. For example, a spherically symmetric Kihara version of the 6–12 LJ potential is

$$\begin{aligned} u(r) &= \infty, \quad r < 2a \\ &= 4\varepsilon\left[\left(\frac{\sigma - 2a}{r - 2a}\right)^{12} - \left(\frac{\sigma - 2a}{r - 2a}\right)^{6}\right], \quad r > \end{aligned} \tag{5.3.12}$$

Such molecules have hard spherical inner cores of diameter $2a$ and a soft outer region in which the interaction depends on the distance $\rho = r - 2a$. The Kihara potential is a way to account somewhat for the fact that the closest parts of the electron clouds about interacting molecules interact more strongly than the most remote parts (that is, if two spherical molecules approach each other, then the electrons distributed between the nuclei interact more strongly than those distributed on the opposite sides of the nuclei). In Table 5.8, we list values of the parameters σ, a, and ε determined from second virial coefficient data. Having one more parameter than the LJ model we expect the Kihara model to give better agreement between the experiment and predictions of B_2 from fitted parameters. That this is the case is illustrated in Figure 5.9.

An example of hard ellipsoids is shown in Figure 5.10. The distance ρ_{12} is, of course, dependent on $\hat{r}_{12}$, $\hat{e}_1$, and $\hat{e}_2$ for a given pair of ellipsoids.

To discuss Kihara's potential for nonspherical molecules, let us introduce the concept of rigid convex bodies. A body is called convex if any line segment whose end points are inside the body lies entirely inside the body. Examples of such bodies are spheres, ellipsoids, rods tetrahedra, triangles, prisms, disks, etc. If $\sigma_{12}(\hat{\mathbf{r}}_{12}, \hat{\mathbf{e}}_1, \hat{\mathbf{e}}_2)$ denotes the distance of separation of the centers of rigid convex bodies in contact with $\hat{\mathbf{r}}_{12}$, the relative orientations of their centers of mass, and $\hat{\mathbf{e}}_1$ and $\hat{\mathbf{e}}_2$, the orientation of each molecule relative to its center of mass, then the hard convex body potential is of the form

$$\begin{aligned} u_{12} &= \infty, \quad r_{12} < \sigma_{12}(\hat{\mathbf{r}}_{12}, \hat{\mathbf{e}}_1, \hat{\mathbf{e}}_2) \\ &= 0, \quad r_{12} > \sigma_{12}(\hat{\mathbf{r}}_{12}, \hat{\mathbf{e}}_1, \hat{\mathbf{e}}_2) \end{aligned} \tag{5.3.13}$$

TABLE 5.8 Parameters for Kihara Potential (Spherical Core) Obtained from Second Virial Coefficient Data

	$a^* \equiv a/\sigma$	σ(Å)	ε/k (K)
Ar	0.121	3.317	146.52
Kr	0.144	3.533	213.73
Ke	0.173	3.880	298.15
CH_4	0.283	3.565	227.13
N_2	0.250	3.526	139.2
O_2	0.308	3.109	194.3
C_2H_6	0.359	3.504	496.69
C_3H_8	0.470	4.611	501.89
CF_4	0.500	4.319	289.7
$C(CH_3)_4$	0.551	5.762	557.75
n-C_4H_{10}	0.661	4.717	701.15
C_6H_6	0.750	5.335	832.0
CO_2	0.615	3.760	424.16
n-C_5H_{12}	0.818	5.029	837.82

Adapted from Prausnitz (1969).

The virial coefficient $B_2^{\alpha\beta}$ for molecules of species α and β with interactions described by eq. (5.3.13) is of the form

$$
\begin{aligned}
B_2^{\alpha\beta} &= -\frac{1}{2}\int \left(e^{-\beta u^{\alpha\beta}} - 1\right) d^3r \frac{d^{2+X_\alpha}e_\alpha d^{2+X_\beta}e_\beta}{(4\pi)^2(2\pi)^{X_\alpha+X_\beta}} \\
&= +\frac{1}{2}\int \left[\int_0^{\sigma^{\alpha\beta}} r^2 dr\right] d^2\hat{r} \frac{d^{2+X_\alpha}e_\alpha d^{2+X_\beta}e_\beta}{(4\pi)^2(2\pi)^{X_\alpha+X_\beta}} \\
&+ \frac{1}{6}\int \left[\sigma^{\alpha\beta}\left(\hat{r}, \mathbf{e}_\alpha, \mathbf{e}_\beta\right)\right] d^2\hat{r} \frac{d^{2+X_\alpha}e_\alpha d^{2+X_\beta}e_\beta}{(4\pi)^2(2\pi)^{X_\alpha+X_\beta}}.
\end{aligned} \tag{5.3.14}
$$

This result is valid for more general rigid molecules than convex ones. Dowling and Davis (1972) have derived the result for arbitons.

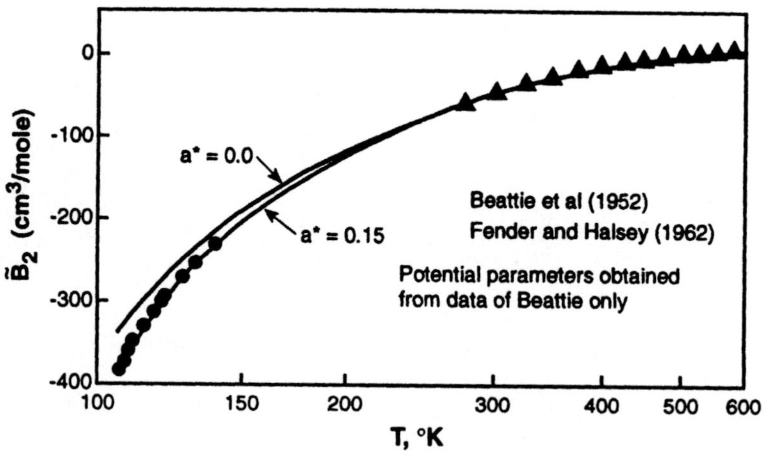

Figure 5.9
Second virial coefficient for krypton. Predictions at low temperature are based on LJ potential ($a^* = 0$) and on Kihara potential. Adapted from Prausnitz, (1969).

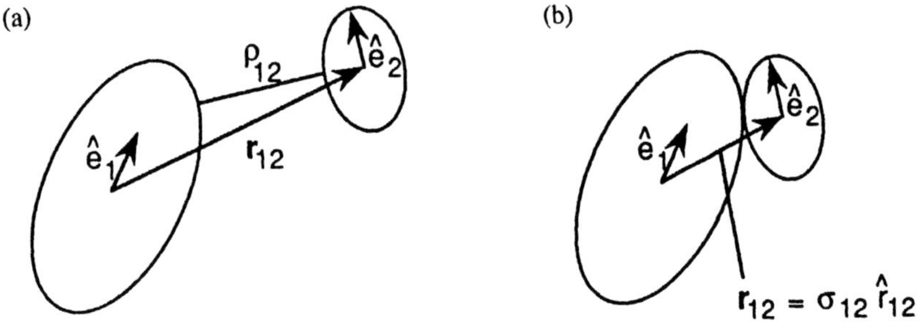

Figure 5.10
Hard ellipsoids.
(a) $r_{12} > \sigma_{12}$, $\rho_{12} > 0$.
(b) $r_{12} = \sigma_{12}$, $\rho_{12} = 0$.

A body is called an arbiton if there exists at least one point in the interior of the body such that any line drawn outward from that point intersects the surface only once. For convex and arbiton systems $2B_2^{\alpha\beta}$ equals the volume excluded from β by α averaged over all relative orientations of the pair.

For a one-component fluid of convex hard bodies, Kihara has reduced B_2 to the form

$$B_2 = v + \frac{M}{4\pi}s, \tag{5.3.15}$$

where v and s are the volume and surface area of the convex body, and

$$M = \frac{1}{2}\int\left(\frac{1}{R_1} + \frac{1}{R_2}\right)ds, \tag{5.3.16}$$

where R_1 and R_2 are the principal radii of curvature (see Chapter 7 for definition of R_i) of the body at the area element ds, which is integrated over all the surface of the body in eq. (5.3.16). In Table 5.9 values of v, s and M are given for several convex objects, and in Table 5.10 convex core assignments made by Kihara are given for several molecules.

As a generalized LJ model, Kihara introduced the potential

$$u(\rho) = u_o\left[\frac{m}{n-m}\left(\frac{\rho_o}{\rho}\right)^n - \frac{n}{n-m}\left(\frac{\rho_o}{\rho}\right)^m\right], \tag{5.3.17}$$

where $\rho(\hat{r}_{12}, \hat{e}_1, \hat{e}_2)$ is the shortest distance between the pair of molecules 1 and 2 and u_o and ρ_o are constants. This potential has the property that $u(\rho) = +\infty$ when the convex cores of molecules touch. For the potential function defined by eq. (5.3.17) for a given convex body, Kihara (1953) obtained the following expression for the second virial coefficient:

$$B_2 = \frac{2\pi}{3}\rho_o^3 F_3\left(\frac{u_o}{kT}\right) + M\rho_o^2 F_2\left(\frac{u_o}{kT}\right) + \left(s + \frac{1}{4\pi}M^2\right)\rho_o F_1\left(\frac{u_o}{kT}\right) + \left(v + \frac{1}{4\pi}Ms\right), \tag{5.3.18}$$

TABLE 5.9 Volume v, Surface Area s, and Surface Integral of Mean Curvature M, Convex Bodies

Core	v	s	M
Sphere (radius a)	$3^{-1}4\pi a^3$	$4\pi a^2$	$4\pi a$
Rectangular (length of each edge l_1, l_2, l_3)	$l_1 l_2 l_3$	$2(l_1 l_1 + l_1 l_3 + l_2 l_3)$	$\pi(l_1 + l_2 + l_3)$
Regular Tetrahedron (length of one edge l)	$6^{-1}2^{-1/2}l^3$	$3^{\frac{1}{2}}l^2$	$6l \tan^{-1}(2^{\frac{1}{2}}$
Regular octahedron	$3^{-1}2^{\frac{1}{2}}l^3$	$3^{\frac{1}{2}}2l^2$	$12l \cos^{-1}(2^{\frac{1}{2}})$
Circular cylinder (length l, radius a)	$\pi a^2 l$	$2\pi a(a+l)$	$\pi(\pi a + l)$
Circular disk (radius a)	0	$2\pi a^2$	$\pi^2 a$
Rectangle (length of each side l_1, l_2)	0	$2l_1 l_2$	$\pi(l_1 + l_2)$
Regular triangle (length of one side l)	0	$2^{-1}3^{1/2}l^2$	$2^{-1}3\pi l$
Regular hexagon (length of one side l)	0	$3^{1/2}3l^2$	$3\pi l$
Thin rod (length l)	0	0	πl

where

$$F_j(z) \equiv \int_o^{\infty} \left[1 - \exp\left(-\frac{m}{n-m}\frac{z}{\xi^n} + \frac{n}{n-m}\frac{z}{\xi^m}\right)\right] d(\xi^j). \tag{5.3.19}$$

TABLE 5.10 Cores of Molecules as Suggested by Kihara

Hydrogen H_2: thin rod connecting 2 H's (length 0.74 Å)		
$v = 0$	$s = 0,$	$M = 1.32$ Å
Nitrogen N_2: thin rod connecting 2 N's (length 1.094 Å)		
$v = 0,$	$s = 0,$	0.3.44 Å
Carbon dioxide CO_2: thin rod connecting 2's (length 2.30 Å)		
$v = 0,$	$s = 0,$	$M = 7.23$ Å
Methane CH_4: regular tetrahedron connecting $4F$'s (length of one edge 2.32 Å)		
$v = 0.670$ Å^3	$s = 9.32$ Å^2	$M = 13.30$ Å
Ethylene C_2H_4: rectangle connecting the mid-point of each C—H bond (length of each side 0.89 Åand 1.95 Å)		
$v = 0,$	$s = 3.47$ Å^2,	$M = 8.92$ Å
Benzene C_6H_6: regular hexagon connecting the mid-point of each C—H bond (length of each side 1.93 Å)		
$v = 0,$	$s = 19.4$ Å^2,	$M = 18.2$ Å

Adapted from Kihara (1953, 1958, 1963).

Values of the functions F_j, $j = 1, 2, 3$, are given in Table 5.11. Using the experimental data given in Table 5.12, and the core assignments of Table 5.10, Kihara determined the values given in Table 5.13 for the parameters ρ_o and u_o for the 6–12 version of eq. (5.3.17). Values of B_2 calculated from Kihara's parameter fit are show in Table 5.12.

5.3.5 Stockmayer Fluid

Consider next polar gases. In addition to the interactions of the non-polar type discussed in the preceding paragraphs, the long-ranged dipole–dipole interaction must be considered. The interaction potential for polar molecules is of the form

$$u_{12} = u_{12}^{(o)} - \frac{\mu_1\mu_2}{r_{12}^3}\left[3\left(\hat{\mathbf{r}}_{12}\cdot\hat{e}_1\right)\left(\hat{\mathbf{r}}_{12}\cdot\hat{e}_2\right) - \hat{e}_1\cdot\hat{e}_2\right], \quad (5.3.20)$$

where $\mu_1\hat{e}_i$ is ith dipole moment vector of the molecule i oriented in the direction $\hat{e}_i$ and $u_{12}^{(o)}$ is the nonpolar part of the interaction potential. If $u_{12}^{(o)}$ is chosen to be the spherically symmetric 6–12 LJ

TABLE 5.11 Functions in Second Virial Coefficient, eq. (5.3.18) for $n = 12$, $m = 6$.

$-\log_{10}(z)$	$F_s(z)$	$F_2(z)$	$F_1(z)$
−0.4	−9.859	−5.211	−1.784
−0.3	−6.138	−3.008	−0.7761
−0.2	−4.003	−1.776	−0.2221
−0.1	−2.673	−1.027	0.1091
0.0	−1.795	−0.5424	0.3198
0.1	−1.189	−0.2151	0.4600
0.2	−0.7587	0.0132	0.5562
0.3	−0.4465	0.1758	0.6234
0.4	−0.2170	0.2930	0.6710
0.5	−0.0469	0.3779	0.7045
0.6	0.0794	0.4392	0.7279
0.7	0.1729	0.4829	0.7436
0.8	0.2415	0.5134	0.7536
0.9	0.2911	0.5336	0.7591
1.0	0.3259	0.5459	0.7611
1.1	0.3493	0.5521	0.7603
1.2	0.3638	0.5534	0.7575
1.3	0.3715	0.5510	0.7528
1.4	0.3737	0.5457	0.7468
1.5	0.3718	0.5381	0.7396
1.6	0.3668	0.5287	0.7320
1.7	0.3594	0.5179	0.7228
1.8	0.3501	0.5062	0.7135
1.9	0.3396	0.4937	0.7038
2.0	0.3281	0.4806	0.6937

Adapted from Kihara (1953, 1958, 1963).

TABLE 5.12 Constants in the Model, eq. (5.3.17), $m = 6, n = 12$.

	$\rho_o(\AA)$	u_o/k (K)
H_2	2.81	39.4
N_2	3.47	124
CO_2	3.36	309
CH_4	1.92	378
CF_4	2.48	372
C_2H_4	2.5	470
C_6H_6	3.4[a]	830

[a] This value of ρ_o is taken from the lattice constant of graphite, because observed values for benzene are not sufficient to determine both ρ_o and u_o.

potential, eq. (5.3.20) is known as the Stockmayer potential. The dipole moment μ_i can be determined from the dielectric constant behavior of a gas of molecules of type i. In Table 5.14 values of the parameters σ, ε, and μ determined from a fit of the second virial coefficient data, are given for the Stockmayer potential for several gases. The second virial coefficient is shown as a function of T and μ in Figure 5.11.

To emphasize the contribution of the dipole–dipole interaction to the second virial coefficient, we have plotted in Figure 5.12 the value of B_2 for water with $\sigma = 2.65$Åand $\varepsilon/k = 380$K for μ equal to zero and μ equal to the correct value of 1.87×10^{-18} esu. The contribution of the dipole–dipole interaction represents a major contribution to B_2 at sufficiently low temperatures (say, $T < 300$ K). In Figure 5.12 we have also plotted values of B_2 predicted by ignoring the dipole–dipole interactions and fitting the 6–12 LJ parameters to second virial coefficient data. These effective LJ parameters, $\sigma = 2.64$ Åand $\varepsilon/k = 809.1$ K give a better fit to the data than one would have expected a priori. Such effective parameters are, however, not expected to be useful in predicting higher virial coefficients or transport properties because the average effect of the dipole–dipole interaction will be different in different quantities.

If it is assumed that the contribution of $u_{12}^{(o)}$ dominates the integrand of B_2 for $0 < r < R$ and is negligible compared to the dipole–dipole interation for $r > R$, then B_2 can be written

$$B_2 = B_2^{(0)}\left[1 - \frac{1}{3}\left(\frac{\mu^2}{R^3kT}\right)^2 - \frac{1}{75}\left(\frac{\mu^2}{R^3kT}\right)^4 + \cdots\right], \quad (5.3.21)$$

in the form where $B_2^{(o)}$ is the virial coefficent of a gas with pair potential $u_{12}^{(o)}$. Equation (5.3.21) was obtained by expanding $e^{-\beta u_{12}^{d,d}}$ in a power series in the dipole–dipole interaction $u^{d,d}$, integrating over the orientations $\hat{e}_1$ and $\hat{e}_2$ and integrating the resulting function of r_{12} from R to ∞. Because the odd terms cancel in this series

TABLE 5.13 Second Virial Coefficient for Nonspherical Molecules, eq. (5.3.18) for 6–12 Potential

	T(K)	B_2(Å^3)	B_2(Å^3)
	Calcd.	Exp.	
	65.1	−30.84	−30.3
	90.0	−9.06	−9.2
	123.0	6.64	4.9
	173	14.64	15.1
H_2 [a]	223	19.66	20.0
	273	22.61	23.2
	323	24.4	25.1
	373	25.69	25.7
	473	27.02	26.0
	143	−100.24	−132
	173	−92.93	−86.0
	223	−46.93	−43.8
	273	−19.72	−17.2
N_2	323	−0.95	−.04
	373	10.15	10.2
	423	17.95	19.1
	473	24.78	25.5
	573	33.54	34.3
	673	39.76	39.0
	273	−241.61	−241
	323	−168.59	−170
	373	−120.57	−119
	423	−84.79	−84.0
CO_2 [b]	473	−58.78	−56.6
	573	−24.12	−22.5
	673	−1.0	−2.6
	773	13.21	10.0
	873	28.61	20.1
	273	−96.49	−89.6
	298	−78.13	−72.1
	323	−58 48	−57.6
CH_4 [c]	348	−47.02	−46.1
	373	−34.98	−35.9
	398	−27.89	−27.2
	423	−20.62	−19.3
	273	−197.52	−184
	323	−121.15	−117
	373	−71.74	−71.6
CF_4 [d]	423	−41.44	−43.2
	523	0.37	−2.1
	573	14.71	15.4
	673	36.75	38.2
	273	−293.84	−280
	298	−232.99	−234
	323	−197.74	−197
C_2H_4 [e]	348	−171.36	−166
	373	−139.59	−142
	398	−122.53	−121
	423	−106.09	−104
	316	−2342.7	−2160
	331	−2157.4	−1890
C_6H_6 [f]	353	−1879.5	−1610
	372	−1694.1	−1520
	398	−1411.5	−1220

[a] The data are from Holborn and Otto (1928) and the values are taken from Fowler and Guggenheim (1939).

[b] MacCormack and Schneider (1950).

[c] Michels and Nederbragt (1936).

[d] MacCormack and Schneider (1951).

[e] Michels and M. Geldermans (1942).

TABLE 5.14 Parameters for Stockmayer's Potential for Polar Fluids

	μ (Debye)	σ (Å)	ε/k (K)
Acetonitrile	3.94	4.38	219
Nitromethane	3.54	4.16	290
Acetaldehyde	2.70	3.68	270
Acetone	2.88	3.67	479
Ethanol	1.70	2.45	620
Chloroform	1.05	2.98	1060
n-Butanol	1.66	2.47	1125
n-Butylamine	0.85	1.58	1020
Methylformate	1.77	2.90	684
n-Proplyformate	1.92	3.06	877
Methylacetate	1.67	2.83	895
Ethylacetate	1.76	2.99	956
Ethyl ether	1.16	3.10	935
Diethyl amine	1.01	2.99	1180
Water	1.87	2.65	380

Adapted from Blanks and Prausnitz (1962).

it converges quite rapidly for $\mu^2/R^3kT < 1$. In the case of water, if we take $R = 4$ Åand $\mu = 1.87 \times 10^{-18}$ esu, then $\mu^2/R^3kT < 1$ for $T \gg 396$ K. Because R is chosen rather arbitrarily, eq. (5.3.21) is only useful to estimate the range of temperatures in which the dipole–dipole interactions will be important. In the special case that $u_{12}^{(o)}$ is the hard sphere potential, however, R will be equal to the hard sphere diameter d and the series of eq. (5.3.21) will be unambiguous. If for water, d is equated to 2.65 Å, then $T = 1018$ K corresponds to a $\mu^2/d^3kT = 1$.

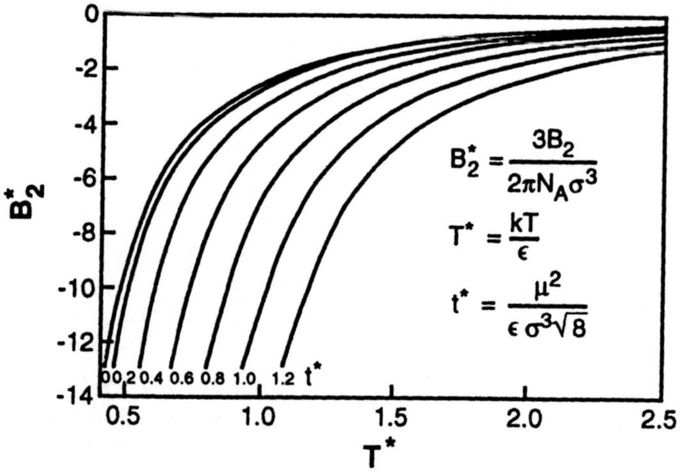

Figure 5.11
Second virial coefficients predicted from the Stockmayer potential for polar molecules. Adapted from Prausnitz (1969).

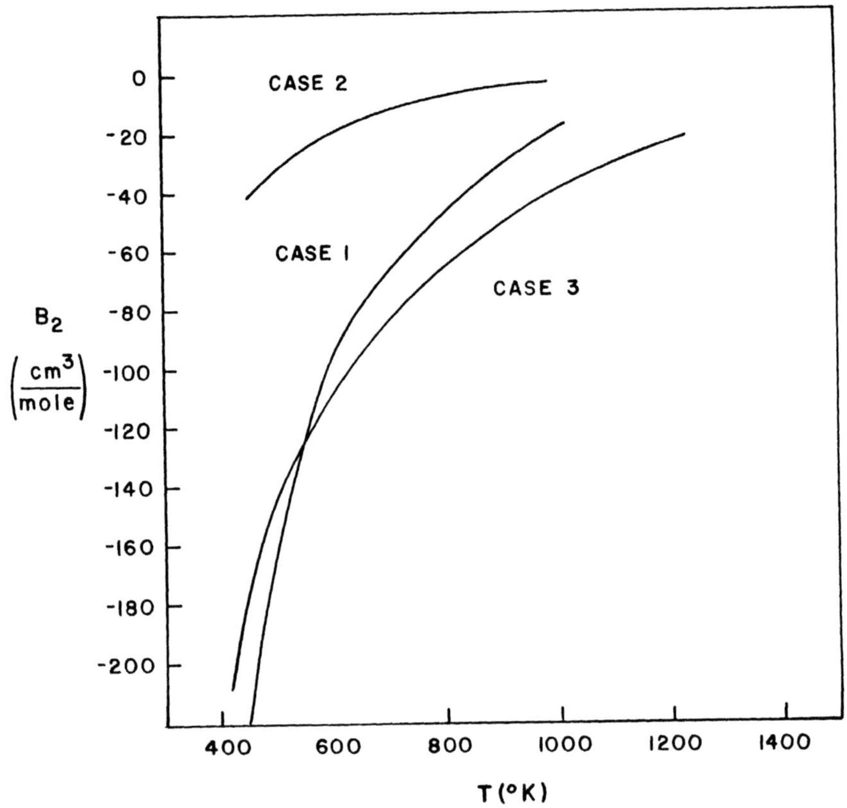

Figure 5.12
Second virial coefficient versus temperature. Case 1: Stockmayer potential for parameter values determined from second virial coefficient data on water; $\varepsilon/k = 380$ K, $\sigma = 2.65$ Åand $\mu = 1.87 \times 10^{-18}$ esu. Case 2: LJ potential; $\varepsilon/k = 380$ K, $\sigma = 2.65$ Å. Case 3: LJ potential for parameter values determined for second virial coefficient data on water; $\varepsilon/k = 809.1$ K, $\sigma = 2.641$ Å.

5.3.6 Ionic Fluid

Dilute plasmas and electrolytes pose a special problem in molecular thermodynamics because of the long-range nature of the Coulombic interaction. The virial expansion does not exist for charged particles, that is, the pressure cannot be expanded in a power series in density. This problem shows up as a divergence in $B_2^{\alpha\beta}$ if one tries to apply the virial coefficient theory. Consider for example charged particles of types α and β interacting with the potential

$$\begin{aligned} u_{\alpha\beta}(r) &= u_{\alpha\beta}^{(o)}(r), \quad r < d_1 \\ &= \frac{e_\alpha e_\beta}{r}, \quad r > d_1, \end{aligned} \tag{5.3.22}$$

where $u_{\alpha\beta}^{(0)}$ is a short-range interaction $\left(|\langle u_{\alpha\beta}^{(o)}\rangle| < \infty\right)$ and e_α the electronic charge on α. Because for sufficiently large r, say $r \geq l$, the conditions $u_{\alpha\beta}^{(o)}(r) \sim 0$ and $e_\alpha e_\beta / rkT \ll 1$ must hold, we can express $B_2^{\alpha\beta}$ for a system described by eq. (5.3.22) in the form

$$B_2^{\alpha\beta} = -\frac{1}{2}\int_{r<l} \left(e^{-\beta u_{\alpha\beta}} - 1\right) - \frac{1}{2}\sum_{\nu=1}^{\infty} \frac{(-\beta e_\alpha e_\beta)^\nu}{\nu!} 4\pi \int_l^\infty r^{-\nu+2} dr. \tag{5.3.23}$$

The first term on the rhs of eq. (5.3.22) is finite, whereas the second term diverges to $+\infty$ or $-\infty$. Thus, $B_2^{\alpha\beta}$ does not exist for charged particles. A divergence would also arise if one tried to compute the Coulombic energy of only the positive charges (or only the negative charges) in a salt crystal. However, owing to electrical screening, the total Coulombic energy of an electrically neutral crystal is finite. Similarly, in a fluid Coulombic screening leads to finite Coulombic energies.

According to the Debye–Hückel theory (Chapter 9), the effect of Coulombic screening in dilute plasmas and electrolytes is to yield a pair correlation function for ions α and β of the form

$$g_2^{\alpha\beta}(r) = e^{-\left(u_{\alpha\beta}^{(o)} + \bar{u}_{\alpha\beta}^{c}\right)/kT} \tag{5.3.24}$$

where

$$\bar{u}_{\alpha\beta}^{c} = \frac{e_\alpha e_\beta}{D_o r} e^{-\kappa r}, \tag{5.3.25}$$

with κ the Debye screening constant defined by the equation

$$\kappa^2 = \frac{4\pi}{D_o kT} \sum_{\alpha=1}^{c} e_\alpha^2 n_\alpha. \tag{5.3.26}$$

D_o is the dielectric constant of the fluid ($D_o = 1$ for a dilute plasma, which is a totally ionized gas, and $D_o \simeq 80$ for water at 25 C) and n_α is the density of ionic species α.

Because the important contribution to $B_2^{\alpha\beta}$ comes from the long-range part of $\bar{u}_{\alpha\beta}^{c}$, where $\bar{u}_{\alpha\beta}^{c}/kT \ll 1$, for very dilute electrolytes, eq. (5.3.29) can be approximated by

$$\begin{aligned} g_2^{\alpha\beta}(r) &\simeq e^{-u_{\alpha\beta}^{(0)}/kT}, \quad 0 < r < \text{range } l \text{ of } u_{\alpha\beta}^{(0)}, \\ &\simeq 1 - \frac{\bar{u}_{\alpha\beta}^{c}}{kT}, \quad r > l. \end{aligned} \tag{5.3.27}$$

Inserting eq. (5.3.27) into the virial equation, eq. (5.3.18), for a c-component plasma and carrying out the integrations, we find

$$\frac{P}{nkT} = 1 + n \sum_{\alpha,\beta} x_\alpha x_\beta B_2^{(0)\alpha\beta} - e^{-\kappa l} \frac{\kappa^3}{n}, \tag{5.3.28}$$

where

$$B_2^{(o),\alpha\beta} \equiv -\frac{1}{2} \int \left[e^{-\beta u_{\alpha\beta}^{(o)}} - 1\right] d^3 r. \tag{5.3.29}$$

For sufficiently low charge concentrations $\kappa l \ll 1$. Thus, because the quantity κ^3/n goes as $n^{\frac{1}{2}}$, we see that P does not obey a power series for a Coulombic system, but rather obeys a series of the form

$$\frac{P}{kT} = n + A_1 n^{3/2} + A_2 n^2 + \cdots \tag{5.3.30}$$

Similarly, a dilute electrolyte virial equation yields the result for eq. (5.3.27)

$$P - P_0 = n_c kT\left[1 + n_c \sum_{\alpha,\beta} x_\alpha^c x_\beta^c B_2^{(o),\alpha\beta} - e^{-\kappa l}\frac{\kappa^3}{n_c}\right], \quad (5.3.31)$$

where P_0 is the pressure of the solvent (gas or liquid or neutral molecules a zero ion concentration), n_c is the total density of ions, and $x_\alpha^c \equiv n_\alpha / n_c$. $P - P_0$ is the osmotic pressure of the electrolyte solutions.

Exercise

Find the form of the fugacity f_α of ionic species α corresponding to the pressure given by eq. (5.3.31).

5.3.7 Empirical Inert Gas Potential

Most interaction potentials used in molecular theory are idealizations or approximations designed to capture the qualitative physics of materials and to predict quantitatively a few properties. However, in two works by Barker and coworkers (1971), (1974), a potential model was introduced and its parameters fit to an extensive array of experimental data for argon, krypton, and xenon. Thus, their potential model is more quantitative than the traditional, simple ones.

Representing the argon potential as the sum of the pair potential to be described below and the Axilrod–Teller triplet potential, eq. (5.3.1), Barker and coworkers obtained an excellent fit between theoretical predictions and experimental data for the second virial coefficient, the vibrational spectrum of argon dimers, molecular beam scattering cross-section determinations, gaseous transport coefficients, and the lattice spacing and cohesive energy of solid argon at 0 K. Furthermore, the thermodynamic properties determined by computer simulations of a liquid of molecules interacting with the argon potential model agree very well with the observed properties of liquid argon. The pair potential introduced for argon is of the form

$$u_{12}(x) = \varepsilon\left\{\sum_{i=0}^{5} A_i (x-1)^i \exp[\alpha(1-x)] - \sum_{j=0}^{2} c_{2j+6}/(\delta + x^{2j6})\right\}, \quad (5.3.32)$$

where

$$x \equiv r_{12}/R_m; \quad \frac{\partial u_{12}}{\partial r_{12}} = 0 \text{ at } r_{12} = R_m; \quad u_{12}(R_m) = -\varepsilon; \quad u_{12}(R_0) = 0. \quad (5.3.33)$$

The parameters entering eq. (5.3.32) are given in Table 5.15. The parameters given in the columns labelled Barker–Pompe and

Bobetic–Barker, respectively, were determined by these investigators by a fit of theory to experiment for thermodynamic and transport properties of dilute argon gas and for the thermodynamic properties of low temperature solid argon. Thus, it is significant that the potential (for either set of parameters) gives good agreement with vibrational data for argon dimers and with molecular beam scattering cross-section data. No other potential model agrees as well with as many separate pieces of experimental data. To further improve agreement between computer simulated liquid properties and measured liquid argon properties, Barker et al. (1971) tried arithmetic averages of u_{BP} and u_{BB} [eq. (4.1.18) with the Barker–Pompe, BP, and Bobetic–Barker, BB, parameters, respectively]. Their best average potential was reported to be

$$u = \frac{3}{4}u_{BB} + \frac{1}{4}u_{BP}, \tag{5.3.34}$$

for which the values of ε, R_m, and R_0 are given in the final column of Table 5.15. The quantity $\varepsilon_{123}/\varepsilon R_m^9$ in Table 5.15 is the reduced value of the Axibrod–Teller parameter, which is $\varepsilon_{123} = 73.2 \times 10^{-84}$ erg cm^9 for argon. The 6–12 LJ potential is compared in Figure 5.13 with eq. (5.3.34). The LJ potential is more shallow at its minimum and is longer ranged than the more accurate potential.

Above 150 K, values of B_2 predicted by the 6–12 LJ model for $\varepsilon/k = 119.8$ K and $\sigma = 3.405$ Åagree with experiment on argon about as well as do values predicted by eq. (5.3.34). Below 150 K, eq. (5.3.33) still agrees well with experiment, whereas the 6–12 LJ model begins to become increasing incorrect, but in error by only about 10% at 100 K. Thus, just for the sake of reproducing second virial coefficients, the added effort in using the more complicated model of Barker and coworkers (1971) rather than the simple LJ

TABLE 5.15 Potential Parameters for Argon

Parameters	*Barker–Pompe*	*Bobetic–Barker*	*Eq. (5.3.34)*
ε/k (K)	147.70	140.235	142.095
R_m (Å)	3.7560	3.7630	3.7612
R_0 (Å)	3.341	3.3666	3.3605
A_0	0.2349	0.29214	0.2718
A_1	−4.7735	−4.41458	−4.5043
A_2	−10.2194	−7.70182	−8.3312
A_3	−5.2905	−31.9293	−25.2696
A_4	0.0	−136.026	−102.020
A_5	0.0	−151.0	−113.25
C_6	1.0698	1.11976	1.1073
C_8	0.1642	0.171551	0.1697
C_{10}	0.0132	0.013748	0.0136
α	12.5	12.5	12.5
δ	0.01	0.01	0.01
$\varepsilon_{123}/\varepsilon R_m^9$	0.02416	0.0250	0.0249

Adapted from Barker et al. (1971).

Figure 5.13
Comparison of 6–12 LJ potential (- - -) Barker–Fisher–Watts potential, eq. (5.3.33) and (—) argon.

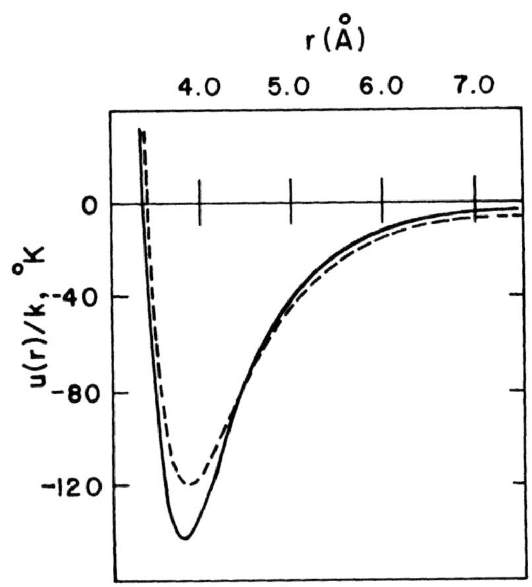

model is difficult to justify. However, it is the goal of a complete molecular theory to obtain the parameters and form of the true potential energy so that all thermodynamic, transport and transient processes may be treated by the same model. It is toward this goal that Barker and colleagues have made significant contributions.

The pair potential proposed by Barker et al. for krypton and xenon is of the form

$$u_{12}^{(x)} = u_{12}^{(A)} + u_{12}^{(B)}, \tag{5.3.35}$$

where $u_{12}^{(A)}$ is defined by eq. (5.3.31) and

$$\begin{aligned} u_{12}^{(B)}(x) &= [P(x-1)^4 + Q(r-1)^5]^{\alpha(1-x)}, \quad x > 1 \\ &= 0 \quad x < 1. \end{aligned} \tag{5.3.36}$$

Two sets of parameters for this potential are given in Table 5.16. The sets are given in Table 5.16. The sets given in columns K2 and X2, corresponding to the case for which $u_{12}^{(B)} \equiv 0$, give the best overall agreement with the widest range of experimental data. The krypton and xenon potential models K2 and X2 appear to be as successful as the argon model given by eq. (5.3.33). Also given in Table 5.16 are reduced values of ε_{123}, the Axilrod–Teller three-body interaction constant. In Figure 5.14, the second virial coefficient predicted by K2 is compared to krypton experimental values. The agreement is very good even down to the lowest temperatures for which data are available (the deviation between the experiment and theory at 100 K being no more than 10% and above 125 K being within experimental error).

To illustrate the importance of three-body interactions, we compare in Figure 5.15 the third virial coefficient of argon computed from the pair potential of Barker et al. [1971; (Eq. (5.3.31)] with and without the Axilrod–Teller interaction [eq. (5.3.1)]. The

TABLE 5.16 Parameters for Krypton and Xenon Potentials for Eq. (4.1.21)

	Krypton (K1)	*Krypton (K2)*	*Xenon (X1)*	*Xenon (X2)*
ε/k (K)	201.3	201.9	293.8	281.0
R_m (Å)	4.008	4.0067	4.355	4.3623
R_0 (Å)	3.580	3.573	3.870	3.890
A_0	0.23526	0.23526	0.18345	0.2402
A_1	−4.78686	−4.78686	−4.6204	−4.8169
A_2	−9.2	−9.2	−27.0	−10.9
A_3	−20.0	−8.0	−58.0	−25.0
A_4	−60.0	−30.0	10.0	−200.0
A_5	−114.0	−205.8	10.0	−200.0
C_6	1.0632	1.0632	1.0052	1.0544
C_8	0.1701	0.1701	0.1590	0.1660
C_{10}	0.0143	0.0143	0.03105	0.0323
α	12.5	12.5	15.5	12.5
δ	0.01	0.01	0.01	0.01
P	0.0	−9.0	0.0	59.3
Q	0.0	68.67	0.0	71.1
d'	...	12.5	...	12.5
$\varepsilon_{123}/\varepsilon R_m^9$	0.0296	0.0296	0.0361	0.0361

Adapted from Barker et al. (1974).

predictions are also compared with experimental values of the third virial coefficient. Addition of three-body interactions substantially improves the predicted results, so much so that one wonders why two-body models do not perform very badly for condensed phase

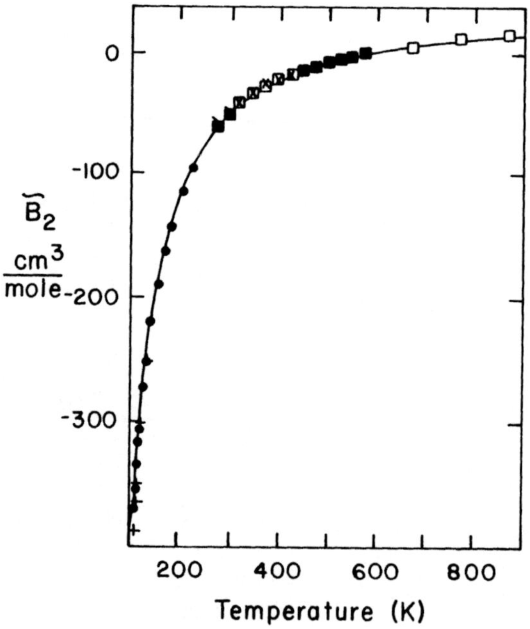

Figure 5.14
Second virial coefficients for krypton; experimental values and those calculated from potential K2 (—). Experimental data: (•), Weir et al.; (×), Trappenniers et al. (+) Fender and Halsey; (□) Whalley and Schneider; (■) Beattie et al. For detailed references see Dymond and Smith, (1969).

Figure 5.15
Predicted and experimental third virial coefficient versus temperature for argon. (- - -) Empirical pair potential, eq. (5.3.31); (· · ·) empirical pair potential plus three-body potential, eq. (5.3.1); (○) experiment. Adapted from Barker et al. (1974).

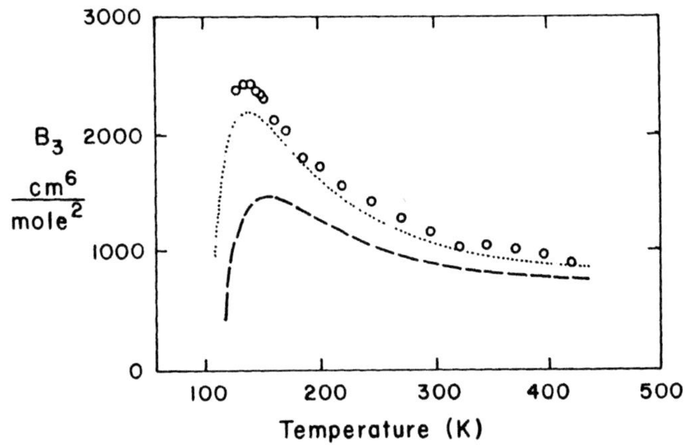

predictions. Barker and coworkers have discussed this point and have concluded that the three body effects partly cancel out in the condensed phase. Additionally, when two-body models are fitted by equating theory and the experiment for thee condensed phases, the many-body effects are incorporated in an average sense in the parameters of the fitted two-body potential.

5.4 Density Expansion of Correlation Functions: One-Component, Simple Fluid

Exercise
The third virial coefficient of a square well fluid is

$$B_3(T) = \frac{b_o^2}{8}[5 - (R^6 - 18R^4 + 32R^2 - 15)x - (2R^6 - 36R^4 + 32R^6 + 18R^2 - 16)x^2 - (6R^6 - 18R^4 + 18R^2 - 6)x^3]$$

for $R \leq 2$, where $R = d_2/d_1$, $x = exp(\beta\epsilon) - 1$ and $b_o = 2\pi d_1^3/3$. Compare the square-well model prediction of B_3 with experiment for argon (Fig. 5.15). For argon $d_1 = 3.16$Å, $d_2/d_1 = 1.85$ and $\epsilon/k = 69.4K$. For krypton and xenon, recommended values for these three parameters are 3.36Å, 1.85, 98.3K and 3.54Å, 1.85 and 126.5K.

To appreciate the basic theory without the burden of notational complexity, let us consider first a one-component classical monatomic fluid whose particles interact via pairwise additive central forces. What we do in the following is derive a set of differential equations for the particle correlation functions and find from this set the virial expansion of the correlation functions.

The νth order distribution function for this fluid is given by

$$p_\nu(\mathbf{r}_1, \ldots, \mathbf{r}_\nu) = \frac{\int \cdots \int e^{-\beta u^N} (d^3r)^{N-\nu}}{Z_N} \tag{5.4.1}$$

where $p_\nu (d^3r)^\nu$ is the probability that particles $1, 2, \ldots, \nu$ are in the volume elements $d^3r_1, \ldots, d^3r_\nu$. We use the notation

$$(d^3r)^\nu \equiv d^3r_1 \cdots d^3r_\nu$$
$$(d^3r)^{N-\nu} \equiv d^3r_{\nu+1} \cdots d^3r_N. \tag{5.4.2}$$

The νth order correlation function is by definition

$$g_\nu = p_\nu / \prod_{j=1}^{\nu} p_1(\mathbf{r}_j). \tag{5.4.3}$$

For homogeneous fluids $p_1 = V^{-1}$ so that

$$g_\nu(\mathbf{r}_1, \ldots, \mathbf{r}_\nu) = \frac{V^\nu}{Z_N} \int \cdots \int e^{-\beta u^N} (d^3 r)^{N-\nu}. \tag{5.4.4}$$

Let $\mathbf{r}_i$ denote the position vector of a particle of the set ν (that is, one of the particles $1, \ldots, \nu$). Taking the gradient of each side of eq. (5.4.4) with respect to $\mathbf{r}_i$, we obtain

$$\nabla_{\mathbf{r}_i} g_\nu = -\beta \frac{V^\nu}{Z_N} \int \cdots \int (\nabla_{\mathbf{r}_i} u^N) e^{-\beta u^N} (d^3 r)^{N-\nu}. \tag{5.4.5}$$

The potential energy can be written as the sum $u^N = u^{\nu, N-\nu} + u^{N-\nu}$, where u_ν is the potential of interaction among the members of the set ν, $u^{N-\nu}$ that among the members of the set $N - \nu$ (particles $\nu + 1, \ldots, N$), and $u^{\nu, N-\nu}$ the potential of interaction between members of ν and of $N - \nu$. For central, pair forces

$$u^\nu = \sum_{i>j=1}^{\nu} u(r_{ij}), \quad u^{\nu,N-\nu} = \sum_{i=1}^{\nu} \sum_{j=\nu+1}^{N} u(r_{ij}), \quad \sum u^{N-\nu} = \sum_{i>j=\nu+1}^{N} u(r_{ij}). \tag{5.4.6}$$

With this division, $\nabla_{\mathbf{r}_i} u^N = \nabla_{\mathbf{r}_i} u^\nu + \nabla_{\mathbf{r}_i} u^{\nu, N-\nu}$, and eq. (5.4.5) can be rearranged to

$$\nabla_{\mathbf{r}_i} g_\nu = -\beta g_\nu \nabla_{\mathbf{r}_i} u^\nu - \prod_{j=\nu+1}^{N} \beta \frac{V^\nu}{Z_N} \int \cdots \int e^{-\beta u^N} \nabla_{\mathbf{r}_i} u(\mathbf{r}_{ij}) (d^3 r)^{N-\nu} \tag{5.4.7}$$

Because the particles are identical the integral on the rhs of this expression will be the same for every particle j in the set $N - \nu$. Thus, we can choose $j = \nu + 1$ as the typical value and replace the sum by multiplication by $N - \nu$, the number of identical terms in the sum. Then eq. (5.4.7) can be rewritten as follows:

$$\nabla_{\mathbf{r}_i} g_\nu = -\beta g_\nu \nabla_{\mathbf{r}_i} u^\nu - \frac{N-\nu}{V} \beta \int \nabla_{\mathbf{r}_i} u(r_{i,\nu+1}) \left[\frac{V^{\nu+1}}{Z_N} \int \cdots \int e^{-\beta u^N} (d^3 r)^{N-(\nu+1)} \right] \tag{5.4.8}$$

By definition the quantity in brackets in this expression is the $\nu + 1$st correlation function $g_{\nu+1}(\mathbf{r}_1, \ldots, \mathbf{r}_{\nu+1})$. Thus, we obtain

$$\nabla_{\mathbf{r}_i} g_\nu = -\beta g_\nu \nabla_{\mathbf{r}_i} u^\nu - \frac{N-\nu}{V} \beta \int \nabla_{\mathbf{r}_i} u(r_{i,\nu+1}) g_{\nu+1} d^3 r_{\nu+1}, \tag{5.4.9}$$

$\nu = 2, 3, \ldots$, a set of equations known as the Yvon–Born–Green (YBG) hierarchy, after its discoverers.

We now seek a solution to eq. (5.4.9) of the form

$$g_\nu = \sum_{m=0}^{\infty} g_\nu^{(m)} V^{-m}, \tag{5.4.10}$$

where $g_\nu^{(m)}$ depends on T and N but not V. Inserting this formula into eq. (5.4.9) and equating coefficients of the same power of V, we find

$$\nabla_{\mathbf{r}_i} g_\nu^{(o)} = -\beta g_\nu^{(o)} \nabla_{\mathbf{r}_i} u^\nu \tag{5.4.11}$$

and, for $m \leq 1$,

$$\nabla_{\mathbf{r}_i} g_\nu^{(m)} + \beta g_\nu^{(m)} \nabla_{\mathbf{r}_i} u^\nu = -\beta(N-\nu)\int \nabla_{\mathbf{r}_i} u(r_{i,\nu+1}) g^{(m-1)} d^3 r_{\nu+1}. \tag{5.4.12}$$

This set of equations can be solved successively to obtain the coefficients $g_\nu^{(m)}$ of the virial expansion of g_ν.

With the boundary condition $g_\nu^{(o)} \to 1$ as particles of the set ν become sufficiently widely separated, eq. (5.4.11) has the solution

$$\nabla_{\mathbf{r}_i} g_\nu^{(1)} + \beta g_\nu^{(1)} \nabla_{\mathbf{r}_i} u^\nu = -\beta(N-\nu)\int \nabla_{\mathbf{r}_i} u(r_{i,\nu+1}) e^{-\beta u^{\nu+1}} d^3 r_{\nu+1}. \tag{5.4.13}$$

For the present purpose we shall not consider higher order terms than $g_\nu^{(1)}$ in the virial expansion of g_ν. This suffices to determine the second and third virial coefficients.

Because $u^{\nu+1} - u^\nu = \sum_{j=1}^{\nu} u(r_{j,\nu+1})$, this expression can be rearranged to

$$\begin{aligned}\nabla_{\mathbf{r}_i}\left[e^{\beta^\nu_u} g_\nu^{(1)}\right] &= -\beta(N-\nu)\int \nabla_{\mathbf{r}_i} u(r_{i,\nu+1}) e^{-\sum_{j=1}^{\nu}\beta u(r_{j,\nu+1})} d^3 r_{\nu+1} \\ &= (N-\nu)\nabla_i \int e^{-\sum_{j=1}^{\nu}\beta u(r_{j,\nu+1})} d^3 r_{\nu+1},\end{aligned} \tag{5.4.14}$$

with the aid of the relation

$$\nabla_{\mathbf{r}_i} e^{-\sum_{j=1}^{\nu}\beta u(r_{j,\nu+1})} = -\beta \nabla_{\mathbf{r}_i} u(r_{i,\nu+1}) e^{-\sum_{j=1}^{\nu}\beta u(r_{j,\nu+1})} \tag{5.4.15}$$

The boundary condition on $g_\nu^{(m)}$, $m > 0$, consistent with the asymptotic property of g_ν ($g_\nu \to 1$ as all ν particles separate) and with the boundary condition imposed on $g_\nu^{(o)}$ is that $g_\nu^{(m)} \to 0$, $m > 0$, as all ν particles separate. Equation (5.4.15) can be integrated to get

$$g_\nu^{(1)} = (N-\nu)e^{-\beta u^\nu}\int e^{-\sum_{j=1}^{\nu}\beta u(r_{j,\nu+1})} d^3 r_{\nu+1} + e^{-\beta u^\nu} C, \tag{5.4.16}$$

where C is a constant of integration. Because $u^\nu \to 0$ and $g_\nu^{(1)} \to 0$ as the particles of the set become widely separated, it follows that

$$C = -\lim (N-\nu)\int e^{-\sum_{j=1}^{\nu}\beta u(r_{j,\nu+1})} d^3 r_{\nu+1}, \tag{5.4.17}$$

where the limit indicated is that members of the set ν are widely separated.

Determination of the final form of C is rather difficult for the general case. Thus, we specialize at this point to the case $\nu = 2$. For fluids having only two body forces g_2 is all one needs for predicting thermodynamic functions.

Consider particles 1 and 2 separated by many multiples of the range of their pair potential, with $\mathbf{r}_1$ lying in volume V_1 far from the boundaries of V_1 and $\mathbf{r}_2$ lying similarly in V_2. Suppose V_1 and V_2 are disjoint volumes such that the total volume $V = V_1 + V_2$. Then, eq. (5.4.17) can be expressed as

$$
\begin{aligned}
C &= -(N-2) \int_{V_1+V_2} e^{-\beta\left[u(r_{13})+u(r_{23})\right]} d^3 r_3 \\
&= -(N-2) \int_{V_2} e^{-\beta u(r_{13})} d^3 r_3 - (N-2) \int_{V_1} e^{-\beta u(r_{23})} d^3 r_3.
\end{aligned}
\tag{5.4.18}
$$

The term $e^{-\beta u(r_{23})} = 1$ is the integral involving V_1 because r_{23} is always larger than the range of the pair potential. For the same reason $e^{-\beta u(r_{13})} = 1$ in the integral involving V_2. Thus,

$$
\begin{aligned}
C &= -(N-2)\left[\int_{V_1} e^{-\beta u(r_{13})} d^3 r_3 + \int_{V_2} e^{-\beta u(r_{23})} d^3 r_3\right] \\
&= -(N-2)\left[\int_{V_1} (e^{-\beta u(r_{13})} - 1) d^3 r_3 + \int_{V_2} (e^{-\beta u(r_{23})} - 1) d^3 r_3 + \int_{V} d^3 r_3\right].
\end{aligned}
\tag{5.4.19}
$$

The second form of C is simply an identity. However, it is useful because the factors $e^{-\beta u(r_{13})} - 1$ and $e^{\beta u(r_{23})} - 1$ are zero except for values of r_{13} and r_{23} with the range of the pair potential. Because $\mathbf{r}_1$ and $\mathbf{r}_2$ are far from the boundaries of V_1 and V_2 then the limits of the integrals over V_1 and V_2 can be taken to be V without introduction of error. Thus,

$$
C = -(N-2) \int_V \left[e^{-\beta u(r_{13})} + e^{-\beta u(r_{23})} - 1\right] d^3 r_3 \tag{5.4.20}
$$

and, consequently, through terms of order V^{-1},

$$
g_2(r_{12}) = e^{-\beta u(r_{12})} \left[1 + n \int_V \gamma(r_{13}) \gamma(r_{23}) d^3 r_3\right], \tag{5.4.21}
$$

where we have neglected 2 in $N - 2$, set $n = N/V$ and used the notation

$$
\gamma(r) = e^{-\beta u(r)} - 1. \tag{5.4.22}
$$

Let us examine the theoretical pair correlation function for a couple of simple models. First consider a hard sphere fluid. With the bipolar coordinate transformation, Figure 5.7, the hard sphere version of eq. (5.4.21) is

$$g_2(r_{12}) = \eta(r_{12} - d)\left[1 + \frac{2\pi n}{r_{12}} \int_0^{\infty} dr_{23} r_{23} \gamma(r_{23}) \int_{r_{12}-r_{23}}^{r_{12}+r_{23}} dr_{13} r_{13} \gamma(r_{13})\right], \tag{5.4.23}$$

where

$$\begin{aligned} \eta(r-d) &= 0, \quad \gamma(r) = -1, \ r < d \\ \eta(r-d) &= 1, \quad \gamma(r) = 0, \ r > d. \end{aligned} \tag{5.4.24}$$

Carrying out the integrations in eq. (5.4.23), we get

$$\begin{aligned} g_2 &= \eta\left(r_{12} - d\right)\left\{1 + \frac{4\pi}{3} nd^3 \left[1 - \frac{3}{4}\left(\frac{r_{12}}{d}\right) + \frac{1}{16}\left(\frac{r_{12}}{d}\right)^3\right]\right\}, \quad 0 < r_{12} < 2d \\ &= 0, \quad r_{12} > 2d. \end{aligned} \tag{5.4.25}$$

Note that the contribution to g_2 of first order in density ($O\,V^{-1}$) is measured by the dimensionless number nd^3, that is a measure of the ratio of the excluded volume of an interacting pair of molecules to the volume per molecule of the system. In general, the contribution to g_2 of order n^s in density will be of the order of $(n V_r)^s$, where V_r is the volume of a sphere with radius equal to the largest range of the components of the potential u^s.

Only the nearest neighbor structure appears in g_2 to first order in density (or V^{-1}). This is shown in Figure 5.16, where the g_2 given by eq. (5.4.22) is compared to a qualitative plot of g_2 for a dense hard sphere fluid.

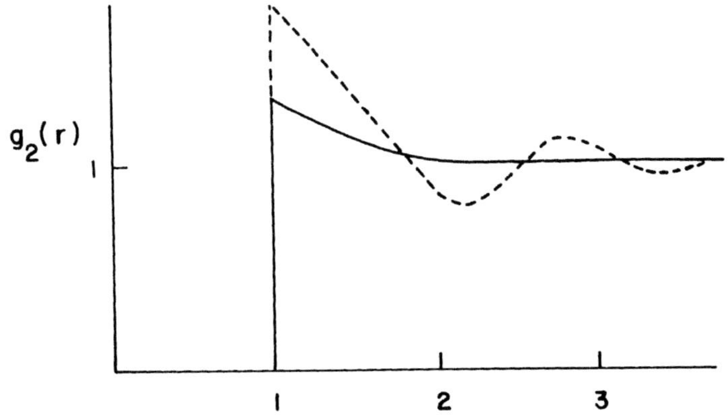

Figure 5.16
Plot of g_2 given by eq. (5.4.24) for $nd^3 = 0.2$ compared to (- - -) qualitative plot of g_2 for a dense hard sphere fluid.

Consider next the 6–12 LJ potential. For this case eq. (5.4.21) can be expressed as

$$g_2(x) = e^{-f(x)/T^*}\left\{1 + \frac{2\pi n\sigma^3}{x}\int_0^\infty dy\, y\gamma(y)\int_{|x-y|}^{x+y} dz\, z\gamma(z)\right\}, \tag{5.4.26}$$

where $x \equiv r_{12}/\sigma,\ T^* \equiv kT/\varepsilon$,

$$\gamma(y) = e^{-f(y)/T^*} - 1 \tag{5.4.27}$$

and

$$f(y) = 4\left(y^{-12} - y^{-6}\right). \tag{5.4.28}$$

The values of g_2 given by eq. (5.4.25) have been computed and plotted in Figure 5.17.

At the density considered, the LJ model, owing to the attractive part of the LJ potential, leads to a sharper nearest neighbor peak than does the hard sphere model. The higher the temperature, the less important is the attractive part of the LJ model and the more the result of the LJ model resembles that of the hard sphere model.

Alternatively to the method we introduced in Section 5.2 to obtain B_2 and B_3, we could determine these quantities directly from eq. (5.4.21). In Chapter 4 we showed that the equation of state for classical particles of the type considered in the present section is

$$P = nkT - \frac{n^2}{6}\int \frac{du(r_{12})}{dr_{12}} r_{12} g_2(r_{12}) d^3 r_{12}. \tag{5.4.29}$$

Inserting eq. (5.4.21) for g_2 in eq. (5.4.28), we identify the second and third virial coefficients to be

$$B_2 = -\frac{1}{6kT}\int e^{-\beta u(r_{12})} r_{12} \frac{du(r_{12})}{dr_{12}} d^3 r_{12} \tag{5.4.30}$$

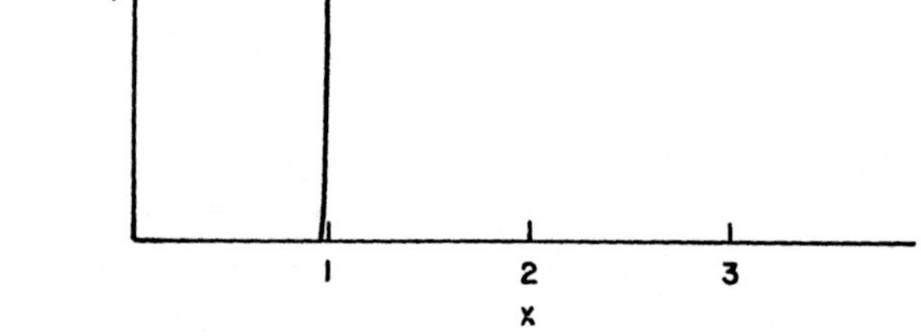

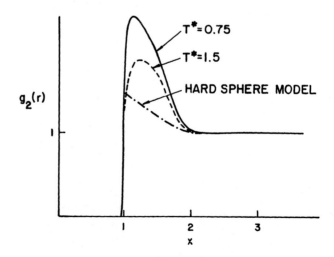

Figure 5.17
Plot of g_2 as predicted by eq. (5.4.25) at $n\sigma^3 = 0.2$ for $T^* = 0.75$ and 1.5 compared with the corresponding hard sphere value of g_2 at $nd^3 = 0.2$. $x \equiv r_{12}/\sigma$ for the LJ model and $\equiv r_{12}/d$ for the hard sphere model.

and

$$B_3 = -\frac{1}{6kT}\int\int r_{12}\frac{du(r_{12})}{dr_{12}}e^{-\beta u(r_{12})}y(r_{13})y(r_{23})d^3r_{12}d^3r_3. \qquad (5.4.31)$$

With the aid of the identity $[dy(r)]/dr = -(1/kT)(du/dr)e^{-\beta u}$ and an integration by parts it is easy to show that eq. (5.4.30) agrees with the formula given at eq. (5.2.29) for B_2. Similarly, but with lengthier manipulations, one can show that eq. (5.4.30) agrees with eq. (5.2.31) for B_3 for the case of no three-body forces.

Supplementary Reading

Gray, C. G. and Gubbins, K. E. 1984. *Theory of Molecular Fluids*, Oxford University Press, Oxford, U.K.

Kestin, J. and Dorfman, J. R. 1971. *A Course in Statistical Thermodynamics*, Academic Press, New York.

Prausnitz, J. M. 1969. *Molecular Thermodynamics of Fluid-Phase Equilibria* , Prentice–Hall, New York.

Prausnitz, J. M., Lichtenthaler, R. M., and Gomes de Azevedo, E. 1986. *Molecular Thermodynamics of Fluid-Phase Equlibria*, Prentice–Hall, New York.

Exercises

1 Given the Stockmayer potential:

$$u(r) = 4\varepsilon\left[\left(\frac{\sigma}{r}\right)^{12} - \left(\frac{\sigma}{r}\right)^{6}\right] - u(\theta, \phi)$$

where

$$u(\theta, \phi) = -\frac{\mu_i\mu_j}{r^3}\left[(2\cos\theta_i\cos\theta_j - \sin\theta_i\sin\theta_j\cos(\phi_i - \phi_j))\right],$$
$$\cos\theta_k = \hat{e}_k\cdot\hat{r}, \quad k = i, j,$$

where $\hat{e}_k$ is a unit vector in the direction of the dipole and the LJ potential:

$$u(r) = 4\varepsilon\left[\left(\frac{\sigma}{r}\right)^{12} - \left(\frac{\sigma}{r}\right)^{6}\right],$$

compare the second virial coefficients B_2 derived from these potentials.

Data for water: $\sigma = 2.65$ Å, $\varepsilon/k = 380$ K, $\mu = 1.87 \times 10^{-18}$ esu.

2 Often second virial coefficient data are fitted to a 6–12 LJ potential even if the molecules are dipolar or non-spherical. In this case σ and ε would represent some angle averaged or "effective" parameters.

For water, the example given in Problem 5.1, effective values of the parameters are

$$\sigma = 1.641\text{Å}, \quad \varepsilon/k = 809.1\text{K}$$

Plot B_2 versus temperature for these parameters and compare the curve with the results obtain from the Stockmayer potential in Problem 5.5.

3 For hard spheres, plot $\mathcal{Z}$ versus n for the approximations

$$\mathcal{Z} = 1 + \sum_{i=1}^{7} n^{i-1} B_i$$

$$\mathcal{Z} = 1 + \sum_{i=1}^{3} n^{i-1} B_i$$

$$\mathcal{Z} = 1 + \sum_{i=1}^{3} \left(\frac{P}{kT}\right)^{i-1} B_i,$$

where $\mathcal{Z} \equiv PV/NkT$. Make the plots for the range $0 \leq nb_o \leq 0.6$.

4 Determine the second and third virial coefficients (B_2 and B_3) of the Redlich–Kwong fluid as a function $\tilde{a}$ and $\tilde{b}$.

5 Consider the pair potentials

$$\begin{aligned} u(r) &= \infty, \quad r < \sigma \\ &= -3\varepsilon\left(\frac{\sigma}{r}\right)^{\nu}, \quad r > \sigma, \end{aligned}$$

where $\nu = 4, 6, 8$.

Compute and plot the reduced second virial coefficient, $B_2/(2\pi\sigma^3/3)$, versus reduced temperature kT/ε for these three potentials. The magnitude of B_2 can be thought of as the volume over which a pair of molecules influence each other. Interpret the differences in B_2 for $\nu = 4, 6, 8$.

6 Using bipolar coordinates evaluate the third virial coefficient for a square-well fluid.

7 Show that the Joule–Thomson inversion temperature obeys the relation

$$\frac{d \ln B_2}{d \ln T} = 1$$

when the pressure can be estimated from the virial expansion keeping terms only through the second virial coefficient.

8 Given the result in Problem 5.5, estimate the Joule–Thomson inversion temperature for the pair potentials in Problem 5.5. Explain the dependence on ν.

9 The activity coefficient γ of a pure fluid can be defined by

$$\mu = \mu^{+}(T) + kT \ln(kT/P_o) + kT \ln(n\gamma),$$

where P_o is a reference pressure. Suppose the fluid is dilute enough that only the second virial coefficient is needed to predict thermodynamic properties. Find the relationship between γ and the second virial coefficient.

10 Using eqs. (5.1.9) and (5.1.8), show that through second order in density the fugacity obeys the equation

$$\ln(f_\alpha/nkT) = 2nB_2 + \frac{1}{2}n^2(B_3 + B_2^2)$$

for one-component fluids. Plot f_α/nkT versus kT/ε for an LJ fluid at density $n\sigma^3 = 0.3$.

11 The Berthelot equation of state is of the form

$$\left(P + \frac{\tilde{a}}{T\tilde{V}^2}\right)\left(\tilde{V} - \tilde{b}\right) = RT.$$

Given that the critical temperature and pressure for normal hexane are 508 K and 29.8 atm, compute the values of $\tilde{a}$ and $\tilde{b}$ for the Berthelot equation.

12 Estimate the fugacity for butane at 10 atm and 400 K using the Peng–Robinson equation of state.

13 Assume argon obeys the LJ interaction potential. Use the formula in Problem 5.10 to esimate the fugacity of argon at 10 atm and 350 K.

14 For benzene plot the experimental virial coefficient, the Kihara prediction for a 6–12 potential [eq. (5.3.18)], and the 6–12 LJ potential.

15 Compute the third virial coefficient B_3 as a function of temperature for argon assuming the interactions are

a) the 6–12 LJ pair additive potential at eq. (2.2.15) and

b) the 6–12 LJ potential plus the three-body potential at eq. (2.2.17).

For argon $\varepsilon = 1.65 \times 10^{-14}$ erg, $\sigma = 3.4$ Å, and $\varepsilon_{123} = 7.32 \times 10^{-83}$ erg-cm^9. Use bipolar coordinates to evaluate the integrals in B_3.

16 Use the theoretical entries in Table 5.12 to construct the universal curve B_2/σ^3 versus kT/ε for a 6–12 LJ fluid.

References

Barker, J. A., Fisher, R. A., and Watts, R. O., 1971. *Molc. Phys.*, **21,** 657.

Barker, J. A., Watts, R. O., Lee, J. K., Schaefer, T. P., and Lee, Y. T. 1974. *J. Chem. Phys.*, **61,** 3081.

Blanks, R. F. and Prausnitz, J. M. 1962. *AIChE J.*, **8,** 86.

Chueh, P. L. and Prausnitz, J. M. 1967. *AIChE J.* **13,** 896.

Curtiss, C. F. in H. Eyring, W. Jost and J. W. Linnet Eds. 1967. *Physical Chemistry,* Vol. II, Academic Press, New York.

Dowling, G. R. and Davis, H. T. 1972. *J. Stat. Phys.*, **4,** 1.

Dymond, J. H. and Smith, E. Z. 1969. *The Virial Coefficients of Gases,* Clarendon, Oxford, UK.

Fowler, R. H. and Guggenheim, E. A. 1939. *Statistical Thermodynamics,* Cambridge University Press, Cambridge, UK, p. 283.

Holborn, L. V. and Otto, J. 1925. *Z. Phys.*, **33,** 1.

Hirschfelder, J. O., McClure, F. T., and Weeks, I. F. 1942. *J. Chem. Phys.*, **10,** 201.

Hirschfelder, J. O., Curtiss, C. F., and Bird, R. B. 1954. *Molecular Theory of Gases and Liquids,* Wiley, New York.

Kihara, T. 1953. *Rev. Mod. Phys.*, **25,** 831; 1958. *Adv. Chem. Phys.*, **1,** 276; 1963. *Adv. Chem. Phys.*, **5,** 147.

Kunz, R. 1968. Ph.D. Thesis, Rensselaer Polytechnic Inst., Troy, NY.

MacCormack, K. E. and Schneider, W. G. 1950. *J. Chem. Phys.*, **18,** 1269.

MacCormack, K. E. and Schneider, W. G. 1951. *J. Chem. Phys.*, **19,** 845.

McGlashan, M. L. and Potter, D. J. B. 1962. *Proc. Roy. Soc. London A,* **267,** 478.

Michels, A. and Geldermans, M. 1942. *Physica,* **9,** 967.

Michels, A. and Nederbragt, G. W. 1936. *Physica,* **3,** 569.

Orentlicher, M. and Prausnitz, J. M. 1967. *Can. J. Chem.*, **45,** 375.

Prausnitz, J. M. 1969. *Molecular Thermodynamics of Fluid-Phase Equilibria,* Prentice–Hall, New York.

6

PHASE EQUILIBRIA

6.1 Thermodynamics of Phase Equilibria

Gibbs' theory of the thermodynamic equilibrium of heterogeneous substances is one of the intellectual marvels in the history of science. The work was totally general, and is as important in today's science of matter as it was when he presented it in the last century.

The conditions of thermodynamic equilibria follow from the second law of thermodynamics that in its simplest form asserts: at equilibrium the entropy S of an isolated system is a maximum. The entropy of a system is a function of its thermodynamic energy U, volume V, and number N_α of each chemical component α. We suppose there are c components. From the maximum entropy principle comes other statements of the second law. Two of the better known of these are (1) in a closed isothermal system of fixed volume the Helmholtz free energy F is a minimum at equilibrium and (2) in a closed, isothermal system at fixed pressure the Gibbs free energy G is a minimum, where

$$F = U - TS \tag{6.1.1}$$

$$G = F + PV = H - TS. \tag{6.1.2}$$

$H(\equiv U + PV)$ is the enthalpy of the system. Another useful statement of the second law is that the grand potential, Ω, of an open, isothermal system of fixed volume is a minimum at equilibrium. The grand potential is defined by

$$\Omega = F - \sum_\alpha \mu_\alpha N_\alpha. \tag{6.1.3}$$

The combination of the first and second laws of thermodynamics yields for a homogeneous system the fundamental equation

$$dU = TdS - PdV + \sum_\alpha \mu_\alpha dN_\alpha, \tag{6.1.4}$$

from which it follows that

$$dH = TdS + VdP + \sum_\alpha \mu_\alpha dN_\alpha, \tag{6.1.5}$$

$$dF = -SdT - PdV + \sum_{\alpha} \mu_{\alpha} N_{\alpha}, \quad (6.1.6)$$

$$dG = -SdT + VdP + \sum_{\alpha} \mu_{\alpha} dN_{\alpha}, \quad (6.1.7)$$

$$d\Omega = -SdT - PdV - \sum_{\alpha} N_{\alpha} d\mu_{\alpha}. \quad (6.1.8)$$

The quantities S, U, V, N_{α}, F, H, G, and Ω are extensive thermodynamic variables, that is, if a homogeneous system of fixed temperature T, pressure P, and chemical potentials μ_{α} is increased k-fold in amount of each component, then the extensive thermodynamic variables become kS, kU, etc. This property allows one to integrate eqs. (6.1.4)–(6.1.8). For example, the integration of eq. (6.1.7) at constant temperature, pressure, and chemical potentials through a k-fold increase in size yields

$$(k-1)G = (k-1)\sum_{\alpha} \mu_{\alpha} N_{\alpha},$$

or

$$G = \sum_{\alpha} \mu_{\alpha} N_{\alpha} \quad (6.1.9)$$

for a homogeneous bulk phase. For multiphase systems there are interfacial contributions not included in eq. (6.1.9). Integration of the other equations at constant intensive variables (T, P, and μ_{α}) similarly yields

$$U = TS - PV + \sum_{\alpha} \mu_{\alpha} N_{\alpha} \quad (6.1.10)$$

$$H = TS + \sum_{\alpha} \mu_{\alpha} N_{\alpha} \quad (6.1.11)$$

$$F = -PV + \sum_{\alpha} \mu_{\alpha} N_{\alpha} \quad (6.1.12)$$

$$\Omega = -PV. \quad (6.1.13)$$

Intensive thermodynamic variables are by definition those variables that are constant throughout a homogeneous phase. They are independent of the amount of phase present.

In an arbitrary infinitesimal variation in the thermodynamic state of the system, then the derivative of U in eq. (6.1.10) yields

$$dU = TdS + SdT - PdV - VdP + \sum_{\alpha} \mu_{\alpha} dN_{\alpha} + \sum_{\alpha} N_{\alpha} d\mu_{\alpha}. \quad (6.1.14)$$

Subtraction of this result from the fundamental eq. (6.1.4) leads to the expression

$$SdT - VdP - \sum_{\alpha} N_{\alpha} d\mu_{\alpha} = 0. \quad (6.1.15)$$

This important equation is known as the Gibbs–Duhem equation. It implies that of the $c+2$ intensive variables T, P, and μ_{α} in a homogeneous phase, only $c+1$ are independent. Thus, for example, if T, $\mu_1, \ldots, \mu_c$ are specified then P is fixed, that is, $P = P(T, \mu_1, \ldots, \mu_c)$. Equivalently it follows that $\mu_\alpha = \mu_\alpha(T, P, \mu_{\beta \neq \alpha})$.

Differentiation of the integrated expression for H, F, G, or Ω and subtraction of the result from the corresponding fundamental equation also yields the Gibbs–Duhem equation. The Gibbs–Duhem equation at first glance appears to give something for nothing. This is not the case, however. In obtaining the integrated expressions in eqs. (6.1.9)–(6.1.13) we assumed that the variables S, U, V, F, H, G, and Ω are homogeneous functions of the first degree in the amounts N_α of the components of a homogeneous system for fixed intensive properties T, P, and μ_α. That is, if E is any one of these extensive variables, it obeys the scaling relation

$$E(T, P, \mu_1, \ldots, \mu_c, kN_1, \ldots, kN_c) = kE(T, P, \mu_1, \ldots, \mu_c, N_1, \ldots, N_c). \qquad (6.1.16)$$

That U, S, and V are homogeneous functions of the first degree in $N_1, \ldots, N_c$ has been verified experimentally for bulk phases that are sufficiently large compared to molecular dimensions. The property follows automatically from the definitions of F, G, H, and Ω once it is established for U, S, and V. The Gibbs–Duhem equation thus results from the "extensivity" of U, S and V, which is a thermodynamic property that is in addition to the fundamental eq. (6.1).

Let us now turn to the problem of equilibrium among phases. To see how the second law yields the conditions of phase equilibria, consider the isolated system depicted in Figure 6.1. Suppose bulk phase 1 is in equilibrium with bulk phase 2. By bulk phase we mean that the phase is large enough that it is homogeneous everywhere except at interfaces with other bulk phases and that the entropy, energy, volume, and number of molecules in the interface are negligible compared to the bulk phase values. The thermodynamics of interfaces will be presented in the next chapter. Consider a small displacement $\delta U^{(i)}$, $\delta V^{(i)}$, $\delta N_\alpha^{(i)}$, $i = 1, 2$, from equilibrium. Because the system is isolated it follows that

$$\begin{aligned} \delta U^{(1)} &= -\delta U^{(2)}, \quad \delta V^{(1)} = -\delta V^{(2)}, \\ \delta N_\alpha^{(1)} &= -\delta N_\alpha^{(2)}, \quad \alpha = 1, \ldots c. \end{aligned} \qquad (6.1.17)$$

The change in entropy corresponding to an arbitrary fluctuation from equilibrium must obey the condition

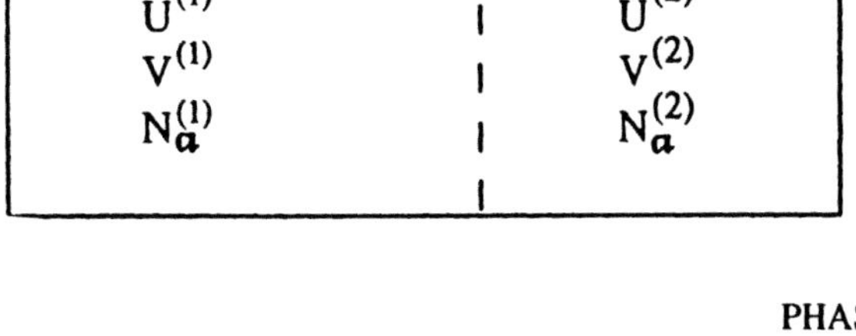

$$U = U^{(1)} + U^{(2)}$$
$$V = V^{(1)} + V^{(2)}$$
$$N_\alpha = N_\alpha^{(1)} + N_\alpha^{(2)}$$

Figure 6.1
Isolated system of fixed energy U, volume V, and number N_α of species α. The interface contributes negligibly to the extensive properties of the two phases if they are sufficiently large.

$$\Delta S = \sum_{i=1}^{2} \left[S^{(i)}(U^{(i)} + \delta U^{(i)}, V^{(i)} + \delta V^{(i)}, \mathbf{N}^{(i)} + \delta \mathbf{N}^{(i)}) - S^{(i)}(U^{(i)}, V^{(i)}, \mathbf{N}^{(i)}) \right] < 0. \tag{6.1.18}$$

With the aid of the Taylor expansion, the isolated system condition in eq. (6.1.18) can for small fluctuations be expressed as

$$\Delta S = \sum_{k=1}^{\infty} \delta^k S < 0, \tag{6.1.19}$$

where

$$\delta S = \sum_{i=1}^{2} (-1)^i \left[\frac{\partial S^{(i)}}{\partial U^{(1)}} \partial U^{(1)} + \frac{\partial S^{(i)}}{\partial V^{(i)}} \delta V^{(1)} + \sum_{\alpha=1}^{c} \frac{\partial S^{(i)}}{\partial N_\alpha^{(i)}} \delta N_\alpha^{(1)} \right] \tag{6.1.20}$$

$$\begin{aligned} \delta^2 S = \sum_{i=1}^{2} \frac{1}{2!} \Bigg[& \frac{\partial^2 S^{(i)}}{\partial U^{(i)2}} (\delta U^{(1)})^2 + \frac{\partial^2 S^{(i)}}{\partial V^{(i)2}} (\delta V^{(1)})^2 \\ & + \sum_{\alpha}^{c} \sum_{\alpha}^{c} \frac{\partial^2 S^{(i)}}{\partial N_\alpha^{(i)} \partial N_\beta^{(i)}} \delta N_\alpha^{(1)} \delta N_\beta^{(1)} + 2 \frac{\partial^2 S^{(i)}}{\partial V^{(i)} \partial U^{(i)}} \delta V^{(1)} \delta U^{(1)} \\ & + 2 \sum_{\alpha}^{c} \frac{\partial^2 S(i)}{\partial N_\alpha^{(i)} \partial U^{(i)}} \delta N_\alpha^{(1)} \delta U^{(1)} + 2 \sum_{\alpha}^{c} \frac{\partial^2 S^{(i)}}{\partial N_\alpha^{(i)} \partial V^{(i)}} \delta N_\alpha^{(1)} \delta V^{(1)} \Bigg], \end{aligned} \tag{6.1.21}$$

etc. for $\delta^k S$, $k > 2$.

From eq. (6.1.4) one can derive the relations

$$\frac{1}{T^{(i)}} = \frac{\partial U^{(i)}}{\partial S^{(i)}}, \quad \frac{P^{(i)}}{T^{(i)}} = -\frac{\partial S^{(i)}}{\partial V^{(i)}}, \quad \text{and} \quad \frac{\mu_\alpha^{(i)}}{T^{(i)}} = \frac{\partial S^{(i)}}{\partial N_\alpha^{(i)}}. \tag{6.1.22}$$

Choosing $\delta V^{(1)} = \delta N_\alpha^{(1)} = 0$ and taking $\delta U^{(1)}$ to be an infinitesimal fluctuation of arbitrary sign, we obtain

$$\Delta S = \left(\sum_{i=1}^{2} \frac{(-1)^i}{T^{(i)}} \right) \delta U^{(1)} + O(\delta U^{(1)})^2. \tag{6.1.23}$$

The only way ΔS can be less than zero for arbitrarily small $\delta U^{(1)}$ of arbitrary sign is that at equilibrium

$$T^{(1)} = T^{(2)}. \tag{6.1.24}$$

Otherwise, the magnitude of $\delta U^{(1)}$ can be chosen sufficiently small that the terms of order $(\delta U^{(1)})^2$ and greater will be negligible and the linear term can be made positive by choosing $\delta U^{(1)}$ to have the same sign as that of $\sum_{i=1}^{2} (-1)^i / T^{(i)}$. Thus, one condition of thermodynamic equilibrium is that the temperature is the same in all parts of the system. Similarly, by choosing $\delta V^{(1)} \neq 0$ and $\delta U^{(1)} = \delta N_\alpha^{(1)} = 0$, we deduce

$$\frac{P^{(1)}}{T^{(1)}} = \frac{P^{(2)}}{T^{(2)}}, \tag{6.1.25}$$

and with $\delta N_\alpha^{(1)} \neq 0, \delta U^{(1)} = \delta V^{(1)} = \delta N_{\beta\neq\alpha}^{(1)} = 0$, we obtain

$$\frac{\mu_\alpha^{(1)}}{T^{(1)}} = \frac{\mu_\alpha^{(2)}}{T^{(2)}}. \tag{6.1.26}$$

Equations (6.1.24)–(6.1.26) represent the conditions of thermal, mechanical, and chemical equilibrium. In summary, according to the second law of thermodynamics, the necessary conditions of thermodynamic equilibrium are that the temperature T, pressure P, and the chemical potentials μ_α, $\alpha = 1, \ldots, \mu$, are the same in all parts of the system. These conditions are independent of whether or not the subparts of the system are the same or different phases. Intensive thermodynamic quantities that are the same in *all coexisting phases* are referred to as *field variables*.

A homogeneous phase is fully specified if its total mass, temperature, pressure, and mole fractions x_i are given. The temperature, pressure, and mole fractions are intensive variables, that is, they are the same throughout a homogeneous bulk phase (mole fractions are, however, not field variables because they are not in general the same in all coexisting bulk phases). Gibbs phase rule incorporates the conditions of thermodynamic equilibrium and prescribes the number of intensive variables that can be independently set in a multiphase system at equilibrium.

To derive the phase rule suppose there are p coexisting phases and c chemical components. The intensive variables to be specified are T, P, and mole fractions $x_\alpha^{(i)}$ of each component α in each phase i. There are $2 + cp$ such variables. However, chemical equilibrium requires that

$$\mu_\alpha^{(1)} = \mu_\alpha^{(2)} \cdots = \mu_\alpha^{(p)}, \quad \alpha = 1, \ldots, c. \tag{6.1.27}$$

Also, since the mole fraction x_α is by definition $x_\alpha = N_\alpha / \sum_{\beta=1}^{c} N_\beta$, it follows that

$$\sum_{\alpha=1}^{c} x_\alpha^{(i)} = 1, \quad i = 1, \ldots, p. \tag{6.1.28}$$

Equation (6.1.27) places $c(p-1)$ constraints on the intensive variables and eq. (6.1.28) imposes another p constraints. Thus, the number f of "degrees of freedom" or intensive variables that can be set independently is the difference between $2 + cp$ and $c(p-1) + p$, namely,

$$f = 2 + c - p. \tag{6.1.29}$$

Equation (6.1.21) is the famous Gibbs phase rule. Consider for example liquid–vapor equilibrium in a one-component system ($c = 1$, $p = 2$). If one intensive parameter, say T, is set, then a unique liquid and vapor pair at a unique vapor pressure must result. On the other hand, if there are two components ($c = 2$, $p = 2$), then a continuous family of liquid–vapor states can occur at a fixed temperature. If both T and P are fixed then a unique liquid and vapor pair results. These examples are illustrated in Figures 6.2 and 6.3. The liquid–vapor coexistence curves (binodals) for methane, butane, octane, and water are shown in Figure 6.2. At a given temperature below the critical point T_c, the coexisting vapor and liquid densities are the densities at the intersections of the constant temperature line with the left and right branches of the binodal. For example, the vapor–liquid "tie-line" for decane at 502 K indicates vapor and liquid densities of 0.04 g/cm^3 and 0.52 g/cm^3. The dashed curves in Figure 6.2 are binodals predicted by the Peng–Robinson (PR) equation of state. We discuss the predictions later.

In a one-component system, the vapor pressure is a function only of temperature along the binodal and thus cannot be varied

Figure 6.2
Liquid–vapor coexistence curves for methane, butane, octane and water. Redrawn from Kuan et al.(1986).

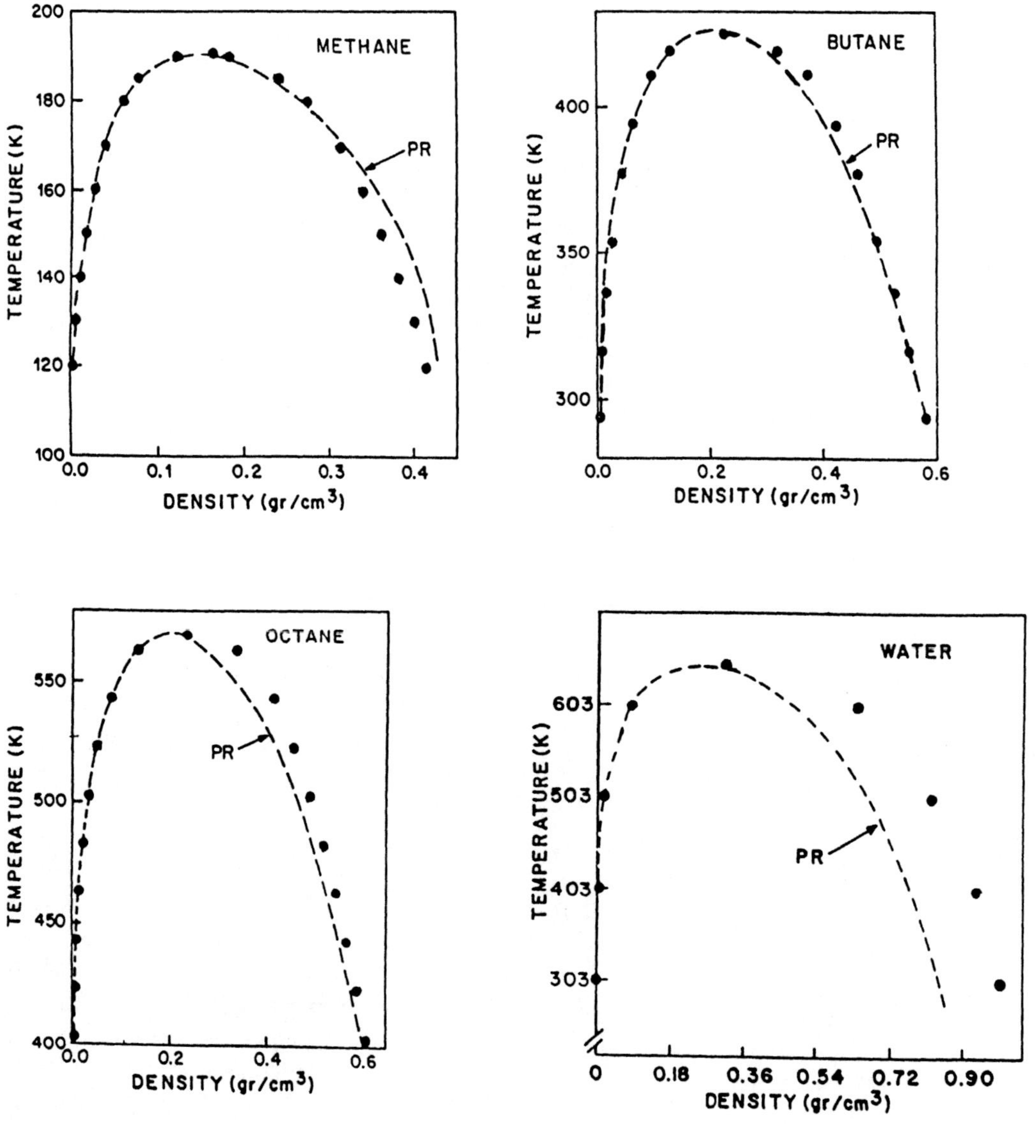

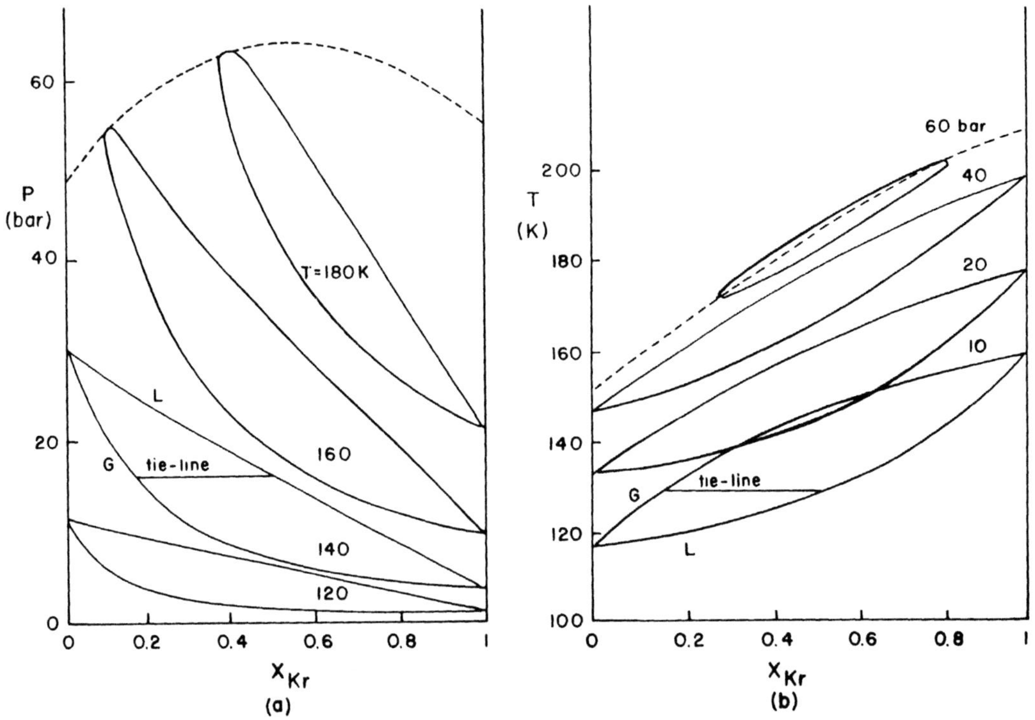

Figure 6.3
Liquid–vapor coexistence curves for argon and krypton mixtures. Redrawn from M. St. Marie, J. Schwedock, and K. E. Gubbins (private communication).

independently of T. On the other hand, for the two-component system, argon and krypton in certain ranges of temperature and pressure, there is a continuous family of binodals. This is illustrated by the pressure–mole fraction and the temperature–mole fraction diagrams in Figure 6.3. At say 140 K there is a liquid–vapor coexistence curve. The coexisting liquid and vapor densities can be read off the $P - x$ diagram by drawing a *constant pressure tie-line* from the vapor branch to the liquid branch of the binodal. For example, at a vapor pressure of 16 bars the mole fraction of argon is about 0.2 in the vapor phase and about 0.5 in the liquid phase. The family of pressure–composition binodals resulting from a scan of temperature are shown in Figure 6.3. Above the critical temperature of pure argon (151 K), the binodal detaches itself from the pure argon axis and ends in a critical pressure P_c. The dashed curve shown in the figure indicates the locus of critical pressures as a function of temperature for the binary system. This is in contrast with a one-component system in which there is a unique liquid–vapor critical point, a contrast again reflecting the phase rule. The family of temperature–composition binodals resulting from a scan of pressure is also illustrated in Figure 6.3.

Let us return to the implications of the second law. The law states that the entropy is a maximum at equilibrium. The equilibrium conditions we have discussed thus far result from the requirement that the first variation, δS, vanish at equilibrium. This is, however, only the necessary condition that S be a maximum. If $\delta^2 S$

is positive for arbitrary fluctuations $\delta U^{(1)}$, $\delta V^{(1)}$, and $\delta N_\alpha^{(1)}$ then the system is not in equilibrium and the state under investigation is not stable to small fluctuations. A multiphase system will be unstable if any one of the phases is unstable to small fluctuations. If $\delta^2 S^{(i)} < 0$ for each phase, it follows that the state of the system corresponds to at least a local maximum in the entropy. Such a state might not be the ultimate equilibrium state because another larger entropy maximum might exist. We say such states are metastable. They may be realized for a short time (e.g., superheated liquid) but will ultimately undergo an irreversible transition to the most stable state.

To learn more about the conditions of stable equilibrium, let us consider the case in which bulk phases 1 and 2 in Figure 6.1 are identical phases. In this case the system can be a single phase divided into equal parts 1 and 2. In this case $S^{(i)} = S/2$ and eq. (6.1.21) reduces to

$$\delta^2 S = \frac{1}{2}\left\{\frac{\partial^2 S}{\partial U^2}(\delta U^{(1)})^2 + \frac{\partial^2 S}{\partial V^2}(\delta V^{(1)})^2 + \sum_{\alpha,\beta}\frac{\partial^2 S}{\partial N_\alpha \partial N_\beta}\delta N_\alpha^{(1)}\delta N_\beta^{(1)} + 2\frac{\partial^2 S}{\partial U \partial V}\delta U^{(1)}\delta V^{(1)} \right.$$
$$\left. 2\sum_\alpha \frac{\partial^2 S}{\partial N_\alpha \partial U}\delta N_\alpha^{(1)}\delta U^{(1)} + 2\sum_\alpha \frac{\partial^2 S}{\partial N_\alpha \partial V}\delta N_\alpha^{(1)}\delta V^{(1)}\right\}. \tag{6.1.30}$$

The entropy maximum principle, eq. (6.1.18), requires $\delta^2 S \leq 0$ for arbitrary fluctuations $\delta U^{(1)}$, $\delta V^{(1)}$, and $\delta N_\alpha^{(1)}$. If $\delta^2 S = 0$ for some fluctuation we have to examine the higher order variations $\delta^k S$ to determine stability. Let us consider only the stability condition $\delta^2 S < 0$. The case $\delta V^{(1)} = \delta N_\alpha^{(1)} = 0$, $\delta U^{(1)} \neq 0$ implies that $\partial^2 S/\partial U^2 < 0$. But from eq. (6.1.14) it follows that $\partial S/\partial U = 1/T$, and so $\partial^2 S/\partial U^2 = (\partial T^{-1}/\partial U)_{\mathbf{N},V} = -T^{-2}(\partial T/\partial U)_{\mathbf{N},V}$. The constant volume specific heat is defined by $C_V = (\partial U/\partial T)_{\mathbf{N},V}$. Thus, according to the second law, stable thermal equilibrium requires that

$$C_V > 0. \tag{6.1.31}$$

Next consider the case $\delta U^{(1)} = \delta N_\alpha^{(1)} = 0$, $\delta V^{(1)} \neq 0$. Because $(\partial S/\partial V)_{U,\mathbf{N}} = P/T$, the second law yields the condition of stable mechanical equilibrium:

$$\left[\frac{\partial(P/T)}{\partial V}\right]_{U,\mathbf{N}} < 0. \tag{6.1.32}$$

Finally, for the case $\delta U^{(1)} = \delta U^{(1)} = 0$, $\delta N_\alpha^{(1)} \neq 0$, there is stable equilibrium if

$$\delta^2 S = \frac{1}{2}\sum_{\alpha,\beta}\frac{\partial^2 S}{\partial N_\alpha \partial N_\beta}\delta N_\alpha^{(1)}\delta N_\beta^{(1)} < 0 \tag{6.1.33}$$

for arbitrary $\delta N_\alpha^{(1)}$. From eq. (6.1.4) we find $(\partial S/\partial N)_{U,V,N_{\beta\neq\alpha}} = -\mu_\alpha/T$. From this it follows that there is stable chemical equilibrium if the symmetric real matrix **A**,

$$\mathbf{A} = [a_{\alpha\beta}], \quad a_{\alpha\beta} = \frac{\partial^2 S}{\partial N_\alpha \partial N_\beta} = -\left[\frac{\partial}{\partial N_\beta}\left(\frac{\mu_\alpha}{T}\right)\right]_{U,V,N_{\alpha\neq\beta}}, \tag{6.1.34}$$

is negative definite, that is, if all of the eigenvalues of **A** are negative.

In the laboratory it is more common to deal with the variables T, V, and N_α than U, V, and N_α. For this reason conditions of a stable equilibrium are frequently derived from the second law statement that the Helmholtz free energy of a closed isothermal system at temperature T is a minimum. According to this condition, for a fluctuation in a homogeneous bulk phase at equilibrium the free energy must vary such that

$$\Delta F = \sum_{i=1}^{2} F(T,\ V^{(i)} + \delta V^{(i)},\ \mathbf{N}^{(i)} + \delta\mathbf{N}^{(i)}) - F(T,\ V,\ \mathbf{N}) \geq 0. \tag{6.1.35}$$

The first variation of F merely recovers the equilibrium conditions $P^{(1)} = P^{(2)}$ and $\mu_\alpha^{(1)} = \mu_\alpha^{(2)}$. (Note that the condition of thermal equilibrium cannot be derived from the minimum Helmholtz free energy version of the second law. This is because one field variable, T, is constrained in this case whereas in the entropy principle no field variables are constrained.) The condition that $\Delta F > 0$ is assured for a homogeneous bulk phase of

$$\delta^2 F = \frac{1}{2}\left\{\frac{\partial^2 F}{\partial V^2}(\delta V^{(1)})^2 + 2\sum_\alpha \frac{\partial^2 F}{\partial V \partial N_\alpha}\delta V^{(1)}\delta N_\alpha^{(1)} + \sum_{\alpha\beta}\frac{\partial^2 F}{\partial N_\alpha \partial N_\beta}\delta N_\alpha^{(1)}\delta N_\beta^{(1)}\right\} > 0 \tag{6.1.36}$$

for arbitrary $\delta V^{(1)}$ and $\delta N_\alpha^{(1)}$ such that not all fluctuations are zero.

The special case $\delta V^{(1)} \neq 0$, $\delta N_\alpha^{(1)} = 0$ implies that the isothermal compressibility is positive, or that

$$\left(\frac{\partial P}{\partial V}\right)_{T,\mathbf{N}} < 0, \tag{6.1.37}$$

which is a more common statement of stable mechanical equilibrium than the one given at eq. (6.1.32). The condition of chemical equilibrium is found by considering $\delta V^{(1)} = 0$ and $\delta N_\alpha^{(1)} \neq 0$. In this case,

$$\delta^2 F = \frac{1}{2}\sum_{\alpha,\beta}\frac{\partial^2 F}{\partial N_\alpha \partial N_\beta}\delta N_\alpha^{(1)}\delta N_\beta^{(1)} > 0. \tag{6.1.38}$$

and so it follows that the bulk phase is stable (or at least metastable) if the matrix

$$\mathbf{A} = [a_{\alpha,\beta}], \quad a_{\alpha\beta} = \frac{\partial^2 F}{\partial N_\alpha \partial N_\beta} = \left(\frac{\partial \mu_\alpha}{\partial N_\beta}\right)_{T,V,N_{\alpha\neq\beta}} \tag{6.1.39}$$

is positive definite, that is, all of its eigenvalues are positive. To obtain the last equality of eq. (6.1.39) we used the relation $\mu_\alpha = (\partial F/\partial N_\alpha)_{\beta\neq\alpha}$ that follows from Eq. (6.1.6). Because one often has equations of state for F as a function of T, V, and N_α (and does not usually have S as a function of U, V, and N_α), eq. (6.1.31) is a more convenient statement of stable chemical equilibrium than is eq. (6.1.33).

In systems having three components or more, one frequently studies phase equilibria at fixed temperature and pressure and chooses mole fractions, $x_\alpha = N_\alpha / \sum N_\beta$, or mass fractions, as the intensive variables to describe the state of the system. In this case it is expedient to use the Gibbs function to investigate chemical stability. At constant T and P a homogeneous phase is locally stable if

$$\delta^2 G = \frac{1}{2}\sum_{\alpha,\beta} \frac{\partial^2 G}{\partial N_\alpha \partial N_\beta} \delta N_\alpha^{(1)} \delta N_\beta^{(1)} \geq 0 \tag{6.1.40}$$

for arbitrary $\delta N_\alpha^{(1)}$ at constant T and P. Unlike the chemical equilibrium stability conditions in terms of S and F [eqs. (6.1.33] and (6.1.38)) there is one composition fluctuation for which $\delta^2 G = 0$ even for a locally stable system. This is the fluctuation in which $\delta N_\alpha^{(1)} = kN_\alpha$, that is, the composition of each component is shifted by the same percent. Noting that the Gibbs–Duhem equation, eq. (6.1.15), for fixed T and P can be expressed as

$$\sum_\alpha \frac{\partial G}{\partial N_\alpha} N_\alpha = 0 \tag{6.1.41}$$

and considering for the fluctuation $\delta N_\alpha^{(1)} = kN_\alpha$, we can rewrite eq. (6.1.40) in the form

$$\delta^2 G = \frac{k^2}{2} \sum_\beta N_\beta \frac{\partial}{\partial N_\beta}\left[\sum_\alpha \frac{\partial G}{\partial N_\alpha} N_\alpha\right]. \tag{6.1.42}$$

Thus, $\delta^2 G = 0$ for this special composition fluctuation. Physically, the fluctuation corresponds to moving the imaginary boundary between parts 1 and 2 in Figure 6.1 without changing the temperature, pressure, or composition in a homogeneous fluid. Thus, the "fluctuation" corresponds to no change at all in the overall system. In view of eqs. (6.1.41) and (6.1.42), it follows that there is stable chemical equilibria at fixed T and P if the matrix $\mathbf{A}$,

$$\mathbf{A} = \left[\frac{\partial^2 G}{\partial N_\alpha \partial N_\beta}\right] = \left[\left(\frac{\partial \mu_\alpha}{\partial N_\beta}\right)_{T,P,N_{\alpha\neq\beta}}\right], \tag{6.1.43}$$

is positive semidefinite with only one zero eigenvalue. The zero eigenvalue corresponds to the trivial concentration fluctuation $\delta N_\alpha^{(1)} = kN_\alpha$ that does not change concentrations in a homogeneous system at fixed T and P.

In Chapter 5 we investigated liquid–vapor phase equilibria of a one-component system. To illustrate the phase equilibria and stability conditions discussed above and to set the stage for discussing multicomponent equilibria, let us reexamine the one-component system. It is convenient to deal with density $n \equiv N/V$ and Helmholtz free energy density $f \equiv F/V$ instead of number of molecules N and free energy F. Because of the extensivity of F and V, f depends only on T and n. The chemical potential and pressure can be computed from

$$\mu = \left(\frac{\partial F}{\partial N}\right)_{T,V} = \left(\frac{\partial (F/V)}{\partial (N/V)}\right)_{T,V} = \left(\frac{\partial f}{\partial n}\right)_T \tag{6.1.44}$$

and

$$\begin{aligned} P &= -\left(\frac{\partial F}{\partial V}\right)_{T,N} = -\left(\frac{\partial (fV)}{\partial V}\right)_{T,N} \\ &= -f - V\left(\frac{\partial f}{\partial V}\right)_{T,N} = -f + n\left(\frac{\partial f}{\partial n}\right)_T. \end{aligned} \tag{6.1.45}$$

According to eqs. (6.1.37) and (6.1.38), a one-component homogeneous fluid is mechanically and chemically stable (or at least metastable) when

$$\left(\frac{\partial^2 F}{\partial V^2}\right)_{T,N} > 0 \quad \left(\frac{\partial^2 F}{\partial N^2}\right)_{T,V} > 0. \tag{6.1.46}$$

But because $(\partial^2 F/\partial V^2)_{T,N} = (n^2/V)\partial^2 f/\partial n^2$ and $(\partial^2 F/\partial N^2)_{T,V} = (1/V)\partial^2 f/\partial n^2$, both inequalities in eq. (6.1.46) reduce to the same thing, namely,

$$\frac{\partial^2 f}{\partial n^2} > 0. \tag{6.1.47}$$

This means that the conditions of mechanical and chemical stability are equivalent in a one-component system. This result is another manifestation of the Gibbs–Duhem equation, that for fixed temperature reads $dP = n\,d\mu$. Using the relations $P = -\partial F/\partial V$, $\mu = \partial F/\partial N$, and $[\partial \mu(T, n)/\partial V]_{T,N} = -(N/V^2)[\partial \mu(T, n)/\partial n]_T = -(N/V)[\partial \mu/\partial N]_{T,V}$, we can differentiate the Gibbs–Duhem equation with respect to V and obtain

$$-\left(\frac{\partial^2 F}{\partial V^2}\right)_{T,N} = -n^2\left(\frac{\partial^2 F}{\partial N^2}\right)_{T,V}. \tag{6.1.48}$$

This again establishes the equivalence of the mechanical and chemical stability conditions for a one-component homogeneous phase.

As a specific example of the equilibrium and stability conditions let us consider the Helmholtz free energy density of the one-component van der Waals fluid we studied in Chapter 5. From eq. (5.3.13), we find for a van der Waals fluid

$$f(T, n) = n\mu^+(T) + nkT \ln[n/(1 - nb)] - n^2 a, \qquad (6.1.49)$$

where $\mu^+(T)$ is the kinetic and internal energy contributions to the free energy that depend on molecular parameters and temperature. The value of μ^+ does not affect phase behavior because it drops out of the equilibrium equations.

Qualitative plots of f versus n are shown in Figure 6.4 for various temperatures. Consider the isotherms for T_1 and T_2. To the left of point A and to the right of point B, the function $f(n)$ has a positive curvature, that is, $\partial^2 f/\partial n^2 > 0$, and so the homogeneous fluid is locally stable (and, therefore, at least metastable) in these ranges of density. At A and B, $\partial^2 f/\partial n^2 = 0$, and between A and B, $\partial^2 f/\partial n^2 < 0$. Thus, homogeneous fluid at any density lying in the range between A and B is not stable because $\delta^2 f = (\partial^2 f/\partial n^2)(\delta n)^2 < 0$ and so the free energy can be lowered by a spontaneous density fluctuation δn. The points A and B denote the limits of local stability of homogeneous fluid and are called spinodal points, as was described in Chapter 5. A single straight line tangent to the T_1 or T_2 isotherm at points G and L locates a

Figure 6.4
Qualitative plots of the Helmholtz free energy versus density of a van der Waals fluid at various temperatures. T_c is the liquid–vapor critical point and $T_1 < T_2 < T_c$.

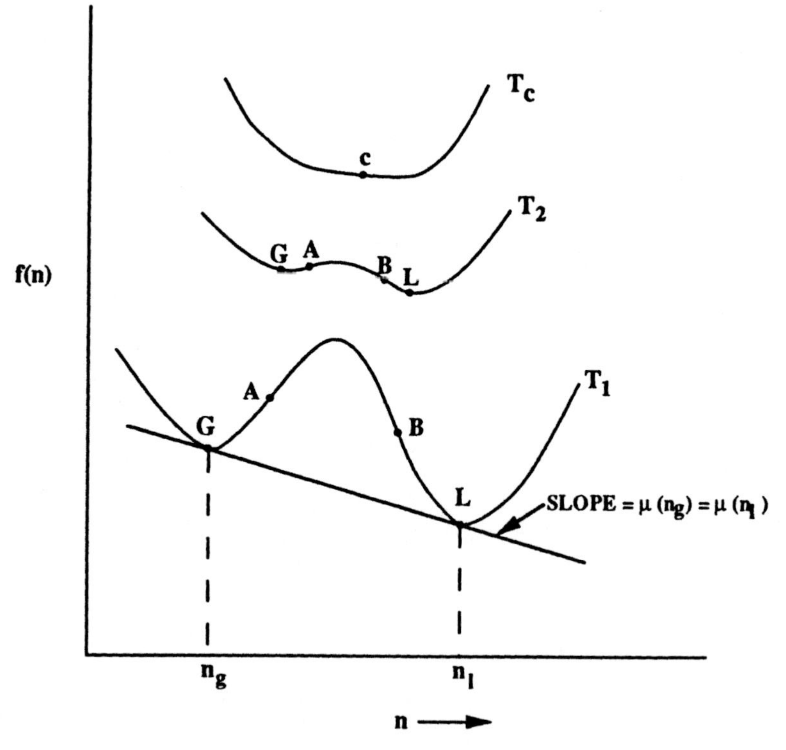

vapor phase at density n_g that has the same chemical potential as a liquid phase at density n_l [the tangent of $f(n)$ is $f'(n) = \mu(n)$ and so $\mu(n_g) = \mu(n_l)$ because points G and L have a common tangent]. From elementary geometry it follows that the points G and L lie on the straight line $y(n) = f(n_g) + (n - n_g)\mu(n_g)$. Because $y(n_l) = f(n_l)$ and $\mu(n_g) = \mu(n_l)$, we deduce that $-f(n_l) + n_l\mu(n_l) = -f(n_g) + n_g\mu(n_g)$, that is, that $P(n_l) = P(n_g)$. Thus, the vapor and liquid phases located by points G and L satisfy the necessary conditions of thermodynamic equilibrium. If n is an overall density lying between n_g and n_l, the free energy of a homogeneous phase would be $f(n)$. Alternatively, the system can split into a vapor phase G having a volume αV and a liquid phase L having a volume $(1-\alpha)V$, where $\alpha n_g + (1-\alpha)n_l = n$ or $\alpha = (n_l - n)/(n_l - n_g)$. The free energy density corresponding to this heterogeneous system is $f_{\text{heter}} = \alpha f(n_g) + (1-\alpha) f(n_l)$. Because $f_{\text{heter}} < f(n)$ for $n_g < n < n_l$, the stable system in the density range between G and L is a vapor phase of density n_g coexisting with a liquid phase at density n_ℓ. The relative volumes of vapor and liquid obey the lever rule $V_g/V_l = (n_l - n)/(n - n_g)$.

As the temperature approaches the critical point T_c, points A and B and points G and L approach each other. At the critical point T_c, the points A, B, G, and L merge. Because A and B are defined by $\partial^2 f/\partial n^2 = 0$, when they come together it follows that $\partial^3 f/\partial n^3 = 0$. But G and L are stable fluids and so at the critical point the spinodal states become stable. Thus, $\partial^4 f/\partial n^4 > 0$ at the critical point. In summary, the spinodal envelope T versus n, which is given by

$$\frac{\partial^2 f}{\partial n^2} = 0, \tag{6.1.50}$$

ends at the critical point T_c, where in addition to eq. (6.1.50) the conditions

$$\frac{\partial^3 f}{\partial n^3} = 0 \quad \text{and} \quad \frac{\partial^4 f}{\partial n^4} > 0 \tag{6.1.51}$$

hold.

In the specific case of a van der Waals fluid

$$\frac{\partial^2 f}{\partial n^2} = \frac{kT}{n(1-nb)^2} - 2a \tag{6.1.52}$$

$$\frac{\partial^3 f}{\partial n^3} = \frac{kT(3nb-1)}{n^2(1-nb)^3} \tag{6.1.53}$$

$$\frac{\partial^4 f}{\partial n^4} = \frac{3bkT}{n^2(1-nb)^3} + \frac{kT(3nb-1)(5nb-2)}{n^3(1-nb)^4}. \tag{6.1.54}$$

At the critical point

$$n_c = \frac{1}{3b}, \quad T_c = \frac{8a}{27bk}, \quad \text{and} \, P_c = \frac{a}{27b^2}. \tag{6.1.55}$$

Also, at the critical point $\partial^4 f/\partial n^4 = 729b^3kT_c/8$, which is positive as required. The graphical construction of the liquid–vapor coexistence states as shown in Figure 6.4 is useful for physical insights. However, it is more accurate for actual computation of the coexistence curve (binodal) to use the equilibrium conditions,

$$\mu(n_g) = \mu(n_l), \qquad P(n_g) = P(n_l), \tag{6.1.56}$$

which can be solved by the Newton–Raphson method outlined in Chapter 5.

The temperature–density liquid–vapor binodal curve, determined by conditions in eq. (6.1.56), and the spinodal curve, determined by the condition in eq. (6.1.50), are plotted in Figure 6.5 for a one-component fluid obeying van der Waals equation of state. Although we have constructed the phase diagram from the van der Waals equation of state, it captures the generic pattern of one-component liquid–vapor equilibria.

Next let us consider a multicomponent fluid. In this case f is a function of T and component densities $n_1, n_2, \ldots, n_c$ and the condition, eq. (6.1.38), of stability of homogeneous fluid can be restated as

$$\delta^2 f = \frac{1}{2}\sum_{\alpha,\beta} \frac{\partial^2 f}{\partial n_\alpha \partial n_\beta}\delta n_\alpha \delta n_\beta > 0. \tag{6.1.57}$$

In stable fluid all of the eigenvalues of the matrix $\mathbf{A} \equiv [\partial^2 f/\partial n_\alpha \partial n_\beta]$ are positive. A spinodal curve corresponds to a curve along which one of the eigenvalues is zero. That is, if we pose at a temperature T the eigenproblem

$$\mathbf{A}\mathbf{v} = \lambda\mathbf{v} \tag{6.1.58}$$

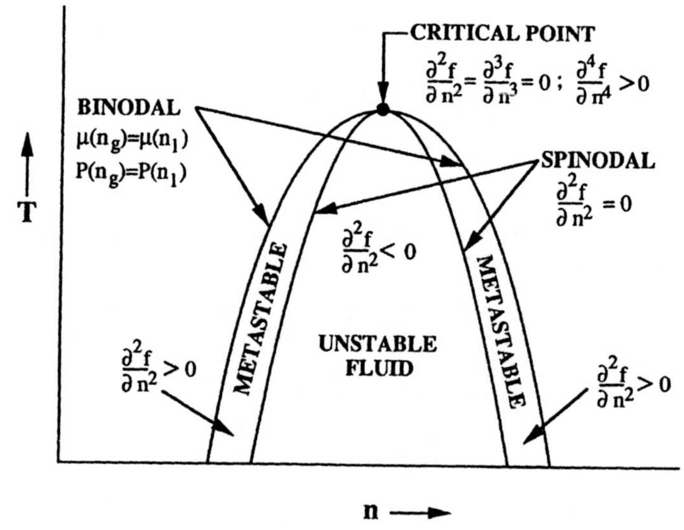

Figure 6.5
Schematic liquid–vapor phase diagram for a one-component fluid. Binodal curve denotes liquid–vapor coexistence densities and spinodal curve denotes the envelope of the region of unstable homogeneous fluid.

where the components $v_1, \ldots, v_c$ of the column vector **v** are eigenvectors of **A**, the spinodal curve will be a path in the c-dimensional density space (if $c = 1$, the spinodal is a pair of points at a fixed temperature) along which one of the eigenvalues of **A** is zero. If the fluctuation $\delta n_\alpha \propto v_\alpha^0$, where v_α^0 is the component of the eigenvector corresponding to $\lambda = 0$, then

$$\delta^2 f \propto \sum_{\alpha,\beta} \frac{\partial^2 f}{\partial n_\alpha \partial n_\beta} v_\alpha^0 v_\beta^0 = \mathbf{v}^{0T} \mathbf{A} \mathbf{v}^0 = 0, \qquad (6.1.59)$$

where $\mathbf{v}^T$ denotes the transpose of **v** and so $\mathbf{v}^T \mathbf{A} \mathbf{v} = \sum_{\alpha,\beta} a_{\alpha\beta} v_\alpha v_\beta$. If for the fluctuation $\delta n_\alpha \propto v_\alpha^0$ the third-order variation obeys the inequality

$$\delta^3 f = \frac{1}{6} \sum_{\alpha,\beta,\gamma} \frac{\partial^3 f}{\partial n_\alpha \partial n_\beta \partial n_\gamma} \delta n_\alpha \delta n_\beta \delta n_\gamma \propto \sum_{\alpha,\beta,\gamma} \frac{\partial^3 f}{\partial n_\alpha \partial n_\beta \partial n_\gamma} v_\alpha^0 v_\beta^0 v_\gamma^0 \neq 0, \qquad (6.1.60)$$

then the curve corresponding to $\lambda = 0$ separates density regions of locally stable and unstable homogeneous fluids. However, at a critical point $\delta^3 f = 0$ and $\delta^4 f > 0$ for $\delta n_\alpha \propto v_\alpha^0$ along the $\lambda = 0$ spinodal curve. One knows that from a critical point a binodal curve of coexisting phases springs. Thus, tracking the spinodal curve to a critical point (i.e., finding where $\delta^3 f = 0$ for $\delta n_\alpha \propto v_\alpha^0$) is one way of locating regions of multiphase equilibria in multicomponent systems. This is sometimes easier than finding the roots of the equilibrium equations $P^{(1)} = P^{(2)}$ and $\mu_\alpha^{(1)} = \mu_\alpha^{(2)}$, $\alpha = 1, \ldots, c$ because root finding is an iterative process whose success depends on having a good first guess.

For a two-component system, the eigenvalue problem posed at eq. (6.1.58) takes on the form

$$\begin{bmatrix} a_{11} & a_{12} \\ a_{12} & a_{22} \end{bmatrix} \begin{bmatrix} v_1 \\ v_2 \end{bmatrix} = \lambda \begin{bmatrix} v_1 \\ v_2 \end{bmatrix}, \qquad (6.1.61)$$

where we have used the symmetry property $a_{12} = a_{21}$ resulting from the fact that $\partial^2 f / \partial n_1 \partial n_2 = \partial^2 f / \partial n_2 \partial n_1$. From linear algebra we know that eq. (6.1.61) has a solution if and only if the determinant of the characteristic matrix $\mathbf{A} - \lambda \mathbf{I}$ is zero, that is,

$$\begin{vmatrix} a_{11} - \lambda & a_{12} \\ a_{12} & a_{22} - \lambda \end{vmatrix} = 0, \qquad (6.1.62)$$

This is a quadratic equation for λ which has two roots λ_1 and λ_2:

$$\lambda_1 = \frac{a_{11} + a_{22} - \sqrt{(a_{11} + a_{22})^2 - 4|\mathbf{A}|}}{4} \quad \text{and} \quad \lambda_2 = \frac{a_{11} + a_{22} + \sqrt{(a_{11} + a_{22})^2 - 4|\mathbf{A}|}}{2} \qquad (6.1.63)$$

where $|\mathbf{A}| = a_{11} a_{22} - a_{12}^2$ is the determinant of **A**. Again it is known from linear algebra that

$$|\mathbf{A}| = \prod_i \lambda_i \quad \sum_i a_{ii} = \sum_i \lambda_i \qquad (6.1.64)$$

for any matrix **A**. Thus, if the determinant $|\mathbf{A}|$ equals zero, at least one eigenvalue is zero.

For a two-component system then, densities n_1 and n_2 lie along a spinodal curve at constant temperature if n_1 and n_2 satisfy the equation

$$\frac{\partial^2 f}{\partial n_1^2}\frac{\partial^2 f}{\partial n_2^2} - \left(\frac{\partial^2 f}{\partial n_1 \partial n_2}\right)^2 = 0. \tag{6.1.65}$$

The spinodal curve can be determined by picking values of n_1 and calculating the corresponding values of n_2 from eq. (6.1.65).

To examine a specific case, let us choose the equation of state for a fluid mixture. The pressure is given by

$$P = \frac{nkT}{1 - nb} - \frac{n^2 a}{1 + 2nb - (nb)^2}, \tag{6.1.66}$$

where

$$b = \sum_{\alpha}^{c} x_\alpha b_\alpha, \quad a = \sum_{\alpha\beta} x_\alpha x_\beta a_{\alpha\beta}. \tag{6.1.67}$$

x_2 is the mole fraction of component α. b_α and $a_{\alpha\alpha}$ are pure fluid parameters computed from

$$b_\alpha = 0.0778 \frac{kT_{c\alpha}}{P_{c\alpha}} \tag{6.1.68}$$

and

$$a_{\alpha\alpha} = 0.45724\left[1 + \kappa(\omega_\alpha)\left(1 - \sqrt{\frac{T}{T_{c\alpha}}}\right)\right]^2 \frac{k^2 T_{c\alpha}^2}{P_{c\alpha}} \tag{6.1.69}$$

with

$$\kappa(\omega_\alpha) = 0.37464 + 1.54226\omega_\alpha - 0.266992\omega_\alpha^2. \tag{6.1.70}$$

ω_α is the acentric factor defined by $\omega_\alpha = -\log(P_\alpha^*/P_{c\alpha}) - 1$. P_α^* the vapor pressure of pure α at $T = 0.8T_{c\alpha}$; $T_{c\alpha}$ and $P_{c\alpha}$ are the critical temperature and pressure of pure α. As discussed in Chapter 5, the equation has been constructed empirically as a quantitative improvement on the van der Waals model. Its best performance is in computing vapor pressures and compositions of coexisting liquid and vapor phases. To complete the model the mixture parameters $a_{\alpha\beta}$ must be specified. This is done by introducing the expression

$$a_{\alpha\beta} = 1 - k_{\alpha\beta}\sqrt{a_{\alpha\alpha} a_{\beta\beta}} \tag{6.1.71}$$

and determining the mixing parameter $k_{\alpha\beta}$ from equation of state data, usually vapor pressure data for binary solutions. Döring and Knapp determined $k_{\alpha\beta}$ for numerous fluid pairs. Also, Ohe (1990) calculated $k_{\alpha\beta}$ parameters for several fluid pairs.

Integrating the expression $dF = -P\,dV$ at constant N and T and using the known form of F in the ideal gas limit, we obtain the following formula for the Helmholtz free energy density of a mixture

$$f(n) = \sum_{\alpha}^{c} n_\alpha \left\{ \mu_\alpha^+(T) + kT \left[\ln \frac{n_\alpha}{1 - nb} + 1 \right] \right\}$$
$$- \frac{na}{2\sqrt{2}b} \ln \left[\frac{1 + (1 + \sqrt{2})nb}{1 + (1 - \sqrt{2}nb} \right]. \qquad (6.1.72)$$

It is important to understand how useful the spinodal curves can be in determining the phase behavior of a multicomponent system. The spinodal curves calculated from eq. (6.1.65) for a solution of carbon dioxide and normal decane at −25°C are shown as dashed curves in Figure 6.6. CO_2 is component 1 and decane is component 2. The critical parameters and acentric factors were taken from Table 4.2 in Chapter 4 and $k_{12} = 0.113$, a value listed by Döring and Knapp. From Figure 6.6a it follows that both pure decane and pure CO_2 have intermediate density ranges in which $\lambda_1 < 0$, and therefore in which a homogeneous fluid is not stable. Thus, there will be a liquid–vapor split along the pure decane and pure CO_2 edges of the density–density phase diagram. The coexistence densities lie outside the region bounded by the spinodal curves. Just by examining the spinodal curves of Figure 6.6a, we can surmise the following things. Somewhere in the region V there is a binodal separating the vapor phase and the $V - L_1$ vapor–liquid coexistence region. Somewhere in the region L_1 there is a binodal separating a vapor–liquid coexistence region. Similarly, somewhere in the region L_2 there is a binodal separating the decane rich phase and the $V - L_2$ vapor–liquid coexistence region. The fact that the spinodal curves intersect the maximum density hypotenuse of the density–density diagram implies the existence of a liquid–liquid ($L_1 - L_2$) coexistence region. We also found that the third-order variation in eq. (6.1.58) vanishes nowhere along the

Figure 6.6
Spinodal curves and phase diagram in a constant temperature density–density phase diagram for carbon dioxide and decane at −25°C. Predicted from PR equation of state.
(a) Spinodal curves. (b) Phase diagram showing spinodals, binodals, tie-lines (compositions of coexisting phases), and the three-phase region as a tie triangle. Adapted from Falls (1982).

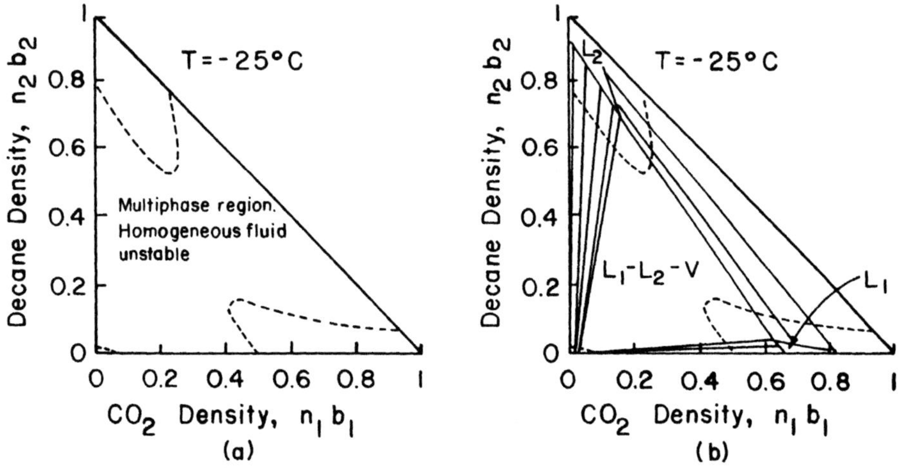

spinodal curves. Thus, there are no critical points for a CO_2-decane mixture at 7°C. From these observations we can conclude without actually computing the coexistence curves that three phase coexistence ($V - L_1 - L_2$) will occur somewhere in the phase diagram because there will be a region in which the $V - L_1$, $V - L_2$, and $L_1 - L_2$ tie-lines overlap. The actual phase diagram computed for the fluid is displayed in Figure 6.6b.

Two other types of spinodal curves are shown qualitatively in Figure 6.7. In Figure 6.7a there is continuous corridor of vapor–liquid equilibria fanning out from the pure component 1 edge to the pure component 2 edge of the density–density phase diagram. Toward the high density hypotenuse of the diagram there is a critical point c that announces the onset of a liquid–liquid binodal (which touches the spinodal at a critical point) in that density region. The liquid–liquid binodal is also implied by the intersections of the spinodal curve with the high density hypotenuse. In Figure 6.7b, however, the pure fluid edges and the high density hypotenuse give no clues as to the existence of multiphase equilibria. On the other hand, the spinodal island with two critical points, reveals the existence of a region of two phase equilibria opening at one critical point (c_1) and closing at another critical point (c_2). The reader is invited to deduce the phase behavior corresponding to the spinodal curves given in Figure 6.7.

For a two-component system the eigenvectors v_1^0, v_2^0 along the spinodal obey the equations

$$\frac{\partial^2 f}{\partial n_1^2} v_1^0 + \frac{\partial^2 f}{\partial n_1 \partial n_2} v_2^0 = 0$$

$$\frac{\partial^2 f}{\partial n_1 \partial n_2} v_1^0 + \frac{\partial^2 f}{\partial n_2^2} v_2^0 = 0. \qquad (6.1.73)$$

Figure 6.7
Schematic plot of spinodal curves in constant temperature density–density phase diagrams. The filled circles denote critical points.

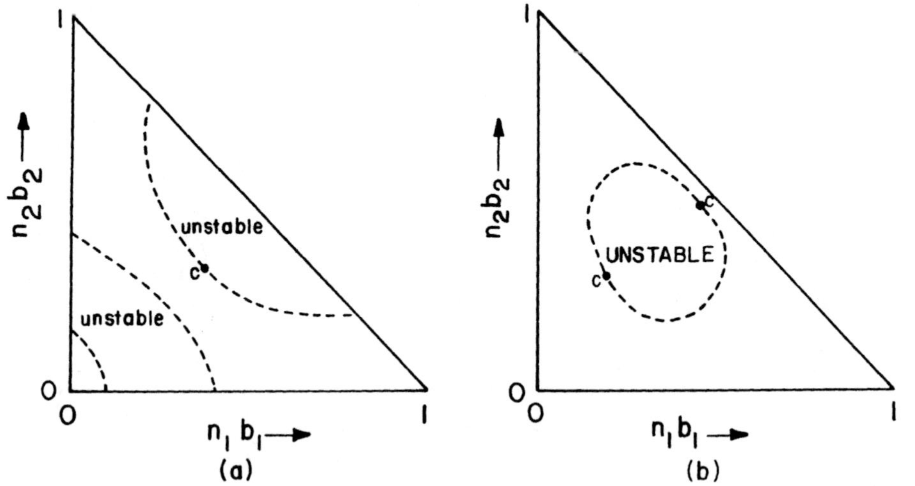

Thus, if we define

$$s \equiv -(\partial^2 f/\partial n_1 \partial n_2)/\partial^2 f/\partial n_1^2 \qquad (6.1.74)$$

we find $v_1^0 = -s v_2^0$, and so we can locate critical points as we compute a spinodal curve by finding points along the curve at which

$$s^3 \frac{\partial^3 f}{\partial n_1^3} + 3s^2 \frac{\partial^3 f}{\partial n_1^2 \partial n_2} + 3s \frac{\partial^3 f}{\partial n_1 \partial n_2^2} + \frac{\partial^3 f}{\partial n_2^3} = 0. \qquad (6.1.75)$$

One such point occurred on a spinodal in Figure 6.7a and two occurred on the spinodal in Figure 6.7b. An isola or island two phase region, as is the case for Figure 6.7b, will always have two critical points.

Let us close this section by noting that for systems having three or more components one often is interested in composition phase diagrams at a given temperature and pressure. These can be done in terms of equations of state for the pressure and the Helmholtz free energy. However, in the case of liquid–liquid equilibria and polymer solutions, there are several useful model equations of state (e.g., the regular solution model or Flory–Huggins model) in which the molar Gibbs free energy $\tilde{g} \equiv G/N$, is given as an explicit function of mole fractions $x_1, \ldots, x_c$ and parameters that depend on temperature and pressure. In this case the chemical potential can be computed as

$$\mu_\alpha = \tilde{g} - \sum_{\beta \neq \alpha}^{c} \left(\frac{\partial \tilde{g}}{\partial x_\beta} \right)_{x_{\gamma \neq \beta}} x_\beta, \qquad (6.1.76)$$

where it is understood that the composition derivatives are taken at constant temperature and pressure. Phase equilibria are then found from the equations $\mu_\alpha(\mathbf{x}^{(1)}) = \mu_\alpha(\mathbf{x}^{(2)})$ and $\sum_\alpha x_\alpha^{(i)} = 1$ for phases $i = 1$ and 2. The homogeneous phase is locally stable if the matrix $\mathbf{A} = [a_{\alpha\beta}]$, $a_{\alpha\beta} = \partial^2 \tilde{g}/\partial x_\alpha \partial x_\beta$, with α and β ranging from 1 to $c - 1$, is positive definite, that is, if

$$\delta^2 \tilde{g} = \frac{1}{2} \sum_{\alpha,\beta=1}^{c-1} \frac{\partial^2 \tilde{g}}{\partial x_\alpha \partial x_\beta} \, \delta x_\alpha \delta x_\beta \qquad (6.1.77)$$

is positive for arbitrary composition fluctuation δx_α. The spinodal region is determined by the condition that **A** have a zero eigenvalue, that is, that

$$|\mathbf{A}| = \begin{vmatrix} \frac{\partial^2 \tilde{g}}{\partial x_1^2} & \frac{\partial^2 \tilde{g}}{\partial x_1 \partial x_2} & \cdots & \frac{\partial^2 \tilde{g}}{\partial x_1 \partial x_{c-1}} \\ \vdots & & \ddots & \\ \frac{\partial^2 \tilde{g}}{\partial x_1 \partial x_{c-1}} & & & \frac{\partial^2 \tilde{g}}{\partial x_{c-1}^2} \end{vmatrix} = 0 \qquad (6.1.78)$$

and a critical point on the spinodal is determined by the conditions

$$\delta^3 \tilde{g} = \frac{1}{6} \sum_{\alpha,\beta,\gamma}^{c-1} \frac{\partial^3 \tilde{g}}{\partial x_\alpha \partial x_\beta \partial x_\gamma} v_\alpha^o v_\beta^o v_\gamma^o = 0 \tag{6.1.79}$$

and $\delta^4 \tilde{g} > 0$ for a fluctuation $\delta \mathbf{x} \propto \mathbf{v}^o$, where $\mathbf{v}^o$ is the eigenvector of $\mathbf{A}$ with zero eigenvalue. In a ternary system $\mathbf{A}$ is a 2×2 matrix and, therefore, the spinodal is a curve in a phase diagram at constant temperature and pressure. Given $\tilde{g}$ a function of T, P, and mole or mass fraction, the mathematics for constructing this ternary phase diagram is exactly the same as that for constructing the constant temperature density–density phase diagram for a binary system. The spinodal becomes a surface in x_1, x_2, x_3 composition space for four components and a hypersurface in $x_1, \ldots, x_{c-1}$ composition space when $c > 4$.

6.2 Some Observed Patterns of Phase Behavior

One of the marvels of nature is the wide variety of phases formed by matter and the intricate patterns of phase equilibria seen as a function of thermodynamic conditions and molecular composition. It is beyond the scope of this text to cover exhaustively all the observed patterns. However, the richness of the subject will be illustrated by several examples drawn from one-, two-, and three-component systems.

6.2.1 One-Component Phase Equilibria

Vapor–liquid equilibria provide the simplest example of phase behavior in a one-component system. We discussed this case at length in Chapter 5 and in the preceding section. One-component systems also form crystalline and liquid crystalline phases. Crystalline phases are solids whose atoms are arranged periodically in three dimensions. Familiar crystalline forms are simple cubic (polonium), body-centered cubic (bcc; e.g., barium, chromium, and sodium), face-centered cubic (fcc; e.g., argon, gold, and silver) hexagonal close packed (e.g., helium, beryllium, and cadminum), and diamond (e.g., carbon, germanium, and silicon).

One way to display a one-component phase diagram is to plot the binodal or envelope of coexisting phases in a temperature- (or pressure-) density diagram as in Figure 6.8. A rather convenient way to summarize the phase diagram of a one-component substance is to plot the pressure as a function of temperature along the path of coexisting phases. Such a phase diagram is shown in Figure 6.8 for water for various ranges of pressure.

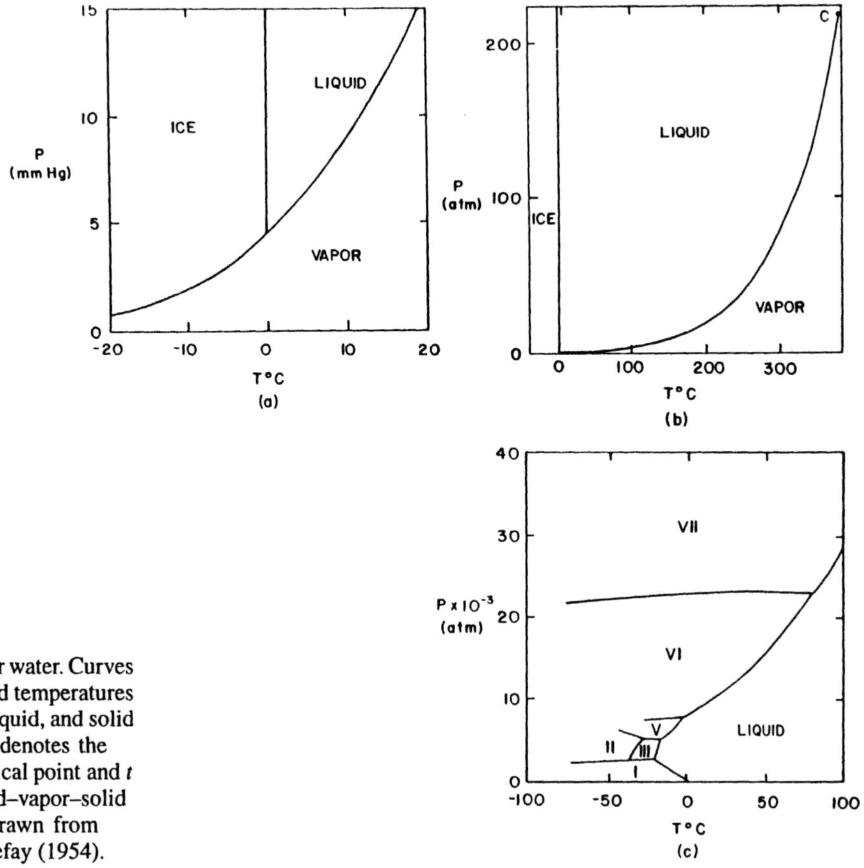

Figure 6.8
Phase diagram for water. Curves give pressures and temperatures at which vapor, liquid, and solid phase coexist. c denotes the liquid–vapor critical point and t denotes the liquid–vapor–solid triple point. Redrawn from Prigogine and Defay (1954).

The liquid–vapor coexistence curve (Fig. 6.8b) begins at the critical point ($T_c = 647.2\text{K}$, $P_c = 217.72$ atm, $\mathcal{P}_c = 0.4$) and with decreasing pressure ends at the triple point (0.0076°C and 4.6 mmHg) at which vapor, liquid, and solid phases are in equilibrium. The triple point is unique because, according to the phase rule, $f = c + 2 - p = 1 + 2 - 3 = 0$, there are no degrees of freedom. Near the triple point the solid–liquid coexistence curve has the unusual property $dP/dT < 0$. Ice having this property is call type I ice. Its structure is hexagonal. At higher pressures, other crystalline forms of ice occur. As many as seven different phases (i.e., different crystalline structures) of ice have been found (Fig. 6.8c).

The negative slope of the liquid–type I ice coexistence curve reflects the fact that the density of the ice phase is smaller than that of the coexisting liquid phase. The relationship between dP/dT and the difference between the densities of coexisting phases is known as the Clausius–Clapeyron equation. Recall that for phase i of a one-component system, the Gibbs–Duhem equation is

$$\tilde{S}^{(i)}dT - \tilde{V}^{(i)}dP + d\mu^{(i)} = 0, \qquad (6.2.1)$$

where $\tilde{S}^{(i)}$ and $\tilde{V}^{(i)}$ are the entropy and volume per molecule in phase i. If phases i and j are in equilibrium, then $\mu^{(i)} = \mu^{(j)}$ and so along the coexistence curve $d\mu^{(i)} = d\mu^{(j)}$. From eq. (6.2.1) it follows that along the $P - T$ coexistence curve

$$\tilde{S}^{(i)} dT - \tilde{V}^{(i)} dP = \tilde{S}^{(i)} dT - \tilde{V}^{(j)} dP,$$

or

$$\frac{dP}{dT} = \frac{\tilde{S}^{(j)} - \tilde{S}^{(i)}}{\tilde{V}^{(j)} - \tilde{V}^{(i)}}. \tag{6.2.2}$$

This is the Clausius–Clapeyron equation. With the aid of the relation $\tilde{H}^{(i)} = T\tilde{S}^{(i)} + \mu^{(i)}$, the Clausius–Clapeyron equation can also be expressed in the form

$$\frac{dP}{dT} = \frac{\tilde{H}^{(j)} - \tilde{H}^{(i)}}{T(\tilde{V}^{(j)} - \tilde{V}^{(i)})} \equiv \frac{\Delta \tilde{H}^{(ij)}}{T \Delta \tilde{V}^{(ij)}}. \tag{6.2.3}$$

Suppose phases i and j denote water and ice, respectively. The heat of melting, $\Delta \tilde{H}^{(sl)}$, of ice I is positive and the density difference $n^{(l)} - n^{(s)}$ is negative. Thus, since $\Delta \tilde{V}^{(sl)} \equiv \tilde{V}^{(l)} - \tilde{V}^{(s)} = (n^{(s)} - n^{(l)})/n^{(s)} n^{(l)}$ it follows from the Clausius–Clapeyron equation that $dP/dT < 0$ along the water–type I ice coexistence curve. The practical importance of the Clausius–Clapeyron equation is that it can be used to compute the enthalpy of phase change from the vapor pressure curve if the phase densities are known. Conversely, if the enthalpy of phase change and the phase densities are known, the vapor pressure curve can be calculated from the Clausius–Clapeyron equation.

If i and j are both condensed phases $\Delta \tilde{H}^{(ij)}$ and $\Delta \tilde{V}^{(ij)}$ in a narrow temperature range can be approximated as constant. In this case, eq. (6.2.3) can be integrated to obtain

$$P = P_0 + \frac{\Delta \tilde{H}_0^{(ij)}}{\Delta \tilde{V}_0^{(ij)}} \ln \frac{T}{T_0}, \tag{6.2.4}$$

where P_0, $\Delta \tilde{H}_0^{(ij)}$, and $\Delta \tilde{V}_0^{(ij)}$ are the vapor pressure and the heat and volume of phase change at temperature T_0. When phase j is a dilute vapor so that $\tilde{V}^{(j)} = kT/P$ and $\tilde{V}^{(j)} - \tilde{V}^{(i)} \simeq \tilde{V}^{(j)}$, integration of the Clausius–Clapeyron equation yields

$$P = P_0 \exp\left[\frac{\Delta \tilde{H}_0^{(ij)}}{k}\left(\frac{1}{T_0} - \frac{1}{T}\right)\right], \tag{6.2.5}$$

where again we have assumed $\Delta \tilde{H}^{(ij)}$ is constant. This assumption of course restricts the range of validity of eq. (6.2.5). Nevertheless, the result provides a good extrapolation formula for estimating the vapor pressure at temperatures not too far from T_0.

Liquid crystals are fluids in which molecules are ordered in at least one direction. Generally liquid crystals are anisotropic,

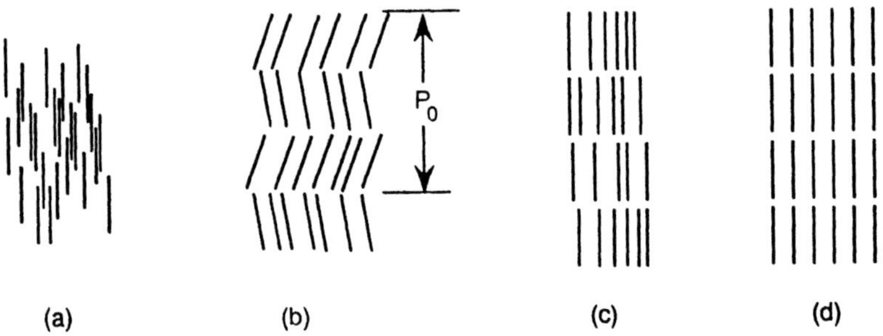

Figure 6.9
Schematic representation of the cross section of (a) nematic, (b) cholesteric, (c) smectic S_A, and (d) smectic S_B liquid crystal structures.

in contrast to ordinary liquids. Examples of one-component liquid crystal structures are shown in Figure 6.9. The structures are classified according to the alignment of the major axis of rodlike molecules, such as 4-methoxybenzylidene-4′-butylaniline

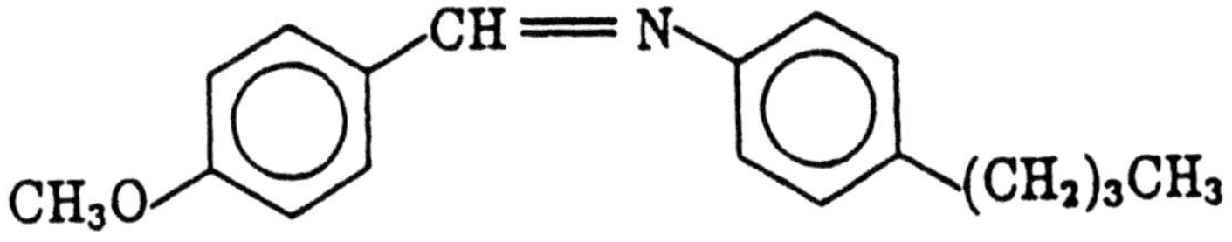

$$\text{CH}_3\text{O}-\!\!\!\bigcirc\!\!\!-\text{CH}=\text{N}-\!\!\!\bigcirc\!\!\!-(\text{CH}_2)_3\text{CH}_3 \qquad (6.2.6)$$

and D (or L)-4-methoxybenzylidene-4-(2-methylbutyl)aniline,

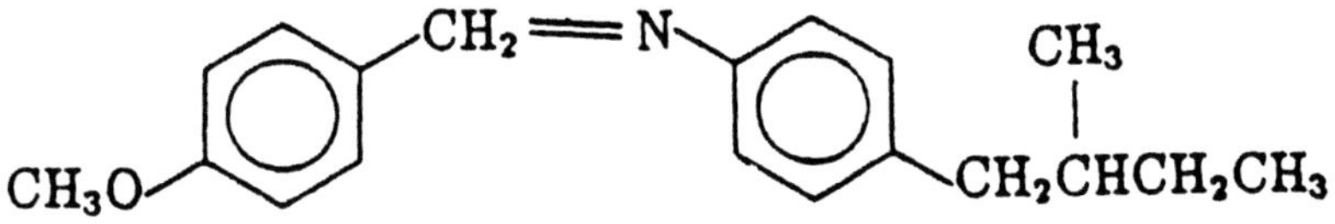

$$\text{CH}_3\text{O}-\!\!\!\bigcirc\!\!\!-\text{CH}_2=\text{N}-\!\!\!\bigcirc\!\!\!-\text{CH}_2\overset{\overset{\displaystyle \text{CH}_3}{|}}{\text{CH}}\text{CH}_2\text{CH}_3 \qquad (6.2.7)$$

In a nematic liquid crystal, the major axis of each of the molecules is aligned in the same direction, but the centers of mass of the molecules are randomly distributed along this direction and in the other two directions. Cholesteric liquid crystals are formed by optically active molecules. The major axis of each of the molecules is aligned in the same direction in successive parallel planes, but the orientation directions in neighboring planes differ by a constant angle. Thus, in advancing normal to these planes the orientation direction rotates as a helix whose pitch P_0 is the distance over which the orientation direction is periodic. Molecules in smectic liquid crystals are in layered or lamellar structures. The structure is periodic in the direction normal to the layers. In each layer the molecules can be randomly distributed (Fig. 6.9c) or periodically arrayed (Fig. 6.9d).

In the range 21–47°C, the molecule at eq. (6.2.6) forms a nematic phase. Below 21°C the substance is crystalline and above 47°C it is an isotropic liquid. In the range 21–24°C the molecule at eq. (6.2.7) forms a cholesteric phase and undergoes a phase transition to an isotropic liquid at 24°C. Because cholesteric liquid crystals are colored (as a result of Bragg scattering of light off the helical structure), the isotropic to cholesteric phase transition is easily identified visually.

6.2.2 Two-Component Phase Equilibria

In this section we illustrate a few of the multitude of interesting patterns of binary phase behavior. Figure 6.10 illustrates three different types of liquid–vapor phase diagrams. Figure 6.10a is a simple case in which the vapor is richer in component A than in component B. More interesting are cases showing azeotropic behavior as shown in Figures 6.10b and 6.10c.

Consider a liquid solution of the type shown in Figure 6.10b that is, a solution exhibiting positive azeotropy. If the liquid is evaporated, the vapor is richer in component 2 if it is below the azeotropic concentration x_{az} and richer in component 1 if it is above x_{az}. Subsequent condensation and distillation of the evaporated vapor moves the vapor concentration toward x_{az}. Similarly, if the liquid composition lies above x_{az}, multistage distillation will generate a vapor decreasing in component 2 until the azeotropic

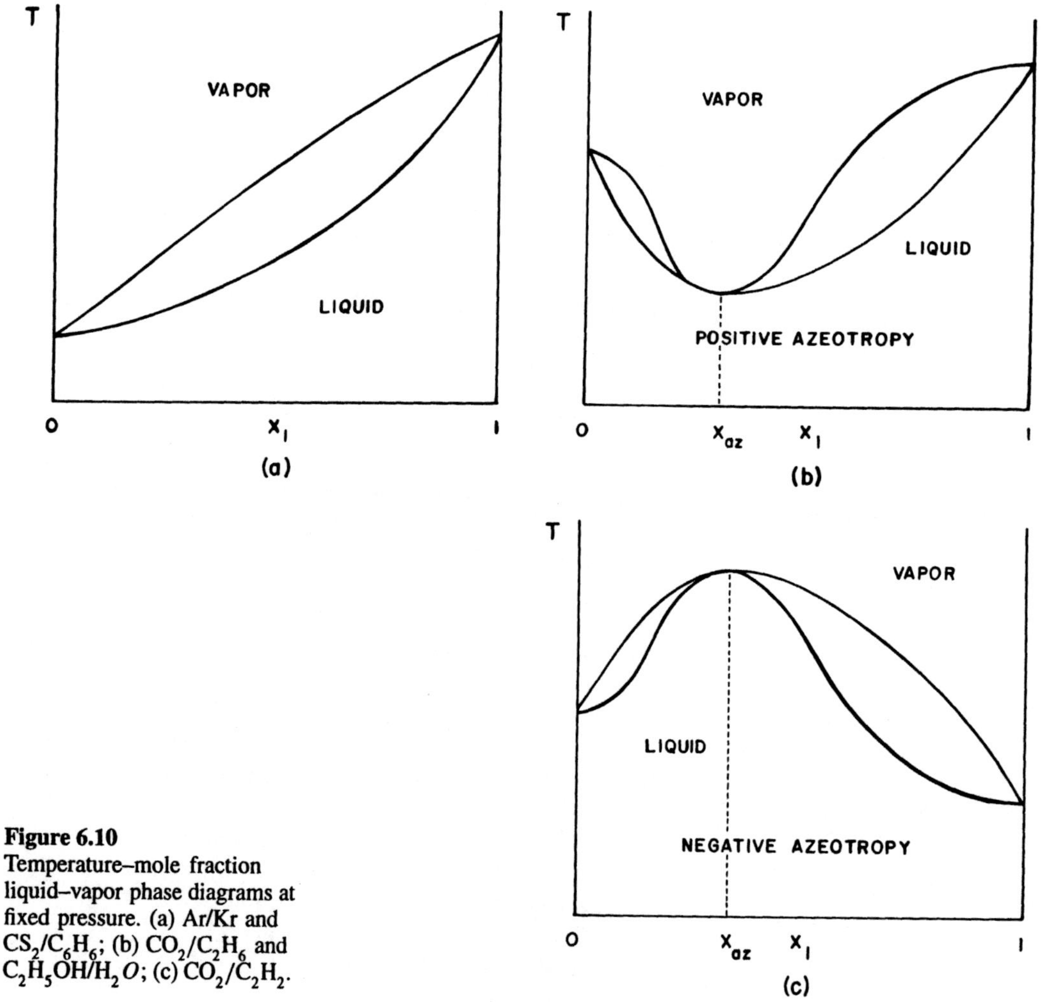

Figure 6.10
Temperature–mole fraction liquid–vapor phase diagrams at fixed pressure. (a) Ar/Kr and CS_2/C_6H_6; (b) CO_2/C_2H_6 and C_2H_5OH/H_2O; (c) CO_2/C_2H_2.

composition is reached. Correspondingly, the mother liquid moves toward the azeotropic composition during distillation. Thus, multistage distillation will always drive the phases toward the azeotropic concentration. A well-known example of this behavior is the water/ethanol mixture. At 1 atm, ethanol can be concentrated by distillation from aqueous solution only up to 95 wt % ethanol, which is the azeotropic composition.

Unlike one-component systems, binary systems can also exhibit liquid–liquid equilibria. For example, normal hexane and nitrobenzene mixtures form coexisting liquid phases below 19°C (Fig. 6.11). The binodal has an *upper critical temperature*, 19°C, above which the liquids are miscible in all proportions. We say upper critical temperature to distinguish it from a *lower critical temperature*, which can occur in liquid–liquid equilibria as illustrated in the phase diagram, Figure 6.12, for water and diethylamine. At an upper critical point two phases become miscible because the contributions of the intermolecular interactions, which favor phase segregation, are overtaken by the entropic contributions (roughly proportional to temperature), which favor mixing or miscibility. Qualitatively, a lower critical point arises from intermolecular association of component 2 into clusters, which can have a very transient existence, that are more energetically compatible with component 1 than are monomers of component 2. As the temperature increases the concentrations or lifetimes of these clusters become sufficiently small that the energetic incompatibility of components 1 and 2 causes a phase split at a lower critical temperature.

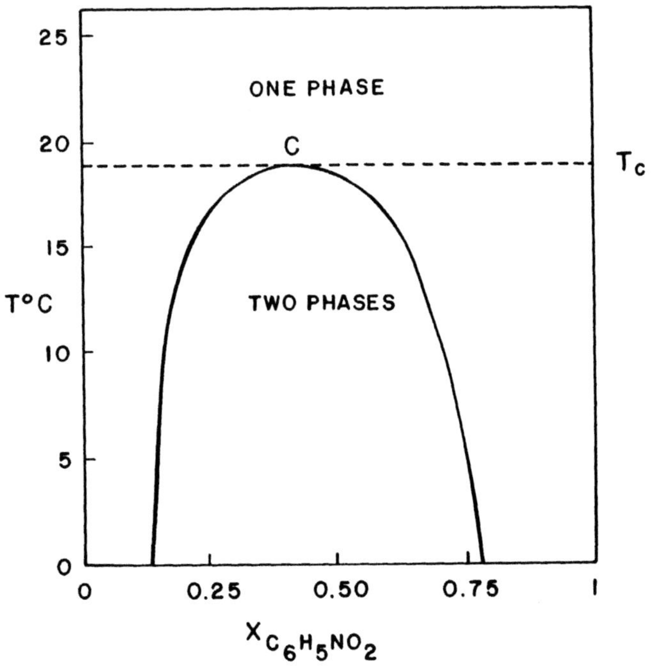

Figure 6.11
Liquid–liquid phase diagram for *n*-hexane and nitrobenzene at atmospheric pressure. Redrawn from Prigogine and Defay (1954).

Figure 6.12
Liquid–liquid phase diagram for water and diethylamine. Redrawn from Prigogine and Defay (1954).

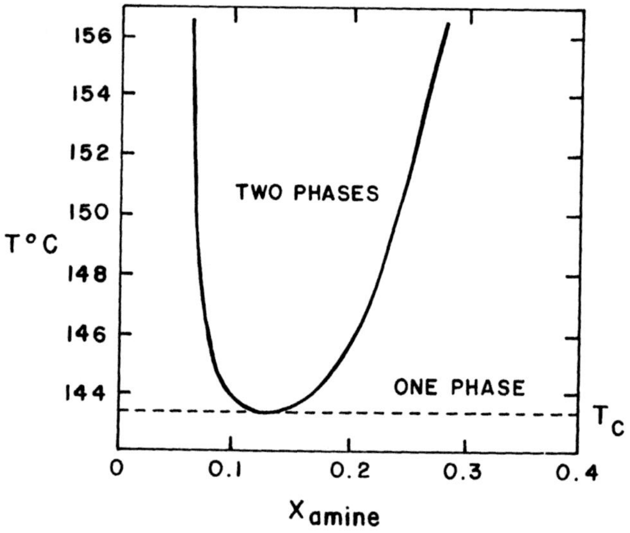

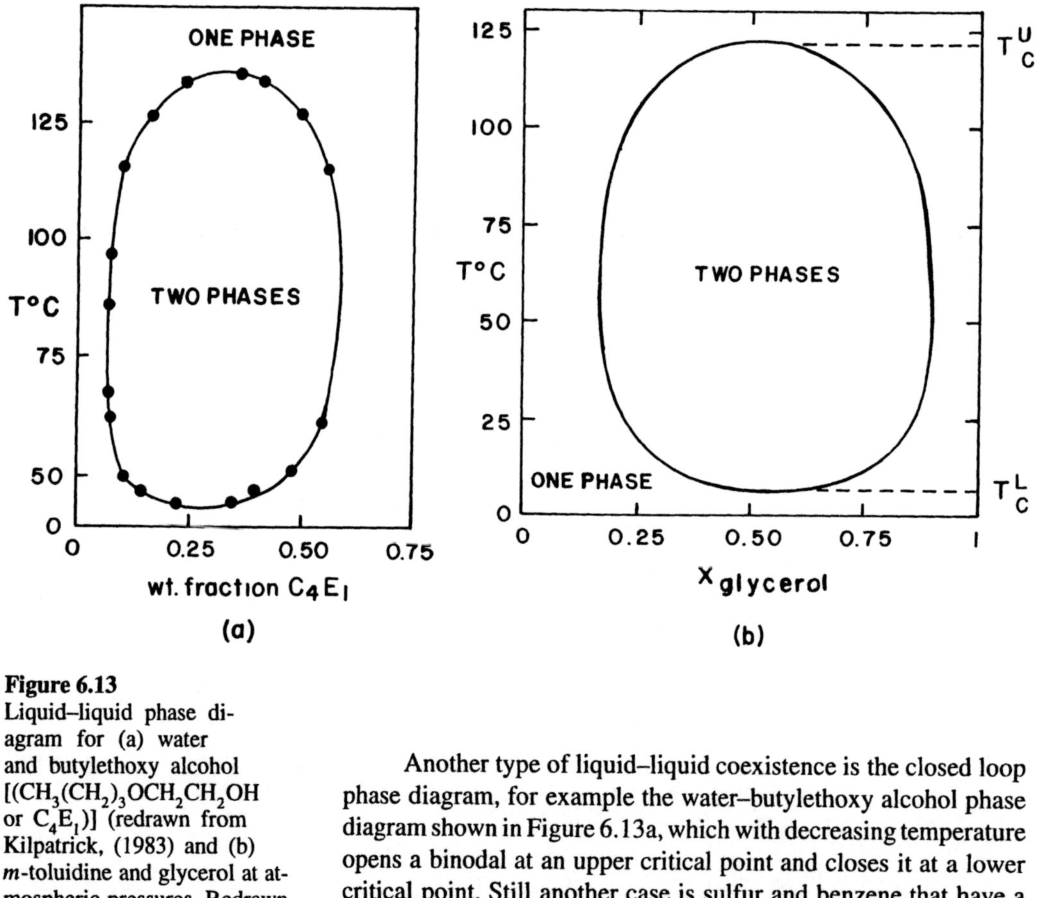

Figure 6.13
Liquid–liquid phase diagram for (a) water and butylethoxy alcohol [$(CH_3(CH_2)_3OCH_2CH_2OH$ or C_4E_1)] (redrawn from Kilpatrick, (1983) and (b) *m*-toluidine and glycerol at atmospheric pressures. Redrawn from Prigogine and Defay Green (1954).

Another type of liquid–liquid coexistence is the closed loop phase diagram, for example the water–butylethoxy alcohol phase diagram shown in Figure 6.13a, which with decreasing temperature opens a binodal at an upper critical point and closes it at a lower critical point. Still another case is sulfur and benzene that have a liquid–liquid coexistence region ending at an upper critical point and above this a liquid–liquid region opening in a lower critical point (Fig. 6.13b). The lower critical point of the sulfur–benzene system is caused by polymerization of sulfur with increasing temperature.

Mixtures of water and carbon dioxide exhibit interesting phase behavior at high temperature and pressure. Three pressure–composition binodals are shown in Figure 6.14. Between 533.15 and 573.15 K the binodal either pinches off into a critical point or a high pressure critical point retracts from above 3000 atm at 533.15 K to about 500 atm at 573.15 K. The solid curves are predictions of a lattice theoretical equation of state. As it is reasonably accurate when compared to experimental data in Figure 6.14, we conclude from its prediction for $T = 553.6$ K that the binodal pinches off into an upper and lower critical point. The dashed curve in Figure 6.14 is the predicted path of the separating critical points.

Surfactants in aqueous solution form association colloids that give rise to fascinating patterns of phase behavior. A surfactant is a molecule that has a water soluble moiety attached to a water insoluble moiety. Examples of surfactants include soaps, e.g.,

$$CH_3(CH_2)_nCOO^-Na^+,$$

Figure 6.14
Pressure–composition phase diagrams of CO_2–water mixtures at different temperatures. Redrawn from Nitsche, et al. (1984).

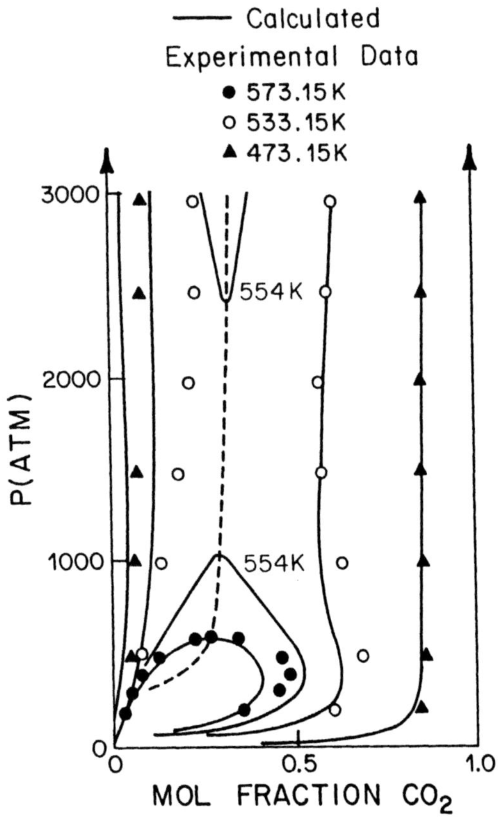

detergents, e.g.,

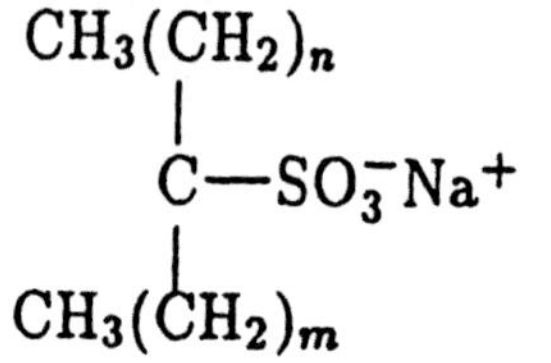

$$\underset{CH_3(CH_2)_m}{\overset{CH_3(CH_2)_n}{C}}-SO_3^- Na^+ \quad \text{or} \quad CH_3(CH_2)_n SO_4^- Na^+$$

foaming and emulsifying agents, e.g.,

$$CH_3(CH_2)_n(OCH_2CH_2)_m OH,$$

and phospholipids, e.g., phosphatidylcholine,

$$\begin{array}{c} CH_2O\overset{O}{\overset{\|}{C}}(CH_2)_{15}CH_3 \\ | \\ HCO\overset{O}{\overset{\|}{C}}(CH_2)_{15}CH_3 \\ | \\ H_2CO\overset{O}{\overset{\|}{P}}OCH_2CH_2\overset{+}{N}(CH_3)_3 \\ | \\ O^- \end{array}$$

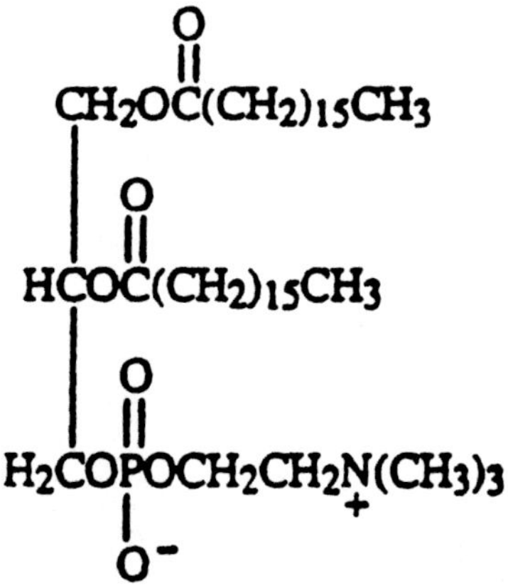

The notation $(CH_2)_n$ indicates a linear chain of n methyl groups and $(OCH_2CH_2)_m$ a linear chain of m ethyl ether groups. It is common in commercial use to find values of n ranging from 6 to 16 and of m ranging from 3 to 8. Surfactants can be anionic, cationic, zwitterionic, and nonionic. The water insoluble part can be a hydrocarbon, a perfluorinated hydrocarbon, silane, a siloxane, or mixtures of these. Surfactants are important ingredients of shampoos, soaps, detergents, repellents, insecticides, paints, hand creams, tooth pastes, sauces, foams, emulsions, fabric softeners, hair conditioners, and the like. And, of course, biological membranes are composed of bilayers of phospholipids.

The special property of surfactants is their tendency to absorb at surfaces or form colloidal aggregates in solution. Because of its water insoluble moiety, a surfactant molecule is not very soluble in water. Consequently, if a little surfactant is added to water it will migrate preferentially to the water–air interface. At the water–air interface, the surfactant forms a monolayer with the water soluble moieties (the "heads") oriented toward the water and the water insoluble moieties (the "tails") oriented toward the air side of the interface. The monolayer lowers the surface tension of the water. At some surfactant concentration the monolayer will become saturated and upon further addition the surfactant must aggregate into micelles or precipitate out as a crystalline or lyotropic (solvent induced) liquid crystalline phase (see Fig. 6.15).

The simplest micelle is a spherical aggregate of surfactant molecules whose water insoluble tails are oriented toward the center of the sphere and whose water soluble heads form the surface of the sphere. There are also surfactants that form disklike, rodlike, or wormlike micelles. At a given temperature, there is a surfactant concentration below which the solution contains only surfactant monomers and above which the additional surfactant aggregates to form micelles. This concentration is called the critical micelle concentration or the cmc. The micellar solution is a stable isotropic phase. The cmc depends on temperature and, because the transition is not a first-order phase transition, the cmc at a given temperature is actually a narrow concentration range rather than a unique concentration. The cmc is illustrated schematically by the dashed line in Figure 6.15.

For the case illustrated by Figure 6.15, depending on temperature and concentration, either crystalline, hexagonal liquid crystalline, or lamellar liquid crystalline phases occur. The hexagonal phase consists of rod micelles whose axes are aligned in a hexagonal array. A lamellar liquid crystal consists of parallel bilayers, stacked periodically. In the association colloids pictured in Figure 6.15, one can think of the surfactant as forming monolayer sheets with a polar (water soluble) side and a nonpolar (water insoluble) side. The different association colloids are then formed from the sheets by closing them onto their hydrocarbon sides to form spheres, disks, or rods or by stacking them in pairs of monolayers with the nonpolar sides facing one another to form a bilayer. The water then resides on the polar sides of the sheet assemblies. Liquid crystals of other symmetries have been identified. Examples include rods on rectangular arrays, spheres on a cubic lattice,

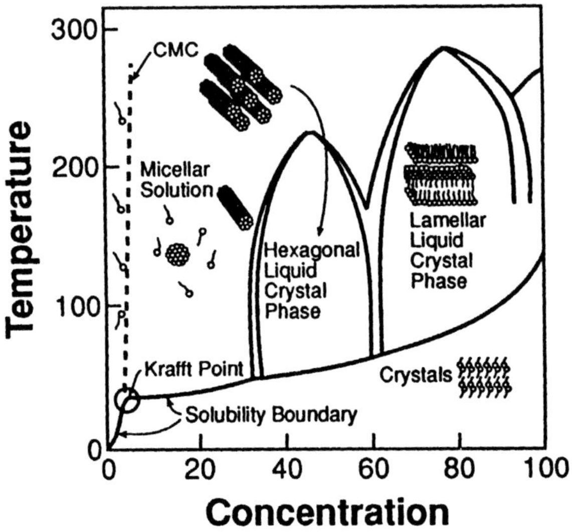

Figure 6.15
Schematic temperature–composition phase diagram of typical single-tail surfactant in water.

and spongelike structures in which surfactant bilayers separate two irregularly connected labyrinths of water.

Figures 6.16 and 6.17 present the temperature–concentration phase diagram of the nonionic surfactant, $CH_3(CH_2)_9(OCH_2CH_2)_4OH$, an ethoxylated alcohol often denoted as $C_{10}E_4$. This system has a lower critical point (at a temperature of about 20.5°C and a $C_{10}E_4$ mass fraction of about 0.02 $C_{10}E_4$) and an upper critical point (at about 296°C and about 0.4 mass fraction of $C_{10}E_4$). The existence of the lower critical point at a very low concentration of solute is common for dilute polymer solutions. Thus, the micelles of $C_{10}E_4$ appear to behave analogously to polymer solutions, as might be expected because the molecular weight of the micelles is on the order of 10^4.

As expected for a surfactant with a normal alkyl tail, hexagonal and lamellar liquid crystalline phases are observed in the water–$C_{10}E_4$ system. There is a region of coexistence between hexagonal and lamellar liquid crystals and between an isotropic micellar phase (ISO-2) and the lamellar liquid crystal phase. The binodal of the micellar and lamellar phases hits the binodal of the low surfactant concentration isotropic phase (ISO-3) and the micellar phase (ISO-2). Three-phase coexistence (ISO-3, ISO-2, and lamellar phases) is the result of this encounter (at about 47°C).

There is another isotropic phase, called the anomalous (ANOM) phase, that occurs between 46 and 59°C. This phase has been observed in many other surfactant solutions and, like the liquid crystalline phases, it is a somewhat generic feature of the phase behavior of association colloids. It is now believed that this anomalous phase is an example of the spongelike phase mentioned above, that is, a phase in which surfactant bilayers divide disordered labyrinths of water-rich regions spanning the phase.

It is interesting to compare the phase diagram of water and $C_{10}E_4$ with that of water and C_4E_1 (Fig. 6.13a), because $C_{10}E_4$ and C_4E_1 belong to the same homologous series. Both binary systems have liquid–liquid coexistence loops opening at a lower critical point and closing at an upper critical point. Thus, there is enough molecular association in each system to cause a phase split at a lower critical point. However, the C_4E_1–water system does not form micelles or liquid crystalline phases, that is, the system does not form association colloids: the molecular association in the C_4E_1–water system is sufficient to result in a lower critical point but is too short lived and is not sufficiently collective to lead to association colloids. Similarly, C_6E_2 does not form association colloids in water, but C_8E_3 does. This ability of an amphiphile, a molecule with a water soluble moiety bonded to a water insoluble moiety, to form association colloids is the property that distinguishes a surfactant, say $C_{10}E_4$ or $C_{12}E_5$, as a special class

Figure 6.16
Temperature–weight fraction phase diagram of water and $CH_3(CH_2)_9(OCH_2CH_2)_3OH$ (3,6,9,12-tetraoxyadocosanol or $C_{10}E_4$). ISO-1, -2, or -3 designate isotropic phases, ANOM anomalous isotropic phase, HEX hexagonal liquid crystalline phase, and LAM lamellar liquid crystalline phase. From Lang and Morgan (1980).

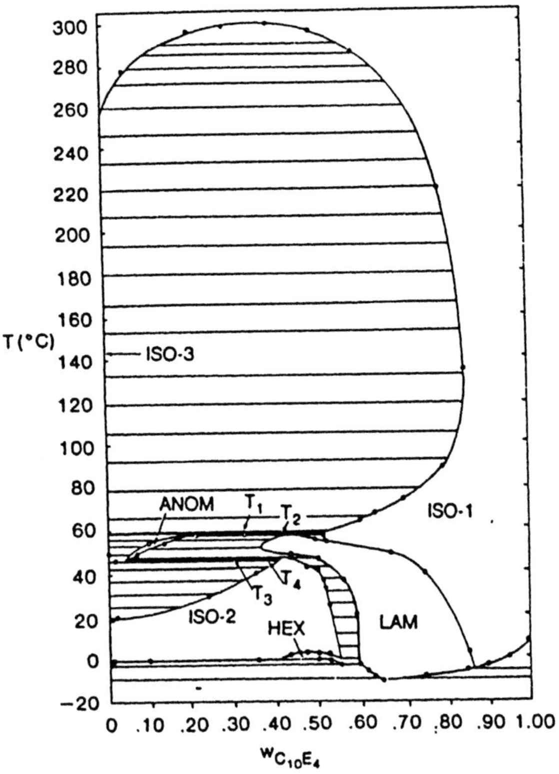

(a)

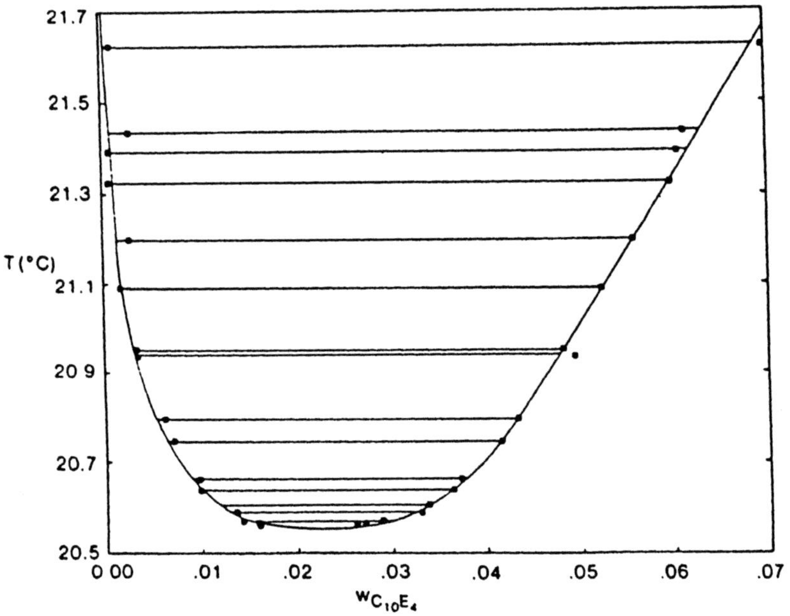

(b)

Figure 6.17
Temperature–weight fraction phase diagram of water and $CH_3(CH_2)_9(OCH_2CH_2)_3OH$ (3,6,9,12-tetraoxyadocosanol or $C_{10}E_4$). ISO-1, -2, or -3 designate isotropic phases, ANOM anomalous isotropic phase, HEX hexagonal liquid crystalline phase, and LAM lamellar liquid crystalline phase. From Lang and Morgan, (1980).

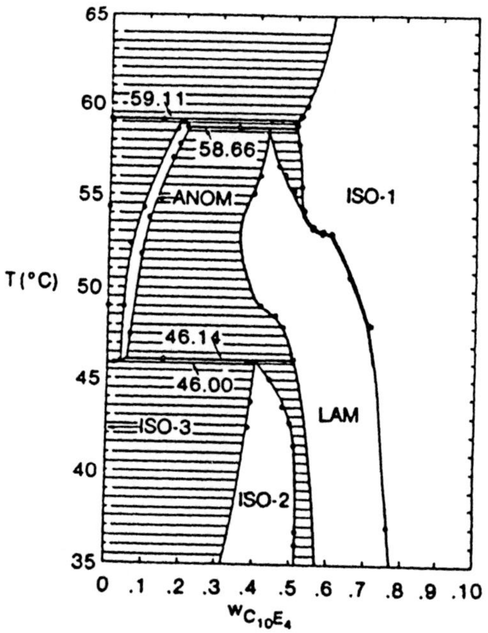

of amphiphiles. Molecules such as C_4E_1, small monohydric alcohols $CH_3(CH_2)_nOH$, $n = 1, 2, \ldots$, aldehydes, amines, etc., are amphiphiles but not surfactants.

Although we have chosen to discuss surfactants in water, it is important to note that surfactants with hydrocarbon tails form in hydrocarbon liquids similar association colloids as those found in water except that the roles of the head and the tail moieties are interchanged.

Binary metal systems also have rich phase behavior owing to their ability to form alloys or solid solutions in variety compositions and crystal structures. The phase diagram of brass, a binary solid solution of copper and zinc, illustrates this richness (see Fig. 6.18). Phases α, β, β', γ, δ, ε and η are the various solid phases of brass. L denotes the liquid phase. Most commercial brasses are α-phase solids, which are solid solutions of zinc in copper. α-brass is a disordered substitutional solid solution, that is, the Zn atoms randomly replace Cu atoms in the fcc Cu crystal. β-brass is a bcc lattice made up of Zn and Cu atoms. At sufficiently low temperatures the atoms of Zn and Cu in this alloy are separately ordered and with increasing temperature they undergo an order–disorder transition to a state in which the Zn and Cu atoms are randomly distributed over the bcc lattice sites. For example, in a β brass with a Cu atom to Zn atom ratio of 1, the Cu atoms primarily occupy the center of the cubic unit cell whereas the Zn atoms primarily occupy the vertices of the cubic unit cell at temperatures lower than 742 K. Above 742 K, the Cu and Zn are found in increasingly

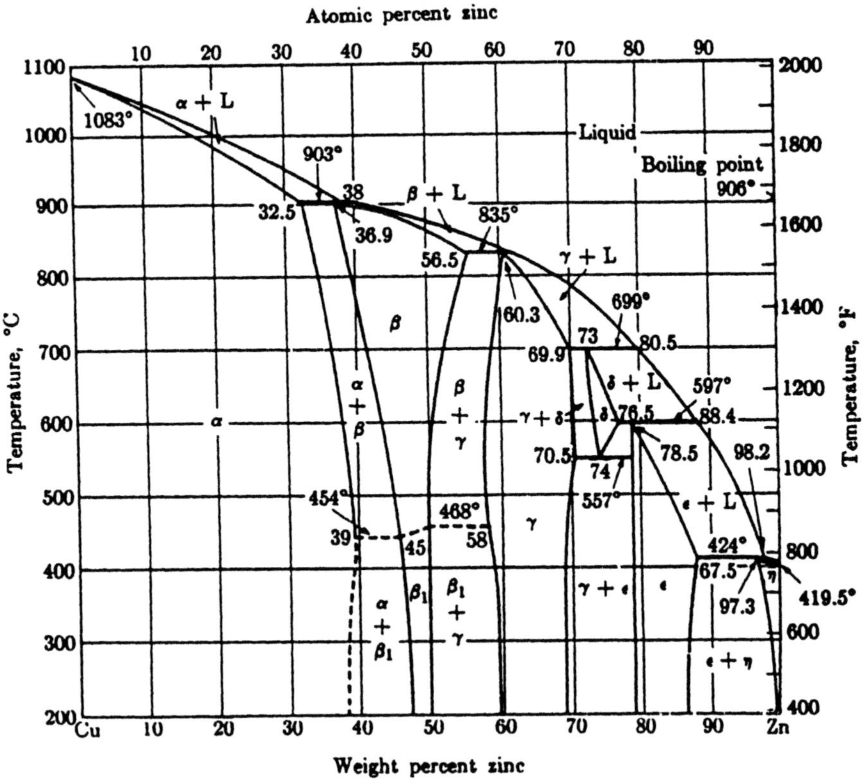

Figure 6.18
The copper–zinc phase diagram adapted from van Vlack (1975).

random distribution on the bcc lattice sites with increasing temperature. Thus, $T = 742$ K is the temperature of an order–disorder transition of the type predicted by the Ising model described in Chapter 4.

The practical importance of alloys is that by combining metals some desired properties of a particular metal (or metals) can be enhanced or the cost of the material may be reduced. Brass and bronze (copper solutions of tin) are stronger and harder than copper, are cheaper than copper, and so are better for making machine parts, vessels and coatings than is copper. Similarly, sterling silver contains 7.4 wt % Cu to harden it without changing its looks or resistance to corrosion.

6.2.3 Three-Component Phase Equilibria

Addition of a third component will obviously lead to more complex phase diagrams. However, knowing the phase behavior of the three binary pairs of a ternary system assists one in guessing what the three-component system will do. In predicting phase splits from equations of state, binary phase splits provide the first iteration of a continuation calculation of two-phase binodals by successively advancing the amount of the third component at fixed temperature and pressure.

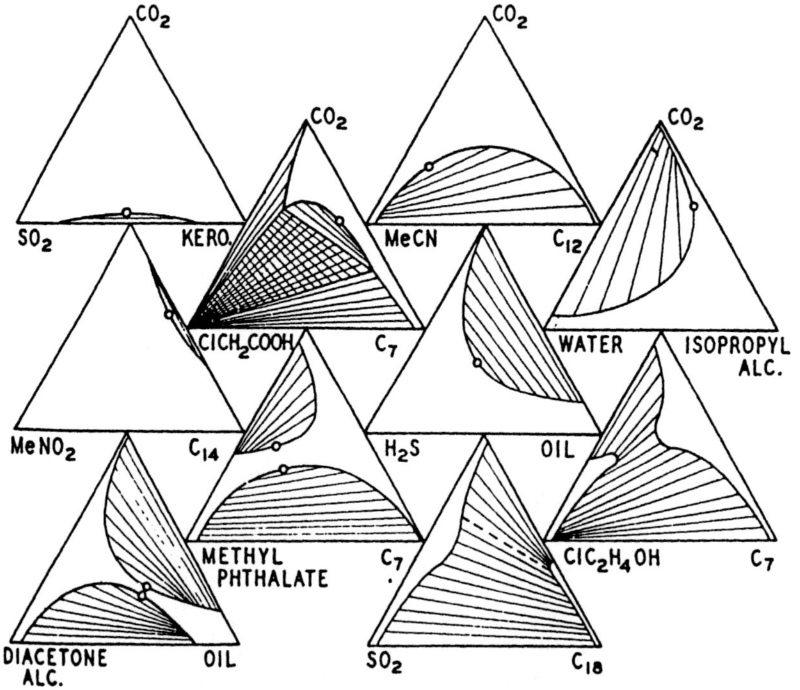

Figure 6.19
Constant temperature and pressure of liquid carbon dioxide with pairs of other liquids. Compositions are given as mass fractions. Adapted from Francis (1963).

In Figure 6.19 are shown 10 ternary constant T and P phase diagrams of liquid carbon dioxide (at elevated pressure) with various liquids. In four of the diagrams, only one binary edge has a two-phase split (the four are carbon dioxide with the pairs SO_2–Kerosene, MeCN–dodecane (C_{12}), $MeNO_2$–tetradecane (C_{14}), and water–isopropanol, respectively). In these cases we know that there will be a binodal curve that begins at the compositions of the two-phase pair on the binary edge and ending in a critical point on the interior of the ternary phase diagram. We know the binodal will exist in the interior of the ternary phase diagram because physically one cannot make a pair of coexisting liquids miscible in all proportions by addition of an infinitesimal amount of a third component. We know that the binodal must end somewhere in a critical point because there are no binodals on the other edges of the ternary phase diagram. Thus, the simplest thing to expect is that the binodal ends in a critical point as do the four cases shown in Figure 6.19. There can, however, be a more complicated pattern as is illustrated by the CO_2—CHOOH—C_{14} system shown in Figure 6.20. In this case, the binodal springing from the CHOOH—C_{14} edge encounters a three-phase region, bordered by two other binodals that end in critical points.

When two edges of a ternary phase diagram have miscibility gaps, a variety of things can happen. Binodals can spring from the edges and end in critical points without encountering one another. This happens in two of the diagrams shown in Figure 6.19. On the other hand the binodals can merge and so a two-phase region can span, without interruption by a critical point, from one edge

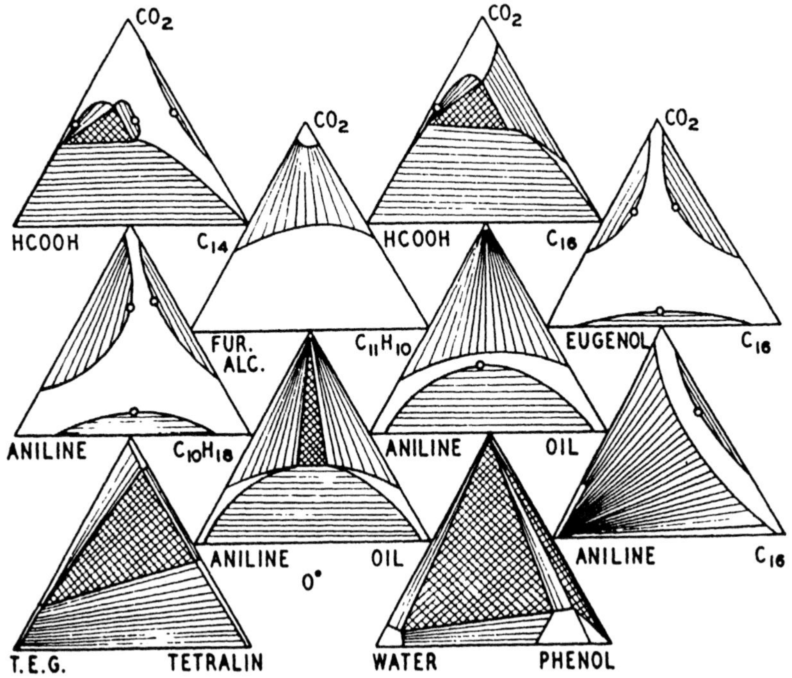

Figure 6.20
Constant temperature and pressure of liquid carbon dioxide with pairs of other liquids. Compositions are given as mass fractions. Adapted from Francis (1963).

to another edge of a ternary phase diagram. Such an edge-to-edge binodal can however be interrupted by a three-phase region (as is the case in the CO_2—$ClCH_2COOH$—C_7 and CO_2—CHOOH—C_{16} systems) two of whose sides merge with the spanning binodal whereas from the third side springs a binodal ending in a critical point.

When the three binary edges of a ternary phase diagram have miscibility gaps even more possibilities exist, as is illustrated by Figure 6.20. The simplest situation is that three binodals spring from the binary edges and end in critical points without encountering one another (note the CO_2–aniline–$C_{10}H_{18}$ and CO_2–eugenol–C_{16} diagrams). Two of the binodals can form an edge-to-edge binodal path while the third binodal ends in a critical point (e.g., CO_2–aniline–C_{16}). Alternatively, the three binodals can encounter a three-phase triangle whose three sides merge with the three binodals (e.g., CO_2–water–phenol).

Most of the phase diagrams shown in Figures 6.19 and 6.20 were determined at about 25°C. However, the phase diagram for the CO_2–aniline–oil system was determined at 25 and 0°C. We see that in lowering the temperature from 25 to 0°C, the binodal with a critical point merges with the binodal spanning from edge-to-edge binodal with the consequence that a three-phase region appears. The sides of the three-phase triangle border the three binodals leading to the binary edges. By varying temperature, then, the type of ternary phase diagram that occurs can be changed. This could be an asset in computing phase diagrams from an equation of state. For example, the binary miscibility gaps could be used as the starting point for computing the binodals in the CO_2–aniline–oil system at 25°C. The binodals of this diagram could than be used as the starting point for computing binodals at successively lower temperatures. Once it is noted that the binodals begin to approach each other, one can guess that a three-phase region is going to occur at some subsequent temperature. The computational algorithm can accordingly be designed to search for such three-phase regions along the temperature path.

In what we have discussed so far miscibility gaps along the binary edges provide the major clues to the structure of the ternary phase diagram and would provide the natural starting points in computing the ternary phase diagram from an equation of state. There are, however, cases in which the binary edges at a give temperature provide no clue to the existence of multiphase regions in the ternary phase diagram. Such a case is illustrated by Figure 6.21, the ternary phase diagram for the liquids acetic acid, *o*-toluidine, and heptane. All three binary pairs of these liquids are miscibile in all proportions. Nevertheless, in the interior of the ternary phase diagram, there is a binodal that opens a two-phase region at a critical point near the acetic acid–C_7 edge and closes it at another critical point near the *o*-toluidine–C_7 edge. The best way to find such an island or isola multiphase region with an equation of state is to compute spinodal curves and search for critical points along the spinodal curves. For the case illustrated by Figure 6.21, there will be two critical points and the spinodal will be a closed curve lying inside the binodal and touching the binodal at each critical point. The critical points and the spinodal curve can be used to guess binodal compositions to initiate a computation of an island binodal.

When oil, water, and a surfactant are mixed, as can be guessed from binary phase diagrams such as Figures 6.16 and 6.17, complex patterns of phase equilibria among microstructured phases occur. As an example we show in Figure 6.22 a sequence of ternary composition phase diagrams for a system of water, normal decane, and

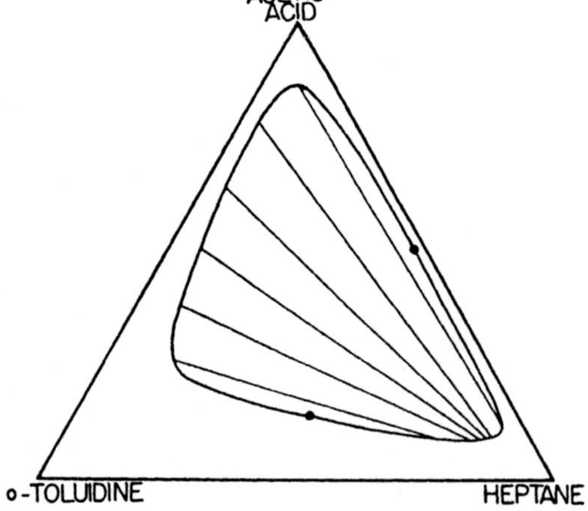

Figure 6.21
Two-phase island in a constant *T* and *P* phase diagram of liquids *o*-toluidine, acetic acid, and *n*-heptane. Compositions in weight percent. Redrawn from Francis (1963).

Figure 6.22
Ternary mass fraction phase diagrams for $CH_3(CH_2)_7(OCH_2CH_2)_3OH$/water/decane mixtures at atmospheric pressure and different temperatures. Adapted from Kilpatrick (1983).

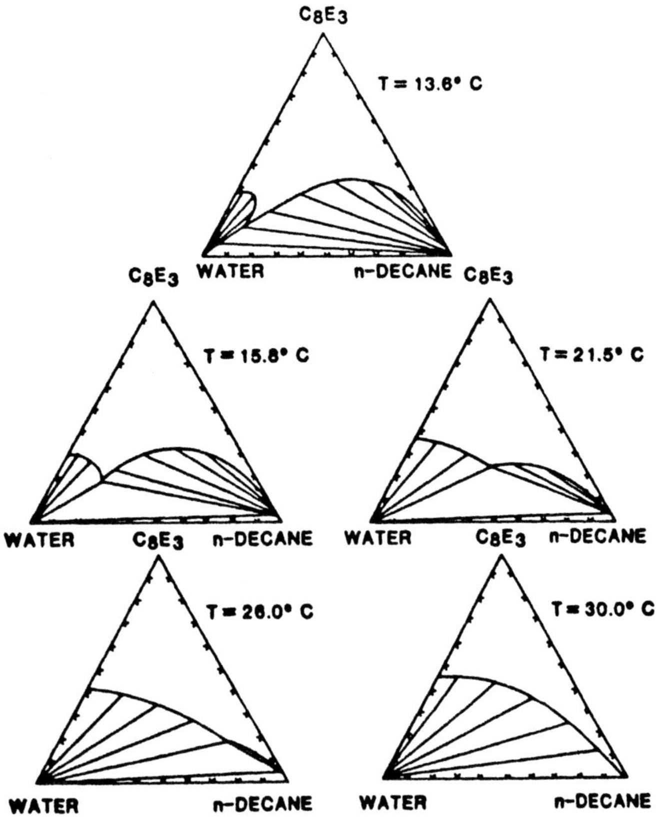

octylpentaethoxy alchohol (C_8E_3) at several temperatures and at atmospheric pressure. Other phase splits occur in the high surfactant part of the phase diagrams but they are not shown in Figure 6.22. Octane and water are almost immiscible at all the temperatures in Figure 6.22. At 30°C a binodal springs from the water–oil edge of the phase diagram and forms a two-phase corridor that ends as a miscibility gap along the water–surfactant edge. The coexisting liquids are isotropic, and the surfactant-rich phase can contain appreciable amounts of both oil and water. This "solubilization" of oil and water is one of the major functions of surfactants in many commercial formulations. An equilibrium, isotropic liquid phase consisting of surfactant and appreciable amounts of oil and water is called a microemulsion. The microstructure of microemulsions has been the subject of intense speculation and investigation in recent years. Later in this section, we will discuss briefly the microstructure of microemulsions. Presently, however, let us continue to examine the patterns of phase behavior in the lower part of the phase diagram. The binodal ends in a critical point in the oil-rich region of the diagram.

As the temperature decreases from 30°C, a third phase, a microemulsion, springs from a tie-line (i.e., a critical end point) of the binodal. At $T = 26$°C, this third phase, although richer in oil, contains an appreciable amount of water. As the temperature continues to decrease the middle phase microemulsion (because oil is lighter than water the surfactant-rich microemulsion phase will literally reside between a lower water-rich and an upper oil-rich phase in a test tube containing the three coexisting phases) becomes increasingly richer in water. Finally, between 15.8 and 13.6°C, the microemulsion and the oil-rich phases merge at a critical end point into a single tie-line of a two-phase binodal. This critical end point is somewhat obscured because at 13.6°C the binodal envelope springing from the water–surfactant edge is virtually touching the binodal envelope springing from the oil–water edge.

The pattern of phase behavior observed in Figure 6.22 is repeated schematically in Figure 6.23 as a temperature–composition prism in which constant temperature cross-sections are ternary phase diagrams. As illustrated in Figure 6.23, at some intermediate temperature the middle phase microemulsion contains equal amounts of water and oil. As the temperature is decreased from this intermediate value, the microemulsion phases merges with the water-rich phase at a critical end point. As the temperature is increased from this intermediate value, the microemulsion phase merges with the water-rich phase at a critical end point. As the temperature is increased from this intermediate value, the microemulsion phase merges with the oil-rich phase as another critical end point. This pattern of two to three to 2-phase splits as a function of temperature turns out to be a generic one, that is, many different surfactants and oils give the same sequence. Moreover, the temperature can be replaced by a variety of field variables in scrolling through the sequence. For example, the pattern is observed in some systems

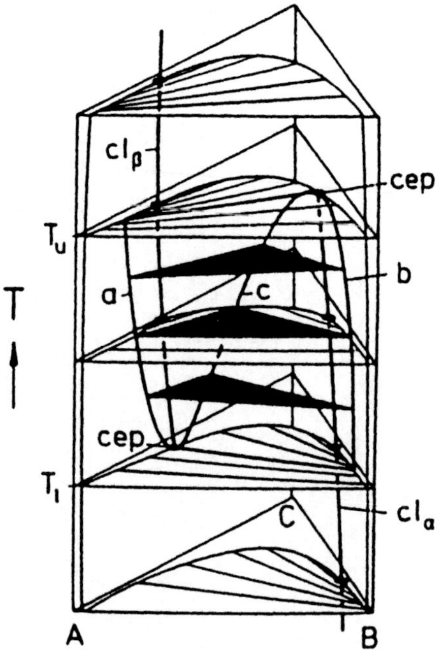

Figure 6.23
A schematic sequence of phase diagrams in which a third phase appears at one end of a binodal tie-line as a critical end point and disappears at the other end of the binodal tie-line as a critical end point.

by varying the activity of a fourth component (salt or alcohol for instance), the pH, or the carbon number in a homologous series of oils.

Kahlweit and coworkers (1990) emphasized that the patterns of microemulsion phase behavior can be anticipated from the binary sides of the temperature–composition ternary phase diagram. The binary sides of the temperature–composition prism are illustrated in Figure 6.24 for a typical nonionic surfactant/oil/water system. Because addition of a little or a third component will not cause a discontinuity of phase behavior, we expect the binodals on the binary sides of the prism will penetrate the prism and will either disappear at critical points or merge with other binodals to form binodal corridors or three-phase equilibria. The critical point T_l is connected by a continuous locus or critical points in the

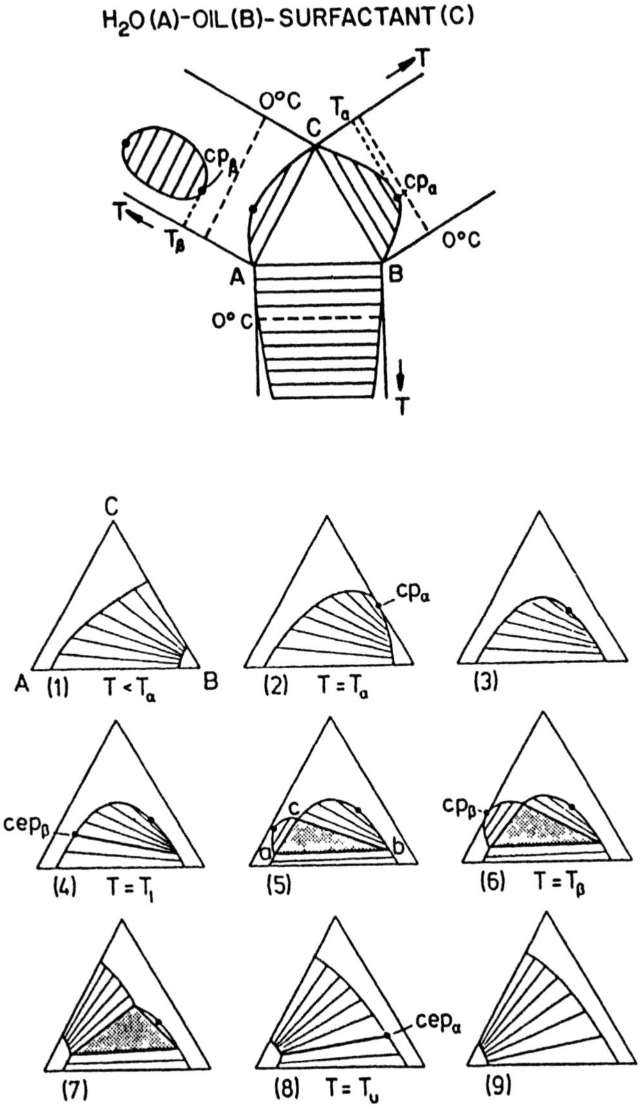

Figure 6.24
Schematic of the evolution of the three-phase triangles with rising temperature in a water, oil, and nonionic surfactant mixture. Redrawn from Kahlweit, et al. (1990).

temperature–composition ternary phase diagram. At the temperature T_l the binodal intrusion from the water–surfactant side of the temperature–composition prism touches the binodal at a critical end point cep$_\beta$) and as temperature is increased the merged binodals split a third phase that becomes increasingly oil-rich until it merges with the oil-rich phase at another critical end point (cep$_\alpha$) at T_u. The critical end point T_u is connected to the critical point T_α by a continuous locus of critical points in the temperature–composition ternary phase diagram.

In the higher surfactant concentration regions of oil/water/surfactant phase diagrams there are various lyotropic (having a solvent) liquid crystalline phases. For example, in the phase diagram shown in Figure 6.25 for hexanol/water/cetyltrimethyl ammonium bromide, there are identified lamellar (neat) and hexagonal liquid crystal phases, micellar and inverted micellar solutions, and midrange microemulsions (middle phase). Numerous micellar and lyotropic liquid crystalline phases have been identified in surfactant solutions. In all of these the surfactant organizes into sheetlike structures that then form periodic arrays of bilayers (lamellar liquid crystal), cylinders (hexagonal liquid crystal), spheres (cubic phase liquid crystal), corrugated bilayers (ripple phase liquid crystal), etc. The sheetlike structures can also form continuous three-dimensional triply periodic surfaces separating bicontinuous regions of oil and water, each of which continuously spans the sample. These are called bicontinuous liquid crystals. Figure 6.26 is a hypothetical oil, water, and surfactant phase diagram. Although the figure is hypothetical, all of the phases indicated they have been found. The microemulsion is shown in the phase diagram as a geometrically disordered bicontinuous phase. The alternatives to a bicontinuous microemulsion are a water-continuous dispersion of oil-swollen globular micelles or an oil-continuous dispersion of water-swollen globular inverted micelles. All three types of microemulsions are now known to exist, the preferred microstructure depending on the chemical components and/or the thermodynamic conditions.

6.3 Peng-Robinson Equation of State

As discussed in Chapter 5, empirical modifications of van der Waal's equation of state greatly improve its value in quantitative predictions of the phase behavior of fluids. One of the most popular modifications is the equation of state developed by Peng and Robinson. The (PR) equation of state and the corresponding free energy density were given at eqs. (6.1.66) and (6.1.72). In the equation of state,

$$P = \frac{nkT}{1 - nb} - \frac{n^2 a}{1 + 2nb - (nb)^2}, \tag{6.3.1}$$

Figure 6.25
Regions of occurrence of various phases in a mixture of hexanol, water and cetyltrimethyl ammonium bromide at room temperature and atmospheric pressure. Adapted from Ekwall *et al.* (1969).

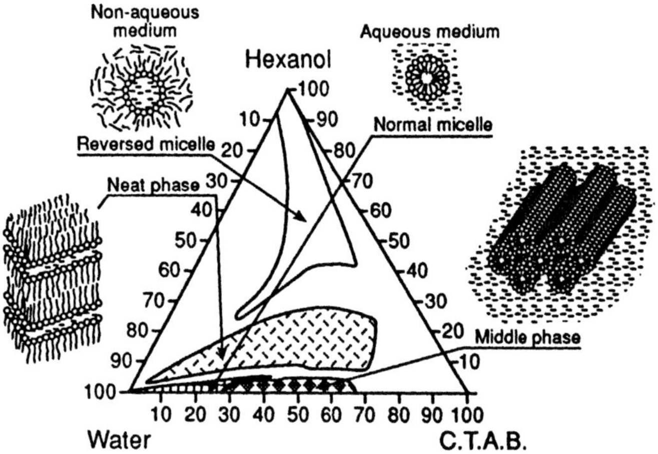

b and a are parameters arising from excluded volume effects and attractive pair interactions. Equation (6.1.67) expresses the usual assumption of semiempirical equations of state of the van der Waals type, namely,

$$b = \sum_{\alpha} x_\alpha b_\alpha \quad \text{and} \quad a = \sum_{\alpha,\beta} x_\alpha x_\beta a_{\alpha\beta}, \tag{6.3.2}$$

Figure 6.26
Hypothetical phase diagram of oil, water, surfactant mixtures. From Davis et al. (1987).

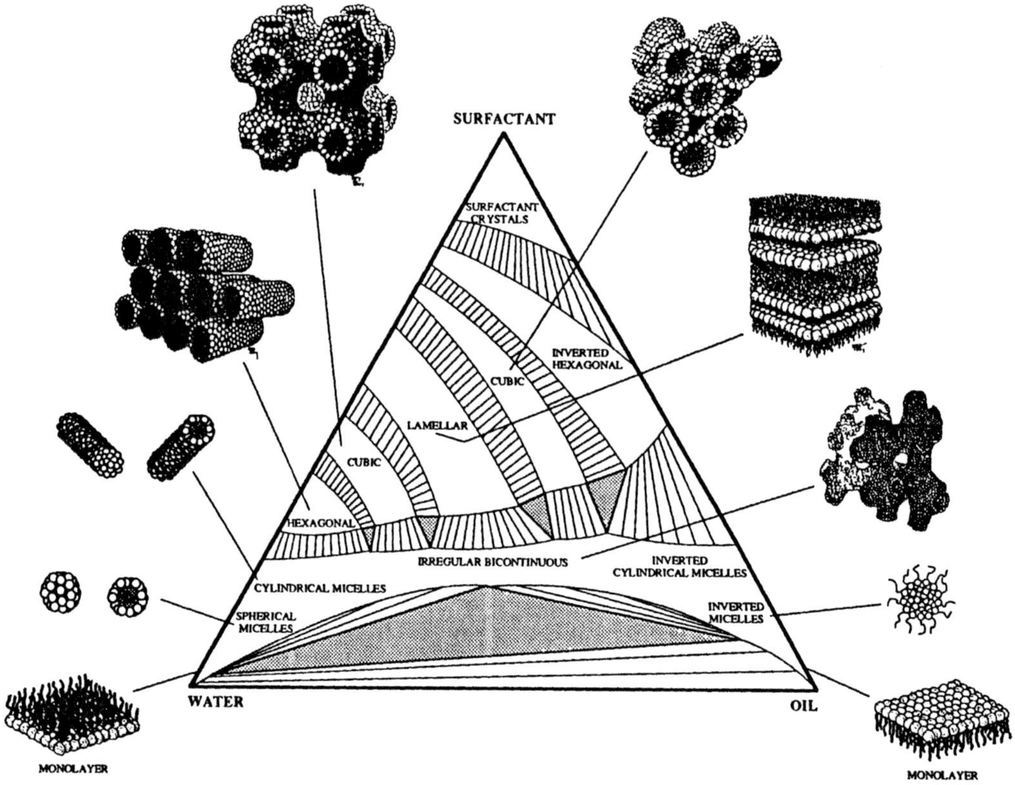

where $x_\alpha = n_\alpha/n$ is the mole fraction of species α and b_α and $a_{\alpha\beta}$ are concentration independent parameters. Consistently with the assumption that b_α and $a_{\alpha\beta}$ are concentration independent, b_α and $a_{\alpha\alpha}$, given by eqs. (6.1.68) and (6.1.69), are chosen to ensure that the PR equation predicts the correct critical temperature and pressure of fluid α and the function $\kappa(w_\alpha)$ at eq. (6.1.70) was introduced to improve the vapor pressure predictions of several hydrocarbon fluids.

In the PR model, the only mixture parameter that enters is $a_{\alpha\beta}$, $\alpha \neq \beta$, the energy parameter representing pair interactions between species α and β. The quantum theory of dispersion forces suggests that $a_{\alpha\beta} \approx \sqrt{a_{\alpha\alpha}a_{\beta\beta}}$ for nonpolar fluids. Thus, a mixing parameter $k_{\alpha\beta}$ is defined by

$$a_{\alpha\beta} = (1 - k_{\alpha\beta})\sqrt{a_{\alpha\alpha}a_{\beta\beta}}, \quad \alpha \neq \beta, \tag{6.3.3}$$

and a fit of the PR equation of state to whatever binary data are available can be used to estimate $k_{\alpha\beta}$. Once the binary $k_{\alpha\beta}$ values are known the PR equation can then be used to study the patterns of phase behavior in multicomponent fluids. In general $k_{\alpha\beta}$ will depend on temperature, although in many cases the temperature dependence is weak. Values of $k_{\alpha\beta}$ are presented in Table 6.1 for a few fluid pairs.

TABLE 6.1 Mixing Parameters k_{12} for Various Fluid Pairs

Fluid Pair	k_{12}	*Temperature T(°C)*
Ammonia–water	−0.26	80–130
Ammonia–water	−0.25	135–150
Butane–decane	0.01	135–240
Butane–methane	0.02	−130–20
Butane–methane	$0.02 + 0.0008(T - 20)$	20–120
Carbon dioxide–butane	0.13	−45–70
Carbon dioxide–decane	0.108	−25–140
Carbon dioxide–decane	0.118	171
Carbon dioxide–decane	0.130	204
Carbon dioxide–decane	0.150	238
Carbon dioxide–octane	0.112	−50–?
Carbon dioxide–methane	0.1	−90–0
Carbon dioxide–propane	0.13	−30–50
Carbon dioxide–water	0.086	300
Carbon dioxide–water	0.130	325
Carbon dioxide–water	0.113	350
Decane–methane	$0.05 + 6.9 \times 10^{-5}$T	70–240
Ethylene–heptane	0.02	−30–75
Ethylene–heptane	$0.02 + 2.66 \times 10^{-5}(T - 75)$	75–195

Smoothed estimated values for k_{12} deduced from values computed by Ohe (1990).

The chemical potential of species α of a PR fluid is obtained by differentiating $f(n)$ in eq. (6.1.72) with respect to n_α at fixed T and $n_{\beta \neq \alpha}$. The result is

$$\mu_\alpha = \mu_\alpha^+(T) + kT \ln\left(\frac{n_\alpha}{1-nb}\right) + \frac{nkTb_\alpha}{1-nb} - \frac{anb_\alpha}{b[1+2nb-(nb)^2]} \tag{6.3.4}$$

$$- \left\{\frac{1}{2\sqrt{2}bn}\frac{\partial}{\partial n_\alpha}[n^2 a(T)] - \frac{ab_\alpha}{2\sqrt{2}b^2}\right\} \ln\left[\frac{1+nb(1+\sqrt{2})}{1+nb(1-\sqrt{2})}\right]. \tag{6.3.5}$$

The conditions for two-phase equilibria are

$$\mu_\alpha(\mathbf{n}^{(1)}) = \mu_\alpha(\mathbf{n}^{(2)}), \quad \alpha = 1, \ldots, c \tag{6.3.6}$$

and

$$P(\mathbf{n}^{(1)}) = P(\mathbf{n}^{(2)}), \tag{6.3.7}$$

where $\mathbf{n}^{(i)} = \{n_1^{(i)}, \ldots, n_c^{(i)}\}$ denotes the composition of phase α. In a two-component system and at a fixed temperature, the phase rule yields $f = c + 2 - p = 2 + 2 - 2 = 2$, which means that only one intensive variable needs to be set to determine the coexisting phases.

6.3.1 Two-Component or Binary Phase Equilibria

To illustrate the strategy of binary phase diagrams from an equation of state, let us compute the pressure–composition phase diagrams of a mixture of carbon dioxide (CO_2) and propane (C_3H_8). At $T = 0°C$, pure liquid C_3H_8 coexists with its vapor at 6 atm and pure liquid CO_2 coexists with its vapor at 34 atm. Thus, the liquid–vapor coexistence curves will spring from the pure C_3H_8 side of the pressure–composition diagram and will end at the pure CO_2 side of the diagram. The first step in computing the phase diagram for C_3H_8 and CO_2 at $T = 0°C$ is then to start near the pure C_3H_8 edge by choosing a vapor pressure P a little higher, say 10% higher, than the vapor pressure of pure C_3H_8. The four densities of the components of the liquid and vapor phases will be determined by the four equations

$$\begin{aligned} \mu_\alpha(\mathbf{n}^{(l)}) - \mu_\alpha(\mathbf{n}^{(v)}) &= 0, \quad \alpha = CO_2, C_3H_8 \\ P(\mathbf{n}^{(l)}) - P &= 0 \\ P(\mathbf{n}^{(v)}) - P &= 0. \end{aligned} \tag{6.3.8}$$

These equations can be solved by the Newton–Raphson (NR) method by guessing the densities $\mathbf{n}^{(l)} = \tilde{\mathbf{n}}^{(l)}$, $\mathbf{n}^{(v)} = \tilde{\mathbf{n}}^{(v)}$ and calculating an improved value from

$$\mathbf{J}(\mathbf{z} - \tilde{\mathbf{z}}) = -\mathbf{f}, \tag{6.3.9}$$

where $f_1 = \mu_{CO_2}(\tilde{\mathbf{n}}^{(l)}) - \mu_{CO_2}(\tilde{\mathbf{n}}^{(v)})$, $f_2 = \mu_{C_3H_8}(\tilde{\mathbf{n}}^{(l)}) - \mu_{C_3H_8}(\tilde{\mathbf{n}}^{(v)})$, $f_3 = P(\tilde{\mathbf{n}}^{(l)}) - P$, $f_4 = P(\tilde{\mathbf{n}}^{(v)}) - P$, $z_1 = n_{CO_2}^{(l)}$, $z_2 = n_{CO_2}^{(v)}$, $z_3 =$

$n^{(l)}_{C_3H_8}$, $z_4 = n^{(v)}_{C_3H_8}$, and the components of the matrix J are $J_{ij} = [\partial f_i/\partial x_j]_{\mathbf{x}=\tilde{\mathbf{x}}}$. Equation (6.3.8) is to be iterated until the latest and next to the latest values of densities obey the inequalities $\sum_\alpha (z_\alpha - \tilde{z}_\alpha)^2 < \varepsilon_{\mathbf{z}}$ and $\sum_\alpha f_\alpha^2 < \varepsilon_{\mathbf{f}}$, where $\varepsilon_{\mathbf{z}}$ and $\varepsilon_{\mathbf{f}}$ are prescribed tolerances.

A good initial guess is that the C_3H_8 densities in the liquid and vapor phases equal those of coexisting pure C_3H_8 at 0°C and that the CO_2 densities equal a few percent, say 10%, of these. If the method fails to converge to a solution with these guesses, then the initial guesses for the CO_2 densities should be changed. Once a converged solution for the densities is found for the given vapor pressure P, the next solution can be found by adding a small increment δP as the first guess in seeking the coexistence densities at $P + \delta P$. The process is continued until the pressure $P = 34$ atm is reached and the pressure–composition phase diagram is determined over the entire binary solution range.

The CO_2—C_3H_8 phase diagrams predicted by the PR equation are compared in Figure 6.27 with experimental results. Note that the agreement is pretty good; this is not unusual for PR predictions of the liquid–vapor equilibria of nonpolar fluids.

The way to use the NR method in changing the temperature of the pressure–composition diagrams is to begin at $T = 0$°C diagram, advance T by a small increment δT, set the vapor pressure P a little higher than the vapor pressure of C_3H_8 at $T = 0°\text{C} + \delta T$, guess the densities to be those of coexisting liquid and vapor at $T = 0$°C and P, and iterate eq. (6.3.8) to obtain the coexistence densities at $T = 0°\text{C} + \delta T$ and P. Then the rest of the diagram for $T = 0°\text{C} + \delta T$ can be computed by advancing P by increments of δP as described above. The process can be continued in temperature to obtain the set of diagrams shown in Figure 6.27. The largest magnitude of increments that can be taken in T and P will be determined by the radius of convergence of the NR method. This is best determined by trial and error once the computation is underway.

Note in Figure 6.27 that the pressure–composition binodal has pulled away from the pure CO_2 axis between 21° and 37.8°C. This is because the critical temperature of pure CO_2 is 31°C. Thus, if one tried to find the liquid–vapor coexistence curve by starting at the pure CO_2 end of the phase diagram at a temperature above 31°C, it would be hard to get a good first guess for the NR method because there is no liquid–vapor equilibrium in pure CO_2 at this temperature. This points up a major strategy in computing phase diagrams for a c-component system: use everything you know about the related $c - 1$ component systems.

The pressure–composition liquid–vapor phase diagram predicted by the PR equation for CO_2 and decane is shown in Figure 6.28. Again the agreement is pretty good.

At lower temperatures than those shown in Figure 6.28, CO_2 and decane exhibit liquid–liquid phase equilibria. The liquid–vapor phase equilibria of pure CO_2 or decane are of no use in

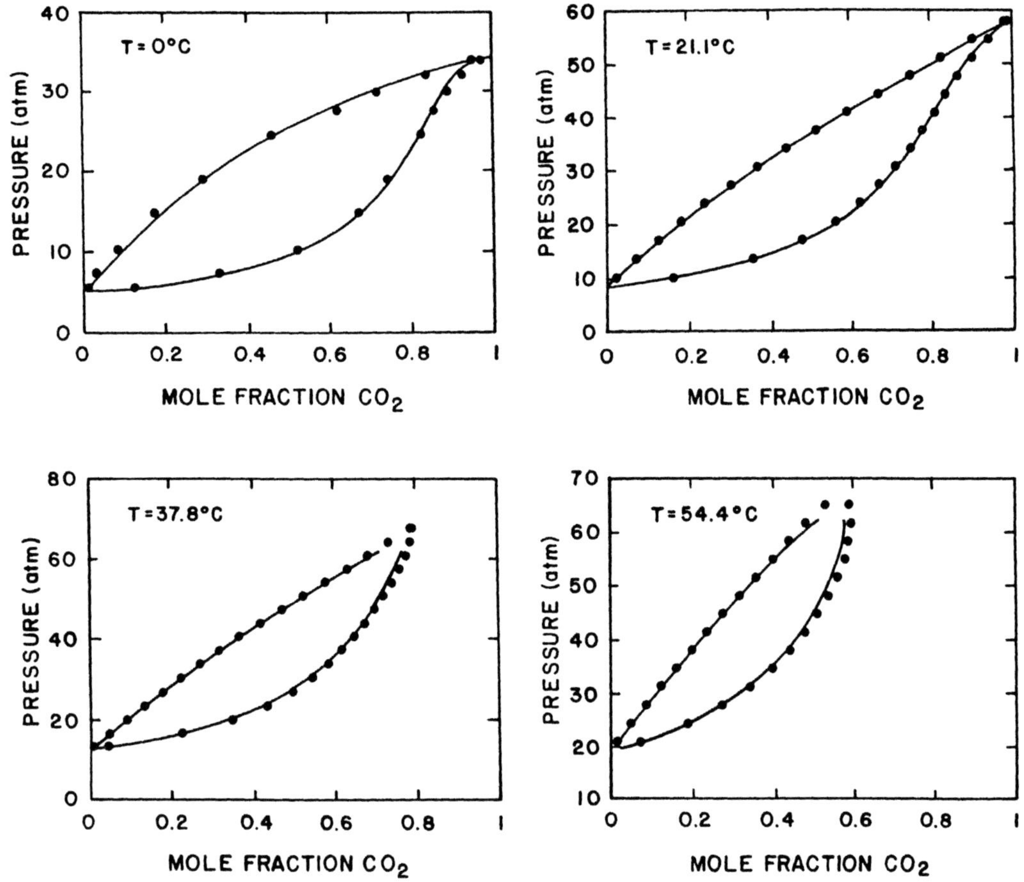

Figure 6.27
Carbon dioxide–propane liquid–vapor phase diagram. (a) $T = 0°\text{C}$, $k_{12} = 0.1518$; (b) $T = 21.1°\text{C}$, $k_{12} = 0.1236$; (c) $T = 37.8°\text{C}$, $k_{12} = 0.1327$; (d) $T = 54.4°\text{C}$, $k_{12} = 0.1390$. Redrawn from Ohe (1990).

identifying the region of liquid–liquid equilibria and so another approach is needed to find the liquid–liquid binodal. As discussed in Section 6.1, the spinodal envelope can be useful in locating phase boundaries. For a two-component system at a fixed temperature, the spinodal curve of n_2 versus n_1 obeys the condition

$$J \equiv \frac{\partial^2 f}{\partial n_1^2}\frac{\partial^2 f}{\partial n_2^2} - \left(\frac{\partial^2 f}{\partial n_1 \partial n_2}\right)^2 = 0. \tag{6.3.10}$$

One way to find the spinodal curve or curves in a density–density phase diagram is to vary n_1 incrementally from 0 to the maximum allowed density, $n_1 b_1 = 1$, and at each increment to plot J versus n_2 to find the solutions to $J = 0$. With this procedure, the PR model will yield the spinodal curves shown in Figure 6.29 for CO_2 and decane at $T = 7°\text{C}$.

Using what we know about the pure decane and CO_2 edges of Figure 6.29, we can find the spinodal curves shown in Figure 6.29 more efficiently than described in the preceding paragraph. We know that there is liquid–vapor equilibrium along each pure

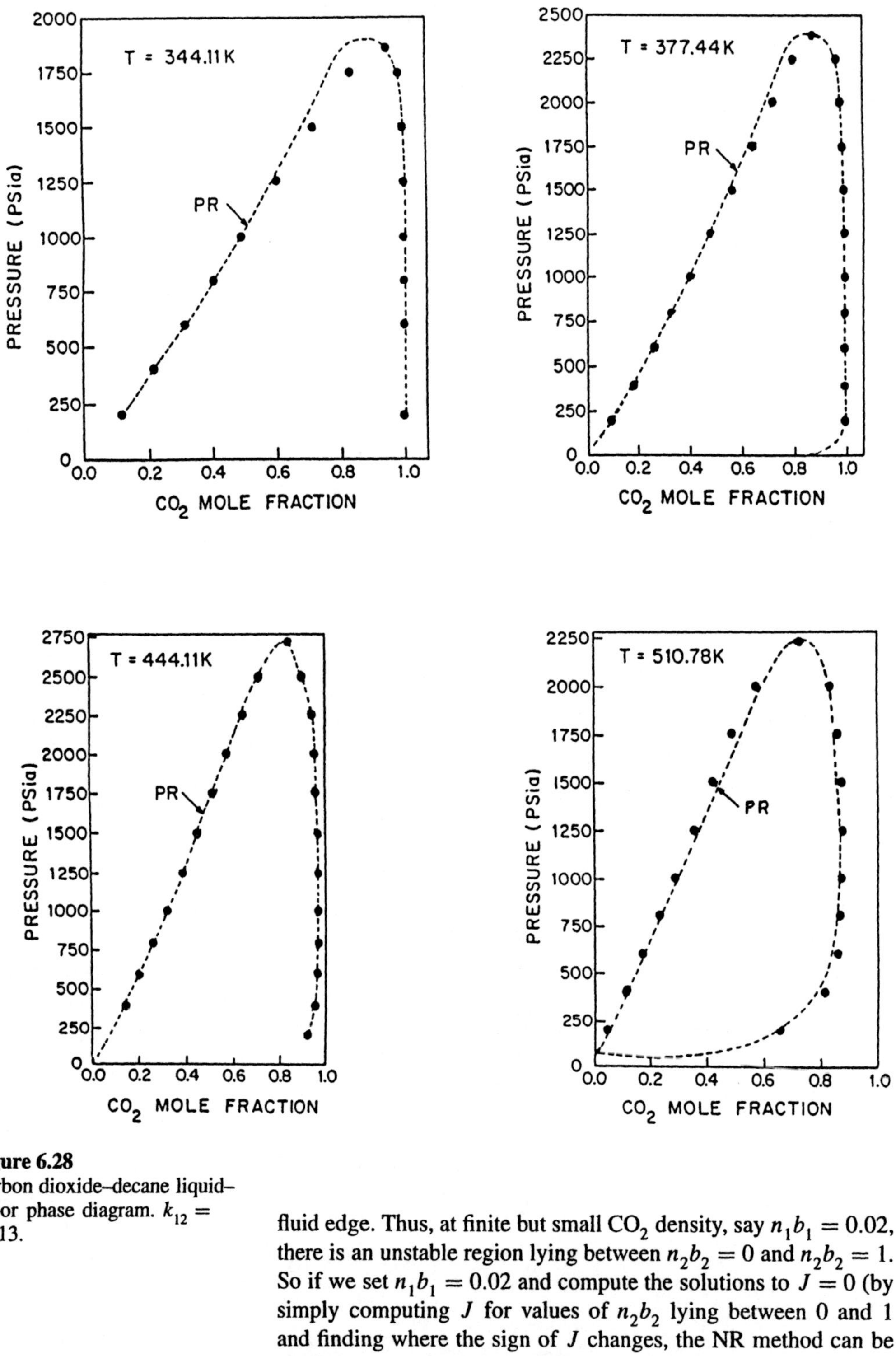

Figure 6.28
Carbon dioxide–decane liquid–vapor phase diagram. $k_{12} = 0.113$.

fluid edge. Thus, at finite but small CO_2 density, say $n_1b_1 = 0.02$, there is an unstable region lying between $n_2b_2 = 0$ and $n_2b_2 = 1$. So if we set $n_1b_1 = 0.02$ and compute the solutions to $J = 0$ (by simply computing J for values of n_2b_2 lying between 0 and 1 and finding where the sign of J changes, the NR method can be used to accurately find the final roots n_2b_2 of $J = 0$), we will find two solutions (one at $n_2b_2 = 0.02$, and one at $n_2b_2 = 0.73$ each of which lie on the spinodal curves. They would be on the same

Figure 6.29
Density–density phase diagram predicted by the PR equation for carbon dioxide at 7°C. Dashed curves are spinodal, solid curves are binodals indicating coexisting liquid–vapor ($L_1 - V$ and $L_2 - V$) and liquid–liquid ($L_1 - L_2$) phase, and straight lines are tie-lines indicating the composition of coexisting phases. Three of these tie-lines form the edges of the tie triangle indicating the region of three-phase coexistence. $k_{12} = 0.113$. Redrawn from Falls (1982).

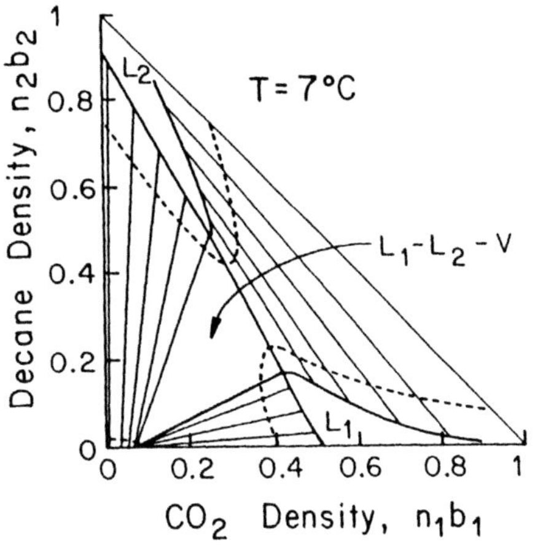

curve if they were on branches that connect at a critical point; this is not the case for the situation illustrated in Figure 6.29. Now that these two points on spinodal curves have been located, the rest of the curves can be computed efficiently. Consider for example, the point $n_1b_1 = 0.02$ and $n_2b_2 = 0.73$. Choose $n_1 = 0.02/b_1 + \delta n_1$ and compute another value of n_2 along the spinodal curve from the method, namely, guess $n_2 = 0.73/b_2$ and compute a new guess from

$$\frac{\partial J}{\partial n_2}(n_1, \tilde{n}_2)(n_2 - \tilde{n}_2) = -J(n_1, \tilde{n}_2). \tag{6.3.11}$$

For the new value of n_1, the new value of n_2 is determined by the NR method by iterating eq. (6.3.10) until the difference between successive estimates, $n_2^{(k+1)} - n_2^{(k)}$, and $J(n_1, n_2^{(k)})$ are as close to zero as some desired accuracy dictates. If one begins on the spinodal at $n_1b_1 = 0$ and $n_2b_2 \simeq 0.73$ and advances n_1 in small increments, then eq. (6.3.10) will yield the spinodal curve until n_1b_1 and n_2b_2 approach the "turning point" at values of about 0.33 and 0.47, respectively. At the turning point $dn_2/dn_1 = \infty$ and so difficulty is encountered in solving eq. (6.3.10) as the turning point is approached.

The origin of the difficulty can be understood by recognizing that $J(n_1, n_2) = 0$ along the spinodal. From this it follows that for incremental steps dn_1 and dn_2 along the spinodal

$$\frac{\partial J}{\partial n_1}dn_1 + \frac{\partial J}{\partial n_2}dn_2 = 0 \tag{6.3.12}$$

Because $dn_1/dn_2 \to 0$ as a turning point where $dn_2/dn_1 \to \infty$ is approached, the derivative $\partial J/\partial n_2 \to 0$ and so eq. (6.1.10) will fail to yield a solution. What is required at such a turning point is that, instead of changing n_1 and computing n_2, we change n_2 and compute n_1 from the NR equation

$$\frac{\partial J}{\partial n_1}(\tilde{n}_1, n_2)(n_1 - \tilde{n}_1) = -J(\tilde{n}_1, n_2), \qquad (6.3.13)$$

because $\partial J/\partial n_1$ is not zero at this turning point. The rest of the spinodal beyond the turning point can be computed by advancing n_2 and finding n_1 by NR iteration of eq. (6.3.12).

The entire spinodal curve beginning at $n_1 b_1 = 0$ and $n_2 b_2 \simeq 0.02$ can be computed from eq. (6.3.10) because along this spinodal n_2 is a monotonic function of n_1 and $|dn_2/dn_1| < \infty$.

Using the fact that there is a miscibility gap along the pure CO_2 edge of Figure 6.29, we know that two spinodals have to spring from the pure CO_2 edge in the liquid density region. One is the low density branch discussed in the preceding paragraph. Taking $n_2 b_2 = 0.02$ and computing $J(n_1, n_2)$ as a function of n_1, we locate a liquid density spinodal at $n_1 b_1 \approx 0.4$. Along this spinodal n_1 can be computed from eq. (6.3.12) by advancing n_2 to the turning point where $dn_1/dn_2 = \infty$ (at $n_1 b_1 \simeq 0.37$ and $n_2 b_2 \simeq 0.17$) and then switching to eq. (6.3.10) and advancing n_1 to compute n_2.

In choosing the next value for changing n_1 or n_2 beyong a turning point, one can usually make a reasonable guess by trial and error without determining an accurate value of the turning point. However, the turning point can be precisely established if desired. For example, suppose a turning point n_1^t, n_2^t for which $dn_2/dn_1 = 0$ has been approached. Its location obeys the equations

$$\begin{aligned} J(n_1^t, n_2^t) &= 0 \\ \frac{\partial J}{\partial n_1}(n_1^t, n_2^t) &= 0, \end{aligned} \qquad (6.3.14)$$

whose equations are

$$\begin{aligned} (n_1^t - \tilde{n}_1)\frac{\partial J(\tilde{n}_1, \tilde{n}_2)}{\partial n_1} + (n_2^t - \tilde{n}_2)\frac{\partial J(\tilde{n}_1, \tilde{n}_2)}{\partial n_2} &= -J(\tilde{n}_1, \tilde{n}_2) \\ (n_1^t - \tilde{n}_1)\frac{\partial^2 J(\tilde{n}_1, \tilde{n}_2)}{\partial n_1^2} + (n_2^t - \tilde{n}_2)\frac{\partial^2 J(\tilde{n}_1, \tilde{n}_2)}{\partial n_1 \partial n_2} &= -\frac{\partial J}{\partial n_1}(\tilde{n}_1, \tilde{n}_2). \end{aligned} \qquad (6.3.15)$$

Iteration of these equations will determine the turning point n_1^t, n_2^t to some desired accuracy.

The density derivatives of J (and even of f in forming J) might be more easily evaluated numerically if the analytical expressions are too cumbersome. The price one pays for this is that the rate of convergence of the NR method is slowed down somewhat, but this is not a problem for low-dimensional problems on modern computers, even small desktop computers.

From the spinodals shown in Figure 6.29 we can deduce that in addition to the liquid–vapor binodals ($L_1 - V$ and $L_2 - V$) springing from the pure fluid axes, there is a miscibility gap lying along the high density edge $n_1b_1 + n_2b_2 = 1$. This means there has to be a liquid–liquid $(L_1 - L_2)$ binodal springing from this high density region. Because the unstable fluid region bounded by the spinodals is a continuous patch extending to all three edges of the triangle in Figure 6.29, we expect the binodals springing from the three edges to intersect. If this happens, three phase equilibrium $(L_1 - L_2 - V)$ will occur somewhere in the central region of the density–density diagram.

The final strategy of the computation of the phase diagram for CO_2 and decane at $T = 7°C$ now will be stated. The liquid–vapor binodals can be computed beginning at the pure fluid edges. For $L_2 - V$ equilibria, we set the CO_2 vapor density to some low value $n_1^{(v)}$ and compute $n_1^{(l)}, n_2^{(v)}, n_2^{(l)}$ from

$$\mu_\alpha(\mathbf{n}^{(v)}) = \mu_\alpha(\mathbf{n}^{(l)}), \quad \alpha = CO_2, \text{decane} \tag{6.3.16}$$

and

$$P(\mathbf{n}^{(v)}) = P(\mathbf{n}^{(l)}). \tag{6.3.17}$$

The initial guess of the densities of decane in the coexisting vapor and liquid phases can be taken as those of the pure decane phases. The initial guess for the density of CO_2 in the liquid phase can be taken to be the same as that set in the gas phase or somewhat higher since the spinodal adjacent to L_2 sweeps into the interior of the phase diagram faster than does the spinodal adjacent to V. Equations (6.3.15) and (6.3.16) can now be solved by the method. The vapor pressure corresponding to the prescribed value of $n_1^{(v)}$ is given by $P(\mathbf{n}^{(v)})$ or $P(\mathbf{n}^{(l)})$. A similar procedure is carried out for the liquid–vapor $(L_1 - V)$ binodal springing from the CO_2 edge of the diagram.

Next consider the liquid–liquid binodals springing from the edge along $n_1b_1 + n_2b_2 = 1$. The recommended procedure is to fix the density of CO_2 in phase L_1 at any desired value, say $n_1^{(l_1)} = 0.9$, to assume that the $L_1 - L_2$ tie-line runs parallel to the $n_1b_1 + n_2b_2 = 1$ edge, and assume that is it $\Delta\%$ longer than the segment of the tie-line lying between the spinodal adjacent to L_1 and L_2. Δ is to be determined by trial and error. One could begin with $\Delta = 10\%$ and increase it or decrease it until the solution to the equilibrium equations

$$\begin{aligned} \mu_\alpha(\mathbf{n}^{(l_1)}), &= \mu_\alpha(\mathbf{n}^{(l_2)}), \quad \alpha = CO_2, \text{decane} \\ P(\mathbf{n}^{(l_1)}) &= P(\mathbf{n}^{(l_2)}) \end{aligned} \tag{6.3.18}$$

converges. Once convergence is obtained for $n_1^{(l_1)}b_1 = 0.9$, $n_1^{(l_1)}$ can be incrementally increased and decreased to compute all the liquid–liquid binodals.

Because three-phase equilibrium is expected in the interior of the phase diagram, the differences between the

$$
\begin{array}{l}
\text{component densities in vapor of the } L_1 - V \text{ and } L_2 - V \text{ binodals,} \\
\text{component densities in liquid } L_1 \text{ of the } L_1 - V \text{ and } L_1 - L_2 \text{ binodals,} \\
\text{component densities in liquid } L_2 \text{ of the } L_2 - V \text{ and } L_1 - L_2 \text{ binodals,} \\
\text{vapor pressures of all three binodals,} \\
\text{and component chemical potentials of all three binodals}
\end{array}
\qquad (6.3.19)
$$

should be tracked. When all these differences become small a three-phase triangle is nearby and the three-phase solution can be computed from the equations

$$
\begin{aligned}
\mu_\alpha(\mathbf{n}^{(l_1)}) &= \mu_\alpha(\mathbf{n}^{(l_2)}) = \mu_\alpha(\mathbf{n}^{(v)}), \ \alpha = 1, 2 \\
P(\mathbf{n}^{(l_1)}) &= P(\mathbf{n}^{l_2)}) = P(\mathbf{n}^{(v)}).
\end{aligned}
\qquad (6.3.20)
$$

The first guess in a solution to these equations can be determined from the binodal values when the above described differences are small. There are six independent equations at eq. (6.3.19) and there are six unknown densities, and so for three-phase equilibrium, in accordance with the phase rule, there are no degrees of freedom left once temperature is set.

Although it is convenient to compute binary phase diagrams in a density–density representation, one can display them in pressure–mole (or mass) fraction phase diagrams. For carbon dioxide and decane at 7°C, the $P - x$ equivalent for Figure 6.29 is shown in Figure 6.30. As temperature and pressure are often the variables controlled in a process, Figure 6.30 is especially useful in practice. However, the diagram suppresses the density information sometimes of value in mass balances. It would be ideal if both phase diagrams were always available.

The predicted liquid–liquid two-phase binodal of the CO_2-decane system ends in an upper liquid–liquid critical point. The upper critical point is shown in Figure 6.31a at 30°C. At 50°C (Fig. 6.31b) the liquid–liquid binodal has disappeared into the liquid–vapor binodal and a liquid–vapor critical point appears instead of a liquid–liquid critical point.

6.3.2 Three-Component or Ternary Phase Behavior

Let us study next a few ternary mixtures of methane, propane, octane, decane, and carbon dioxide. The vapor pressure curves predicted by the PR equation for pure propane, decane, and CO_2 are compared with the experiment in Figure 6.32. Predicted densities of the coexisting liquid and vapor phases are compared in Figure 6.33 with the experiment for the same compounds. The corresponding comparison was given in Figure 6.2 for methane and octane. For these nonpolar fluids the PR equation clearly performs quite well.

Figure 6.30
Pressure–composition phase diagram for CO_2 and decane at 7°C predicted by the PR equation. Redrawn from Falls (1982).

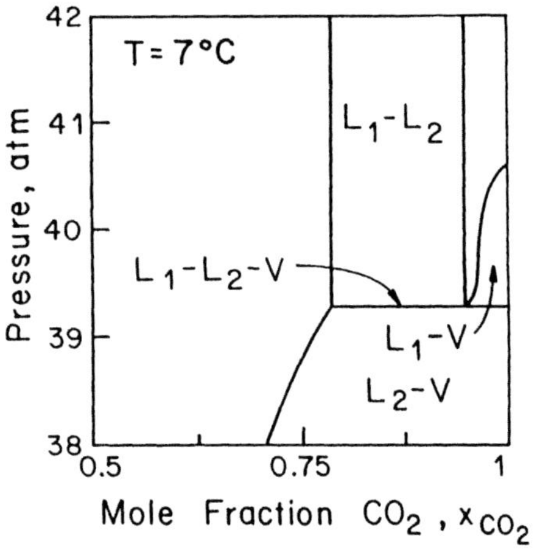

To compute the ternary phase diagram it is convenient to set the temperature T and pressure P and to seek two or three phase equilibria by solving the equations

$$\begin{aligned} \mu_\alpha(\mathbf{n}^{(1)}) &= \mu_\alpha(\mathbf{n}^{(2)}), \quad \alpha = 1, 2, 3 \\ P(\mathbf{n}^{(1)}) &= P(\mathbf{n}^{(2)}) = P \end{aligned} \tag{6.3.21}$$

or

$$\begin{aligned} \mu_\alpha(\mathbf{n}^{(1)}) &= \mu_\alpha(\mathbf{n}^{(2)}) = \mu_\alpha(\mathbf{n}^{(3)}), \quad \alpha = 1, 2, 3 \\ P(\mathbf{n}^{(1)}) &= P(\mathbf{n}^{(2)}) = P(\mathbf{n}^{(3)}) = P. \end{aligned} \tag{6.3.22}$$

Figure 6.31
Pressure–mole fraction phase diagrams predicted by the PR equation for carbon dioxide and decane mixtures. Redrawn from Quinones et al. (1991).

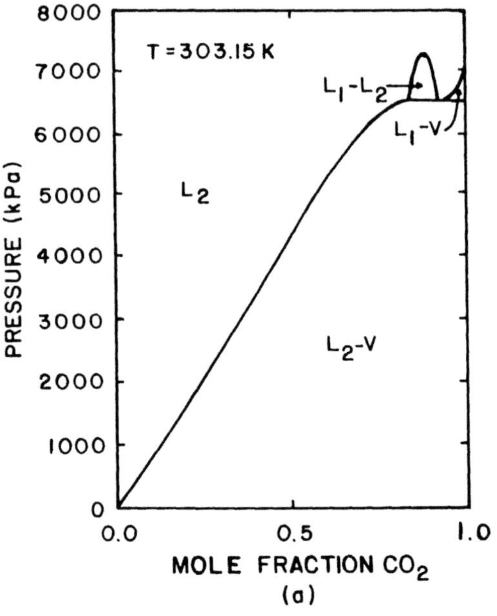

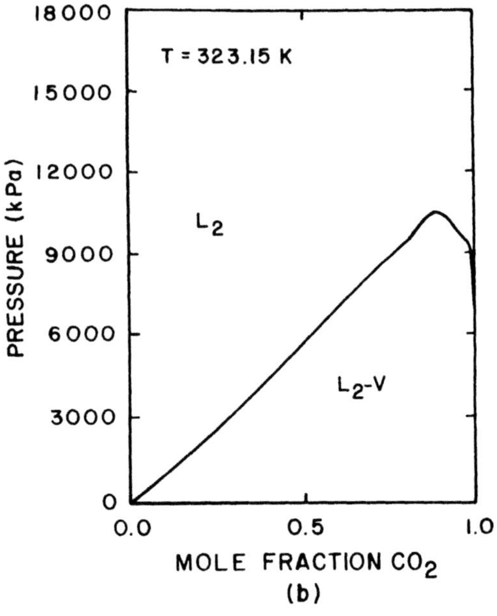

In the case of the two-phase equilibrium there are five equations and six unknown component densities. Thus, the density (or mole fraction) of one component must be specified in solving eq. (6.3.23) for a two-phase equilibrium state. For three-phase equilibrium, there are nine equations and nine unknown component densities and so none of the component compositions can be preset in solving eq. (6.3.20).

We have, of course, just repeated the phase rule, $f = 2 + c - p$, which tells us that $f = 3$ for two-phase equilibria and $f = 2$ for three-phase equilibria in a ternary system. Thus, setting T and P leaves one and zero degrees of freedom for $p = 2$ and 3, respectively.

Two ternary phase diagrams predicted for propane/decane/CO_2 mixtures are shown in Figure 6.34. The compositions are given in mole fractions. At the conditions considered in Figure 6.34a we know from Figure 6.32 that all three pure components are liquids and from Figure 6.30 that along the CO_2–decane edge of the ternary composition triangle there is a liquid–liquid miscibility gap (coexisting L_1 and L_2 phases). Thus, the two-phase binodal in Figure 6.34a is easily constructed by incrementally increasing the composition of propane beginning at zero and solving eq. (6.3.20) by the NR method taking initial guesses from the known CO_2–decane densities in the L_1 and L_2 phases. The binodal ends at a critical point. Because propane is totally miscible with CO_2 and decane at the conditions of Figure 6.34a, no binodals spring from the propane–CO_2 or decane–CO_2 binary edges of the ternary phase diagram. Thus, a reasonable guess is that no multiphase splits occur

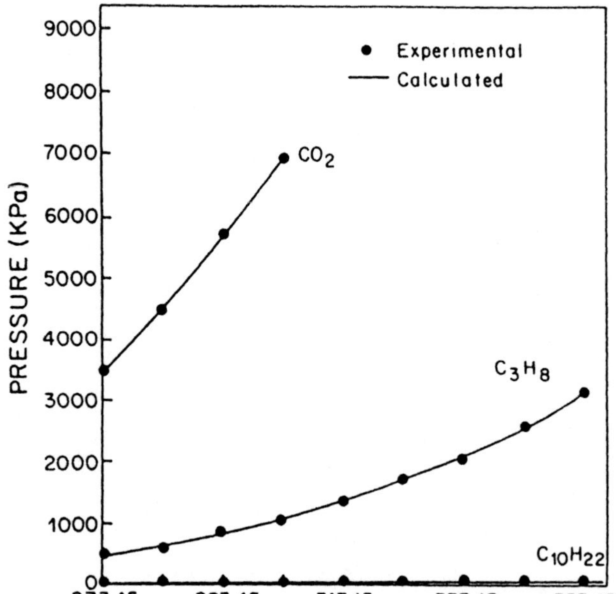

Figure 6.32
Vapor–pressure curves for pure propane (C_3H_8), decane ($C_{10}H_{22}$), and carbon dioxide (CO_2). Curves were computed from the PR equation of state. Adapted from Sahimi et al. (1985).

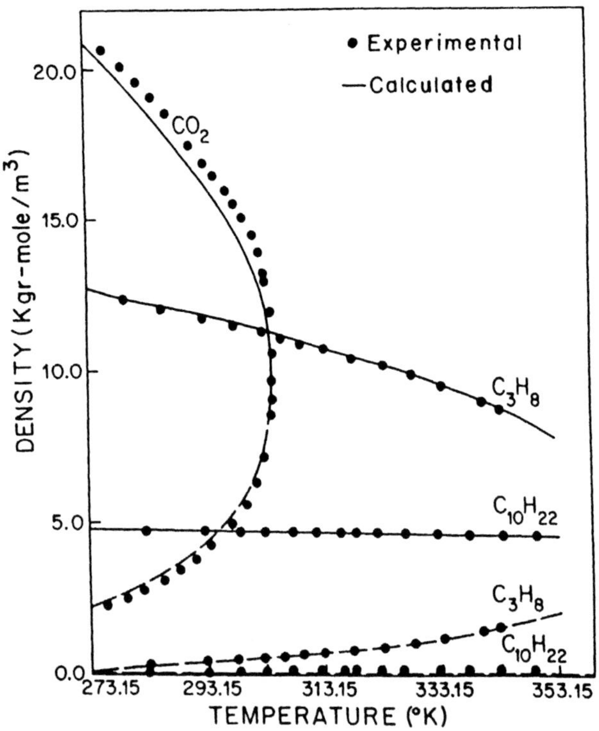

Figure 6.33
Liquid–vapor coexistence densities for pure propane, decane, and carbon dioxide. Curves were computed from the PR equation of state. Adapted from Sahimi, et al.(1985).

other than the binodal shown in Figure 6.34a. To be certain that this is true one has to prove that no spinodal regions exist outside the binodal region shown. In other words, one has to find that the determinant of the matrix **A**,

$$\mathbf{A} = \left[\frac{\partial^2 f}{\partial n_\alpha \partial n_\beta} \right], \quad \alpha, \beta = 1, 2, 3, \tag{6.3.23}$$

has the same sign (which is positive in a stable phase) everywhere outside the binodal region. When a determinant changes signs one of its eigenvalues changes sign, which is the condition of passing

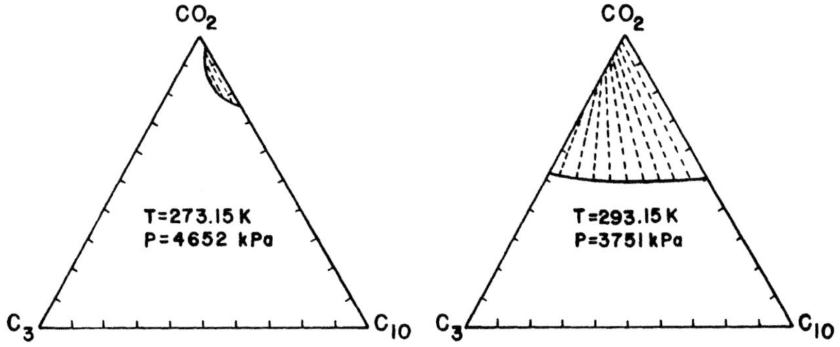

Figure 6.34
Ternary phase diagrams predicted for propane, decane, and CO_2 mixtures. Compositions are expressed in mole fractions. The temperatures and pressures are (a) 273.15 K and 4652 kPa and (b) 293.15 K and 3751 kPa. Redrawn from Quinones et al. (1991).

from a stable to an unstable region of fluid. Already at three components the task of finding the spinodal regions becomes a bit tedious, but to be sure no isolas [multiphase regions in a c-component system not springing from $(c - 1)$-component systems] occur the spinodal regions must be mapped out.

The ternary phase diagram shown in Figure 6.34b is quite different from that in Figure 6.34a. In Figure 6.34a the pure fluids and all mixtures are in the liquid state. In Figure 6.34b, pure CO_2 is a vapor, whereas pure propane and decane are liquids. In this case, there are liquid–vapor miscibility gaps along the CO_2–propane edge and the CO_2–decane edge of the ternary phase diagram. In principle, the binodals springing from these edges can end in critical points. However, they do not. The solution of eq. (6.3.20) beginning either at the CO_2–propane or CO_2–decane edge leads to a continuous binodal region spanning from one edge to the other. Again, because liquid propane and decane are totally miscible at the conditions of Figure 6.34b, we do not expect phase splits other than the two-phase region shown in Figure 6.34b. However, as explained above, to establish this as a fact one must show that the determinant of **A** does not change signs outside the two-phase region shown.

Ternary phase diagrams predicted for methane, octane, and CO_2 at a couple of different conditions are shown in Figure 6.35. In Figure 6.35a, there are liquid–liquid miscibility gaps along the CO_2–octane edge and liquid–vapor miscibility gaps along the binary edges of methane with octane or CO_2. Thus, two-phase binodals emanating from all three edges can be computed by solving eq. (6.3.20) as described above. However, because the tie-lines of these three binodals may come together to form a three-phase tie triangle, we also compute the differences described at eq. (6.3.18), and when these differences are sufficiently small we switch to eq. (6.3.20) and search for three-phase equilibria. In Figure 6.35a there is a three-phase equilibrium region where liquid–liquid-vapor phases coexist. As the pressure is increased beyond the pressure of Figure 6.35a, the octane-rich phase of the three-phase triangle moves in composition toward the CO_2–methane edge of the phase diagram and the CO_2-rich phase moves toward higher methane compositions. Somewhere between the pressure of Figure 6.35a and that of Figure 6.35b the octane-rich phase and the CO_2-rich phases merge with each other to form a two-phase tie-line that is also a critical point of the two-phase liquid–liquid binodal springing from the CO_2–octane edge of the phase diagram. With a further increase of pressure the liquid–liquid and liquid–vapor binodals pull away from one another leaving a large one-phase corridor spreading from the pure octane corner of the ternary phase diagram to the CO_2–methane binary edge of the diagram.

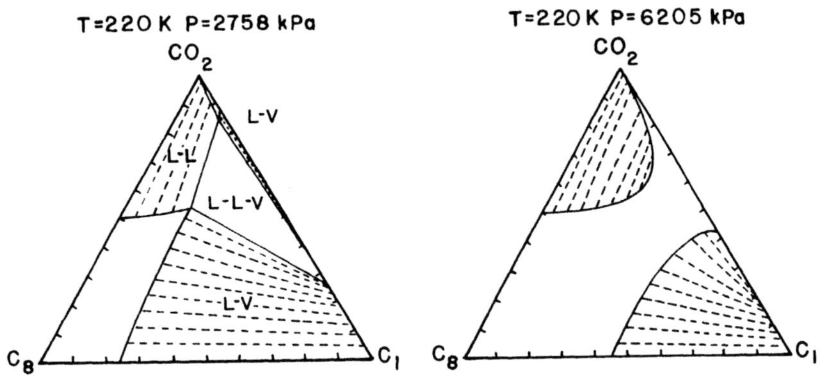

Figure 6.35
Ternary phase diagrams predicted for methane, octane, and CO_2 mixtures. Compositions are expressed in mole fractions. Temperatures and pressures are (a) 220 K and 2758 kPa and (b) 220 K and 6205 kPa. Redrawn from Quinones et al. (1991).

6.3.3 Four-Component or Quaternary Phase Behavior

To construct quaternary phase diagrams the basic strategy is similar to that for constructing binary and ternary diagrams. The first step is to find all the ternary composition phase diagrams. These will form the faces of a quaternary tetrahedral phase diagram. Plainly, the difficulty of finding isola (or spinodal regions not touching the ternary faces) will be harder than for ternary systems. We will not pursue the problem further here but as an example we show in Figure 6.36 the quaternary phase diagram predicted for a CO_2, methane, butane, and decane system.

Often in dealing with condensed phases one introduces models in which the effect of pressure on phase equilibria is ignored. The examples discussed in this section bear witness to the fact that this is not always a good approximation. However, there are numerous examples for which it is and in the next section two of these models, a regular solution and a Flory–Huggins solution, will be examined. Their advantage lies in the fact that the degrees of freedom are reduced by one. Thus, for example, construction

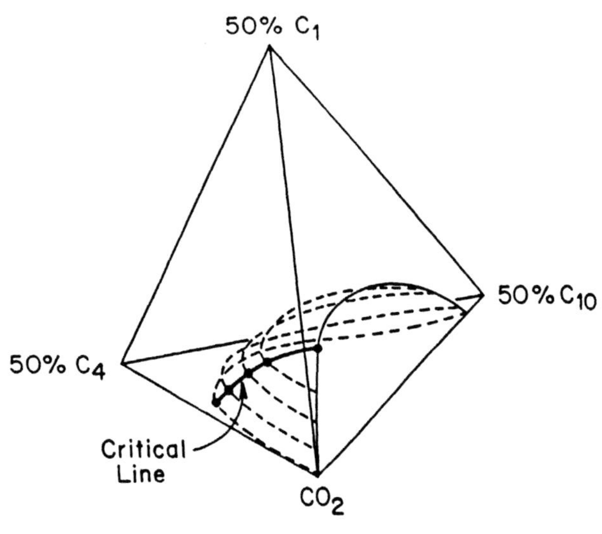

Figure 6.36
Quaternary tetrahedron of a CO_2–methane–butane–decane mixture at 344 K and 150 psia showing the liquid–vapor binodal surface and its line of critical points.

of quaternary phase diagrams becomes mathematically equivalent to construction of a ternary phase diagram with a full equation of state.

Example

If pure water and an aqueous sugar solution are separated by a rigid membrane permeable to water but not to sugar, the pressure P_s on the solution side is greater by amount P_{os} than the pressure P_w on the pure water side. P_{os} is called the osmotic pressure. The water on each side of the membrane is at chemical equilibrium i.e., $\mu_w^0(P_w, T) = \mu_w(P_s, T, x_w)$, where x_w is the mole fraction of water in the solution. Assume that the sugar concentration is low enough for the water to behave as an ideal solution. Then $\mu_w(P_s, T, x_w) = \mu_w^o(P_s, T) + kT\ell n x_w$. But $\mu_w^o(P_s, T) = \mu_w^o(P_w, T) + \int_{P_w}^{P_s} \tilde{V}_w dP$. $\tilde{V}_w$ is the molecular volume of pure water and is insensitive to pressure in the typical osmotic range. This fact leads to the result

$$P_{os}\tilde{V}_w \simeq -kT\ell n x_w$$

for dilute solutions of sugar (or any other component for which a semipermeable membrane can be found. At 303K, the osmotic pressure of sugar solution is 2.53×10^5 and 5.17×10^5 Pa for 0.1 and 0.2 molar sugar solutions, respectively. If instead of an ideal solution, it is a regular solution, the osmotic pressure obeys the equation

$$P_{os}\tilde{V}_w = -kT\ell n x_w \gamma_w$$

where

$$\gamma_w = exp[(1 - x_w)^2 a/kT].$$

6.4 Lattice Theory of Solutions: Regular Solutions

Many of the patterns of phase equilibria of solutions can be understood with the aid of regular solution theory. In its simplest form, the theory is based on the assumption that molecules of species 1 or 2 occupy lattice sites of identical volume v_o. It is assumed that the lattice is regular (periodic) and each site has z nearest neighbors. The total number of sites equals the total number $N\ (= N_1 + N_2)$ of molecules in the solution. The volume of the system is $V = Nv_o$ and cannot be varied independently of the total number of molecules. Thus, there is no equation of state for the pressure of a regular solution, and so a phase can be meaningfully approximated as a regular solution only if it is a condensed phase at a relatively low pressure, that is, such that $PV \ll NkT$. In this case the Gibbs and Helmholtz free energies are approximately the same.

According to the theory, there are only nearest neighbor pair interactions. The pair interaction energy between neighboring species i and j is w_{ij}. In a binary system there are interactions w_{11}, w_{22} and w_{12}. For a regular solution one makes the so-called Bragg–Williams approximation, namely, it is assumed that the molecules are randomly distributed on the lattice sites. If the molecules are randomly distributed on the lattice sites then

$$\frac{N_\alpha}{N} = \text{the probability that species } \alpha \text{ occupies a given site}$$

$$\frac{zN_\beta}{N} = \text{the probable number of nearest neighbors of a given site that are molecules of species } \beta.$$

The average energy of interaction between a given site and its nearest neighbors is then

$$\sum_{\alpha,\beta=1}^{2} \frac{N_\alpha}{N}\frac{zN_\beta}{N} w_{\alpha\beta}$$

and so the average energy of interaction of the system is half (in order to not count pairs twice) of N times this energy is

$$W = \frac{zN}{2}\sum_{\alpha,\beta=1}^{2} x_\alpha x_\beta w_{\alpha\beta} = \frac{zN}{2}[x_1^2 w_{11} + x_2^2 w_{22} + 2x_1 x_2 w_{12}]$$
$$= x_1 x_2 Na + \frac{z}{2}N_1 w_{11} + \frac{z}{2}N_2 w_{22}, \qquad (6.4.1)$$

where $x_\alpha = N_\alpha/N$ and

$$a \equiv -\frac{z}{2}[w_{11} + w_{22} - 2w_{12}]. \tag{6.4.2}$$

The partition function for the regular solution is then

$$Q_N = [q_1^I(T)]^{N_1}[q_2^I(T)]^{N_2}\frac{N!}{N_1!N_2!}e^{-W/kT}, \tag{6.4.3}$$

$$= [q_1^I(T)e^{-zw_{11}/2kT}]^{N_1}[q_2^I(T)e^{-zw_{22}/2kT}]^{N_2}\frac{N!}{N_1!N_2!}e^{-x_1x_2Na} \tag{6.4.4}$$

where $q_\alpha^I(T)$ is the internal energy contribution to the partition function of a molecule of species α and $N!/N_1!N_2!$ is the number of ways a lattice of N sites can be randomly occupied by N_1 molecules of species 1 and N_2 of species 2 (here $N = N_1 + N_2$).

The free energy, entropy, and energy of a two-component regular solution are given by

$$F = -\sum_{\alpha=1}^{2}\left[N_\alpha kT\ln q_\alpha^I(T) - \frac{1}{2}N_\alpha z w_{\alpha\alpha}\right] + kT\sum_{\alpha=1}^{2}N_\alpha \ln x_\alpha + x_1x_2Na \tag{6.4.5}$$

$$S = \sum_\alpha\left[N_\alpha k\ln q_\alpha^I(T) + N_\alpha kT\frac{\partial \ln q_\alpha^I}{\partial T}\right] - \sum_{alpha=1}^{2}N_\alpha k\ln x_\alpha \tag{6.4.6}$$

$$U = \sum_\alpha\left[N_\alpha kT^2\frac{\partial \ln q_\alpha^I}{\partial T} + \frac{1}{2}N_\alpha z w_{\alpha\alpha}\right] + x_1x_2Na. \tag{6.4.7}$$

The chemical potential of species α is

$$\mu_\alpha = -kT\ln q_\alpha^I + \frac{z}{2}w_{\alpha\alpha} + kT\ln x_\alpha + (1 - x_\alpha)^2 a \tag{6.4.8}$$

$$= \mu_\alpha^p + kT\ln x_\alpha + (1 - x_\alpha)^2 a \tag{6.4.9}$$

where μ_α^p is the chemical potential of a fluid of pure α.

Frequently solutions are characterized by their thermodynamic mixing properties. If E is an extensive variable, then the mixing part of E for a binary solution is defined to be

$$\Delta E_m \equiv E(T, P, N_1, N_2) - E(T, P, N_1) - E(T, P, N_2), \tag{6.4.10}$$

that is, ΔE_m is an excess of E in a solution of N_1 particles of the species 1 and N_2 of species 2 over the sum of the values of E for a pure fluid of N_1 particles of species 1 and a pure fluid of N_2 particles of species 2. The mixing is done at constant temperature T and pressure P. Because P is not a thermodynamic variable in regular solutions, its constancy can be ignored here. The quantities F, S, and U for the pure fluid state can be obtained from eqs. (6.4.5)–(6.4.7) by setting $N = N_\alpha$ so that $N_\beta = x_\beta = 0$ and $x_\alpha = 1$. The regular solution mixing properties can then be shown to be

$$\Delta F_m = kT\sum_{\alpha=1}^{2}N_\alpha \ln x_\alpha + x_1x_2Na \tag{6.4.11}$$

$$\Delta S_m = -k \sum_{\alpha=1}^{2} N_\alpha \ln x_\alpha \tag{6.4.12}$$

$$\Delta U_m = x_1 x_2 N a. \tag{6.4.13}$$

The entropy of mixing of a regular solution is the same as that of an ideal solution. There is, however, a nonideal contribution to the energy of mixing of a regular solution.

The generalization to a c-component regular solution is straightforward. The result of the Bragg–Williams approximation is

$$W = z\frac{N}{2} \sum_{\alpha,\beta=1}^{c} x_\alpha x_\beta w_{\alpha\beta}$$

$$= -\sum_{\alpha,\beta=1}^{c} \frac{z}{2} N x_\alpha x_\beta (w_{\alpha\alpha} - w_{\alpha\beta}) + \sum_{\alpha=1}^{c} \frac{z}{2} N_\alpha w_{\alpha\alpha}$$

$$= \frac{1}{2} \sum_{\alpha,\beta=1}^{c} N x_\alpha x_\beta a_{\alpha\beta} + \sum_{\alpha=1}^{c} \frac{z}{2} N_\alpha w_{\alpha\alpha}, \tag{6.4.14}$$

where

$$a_{\alpha\beta} \equiv -\frac{z}{2}(w_{\alpha\alpha} + w_{\beta\beta} - 2w_{\alpha\beta}). \tag{6.4.15}$$

The corresponding thermodynamic functions are

$$F = -\sum_{\alpha=1}^{c} \left[N_\alpha kT \ln q_\alpha^I - \frac{1}{2} N_\alpha z w_{\alpha\alpha} \right] + kT \sum_{\alpha=1}^{c} N_\alpha \ln x_\alpha + \frac{1}{2} \sum_{\alpha,\beta=1}^{c} N x_\alpha x_\beta a_{\alpha\beta}, \tag{6.4.16}$$

$$S = \sum_{\alpha=1}^{c} \left[N_\alpha k \ln q_\alpha^I + N_\alpha kT \frac{\partial \ln q_\alpha^I}{\partial T} \right] - \sum_{\alpha=1}^{c} N_\alpha k \ln x_\alpha, \tag{6.4.17}$$

$$U = \sum_{\alpha=1}^{c} \left[N_\alpha kT^2 \frac{\partial \ln q_\alpha^I}{\partial T} + \frac{1}{2} N_\alpha z w_{\alpha\alpha} \right] + \frac{1}{2} \sum_{\alpha,\beta=1}^{c} N x_\alpha x_\beta a_{\alpha\beta} \tag{6.4.18}$$

and

$$\mu_\alpha = -kT \ln q_\alpha^I + \frac{z}{2} w_{\alpha\alpha} + \sum_{\beta=1}^{c} x_\beta a_{\alpha\beta} - \sum_{\beta,\gamma=1}^{c} x_\beta x_\gamma a_{\beta\gamma}. \tag{6.4.19}$$

Note that in $a_{\alpha\alpha} = 0$ that the sign of $a_{\alpha\beta}$ depends on the magnitude of the pair interaction between species α and β relative to the arithmetic mean of the pair interactions of species α with itself and β with itself. In general $a_{\alpha\beta}$ can be positive or negative.

6.4.1 Binary Phase Equilibria

Regular solution theory is often useful for predicting the vapor pressure or boiling point of a binary solution when the vapor is dilute enough to behave as an ideal gas. The chemical potential of species α in a dilute gas is

$$\mu_\alpha = -kT \ln q_\alpha^I + kT \ln y_\alpha P, \tag{6.4.20}$$

where y_α is the mole fraction of α and P is the pressure of the gas. At equilibrium between vapor and liquid phases, the chemical potentials in eqs. (6.4.8) and (6.4.19) are equal, that is,

$$kT \ln y_\alpha P = \frac{z}{2} w_{\alpha\alpha} + kT \ln x_\alpha + a(1 - x_\alpha)^2, \qquad (6.4.21)$$

for $\alpha = 1$ and 2. If $P_\alpha^s(T)$ is the vapor pressure of pure α at temperature T, it follows from eq. (6.4.20) when $y_\alpha = x_\alpha = 1$ that $zw_{\alpha\alpha}/2 = kT \ln P_\alpha^s(T)$, and so

$$y_\alpha P / x_\alpha P_\alpha^s(T) = \exp\left[\frac{a}{kT}(1 - x_\alpha)^2\right]. \qquad (6.4.22)$$

For an ideal solution ($w = 0$), eq. (6.4.21) reduces to Raoult's law, $y_\alpha P = x_\alpha P_\alpha^s(T)$.

Ethyl ether and acetone behave as a regular solution with $a/k \simeq 226$ K (Porter, 1920). The partial pressures, $P_\alpha = y_\alpha P$, and total vapor pressure P predicted by eq. (6.4.21) for a binary solution of ether and acetone are shown in Figure 6.37. Because a is positive, the partial pressures lie above those of an ideal solution (dashed lines in Fig. 6.37, predicted from Raoult's law).

By combining eq. (6.4.21) and the approximate Clausius–Clapeyron expression, eq. (6.2.5),

$$P_\alpha^s(T) = P \exp\left[\frac{\Delta \tilde{H}_\alpha^v}{k}\left(\frac{1}{T_\alpha} - \frac{1}{T}\right)\right], \qquad (6.4.23)$$

where $\Delta \tilde{H}_\alpha^v$ is the heat of vaporization of pure liquid α and T_α is the boiling point of pure α at pressure P, we obtain the following expression for the boiling point T of a regular solution at pressure P:

$$x_2\{\exp[h_1 + h_2 + a(1 - x_2)^2/kT] - \exp[ax_2^2/kT]\} + \exp(h_1) - \exp(ax_2^2/kT) = 0, \qquad (6.4.24)$$

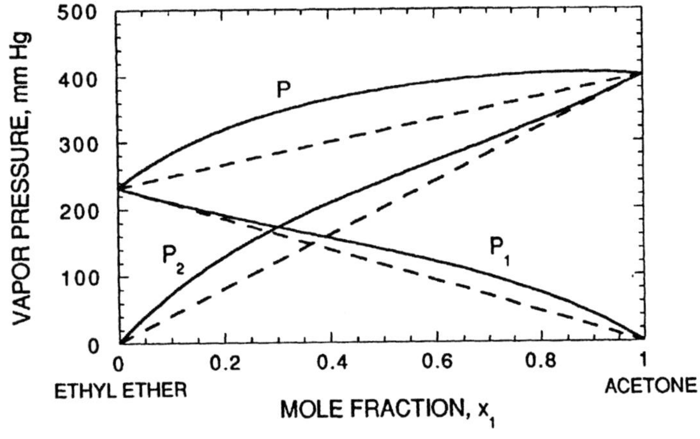

Figure 6.37
Vapor pressure P and partial pressures P_1 and P_2 of a solution of ethyl ether and acetone versus liquid composition at 40°C. Solid curves predicted for a regular solution and dashed lines are for an ideal solution.

where

$$h_\alpha = \frac{\Delta \tilde{H}_\alpha^v}{k}\left(\frac{1}{T_\alpha} - \frac{1}{T}\right). \qquad (6.4.25)$$

The normal boiling points (T_α at $P = 1$ atm) of acetone and diethyl ether are $T_1 = 329.8$K and $T_2 = 308.0$K, and their heats of vaporization are $\Delta\tilde{H}_1^v/k = 4086$K and $\Delta\tilde{H}_2^v/k = 3150$K, respectively. With these values and the Porter's estimated regular solution parameter, $a/k = 226$K, eq. (6.4.23) predicts the boiling point curve of the ether-acetone solution very accurately. The results are shown in Table 6.2.

Let us next consider the implications of regular solution theory for azeotropy. If at a given temperature and pressure the composition of a liquid is the same as that of its coexisting vapor (i.e., if $x_\alpha = y_\alpha$), we say the system is an azeotrope or that it has reached an azeotropic point in the phase diagram. The temperature–composition phase diagrams of binary systems exhibiting positive and negative azeotropy were illustrated in Figures 6.9b and 6.9c. The condition for an azeotrope to exist according to eq. (6.4.21) is that the equations

$$\frac{P}{P_\alpha^s(T)} = \exp\left[\frac{a(1-x_\alpha)^2}{kT}\right], \quad \alpha = 1, 2 \qquad (6.4.26)$$

have a solution for x_α lying between 0 and 1. Taking the ratio of eq. (6.4.25) for $\alpha = 1$ and 2 and solving the expression for x_2 yields

$$x_2 = \frac{1}{2} + \frac{kT}{2a}\ln\left[\frac{P_2^s(T)}{P_1^s(T)}\right]. \qquad (6.4.27)$$

The condition that x_2 lie between 0 and 1 is that

$$-1 < \frac{kT}{2a}\ln\left[\frac{P_2^s(T)}{P_1^s(T)}\right] < 1 \quad \text{or} \quad \frac{2|a|}{kT} > \left|\ln\left[\frac{P_2^s(T)}{{}^s_1(T)}\right]\right|. \qquad (6.4.28)$$

TABLE 6.2 Boiling Point Curve Composition of Diethyl Ether-Acetone Solution

T (K)	x_2 [Eq. 6.4.23]	x_2 (Exp.)
308	1	1
313	0.45	0.45
317	0.28	0.28
321	0.16	0.16
325	0.007	0.007
329.8	0	0

Component 2 is ether. Adapted from Prigogine and R. Defay (1962).

Equation (6.4.27) implies that if there is a temperature T' at which the vapor pressure curves of 1 and 2 cross ($P_1^s(T') = P_2^2(T')$), then the system will be an azeotrope for any amount of nonideality, that is, for $|a|$ having any nonzero value. The general rule for azeotropy in a regular solution is that the vapor pressure curves $P_2^s(T)$ and $P_1^s(T)$ must pass close enough to each other that the magnitude of $(kT'/2)\ln[P_L^s(T')/P_1^{s^2}(T')]$ be smaller than $|a|$.

Let us next consider phase equilibria between condensed phases behaving as regular solutions. Because the regular solution free energy is an explicit function of mole fractions of the components, it is convenient to describe the phase stability and critical points of regular solutions in terms of the Gibbs free energy density $\tilde{g} = G/N$. The spinodals and critical points are governed by eqs. (6.1.75) and (6.1.76). For a binary solution the binodal obeys the equation

$$\frac{\partial^2 \tilde{g}}{\partial x_1^2} = 0 \tag{6.4.29}$$

and critical points obey eq. (6.4.28) and

$$\frac{\partial^3 \tilde{g}}{\partial x_1^3} = 0, \quad \frac{\partial^4 \tilde{g}}{\partial x_1^4} \neq 0. \tag{6.4.30}$$

Because $G = PV + F$ and $V = Nv$, v the volume of a lattice site, and because the derivatives are taken at constant T and P, $\tilde{g}$ can be replaced by F/N in eqs. (6.4.28) and (6.4.29). Thus, from eq. (6.4.5), we find that

$$\frac{\partial^2 \tilde{g}}{\partial x_1^2} = \frac{kT}{x_1} + \frac{kT}{1 - x_1} - 2a \tag{6.4.31}$$

and

$$\frac{\partial^3 \tilde{g}}{\partial x_1^3} = -\frac{kT}{x_1} + \frac{kT}{1 - x_1}. \tag{6.4.32}$$

From these expressions it follows that the spinodal curve obeys

$$T = (2a/k)x_1(1 - x_1) \tag{6.4.33}$$

and there is a critical point at

$$x_{1c} = 1/2, \quad T_c = a/2k. \tag{6.4.34}$$

We see that the interaction parameter a must be positive for two-phase equilibria to occur between regular solutions. The liquid–liquid binodal compositions at a given T obey the two equilibrium conditions,

$$\mu_\alpha(\mathbf{x}^{(1)}) = \mu_\alpha(\mathbf{x}^{(2)}), \quad \alpha = 1, 2, \tag{6.4.35}$$

which obey the symmetry condition $x_1^{(1)} = x_2^{(2)}$, and so equilibrium compositions can be computed from the equation for $\alpha = 1$ (or 2). The result is

$$T = (a/k)(1 - 2x_1^{(1)}) / \ln[(1 - x_1^{(1)})/x_1^{(1)}]. \tag{6.4.36}$$

The binodal spinodal and unstable region of the binary system is shown in Figure 6.38.

The regular solution model predicts an upper critical point only. More sophisticated lattice solution models must be introduced to include the physics responsible for lower critical point behavior. Such models will be discussed in a later section.

6.4.2 Ternary Phase Behavior

In a three-component system

$$\tilde{g} = \sum_{\alpha} x_\alpha \mu_\alpha^P(T, P) + kT \sum_{\alpha=1}^{3} x_\alpha \ln x_\alpha + a_{12}x_1x_2 + a_{13}x_1x_3 + a_{23}x_2x_3. \tag{6.4.37}$$

At fixed T and P the spinodal curves in a ternary composition phase diagram are determined by the equation

$$\frac{\partial^2 \tilde{g}}{\partial x_1^2}\frac{\partial^2 \tilde{g}}{\partial x_2^2} - \left(\frac{\partial^2 \tilde{g}}{\partial x_1 \partial x_2}\right)^2 = 0. \tag{6.4.38}$$

By differentiating eq. (6.4.36), noting that $x_3 = 1 - x_1 - x_2$, we get

$$\frac{\partial^2 \tilde{g}}{\partial x_1^2} = -2a_{13} + kT\left[\frac{1}{x_1} + \frac{1}{x_3}\right] \tag{6.4.39}$$

$$\frac{\partial^2 \tilde{g}}{\partial x_2^2} = -2a_{23} + kT\left[\frac{1}{x_2} + \frac{1}{x_3}\right] \tag{6.4.40}$$

$$\frac{\partial^2 \tilde{g}}{\partial x_1 \partial x_2} = a_{12} - a_{13} - a_{23} + \frac{kT}{x_3}. \tag{6.4.41}$$

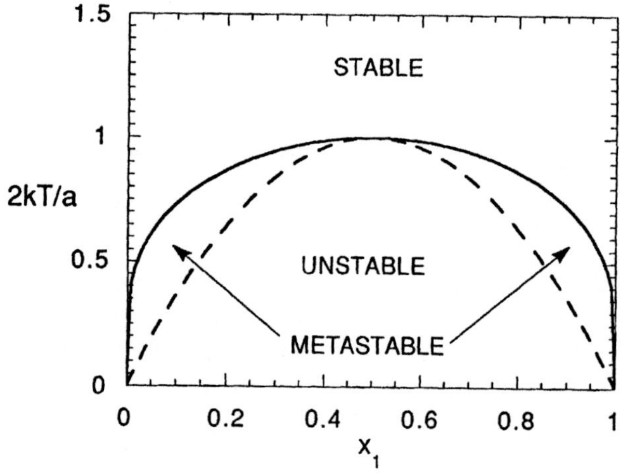

Figure 6.38
Phase diagram of a binary regular solution. The solid curve is the binodal and the dashed curve is the spinodal.

Insertion of eq. (6.4.38) into (6.4.37) yields for the spinodal curves the equation

$$Jx_1x_2x_3 + 2(a_{12}x_1x_2 + a_{13}x_1x_3 + a_{23}x_2x_3)/kT - 1 = 0, \qquad (6.4.42)$$

where

$$J = (a_{12}^2 + a_{13}^2 + a_{23}^2 - 2a_{12}a_{13} - 2a_{12}a_{23} - 2a_{13}a_{23})/(kT)^2. \qquad (6.4.43)$$

To search for spinodal curves for a given set of interaction parameters, one simply sets, say x_2 and computes x_1 from eq. (6.4.39).

If δx_1 and δx_2 are variations of composition along a spinodal curve, we know that

$$\delta^2 \tilde{g} = \frac{1}{2}\left[\frac{\partial^2 \tilde{g}}{\partial x_1^2}\delta x_1^2 + 2\frac{\partial^2 \tilde{g}}{\partial x_1 \partial x_2}\delta x_1 \delta x_2 + \frac{\partial^2 \tilde{g}}{\partial x_2^2}\delta x_2^2\right] = 0. \qquad (6.4.44)$$

Moreover, if a critical point lies on the spinodal, then

$$\delta^3 \tilde{g} = \frac{1}{6}\left[\frac{\partial^3 \tilde{g}}{\partial x_1^3}\delta x_1^3 + 3\frac{\partial^3 \tilde{g}}{\partial x_1^2 \partial x_2}\delta x_1^2 \delta x_2 + 3\frac{\partial^3 \tilde{g}}{\partial x_1 \partial x_2^2}\delta x_1 \delta x_2^2 + \frac{\partial^3 \tilde{g}}{\partial x_2^3}\delta x_2^3\right] = 0 \qquad (6.4.45)$$

at this point. From eqs. (6.4.37) and (6.4.41) it follows that along the spinodal

$$s \equiv \frac{\delta x_2}{\delta x_1} = -\frac{\partial^2 \tilde{g}/\partial x_1^2}{\partial^2 \tilde{g}/\partial x_1 \partial x_2}, \qquad (6.4.46)$$

and so a critical point along the spinodal obeys

$$\frac{\partial^3 \tilde{g}}{\partial x_1^3} + 3s\frac{\partial^3 \tilde{g}}{\partial x_1^2 \partial x_2} + 3s^2\frac{\partial^3 \tilde{g}}{\partial x_1 \partial x_2^2} + s^3\frac{\partial^3 \tilde{g}}{\partial x_2^3} = 0 \qquad (6.4.47)$$

in addition to eq. (6.4.40). The third derivatives of $\tilde{g}$ are

$$\frac{\partial^3 \tilde{g}}{\partial x_1^3} = kT\left[\frac{1}{x_3^2} - \frac{1}{x_1^2}\right] \qquad (6.4.48)$$

$$\frac{\partial^3 \tilde{g}}{\partial x_2^3} = kT\left[\frac{1}{x_3^2} - \frac{1}{x_2^2}\right] \qquad (6.4.49)$$

$$\frac{\partial^3 \tilde{g}}{\partial x_1^2 \partial x_2} = \frac{\partial^3 \tilde{g}}{\partial x_1 \partial x_2^3} = \frac{kT}{x_3^2}. \qquad (6.4.50)$$

As we saw in applications of the PR equation of state when miscibility gaps lie along the binary edges of a ternary phase diagram, then the interior binodals are easily constructed by continuation from the edges. If, however, a multiphase region is isolated from the edges, then the spinodal curves and its critical points are the best starting point for the construction of binodals. As an example of this situation, consider a ternary system with

$$a_{12} = -10kT, \quad a_{13} = a_{23} = 0. \qquad (6.4.51)$$

Because $a_{\alpha\beta} \leq 0$ there are no miscibility gaps along the binary edges of the phase diagram. On the other hand, the spinodal equation,

$$100x_1x_2x_3 - 20x_1x_2 - 1 = 0, \qquad (6.4.52)$$

admits a solution of x_2 versus x_1. The spinodal is shown as a closed dash curve in Figure 6.39. At compositions inside the curve a homogeneous phase is unstable and outside the curve a homogeneous phase is stable or unstable. The curve is symmetrical in x_1 and x_2, as is eq. (6.4.47), and two critical points (filled circles) occur on the left and right branches of the curve.

The closed unstable region in Figure 6.39 guarantees the existence of an isolated region (not reaching the *binary* edges) of two-phase equilibria. The equilibrium conditions, $\mu_\alpha(\mathbf{x}^{(1)}) = \mu_\alpha(\mathbf{x}^{(2)})$, $\alpha = 1, 2$, can be solved easily by the NR method using the critical points and spinodal as a guide to a first guess and then using continuation to calculate the entire binodal. The binodal and a few tie-lines are shown in Figure 6.39.

It is interesting to consider the physics of the origin of the isola in Figure 6.39. The reason for a binary phase split of a regular solution is that $a_{\alpha\beta}/kT > 2$, that is, that the attraction between molecules of types α and β is sufficiently small compared to the mean of the pair attractions of the species α with itself and species β with itself. Phase separation occurs because α and β prefer to be with their own species. However, in the ternary example just examined, $a_{13} = a_{23} = 0$ indicating 1 and 2 have no preferences relative to species 3 and $a_{12}/kT = -100$ indicating that species 1 and 2 have a strong preference for each other. Thus, in a binary system 1 and 2 will be miscible in all proportions. In a ternary system species 1

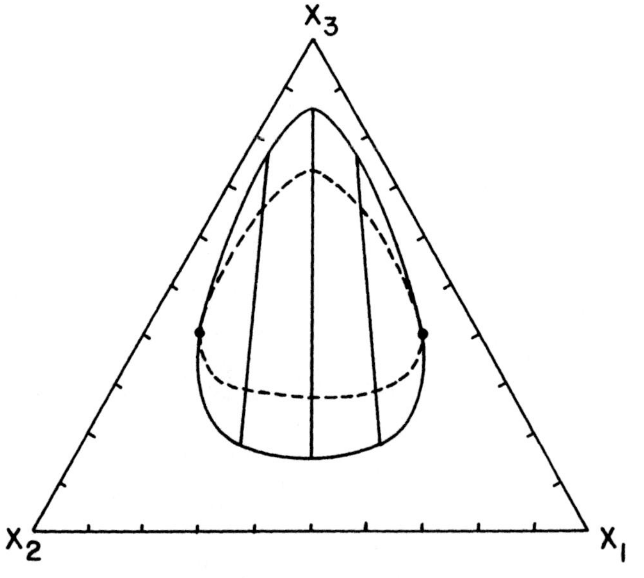

Figure 6.39
Ternary phase diagram for a regular solution with $a_{12}/kT = -10$, $a_{13} = a_{23} = 0$. Critical points (filled circles) are at mole fractions $x_1 = 0.5$ and $x_2 = 0.1$ and $x_1 = 0.1$ and $x_2 = 0.5$. Binodal shown as a solid curve with tie-lines and spinodal shown as a dashed curve. Redrawn from Meijering (1951).

and 2 prefer each other so strongly relative to species 3 that this energetic preference splits a species 1 and 2 rich phase from a species 3 rich phase when the 12 pair probability, x_1x_2, is sufficiently large compared to the 13 or 23 probabilities, x_1x_3 or x_2x_3.

Acetic acid and the weak base toluidine have a strong affinity to each other compared to their attraction to heptane. Thus, the observed isola reported in Figure 6.21 is a real example of the case studied here with regular solution theory.

Meijering (1950, 1951) published a rather extensive study of the patterns of ternary phase behavior predicted by regular solution theory. He categorized different cases as:

1 all $a_{\alpha\beta}$ negative, no phase splits in binary solutions;

2 two $a_{\alpha\beta}$ negative, one positive, miscibility gap along one binary edge;

3 one $a_{\alpha\beta}$ negative, two positive, miscibility gaps along two binary edges;

4 all $a_{\alpha\beta}$ positive, miscibility gaps along all binary edges.

We will not discuss all of the subtleties explored by Meijering, but several of the special cases he considered are shown in Figure 6.40. It is plain to see that by judicious choice of parameters $a_{\alpha\beta}$ something approaching the richness of the observed patterns of ternary phase behavior shown in Figures 6.19–6.21 can be predicted by regular solution theory.

In our derivation of the thermodynamic functions of a regular solution we imagined that all molecules occupied the same volume, namely the volume v_s of a lattice site. This is of course generally not the case. Hildebrand and Scatchard independently introduced a variation on the model to improve its quantitative performance. They introduced the quantity $c_{\alpha\beta}$ as the interaction energy between species α and β per unit volume ($c_{\alpha\beta} = c_{\beta\alpha}$). If it is then assumed that species α and β are randomly distributed throughout the volume V, the average interaction energy per unit volume is

$$U^E = \frac{1}{2}\sum_{\alpha,\beta} V c_{\alpha\beta} \cdot \frac{V_\alpha}{V} \cdot \frac{V_\beta}{V}. \tag{6.4.53}$$

If v_α is the lattice volume occupied by a molecule of species α, then $V_\alpha = N_\alpha v_\alpha$ and $V = \sum_\alpha N_\alpha v_\alpha$ and eq. (6.4.48) becomes

$$U^E = N \sum_{\alpha,\beta} \frac{x_\alpha x_\beta v_\alpha v_\beta c_{\alpha\beta}}{\sum_\alpha x_\alpha v_\alpha}. \tag{6.4.54}$$

Figure 6.40
Some ternary phase diagrams predicted by regular solution theory for the parameters a_{12}/kT, a_{13}/kT, a_{23}/kT set equal to (a) 0, 3, −3; (b) 0, 2.5, 2.5; (c) 0.5, 2.5, 2.25; (d) −0.5, 2.5, 2.25; (e) 3.5, 2.5, 2.67; (f) 2.57, 2.57, 2.57; (g) 2.62, 2.62, 2.62. Redrawn from Meijering (1951).

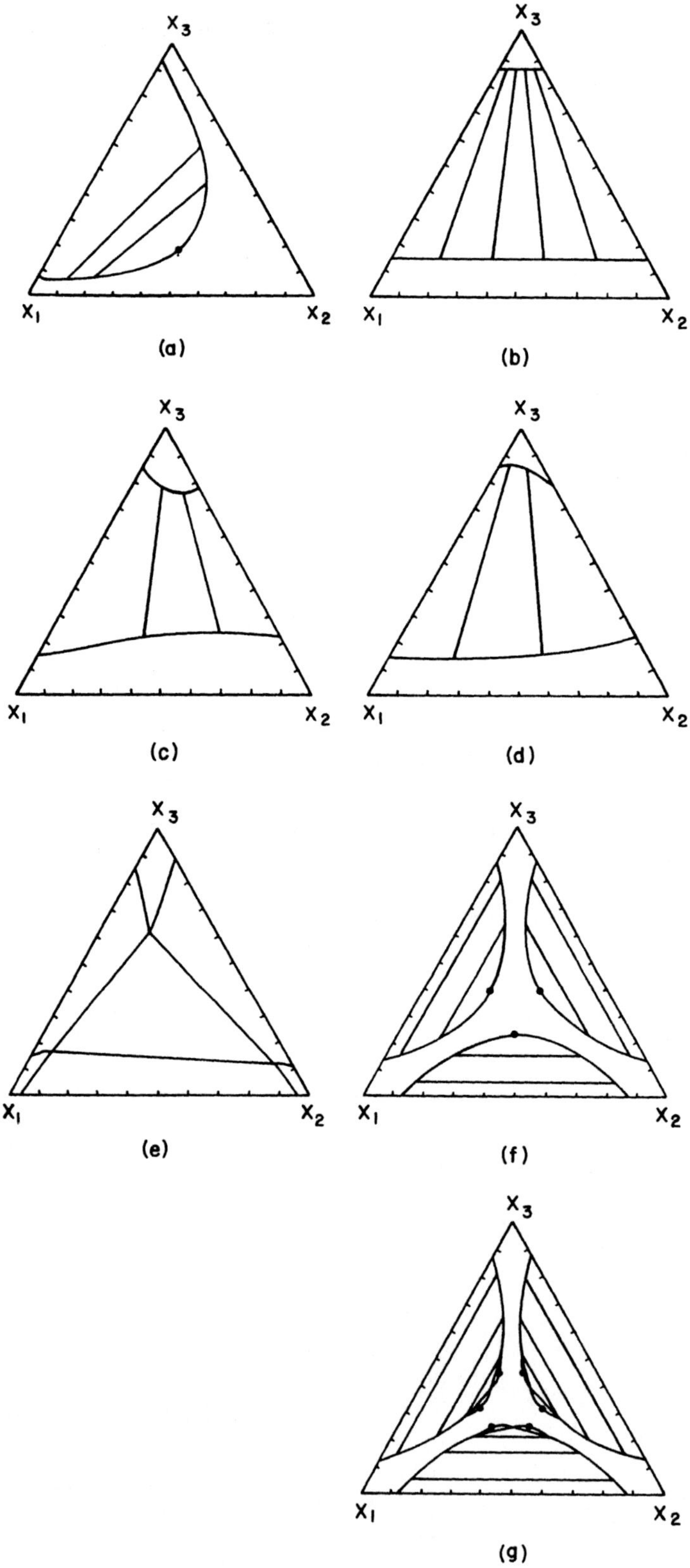

Hildebrand and Scatchard assumed that the entropy of mixing is that of an ideal solution and so the molecular Gibbs free energy, $\tilde{g}$, of the solution is

$$\tilde{g} = \sum_{\alpha} x_\alpha \mu_\alpha^P(T, P) + \sum_{\alpha} kT x_\alpha \ln x_\alpha + \frac{1}{2} \sum_{\alpha,\beta} \frac{x_\alpha x_\beta v_\alpha v_\beta c_{\alpha\beta}}{\sum_\alpha x_\alpha v_\alpha} - \sum_{\alpha} x_\alpha v_\alpha c_{\alpha\alpha} \tag{6.4.55}$$

and the chemical potential of species α is

$$\mu_\alpha = \mu_\alpha^P(T, P) + kT \ln x_\alpha + v_\alpha \sum_{\beta} \Phi_\beta c_{\alpha\beta} - \frac{1}{2} v_\alpha c_{\alpha\alpha} - \frac{v_\alpha}{2} \sum_{\beta,\gamma} \Phi_\beta \Phi_\gamma c_{\beta\gamma}, \tag{6.4.56}$$

where Φ_β is the volume fraction, namely,

$$\Phi_\beta \equiv \frac{N_\beta v_\beta}{\sum_\gamma N_\gamma v_\gamma}. \tag{6.4.57}$$

In a pure fluid the quantity $c_{\alpha\alpha}$ is well approximated by

$$c_{\alpha\alpha} = \frac{\Delta \tilde{u}_\alpha^v}{v_\alpha}, \tag{6.4.58}$$

where $\Delta \tilde{u}_\alpha^v$ is the molecular energy of vaporization of species α per unit volume. With the mixing rule

$$c_{\alpha\beta} = (c_{\alpha\alpha} c_{\beta\beta})^{1/2} \tag{6.4.59}$$

and the definition of the so-called solubility parameter δ_α,

$$\delta_\alpha \equiv (\Delta \tilde{u}_\alpha^v / v_\alpha)^{1/2}, \tag{6.4.60}$$

the chemical potential of species α can be expressed as

$$\mu_\alpha = \mu_\alpha^P(T, P) + kT \ln x_\alpha + v_\alpha (\delta_\alpha - \bar{\delta})^2, \tag{6.4.61}$$

where

$$\bar{\delta} = \sum_{\alpha} \Phi_\alpha \delta_\alpha. \tag{6.4.62}$$

In a binary system of species α and β,

$$\mu_\alpha = \mu_\alpha^P(T, P) + kT \ln x_\alpha + v_\alpha \Phi_\alpha (\delta_\alpha - \delta_\beta)^2. \tag{6.4.63}$$

The molar volumes and solubility parameters of several nonpolar fluids are given in Table 6.3. The theory with the simple mixing rule, eq. (6.4.54), and the solubility parameter assumption, eq. (6.5.55), yields reasonably good approximations (say 10–20% error) of compositions of liquid–vapor equilibria of many nonpolar mixtures (McGlashan, 1962). However, in the case of condensed phase equilibria, deviations from ideality are large and regular solution theory is more often of qualitative value than quantitative.

Exercise

Find the formula for the critical point of binary liquid solutions from the Hildebrand-Scatchard regular solution theory. Do any liquid pairs listed in Table 6.3 exhibit liquid-liquid phase splits at 25°C? If so determine the binodal and spinodal curves for the least miscible pair. Assume that the molar volumes and solubility parameters are independent of temperature.

TABLE 6.3 Molar Liquid Volumes and Solubility Parameters of Some Nonpolar Liquids

Liquids	*v (ml/g-mol)*	*δ (cal/ml)$^{1/2}$*
90 K		
Nitrogen	38.1	5.3
Carbon monoxide	37.1	5.7
Argon	29.0	6.8
Oxygen	28.0	7.2
Methane	35.3	7.4
Carbon tetrafluoride	46.0	8.3
Ethane	45.7	9.5
25°C		
Perfluoro-*n*-heptane	226	6.0
Neopentane	122	6.2
Isopentane	117	6.8
n-Pentane	116	7.1
n-Hexane	132	7.3
1-Hexene	126	7.3
n-Octane	164	7.5
n-Hexadecane	294	8.0
Cyclohexane	109	8.2
Carbon tetrachloride	97	8.6
Ethyl Benzene	123	8.8
Toluene	107	8.9
Benzene	89	9.2
Styrene	116	9.3
Tetrachloroethylene	103	9.3
Carbon disulfide	61	10.0
Bromine	51	11.5
Methanol	40.7	14.5
Water	18.07	23.4

Adapted from Prausnitz (1969).

Exercise
Consider the liquids at 25°C in Table 6.3. According to the Scatchard-Hildebrand regular solution theory, which of the liquids are miscible with water? With methanol?

6.5 Lattice Theory of Solutions: Quasichemical Approximation

Bethe, in studying the order-disorder transition in alloys, and Guggenheim, in studying fluid–fluid equilibria, introduced the quasichemical approximation, which is an improvement on the Bragg–Williams approximation used in the regular solution model.

Consider a lattice of coordination z occupied by N_1 molecules of species 1 and N_2 of species 2. The number of lattice sites is $N = N_1 + N_2$. Assume only nearest neighbor interactions. The number of nearest neighbor sites is $zN/2$ and in a given configuration there will be $N_{\alpha\beta}$ distinct pairs of nearest neighbors of molecules of species α and β. Of the zN_1 pairs of neighboring sites to molecules of species 1, $2N_{11}$ are occupied by species 1 and N_{12} are occupied by species 2, that is,

$$zN_1 = 2N_{11} + N_{12}. \tag{6.5.1}$$

Similarly,

$$zN_2 = 2N_{22} + N_{12}. \tag{6.5.2}$$

The reason $zN_{\alpha\alpha}$ enters the balance represented by these equations and not $N_{\alpha\alpha}$ is that $N_{\alpha\alpha}$ denotes distinct pairs whereas zN_α counts twice a given neighboring $\alpha\alpha$ pair. There are of course $zN/2$ distinct nearest neighbor pairs of sites and the sum of eqs. (6.5.1) and (6.5.2) gives $zN/2 = N_{11} + N_{22} + N_{12}$, which is the correct accounting of distinct pairs of nearest neighbor molecular occupancy.

The energy of a given molecular configuration is

$$W = N_{11}w_{11} + N_{22}w_{22} + N_{12}w_{12} \tag{6.5.3}$$

and if $\nu(N_1, N_2, N_{12})$ denotes the number of possible configurations in which N_1 and N_2 molecules of species 1 and 2 are arranged such that there are N_{12} nearest neighbor pairs of unlike species, the canonical ensemble partition function is

$$Q_N = [q_1^I(T)e^{-zw_{11}/2kT}]^{N_1}[q_2^I(T)e^{-zw_{22}/2kT}]^{N_2} \sum_{N_{12}} \nu(N_1, N_2, N_{12})e^{N_{12}w/2kT}, \tag{6.5.4}$$

where

$$w = w_{11} + w_{22} - 2w_{12}. \tag{6.5.5}$$

The summation in eq. (6.5.4) is over every possible configuration N_{12} consistent with eqs. (6.5.1) and (6.5.2). Note that $\nu(N_1, N_2, N_{12})$ obeys the condition

$$\sum_{N_{12}} \nu(N_1, N_2, N_{12}) = \frac{N!}{N_1!N_2!}, \tag{6.5.6}$$

because this is how many distinct ways a lattice of N sites can be occupied by N_1 molecules of type 1 and N_2 of type 2.

Because we are interested in bulk properties, N_1 and N_2 are extremely large numbers, and so the sum in eq. (6.5.4) can be approximated by its largest term, that is,

$$Q_N \simeq [q_1^I e^{-zw_{11}/2kT}]^{N_1}[q_2^I e^{-zw_{22}/2kT}]^{N_2}\nu^* e^{N_{12}^* w/2kT} \tag{6.5.7}$$

where N_{12}^* is determined by maximum condition

$$\frac{\partial}{\partial N_{12}} \ln[\nu(N_1, N_2, N_{12})e^{N_{12}w/2kT}] = 0 \quad \text{at} \quad N_{12}^*. \tag{6.5.8}$$

In the quasichemical theory one assumes that pairs of nearest neighbor sites are independent of each other. Although the assumption yields the exact result for a one-dimensional lattice, it is an approximation for two- or three-dimensional lattices. For example, if sites a, b, c, and d form a nearest neighbor square with a and b occupied by 1, and c and d by 2 for two 12 bonds, then 11 22 bonds must occur. The approximation overestimates the number of configurations possible. Nevertheless, quasichemical theory does represent an improvement over regular solution theory.

A given pair of nearest neighbor sites can be occupied four ways: 11, 22, 12, and 21. There are N_{11}, N_{22}, $N_{12}/2$, and $N_{12}/2$ of such pairs and there are $zN/2$ total pairs of nearest neighbor sites. The number of ways to occupy the $zN/2$ pairs of sites if the pairs $\alpha\beta$ are independent is

$$\omega = \frac{(zN/2)!}{N_{11}!N_{22}!(N_{12}/2)!(N_{12}/2)!}, \tag{6.5.9}$$

because $(zN/2)!$ overcounts distinct configurations of 11 pairs by $N_{11}!$, 22 by $N_{22}!$, etc. According to the approximation, ν is proportional to this number, that is,

$$\nu(N_1, N_2, N_{12}) = \alpha \frac{(zN/2)!}{N_{11}!N_{22}![(N_{12}/2)!]^2}. \tag{6.5.10}$$

The proportionality constant assures the condition at eq. (6.5.6) and thus is a correction for the configuration overcounting implicit in the quasichemical approximation.

From eqs. (6.5.1) and (6.5.2) it follows that $\partial N_{11}/\partial N_{12} = \partial N_{22}/\partial N_{12} = -1/2$, and so insertion of eq. (6.5.10) into eq. (6.5.8) yields

$$\frac{1}{2}\ln N_{11}^* + \frac{1}{2}\ln N_{22}^* - \ln(N_{12}^*/2) + w/2kT = 0 \tag{6.5.11}$$

or

$$\frac{N_{11}^* N_{22}^*}{(N_{12}^*/2)^2} = e^{-w/kT}. \tag{6.5.12}$$

Equation (6.5.12) is the reason for the name quasichemical approximation, because it evokes the image of an equilibrium between "pair species"

$$\frac{1}{2}[12] + \frac{1}{2}[21] \rightleftharpoons [11] + [22] \tag{6.5.13}$$

that according to the law of mass action would obey eq. (6.5.12) with an equilibrium constant $K = \exp[-w/kT]$. A positive value of w favors 12 pairs and negative value favors 11 and 22 pairs. To determine α, the sum in eq. (6.5.6) is replaced by its maximum term, $\alpha w_{max} \simeq N!/N_1!N_2!$. The condition $\partial\omega\partial N_{12}$, yields $\omega_{max} = (N!/N_1!N_2!)$, and so

$$\alpha = \left(\frac{N!}{N_1!N_2!}\right)^{1-z}. \tag{6.5.14}$$

Combination of the appropriate equations leads to the following quasichemical approximation to the partition function

$$Q_N = [q_1^I e^{zw_{11}/2kT}]^{N_1}[q_2^I e^{-zw_{22}/2kT}]^{N_2} \left(\frac{N!}{N_1!N_2!}\right)^{1-z} \frac{(zN/2)!\exp(N_{12}^* w/2kT)}{N_{11}^* N_{22}^*![(N_{12}^*/2)!]^2}, \qquad (6.5.15)$$

where N_{12}^* obeys the equation

$$N_{12}^* = (zN/2)\left[1 - \frac{\sqrt{1 - 4x_1x_2 1 - e^{-w/kT})}}{1 - e^{-w/kT}}\right], \qquad (6.5.16)$$

and $N_{11}^* = (zN_1 - N_{12}^*)/2$, $N_{22}^* = (zN_2 - N_{12}^*)/2$.

The chemical potential of species 1 can be computed from

$$\mu_1 = -kT\left(\frac{\partial \ln Q_N}{\partial N_1}\right)_{N_2} = -kT\left(\frac{\partial \ln Q_N}{\partial N_1}\right)_{N_2,N_{12}^*} - kT\left(\frac{\partial \ln Q_N}{\partial N_{12}^*}\right)_{N_1,N_2}\left(\frac{\partial N_{12}^*}{\partial N_1}\right)_{N_2}$$

$$= -kT\left(\frac{\partial \ln Q_N}{\partial N_1}\right)_{N_2,N_{12}^*}. \qquad (6.5.17)$$

The term involving $\partial \ln Q_N/\partial N_{12}^*$ equals 0 because of the condition at eq. (6.5.8). The quasichemical approximation for μ_α for a binary solution is easily shown from eq. (6.5.17) to be

$$\mu_\alpha = \mu_\alpha^P + kT\ln x_\alpha + \frac{z}{2}kT\ln\left[\frac{t - 1 + 2x_\alpha}{x_\alpha(1+t)}\right], \qquad (6.5.18)$$

where $t \equiv \sqrt{1 - 4x_1x_2(1 - e^{-w/kT})}$.

μ_α^P is the chemical potential of pure component $\alpha(\mu_\alpha^P = -kT\ln q_\alpha^I + zw_{\alpha\alpha}/2)$.

The thermodynamic mixing properties [eq. (6.4.9)] of the quasichemical model solution are

$$\Delta F_m = \Delta G_m = \sum_{\alpha=1}^{2} N_\alpha kT\ln x_\alpha + \frac{z}{2}\sum_{\alpha=1}^{2} N_\alpha kT\ln\left[\frac{t - 1 + 2x_\alpha}{(t+1)x_\alpha}\right] \qquad (6.5.19)$$

$$\Delta U_m = -\frac{Nzwx_1x_2}{t+1} \qquad (6.5.20)$$

and

$$\Delta S_m = -\sum_{\alpha=1}^{2} N_\alpha k\ln x_\alpha - \frac{Nzwx_1x_2}{T(t+1)} - \frac{z}{2}\sum_{\alpha=1}^{2} N_\alpha k\ln\left[\frac{t - 1 + 2x_\alpha}{(t+1)x_\alpha}\right]. \qquad (6.5.21)$$

If these functions are expanded in a Taylor series about $w/kT = 0$, the first few terms are

$$\frac{\Delta G_m}{NkT} = \sum_{\alpha=1}^{2} x_\alpha \ln x_\alpha - x_1x_2\left(\frac{zw}{2kT}\right) - \frac{x_1^2x_2^2}{z}\left(\frac{zw}{2kT}\right)^2 + \cdots \qquad (6.5.22)$$

$$\frac{\Delta U_m}{NkT} = -x_1x_2\left(\frac{zw}{2kT}\right) - \frac{2x_1^2x_2^2}{z}\left(\frac{zw}{2kT}\right)^2 + \cdots \qquad (6.5.23)$$

$$\frac{\Delta S_m}{Nk} = -\sum_{\alpha=1}^{2} x_\alpha \ln x_\alpha - \frac{x_1^2x_2^2}{z}\left(\frac{zw}{2kT}\right)^2 + \cdots. \qquad (6.5.24)$$

For $w = 0$, these quantities are those of an ideal solution. To first order in w/kT, the quantities correspond to a regular solution, that is, to the random mixing model. Thus, the terms of second order in w/kT represent estimated corrections to random mixing. To get an idea of the magnitude of the corrections, consider the case $zw/2kT = -2$ and $x_1 = x_2 = 1/2$, that corresponds to a regular solution at its critical point (note that $zw/2 = -a$ in the regular solution notation). A reasonable value for z is 6. For these parameters

$$\frac{\Delta G_m}{NkT} - \sum_\alpha x_\alpha \ln x_\alpha = \frac{1}{2} - \frac{1}{24} \tag{6.5.25}$$

$$\frac{\Delta U_m}{NkT} = \frac{1}{2} - \frac{1}{12} \tag{6.5.26}$$

and

$$\frac{\Delta S_m}{NkT} + \sum_\alpha x_\alpha \ln x_\alpha = -\frac{1}{24}, \tag{6.5.27}$$

indicating corrections to the random mixing results that are on the order of 10–20%.

The liquid–liquid binodal of a binary solution can be computed from the equilibrium condition $\mu_1(x_1^{(1)} = \mu_1(x_1^{(2)} = 1 - x_1^{(1)})$. The binodals predicted by the quasichemical solution model and the regular solution model are compared in Figure 6.41 for the case $z = 6$. The concentration at the critical point of both models is $x_{1c} = 1/2$ but the critical temperature T_c of the quasichemical model equals $-1.233(w/k)$ whereas that of the regular solution is $-1.5(w/k)$. Thus, the molecular correlations or deviations from random site occupancy tend to lower the critical temperature of the system. Also, near the critical point the binodal of the quasichemical model is flatter than that of the regular solution model. Observed binodals are generally flatter than those predicted by the regular solution model and so the quasichemical approximation is an improvement over the regular solution model.

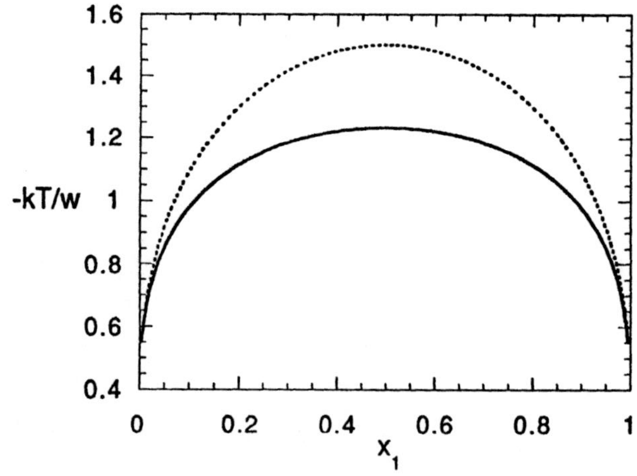

Figure 6.41
Liquid–liquid phase diagram predicted by the quasichemical model (solid curve) and the regular solution model (dotted curve). $z = 6$.

6.6 Lattice Theory of Solutions: Directed Bond and Decorated Lattice Models

In Section 6.2 we saw that certain pairs of liquids, e.g., *m*-toludine and water, have upper and lower critical points and have a closed loop temperature–composition phase diagram. Neither the regular solution nor the quasichemical approximation predict the existence of lower solution critical temperatures (LSCT) in binary systems. Hirshfelder, et al. (1937) suggested that the LCST was due to directional interactions such as hydrogen bonding that interfere with the free rotation of molecules. The average interaction energy between freely rotating molecules and those held by strong, directed bonds would be quite different. We shall see in what follows that Hirschfelder et al.'s idea provides an understanding of the LSCT.

Recall that for the regular solution, the energy of mixing in a binary fluid is $\Delta U_m = x_1 x_2 N a$. If $a/kT > 2$, then the binary system has a concentration region of liquid–liquid coexistence. If a is a function of temperature that increases faster with temperature than T itself, it is possible that

$$\begin{aligned} a/kT &< 2, \quad T < T_1 \\ &= 2, \quad T = T_1 \\ &> 2, \quad T > T_1. \end{aligned} \tag{6.6.1}$$

In this case T_1 would be a lower critical temperature, below which the liquids would be miscible in all proportions and above which a two-phase split would occur. Consider now how directed interactions could lead to such a pattern in the interaction parameter a. We assume the like pair interactions are w_{11} and w_{22}, but that the unlike pair interactions depend on the orientation of a particle of type 1 relative to one of type 2. The molecules are imagined to have z interaction points, each point acting along the line connecting nearest neighbor lattice sites. $z/2$ of the points of each molecule are "bonding points." When a bonding point of a molecule of species 1 is the nearest neighbor of a bonding point of a molecule of species 2 the interaction energy is $w_{12}^{(1)}$. Otherwise the interaction energy is $w_{12}^{(2)}$. The situation is illustrated in Figure 6.42.

If, for the model just described, we assume that nearest neighbors are randomly distributed but that the orientation of points between unlike molecules is weighted by the Boltzmann factor $\exp(-w_{12}^{(i)}/kT)$ of the directed pair interaction, then the average total interaction energy W of the system will be

$$W = N\left\{x_1^2\frac{z}{2}w_{11} + x_2^2\frac{z}{2}w_{22} + x_1x_2 z\left[\frac{w_{12}^{(1)}e^{-\beta w_{12}^{(1)}} + w_{12}^{(2)}e^{-\beta w_{12}^{(2)}}}{e^{-\beta w_{12}^{(1)}} + e^{-\beta w_{12}^{(2)}}}\right]\right\}. \tag{6.6.2}$$

Comparison of eq. (6.6.2) with eq. (6.4.1), we see that this approximation yields thermodynamic functions of precisely the same form

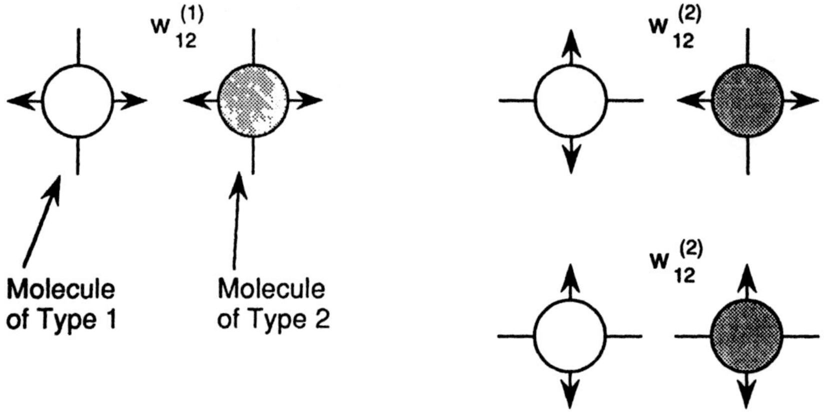

Figure 6.42
Schematic of directional pair interactions. When the bonding points (arrow) are pointing toward one another the pair interaction energy is $w_{12}^{(1)}$, otherwise it is $w_{12}^{(2)}$.

as those of the regular solution except that the parameter a is of the form

$$a = \frac{a^{(1)}e^{E/kT} + a^{(2)}}{e^{E/kT} + 1}, \tag{6.6.3}$$

with

$$E \equiv w_{12}^{(2)} - w_{12}^{(1)}, \quad a^{(i)} \equiv \frac{z}{2}[2w_{12}^{(i)} - w_{11} - w_{22}]. \tag{6.6.4}$$

If the bonding interaction is negative and if $|w_{12}^{(1)}| > |w_{12}^{(2)}|$, then $E > 0$ and the parameter a ranges from $a^{(1)}$ to $[a^{(1)} + a^{(2)}]/2$ as T goes from zero to infinity. To proceed further with the model, we must specify the parameters $a^{(1)}$, $a^{(2)}$, and E. For illustration we have chosen

$$a/k = 1100\text{K}\left[\frac{4 - \exp(245\text{K}/T)}{1 + \exp(245\text{K}/T)}\right]. \tag{6.6.5}$$

Values of a/kT computed from eq. (6.6.5) for various temperatures are given in Table 6.4. Because at a critical point $a/kT_c = 2$ and in the two-phase region $a/kT > 2$, it follows from the entries in Table 6.4 that a solution obeying eq. (6.6.5) will have upper and lower critical temperatures bounding a closed loop binodal.

TABLE 6.4 Values of a/kT for Various Temperatures

T (K)	a/kT
250	1.60
300	1.95
350	2.07
400	2.08
450	2.04
500	1.98

Because the present model is equivalent to a regular solution with a temperature dependent interaction parameter a, the chemical potential of species α is given by eq. (6.4.8) and the liquid–liquid compositions can be computed from

$$kT/a(T) = (1 - 2x_1^{(1)}) \ln[(1 - x_1^{(1)})/x_1^{(1)}], \tag{6.6.6}$$

which is the same as eq. (6.4.35) except that $a(T)$ depends on T. The liquid–liquid binodal is thus calculated by setting $x_1^{(1)}$, determining $kT/a(T)$ from eq. (6.6.6), and finding T from eq. (6.6.5). The resulting phase diagram is shown in Figure 6.43. The lower critical point occurs at $x_{1c} = 1/2$ and $T_c^L = 313.5$K and the upper critical point occurs at $x_{1c} = 1/2$ and $T_c^u = 484.7$K.

The origin of the closed loop binodal for the model developed here can be understood by examination of eqs. (6.6.3) and (6.6.4). The unlike particle pair interaction $w_{12}^{(1)}$ is more strongly attractive than the mean of the like particle interaction, $(w_{11} + w_{22})/2$, and so $a^{(1)}$ is negative. This directed bond interaction thus favors miscibility of two liquids. The unlike particle pair interaction $w_{12}^{(2)}$ is less strongly attractive than the like particle mean, $(w_{11} + w_{22}/2)$, and so $a^{(2)}$ is positive. This favors a phase split. Low temperature favors configurations maximizing pair interactions of the type $w_{12}^{(1)}$ [note that $E > 0$ in the Boltzmann factor $\exp(E/kT)$], and so low temperature favors miscibility. With increasing temperature, configurations with more $w_{12}^{(2)}$ interactions increase in number, resulting in an increase in a and thus a phase split at T_c^L. With further increase in temperature, a approaches the limit $(a^{(1)} + a^{(2)})/2$, that is, all configurations approach equal likelihood. This means the ratio a/kT must, with increasing T, decrease eventually to 2 at an upper critical temperature, T_c^u, above which the entropy of mixing dominates the energy of phase segregation, and so the liquids are again miscible in all proportions.

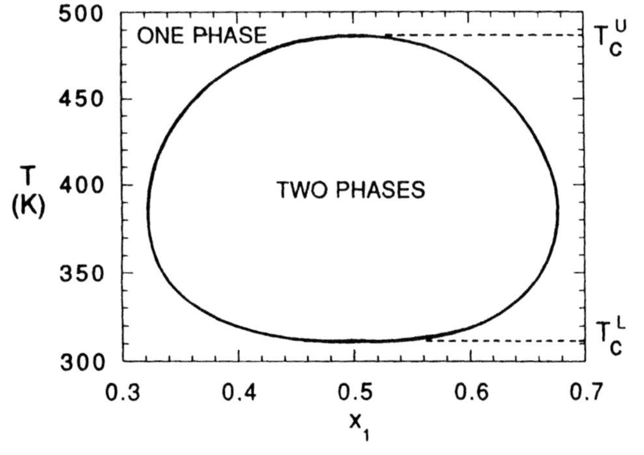

Figure 6.43
Closed loop liquid–liquid phase diagram predicted by the directed bond model using the parameters in eq. (6.6.5). The LSCT and USCT are 313.5 K and 484.7 K and the critical composition is $x_{1c} = 0.5$ in each case.

The model detailed above was inspired by a more sophisticated one introduced by J. Barker and Fock (1953) using a quasichemical approximation. They assumed that each molecule on a lattice of coordination z has z contact points (in the preceding paragraphs we referred to these as interaction points). One of these they called a q-type contact point, and the $z-1$ others they called p-type contact points. Between pairs of like molecules all the contact points interact alike. Between p-type contacts of an unlike pair the interaction was assigned a positive value u_1 and between p and q-type and between q and q-type contacts of unlike pairs the interaction was assigned a negative value u_2. The quantities u_1 and u_2 actually represent the difference between the pair interaction and the arithmetic mean of like pair interactions.

With their assumptions and the quasichemical approximation, Barker and Fock found for the chemical potential of species

$$\mu_1 = \mu_1^P + kT[\ln x_1 + (z-1)\ln(\chi_1/x_1\chi_1') + \ln(\Psi_1/x_1\Psi_1')] \qquad (6.6.7)$$

and

$$\mu_2 = \mu_2^P + kT[\ln x_2 + (z-1)\ln(\chi_2/x_2\chi_2') + \ln(\Psi_2/x_2\Psi_2')], \qquad (6.6.8)$$

where the χ's and x's obey the equations

$$\begin{aligned} \chi_1(x_1 + \psi_1 + e^{-\beta u_1}\chi_2 + e^{-\beta u_2}\psi_2) - \frac{1}{2}(z-1)x_1 &= 0 \\ \psi_1(\chi_1 + \psi_1 + e^{-\beta u_2}\chi_2 + e^{-\beta u_2}\psi_2) - \frac{1}{2}x_1 &= 0 \\ \chi_2(e^{-\beta u_1}\chi_1 + e^{-\beta u_2}\psi_1 + \chi_2 + \psi_2) - \frac{1}{2}(z-1)x_2 &= 0 \\ \psi_2(e^{-\beta u_2}\chi_1 + e^{-\beta u_2}\psi_1 + \chi_2 + \psi_2) - \frac{1}{2}x_2 &= 0, \end{aligned} \qquad (6.6.9)$$

with $\chi_i' = \chi_i(x_i = 1)$ and $\chi_i' = \psi_i(x_i = 1)$. The model predicts a closed loop binodal for certain values of u_1 and u_2. Two cases are shown in Figure 6.44.

The qualitative situation is again that the lower critical point occurs because the strong pair attraction (u_2) between the special contact points causes the liquids to be miscible in all proportions at temperatures below T_c^L. Above T_c^L but below T_c^u, enough repulsive bonds (u_1) are formed for the energy or mixing to drive a phase split; and above T_c^u the entropy of mixing overrides the energy of mixing and again the liquids are miscible in all proportions.

Wheeler (1975) introduced a decorated-lattice model that captures the essence of the Barker–Fock model but can be mapped into an Ising model whose phase behavior has been established accurately by a combination of theory and computer simulations. Wheeler first defines a primary lattice and midway between every pair of sites of the primary lattice he locates a secondary site, the collection of which defined the secondary (or decorating) lattice. Every site of the primary and the secondary lattice is occupied by a molecule of either species 1 and 2. The molecules have ω

Figure 6.44
Liquid–liquid coexistence curves predicted by the Barker–Fock model for $z = 6$ and $-u_2/u_1 = 0.333$ (solid curve) and 0.620 (dashed curve). Redrawn from Barker and Fock (1953).

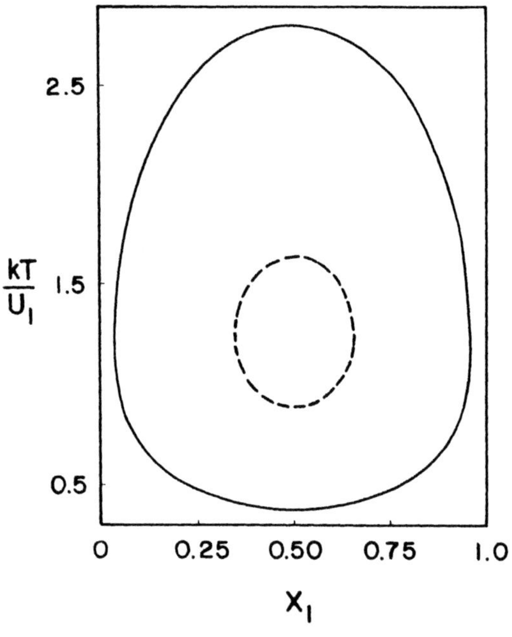

contact points (not necessarily the same number as the lattice coordination number) that can be thought of as the orientations of the molecules on secondary lattice sites. One of the contact points is assumed to be special. The only interaction energies in the model are between particles on adjacent primary–secondary sites. Interactions between like species are assumed to be zero. The 1–2 interaction depends upon the orientation of the molecule in the secondary cell. If the 1–2 bond involves the special contact point of the molecule on the secondary site, the interaction energy u_z is negative; otherwise the energy u_1 is positive. Thus, a molecule of type α on a secondary site has one orientation with energy of interaction $u_2 < 0$ with an unlike molecule β on an adjacent primary site and $\omega - 1$ orientations with energy of interaction $u_1 > 0$. Each secondary side has two adjacent primary neighbors. With one like and one unlike primary neighbors there is one orientation with energy $u_2 < 0$ and $\omega - 1$ orientations with energy $u_1 > 0$. If both primary neighbors are unlike molecules, then there are two orientations with energy $u_1 + u_2$ and $\omega - 2$ orientations with energy $2u_1$.

Wheeler proved that the decorated-lattice model can be transformed exactly into the spin$-1/2$ Ising model for which formulas yielding accurate prediction of phase behavior are available. Predictions of the model are compared with the experiment in Figure 6.45. Because a lattice theory does not account for differences in molecular volumes, Anderson and Wheeler (1978) presented the experimental data in terms of the composition variable

$$\phi_1 = \frac{\alpha x_1}{\alpha x_1 + x_2} \qquad (6.6.10)$$

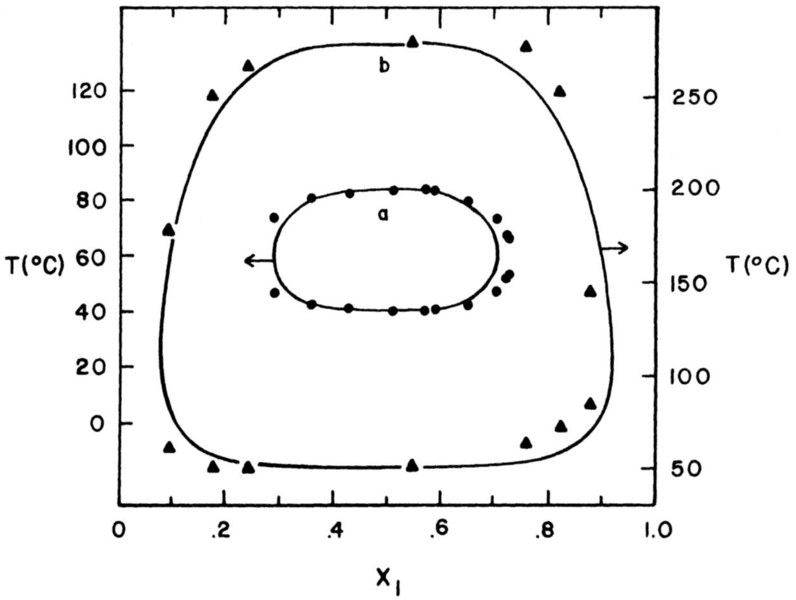

Figure 6.45
Liquid–liquid coexistence curves for the mixtures (a) glycerol and guaiacol and (b) glycerol and benzylethylamine. Points are experimental results and curves are predictions of the decorated lattice model with $w = 5000$. Compositions are mole fraction x_1 (for theory) and ϕ_1 of eq. (6.6.10) (for experimental data). Glycerol is component 2. Redrawn from Anderson and Wheeler (1978).

where α was set equal to 0.86 for the glycerol–benzylethylamine mixture and 1.12 for the glycerol–quaiacol mixture. In the first case ϕ is the volume fraction (the choice often used to improve the regular solution model as described in Section 6.5), and in the second case ϕ is a measure of composition lying between the mass and mole fraction.

One can consider α as another parameter of the model added to improve agreement with the experiment (the other parameters are z, w, u_1 and u_2). The agreement between the experiment and the model is indeed excellent compared to that achieved by mean field theories such as the regular solution quasichemical approximation. The most important aspect of the agreement is that the predicted binodals are wider and flatter at the top and bottom than those predicted by mean fluid theories. Part of the reason for this is that the Ising model equations that Anderson and Wheeler used obey the correct near-critical scaling laws. In particular, near the critical temperature the mole fractions of species α in phase 1 and 2 obey the asymptotic formula

$$x_\alpha^{(1)} - x_\alpha^{(2)} \propto |T_c - T|^\beta. \tag{6.6.11}$$

The correct (and universal) value of the critical exponent β in three dimensions is $\beta \simeq 0.34$, whereas mean field theories yield $\beta = 0.5$.

6.7 Lattice Theory of Solutions: Flory–Huggins Theory of Polymer Solutions

The two-phase binodal of a binary regular solution is symmetric, the mole of volume fraction of component 1 in one phase being the same as that of component 2 in the other phase. When the two components have low molecular weights, the regular solution model gives a reasonably accurate qualitative prediction. In a solution of polymer in a small molecule solvent the situation is quite different owing to entropic effects. In 1941 the Flory–Huggins (FH) theory provided an explanation of the special properties of polymer solutions.

Flory and Huggins used the lattice model illustrated in Figure 6.46. A small molecule, represented by an open circle, occupies one site in the lattice, whereas a polymer molecule, represented by connected filled circles, occupies ν sites, where

$$\nu = \tilde{V}_2/\tilde{V}_1. \tag{6.7.1}$$

$\tilde{V}_1$ is the molar volume of the solvent molecules and $\tilde{V}_2$ is the molar volume of the polymer molecules. Thus, each filled circle represents a polymer segment. In general ν is proportional to but not equal to the molecular weight of the polymer. In the model there are no vacant sites, which means that as with regular solutions there is no volume of mixing.

The FH theory assumes only nearest neighbor interactions and assigns the value $-\varepsilon_{11}$ to a neighboring solvent pair, $-\varepsilon_{22}$ to a pair of neighboring polymer segments and $-\varepsilon_{12}$ to solvent molecule–polymer segment neighbors. The energy between bonded polymer segments should not be included in computing the thermodynamic energy. However, the FH theory ignores the distinction between bonded and nonbonded segments on the grounds that the coordination number z (average number of nearest neightors) is of the order of 10 whereas at most two segments can be bonded to a given segment. If a mixture has N_1 solvent molecules and N_2

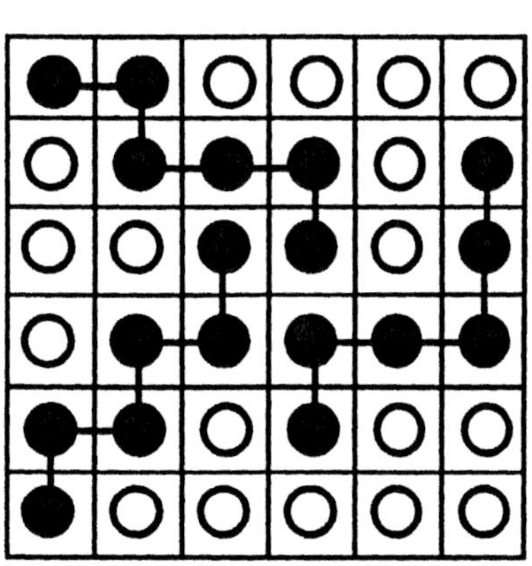

Figure 6.46
Two-dimensional illustration of a lattice decorated with small molecules and polymer molecules.

polymer molecules, then the fractions of sites occupied by solvent and polymer segments

$$\Phi_1 = \frac{N_1}{N_s}, \quad \Phi_2 = 1 - \Phi = \frac{\nu N_2}{N_s}, \tag{6.7.2}$$

where $N_s = N_1 + \nu N_2$ is the number of sites in the lattice. Plainly, Φ and Φ_2 are volume fractions, not mole fractions. If solvent molecules and polymer segments are assumed to be randomly distributed on the N_s sites, the

$$\frac{N_1}{N_s} \quad \frac{\nu N_2}{N_s} = \Phi_1 \quad \Phi_2 =$$ the probability that a site is occupied by solvent or polymer segment.

and

$$\frac{zN_1}{N_s} \quad \frac{z\nu N_2}{N_s} = z\Phi_1 \quad z\Phi_2 =$$ the probable number of nearest neighbors of a site that are solvent or polymer segment.

Corresponding to these probabilities, the average energy of interaction for the FH model is

$$W = -\frac{1}{2} N_s \sum_{\alpha,\beta=1}^{2} \varepsilon_{\alpha\beta} \Phi_\alpha z \Phi_\beta. \tag{6.7.3}$$

For a pure solvent of N_1 molecules,

$$W_1 = W(N_2 = 0, N_s = N_1) = -\frac{1}{2} N_1 z \varepsilon_{11}, \tag{6.7.4}$$

and for a pure polymer system of N_2 molecules,

$$W_2 = W(N_1 = 0, \quad N_s = \nu N_2) = -\frac{1}{2} \nu N_2 z \varepsilon_{22}. \tag{6.7.5}$$

Thus, the FH theory gives for the energy of mixing ($W - W_1 - W_2$)

$$\Delta U_m = (N_1 + \nu N_2) z (-\varepsilon_{12} + \frac{1}{2}\varepsilon_{11} + \frac{1}{2}\varepsilon_{22}) \Phi_1 \Phi_2, \tag{6.7.6}$$

which is similar in form to the regular solution energy of mixing. It is, in fact, the same as the Hildebrand–Scatchard energy of mixing and, in the special case that $\nu = 1$, Φ_α becomes the mole fraction and eq. (6.7.6) reduces to the simple regular solution model.

The partition function of the FH theory (under the mean field approximation used above to calculate W) is

$$Q(N_1, N_2, \nu, T) = \left[q_1^I(T)\right]^{N_1} \left[q_2^I(T)\right]^{N_2} \Omega(N_1, N_2, \nu) e^{-W/kT} \tag{6.7.7}$$

where $\Omega(N_1, N_2, \nu)$ denotes the number of arrangements of solvent and polymer molecules on the lattice. For a simple regular solution the exact result is $\Omega(N_1, N_2, 1) = (N_1 + N_2)!/N_1!N_2!$. The requirement that no site can be occupied by more than one solvent molecule or polymer segment, prevents the exact evaluation of Ω when polymers are involved. Without the restriction the number of arrangements can be computed as follows: there are N_2 positions to place the first segment of each polymer chain, z ways of placing the second, $z - 1$ of placing each subsequent segment (0° bond angles are not allowed), and N_2 positions for each solvent molecule. For this unconstrained problem the number of lattice arrangements is thus given by

$$\Omega^*(N_1, N_2, \nu) = \frac{N_s^{N_2}\left[z(z-1)^{\nu-2}\right]^{N_2}}{N_2!}\frac{N_s^{N_1}}{N_1!}. \tag{6.7.8}$$

Ω^* is of course a serious overestimate of Ω. The larger the molecular weight (i.e., the larger ν), the larger the overestimate caused by ignoring the constraint of single site occupancy.

In the FH theory the correction factor, defined by

$$\Theta \equiv \frac{\Omega(N_1, N_2, \nu)}{\Omega^*(N_1, N_2, \nu)}, \tag{6.7.9}$$

is approximated by assuming that it depends only on the number of polymer segments present, not on the degree to which they are linked into polymer chains. With this approximation

$$\begin{aligned}\Theta &\simeq \frac{\Omega(N_1, \nu N_2, 1)}{\Omega^*(N_1, \nu N_2, 1)} = \frac{N_s!/N_1!(\nu N_2)!}{N_s^{N_1}N_s^{\nu N_2}/N_1!(\nu N_2)!} \\ &= \frac{N_s!}{N_s^{N_s}} \approx e^{-N_s},\end{aligned} \tag{6.7.10}$$

where Stirling's approximation, $N_s! \approx (N_s/e)^{N_s}$, has been used to obtain the final result. Combining eqs. (6.7.8)–(6.7.10), we obtain

$$\Omega(N_1, N_2, \nu) \simeq \frac{N_s^{N_1+N+2}\left[z(z-1)^{\nu-2}\right]^{N_2}}{N_1!N_2!}e^{N_s} \tag{6.7.11}$$

for the mixture and

$$\Omega(N_1, 0, 1) \approx \frac{N_1^{N_1}e^{-N_1}}{N_1!} \tag{6.7.12}$$

and

$$\Omega(0, N_2, \nu) \simeq \frac{(\nu N_2)^{N_2}\left[z(z-1)^{\nu-2}\right]^{N_2}e^{-\nu N_2}}{N_2!}. \tag{6.7.13}$$

The entropy of mixing for the model is given by

$$\Delta S_m = k\ln\left[\frac{\Omega(N_1, N_2, \nu)}{\Omega(N_1, 0, 1)\Omega(0, N_2, \nu)}\right], \tag{6.7.14}$$

and so, for the FH theory,

$$\begin{aligned}\Delta S_{m'} &= -k\left[N_1\ln(N_1/N_s) + N_2\ln(\nu N_2/N_s)\right] \\ &= -N_s k\left(\Phi_1\ln\Phi_1 + \frac{\Phi_2}{\nu}\ln\Phi_2\right).\end{aligned} \tag{6.7.15}$$

The entropy predicted by the FH theory is very different from the regular solution counterpart because of the factor of ν^{-1} in the polymer term. As we shall see below, this factor will turn out to have profound effects on phase equilibria.

The free energy of mixing (Gibbs or Helmholtz because the model has zero volume of mixing) of the FH theory is

$$\Delta F_m = N_s kT\left(\Phi_1 \ln \Phi_1 + \frac{\Phi_2}{\nu} \ln \Phi_2 + \chi \Phi_1 \Phi_2\right), \tag{6.7.16}$$

where the so-called FH parameter is

$$\chi \equiv \frac{z}{kT}\left(\frac{1}{2}\varepsilon_{11} + \frac{1}{2}\varepsilon_{22} - \varepsilon_{12}\right). \tag{6.7.17}$$

χ can be positive or negative, but, if the components are not too polar and do not have strong hydrogen bonds, it is usually positive.

Differentiation of eq. (6.7.16) with respect to N_1 and N_2, respectively, gives for the chemical potentials of solvent molecules and polymer molecules

$$\mu_1 = \mu_1^p(T) + kT \ln \Phi_1 + kT\left(1 - \frac{1}{\nu}\right)\Phi_2 + kT\chi\Phi_2^2 \tag{6.7.18}$$

and

$$\mu_2 = \mu_2^p(T) + kT \ln \Phi_2 + kT(1 - \nu)\Phi_1 + \nu kT\chi\Phi_1^2, \tag{6.7.19}$$

where $\mu_i^p(T)$ is the chemical potential of pure i in its reference state.

The FH chemical potential contains an ideal mixing contribution, an excluded volume contribution (from the single site occupancy constraint), and an energy of mixing contribution. The energy contribution of a solvent molecule goes as $kT\chi$ whereas for a polymer molecule it goes as $\nu kT\chi$ because each polymer has ν polymer segment–solvent molecule interactions. For a large molecular weight polymer the excluded volume contribution to the solvent molecule is almost independent of ν, whereas there is a large negative contribution ($\sim -\nu kT\Phi_1$) to the polymer molecule. This contribution favors dilute polymeric solutions (large Φ_1) in phase equilibria. In fact, at the critical point, where

$$0 = \frac{d^2 \Delta F_m}{d\Phi_1^2} = \frac{1}{\nu\Phi_{2c}} - 2\chi_c \tag{6.7.20}$$

and

$$0 = \frac{d^3 \Delta F_m}{d\Phi_1^3} = -\frac{1}{\Phi_{1c}^2} + \frac{1}{\nu\Phi_{2c}^2}, \tag{6.7.21}$$

the FH theory predicts

$$\Phi_{2c} = \frac{1}{1+\sqrt{\nu}}, \quad \chi_c = \frac{1}{2}\left(1 + \frac{1}{\sqrt{\nu}}\right)^2. \tag{6.7.22}$$

For $\nu = 1000$, not an unusual value, the critical polymer concentration is about 3% by volume and $\chi_c \simeq 1/2$. This means that just below the critical point the two coexisting phases are nearly pure

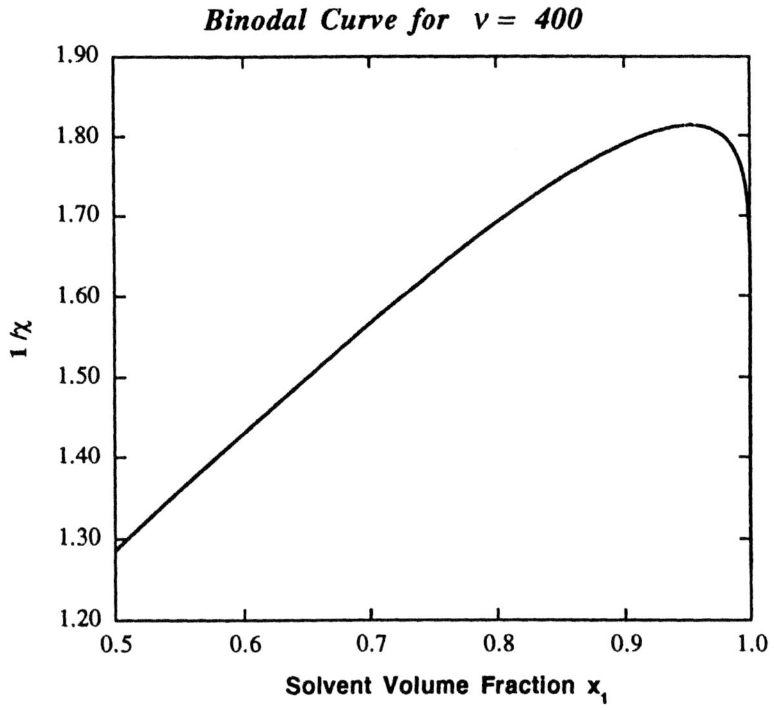

Figure 6.47
The two-phase coexistence curve predicted by the FH theory of a polymer solution.

water! Thus, the phase behavior of polymer solutions is qualitatively very different from that of small molecule mixtures. The reason for this is the large contribution to the entropy of mixing from polymer excluded volume effects.

The two-phase binodal of an FH solution can be computed as usual from the conditions of equilibrium, μ_i (phase 1 composition) $= \mu_2$ (phase 2 composition), $i = 1, 2$. The binodal for the case $\nu = 400$ is shown in Figure 6.47.

The binodal curves of a solvent–polymer solution approaches the pure solvent side of the phase diagram as ν increases. Compare Figure 6.47 with the regular solution binodal given in Section 6.4.

6.8 Lattice Theory of Solutions: Association Colloids

As indicated in the phase diagrams shown in Figures 6.15, 6.25, and 6.26, surfactants (amphiphiles) in solution form supramolecular aggregates (association colloids) in aqueous and/or oleic solutions. Understanding the molecular origins of the phase and microstructural behavior of these systems is deemed sufficiently important and fascinating to justify devotion of an entire chapter to the subject. The interested reader should advance directly to Chapter 14 in which Michael Schick introduces a lattice model that captures many of the important features of micellar solutions, microemulsions, and liquid crystals.

Supplementary Reading

Callen, H. B. 1986. *Thermodynamics and an Introduction to Thermostatics*, Wiley, New York.

Gibbs, J. W. 1876. *Trans. Connecticut Acad.*, **3**, 108; 1878. *Trans. Connecticut Acad.* **3**, 343.

Guggenheim, E. A. 1957. *Thermodynamics; an Advanced Treatment for Chemists and Physicists*, Interscience.

Prausnitz, J. M. 1969. *Molecular Thermodynamics of Fluid-Phase Equilibria*, Prentice–Hall, New York.

Robertson, H. S. 1993. *Statistical Thermophysics*, Prentice–Hall, New York.

Rowlinson, J. S. 1969. *Liquids and Liquid Mixtures*, Butterworths, London.

Sandler, S. I. 1989. *Chemical Engineering Thermodynamics*, Wiley, New York.

Exercises

1 Compute the $T - n$ spinodal curves for a one-component PR fluid for the cases $\omega = 0$ and 0.2. Plot the curves using the units b/a for T and b^{-1} for n.

2 Consider a decane–carbon dioxide mixture at $-15°C$. Assume the mixture obeys the PR equation of state.

a) Compute the spinodal curves and plot them in an $n_1 - n_2$ diagram.

b) Determine and plot the $n_1 - n_2$ phase diagram. *Hint:* See Figure 6.6.

3 Compute and plot the ternary phase diagram for a CO_2–decane–octane mixture at $-25°C$ and 1 atm. Assume that the PR equation of state is valid for the mixture.

4 Consider the regular solution depicted in Figure 6.39. Compute and compare the ternary phase diagrams for $a_{12}/kT = -10, -9, -5, -2$.

5 Compute and compare the binary liquid–liquid phase diagram predicted by the quasichemical model and the regular solution model for $z = 4, 6, 8$, and 12.

6 Consider the FH theory of polymer solutions.

a) Compute the chemical potentials of the solvent and polymer in the solution at various temperatures.

b) For temperatures above the consolute (critical) point T_c, only one phase solutions can exist. Above T_c,

$$\left(\frac{\partial \mu_1}{\partial x_1}\right)_{T,P} > 0$$

for all solvent mole fractions x_1. The spinodal curve T versus x_1 is defined as the limit of stability of a one-phase solution, that is, T versus x_1 is determined from the condition

$$\left(\frac{\partial \mu_1}{\partial x_1}\right)_{T,P} = 0.$$

Determine the formula for T versus x_1 along the spinodal curve for the polymer solution.

c) At the critical point the conditions

$$\left(\frac{\partial \mu_1}{\partial x_1}\right)_{T,P} = \left(\frac{\partial^2 \mu_1}{\partial^2 x_1}\right)_{T,P} = 0$$

Calculate the critical temperature and mole fraction for a polymer solution for which $\nu = 1000$.

d) Give qualitative plots of the T versus x_1 phase diagram for $\nu = 1000$. Show the one-phase region, the binodal, the spinodal, and the unstable region.

e) Plot the binodal for $\nu = 1000$ and $T = 0.9T_c$.

References

Anderson, G. R. and Wheeler, G. C. 1978. *J. Chem. Phys.*, **69**, 2082.

Barker, J. A. and Fock, W. 1953. *Discussions of the Faraday Society*, **15**, 188.

Davis, H. T., Bodet, J. F., Scriven, L. E., and Miller, W. G. in 1987. *Physics of Amphiphilic Layers*, D. Langevin and J. Meunier, Eds., Springer–Verlag, New York.

Döring, R. and Knapp, H. 1980. *Phase Equilibria and Fluid Properties in the Chemical Industry*, DECHEMA, Eds. S. I. Sandler and H. Knapp, Frankfurt, Part I, pp. 34–38.

Ekwall, P., Mandell, L., and Fontell, K. 1969. *J. Colloid and Interface Sci.* **29**, 639.

Falls, A. H. 1982. Ph.D. Thesis, University of Minnesota at Minneapolis, MN.

Francis, A. W. 1963. *Liquid–Liquid Equilibriums*, Interscience Publishers, New York.

Hirshfelder, J., Stevenson, D., and Eyring, H. 1937. *J. Chem. Phys.*, **5** 896.

Kahlweit, M., Strey, R., and Busse, G. 1990. *J. Phy. Chem.*, **94**, 3881.

Kilpatrick, P. K. 1983. Ph.D. Thesis, University of Minnesota at Minneapolis, MN.

Kuan, D. Y., Kilpatrick, P. K., Sahimi, M., Scriven, L. E., and Davis, H. T. 1986. *SPE Reservior Eng.*, 61–72.

McGlashan, M. L. 1962. *Chem. Soc. (London), Ann. Rep.*, **59**, 73.

Meijering, J. L. 1950. *Philips Res. Rep.*, **5**, 333; 1951. *Philips Res. Rep.*, **6**, 183.

Nitsche, J. M., Teletzke, G. F., Schriven, L. E., and Davis, H. T. 1984. *Fluid Phase Equilibria*, **17**, 243.

Ohe, S. 1990. *Vapor-Pressure Equilibrium Data at High Pressure*, Elsevier/Kodansha, Amsterdam/Tokyo.

Prausnitz, J. M. 1969. *Molecular Thermodynamics of Fluid-Phase Equilibria*, Prentice–Hall, New York.

Porter, W. 1920. *Trans. Faraday Soc.*, **16**, 336.

Prigogine, I., and Defay, R. 1962. *Chemical Thermodynamics*, Wiley, New York.

Prigogine, I., and Defay, R. 1954. *Chemical Thermodynamics*, Longmans Green and Co., London.

Quinones, S. E., Blackburn, M. B., Schriven, L. E., and Davis, H. T. 1991. *SPE Reservoir Eng.*, **February**, 33–36.

Sahimi, M., Davis, H. T., and Scriven, L. E. 1985. *SPE J.*, **April**, 235–254.

van Vlack, L. H. 1975. *Elements of Materials Science and Engineering*, 3rd ed., Addison–Wesley, Reading, MA.

Wheeler, J. E. 1975. *J. Chem. Phys.*, **62**, 433.

7

CAPILLARITY AND INTERFACIAL THERMODYNAMICS

7.1 Manifestations of Surface or Interfacial Tension

To the naked eye the interfacial zone between two fluid phases, liquid–vapor or liquid–liquid phases, appears to be a surface of zero thickness. In many ways this surface or interface acts like a membrane under tension. The most common observation of the effect of this surface or interfacial tension is the rise or depression of a liquid in a vertically held capillary tube partially penetrating the surface of a liquid exposed to air. If the angle of contact θ that the air–liquid (meniscus) interface makes with the capillary wall is less than 90°, the liquid will rise in the tube to some equilibrium height and if the angle is greater than 90°, the liquid in the capillary will be lower than the liquid–vapor surface at equilibrium. As illustrated in Figure 7.1, water in glass and mercury in glass are well-known examples of capillary rise and capillary depression. Also illustrated in Figure 7.1 are suspended columns of liquid and liquid droplets sitting on or hanging from a solid surface, all being situations influenced by interfacial tension.

Combined with the theory of capillary hydrostatics all of the phenomena in Figure 7.1 can be used to measure the tension γ of an interface. For instance, consider the case in which a liquid rises in a capillary to an equilibrium height h above the air–liquid surface. If the capillary is circular in cross-section and if its radius r is not too large, then the meniscus will be approximately a cap of a hemisphere of radius $R = r/\cos\theta$. If the height h is large compared to r, the following force balance on the contents of the capillary above the liquid level can be made:

$$-P_1\pi r^2 + 2\pi r\gamma\cos\theta + P_2\pi r^2 = \rho_1\pi r^2 hg, \tag{7.1.1}$$

where g is the acceleration of gravity and ρ_1 is the density of liquid 1. But the difference $P_2 - P_1$ is equal to $\rho_2$gh (from applying

Figure 7.1
Phenomena affected by interfacial tension.

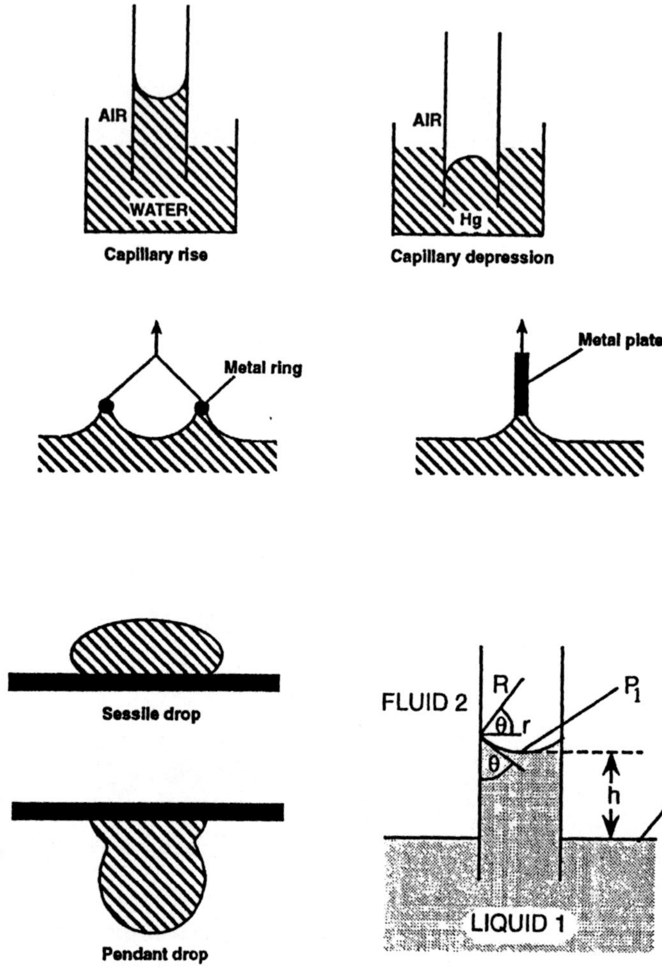

the hydrostatic equation $dP/dz = -\rho_2 g$ to fluid 2 with neglect of compressibility over the distance h). Thus we find

$$h = \frac{2\gamma \cos\theta}{(\rho_1 - \rho_2)gr}. \tag{7.1.2}$$

eq. (7.1.2) is valid for capillary rise ($0° \le \theta \le 90°$) or depression ($90° \le \theta \le 180°$). By determining the contact angle and h, one can use eq. (7.1.2) to estimate the interfacial tension between fluid phases.

Example

If the air–water surface tension is 72 dyn/cm and if $\theta \simeq 0$ for a particular glass capillary, how high would water rise in a 2 mm diameter capillary? $\cos\theta = 1$, $\rho_1 - \rho_2 \simeq 1$ g/cm^3, $g = 980$cm/s^2

$$\text{Ans}: \quad h = \frac{2 \times 72}{1 \times 980 \times 0.1} = 1.47\text{cm}.$$

The tension of an interface can be seen at work in the simple experiment depicted in Figure 7.2. If a drop of detergent solution is deposited on the surface of water it will quickly spread with a consequent reduction of the air–water surface tension. A loop of waxed string will float on a clear water surface in whatever configuration it is placed because the surface tension is the same on each side of the string. When a drop of detergent solution is put inside the loop it reduces the tension there with the result that the unbalanced external tension quickly pulls the loop into a circle.

Values of surface tensions (liquid–vapor) and interfacial tensions (liquid–liquid) are listed in Tables 7.1 and 7.2 for several systems. Water has a higher surface tension than most organic liquids at room temperature. Liquid metals because of electronic effects, have very high surface tensions at their boiling points. All surface and interfacial tensions approach zero as a critical point is approached.

Several simple empirical or theoretical formulas have beeen developed to estimate surface and interfacial tensions. Guggenheim (1945) proposed for the liquid–vapor surface tension the empirical expression

$$\gamma = \gamma_o \left(1 - \frac{T}{T_c}\right)^n \tag{7.1.3}$$

where γ_o and n are constants and T_c is the critical temperature. Good fits (2 or 3% error) to benzene and water data are achieved

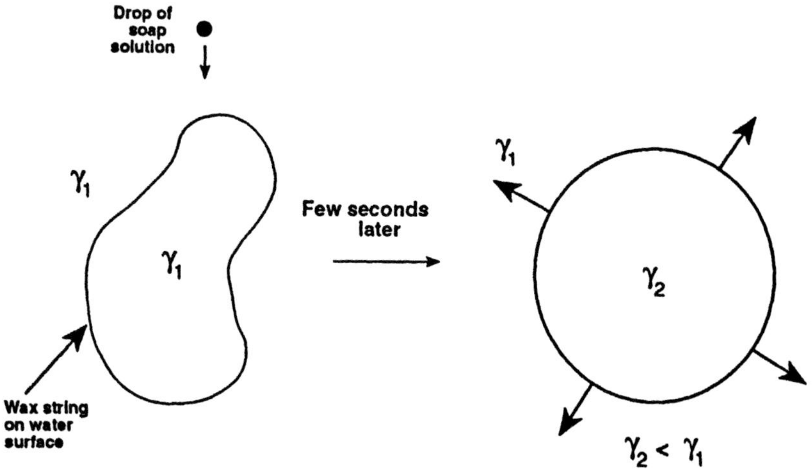

Figure 7.2
A simple experiment to observe interfacial tension.

TABLE 7.1 Observed Interfacial Tensions for Various Substances

	γ *(dyn/cm)*	*Temperature,* T *(°C)*
	Liquid–Vapor Interface	
Water	72.8	20
Methylene iodide	67.0	20
Dimethyl sulfoxide	43.5	20
Nitromethane	32.7	20
Benzene	28.9	20
Toluene	28.5	20
Methanol	22.5	20
Ethanol	22.4	20
n-Heptane	20.1	20
n-Octane	21.8	20
Ether	20.1	25
Carbon tetrachloride	27.0	20
Perfluoroheptane	13.2	20
Perfluoropentane	9.9	20
Chlorine	25.6	-30
Argon	11.9	-183
Sodium	198.0	130
Mercury	486.5	20
Tin	544.0	332
Copper	1300.0	1083
Iron	1880.0	1535
	Liquid–Liquid Interface	
Water/mercury	425.0	20
Water/hexane	51.1	20
Water/heptane	50.2	20
Water/octane	50.8	20
Water carbon tetrachloride	45.0	20
Water/benzene	35.0	20
Benzene/mercury	357.0	20
Water/*n*-octyl alcohol	8.5	20
Water/*n*-butyl alcohol	1.8	20

Adapted from Weast (1972–1973).

with $\gamma_o = 102.8$ dyn/cm, $n = 1.44$, and $\gamma_o = 136.9$ dyn/cm, $n = 1.09$, respectively. Sufficiently near the critical point $n \simeq 1.3$ for all fluids. This universality will be discussed in subsequent chapters.

Davis and Scriven (1976) used a lattice model to derive the formula

$$\gamma = \frac{1}{8n_l^{1/3}}\left(\frac{\partial U_l}{\partial V}\right)_{T,N} \tag{7.1.4}$$

for the surface tension of pure liquids at low vapor pressure. n_l and U_l are the density and internal energy of the liquid phase. For some 15 hydrocarbon and slightly polar liquids eq. (7.1.4) estimates the tensions to within 10 or 15% of the observed values.

TABLE 7.2 Observed Interfacial Tensions for Various Substances

T (K)	T/T_c	γ (dyn/cm)	T (K)	T/T_c	γ (dyn/cm)
	Argon[a]			Carbon Tetrachloride[b]	
84.5	0.56	16.10	293.2	0.53	29.95
99.6	0.66	10.91	373.2	0.67	17.26
114.7	0.76	6.60	473.2	0.85	6.53
129.7	0.86	2.99			
144.8	0.96	0.47			
150.1	0.995	0.028			
	Carbon Dioxide[a]			Benzene[b]	
273.2	0.80	10.46	269.8	0.48	40.1
252.3	0.83	8.06	320.4	0.57	30.7
264.6	0.87	5.86	371.0	0.66	22.0
273.7	0.90	3.90	421.5	0.75	14.15
282.9	0.93	2.21	472.1	0.84	7.41
289.0	0.95	1.49	522.7	0.93	2.23
292.0	0.96	0.86			
298.1	0.98	0.36			
303.2	0.997	0.03			
	Propane[a]			Water[b]	
185.0	0.50	19.08	265	0.41	77
203.5	0.55	16.28	295	0.46	73
222.0	0.60	13.59	323	0.50	68
244.2	0.66	11.08	373	0.58	59
262.7	0.71	8.64	400	0.62	54
284.9	0.77	6.41	460	0.71	40
			500	0.77	31
			590	0.91	10

[a] Adapted from Jasper (1972).

[b] Adapted from Washburn (1928).

Winterfeld et al. (1978) have found that at low vapor pressures, the liquid–vapor surface tension of many multicomponent liquids is approximated fairly well (to within 10 or 15%) by the formula

$$\gamma = \sum_{\alpha,\beta} \frac{x_\alpha x_\beta n^2}{n_\alpha^o n_\beta^o} \sqrt{\gamma_\alpha^o \gamma_\beta^o}, \tag{7.1.5}$$

where $1/n = \sum_\alpha x_\alpha / n_\gamma^o$, and x_α is the mole fraction of component α, γ^o is the surface tension of pure α, and n_α^o is the liquid density of pure α.

For an immiscible pair of liquids or a solid and a liquid, Girifalco and Good (1957) approximated the liquid–liquid interfacial tension by the equation

$$\gamma = \gamma_1^o + \gamma_2^o - 2\Phi\sqrt{\gamma_1^o \gamma_2^o}, \tag{7.1.6}$$

where

$$\Phi = \frac{4(n_1^o n_2^o)^{1/3}}{[(n_1^o)^{1/3} + (n_2^o)^{1/3}]^2} \tag{7.1.7}$$

and γ_α^o and n_α^o are the surface tension and density of pure liquid or solid α. For numerous nonpolmeric systems Φ lies between 0.5 and 1.2. Examples of the performance of eq. (7.1.6) are benzene and $(C_4F_9)O$ for which $\Phi = 0.97$, $\gamma_1^o = 28.9$ dyn/cm, $\gamma_2^o = 12.2$ dyn/cm, and n-heptane and $(C_2F_9)_3N$ for which $\Phi = 0.99$, $\gamma_1^o = 20.4$ dyn/cm, and $\gamma_2^o = 16.8$ dyn/cm. The liquid–liquid interfacial tensions predicted by eq. (7.1.6) for these systems are $\gamma = 4.7$ and 0.6 dyn/cm. Experimentally the corresponding tensions are 5.7 and 1.6 dyn/cm.

Fowkes (1964) improved upon the Girifalco and Good model by arguing that in strongly polar or metallic liquids the nonpolar dispersion forces must be treated differently from the hydrogen bonding or metallic forces. For example, for pure water and pure mercury he decomposes the surface tension as follows:

$$\gamma_{H_2O}^o = \gamma_{H_2O}^{o,h} + \gamma_{H_2O}^{o,d}; \quad \gamma_{Hg}^o = \gamma_{Hg}^{o,m} + \gamma_{Hg}^{o,d}, \tag{7.1.8}$$

where $\gamma^{o,d}$ is the contribution from nonpolar dispersion forces, $\gamma^{o,h}$ from hydrogen bonding and $\gamma^{o,m}$ from metallic forces.

Fowkes assumed that interaction between immiscible liquid phases 1 and 2 involve only dispersion forces and recommended that the liquid–liquid (or solid–liquid) interfacial tension be approximated by

$$\gamma = \gamma_1^o + \gamma_2^o - 2\sqrt{\gamma_1^{o,d}\gamma_2^{o,d}}. \tag{7.1.9}$$

For a nonpolar fluid, $\gamma_2^o = \gamma_2^{o,d}$ and so, according to the Fowkes theory, one can use eq. (7.1.9) and measurements of the liquid–liquid interfacial tension of a metallic or hydrogen bonded liquid against a nonpolar liquid to estimate $\gamma^{o,d}$ for the metallic or hydrogen bonded liquid. Using hydrocarbons at 20°C against mercury, Fowkes found

$$\gamma_{Hg}^{o,d} \approx 200 \text{dyn/cm}. \tag{7.1.10}$$

Hydrocarbons against water yields

$$\gamma_{H_2O}^{o,d} \approx 21.8 \text{dyn/cm}. \tag{7.1.11}$$

As a test of the theory, eq. (7.1.9) can be used to predict [with the aid of the values given at eqs. (7.1.10) and (7.1.11)] the interfacial tension of water against mercury. The prediction is $\gamma = 427$ dyn/cm compared to the experimental value 426 dyn/cm.

The theoretical basis for eqs. (7.1.4)–(7.1.6) will be discussed in Chapter 8.

7.2 The Young–Laplace and Kelvin Equations

An important general consequence of interfacial tension can be deduced from the capillary tube experiment: *there is a pressure jump* ΔP *across a curved interface.* Consider the case of capillary rise in Figure 7.1. The pressure P_1' just below the meniscus (interface) is related to P_2 by

$$P_1' = P_2 - \rho_1 g h \tag{7.2.1}$$

according to the equation hydrostatics. Combining this result with eq. (7.1.1) and defining $\Delta P = P_1 - P_1'$, we obtain

$$\Delta P = \frac{2\gamma}{R} \tag{7.2.2}$$

where $R = r/\cos\theta$. Equations (7.2.2) and (7.2.3) are special cases of the Young–Laplace equation. From this equation it follows that the pressure is greater on the concave side of the interface: $P_1 \geq P_1'$ for capillary rise and $P_1' \geq P_1$ for capillary depression.

Because we are considering here capillary tubes small enough that the meniscus is a hemisphere (how small this must be will be considered later), we can already conclude that the pressure in a spherical drop or bubble is greater than that outside the drop or bubble. It is instructive, however, to derive the YL equation for a spherical droplet by a different argument. For this we consider a droplet of a radius R small enough to neglect the effect of gravity on it. The pressure in the bulk phase inside the droplet is P_I and that in the bulk phase outside the droplet is P_0 as shown in Figure 7.3. The desired equation comes from a force balance on the top hemisphere of the droplet. The downward force exerted on the top of the hemisphere by the external phase is $P_0 \pi R^2$, the upward force exerted on the bottom of the hemisphere is $P_I \pi R^2$. The downward force of tension exerted on the perimeter of the hemisphere

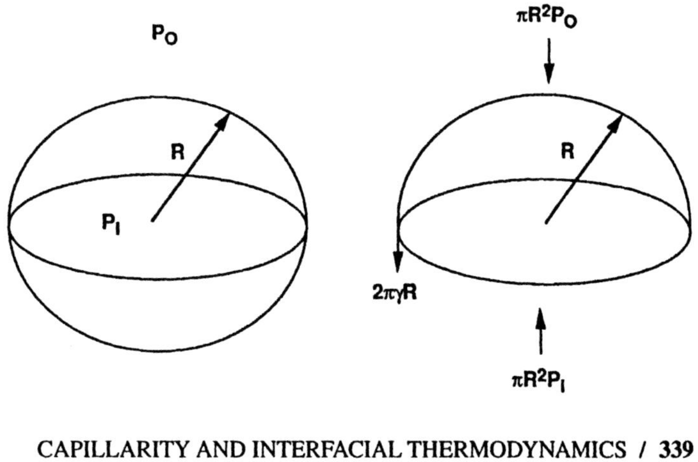

Figure 7.3
Spherical drop or bubble.

is $2\pi R\gamma$. The balance of these forces, $P_0\pi R^2 + 2\pi R\gamma = P_I\pi R^2$, yields again the YL equation

$$\Delta P = \frac{2\gamma}{R} \qquad (7.2.3)$$

Example
Two soap bubbles of radius $R_1 = 20\mu m$ and $R_2 = 200\mu m$ are joined by a thin tube. The pressure in bubble 1 is greater than that in bubble 2 by the amount $2\gamma(1/R_1 - 1/R_2)$. Consequently the small bubble collapses and the large bubble grows. Similarly, small droplets or bubbles in emulsions or foams tend to feed large droplets or bubbles because of the capillary pressure differences.

for the pressure jump $\Delta P = P_I - P_0$ across a spherical meniscus. Equation (7.2.4) is equally valid for bubbles or droplets.

A similar force balance on a long cylindrical droplet or bubble yields $2L\gamma = 2RL\Delta P$, or

$$\Delta P = \frac{\gamma}{R}, \qquad (7.2.4)$$

where R and L are the radius and length of the cylinder.

The results for the sphere and cylinder are special cases of the YL equation. In general any point on a surface possesses two principal radii of curvature R_1 and R_2. The radius R_i is the distance away from the point on the surface that the center of a tangent circle of the same curvature as the surface must be placed. Depending on the shape of the surface R_1 and R_2 may or may not have the same signs (see Figure 7.4). A force balance on an element of surface having arbitrary curvature yields the general YL equation

$$\Delta P = 2H\gamma \qquad (7.2.5)$$

where H is the mean curvature defined by

$$H \equiv \frac{1}{2}\left(\frac{1}{R_1} + \frac{1}{R_2}\right). \qquad (7.2.6)$$

A summary of special cases is shown in Figure 7.5.

Figure 7.4
Elements of a surface whose principal radii of curvature have (a) the same sign and (b) opposite signs.

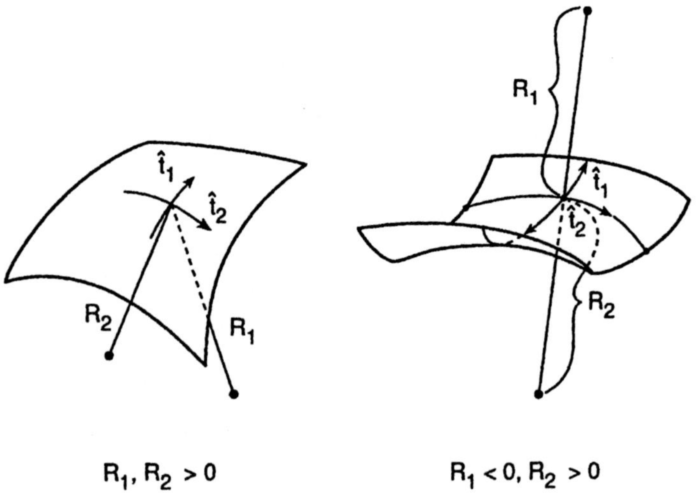

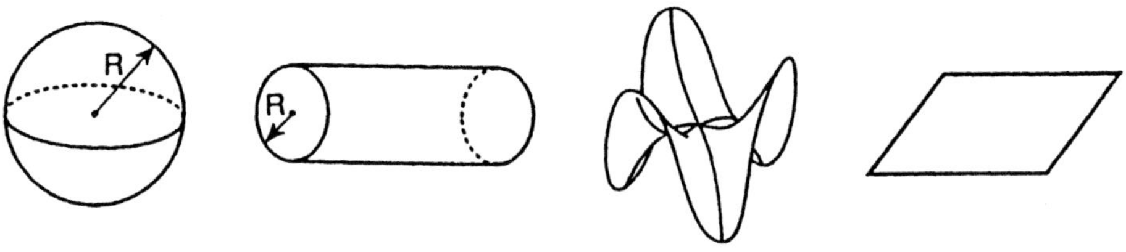

Sphere $R_1 = R_2 = R$ $\Delta P = 2\gamma/R$

Cylinder $R_1 = R$, $R_2 = \infty$ $\Delta P = \gamma/R$

Monkey Saddle $R_1 = -R_2$ $\Delta P = 0$

Plane $R_1 = R_2 = \infty$ $\Delta P = 0$

Figure 7.5
The YL equation for some highly symmetric surfaces.

Example

a) What is the pressure in a 1000 Å air bubble in water at 20°C and 1 atm pressure?

Ans: $\gamma = 72$dyn/cm, $\Delta P = 2\gamma/R = 14.4$atm
and so $P_l = 15.4$atm

b) What is the pressure in a 1 μm drop of water in air at 20°C and 1 atm pressure?

Ans: $\Delta P = 1.44$atm and so $P_l = 2.44$atm.

The Young-Laplace (YL) equation is derived from a force balance and as such represents mechanical equilibrium. It is valid for the interface between any two coexisting fluid phases, pure or multicomponent. Additional conditions of thermodynamic equilibrium are that the chemical potential μ_i of species i is the same in every phase. A very useful formula, known as Kelvin's equation, can be derived for the one-component liquid–vapor interface by combining the YL equation with the condition of chemical equilibrium.

Consider a liquid phase separated from a vapor phase by an interface of mean curvature H. The chemical potential μ_l and the pressure P_l in the liquid phase obey the Gibbs–Duhem equation for a constant temperature (T) process

$$d\mu_l = \frac{1}{n_l} dP_l, \tag{7.2.7}$$

where n_l is the density of the liquid. Assume that the liquid is incompressible and integrate eq. (7.2.7) from P_{sat} to P_l to obtain

$$\mu_l(T, P_l) = \mu_l(T, P_{sat}) + \frac{1}{n_l}(P_l - P_{sat}). \tag{7.2.8}$$

By P_{sat} we denote the saturated vapor pressure, which is the coexistence pressure of a liquid at equilibrium with its vapor across a

Figure 7.6
Liquid and vapor phases coexisting across an interface of mean curvature H.

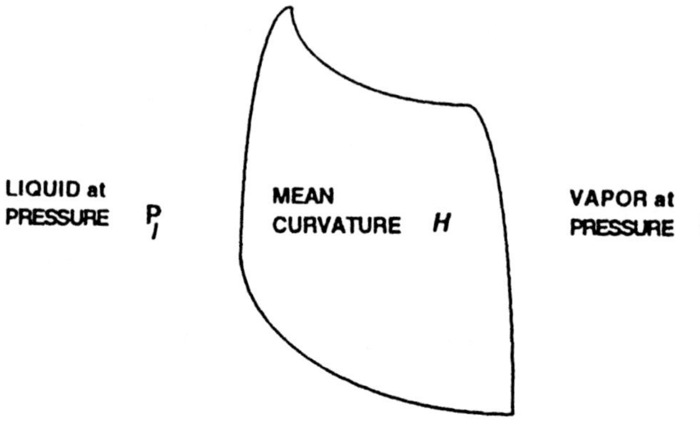

planar interface (for which $H = 0$). We make the further assumption that the vapor behaves as an ideal gas. Then, beecause the liquid is in chemical equilibrium with the vapor at pressure P,

$$\mu_l(T, P) = \mu_v(T, P) = \mu^+(T) + kT \ln P, \tag{7.2.9}$$

where $\mu^+(T)$ is the chemical potential of the ideal gas in the standard state. The condition of chemical equilibrium of a liquid and vapor at a planar interface is

$$\mu_l(T, P_{sat}) = \mu_v(T, P_{sat}) = \mu^+(T) + kT \ln P_{sat}, \tag{7.2.10}$$

Equations (7.2.9)–(7.2.10) combine to yield

$$kT \ln \frac{P}{P_{sat}} = \frac{1}{n_l}(P_l - P_{sat}). \tag{7.2.11}$$

According to the YL equation $P_l = P + 2H\gamma$, and so P_l can be eliminated from eq. (7.2.11) to obtain

$$kT \ln \frac{P}{P_{sat}} = \frac{2H\gamma}{n_l} + \frac{P - P_{sat}}{n_l}. \tag{7.2.12}$$

In most practical situations $(P - P_{sat})/n_l kT \ll 1$ and $2H\gamma/n_l kT < 1$. Under these conditions the second term on the right-hand side (rhs) of eq. (7.2.12) can be neglected to obtain *Kelvin's equation*

$$P = P_{sat} \exp\left[\frac{2H\gamma}{n_l kT}\right], \tag{7.2.13}$$

an important equation of surface chemistry.

For a liquid drop $2H = 2/R$ because the liquid is on the concave side of the meniscus. Thus, it follows from Kelvin's equation, $P = P_{sat} \exp(2\gamma/n_l kTR)$, that the vapor pressure of a liquid drop is greater than the saturated vapor pressure. On the other hand, for a bubble $2H = -2/R$, because the liquid is on the convex side of the meniscus and so Kelvin's equation, $P = P_{sat} \exp(-2\gamma/n_l kTR)$, implies the vapor pressure in a bubble is less than the saturated vapor pressure.

Example

Calculate P from (a) eq. (7.2.13) and (b) eq. (7.2.14) for a spherical liquid drop of radius $R = 100$ Å at $T = 298$ K, $n_l = 2 \times 10^{22}$ molecules/cm^3, $\gamma = 50$ dyn/cm, and $P_{sat} = 10^6$ dyn/cm.

Ans: For a drop $2H = 2/R$. (a) P = 1.12946×10^6 dyn/cm^2 and (b) $P = 1.12928 \times 10^6$ dyn/cm^2, less than 0.02% difference.

Example

For water at 300K, P/P_{sat} is about 1.001 if $R = 10^4$ Å and is about 1.114 if $R = 100$ Å. $\Delta P = P_l - P$ is about 1.4 atm if $R = 10^4$ Å and is about 140 atm if $R = 100$ Å. One wonders if the high pressure inside small liquid drops could cause solidification.

Exercise

The pore radius distribution for a porous silica solid was found to be

$$\hat{v}(R) = \frac{2R}{\sigma^2} e^{-R^2/\sigma^2}$$

where $\sigma = 25$Å. The saturated vapor pressure of water between 273.15 and 373.15 K can be estimated to within a few percent from the expression $\log P = 9.0725 - 2262/T$ where P is in kPa and T in K. Consider water at 25° K where the surface tension is 72 dyn/cm and the liquid density is 1 gm/cm^3. Plot the liquid water saturation in the pore space versus pressure of water vapor if the contact angle of water is 0°.

An important application of Kelvin's equation is the determination of the distribution of pore sizes in a porous solid. Consider a liquid in a cylindrical pore as illustrated in Figure 7.7. If the liquid has a contact angle $\theta = 0°$, then the liquid–vapor meniscus is a hemisphere of radius R. Because the liquid is on the convex side of the meniscus, $2H = -2/R$, and so according to Kelvin's equation, the pressure $P(R)$ of vapor in equilibrium with the liquid in the pore of radius R is given by $P(R) = P_{sat} \exp[-2\gamma/n_l kTR]$. Thus, if a porous sample having many pores of different radii is placed in an environment with vapor pressure controlled at $P(R)$, then liquid will condense in all of the pores with radius less than or equal to R. By measuring the mass M_l of the condensed liquid, we can calculate the total volume $V(R)$ of pores having radius less than or equal to R: $V(R) = M_l/\rho_l$, where ρ_l is the mass density of the liquid. The radius R is computed from the Kelvin formula

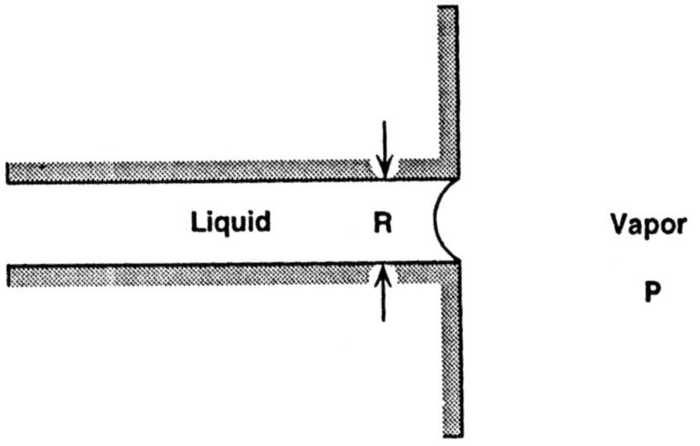

Figure 7.7
Capillary condensation in cylindrical pore of radius R. Contact angle $\theta = 0°$.

Figure 7.8
Schematic of the volume distribution of pores in a disordered porous medium.

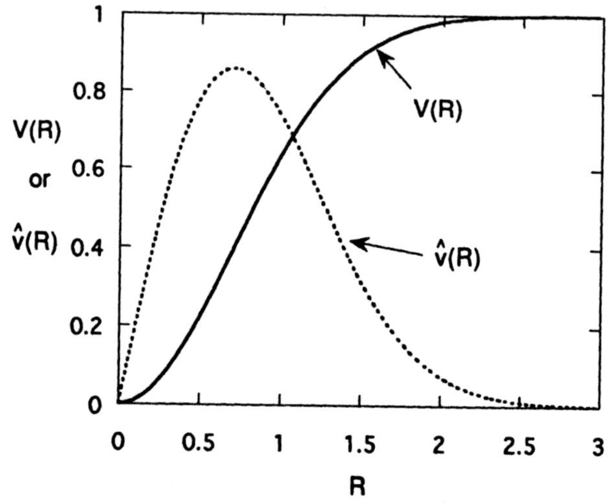

$R = 2\gamma/\{n_l kT \ln[P_{sat}/P(R)]\}$. By varying the pressure of the vapor, we can calculate $V(R)$ versus R. From this we obtain the volume distribution of porosity, defined by

$$\hat{v}(R) = \frac{dV(R)}{dR}. \tag{7.2.14}$$

$\hat{v}(R)dR$ is the volume of pores in the sample having a radius between R and $R + dR$. The distribution of porosity is needed for understanding catalytic activity, separations, and multiphase displacement processes in porous media. Data typical of a disordered porous solid are illustrated in Figure 7.8.

7.3 Capillary Hydrostatics

The shape of the meniscus between two fluid phases will depend not only on the interfacial tension but also on the presence of body forces such as gravitational, centrifugal, electrical, or magnetic forces. For example, consider the drop in Figure 7.9 that is sitting on a flat solid (xy plane) in the presence of gravity (z direction). The height $h(x, y)$ of the meniscus above the solid varies with position in the xy plane. In each of the fluid phases the pressure obeys the equation of hydrostatics, that is,

$$\nabla P_i = \rho_i \mathbf{g}, \tag{7.3.1}$$

where ∇P_i is the gradient of P_i. Because $\mathbf{g} = -g\hat{\mathbf{k}}$, $\hat{\mathbf{k}}$ a unit vector along the z axis, eq. (7.3.1) reduces to the scalar equation $dP_i/dz = -\rho_i g$ that can be integrated to give

$$P_i(z) - P_i(z_o) = -\rho_i g(z - z_o), \quad i = 1, 2, \tag{7.3.2}$$

if the effect of gravity on density is ignored.

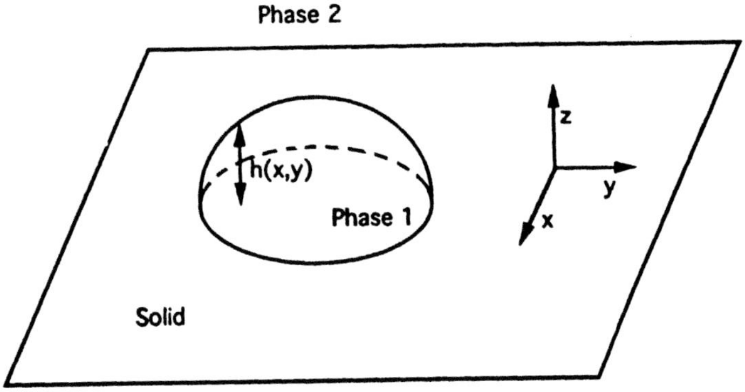

Figure 7.9
Droplet of phase 1 sitting on a flat solid in the presence of gravity.

The difference between the pressure $P_1(z = h(x, y) - \varepsilon)$ just below the meniscus and the pressure $P_2(z = h(x, y) + \varepsilon)$ just above the meniscus is equal to $2H\gamma$ according to the YL equation. Combining this jump condition with eq. (7.3.2) evaluated at $z = h(x, y)$, we obtain the equation of capillary hydrostatics,

$$2\gamma H = -\Delta\rho g h(x, y) + \Delta P_o, \tag{7.3.3}$$

for the meniscus of the drop in the presence of gravity. We have introduced the notation $\Delta\rho \equiv \rho_1 - \rho_2$. Because $\Delta P_o \equiv P_1(z_o) - P_2(z_o)$ is a fixed reference pressure difference, the fact that meniscus height depends on position x, y implies that the mean curvature also depends on position x, y. The basic problem of capillary hydrostatics is to determine how h depends on x, y. Equation (7.3.3) is not restricted to droplets on flat solids. The hydrostatic pressure in any fluid phase in the presence of gravity can be expressed in the form given at eq. (7.3.2), where $h(x, y)$ is the height of the meniscus from an xy plane chosen at some elevation z_o, and so with the YL jump condition $P_1(h(x, y) + \varepsilon) - P_2(h(x, y) - \varepsilon) = 2\gamma H$ one finds eq. (7.3.3) for the general case.

To investigate how the equation of capillary hydrostatics determines the shape of the meniscus let us first consider a meniscus with translational symmetry, that is, h varies in only the x direction as illustrated in Figure 7.10. The vectors $\mathbf{x}$ and $\mathbf{x} + d\mathbf{x}$ denote two

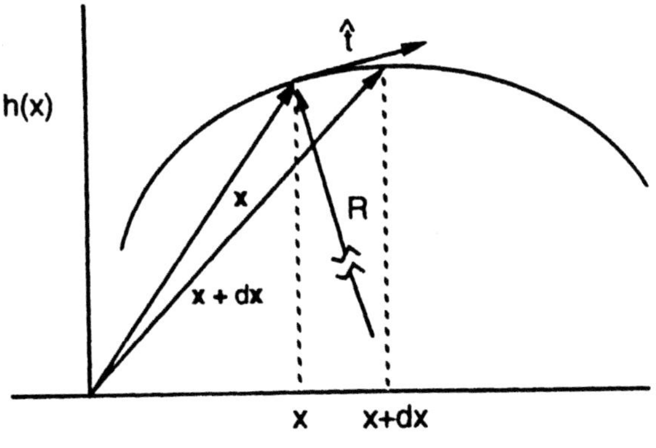

Figure 7.10
The height $h(x)$ of a translationally symmetric meniscus as a function of position in the x direction.

points on the meniscus curve separated by the infinitesimal $d\mathbf{x}$. The arc length between the two points is $ds = \sqrt{(d\mathbf{x})^2}$. The unit tangent vector $\hat{t}$ at $\mathbf{x}$ is given by

$$\hat{t} = \frac{d\mathbf{x}}{ds} \tag{7.3.4}$$

according to the elementary differential geometry of a curve. The radius of curvature R at the point $\mathbf{x}$ can be computed by imagining that the point $\mathbf{x} + d\mathbf{x}$ is reached from the point $\mathbf{x}$ by an infinitesimal rotation $d\theta$ about the radius of curvature R. In this case $ds = |R| d\theta$. Because the tangent vector $\hat{t}$ is orthogonal to the direction of the radius of curvature R, it follows that the angle between $\hat{t}$ and $\hat{t} + d\hat{t}$ is also $d\theta$. Thus, $|d\hat{t}| = |\hat{t}| d\theta = d\theta$, or

$$\frac{1}{|R|} = \left| \frac{d\hat{t}}{ds} \right|, \tag{7.3.5}$$

another general result from elementary differential geometry.

For a translationally symmetric meniscus $d\mathbf{x} = dx\hat{\imath} + dh\hat{\jmath}$, $\hat{\imath}$ and $\hat{\jmath}$ being Cartesian unit vectors, and so

$$ds = \sqrt{(dx)^2 + (dh)^2} = \sqrt{1 + h_x^2}\,dx, \tag{7.3.6}$$

where $h_x \equiv dh/dx$. From this and eq. (7.3.4) it follows that

$$\hat{t} = \frac{\hat{\imath} + h_x \hat{\jmath}}{\sqrt{1 + h_x^2}}. \tag{7.3.7}$$

Because

$$d\hat{t}/dx = [-h_x\hat{\imath} + \hat{\jmath}]h_{xx}/(1 + h_x^2)^{3/2} \quad \text{and} \quad d\hat{t}/ds = d\hat{t}/dx\sqrt{1 + h_x^2}, \tag{7.3.8}$$

we find from eq. (7.3.5) that

$$\frac{1}{R} = \pm \frac{h_{xx}}{[1 + h_x^2]^{3/2}}, \tag{7.3.9}$$

where $h_{xx} = d^2h/dx^2$. Thus, because for a translationally symmetric meniscus $2H = 1/R$, the above equation of capillary hydrostatics becomes

$$-\frac{\gamma h_{xx}}{[1 + h_x^2]^{3/2}} = -\Delta\rho g h + \Delta P_0, \tag{7.3.10}$$

a nonlinear second-order differential equation for the meniscus position h. We have chosen the negative sign for the curvature sign so that the curvature R will be positive when phase 1 lies on the concave side of the meniscus.

Multiplying eq. (7.3.9) by dh/dx, we obtain

$$\gamma \frac{d}{dx}\left[1 + \left(\frac{dh}{dx}\right)^2\right]^{-1/2} = \frac{d\psi}{dx}, \tag{7.3.11}$$

where

$$\psi(h) \equiv -\frac{1}{2}\Delta\rho g h^2 + h\Delta P_0. \qquad (7.3.12)$$

Equation (7.3.10) can be integrated and rearranged to yield

$$\frac{dh}{[\gamma^2/(\psi + K)^2 - 1]^{1/2}} = \pm dx, \qquad (7.3.13)$$

where K is the constant of integration and the sign on the rhs will be determined by the physical situation. To gain some insight into the behavior of a translationally symmetric meniscus in the presence of gravity, consider the two situations depicted in Figure 7.11. A common laboratory situation approximating Figure 7.11a is water in a beaker whose radius is large compared to the distance away form the beaker at which the meniscus becomes approximately flat. Figure 7.11b can be created easily by dipping in water a pair of parallel microscope slides held a small fixed distance $2r$ apart.

For Figure 7.11a, $\Delta P_0 = 0$ because the meniscus is flat far from the wall. We can choose a coordinate system such that $h(x = \infty) = 0$. This boundary condition yields $K = \gamma$. The other boundary condition is that the meniscus makes a contact angle θ at $x = 0$, that is, $[dh/dx]_{x=0} = -\cot\theta$. Inserted into eq. (7.3.12), this boundary condition gives

$$h(x = 0) = \text{sgn}(90° - \theta)\left(\frac{2\gamma}{\Delta\rho g}\right)^{1/2}[1 - (1 + \cot^2\theta)^{-1/2}]^{1/2}, \qquad (7.3.14)$$

where $\text{sgn}x = 1$ when $x > 0$ and $\text{sgn}x = -1$ when $x < 0$. Thus, the positive sign in the case of capillary rise and the negative sign is chosen in the case of capillary depression. The meniscus height $h(x)$ in Figure 7.11 a can then be computed from

$$\int_{h(x)}^{h(0)} \frac{dh}{[\gamma^2/(\gamma - \Delta\rho g h^2/2)^2 - 1]^{1/2}} = x\ \text{sgn}(90° - \theta). \qquad (7.3.15)$$

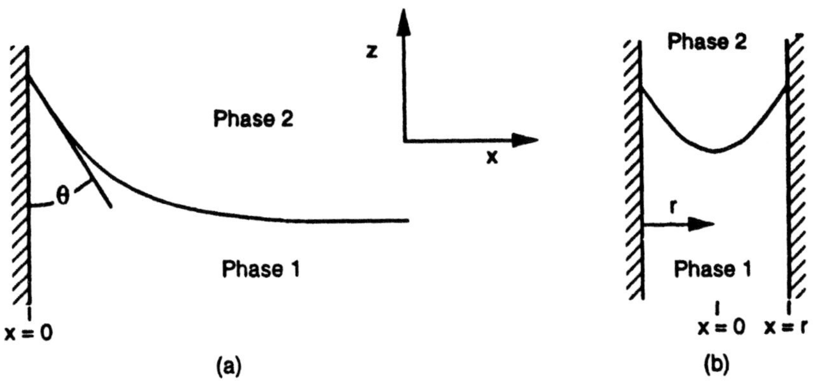

Figure 7.11
(a) Fluids against a flat slab and (b) fluids contained between slabs separated by the distance $2r$.

From eq. (7.3.13), we see that the maximum capillary rise ($\theta = 0°$) is $h(0) = (2\gamma/\Delta\rho g)^{1/2}$ and the minimum capillary depression ($\theta = 180°$) is $h(0) = -(2\gamma/\Delta\rho g)^{1/2}$. Also, because the only characteristic length in eq. (7.3.14) is

$$x_c = (\gamma/\Delta\rho g)^{1/2}, \tag{7.3.16}$$

it follows that the meniscus height $h(x)$ approaches zero on a length scale of the order of x_c. This is easy to see when θ lies between 55° and 125°. In this situation, eq. (7.3.9) can be linearized, that is, $h_{xx}/(1+h_x^2)^{1/2} \simeq h_{xx}$, and the solution to the linear equation is

$$h(x) = x_c \cot\theta e^{-x/x_c}. \tag{7.3.17}$$

The meniscus height for this range of θ's thus goes to zero exponentially with increasing x/x_c.

Example

a) What are the values of x_c for the air–water and air–mercury interfaces at 20°C? If $\theta = 55°$ for water and 125° C for mercury against a particular solid what are the maximum capillary rise and depression for the air–liquid interfaces.
Ans: $x_{c,H_2O} = \sqrt{\frac{72}{1\times 980}} = 0.27\,\text{cm}, x_{c,Hg} = \sqrt{\frac{480}{13\times 980}} = 0.19$ cm.

b) From exact eq. (7.3.13): $h_{H_2O}(0) = 0.165$ cm and $h_{Hg}(0) = -0.11$ cm. From linear eq. (1.3.17): $h_{H_2O}(0) = 0.19$ cm and $h_{Hg}(0) = -0.13$ cm

From these sample calculations it follows that the meniscus of fluids such as water or mercury would be flat in a typical laboratory beaker except in a region a few millimeters from the beaker wall. The extent of this region is determined by the characteristic length x_c.

In a narrow pore typified in Figure 7.11b, the meniscus will be curved everywhere and so ΔP_0 is not zero. In this case it is convenient to choose $x = 0$ at the midpoint plane between the slabs and to place the datum for the z axis such that $h(0) = 0$. The boundary conditions for this problem are

$$\frac{dh}{dx} = 0 \quad \text{at} \quad x = 0 \quad \text{and} \quad \frac{dh}{dx} = \cot\theta \quad at \quad x = r. \tag{7.3.18}$$

From the boundary conditions it follows that $K = \gamma$ and

$$h(x = r) = \frac{\Delta P_0}{\Delta\rho g}\left\{1 \pm \text{sgn}(\theta - 90°)\left[1 + \frac{2\Delta\rho g\gamma}{(\Delta P_0)^2}[1 - (1 + \cot^2\theta)^{-1/2}]\right]\right\}^{1/2}. \tag{7.3.19}$$

The meniscus height for this case is given by

$$\int_0^{h(x)} \frac{dh}{\{\gamma^2/[\gamma + h\Delta P_0 - \Delta\rho g h^2/2]^2 - 1\}^{1/2}} = x\,\mathrm{sgn}(90^\circ - \theta). \tag{7.3.20}$$

Evaluation of eq. (7.3.20) is not straightforward because the "capillary pressure" ΔP_0 is not known. What one has to do is to solve eq. (7.3.20) for $x = r$ simultaneously with eq. (7.3.19). This procedure, for which some iterative scheme must be used, will give $h(x)$ and ΔP_0 for given γ, $\Delta\rho$ and r.

For a simpler formula for $h(x)$, let us again assume $55^\circ < \theta < 125^\circ$ so that eq. (7.3.9) can be linearized. The solution to the linear equation for the boundary condition at eq. (7.3.17) is

$$h = \frac{\Delta P_0}{\Delta\rho g} + \frac{x_c \cot\theta}{e^{r/x_c} - e^{-r/x_c}}\left(e^{x/x_c} + e^{-x/x_c}\right). \tag{7.3.21}$$

When the datum for h is chosen to be $h(0) = 0$, then the capillary pressure is found from eq. (7.3.20) to be

$$\Delta P_0 = -\frac{2\Delta\rho g x_c \cot\theta}{e^{r/x_c} - e^{-r/x_c}} \tag{7.3.22}$$

and so

$$h = \frac{x_c \cot\theta}{e^{r/x_c} - e^{-r/x_c}}[e^{x/x_c} + e^{-x/x_c} - 2]. \tag{7.3.23}$$

Equation (7.3.22) exposes for the special case of translationally symmetric meniscus a general property of the meniscus shape in small capillaries: if the width or diameter of the capillary is sufficiently small that $r \ll x_c$, then the mean curvature of the mensicus is approximately constant and is independent of gravity. For a linear approximation to the case considered here $R^{-1} = -d^2h/dx^2 = -(\cot\theta/r)[1 + O(r/x_c)]$ and $\Delta P_0 = -\gamma\cot\theta/r$, and so as expected we recovered the YL equation $\Delta P_0 = \gamma/R$ for a cylindrical meniscus in the absence of gravity. The physical origin of this result is that in a narrow capillary of width or diameter r the vertical extent of the mensicus, $h(r) - h(0)$, will be of the order of r and if this is sufficiently small, gravity will have a negligible effect on the shape of the meniscus. The general result can be seen at eq. (7.3.3). In a sufficiently small capillary, the capillary pressure jump ΔP_0 across the curved meniscus becomes large compared to $h_{\max} - h_{\min}$ and so the gravitational term in eq. (7.3.3) becomes negligible compared to ΔP_0, the result being that eq. (7.3.3) reduces to the YL equation $2\gamma H = \Delta P_0$ over the entire meniscus.

Let us close this section with an exploration of menisci on which the effect of gravity is negligible. For the translationally symmetric meniscus, the YL equation is

$$-\frac{1}{[1 + (dh/dx)^2]^{3/2}}\frac{d^2h}{dx^2} = \frac{\Delta P_0}{\gamma}. \tag{7.3.24}$$

Integrating this equation by the same trick used above, solving for dh/dx, and integrating again we obtain

$$h = b + \sqrt{\left(\frac{\gamma}{\Delta P_0}\right)^2 - (x-a)^2}, \tag{7.3.25}$$

where a, b, and ΔP_0 must be determined by boundary conditions and the choice of the position x_o at which h is zero.

From eq. (7.3.23) it follows that with the neglect of gravity a translationally symmetric meniscus is always a piece of a cylinder [centered at (a, b)]. In the case of a free standing cylinder of fluid of radius R, we know from the YL equation that $R^2 = h^2 + x^2$, and so $a = b$ and $\Delta P_0 = \gamma/R$. For a meniscus lying between two parallel slabs and having a contact angle θ_1, at the left wall and θ_2 at the right wall, the boundary conditions are

$$\left.\frac{dh}{dx}\right|_{x=0} = -\cot\theta_1 = \frac{a}{\sqrt{(\gamma/\Delta P_0)^2 - a^2}} \tag{7.3.26}$$

and

$$\left.\frac{dh}{dx}\right|_{x=2r} = \cot\theta_2 = \frac{-(2r-a)}{\sqrt{(\gamma/\Delta P_0)^2 - (2r-a)^2}} \tag{7.3.27}$$

where $2r$ is the separation between slabs.

A third condition is that $h = 0$ at $x = r$, or

$$b = -\sqrt{\left(\frac{\gamma}{\Delta P_0}\right)^2 - (r-a)^2}. \tag{7.3.28}$$

Simultaneous solution of eqs. (7.3.25)–(7.3.27) yields

$$a = \frac{2r\cos\theta_1}{\cos\theta_1 + \cos\theta_2}, \quad b = \frac{a}{\cos\theta_1}\sqrt{1 - \left(\frac{\cos\theta_1 - \cos\theta_2}{2}\right)^2}, \qquad \frac{\Delta P_0}{\gamma} = \frac{\cos\theta_1 + \cos\theta_2}{2r}. \tag{7.3.29}$$

Because $\Delta P_0/\gamma = 2H$, we see from eq. (7.3.28) that the radius of curvature of the cylinder of which the meniscus is a piece

$$R = 2r/(\cos\theta_1 + \cos\theta_2). \tag{7.3.30}$$

If $\theta_1 = \theta_2$, $R = r/cos\theta_1$, a result mentioned earlier.

Several special cases are illustrated in Figure 7.12. For $\theta_1 = \theta_2 = 0°$ or $180°$ the meniscus is a hemicylinder of radius of curvature $R = r$ or $-r$. The curvature is zero (or $R = \pm\infty$) in the limit $\theta_1 = \theta_2 \pm 90°$, that is, the meniscus is a piece of an inclined plane.

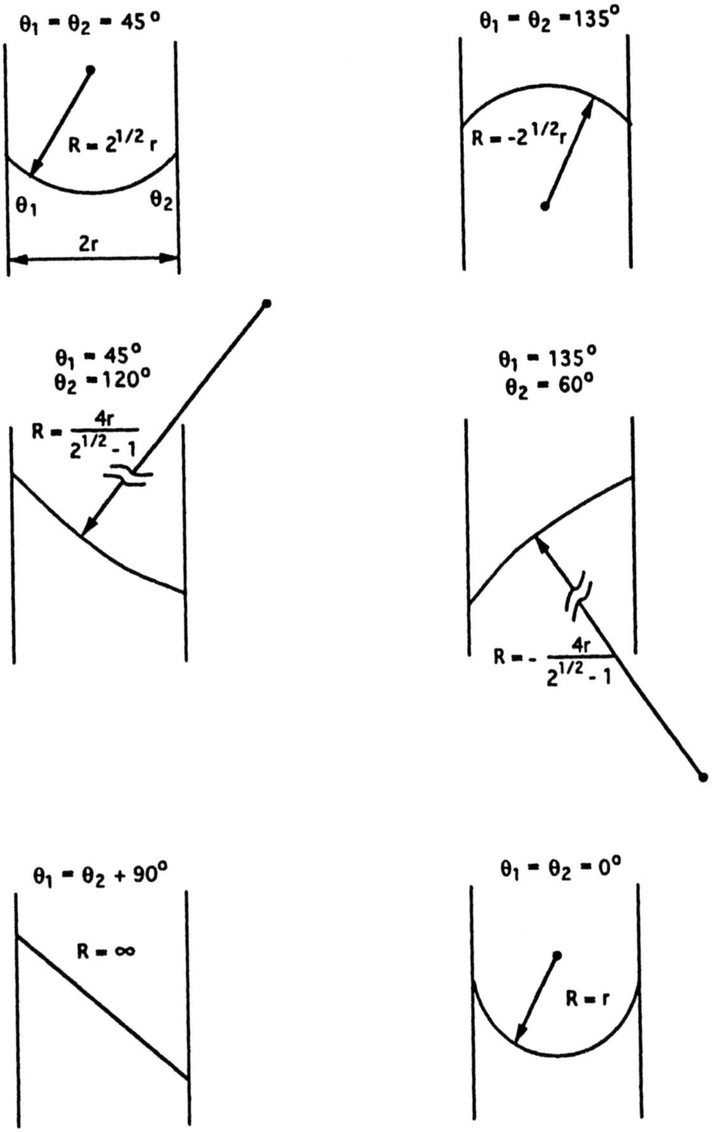

Figure 7.12
Shapes of menisci between two parallel slabs for various contact angles when gravity is negligible.

The relationship between the mean curvature of an arbitrary surface and its height $h(x, y)$ above some xy plane is

$$2H = -\frac{(1 + h_y^2)h_{xx} - 2h_x h_y h_{xy} + (1 + h_x^2)h_{yy}}{(1 + h_x^2 + h_y^2)^{3/2}} \quad (7.3.31)$$

where $h_x \equiv \partial h/\partial x$, $h_{xy} = \partial^2 h/\partial x \partial y$, etc. The derivation of this equation is rather tedious and will not be reproduced here. It can be found in texts on differential geometry. Insertion of eq. (7.3.33) into the equation of capillary hydrostatics, eq. (7.3.3), yields a second-order nonlinear partial differential equation for $h(x, y)$.

In the case of an axisymmetric meniscus, that is, $h = h(\rho)$, $\rho = \sqrt{x^2 + y^2}$, q. (7.3.31) can be simplified [using $\partial h/\partial x =$

$(\partial\rho/\partial x)dh/d\rho = (x/\rho)dh/d\rho$, $\partial h/\partial y = (y/\rho)dh/d\rho$, etc.] to

$$2H = -\left[\frac{h_{\rho\rho}}{(1+h_\rho^2)^{3/2}} + \frac{h_\rho}{\rho(1+h_\rho^2)^{1/2}}\right], \qquad (7.3.32)$$

where $h_\rho \equiv dh/d\rho$ and $h_{\rho\rho} \equiv d^2h/d\rho^2$, and eq. (7.3.3) becomes

$$-\left[\frac{h_{\rho\rho}}{(1+h_\rho^2)^{3/2}} + \frac{h_\rho}{\rho(1+h_\rho^2)^{1/2}}\right] = -\frac{\Delta\rho g h}{\gamma} + \frac{\Delta P_0}{\gamma}. \qquad (7.3.33)$$

With appropriate boundary conditions (which must be axisymmetric) this equation yields the meniscus shape for interfaces in vertical cylindrical capillary tubes, drops sitting on or hanging from solids flat in the xy plane or sitting on or hanging from capillary tubes.

When the net vertical extent of a meniscus is small compared to x_c, the gravitational term, $\Delta\rho g h/\gamma$, in eq. (7.3.33) is negligible. In this case the axisymmetric equations admits the solution $h^2 + \rho^2 = R^2$. Thus, the meniscus is a piece of sphere of radius of curvature R and, of course, $\Delta P_0 = 2\gamma/R$. A few examples of such a solution are shown in Fig. 7.13 for a drop of a fixed volume but on solids with which it has different contact angles.

7.4 Interfacial Thermodynamics

This section will be restricted to planar interfaces. Its purpose is to set the stage for later developing a molecular theory of interfacial structure and stress. For interfaces whose radii of curvature are large in magnitude compared to interfacial thickness (of order of 10–100 Å usually) the macroscopic theory of capillarity developed in the preceding two sections suffices to handle curved interfaces.

To construct the thermodynamics of planar fluid interfaces, let us consider the system shown in Figure 7.14. The system is a closed, two-phase system with a planar interface of area A normal to the direction of the z axis. Except near critical or consolute points, the interfacial region (indicated by a straight line in the figure) is narrow in thickness (i.e., in extent along the z axis) compared to the thicknesses of phases 1 and 2. The system can contain an arbitrary

Figure 7.13
Contact angle convention. Sessile droplets of fixed volume V_D but different contact angles when gravity is negligible. The radius R of curvature for a piece of sphere making a contact angle θ with a flat solid surface is $R = \{3V_D/[2\pi(1+\cos\theta)]\}^{1/3}$. Thus in the contact angle range $0° \geq \theta \geq 90°$, the droplet radius of curvature is in the range $(3V_D/4\pi)^{1/3} \leq R \leq (3V_D/2\pi)^{1/3}$ whereas for $90° \geq \theta \geq 180°$ the range is $(3V_D/2\pi)^{1/3} \leq R \leq \infty$.

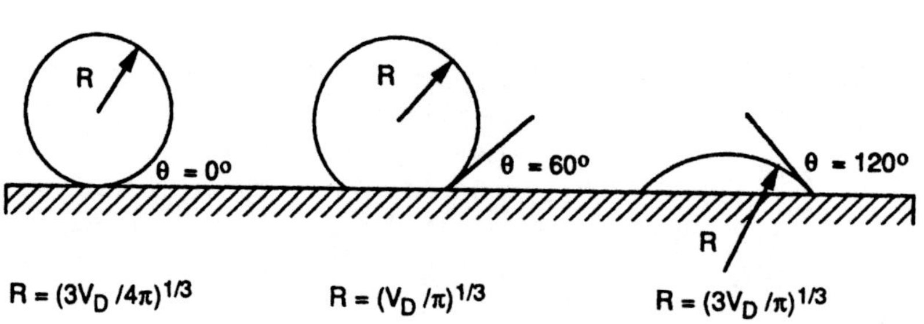

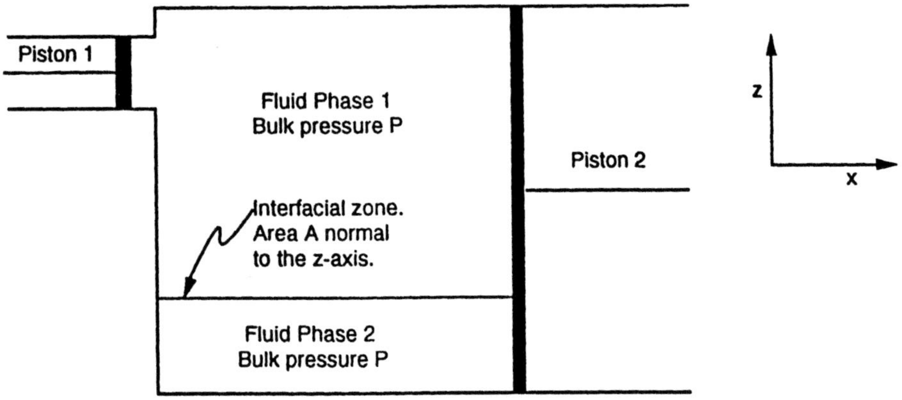

Figure 7.14
Illustration of closed, two-phase fluid system with planar interface.

number of components and can be a vapor–liquid or a liquid–liquid system. Movement of piston 1 changes the volume V of the system and without changing the interfacial area A. Movement of piston 2 changes A and V.

As we saw in the study of capillary hydrostatics, the shape of an interface near a solid surface or in a capillary tube is the result of competition between interfacial tension and gravity. However, if the interface spans a distance large compared to x_c in all directions in the interfacial plane, then the interface will be flat everywhere except very near the solid walls. Even there, for a thought experiment, we can choose a contact angle of $\theta = 90°$ and so the interface is flat everywhere. Then, because the effect of gravity on density is negligible in a laboratory sized system, the only role it plays in the system illustrated in Figure 7.14 is to cause the lower density phase to be above the higher density phase. Thus, we will ignore gravity in discussing the thermodynamics of planar interfaces in this section.

With the neglect of gravity, a force balance on an infinitesimal cube centered on the point x, y, z in the system gives the following results for a static fluid at equilibrium:

$$\frac{\partial P_{xx}(x, y, z)}{\partial x} = \frac{\partial P_{yy}(x, y, z)}{\partial y} = \frac{\partial P_{zz}(x, y, z)}{\partial z} = 0. \qquad (7.4.1)$$

where P_{xx}, P_{yy}, and P_{zz} are the pressures exerted on surfaces normal to the x, y, and z axes, respectively. In a planar system of the type shown in Figure 7.14, the fluid properties are isotropic in the xy plane, that is, the local density and pressure depend only on z and not on position in the xy plane. Thus, P_{xx}, P_{yy}, and P_{zz} are functions only of z. Also, because of the condition of isotropy in the xy plane, the pressure exerted on a surface normal to the y axis is the same as that exerted on a surface normal to the x axis, that is, $P_{xx} = P_{yy}$. Equation (7.4.1) can then be written as

$$\frac{\partial P_{zz}(z)}{\partial z} = 0 \quad \text{or} \quad P_{zz} = \text{constant} \equiv P_N \qquad (7.4.2)$$

and

$$P_{xx} = P_{yy} \equiv P_T(z). \tag{7.4.3}$$

P_N denotes the pressure on a surface whose normal is parallel to the normal of the interface and according to eq. (7.4.2) is independent of z. That P_N is constant across the interface is consistent with the YL equation. P_T is the pressure on a surface at z whose normal is perpendicular to the normal of the interface. Outside the interfacial region the fluids are isotropic (neglecting the effect of gravity on the bulk properties), and $P_T(z) = P_N =$ constant for values of z sufficiently far from the interfacial region (far from the interfacial region can be a distance of as little as some 10–100 Å). Because P_N is independent of z we conclude that P_T and P_N are equal to the same bulk value, call it P, in phases 1 and 2 sufficiently far from the interfacial region.

In summary then, in a two-phase system with a planar interface, the conditions of hydrostatic equilibrium are

$$P_N = P \text{ everywhere}$$

and

$$P_T(z) = P \text{ sufficiently far from the interfacial region}$$

These conditions imply that P_T does not have to equal P_N in the interfacial region. They do not prove that P_T must be different from P_N anywhere. It is, however, an experimental fact that $P_T \neq P_N$ in the interfacial region. In fact, the difference between P_T and P_N gives rise to the interfacial tension. Imagine a measurement of the force exerted on a rectangular surface element of area lw with a normal lying in the zy plane (i.e., a normal perpendicular to the normal of the interface). Let l denote the length of the element in the z direction and w the width in a direction perpendicular to the yz plane. Suppose l is large compared to the thickness of the interfacial region so that if the element is centered on the interfacial region, the top edge of the element will lie in a bulk phase 1 and the bottom will lie in bulk phase 2. The force exerted on a face of the element when it is centered across the interfacial region is

$$-\int_{-l/2}^{l/2} P_T(z) w dz. \tag{7.4.4}$$

When the element is sufficiently far from the interfacial region the force exerted on a face of the element is

$$-Pwl \equiv -\int_{-l/2}^{l/2} Pwdz. \tag{7.4.5}$$

We define $-\gamma w$, the reduction in the force on the element caused by the interface, as the difference between eqs. (7.4.5) and (7.4.4). Thus

$$\gamma w = w \int_{-l/2}^{l/2} [P - P_T(z)] dz. \qquad (7.4.6)$$

As long as the length l of the test surface element is longer than the thickness of the interfacial region, then $P_T(z) - P = 0$ for $|z| > l/2$ so that the limits in eq. (7.4.6) can be extended to infinity without error. In this case, we find

$$\gamma = \int_{-\infty}^{\infty} [P - P_T(z)] dz. \qquad (7.4.7)$$

The quantity γ is of course the interfacial or surface tension, and is seen here to arise from excess tangential stress in the interfacial zone.

Next let us examine the thermodynamic connections of γ. Consider the work associated with a reversible expansion process in which piston 1 changes by amount dV_1 and piston 2 by amount dV_2 such that the net volume change of the system is zero, that is, $dV_1 = -dV_2$. Note that $dV_2 = L_x L_z dy \equiv \int_{-L_z/2}^{L_z/2} dz L_x dy$, where L_x and L_z are the x and z dimensions of piston 2 and dy is the displacement of the piston. The work involved in the process is

$$dW = -P dV_1 - \left[\int_{-L_z/2}^{L_z/2} P_T(z) L_x dz \right] dy = \left\{ \int_{-L_z/2}^{L_z/2} [P - P_T(z)] dz \right\} L_x dy = \gamma dA. \qquad (7.4.8)$$

Thus, for a planar system, the work of a reversible constant volume process that changes the interfacial surface area by dA is equal to the product of the surface tension γ and the change of area dA.

In the reversible process involving arbitrary dV_1 and dV_2, the work can be expressed in the form

$$dW = -P dV + \gamma dA, \qquad (7.4.9)$$

where $dV = dV_1 + dV_2$. If a closed system undergoes a process in which heat dQ is transferred to the system and work dW is done by the system, then the first law of thermodynamics states

$$(dU)_{\mathbf{N}} = dQ + dW, \qquad (7.4.10)$$

where the subscript **N** on $(dU)_{\mathbf{N}}$ indicates that the system is closed to molecular transfer. If the process is reversible, $dQ = T dS$, and dW is given by eq. (7.4.9) for systems with a planar interface, and so

$$(dU)_{\mathbf{N}} = T dS - P dV + \gamma dA. \qquad (7.4.11)$$

With the aid of the definitions of enthalpy ($H = U + PV$), Helmholtz free energy ($F = U - TS$), and Gibbs free energy ($G = F + PV$), we can derive from eq. (7.4.11) the following equations,

$$(dH)_{\mathbf{N}} = TdS + VdP + \gamma dA \tag{7.4.12}$$

$$(dF)_{\mathbf{N}} = -SdT - PdV + \gamma dA \tag{7.4.13}$$

$$(dG)_{\mathbf{N}} = -SdT + VdP + \gamma dA \tag{7.4.14}$$

that describe the thermodynamics of a planar interface in terms of four alternative energy functions.

From eqs. (7.4.11)–(7.4.14) it follows that the tension can be computed as an appropriate area derivative of any of the four energy quantities, namely,

$$\gamma = \left(\frac{\partial U}{\partial A}\right)_{\mathbf{N},S,V} = \left(\frac{\partial H}{\partial A}\right)_{\mathbf{N},S,P} = \left(\frac{\partial F}{\partial A}\right)_{\mathbf{N},V,T} = \left(\frac{\partial G}{\partial A}\right)_{\mathbf{N},P,T}. \tag{7.4.15}$$

In terms of these equations we view the interfacial tension as the excess of a thermodynamic energy quantity per unit interfacial area. On the other hand, eq. (7.4.7) relates the tension to the tangential stress in the interfacial zone. Although eq. (7.4.7) is perhaps more appealing from the physical point of view, the free energy formulas for γ have proven more fruitful in the formulations of a statistical mechanical theory of interfaces.

Consider next adsorption at interfaces. For this, it is convenient to allow mass transfer in the reversible work experiment used in deriving eq. (7.4.11). If an open system undergoes a reversible process in which heat $dQ = TdS$ is transferred to the system, work $dW = -PdV + \gamma dA$ is done, and a transfer dN_i of each component i occurs, then to eq. (7.4.11) we must add the energy $\sum_i \mu_i dN_i$ brought to the system by addition of components. μ_i is the chemical potential of component i. Thus, the appropriate open system generalization of the internal energy equation is

$$dU = TdS - PdV + \gamma dA + \sum_i \mu_i dN_i, \tag{7.4.16}$$

and for the other energy quantities the generalizations are

$$dH = TdS + VdP + \gamma dA + \sum_i \mu_i dN_i \tag{7.4.17}$$

$$dF = -SdT - PdV + \gamma dA + \sum_i \mu_i dN_i \tag{7.4.18}$$

$$dG = -SdT + VdP + \gamma dA + \sum_i \mu_i dN_i \tag{7.4.19}$$

The above equations can be integrated at constant intensive variables T, P, γ, μ_i to yield

$$U = TS - PV + \gamma A + \sum_i \mu_i N_i$$

$$H = TS + \gamma A + \sum_i \mu_i N_i$$

$$F = -PV + \gamma A + \sum_i \mu_i N_i \qquad (7.4.20)$$

$$G = \gamma A + \sum_i \mu_i N_i.$$

Differentiation of the expression for U at eq. (7.4.20) and subtraction of the result from eq. (7.4.16) leads to the Gibbs–Duhem equation,

$$0 = SdT - VdP + Ad\gamma + \sum_{i=1}^{C} N_i d\mu_i, \qquad (7.4.21)$$

for systems having a planar interface. Differentiation of any other of the energy quantities at eq. (7.4.20) and subtraction from the previous equation for that quantity give the same equation. Physically, the Gibbs–Duhem equation asserts that only $C + 2$ of the $C + 3$ intensive variables $T, P, \gamma, N_1, \ldots, N_c$ can vary independently if a system is to remain at thermodynamic equilibrium.

For the purposes of interfacial thermodynamics, it is convenient to eliminate bulk extensive quantities from eq. (7.4.21). To do this we use the Gibbsian device of replacing the real system by a hypothetical one of the same volume as indicated in Figure 7.15. In a real system the density n_i of each component i varies continuously across the interfacial zone from its bulk value $n_i^{(1)}$ in phase 1 to its bulk value $n_i^{(2)}$ in phase 2 as illustrated by the curve for $n_i(z)$ in Figure 7.15. In the hypothetical system, we imagine

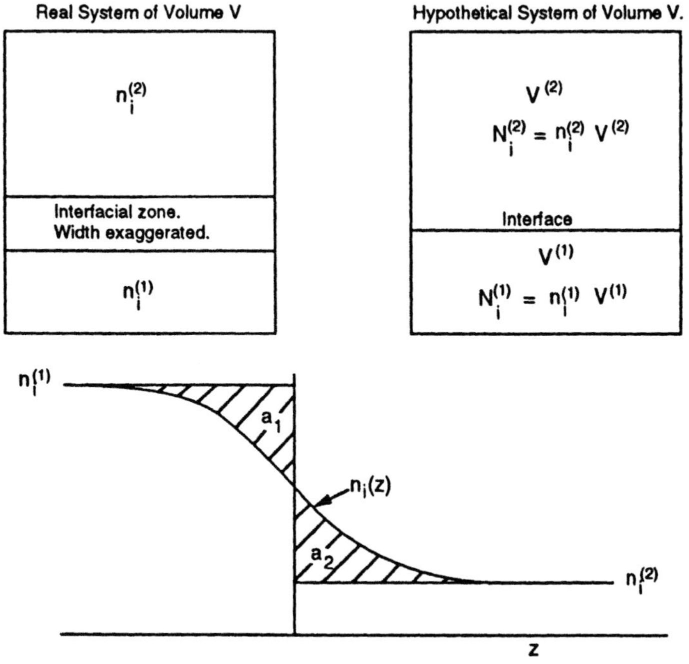

Figure 7.15
Schematic of real system and hypothetical system in which the interfacial zone is replaced by a mathematical dividing surface.

bulk phase 1 and 2 separated by a mathematical dividing surface ($z = 0$) where component densities change discontinuously from bulk value $n_i^{(1)}$ to $n_i^{(2)}$.

We are free to choose the hypothetical system to have the same volume as the actual system. However, the number of molecules of each component and the total entropy of the two systems will differ to the extent that the actual interface differs from the hypothetical one. The *surface excess density* Γ_i of component i is defined as

$$\Gamma_i \equiv (N_i - N_i^{(1)} - N_i^{(2)})/A. \tag{7.4.22}$$

This quantity equals the difference between area a_1 and area a_2 shown in Figure 7.15. In later chapters where we develop theories for computing $n_i(z)$, Γ_i affords a way to quantitate the excess or deficit of component i in the interfacial zone. Similarly, the *surface excess entropy* is defined as

$$S^{ex} \equiv (S - S^{(1)} - S^{(2)})/A. \tag{7.4.23}$$

The Gibbs–Duhem equation for bulk phase α is

$$0 = S^{(\alpha)}dT - V^{(\alpha)}dP + \sum_i N_i^{(\alpha)}d\mu_i. \tag{7.4.24}$$

If this equation is subtracted from eq. (7.4.21) for each bulk phase, $\alpha = 1$ and 2, the result is

$$0 = -S^{ex}dT + d\gamma + \sum_i \Gamma_i d\mu_i, \tag{7.4.25}$$

whose isothermal form,

$$d\gamma = -\sum_i \Gamma_i d\mu_i, \tag{7.4.26}$$

is known as the *Gibbs adsorption equation.*

As the reader might have noticed, the position of the mathematical dividing surface in the hypothetical system is arbitrary. Clearly for Γ_i to have the meaning ascribed to it the position ought to be in the interfacial zone. Beyond this its position is a matter of convention. It is common to choose it such that $\Gamma_1 = 0$ where Γ_1 is the surface excess density of the major component.

The Gibbs isotherm equation is often used to measure the degree of surface activity of a surfactancy at a water–air, water–oil or oil–air interface. A surfactant is a molecule consisting of a hydrophilic or water soluble moiety connected to a hydrophobic or water insoluble moiety. Examples of surfactant molecules are given below:

Sodium dodecylsulfate (SDS): $CH_3\,(CH_2)_{11}\,SO_4^-\,Na^+$

Sodium palmitate: $(CH_3)(CH_2)_{14}COO^-Na^+$

Dodecylpentaethoxy alcohol: $CH_3(CH_2)_{11}\,(OCH_2CH_2)_5OH$.

When a small amount of a surfactant is added to water (or oil) it will tend to accumulate preferentially at the interface with the hydrophobic (or hydrophilic) part oriented away from the bulk phase as illustrated in Figure 7.16.

Surfactant is frequently present in an aqueous solution at low concentration, in which case the ideal solution approximation is

$$d\mu_i = R_g T d \ln C_i, \tag{7.4.27}$$

where μ_i is energy per mole, R_g the gas constant, and C_i the molar concentration. If the Gibbs dividing surface is chosen so that $\Gamma_1 = 0$, component 1 being water, then in the low concentration regime

$$d\gamma = -\sum_{i=2}^{c} \Gamma_i R_g T d \ln C_i. \tag{7.4.28}$$

For a one-component nonionic surfactant, this expression becomes

$$d\gamma = -\Gamma_s R_g T d \ln C_s, \tag{7.4.29}$$

and so the tension versus logarithm of surfactant concentration can be measured and the surface excess density of the surfactant can be computed from

$$\Gamma_s = -\frac{d\gamma}{R_g T d \ln C_s}. \tag{7.4.30}$$

If the surfactant is ionic, then electroneutrality has to be accounted for. For example, consider a surfactant that is totally dissociated in solution and having a cation valence z_+ and anion valence z_-. Then the number of surfactant cations and anions, N_{s+} and N_{s-}, cannot be varied independently, but rather must obey the electroneutrality condition

$$dN_s = \frac{1}{z_+} dN_{s+} = \frac{1}{|z_-|} dN_{s-}, \tag{7.4.31}$$

where N_s is the number of surfactant molecules.

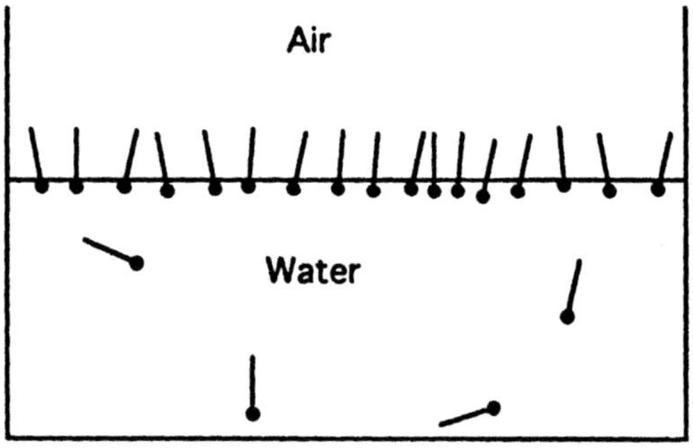

Figure 7.16
Illustration of surface activity of a surfactant (—) denotes water insoluble tail and (•) water soluble head of surfactant molecule at an air–water interface.

Thus, eqs. (7.4.19), (7.4.21), and (7.4.26) become

$$dG = -SdT + VdP + \gamma dA + \mu_1 dN_1 + (z_+\mu_{s+} + |z_-|\mu_{s-})dN_s, \quad (7.4.32)$$
$$0 = SdT - VdP + Ad\gamma + N_1 d\mu_1 + N_s d(z_+\mu_{s+} + |z_-|\mu_{s-}), \quad (7.4.33)$$

and

$$d\gamma = -\Gamma_1 d\mu_1 - \Gamma_s d(z_+\mu_{s+} + |z_-|\mu_{s-}). \quad (7.4.34)$$

If we again choose $\Gamma_1 = 0$ and consider very low surfactant concentration, we find from eq. (7.4.34)

$$\begin{aligned} d\gamma &= -\Gamma_s R_g T(z_+ d\ln C_{s+} + |z_-| d\ln C_{s-}) \\ &= -\Gamma_s R_g T(z_+ + |z_-|) d\ln C_s. \end{aligned} \quad (7.4.35)$$

If one adds an excess of electrolyte containing the counterion of the surfactant, say add NaCl to aqueous SDS, then changing surfactant concentration a small amount does not change the chemical potentials of Na^+ and Cl^- ions appreciably. In this case,

$$d\gamma \simeq -\Gamma_s RT|z_\nu| d\ln C_s, \quad (7.4.36)$$

where z_ν is the valence of the surfactant ion (the dodecylsulfate ion in the case just mentioned). We assume in eg. (7.4.36) that surfactant is added at fixed NaCl concentration.

As an example, let us consider the surface tension data given in Table 7.3 for SDS. In the case of surfactant in pure water, eq. (7.4.35) yields

$$\Gamma_{SDS} = 2.3 \times 10^{-10} \text{gmol/cm}^2. \quad (7.4.37)$$

For surfactant in tenth molar NaCl brine eq. (7.4.36) gives

$$\Gamma_{SDS} = 3.11 \times 10^{-10} \text{gmol/cm}^2. \quad (7.4.38)$$

At a bulk concentration of 3.47×10^{-3}M in the pure water case, a layer of bulk solution 0.66 μm thick would have to be depleted to form the surface excess layer. At a bulk concentration of 1.73×10^{-4}M in the case of 0.1M NaCl brine, a layer of 18 μm would have to be depleted.

TABLE 7.3 Interfacial Tension γ of Decane Against Water as a Function of Molar Concentration C_{SDS} of the Anion, Dodecylsulfate, of SDS. ($T = 20°C$)

Pure Water		*0.1 M NaCl Brine*	
γ (dyn/cm)	C_{SDS}(M)	γ (dyn/cm)	C_{SDS}(M)
28.3	1.74×10^{-3}	27.5	8.67×10^{-5}
20.8	3.47×10^{-3}	22.7	1.73×10^{-4}
15.3	5.21×10^{-3}	17.4	3.46×10^{-4}

The area per molecule adsorbed in the interfacial zone is defined by $\sum_i = 1/\Gamma_i$. $\Gamma_{SDS} = 72.2$ Å^2/molecule and 53.4 Å^2/molecule for the cases of adsorption of SDS at the pure water–decane and 0.1 M NaCl brine–decane interfaces, respectively. We observe that there is a high population of surfactant at the interface in both cases (average separations of about 8.5 and 7.3Å, respectively.) A higher population is allowed when NaCl is present because of the Coulombic screening of the electrolyte.

As a second example of the use of the Gibbs adsorption equation, consider the data given in Figure 7.17. This is a typical plot of the surface tension versus concentration of surfactant in water. The breakpoint in the plot is called the critical micelle concentration (cmc). Below this concentration the surfactant is either adsorbed at the air–water interface or is monomolecularly dispersed in solution. Above the cmc the monomolecular concentration remains nearly constant and the excess surfactant self-assembles into micellar aggregates as depicted in the preceding chapter. Because at low concentration $d\mu_s \propto d \ln C_s^m$, and $d\gamma \propto -d\mu_s$, where C_s^m is the monomolecular concentration, the surface tension decreases with increasing surfactant concentration below the cmc and remains almost constant with increasing concentration above the cmc because $d\mu_s \sim 0$ above the cmc. Thus, the Gibbs isotherm is useful for measuring the surface excess density as well as the cmc.

The surfactant studied in Figure 7.17 is the following phenylene diacrylate bromide:

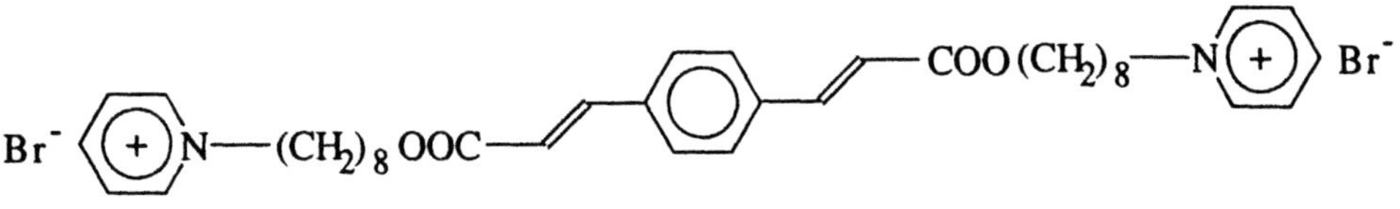

(7.4.39)

The surface tension at the cmc is 41 dyn/cm and the cmc = 0.015 mol/L. The surface excess concentration, computed from the slope

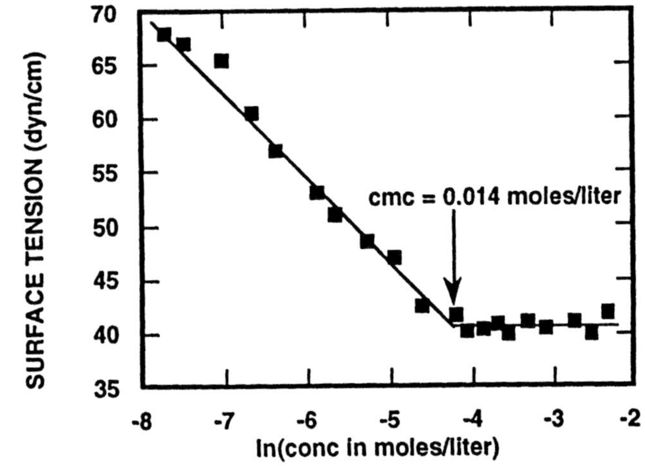

Figure 7.17
Surface tension of aqueous solution of the surfactant shown at eq. (7.4.39) at 23°C. Mao et al. (1973).

of the linearly decreasing part of the plot in Figure 7.17 using eq. (7.4.35) with $z_+ = 2$ and $|z_-| = 1$ is

$$\Gamma_s = 8.45 \times 10^{-8}\text{gmol/cm}^2 \quad \text{or} \quad \Sigma_s = 151\text{Å}^2/\text{molecule}. \tag{7.4.40}$$

The fact that Σ_s for the molecule shown at eq. (7.4.39) is about twice that of SDS indicates that both ionic ends of the former molecule are embedded in the aqueous surface at the air–water interface.

Note that the slope of γ versus $\ln C_s$ below the cmc remains nearly constant in Figure 7.17 but the concentration C_s changes by a factor of 50. This is not unusual behavior for a surfactant solution and it indicates that the surface excess concentration is nearly independent of bulk surfactant concentration as the cmc is approached. Qualitatively, one can understand this behavior in terms of the Langmuir isotherm. According to this isotherm

$$\Gamma_s = \frac{aC_s}{1 + bC_s}. \tag{7.4.41}$$

At sufficiently large C_s, Γ_s is approximately equal to its saturation value $\Gamma_s^s = a/b$. Near saturation,

$$\Gamma_s \simeq \Gamma_s^s[1 - \frac{1}{bC_s} + \mathcal{O}(bC_s)^{-2}]. \tag{7.4.42}$$

Suppose $bC_s = 50$. Then $\Gamma_s \approx 0.98\Gamma_s^s$ and increasing bC_s to 500 changes Γ_s very little (to $\Gamma_s \simeq 0.9998\Gamma_s^s$), whereas the chemical potential change is substantial ($\Delta\mu_s \approx -R_g T \ln 10$). If Γ_s is not near its saturation value in approaching the cmc, the plot in Figure 7.17 would be nonlinear below the cmc making it harder to extract a value of the cmc.

Example

In a binary liquid solution, a solute often obeys the Langmuir isotherm,

$$\Gamma = \frac{\Gamma_o C}{b + C}$$

at a liquid-vapor interface. At low solute concentration C, the solution is ideal and so

$$d\mu = RT d\ell n C.$$

According to the Gibbs isotherm

$$d\gamma = -\Gamma d\mu.$$

Putting all this together we find the Szyszkowski equation,

$$\Pi = \Gamma_o RT\ell n(1 + C/b),$$

where the spreading pressure $\Pi = \gamma_o - \gamma$, the difference between the surface tension of the pure solvent and the surface tension of the solution at concentration C.

7.5 Thermodynamics of Wetting

Suppose phases 1, 2 and 3 are in thermodynamic equilibrium. When a small drop of phase 2 is placed at the interface between phase 1 and 3 it will either remain as a droplet (whose shape is determined by capillary hydrostatics) or it will spread spontaneously to form a layer of phase 2 between phases 1 and 3 (assuming the interfacial area between phases 1 and 2 is not so large that the layer becomes molecularly thin). The layering of phase 2 is said to completely wet the interface between phases 1 and 3. The situations are illustrated in Figure 7.18.

If the interfacial area between phases α and β is $A_{\alpha\beta}$, then the Gibbs free energy of a system of coexisting phases is $G_1 = \sum_{\alpha\beta} \gamma_{\alpha\beta} A_{\alpha\beta} + \sum_i \mu_i N_i$, where $\gamma_{\alpha\beta}$ is the surface interfacial excess free energy per unit area of interface (we assume layers are thick enough and radii of curvature large enough in magnitude for $\gamma_{\alpha\beta}$ to be that of a planar interface). If α and β are fluids, $\gamma_{\alpha\beta}$ is the interfacial

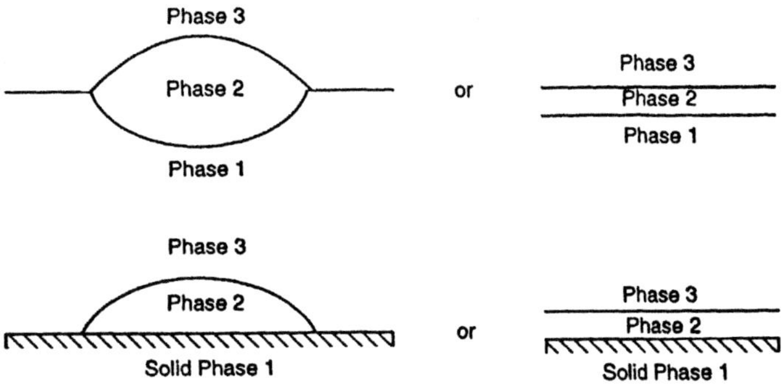

Figure 7.18
A droplet of phase 2 at the interface of phases 1 and 3 will either form a droplet whose shape is determined by capillary hydrostatics or will spontaneously spread to form a thin layer of phase 2 completely wetting the interface between phase 1 and 3.

tension. However, if α or β is a solid, $\gamma_{\alpha\beta}$ is best thought of as the surface excess free energy of the interface, because the lack of deformability of the solid surface prevents the manifestations of interfacial tension available to fluid–fluid interfaces.

In any case, to investigate whether a droplet of phase 2 will spontaneously spread and wet the 13 interface, consider a droplet whose edge makes a small thin wedge of phase 2 intruding between phases 1 and 3 as shown in Figure 7.19. The free energy difference before and after the wedge formation is

$$G(\text{before}) - G(\text{after}) = [\gamma_{13} - (\gamma_{12} + \gamma_{23})]dA = w\,dA, \quad (7.5.1)$$

where the *wetting or spreading coefficient* w is defined by

$$w \equiv \gamma_{13} - (\gamma_{12} + \gamma_{23}). \quad (7.5.2)$$

According to the second law of thermodynamics the Gibbs free energy of a fixed **N**, T, P system is a minimum at equilibrium. Thus, whether a droplet of 2 will spontaneously spread at the 13 interface depends on the sign of the wetting coefficient w. In particular,

1 if $w \geq 0$, a droplet of 2 spreads spontaneously, completely wetting the 13 interface and

2 if $w < 0$, a droplet of 2 remains as a lens in a configuration of capillary hydrostatic equilibrium

Later we show that $w > 0$ is possible for macroscopic systems only if the three phases are not in mutual equilibrium.

An example of a nonwetting droplet is hexadecane (phase 2) at an air–water (phases 3 and 1) interface. These three phases are virtually totally immiscible, and so the interfacial tensions can

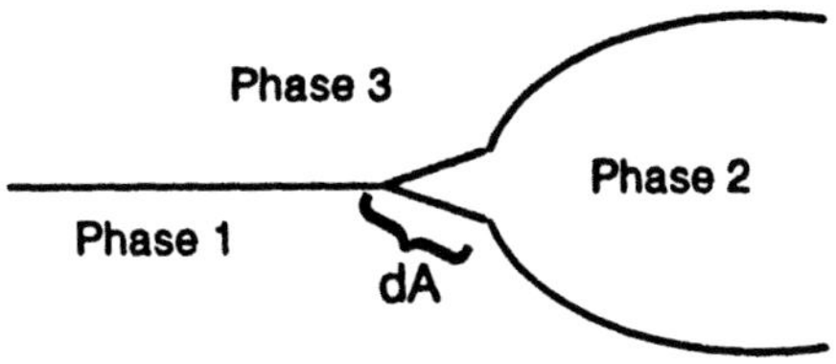

Figure 7.19
Droplet of phase 2 with a small thin wedge intruding between phases 1 and 3.

be approximated as those measured for each pair of fluids in this absence of the third phase. In this case,

$$w = 72.8 - (30.0 + 52.1) = -9.3\text{dyn/cm},$$

and so a hexadecane droplet will not wet on an air–water interface. Another interesting example is a droplet of benzene at an air–water interface. If one puts a drop of pure benzene on the air–water surface, the spreading coefficient is initially

$$w_{initial} = 72.8 - (28.8 + 35.0) = 8.9\text{dyn/cm},$$

indicating that the droplet of benzene will initially spread. However, benzene is partially soluble in water. When the three phases are fully equilibrated

$$w_{equil} = 62.4 - (28.8 + 35.0) = -1.4\text{dyn/cm},$$

indicating that at three-phase equilibrium a droplet of the benzene phases will not wet the air–water interface. Thus, a drop of pure benzene initially spreads on the air–water interface, but as equilibration occurs the wettability reverses and the droplet retracts to a lenslike equilibrium. The same situation occurs with n-hexanol at the air–water interface:

$$w_{\text{initial}} = 72.8 - (24.8 + 6.8) = 41.2\text{dyn/cm}$$
$$w_{\text{equil}} = 28.8 - (24.7 + 6.8) = -3.0\text{dyn/cm}.$$

An example of a liquid that does completely wet the air–water interface is n-octanol ($w = 35.7$ dyn/cm). Adamson (1990) lists wetting (spreading) coefficients for several three-phase fluid systems.

Examples of complete wetting when phase 1 is solid are water on clean glass and on clean metal in the presence of air and hydrocarbon liquids on clean metal in air. Normal hexane wets Teflon but normal alkanes of larger molecular weight than n-octane do not wet Teflon.

Let us examine the question of whether, if phases α, β, and γ are at equilibrium, the inequality

$$\gamma_{\alpha\gamma} - (\gamma_{\alpha\beta} + \gamma_{\beta\gamma}) > 0 \qquad (7.5.3)$$

is possible. Consider the grand potential

$$\Omega \equiv F - \sum_i \mu_i N_i. \qquad (7.5.4)$$

At thermodynamic equilibrium, Ω is a minimum for a system in contact with a bath at constant T, V, $\mu_1, \ldots, \mu_c$. Consider a system with planar interface between phases α and γ. The grand potential for the system is

$$\Omega = -PV + \gamma_{\alpha\gamma} A. \qquad (7.5.5)$$

By exchanging enough material with the bath to form a layer of phase β of near interfacial thickness, the system can achieve the grand potential

$$\Omega' = -PV + (\gamma_{\alpha\beta} + \gamma_{\beta\gamma})A, \tag{7.5.6}$$

$$\Omega' - \Omega = (\gamma_{\alpha\beta} + \gamma_{\beta\gamma} - \gamma_{\alpha\gamma})A. \tag{7.5.7}$$

A microscopic layer of β phase will be recruited spontaneously to the $\alpha\gamma$ interface if the inequality of eq. (7.5.3) would otherwise hold. Therefore, if phases α, β, and γ are in equilibrium and β is a wetting phase, the tension measured between α and β will always obey the equality

$$\gamma_{\alpha\gamma} - (\gamma_{\alpha\beta} + \gamma_{\beta\gamma}) = 0. \tag{7.5.8}$$

Thus, we arrive at a generalization of Antonov's rule: if phases α, β, and γ are in equilibrium,

$$\gamma_{\alpha\gamma} - (\gamma_{\alpha,\beta} + \gamma_{\beta,\gamma}) \leq 0. \tag{7.5.9}$$

When the inequality holds, phase β does not wet the $\alpha\gamma$ interface, but when the equality holds β wets the $\alpha\gamma$ interface. Of course, the wetting phase may not be the droplet phase, as is illustrated in Figure 7.20.

When a droplet is nonwetting, its configuration is determined by the equation of capillary hydrostatics whose solution depends on boundary conditions that are often related to contact angles (Fig. 7.21). Shown in Figure 7.21 is a blow-up of the menisci sufficiently near the three-phase contact line that the gravitational bending of the menisci is negligible.

In the case of three fluid phases, a force balance in the direction orthogonal to the direction of the ij meniscus yields $\gamma_{ik}\sin(\pi - \alpha_k) = \gamma_{ij}\sin(\pi - \alpha_j)$, or $\gamma_{ik}\sin\alpha_k = \gamma_{ij}\sin\alpha_j$. The relationships can be summarized as

$$\frac{\gamma_{12}}{\sin\alpha_3} = \frac{\gamma_{13}}{\sin\alpha_2} = \frac{\gamma_{23}}{\sin\alpha_1} \tag{7.5.10}$$

Because $\alpha_1 + \alpha_2 + \alpha_3 = 360°$, one can use eq. (7.5.10) to compute from interfacial tensions the contact angles α_i needed for boundary conditions in capillary hydrostatics.

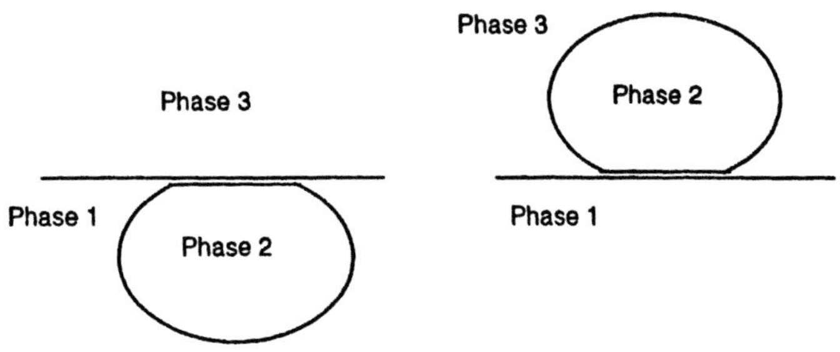

Figure 7.20
Cases in which (a) phase 1 is the wetting phase and (b) in which phase 3 is the wetting phase.

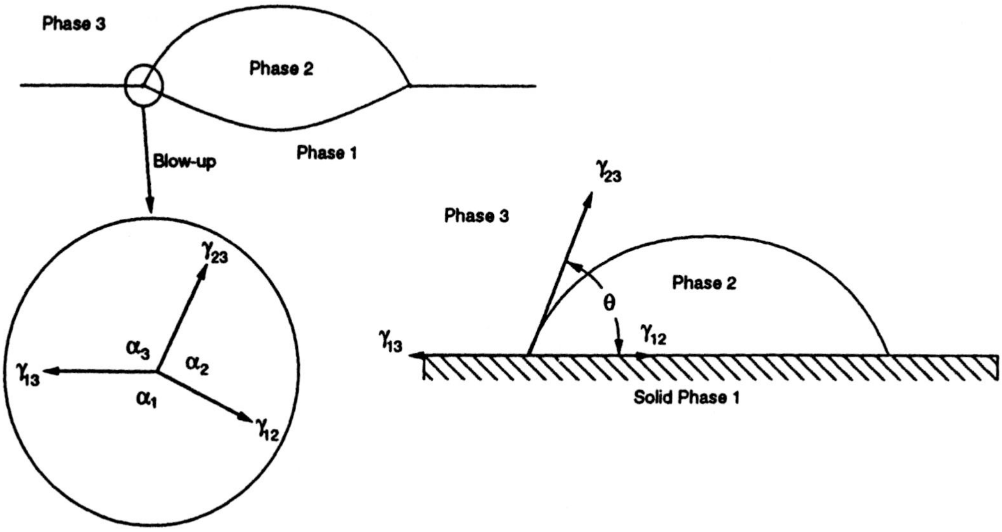

Figure 7.21
Illustration of tension balances in the region near the three-phase contact line (into the page).

When phase 1 is a flat solid as shown in Figure 7.21(b), the force balance yields

$$\gamma_{13} = \gamma_{12} + \gamma_{23} \cos\theta, \tag{7.5.11}$$

which is known as *Young's equation*. Fluid–solid tensions [γ_{12} and γ_{13} in eq. (7.5.11)] cannot be measured, and so one typically measures γ_{23} and the contact angle in separate experiments and uses eq. (7.5.11) to express the surface excess free energy density of phases 1 and 3 relative to γ_{12}. Some convenient reference fluid 2 could be chosen to compare among solids or different choices of phase 3.

Often by varying conditions, say temperature, the activity of an added component, the carbon number of an alkane in a homologous series of oils, etc., one can convert a nonwetting phase to a wetting phase. The point in a parameter sequence at which this occurs is called the wetting transition point. If phase 1 is a solid the contact angle θ approaches 0° or 180° at the wetting transition. If all three phases are fluids and phase i is the wetting phase, then α_i approaches 0° as the wetting transition is approached.

In a study that has proven very useful in the design of products of prescribed wettability, Zisman found that a drop of low vapor pressure liquid will spread on a given solid in the presence of air if the liquid–vapor surface tension γ_{LV} is less than a value γ_c characteristic of the solid (γ_c is only weakly dependent on the liquid). Some of his results for hydrocarbons on polymer substrates are shown in Figure 7.22. The importance of Zisman's discovery is that desired wettability of fabrics, papers, containers, and the like can be obtained by applying a polymer coating to or chemically modifying the surface of the product based on known values of γ_c.

Zisman's discovery can be understood in terms of the theories of Girifalco and Good (1957) and Fowkes (1962, 1964) discussed

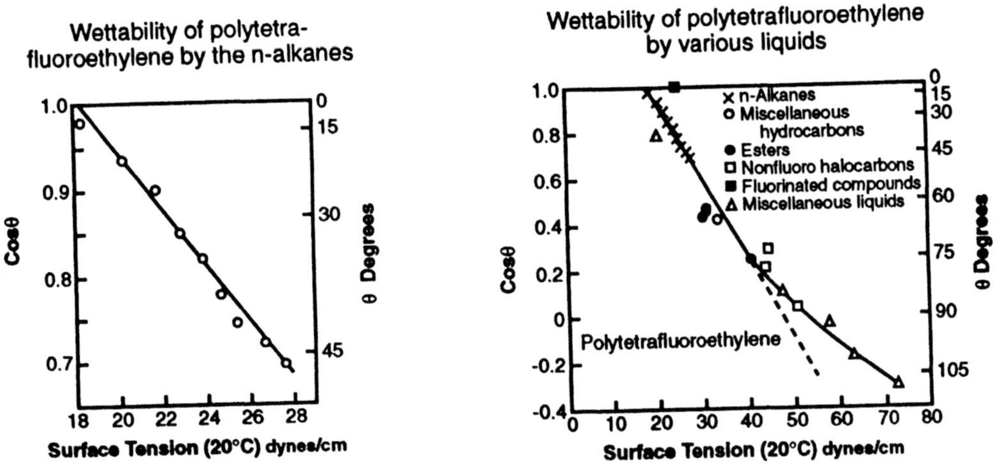

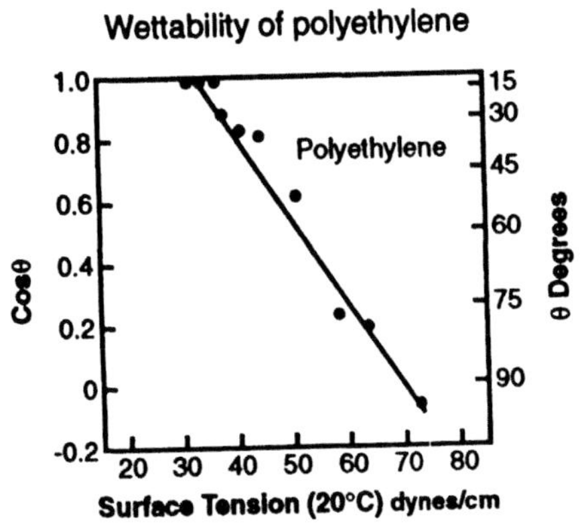

Critical Surface Tensions of Various Polymeric Solids

Polymeric Solid	γ_c, Dynes/Cm. at 20°C.
Polymethacrylic ester of ϕ'-octanol	10.6
Polyhexafluoropropylene	16.2
Polytetrafluoroethylene	18.5
Polytrifluoroethylene	22
Poly(vinylidene fluoride)	25
Poly(vinyl fluoride)	28
Polyethylene	31
Polytrifluorochloroethylene	31
Polystyrene	33
Poly(vinyl alcohol)	37
Poly(methyl methacrylate)	39
Poly(vinyl chloride)	39
Poly(vinylidene chloride)	40
Poly(ethylene terephthalate)	43
Poly(hexamethylene adipamide)	46

Figure 7.22
Zisman's contact angle studies (1964) of liquid drops on polymer substrates.

in Section 7.1. For a liquid drop on an immiscible solid or liquid (phase 1) in the presence of vapor (phase 3), the model of Girifalco and Good for the spreading coefficient

$$W = 2(\Phi\sqrt{\gamma_1^o\gamma_2^o} - \gamma_2^o). \tag{7.5.12}$$

According to the Fowkes theory

$$W = 2(\sqrt{\gamma_1^{o,d}\gamma_2^{o,d}} - \gamma_2^o). \tag{7.5.13}$$

If the liquid drop is a normal alkane, it has only dispersion forces and so $\gamma_2^{o,d} = \gamma_2^o$.

The wetting transition occurs at $w = 0$. If we consider the spreading of alkanes on water at 20° and the surface tension γ_2^0 of the alkanes at which $w = 0$ according to the Girifalco–Good model is (the surface tension of water is 72.8 dyn/cm)

$$\gamma_2^o \simeq \Phi^2 72.8\text{dyn/cm}. \tag{7.5.14}$$

Because $\Phi \approx 1$, the Girifalco–Good model estimates an oil spreading tension that is too high.

Because $\gamma_{H_2O}^{o,d} \simeq 21.8$ dyn/cm and $\gamma_{oil}^{o,c} \simeq \gamma_{oil}^{o}$, the Fowkes theory predicts that the wetting transition occurs at

$$\gamma_{H_2O}^{o} = \gamma_{oil}^{o,d} = 21.8\text{dyn/cm}. \tag{7.5.15}$$

Thus, hydrocarbons with surface tensions less than 21.8 dyn/cm (heptane, hexane, etc.) should spread on water and those of higher molecular weight (octane, nonane, etc.) should not. This prediction has been verified and indicates that Fowkes' approximation is better than that of Girifalco and Good.

For hydrocarbons on mercury, Fowkes theory predicts that the wetting transition occurs at $\gamma_{oil} = \gamma_{Hg}^{o,d} = 200$ dyn/cm, implying what is well known, namely that all hydrocarbon liquids spread readily on clean liquid mercury surfaces. By contrast, for water Fowkes formula predicts a spreading coefficient of $w = -13.5$ dyn/cm, that is, water will not spread on clean mercury. This also agrees with observation.

Fowkes theory provides the basis for understanding Zisman's observation. From eq. (7.5.13) it follows that for nonpolar fluids on solids Zisman's $\gamma_c = \gamma_1^{o,d}$. For example, from the data on n-alkanes on Teflon $\gamma_c = \gamma_{Teflon}^{o,d} = 18.5$ dyn/cm.

If phase 1 is a solid, Young's equation, (7.5.11), is valid. Combining this equation and Fowkes approximation, (7.1.9), we find for the contact angle

$$\cos\theta = 2\sqrt{\frac{\gamma_1^{o,d}\gamma_2^{0,d}}{(\gamma_2^o)^2}} - 1. \tag{7.5.16}$$

For nonpolar drops $\gamma_2^o = \gamma_2^{o,d}$ and

$$\cos\theta = 2\sqrt{\frac{\gamma_1^{o,d}}{\gamma_2^o}} - 1. \tag{7.5.17}$$

Because $\gamma_2^{o,d} < \gamma_2^o$ for hydrogen bonded fluids, thee above two equations would explain the smaller slope of $\cos\theta$ versus γ_2^o for the miscellaneous liquids on Teflon (Fig. 7.22) if these liquids are hydrogen bonded.

An interesting spreading situation, known as "critical-wetting," occurs when two of the three phases are near critical. J. W. Cahn (1977) argued that as a pair of phases, say 2 and 3, approach sufficiently close to a critical point, then either phase 2 will become a spreading phase between 1 and 3 or 3 will become a spreading phase between 1 and 2. The theory is based on the universal scaling law of tension near a critical point according to which the tension between 2 and 3 approaches zero as

$$\gamma_{23} = A\left|1 - \frac{a}{a_c}\right|^{1.3} \tag{7.5.18}$$

as the field variable a (temperature, pressure, chemical potential, carbon number in homologous series, etc.) approaches the critical value a_c. A is a scale factor depending on the particular fluid pair. The difference $\gamma_{12} - \gamma_{13}$ approaches zero as phases 2 and 3 approach their critical point. According to the theory of critical phenomena, the difference is expected to obey the universal scale law

$$\gamma_{12} - \gamma_{13} = B \left| 1 - \frac{a}{a_c} \right|^{\beta_1} \tag{7.5.19}$$

for field variable "a" sufficiently close to its critical point "a_c." B is a scale factor which can be either positive or negative. Cahn hypothesized that $\beta_1 = 0.34$, the value of the composition and density critical exponent. This is a reasonable expectation because the tension difference approaches zero because component densities in each phase become identical at the critical point. However, it has recently been found (Turkevich, 1995) that

$$\beta_1 = 0.6.$$

Although this is larger than expected by Cahn, his argument still holds. That is, beginning from a nonwetting situation (for which $\gamma_{23} > |\gamma_{12} - \gamma_{13}|$), as a approaches a_c the quantity γ_{23} approaches zero faster than does $|\gamma_{12} - \gamma_{13}|$. Thus it follows that at some field variable, a_w lying a finite distance from a_c, the equality

$$\gamma_{23} = \gamma_{13} - \gamma_{12} \tag{7.5.20}$$

will occur and the droplet of 2 will perfectly wet the interface. The more general conclusion is that eqs. (7.5.12) and (7.5.13) imply the existence of an $a_w \neq a_c$ such that $\gamma_{23} = \pm(\gamma_{13} - \gamma_{12})$. When the sign is positive, phase 2 is the wetting phase. With the negative sign, 3 is the wetting phase. Although more rigorous scaling theory and experiment showed that the power of 0.34 assumed by Cahn is wrong, the value 0.6 is certainly less than the exponent (≈ 1.3) in eq. (7.5.6), and so Cahn's conclusion was correct.

The conclusion of the theory is that a wetting (spreading) phase will always occur if two of the fluids are near enough to a critical point. Moldover and Cahn (1980) verified this theory for a mixture of methanol, cyclohexane, and water (two phases) in the presence of air. The pertinent phase diagrams are shown in Figure 7.23.

Water is added to move away from the methanol–cyclohexane critical point with the result that the wetting film of the cyclohexane-rich phase dewets the air–methanol-rich phase interface (Figure 7.24). The cyclohexane-rich phase completely wets the cuvette–methanol-rich phase interface at water concentrations they studied.

As noted by Moldover and Cahn, their finding suggests the strategy of inducing wetting or nonwetting as desired. By addition of another chemical component, the system can be driven toward

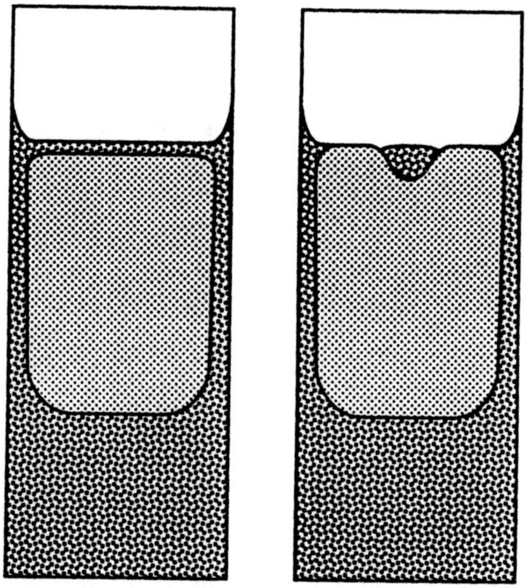

Figure 7.23
Wetting transition of a liquid drop at a vapor–liquid interface. Sketch of cuvettes containing two liquid phases and vapor. (Left) The lower (CH_3OH-rich) phase intrudes between the middle (C_6H_{12}-rich) phase and the cuvette wall. The lower phase also intrudes between the middle phase and the vapor. (Right) Water has been added to the solution. The lower phase still intrudes between the cuvette wall and the middle phase. The lower phase no longer intrudes between the middle phase and the vapor. Instead, there is a line of three-fluid phase contact near the cuvette wall and a line of three-fluid phase contact surrounding a lenticular drop of the lower phase suspended at the vapor interface. Redrawn from Moldover and Cahn, (1980).

or away from the critical point and therefore one can attain wetting or dewetting.

7.6 Thin Films

When a material exists as a thin film between two other bulk phases, the pressure P_f in the film can differ from the pressure P_B the material has in its adjoining bulk phase. Two experiments that can measure this phenomenon are illustrated in Figure 7.25. The difference between P_f and P_B has been named the disjoining pressure by Deryagin and is denoted by

$$\Pi \equiv P_f - P_B \tag{7.6.1}$$

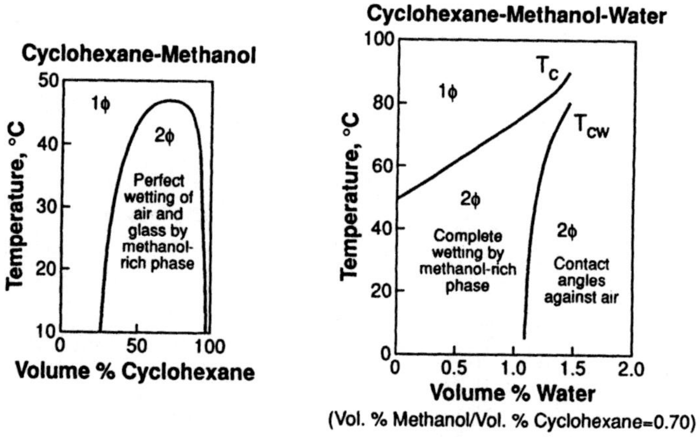

Figure 7.24
Liquid–liquid binodal of cyclohexane–methanol solution. The right panel shows the effect adding water has on the wetting behavior of this solution.

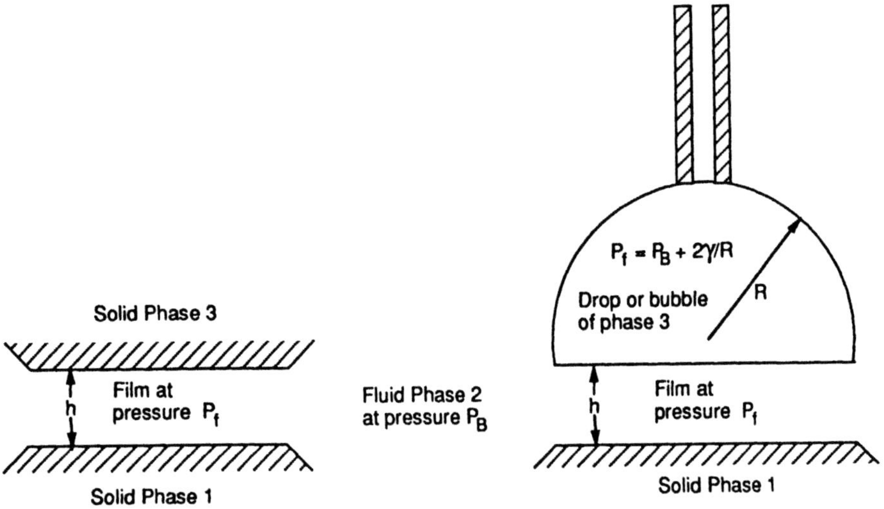

Figure 7.25
Thin planar films of material confined by phases 1 and 3 and at chemical equilibrium with bulk phase 2.

The concept of "disjoining" is that to squeeze a film to thickness h, an excess pressure of $\Pi(h)$ must be applied to offset the tendency of the film phase to separate or disjoin the confining phases. If Π is negative, the film material wants to retreat from the region between the confining phase. Thus, if $\Pi > 0$, the film material is wetting, and if $\Pi < 0$, it is nonwetting.

The dependence of Π on h depends on the molecular components of the film and phases. Also, there is a "thick thin-film" domain in which the interfacial tensions 12 and 23 between the film and adjoining phases are the same as those of bulk phase 2 against phases 1 and 3, respectively. Phases behave as thick thin-films when h is a few times greater than the thickness of the interfacial zones between phases 1 and 2 and 2 and 3. For nonpolymeric fluids not too close to a critical point, a film typically behaves as a thick thin-film when h is greater than about 100 Å. The disjoining pressure of thin thin-films (which are just adsorbed submonolayers in limit of ultimate thinness) is a complicated oscillatory function of h if the adjoining phases are solids. This behavior is due to the formation of fluid layers between confining solids. Layers are favored (pressure minima) at thickness h allowing an integral number of them.

The disjoining pressure of thick thin-films of nonpolar molecules interacting with adjoining phases via nonpolar pair potentials depends on intermolecular separation as

$$u_{ij} = -e_{ij}/s^6$$

is of the form

$$\Pi(h) = -A_H/6\pi h^3 \tag{7.6.2}$$

where the Hamaker constant is given by

$$A_H = \pi^2(n_1 n_3 e_{13} + n_2^2 3_{22} - n_1 n_2 e_{12} - n_2 n_3 e_{23}). \tag{7.6.3}$$

n_i denotes the density of material i. Equation (7.6.2) has been shown to be valid, for example, for alkanes and liquid helium on glass or quartz in the presence of their vapors; A_H can be negative or positive. In the case of octane on quartz

$$\Pi(h) = \frac{0.9 \times 10^{-14}\text{erg}}{h^3} \tag{7.6.4}$$

for $h < 1$ μm. Owing to retardation effects, nonpolar attractive interactions go as $u_{ij} \propto -s^{-7}$ at large separations. The effect of this on the disjoining pressure is that Π is proportional to h^{-4} when $h > 1$ μm.

Water is a polar, hydrogen bonded fluid. Its measured disjoining pressure on quartz can be approximated by

$$\begin{aligned} \Pi(h) &= \frac{41\text{dyn/cm}}{h}, \quad h < 800\text{Å} \\ &= \frac{2 \times 10^{-7}\text{dyn}}{h^2}, \quad h > 1200\text{Å}. \end{aligned} \tag{7.6.5}$$

These formulas were measured in different experiments that did not overlap in the range of thicknesses. Below some tens of angstroms, eq. (7.6.5) fails.

If electrolyte is present in the film, there is a double layer contribution to the disjoining pressure. At low ionic concentrations, this contribution is

$$\Pi_{\text{el}} = 64 C_i R_g T \chi e^{-\kappa h}, \tag{7.6.6}$$

where C_i is ion concentration in bulk phase, κ^{-1} the Debye double layer thickness, and χ a parameter characteristic of the adjoining phases.

At small film thickness, as mentioned above, pressure of fluid confined between solids becomes an oscillatory function of h as illustrated in Figure 7.26. The interactions and fluid rearrangements responsible for the behavior depicted in this figure are often referred to as solvation forces. Note that length scale in Figure 7.26 is the molecular diameter of the fluid particles, and so on the

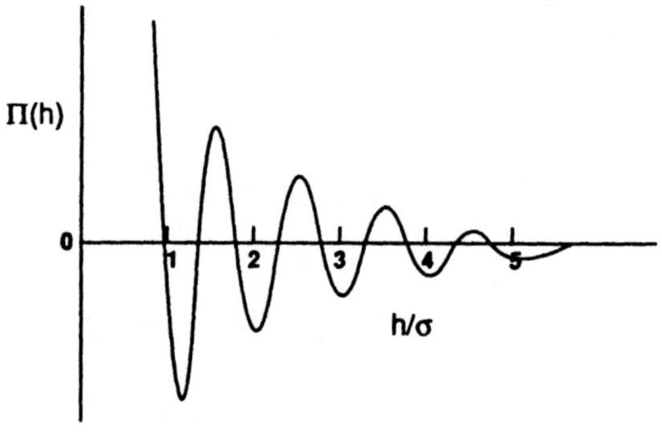

Figure 7.26
Disjoining pressure of a film confined between flat solid surfaces as a function of distance of separation of h. Distance is given in units of diameter of the fluid molecules.

coarse scale on which Eqs. (7.6.4)–(7.6.6) are observed, the short-range behavior shown in Figure 7.26 might appear to be a nearly vertical, strongly positive, disjoining pressure. In his summary of the various types of expected disjoining pressures, Lifshitz has indeed collapsed the short-range solvation force behavior into a steep monotonic repulsion at molecular distances. Lifshitz' (1956) catalog of disjoining pressures is shown in Figure 7.27. These result from the assumption that $\Pi(h) = \Pi_s + \Pi_{el} + A/h^3 + B/h^4$ and considering various signs for A and B, zero ionic concentration ($\Pi_{el} = 0$), nonzero ionic concentration ($\Pi_{el} > 0$), and that the oscillatory structure of the solvation force contribution, Π_s, is absent.

For describing the meniscus shape of fluids A and B in contact with a solid on which a film can form, one derives from the YL equation the augmented YL equation. Ignore for the moment the effect of gravity. The YL equation still holds for the pressure jump across the interface, that is, $\Delta P = P_f - P_A = 2\gamma H$. Thus, because $P_f = P_B + \Pi(h)$, the YL equation becomes the augmented YL equation

$$2\gamma H(h) - \Pi(h) = P_B - P_A. \tag{7.6.7}$$

In the presence of gravity, the equation of hydrostatics gives for the pressure in phase i

$$P_i = P_{io} + \rho_i \mathbf{g} \cdot (\mathbf{r} - \mathbf{r}_o), \tag{7.6.8}$$

where P_{io} is the pressure at some reference point $\mathbf{r}_0$ and h is the distance of the meniscus from the solid surface, whose unit normal is denoted as $\hat{e}_z$. The direction of the acceleration gravity is not

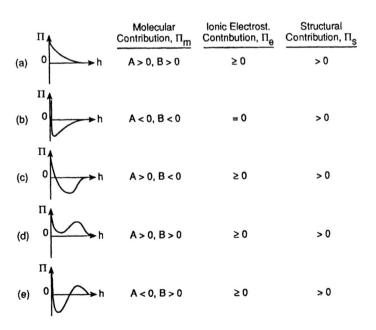

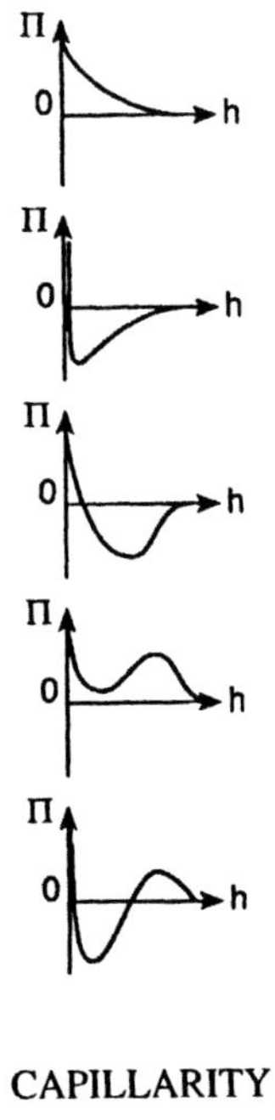

	Molecular Contribution, Π_m	Ionic Electrost. Contribution, Π_e	Structural Contribution, Π_s
(a)	$A > 0, B > 0$	≥ 0	> 0
(b)	$A < 0, B < 0$	$= 0$	> 0
(c)	$A > 0, B < 0$	≥ 0	> 0
(d)	$A > 0, B > 0$	≥ 0	> 0
(e)	$A < 0, B > 0$	≥ 0	> 0

Figure 7.27
Lifshitz (1956) catalogue for dependence of disjoining pressure on film thickness.

necessarily paralleled to $\hat{e}_z$. If $\mathbf{r}_0$ is chosen as a point x_0, y_0, and $z_0 = 0$ on the solid, then the meniscus position at a given value of x and y is $\mathbf{r} = \mathbf{r}_0 + (x - x_0)\hat{e}_x + (y - y_0)\hat{e}_y + h\hat{e}_z$, and so the pressure there is

$$P_i = P_{io} + \rho_i[(x - x_0)\mathbf{g} \cdot \hat{e}_x + (y - y_0)\mathbf{g} \cdot \hat{e}_y + h\mathbf{g} \cdot \hat{e}_z], \quad (7.6.9)$$

where $\hat{e}_x$ and $\hat{e}_y$ are orthogonal unit vectors lying in the plane of the solid surface. Substituting eq. (7.6.9) for the pressures in eq. (7.6.7), we obtain the augmented equation of capillary hydrostatics, namely,

$$2\gamma H(h) - \Pi(h) - \Delta\rho[(x - x_0)\mathbf{g} \cdot \hat{e}_x + (y - y_0)\mathbf{g} \cdot \hat{e}_y + h\mathbf{g} \cdot \hat{e}_z] = \Delta P_o, \quad (7.6.10)$$

where $\Delta P_0 \equiv P_B(\mathbf{r}_o) - P_A(\mathbf{r}_o)$ and $\Delta\rho \equiv \rho_2 - \rho_3$. Equation (7.6.10), plus appropriate boundary conditions, determines the shape of a meniscus of phase 2 in equilibrium with its thin film. On a flat solid whose normal is vertical ($g = -g_{e_z}$ a thin film of z in equilibrium with bulk fluid z is illustrated in Figure 7.28. The boundary conditions in this case depend on the volume of phase 2.

If the solid is a vertical flat wall ($\mathbf{g} \cdot \hat{e}_x = -g$, $\mathbf{g} \cdot \hat{e}_y = \mathbf{g} \cdot \hat{e}_z = 0$) and if the meniscus is translationally symmetric and becomes flat at $x_o = 0$ far from the wall (i.e., as $h \to \infty$), then $\Delta P_o = 0$ and eq. (7.6.10) becomes

$$\frac{+\gamma h_{xx}}{(1 + h_x^2)^{3/2}} + \Delta\rho g x - \Pi(h) = 0. \quad (7.6.11)$$

Here x denotes the vertical height in the direction of gravity and $h(x)$ is the distance of the meniscus from the wall at position x. High enough up the wall, the meniscus will be sufficiently flat for the curvature term to be negligible. In this case, for the disjoining potential $\Pi(h) = A/h^3$, we obtain the asymptotic solution

$$h \simeq (A/\Delta\rho g x)^{1/3}. \quad (7.6.12)$$

For octane on quartz in the presence of air $A = 0.9 \times 10^{-14}$ erg and $\Delta\rho = 0.7$ g/cm^3. Thus the thickness of a film of octane on quartz will be $h \simeq 2.36 \times 10^{-6}$ cm $(1 \text{ cm}/x)^{1/3}$ in the asymptotic region. This predicts that the film is 236 Å thick 1 cm above the flat surface at x_o and 51 Å thick 1 m above x_o. On a flat horizontal solid (i.e., a

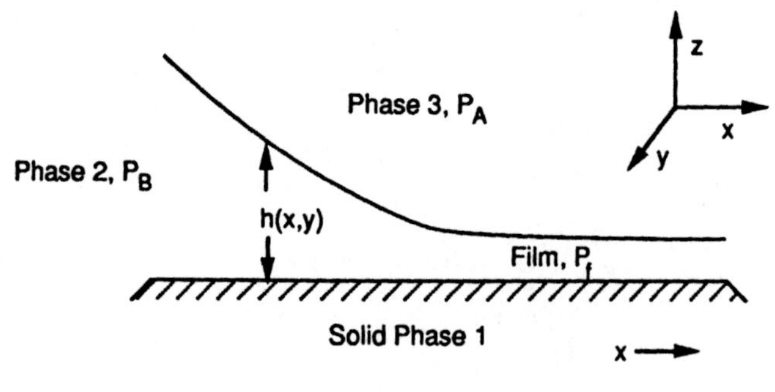

Figure 7.28
Meniscus of fluid 2 in equilibrium with its thin film between fluid phase 3 and solid phase 2.

solid whose normal is in the direction of gravity), certain disjoining pressures allow planar thin films and others allow coexistence of thin films of different thickness. For a horizontal flat solid, eq. (7.6.10) for a translationally symmetric system becomes

$$\frac{-\gamma h_{xx}}{(1+h_x^2)^{3/2}} - \Pi(h) + \Delta\rho g h = \Delta P_o. \tag{7.6.13}$$

Multiplying this equation by h_x, we obtain

$$\gamma \frac{d}{dx}(1+h_x^2)^{-1/2} = \frac{dh}{dx}\mathcal{P}(h) \tag{7.6.14}$$

where

$$\mathcal{P}(h) \equiv \Delta P_o - \Delta\rho g h + \Pi(h). \tag{7.6.15}$$

Defining

$$\mathcal{A}(h) \equiv \int_h^{ho} \mathcal{P}(h')dh' \tag{7.6.16}$$

we can write eq. (7.6.14) in the form

$$\gamma \frac{d}{dx}(1+h_x^2)^{-1/2} = \frac{-d\mathcal{A}(h)}{dx} \tag{7.6.17}$$

that integrates to

$$\gamma(1+h_x^2)^{-1/2} = -\mathcal{A}(h) + K. \tag{7.6.18}$$

This result can be rearranged to yield

$$\frac{dh}{\{[\gamma/(K-\mathcal{A}(h))]^2 - 1\}^{1/2}} = dx. \tag{7.6.19}$$

To seek coexisting flat films, we require that $h(x) \to h_1$, as $x \to -\infty$ and $h(x) \to h_2$ as $x \to \infty$. Equations (7.6.13) and (7.6.18) and the asymptotic boundary conditions imply that

$$\mathcal{P}(h_1) = \mathcal{P}(h_2) = 0 \tag{7.6.20}$$

and

$$\mathcal{A}(h_1) = \mathcal{A}(h_2) \quad \text{or} \quad \int_{h1}^{h2} \mathcal{P}(h)dh = 0. \tag{7.6.21}$$

For these equations to have a nontrivial solution (i.e., one for which $h_1 \neq h_2$), the quantity $\mathcal{P}(h)$ must be nonmonotonic. Disjoining pressures (a)–(c) in Figure 7.27 do not admit coexisting thin films because eq. (7.6.21) cannot be satisfied. The condition can be satisfied by (d) and (e). As illustrated in Figure 7.29, eq. (7.6.21) represents an "equal area tie-line" construction. Disjoining pressures of the types (d) and (e) in Figure 7.27 will admit coexisting thin films if the areas a_1 and a_2 shown in Figure 7.29 can be chosen to be equal.

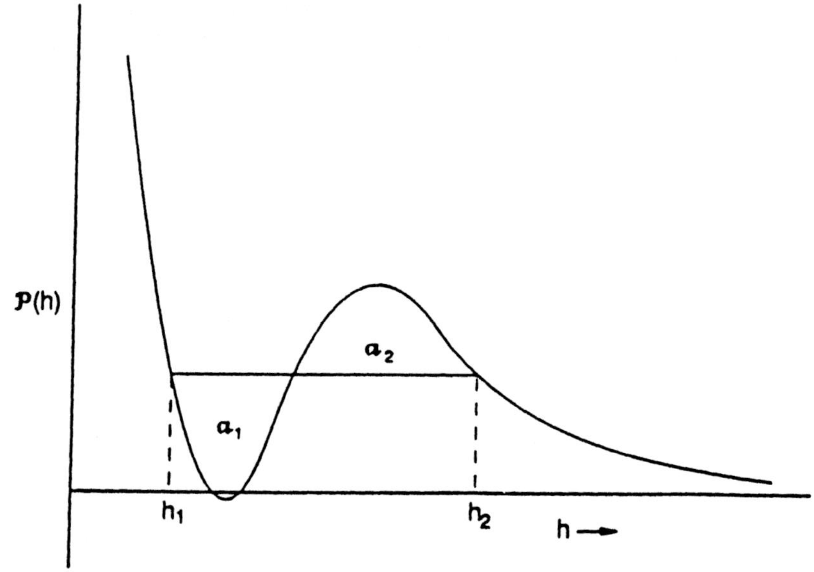

Figure 7.29
Equal area tie-line construction of coexisting thin films.

If coexisting flat thin films exist, one can choose $\Delta P_o = 0$, $h_o = h_1$, $\mathcal{A}(h_1) = 0$ and $K = \mathcal{A}(h_1) + \gamma = \gamma$. The film thickness profile can then be computed from

$$\int_{(h_1+h_2)/2}^{h(x)} \frac{dh}{\{[\gamma/(\gamma - \mathcal{A}(h))]^2 - 1\}^{1/2}} = x \tag{7.6.22}$$

where the origin $x = 0$ of the coordinate system has been chosen so that $h(0) = (h_1 + h_2)/2$ as indicated in Figure 7.30. The detailed structure of the meniscus in the interfilm region between the coexisting films depends primarily on the shape of the disjoining pressure because the effect of gravity will usually be rather small in thin films. For example, for a 1000 Å thick film of octane on quartz, $\Pi(h) = 9\ \text{dyn/cm}^2$, whereas $\Delta\rho g h = 6.9 \times 10^{-3}\ \text{dyn/cm}^2$.

Flat thin film solutions of arbitrary thickness exist only for (a) of those shown in Figure 7.27. The reason is that the film is mechanically unstable when $\partial\mathcal{P}/\partial h > 0$, because in this case, the film could thin spontaneously with a resulting reduction in film pressure. This is analogous to bulk fluids that are unstable when the compressibility is negative, that is, when $\partial P/\partial V > 0$.

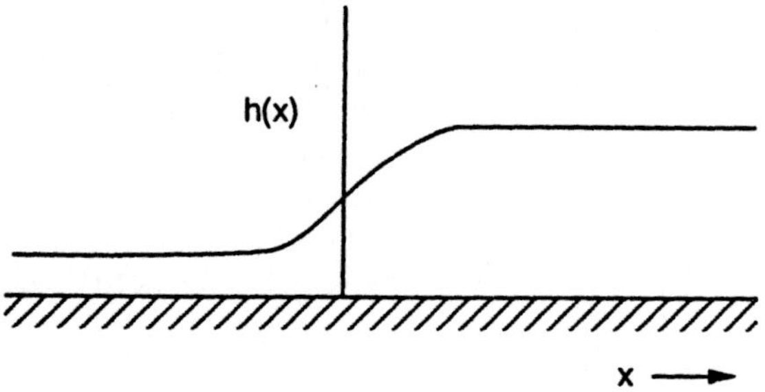

Figure 7.30
Coexisting thin films.

Let us close this section by noting that a "disjoining potential" is sometimes used instead of the disjoining pressure in describing one-component thin films. The disjoining pressure is defined by $\Pi(h) = P_f(h) - P(h = \infty)$, where $P_f(h)$ is the pressure of the film of thickness h in chemical equilibrium with bulk phase at pressure $P(h = \infty)$. A disjoining potential $\Delta\mu_2(h)$ is defined as the difference between the chemical potential of the film and the chemical potential of bulk phase at the film pressure P_f, that is,

$$\Delta\mu_2(h) = \mu_2(P_f, h) - \mu_2(P_f, h = \infty). \tag{7.6.23}$$

Bulk phase 2 obeys the Gibbs–Duhem equation

$$d\mu_2 = \frac{1}{n_2} dP \tag{7.6.24}$$

that can be integrated to obtain

$$\mu_2(P_f, h = \infty) = \mu_2(P, h = \infty) + \int_P^{P+\Pi} \frac{1}{n_2} dP. \tag{7.6.25}$$

Because $\mu_2(P_f, h) = \mu_2(P, h = \infty)$ at chemical equilibrium, it follows from eqs. (7.6.23) and (7.6.25) that

$$\Delta\mu_2(h) = -\int_P^{P+\Pi(h)} \frac{1}{n_2} dP. \tag{7.6.26}$$

For an incompressible film

$$\Delta\mu_2(h) = -\frac{1}{n_2}\Pi(h) \tag{7.6.27}$$

in which case $\Delta\mu_2(h)$ and $-\Pi(h)$ differ only by a constant factor.

Supplementary Reading

Davis, H. T. and Scriven, L. E. 1982. Adv. Chem. Phys., **49**, 357.

Evans, D. F. and Wennerström, H. 1994. *The Colloidal Domain. Where Physics, Chemistry, Biology and Technology Meet,* VCH Publishers, New York.

Derjaguin, B. V., Churaev, N. V., and Muller, V. M. in V. I. Kisin, Trans., J. A. Kitchner, Ed. *Surface Forces,* 1989. Consultants Bureau, New York.

Henderson, D., Ed. 1992. *Fundamentals of Inhomogeneous Fluid,* Dekker, New York.

Hunter, R. J. 1993. *Introduction to Modern Colloid Science,* Oxford University Press, New York.

Israelachvili, J. N. 1991. *Intermolecular and Surface Forces,* Academic Press, New York.

Mahanty, J. and Ninham, B. W. 1976. *Dispersion Forces,* Academic Press, New York.

Safran, S. A. 1994. *Statistical Thermodynamics of Surfaces, Interfaces and Membranes,* Addison-Wesley, Reading, MA.

Exercises

1 The contact angle of tin on tungsten is 0°, calculate

a) the capillary rise of tin in the presence of its vapor in a tungsten capillary of radius 0.07 cm and

b) the meniscus profile of a large pool of tin against a wide vertical flat plate of tungsten.

2 Assume that the droplets in Figure 7.13 are water in the presence of air at room temperature. Calculate the pressure inside the droplet for each of the cases shown in the figure.

3 Suppose that the volume distribution of porosity of a porous solid is

$$\hat{v}(R) = \frac{2R}{\sigma^2} e^{-R^2/\sigma^2},$$

where $\sigma = 500$ Å. Consider water at room temperature. Assume water has a 0° contact angle with the solid in the presence of its vapor. Define the saturation S of water in the porous solid as the fraction of porosity filled with water. Calculate the saturation S as a function of the pressure of water vapor at room temperature.

4 Consider a vertical flat quartz plate whose bottom edge is immersed in octane at 20°C. Solve eq. (7.6.11) for the meniscus h versus the distance x from the surface of the plate.

5 The excess free energy of a two-phase fluid system with meniscus at postion $h(x, y)$ above the xy plane given by

$$\Delta F(h) = \gamma \int \sqrt{1 + h_x^2 + h_y^2} dxdy - \Delta P \int \eta(h(x, y) - z) dxdydz,$$

where $h_x = \partial h/\partial x$, $h_y = \partial h/\partial y$ and $\eta(\zeta)$ is a step function that is zero when $\zeta < 0$ and one when $\zeta > 0$. Gravity is assumed to be negligible.
The meniscus takes the equilibrium shape $h(x, y)$ that minimizes ΔF. Thus,

$$\frac{d\Delta F}{d\varepsilon}(h + \varepsilon v) = 0 \quad \text{at} \quad \varepsilon = 0, (A)$$

where $v(x, y)$ is an arbitrary function that vanishes at the boundary of the meniscus. Prove that condition (A) yields the YL equation

$$\Delta P = 2H\gamma,$$

where

$$2H = \text{Equation(7.3.31)}.$$

6 Suppose water on glass has the disjoining pressure given by eq. (7.6.5). Using either the method of finite differences or finite elements, solve eq. (7.6.11) for the film thickness versus height for water rising on a vertical flat glass surface.

7 A pair of glass disks 10 cm in diameter are held 1 mm apart and the space between them is filled with water as indicated below:

GLASS PLATE
WATER
GLASS PLATE

The system is in air at atmospheric pressure. Estimate the force needed to separate the plates. The contact angle that the water/air meniscus makes with the glass surface is 0° and the surface tension of the water is 72 dyn/cm.

8 Consider a drop of liquid phase A resting on a flat solid in the presence of air. Suppose the contact angle that the air–liquid meniscus makes with the solid is 90°. Neglect gravity and assume that the drop volume is 4×10^{-3} mL.

a) What is the radius of the drop?

b) Sketch the drop.

c) Sketch the drop qualitatively if gravity is present.

9 Consider the height $h(x)$ of the air–water meniscus between two vertical parallel plates contacting a pool of water. The equation of capillary hydrostatics for this case is

$$\frac{-\gamma h_{xx}}{(1+h_x^2)^{3/2}} = -\Delta\rho g h + \Delta P_o. \qquad (*)$$

Suppose that the contact angle of the meniscus at the wall is 45° and that the gap between the plates is 2 mm. For water $\gamma = 72$ dyn/cm and $\Delta\rho = 1$ g/cm^3.

a) Linearize eq. (*) and solve the resulting equation. Chose the datum for h such that $h(x = 1 \text{ mm}) = 0$.

b) Calculate $h(0)$, $h(2\text{mm})$, and ΔP_o.

c) Calculate how far the minimum point $h(x = 1\text{mm})$ of the liquid meniscus lies above the level of the pool of water. Sketch the system qualitatively.

References

Adamson, A. W. 1990. *Physical Chemistry of Surfaces,* Wiley, New York.

Cahn, J. W. 1977. *J. Chem. Phys.*, **66,** 3667.

Davis, H. T. and Scriven, L. E. 1976. *J. Phys. Chem.*, **80,** 2805.

Fowkes, F. M. 1962. *J. Phys. Chem.*, **66,** 1863; 1964. *Ind. Eng. Chem.*, **56,** 40.

Girifalco, L. A. and Good, R. J. 1957. *J. Phys. Chem.*, **61,** 904.

Guggenheim, E. A. 1945. *J. Chem. Phys.*, **13,** 253.

Jasper, J. J. 1972. *J. Phys. Chem.*, **1,** 841.

Lifshitz, E. M. 1956. *Soviet Physics JETP,* **2,** 480.

Mao, G., Tsao, Y., Tirell, M., Davis, H. T., Hessel, V., and Ringsdorf, H. 1993. **9,** 3461.

Moldover, M. R. and Cahn, J. W. 1980. *Science,* **207,** 1073.

Turkevich, L. Private communication, 1995.

Washburn, E. W., Ed. 1928. *International Critical Tables,* Vol. IV, McGraw–Hill, New York.

Weast, R. C., editor. *Handbook of Chemistry and Physics.* 1972–1973. 53rd ed., CRC Press, New York.

Winterfeld, P., Scriven, L. E., and Davis, H. T. 1978. *AIChE J.,* **24,** 1010.

Zisman, W. A. 1964. *Contact Angle, Wettability, and Adhesion,* Advances in Chemistry, Series 43, ACS, Washington, D. C.

8

STRUCTURE AND STRESS IN SIMPLE FLUIDS AND THEIR MIXTURES: APPLICATIONS TO INTERFACES

8.1 Density Distribution Functions: The Yvon–Born–Green Hierarchy

To understand the role of molecular interactions in the equilibrium behavior of fluids, it is useful to investigate density distribution functions. We will restrict our attention to fluids of particles interacting via pairwise additive, centrally symmetric forces in the presence of conservative external forces. Thus, in a system of N identical particles, the total potential energy u^N is

$$u^N(\mathbf{r}_1, \ldots, \mathbf{r}_N) = \sum_{i>j=1}^{N} u(r_{ij}) + \sum_{i=1}^{N} u^e(\mathbf{r}_i), \tag{8.1.1}$$

where $u(r_{ij})$ is the potential energy between the pair of particles i and j and $u^e(\mathbf{r}_i)$ is the potential of the external force on particle i. Typical external forces are gravitational, centrifugal, or electrical. More interesting in this text, however, is the case in which the external potential arises from interaction between the fluid phase and an adjoining solid.

Recall from Chapter 4 that the s-particle configuration probability distribution function for a canonical ensemble of N particles is given by

$$\begin{aligned}\mathcal{P}^{(s,N)}(\mathbf{r}_1, \ldots, \mathbf{r}_s) &= \int_V \cdots \int e^{-\beta u^N} (d^3r)^{N-s} \Big/ \int_V \cdots \int e^{-\beta u^N} (d^3r)^N \\ &= \int_V \cdots \int e^{-\beta u^N} (d^3r)^{N-s} / Z_N.\end{aligned} \tag{8.1.2}$$

It is sometimes convenient to replace the probability distribution function by the density distribution function

$$n^{(s,N)}(\mathbf{r}_1, \ldots, \mathbf{r}_s) \equiv \frac{N!}{(N-s)!}\mathcal{P}^{(s,N)}(\mathbf{r}_1, \ldots, \mathbf{r}_s), \qquad (8.1.3)$$

ordinary number density of the fluid at position $\mathbf{r}_1$ [later when no confusion is caused we identify the number density by the simpler notation $n(\mathbf{r}_1)$ used earlier in the text]. The quantity $\mathcal{P}^{(s,N)}$ was represented by the simpler notation $\mathcal{P}_s$ in Chapter 5. The reason for the complication of added notation here is to distinguish between canonical and grand canonical ensemble quantities and to reserve subscripts for labeling molecular species.

For an open system, the canonical ensemble distribution functions must be replaced by the corresponding grand canonical ensemble quantities, namely,

$$\mathcal{P}^{(s)} = \sum_{N=s}^{\infty} \mathcal{P}^{(s,N)} p(N) = \sum_{N=s}^{\infty} \frac{\mathcal{P}^{(s,N)} Q_N e^{\beta N \mu}}{\Xi} \qquad (8.1.4)$$

and

$$n^{(s)} = \sum_{N=s}^{\infty} \frac{n^{(s,N)} Q_N e^{\beta N \mu}}{\Xi}, \qquad (8.1.5)$$

where Q_N and Ξ are the partition functions of the canonical and grand canonical ensembles, respectively.

By differentiating eq. (8.1.2) with respect to the position one of the particles 1, 2, $\ldots$, s, say of particle i, we obtain an integrodifferential equation that leads to what is known as an Yvon–Born–Green equation. In particular,

$$\begin{aligned}\nabla_{\mathbf{r}_i} \mathcal{P}^{(s,N)} &= -\beta \left[\nabla_{\mathbf{r}_i} u^e(\mathbf{r}_i) + \sum_{j \neq i}^{s} \nabla_{\mathbf{r}_i} u(r_{ij}) \right] \mathcal{P}^{(s,N)} \\ &\quad - \beta \sum_{j=s+1}^{N} \int \nabla_{\mathbf{r}_i} u(\mathbf{r}_{ij}) \left[\int \cdots \int e^{-\beta u^N} (d^3 r)^{N-s-1} / Z_N \right] d^3 r_j .\end{aligned} \qquad (8.1.6)$$

Using the definition of $\mathcal{P}^{(s+1,N)}$ and noting that every term in the sum $\sum_{j=s+1}^{N}$ on the right-hand side (rhs) of eq. (8.1.6) is identical, we can reduce eq. (8.1.6) to the form

$$\begin{aligned}-kT \nabla_{\mathbf{r}_i} \mathcal{P}^{(s,N)} &= \left[\nabla_{\mathbf{r}_i} u^e(\mathbf{r}_i) + \sum_{j \neq i}^{s} \nabla_{\mathbf{r}_i} u(r_{ij}) \right] \mathcal{P}^{(s,N)} \\ &\quad (N-s) \int \nabla_{\mathbf{r}_i} u(r_{i,s+1}) \mathcal{P}^{(s+1,N)} d^3 r_{s+1} .\end{aligned} \qquad (8.1.7)$$

The equation is usually expressed in terms of density distributions instead of probability distributions. To do this, we multiply by

$N!/(N-s)!$, use the fact $[N!/(N-s)!](N-s) = N!/(N-s-1)!$, apply the definition at eq. (8.1.3), and find

$$-kT\nabla_{\mathbf{r}_i} n^{(s,N)} = \left[\nabla_{\mathbf{r}_1} u^e(\mathbf{r}_1) + \sum_{j\neq i}^{s} \nabla_{\mathbf{r}_i} u(r_{ij})\right] n^{(s,N)} + \int \nabla_{\mathbf{r}_i} u(\mathbf{r}_{i,s+1}) n^{(s+1,N)} d^3 r_{s+1}. \tag{8.1.8}$$

For $s = 1, 2, \ldots, N-1$, eq. (8.1.8) represents the YBG hierarchy of integrodifferential equations of the canonical ensemble (i.e., closed isothermal systems). Averaging eq. (8.1.8) as indicated at eq. (8.1.5) converts the hierarchy to the set

$$-kT\nabla_{\mathbf{r}_i} n^{(s)} = \left[\nabla_{\mathbf{r}_i} u^e(\mathbf{r}_i) + \sum_{j\neq i}^{s} \nabla_{\mathbf{r}_i} u(r_{ij})\right] n^{(s)} + \int \nabla_{\mathbf{r}_i} u(r_{i,s+1}) n^{(s+1)} d^3 r_{s+1}, \tag{8.1.9}$$

which are the YBG equations of the grand canonical ensemble (i.e., open isothermal systems).

The results obtained above for a one-component fluid generalize in a straightforward way to a solution containing ν different chemical species. The s-body probability distribution function is represented by $\mathcal{P}^{(s,\mathbf{N})}_{\{\alpha\}}(\mathbf{r}_1, \ldots, \mathbf{r}_s)$, where bold $\mathbf{N}$ denotes the set of species numbers $\{N_1, \ldots, N_\nu\}$, with $N = \sum_{\alpha=1}^{\nu} N_\alpha$, and $\{\alpha\}$ denotes the species labels of the set of particles at $\mathbf{r}_1, \mathbf{r}_2, \ldots, \mathbf{r}_s$. In general, none, some or all of the species α_i can be different. The multicomponent density distribution function is defined by the equation

$$n^{(s,\mathbf{N})}_{\{\alpha\}}(\mathbf{r}_1, \ldots, \mathbf{r}_s) = \frac{\prod_{i=1}^{s} N_{\alpha_i}!}{\prod_{i=1}^{s}(N_{\alpha_i} - s_{\alpha_i})!} \mathcal{P}^{(s,\mathbf{N})}_{\{\alpha\}}(\mathbf{r}_1, \ldots, \mathbf{r}_s), \tag{8.1.10}$$

where N_{α_i} is the total number of particles of species α_i in the system and s_{α_i} equals the number of the particles in the set of s particles that are of type α_i. The product Π is over the different particles represented in the set $\{\alpha\}$. For example, if $s = 2$ and one particle is of type α and the other of type β, then eq. (8.1.10) reads

$$n^{(2,N)}_{\alpha\beta}(\mathbf{r}_1, \mathbf{r}_2) = \frac{N_\alpha! N_\beta!}{(N_\alpha - 1)!(N_\beta - 1)!} \mathcal{P}^{(2,N)}_{\alpha\beta}(\mathbf{r}_1, \mathbf{r}_2) = N_\alpha N_\beta \mathcal{P}^{(2,N)}_{\alpha\beta}(\mathbf{r}_1, \mathbf{r}_2). \tag{8.1.11}$$

The grand canonical ensemble s-body density distribution function in the ν-component system is

$$n^{(s)}_{\{\alpha\}} = \sum_{N_1=0}^{\infty} \cdots \sum_{N_\nu=0}^{\infty} \frac{n^{(s,\mathbf{N})}_{\{\alpha\}} Q_{\mathbf{N}} e^{\beta \mathbf{N}\cdot\mu}}{\Xi}, \quad \sum_{\alpha=1}^{\nu} N_\alpha \geq s, \tag{8.1.12}$$

where $\mathbf{N} \cdot \boldsymbol{\mu} = \sum_{\alpha=1}^{\nu} N_\alpha \mu_\alpha$ and

$$\Xi = \sum_{N_1=0}^{\infty} \cdots \sum_{N_\nu=0}^{\infty} Q_{\mathbf{N}} e^{\beta \mathbf{N}\cdot\boldsymbol{\mu}} \tag{8.1.13}$$

with $Q_{\mathbf{N}}$ the canonical ensemble partition function of the system.

With the notation given above, the YBG equations for a ν-component system of particles interacting via central, pair additive forces are

$$-kT\nabla_{\mathbf{r}_i} n_{\{\alpha\}}^{(s,\mathbf{N})} = \left[\nabla_{\mathbf{r}_i} u_{\alpha_i}^{e}(\mathbf{r}_i) + \sum_{j\neq i}^{s} \nabla_{\mathbf{r}_i} u_{\alpha_i\alpha_j}(r_{ij})\right] n_{\{\alpha\}}^{(s,N)} + \sum_{\beta=1}^{\nu} \int \nabla_{\mathbf{r}_i} u_{\alpha_i\beta}(\mathbf{r}_{i,s+1}) n_{\{\alpha\}}^{(s+1,\mathbf{N})} d^3 r_{s+1} \tag{8.1.14}$$

for a canonical ensemble and

$$-kT\nabla_{\mathbf{r}_i} n_{\{\alpha\}}^{(s)} = \left[\nabla_{\mathbf{r}_i} u_{\alpha_1}^{e}(\mathbf{r}_i) + \sum_{j\neq i}^{s} \nabla_{\mathbf{r}_i} u_{\alpha_i\alpha_j}(r_{ij})\right] n_{\{\alpha\}}^{(s)} + \sum_{\beta=1}^{\nu} \int \nabla_{\mathbf{r}_i} u_{\alpha_i\beta}(r_{i,s+1}) n_{\{\alpha\}}^{(s+1)} d^3 r_{s+1} \tag{8.1.15}$$

for a grand ensemble.

As an example of eq. (8.1.15), suppose that there are two components and that $s = 1$. Then the density $n_\alpha^{(1)}$ of component α obeys the equation

$$-kT\nabla_{\mathbf{r}} n_\alpha^{(1)} = \nabla_{\mathbf{r}} u_\alpha^{e}(\mathbf{r}) n_\alpha^{(1)}(\mathbf{r}) + \int \nabla_{\mathbf{r}} u_{\alpha 1}(|\mathbf{r}-\mathbf{r}'|) n_{\alpha 1}^{(2)}(\mathbf{r},\mathbf{r}') d^3 r' + \int \nabla_{\mathbf{r}} u_{\alpha 2}(|\mathbf{r}-\mathbf{r}'|) n_{\alpha 2}^{(2)}(\mathbf{r},\mathbf{r}') d^3 r'. \tag{8.1.16}$$

Because $\alpha = 1$ and 2 it follows that there are two YBG equations relating the densities $n_1^{(1)}$ and $n_2^{(1)}$ to the three doublet density distribution functions $n_{11}^{(2)}$ and $n_{22}^{(2)}$ and $n_{12}^{(2)}$.

Because in the equation for $n^{(s)}$ the density distribution $n^{(s+1)}$ appears, the YBG equations are useful for computing distribution functions only if some approximation is introduced to truncate the hierarchy. The development of such approximations has been an area of active interest for several decades.

One of the simplest and oldest such approximations is the so-called superposition approximation, for which the motivation necessitates introduction of the concept of the potential of mean force. In a one-component canonical ensemble, the s-particle potential of mean force is defined by

$$e^{-\beta w^{(s,N)}(\mathbf{r}_1,\ldots,\mathbf{r}_s)} = V^s \mathcal{P}^{(s,N)}(\mathbf{r}_1,\ldots,\mathbf{r}_s). \tag{8.1.17}$$

With this definition, it can be shown easily that

$$\nabla_{\mathbf{r}_i} w^{(s,N)} = \overline{\nabla_{\mathbf{r}_i} u^N}^s, \quad (8.1.18)$$

where i is any one of the set of particles $\mathbf{r}_1, \ldots, \mathbf{r}_s$ and the mean force $\nabla_{\mathbf{r}_i} u^N$ on particle i is defined by

$$\overline{\nabla_{\mathbf{r}_i} u^N}^s = \frac{\int \cdots \int (\nabla_{r_i} u^N) e^{-\beta u^N} (d^3 r)^{N-s}}{\int \cdots \int e^{-\beta u^N} (d^3 r)^{N-s}}. \quad (8.1.19)$$

For a given configuration $\mathbf{r}_1, \ldots, \mathbf{r}_N$, the force exerted on molecule i is $\mathbf{F}_i = -\nabla_{\mathbf{r}_i} u^N$ and so $-\nabla_{\mathbf{r}_i} \omega^{(s,N)}$ is the mean value of this force averaged over all configurations $\mathbf{r}_{s+1}, \ldots, \mathbf{r}_N$ of the $N - s$ particles while holding the members of the set s fixed at positions $\mathbf{r}_1, \ldots, \mathbf{r}_s$.

If eqs. (8.1.17) and (8.1.7) are combined the following equation is obtained for the potential of mean force:

$$\nabla_{\mathbf{r}_i} w^{(s,N)} = \nabla_{\mathbf{r}_i} u^s + \frac{N-s}{V} \int \nabla_{\mathbf{r}_i} u(\mathbf{r}_{i,s+1}) e^{-\beta[w^{(s+1,N)} - w^{(s,N)}]} d^3 r_{s+1}, \quad (8.1.20)$$

where $u^s \left(\equiv \sum_{j>k=1}^{s} u(r_{jk}) + \sum_{j=1}^{s} u^e(\mathbf{r}_1)\right)$ is the potential of interaction of the set of particles 1, 2, ..., s. For $s \ll N$ and in the dilute gas limit,that is, $N/V \to 0$, the solution to eq. (8.1.20) is simply

$$w^{(s,N)} = u^s + \text{constant}, \quad (8.1.21)$$

or

$$\mathcal{P}^{(s,N)} = \alpha_s e^{-\beta u^s} \quad (8.1.22)$$

where α_s is a constant. In this case,

$$\mathcal{P}^{(1,N)}(\mathbf{r}_1) = \alpha_1 e^{-\beta u^e(\mathbf{r}_1)} \quad (8.1.23)$$

$$\mathcal{P}^{(2,N)}(\mathbf{r}_1, \mathbf{r}_2) = \alpha_2 e^{-\beta[u(r_{12}) + u^e(\mathbf{r}_1) + u^e(\mathbf{r}_2)]} \quad (8.1.24)$$

$$\mathcal{P}^{(3,N)}(\mathbf{r}_1, \mathbf{r}_2, \mathbf{r}_3) = \alpha_3 e^{-\beta[u(r_{12}) + u(r_{13}) + u(r_{23}) + u^e(\mathbf{r}_1) + u^e(\mathbf{r}_2) + u^e(\mathbf{r}_3)]}. \quad (8.1.25)$$

If we define the s-body correlation functions as

$$g^{(s,N)} \equiv \frac{\mathcal{P}^{(s,N)}}{\pi_{i=1}^{s} \mathcal{P}^{(1,N)}(\mathbf{r}_i)}, \quad (8.1.26)$$

it follows that in the dilute gas limit

$$g^{(2,N)}(r_{12}) = e^{-\beta u(r_{12})} \quad (8.1.27)$$

and

$$g^{(3,N)}(\mathbf{r}_1, \mathbf{r}_2, \mathbf{r}_3) = e^{-\beta[u(r_{12}) + u(r_{13}) + u(r_{23})]}$$
$$= g^{(2,N)}(r_{12}) g^{(2,N)}(r_{13}) g^{(2,N)}(r_{23}). \quad (8.1.28)$$

The approximation that the triplet correlation function equals the product of the pair correlation functions of the three pairs, as in eq. (8.1.28), is called the superposition approximation. In the dilute gas limit it is a rigorous result of the fact that the three-body potential is the superposition of the pair potentials. In analogy with this, if the s-body potential of mean force $\omega^{(s,N)}$ assumed to preserve the same additivity as u^s, which is

$$\omega^{(s,N)}(\mathbf{r}_1, \ldots, \mathbf{r}_s) - \sum_{i=1}^{s} \omega^{(1,N)}(\mathbf{r}_i)$$
$$= \sum_{i>j=1}^{s} [\omega^{(2,N)}(\mathbf{r}_i, \mathbf{r}_j) - \omega^{(1,N)}(\mathbf{r}_i) - \omega^{(1,N)}(\mathbf{r}_j)], \tag{8.1.29}$$

then the s-body correlation function obeys the superposition principle

$$g^{(s,N)}(\mathbf{r}_1, \ldots, \mathbf{r}_s) = \prod_{i>j=1}^{s} g^{(2,N)}(\mathbf{r}_i, \mathbf{r}_j), \quad s \geq 2. \tag{8.1.30}$$

As shown above, this relationship is valid for simple fluids in the limit of sufficiently low density. It is also exact for a one-dimensional fluid of hard-rod fluids at arbitrary density and composition.

In terms of density distribution functions in the grand canonical ensemble, it is convenient to define the s-body correlation function as

$$g^{(s)}(\mathbf{r}_1, \ldots, \mathbf{r}_s) = \frac{n^{(s)}(\mathbf{r}_1, \ldots, \mathbf{r}_s)}{\pi_{i=1}^{s} n^{(1)}(\mathbf{r}_i)}, \tag{8.1.31}$$

and state the superposition approximation as

$$g^{(s)}(\mathbf{r}_1, \ldots, \mathbf{r}_s) = \pi_{i>j=1}^{s} g^{(2)}(\mathbf{r}_t, \mathbf{r}_j) \tag{8.1.32}$$

Using this superposition approximation, we can express the first two grand canonical YBG equations as

$$-kT\nabla_{\mathbf{r}} n^{(1)}(\mathbf{r}) = \nabla_{\mathbf{r}} u^e(\mathbf{r}) n^{(1)}(\mathbf{r}) + \int \nabla_{\mathbf{r}} u(|\mathbf{r} - \mathbf{r}'|) n^{(2)}(\mathbf{r}, \mathbf{r}') d^3r' \tag{8.1.33}$$

and

$$-kT\nabla_{\mathbf{r}} n^{(2)}(\mathbf{r}, \mathbf{r}') = [\nabla_{\mathbf{r}} u^e(\mathbf{r}) + \nabla_{\mathbf{r}} u(|\mathbf{r} - \mathbf{r}'|)] n^{(2)}(\mathbf{r}, \mathbf{r}')$$
$$+ \int \frac{\int \nabla_{\mathbf{r}} u(|\mathbf{r} - \mathbf{r}''|) n^{(2)}(\mathbf{r}, \mathbf{r}') n^{(2)}(\mathbf{r}', \mathbf{r}'') n^{(2)}(\mathbf{r}, \mathbf{r}'')}{n^{(1)}(\mathbf{r}) n^{(1)}(\mathbf{r}') n^{(1)}(\mathbf{r}'')} d^3r''. \tag{8.1.34}$$

These two integrodifferential equations form a closed nonlinear system that can be solved for $n^{(1)}$ and $n^{(2)}$. Thus, the superposition approximation accomplishes the desired closure of the YBG hierarchy.

Equations (8.1.33) and (8.1.34) have not been studied a great deal for an inhomogeneous system. However, they have been studied extensively for a homogeneous fluid, the first studies going back to Kirkwood and coworkers (1949). In a homogeneous fluid, $u^e(\mathbf{r}) = 0$, $n^{(1)}$ is constant, and $n^{(2)}(\mathbf{r}, \mathbf{r}') = [n^{(1)}]^2 g^{(2)}(|\mathbf{r} - \mathbf{r}'|)$. Thus, eq. (8.1.33) drops out and eq. (8.1.34) simplifies to

$$
\begin{aligned}
&-kT\nabla_{\mathbf{r}} \ln g^{(2)}(|\mathbf{r} - \mathbf{r}'|) = \nabla_{\mathbf{r}} u(|\mathbf{r} - \mathbf{r}'|) \\
&n \int \nabla_{\mathbf{r}} u(|\mathbf{r} - \mathbf{r}''|) g^{(2)}(|\mathbf{r} - \mathbf{r}''|) g^{(2)}(|\mathbf{r}' - \mathbf{r}''|) d^3 r''.
\end{aligned}
\tag{8.1.35}
$$

With the coordinate transformations $\mathbf{s} = (\mathbf{r} - \mathbf{r}')$ and $\mathbf{s}' = \mathbf{r} - \mathbf{r}''$, eq. (8.1.35) can be rearranged to

$$
-kT \frac{d}{ds} \ln g^{(2)}(s) = u'(s) + n \int \frac{\mathbf{s} \cdot \mathbf{s}'}{ss'} u'(s') g^{(2)}(s') g^{(2)}(|\mathbf{s} - \mathbf{s}'|) d^3 s'. \tag{8.1.36}
$$

Equation (8.1.36) has been solved by a variety of numerical techniques. An early solution by Kirkwood, et al. (1952) using a Lennard–Jones potential is compared in Figure 8.1 with the correlation function deduced for argon from X-ray scattering data. Clearly, the superposition approximation captures the qualitative behavior of $g^{(2)}$. The most modern solution to eq. (8.1.35) is that of Kerins et al. (1986) and is based on finite element methods. In a later chapter on homogeneous fluid structure we will return to the superposition approximation along with other more recent approximations to the pair correlation function.

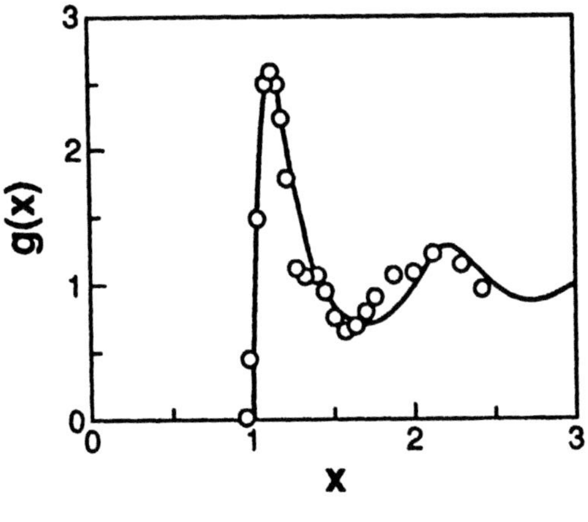

Figure 8.1
Pair correlation function. Solid line is superposition approximation (Kirkwood, et al., 1952) and points are from X-ray data (Eisenreich and Gingrich, 1942). $x = r/\sigma$, where r is the intermolecular separation and σ is the Lennard–Jones diameter of argon.

8.2 Pressure Tensor and Fluid Mechanical Equilibrium: Continuum Mechanics

As we discovered in Chapter 7, the pressure in an inhomogeneous material is not isotropic. It is in fact a local tensor function that measures the force exerted by matter on one side of a unit area on matter on the other side of the same unit area. Let us denote by $\mathbf{t}_{\hat{\varepsilon}}$dA the force exerted through the area dA by the material above the element ($\hat{\varepsilon}$ is the unit normal pointing above the element). The quantity $\mathbf{t}_{\hat{\varepsilon}}$ is called the traction vector. It has units of force per unit area and depends on the location $\mathbf{r}$ and orientation $\hat{\varepsilon}$ of dA in the material. It is, however, closely related to the pressure tensor $\mathbf{P}$ which is a function of location $\mathbf{r}$ but not orientation of dA. Thus, $\mathbf{P}$ is a material property.

To relate the traction vector to the pressure tensor we use the force balance on a Cauchy tetrahedron as shown in Figure 8.2. The center of the infinitesimal tetrahedron is located at $\mathbf{r}$ and $\hat{e}_1$, $\hat{e}_2$, and $\hat{e}_3$ are a set of Cartesian unit vectors ($\hat{e}_i \cdot \hat{e}_j = \delta_{ij}$). Accordingly, the area dA_i of the face of the tetrahedron whose normal is $-\hat{e}_i$ satisfies the relationship $dA_i = -\hat{e}_i \cdot \hat{\varepsilon} dA$. Newton's second law can be applied to the tetrahedron to obtain

$$(\rho dV)\frac{d\mathbf{U}}{dt} = \mathbf{t}_{\hat{\varepsilon}} dA - \sum_{i=1}^{3} \mathbf{t}_{\hat{e}_i} \hat{e}_i \cdot \hat{\varepsilon} dA + (\rho dV)\hat{\mathbf{F}}_e, \qquad (8.2.1)$$

where dV is the volume of the tetrahedron, $\mathbf{U}$, and ρ mass average velocity and average mass density of the material in the tetrahedron, and $\hat{\mathbf{F}}^e$ the external force per unit mass exerted on the contents of

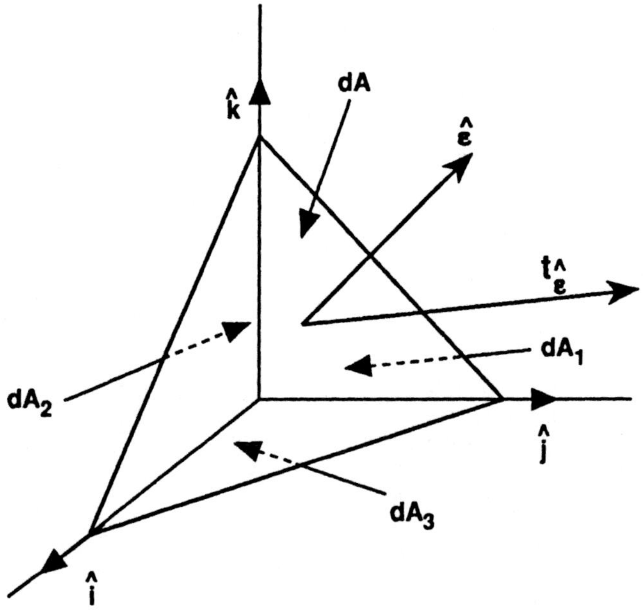

Figure 8.2
Cauchy tetrahedron for resolving the traction force $\mathbf{t}_{\hat{\varepsilon}}$ on dA.

the tetrahedron. Dividing eq. (8.2.1) by dA and passing to the limit $dA = 0$, noting that $dV/dA = 0$ in this limit, we obtain

$$\mathbf{t}_{\hat{\varepsilon}} = -\hat{\varepsilon} \cdot \mathbf{P}, \tag{8.2.2}$$

where the pressure tensor,

$$\mathbf{P} \equiv -\hat{e}_1 \mathbf{t}_{e_1} - \hat{e}_2 \mathbf{t}_{e_2} - \hat{e}_3 \mathbf{t}_{e_3}, \tag{8.2.3}$$

is a possible function of position $\mathbf{r}$ but is independent of $\hat{\varepsilon}$ since the Cartesian coordinate system, $\hat{e}_1$, $\hat{e}_2$, $\hat{e}_3$, is arbitrary.

In an isotropic fluid, the force exerted on a surface dA is $-\hat{\varepsilon} P dA$. Thus, in an isotropic fluid the pressure tensor is $P\mathbf{I}$, where $\mathbf{I}$ is the unit tensor,

$$\mathbf{I} = \sum_{i=1}^{3} \hat{e}_i \hat{e}_i, \tag{8.2.4}$$

having the property $\mathbf{I} \cdot \mathbf{a} = \mathbf{a}$ for arbitrary vector $\mathbf{a}$. In general the pressure tensor can be expressed as

$$\mathbf{P} = \sum_{i,j=1}^{3} P_{ij} \hat{e}_i \hat{e}_j, \tag{8.2.5}$$

where P_{ij} are the Cartesian components $\hat{e}_i \cdot \mathbf{P} \cdot \hat{e}_j$. It is usually presumed that $\mathbf{P}$ is symmetric, that is, that $P_{ij} = P_{ji}$. For the simple systems considered in this chapter the symmetry property will be proved in what follows.

If the material is a fluid, the momentum balance on an arbitrary fixed volume V with boundary A is

$$\frac{d}{dt} \int_V \rho \mathbf{u} dV = \int_A [-\rho \mathbf{u}(\hat{\varepsilon} \cdot \mathbf{u}) - \hat{\varepsilon} \cdot \mathbf{P}] dA + \int_V \rho \hat{\mathbf{F}}^e dV. \tag{8.2.6}$$

The accumulation of momentum in the volume V results from flow across the boundary, traction forces on the boundary, and external forces on the contents of V. The mass balance on the fixed volume V yields

$$\frac{d}{dt} \int_V \rho dV + \int (-\rho \hat{\varepsilon} \cdot \mathbf{u}) dA. \tag{8.2.7}$$

Using the property $d/dt \int_V f dV = \int (\partial f/\partial t) dV$ and the divergence theorem $\int_A \hat{\varepsilon} \cdot \mathbf{f} dA = \int_V \nabla \cdot \mathbf{f} dV$, we deduce from eq. (8.2.7) the mass continuity equation

$$\frac{\partial \rho}{\partial t} = -\nabla \cdot (\rho \mathbf{u}), \tag{8.2.8}$$

and from eq. (8.2.6) the equation of motion

$$\rho \frac{\partial \mathbf{u}}{\partial t} + \rho \mathbf{u} \cdot \nabla \mathbf{u} = -\nabla \mathbf{P} + \rho \hat{\mathbf{F}}^e. \tag{8.2.9}$$

At equilibrium $\mathbf{u} = 0$ and eq. (8.2.9) reduces to the equation of hydrostatics

$$0 = -\nabla \cdot \mathbf{P} + \rho \mathbf{F}^e, \tag{8.2.10}$$

which in the special case of an isotropic fluid, $\mathbf{P} = P\mathbf{I}$, has the more familiar form

$$0 = -\nabla P + \rho \mathbf{F}^e \tag{8.2.11}$$

presented in Chapter 1.

As we discover in the next section, for a one-component fluid $\mathbf{P}$ is a functional of the density distribution $n(\mathbf{r})$. Thus, the density distribution at equilibrium can be determined by solving the equation of hydrostatics. For a multicomponent fluid, the overall equation of hydrostatics is not sufficient for determining the density distributions of all components. What must be solved are the equations of hydrostatics of the separate species, namely,

$$0 = -\nabla \cdot \mathbf{P}_\alpha + \mathbf{B}_\alpha + \rho_\alpha \hat{\mathbf{F}}^e_\alpha, \tag{8.2.12}$$

where $\mathbf{P}_\alpha$ and ρ_α are the pressure tensor and mass density of species α, $\mathbf{B}_\alpha$ the density of asymmetric force on α arising from the other species, and $\hat{\mathbf{F}}^e_\alpha$ the external force per unit mass on species α. The total pressure tensor $\mathbf{P}$ equals $\sum_\alpha \mathbf{P}_\alpha$ and the total asymmetric force must be zero, that is, $\sum_\alpha \mathbf{B}_\alpha = 0$, to insure that the sum of eq. (8.2.12) over species α yields the overall equation of hydrostatic, eq. (8.2.10). The particular forms of $\mathbf{P}_\alpha$ and $\mathbf{B}_\alpha$ cannot be determined from continuum mechanics and so to make further progress in studying inhomogeneous fluids, such as interfaces and thin films, one has to turn to molecular theory. The molecular theoretical fundamentals of the pressure tensor are laid out in the next section for the simple classical fluids of interest in this chapter.

8.3 Pressure Tensor and Fluid Mechanical Equilibrium: Molecular Theory

For classical molecules interacting in a pair additive, centrally symmetric forces the the formula for the intermolecular contribution to the pressure tensor can be found by the method introduced by Irving and Kirkwood (1950).

Consider first a one-component system in which the fluid density is $n^{(1)}(\mathbf{r})$ and the doublet density is $n^{(2)}(\mathbf{r}, \mathbf{r}')$. At equilibrium these are the one- and two-body density distribution functions defined at eq. (8.1.10) or (8.1.12). In this case $n^{(1)}(\mathbf{r})d^3r$ represents the probable number of particles lying between $\mathbf{r}$ and $\mathbf{r} + d\mathbf{r}$ (i.e., lying in the volume element d^3r). Similarly, $n^{(2)}(\mathbf{r}, \mathbf{r}')d^3rd^3r'$ is to be interpreted as the probable number of pairs of particles found such that one of the pair lies between $\mathbf{r}$ and $\mathbf{r} + d\mathbf{r}$ and the other lies between $\mathbf{r}'$ and $\mathbf{r}' + d\mathbf{r}'$. To determine $\mathbf{P}^u$, the intermolecular part of the pressure tensor, imagine the situation illustrated in Figure 8.3.

Particles 1 and 2 can exert a force on each other across an element of area dA located at $\mathbf{r}$ and having a unit normal $\hat{\varepsilon}$. Suppose particle 1 is at the position $\mathbf{r}'$ and particle 2 is at $\mathbf{r}' + s$. Particles 1 and 2 will exert a force on each other across dA only if their connecting vector s intersects the surface at some distance $\eta s (0 \leq \eta \leq 1)$ along s. Then the resulting force on the positive $\hat{\varepsilon}$ side of dA is $(\mathbf{s}/s)(du/ds)$, where $u(s)$ is the pair potential between particles 1 and 2. Particle 1 lies in the volume $dV = dA\hat{\varepsilon} \cdot \mathbf{s}d\eta$ and particle 2 lies in the volume d^3s fixed on s. The probable number of pairs in the configurations indicated in Figure 8.3 is $n^{(2)}(\mathbf{r}', \mathbf{r}' + \mathbf{s})dA\hat{\varepsilon} \cdot \mathbf{s}d\eta d^3s$ and the probable force they exert across dA is

$$\frac{\mathbf{s}}{s}\frac{du}{ds}n^{(2)}(\mathbf{r}', \mathbf{r}' + \mathbf{s})dA\hat{\varepsilon} \cdot \mathbf{s}d\eta d^3s. \tag{8.3.1}$$

The total force exerted across dA is obtained by integrating eq. (8.3.1) over $d\eta$ and d^3s with the constraints $0 \leq \eta \leq 1$ and $\mathbf{s} \cdot \hat{\varepsilon} > 0$ to insure that the interacting pairs lie on opposite sides of dA. The resulting traction force is

$$\mathbf{t}_{\varepsilon}^{u}dA = dA\hat{\varepsilon} \cdot \int_{\mathbf{s}\cdot\hat{\varepsilon}>0}\int_0^1 \frac{\mathbf{ss}}{s}\frac{du}{ds}n^{(2)}(\mathbf{r} - \eta\mathbf{s}, \mathbf{r} - \eta\mathbf{s} + \mathbf{s})d\eta d^3s. \tag{8.3.2}$$

We have used the relationship $\mathbf{r}' = \mathbf{r} - \eta\mathbf{s}$ to eliminate $\mathbf{r}'$ from this expression. If we introduce the coordinate transformations $\mathbf{s}' =$

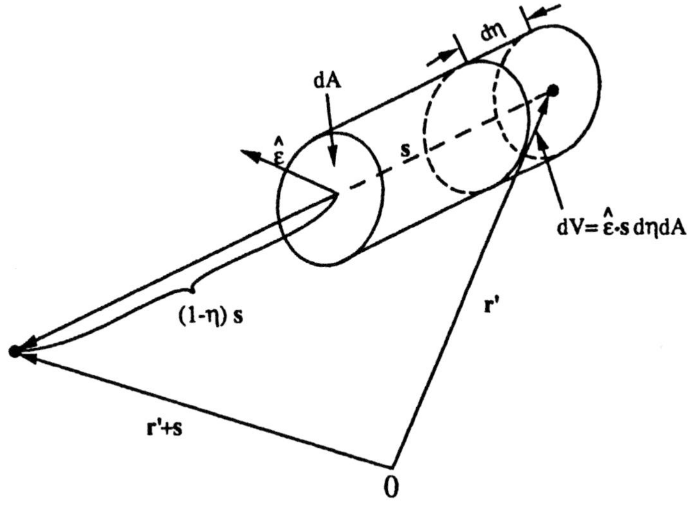

Figure 8.3
Particles at $\mathbf{r}'$ and $\mathbf{r}' + \mathbf{s}$ exert a force on each other across the area element dA with unit normal $\hat{\varepsilon}$ and located at $\mathbf{r}$.

$-\mathbf{s}$ and $\eta' = 1 - \eta$ and use the symmetry condition, $n^{(2)}(\mathbf{r}, \mathbf{r}') = n^{(2)}(\mathbf{r}', \mathbf{r})$, we can transform eq. (8.3.2) to the form

$$\mathbf{t}^u_\varepsilon dA = dA\hat{\varepsilon} \cdot \int_{\mathbf{s}'\cdot\hat{\varepsilon}<0} \int_0^1 \frac{\mathbf{s}'\mathbf{s}'}{s'}\frac{du}{ds'} n^{(2)}(\mathbf{r} - \eta'\mathbf{s}', \mathbf{r} - \eta'\mathbf{s}' + \mathbf{s}')d\eta' d^3s'. \quad (8.3.3)$$

Because $\mathbf{s}'$ and η' are dummy variables the primes can be dropped in eq. (8.3.3). Thus, the rhs of eq. (8.3.3) differs from that of eq. (8.3.2) only in that the constraint on $\mathbf{s}$ in the former is $\mathbf{s} \cdot \hat{\varepsilon} > 0$ and in the latter is $\mathbf{s} \cdot \hat{\varepsilon} < 0$. Adding eqs. (8.3.2) and (8.3.3) and dividing by 2 and recalling that $\mathbf{t}^u_\varepsilon dA = -\varepsilon \cdot \mathbf{P}^u dA$,

$$\mathbf{P}^u = -\frac{1}{2} \int \int_0^1 \frac{\mathbf{s}\mathbf{s}}{s}\frac{du}{ds} n^{(2)}(\mathbf{r} - \eta\mathbf{s}, \mathbf{r} - \eta\mathbf{s} + \mathbf{s})d\eta d^3s \quad (8.3.4)$$

for the intermolecular potential contribution to the pressure tensor. Note that as found from a continuum mechanics argument the pressure tensor is independent of $\hat{\varepsilon}$. It is a material property depending only on the position $\mathbf{r}$ at which we place dA to measure it. Equation (8.3.4) is valid even for a nonequilibrium fluid, although in this case $n^{(2)}$ cannot be determined from equilibrium ensemble theory.

Recall from Chapter 1 that there is a contribution to pressure from the kinetic motion of molecules into element dA. For this contribution we will consider only the equilibrium situation. Denote by $f(\mathbf{r}, \mathbf{v})d^3r d^3\mathbf{v}$ the probability that a particle lies between $\mathbf{r}$ and $\mathbf{r} + d\mathbf{r}$ and has a velocity between $\mathbf{v}$ and $\mathbf{v} + d\mathbf{v}$. At equilibrium we know from canonical ensemble theory of

$$\begin{aligned} f(\mathbf{r}, \mathbf{v}) &= \frac{N \int e^{-\beta H}(d^3r d^3v)^{N-1}}{\int e^{-\beta H}(d^3r d^3v)^N} \\ &= \phi(v) n^{(1, N)}(\mathbf{r}), \end{aligned} \quad (8.3.5)$$

where $\phi(v)$ is the Maxwell velocity distribution function

$$\phi(v) = \left(\frac{kT}{2\pi m}\right)^{3/2} e^{-mv^2/2kT}. \quad (8.3.6)$$

In the grand canonical ensemble, $f(\mathbf{r}, \mathbf{v}) = \phi(v)n^{(1)}(\mathbf{r})$. Because it will lead to no confusion in this chapter, from this point on we will drop the superscript on the singlet density distribution and simply denote density by $n(\mathbf{r})$. Thus, $f(\mathbf{r}, \mathbf{v}) = \phi(\mathbf{v})n(\mathbf{r})$.

If we consider an area element dA at $\mathbf{r}$ with normal $\hat{\varepsilon}$, then the number of particles with velocity $\mathbf{v}$ that will impinge on this area during the short time δt will be the number of such particles lying in the cylinder shown in Figure 8.4. Thus, the momentum transferred across dA by kinetic motion during δt will be $m\mathbf{v}$ times the number of particles having the velocity $\mathbf{v}$ and lying in the volume $\delta V = |\hat{\varepsilon} \cdot \mathbf{v}|\delta t dA$. The probable number of particles

Figure 8.4
Elements of volume containing particles of velocity **v** that will impinge on area dA within the time interval δt.

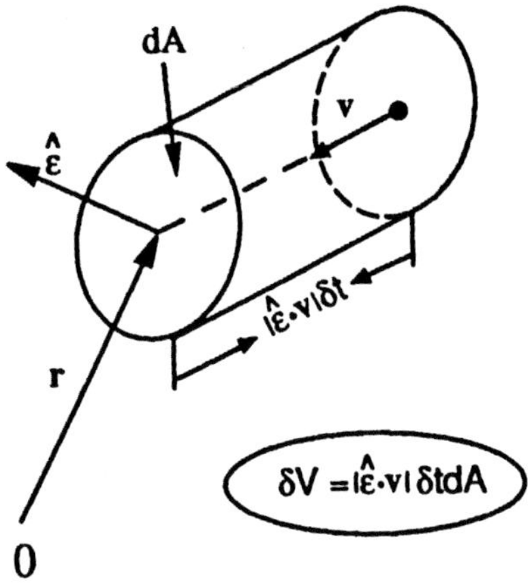

having velocity between $\mathbf{v}$ and $\mathbf{v}+d\mathbf{v}$ and at some position $\mathbf{r}$ in δV is $f(\mathbf{r},\,\mathbf{v})d^3v|\hat{\varepsilon}\cdot\mathbf{v}|\delta t\,dA$, and so the average momentum transferred across dA during the time δt by particle motion is

$$\int_0^{\delta t}\int_{\hat{\varepsilon}\cdot\mathbf{v}<0} m\mathbf{v}f(\mathbf{r},\,\mathbf{v})d^3v|\hat{\varepsilon}\cdot\mathbf{v}|dt\,dA$$
$$-\int_0^{\delta t}\int_{\hat{\varepsilon}\cdot\mathbf{v}>0} m\mathbf{v}f(\mathbf{r},\,\mathbf{v})d^3v|\hat{\varepsilon}\cdot\mathbf{v}|dt\,dA. \tag{8.3.7}$$

The first term in eq. (8.3.8) represents the momentum transferred into dA (opposite the direction of the normal vector $\hat{\varepsilon}$) and the second term represents the momentum out of dA, thus the negative sign. Becausee force is the momentum transfer per unit time, we compute the kinetic part of the traction force $\mathbf{t}_{\hat{\varepsilon}}^K dA$ by dividing eq. (8.3.8) by δt and passing to the limit $\delta t = 0$. Using the fact that $|\mathbf{v}\cdot\hat{\varepsilon}| = -\mathbf{v}\cdot\hat{\varepsilon}$ when $\mathbf{v}\cdot\hat{\varepsilon} < 0$, we can combine the two terms in eq. (8.3.8) and obtain

$$\mathbf{t}_{\hat{\varepsilon}}^K dA = -dA\hat{\varepsilon}\cdot\int m\mathbf{v}\mathbf{v}f(\mathbf{r},\,\mathbf{v})d^3v. \tag{8.3.8}$$

The connection between the traction vector and pressure tensor, $\mathbf{t}_{\hat{\varepsilon}}^K = -\hat{\varepsilon}\cdot\mathbf{P}^K$, yields for the kinetic part of pressure tensor

$$\mathbf{P}^K = \int m\mathbf{v}\mathbf{v}f(\mathbf{r},\,\mathbf{v})d^3v$$
$$= \left[\int m\mathbf{v}\mathbf{v}\phi(\mathbf{v})d^3v\right]n(\mathbf{r}) \tag{8.3.9}$$

Becausee $\phi(\mathbf{v})$ is the Gaussian distribution, $\int v_i v_j \phi(\mathbf{v})d^3v = 0$ when $i \neq j$ and $\int v_i^2\phi(\mathbf{v})d^3v = (1/3)\int v^2\phi(\mathbf{v})d^3v = kT/m$, where v_1, v_2, and v_3 are Cartesian coordinates of $\mathbf{v}$. Thus, at equilibrium the kinetic part of the pressure tensor is isotropic and the same as that of an ideal gas, that is,

$$\mathbf{P}^K = n(\mathbf{r})kT\mathbf{I}, \tag{8.3.10}$$

where $\mathbf{I}$ is again the unit tensor.

Equations (8.3.11) and (8.3.4) combine to give

$$\mathbf{P} = n(\mathbf{r})kT\mathbf{I} - \frac{1}{2}\int\int_0^1 \frac{\mathbf{ss}}{s}\frac{du}{ds}n^{(2)}(\mathbf{r}-\eta\mathbf{s}, \mathbf{r}-\eta\mathbf{s}+\mathbf{s})d\eta d^3s \tag{8.3.11}$$

for the pressure tensor of a classical mechanical inhomogeneous fluid of simple particles at equilibrium. Because $\mathbf{I}$ and $\mathbf{ss}$ are symmetric tensors, it follows from eq. (8.3.12) that the pressure tensor is symmetric.

In the special case of a homogeneous fluid at equilibrium $n(\mathbf{r})$ is a constant n and $n^{(2)}(\mathbf{r}, \mathbf{r}') = n^2 g^{(2)}(|\mathbf{r}-\mathbf{r}'|)$. In this case, because

$$\int \frac{\mathbf{ss}}{s}\frac{du(s)}{ds}g^{(2)}(s)d^3s = \frac{\mathbf{I}}{3}\int s\frac{du}{ds}g^{(2)}(s)d^3s. \tag{8.3.12}$$

Equation (8.3.12) reduces to $\mathbf{P} = P\mathbf{I}$, where the pressure P obeys the virial equation of state derived in Chapter 5, namely,

$$P = nkT - \frac{n^2}{6}\int s\frac{du}{ds}g^{(2)}(s)d^3s. \tag{8.3.13}$$

The multicomponent version of eq. (8.3.4) can be derived similarly and is

$$\mathbf{P}^u = -\sum_{\alpha,\beta}\frac{1}{2}\int\int_0^1 \frac{\mathbf{ss}}{s}\frac{du_{\alpha\beta}}{ds}n_{\alpha\beta}^{(2)}(\mathbf{r}-\eta\mathbf{s}, \mathbf{r}-\eta\mathbf{s}+\mathbf{s})d\eta d^3s \tag{8.3.14}$$

where the sums α and β are over the different species. It is also straightforward to show that

$$\mathbf{P}^K = \sum_\alpha n_\alpha(\mathbf{r})kT\mathbf{I}, \tag{8.3.15}$$

and so the pressure tensor for an inhomogeneous multicomponent fluid at equilibrium is

$$\mathbf{P} = \sum_\alpha n_\alpha(\mathbf{r})kT\mathbf{I} - \sum_{\alpha,\beta}\frac{1}{2}\int\int_0^1 \frac{\mathbf{ss}}{s}\frac{du_{\alpha\beta}}{ds}n_{\alpha\beta}^{(2)}(\mathbf{r}-\eta\mathbf{s},\ \mathbf{r}-\eta\mathbf{s}+\mathbf{s})d\eta d^3s. \tag{8.3.16}$$

Again we see that the pressure tensor is symmetric.

From continuum mechanics, we know that the pressure tensor obeys the equation of hydrostatics [eq. (8.2.10)] at mechanical equilibrium. This equation turns out to be equivalent to the YBG equation for the singlet density distribution (i.e., for the density). To see this consider first the one-component YBG equation, namely, eq. (8.1.9) for $s = 1$. It can written as

$$kT\nabla n(\mathbf{r}) + \int \nabla u(|\mathbf{r} - \mathbf{r}'|)n^{(2)}(\mathbf{r}, \mathbf{r}')d^3r' = -n(\mathbf{r})\nabla u^e(\mathbf{r}), \tag{8.3.17}$$

where we have suppressed the superscript on the singlet density distribution and have denoted $\nabla_{\mathbf{r}}$ as ∇. The equation of hydrostatics is

$$\nabla\mathbf{P} = -\rho\nabla\hat{u}^e(\mathbf{r}), \tag{8.3.18}$$

where ρ is the mass density and $\hat{u}^e(\mathbf{r})$ is the external potential energy per unit mass. What we want to demonstrate is that the insertion of the formula given for $\mathbf{P}$ at eq. (8.3.12) into eq. (8.3.19) gives the YBG eq. (8.3.18). Because the mass density obeys $\rho = mn$ and external potential energy per unit mass obeys $\hat{u}^e(\mathbf{r}) = u^e(\mathbf{r})/m$ it follows that $\rho\nabla\hat{u}^e(\mathbf{r}) = n\nabla u^e(\mathbf{r})$. Thus, the external force terms in eqs. (8.3.18) and (8.3.19) are the same. Similarly, because $\nabla \cdot \mathbf{P}^K = kT\nabla n(\mathbf{r})$, it follows that all we have to prove is that $\nabla \cdot \mathbf{P}^u$ is equal to the second term on the left-hand side (lhs) of eq. (8.3.18).

Note that $\nabla u(|\mathbf{r} - \mathbf{r}'|) = [(\mathbf{r} - \mathbf{r}')/|\mathbf{r} - \mathbf{r}'|]u'(|\mathbf{r} - \mathbf{r}'|)$, where u' denotes the derivative of u with respect to its argument. With this result and the coordinate transformation $\mathbf{r},\ \mathbf{r}' \to \mathbf{r},\ \mathbf{s} = \mathbf{r}' - \mathbf{r}$, we find the equation

$$\int \nabla u(|\mathbf{r} - \mathbf{r}'|)n^{(2)}(\mathbf{r}, \mathbf{r}')d^3r' = -\int \frac{\mathbf{s}}{s}u'(s)n^{(2)}(\mathbf{r}, \mathbf{r} + \mathbf{s})d^3s,$$

and with the transformation $\mathbf{s} \to -\mathbf{s}$ this becomes

$$\int \nabla u(|\mathbf{r} - \mathbf{r}'|)n^{(2)}(\mathbf{r}, \mathbf{r}')d^3\mathbf{r}' = -\int \frac{\mathbf{s}}{s}u'(s)n^{(2)}(\mathbf{r}, \mathbf{r} - \mathbf{s})d^3s.$$

Adding these two equations, we obtain

$$\int \nabla u(|\mathbf{r} - \mathbf{r}'|)n^{(2)}(\mathbf{r}, \mathbf{r}')d^3\mathbf{r}' = -\frac{1}{2}\int \frac{\mathbf{s}}{s}u'(s)[n^{(2)}(\mathbf{r}, \mathbf{r} + \mathbf{s}) - n^{(2)}(\mathbf{r}, \mathbf{r} - \mathbf{s})]d^3s. \tag{8.3.19}$$

Consider next

$$\nabla \cdot \mathbf{P}^u = -\frac{1}{2}\int \int_0^1 \frac{\mathbf{s}}{s}u'(s)\mathbf{s} \cdot \nabla n^{(2)}(\mathbf{r} - \eta\mathbf{s}, \mathbf{r} - \eta\mathbf{s} + \mathbf{s})d\eta d^3s. \tag{8.3.20}$$

To take the gradient of $n^{(2)}$ with respect to $\mathbf{r}$, it is convenient to choose a Cartesian coordinate system in which one axis lies along $\mathbf{s}$. In this case

$$\begin{aligned}\mathbf{s} \cdot \nabla n^{(2)}(\mathbf{r} - \eta\mathbf{s}, \mathbf{r} - \eta\mathbf{s} + \mathbf{s}) &= s\lim_{h\to 0}\frac{[n^{(2)}(\mathbf{r} + (h - \eta)\mathbf{s}, \mathbf{r} + (h - \eta)\mathbf{s} + \mathbf{s}) - n^{(2)}(\mathbf{r} - \eta\mathbf{s}, \mathbf{r} - \eta s + \mathbf{s})]}{hs} \\ &= -\frac{\partial n^{(2)}}{\partial\eta}(\mathbf{r} - \eta\mathbf{s}, \mathbf{r} - \eta\mathbf{s} + \mathbf{s}),\end{aligned} \tag{8.3.21}$$

When this result is inserted into eq. (8.3.21) the integration over $d\eta$ can be carried out analytically to yield

$$\nabla \cdot \mathbf{P}^u = -\frac{1}{2}\int \frac{\mathbf{s}}{s}u'(s)[n^{(2)}(\mathbf{r},\ \mathbf{r}+\mathbf{s}) - n^{(2)}(\mathbf{r}-\mathbf{s},\ \mathbf{r})]d^3s. \qquad (8.3.22)$$

For a one-component system $n^{(2)}(\mathbf{r},\ \mathbf{r}-\mathbf{s}) = n^{(2)}(\mathbf{r}-\mathbf{s},\ \mathbf{r})$, because the probability that particle 1 is at $\mathbf{r}$ and particle 2 is at $\mathbf{r}-\mathbf{s}$ is the same as that of 2 at $\mathbf{r}$ and 1 at $\mathbf{r}-\mathbf{s}$ because the particles are identical. Thus, we have shown that $\nabla \cdot \mathbf{P}^u$ is identical to the rhs of eq. (8.3.20), which completes the proof that the YBG equation for the density distribution is in fact equivalent to the equation of hydrostatics, that is, the YBG equation is the condition of mechanical equilibrium.

Consider next the multicomponent YBG equation for the density distribution:

$$kT\nabla n_\alpha(\mathbf{r}) + \sum_\beta \int \nabla u_{\alpha\beta}(|\mathbf{r}-\mathbf{r}'|)n^{(2)}_{\alpha\beta}(\mathbf{r},\ \mathbf{r}')d^3r' = -n_\alpha(\mathbf{r})\nabla u^e_\alpha(\mathbf{r}). \qquad (8.3.23)$$

We want to relate this YBG equation to a hydrostatic equation for species α (i.e., to the momentum balance on species α at equilibrium). Let us first decompose the pressure tensor into its species contributions, namely,

$$\mathbf{P} = \sum_\alpha \mathbf{P}_\alpha, \qquad (8.3.24)$$

where

$$\mathbf{P}_\alpha = n_\alpha kT\mathbf{I} - \sum_\beta \frac{1}{2}\int\int_0^1 \frac{\mathbf{ss}}{s}u'_{\alpha\beta}(s)n^{(2)}_{\alpha\beta}(\mathbf{r}-\eta\mathbf{s},\ \mathbf{r}-\eta\mathbf{s}+\mathbf{s})d\eta d^3s. \qquad (8.3.25)$$

Then, using the relationship derived at eq. (8.3.22) we obtain

$$\nabla\mathbf{P}_\alpha = kT\nabla_\mathbf{r}n_\alpha(\mathbf{r}) - \sum_\beta \frac{1}{2}\int \frac{\mathbf{s}}{s}u'_{\alpha\beta}(s)[n^{(2)}_{\alpha\beta}(\mathbf{r},\ \mathbf{r}+\mathbf{s}) - n^{(2)}_{\alpha\beta}(\mathbf{r}-\mathbf{s},\ \mathbf{r})]d^3s. \qquad (8.3.26)$$

Similarly, the coordinate transformations leading to eq. (8.3.20) for the one-component fluid yield for a multicomponent fluid the result

$$\sum_\beta \int \nabla u_{\alpha\beta}(|\mathbf{r}-\mathbf{r}'|)n^{(2)}_{\alpha\beta}(\mathbf{r},\ \mathbf{r}')d^3r' = -\sum_\beta \frac{1}{2}\int \frac{\mathbf{s}}{s}u'_{\alpha\beta}(s)[n^{(2)}_{\alpha\beta}(\mathbf{r},\ \mathbf{r}+\mathbf{s}) - n^{(2)}_{\alpha\beta}(\mathbf{r},\ \mathbf{r}-\mathbf{s})]d^3s. \qquad (8.3.27)$$

Substitution of eq. (8.3.28) into the YBG equation and combining the result with eq. (8.3.27), we obtain the equation of hydrostatic equilibrium for species α, namely,

$$\nabla \cdot \mathbf{P}_\alpha - \mathbf{B}_\alpha = -n_\alpha \nabla u^e_\alpha, \qquad (8.3.28)$$

where

$$\mathbf{B}_\alpha \equiv \sum_\beta \frac{1}{2} \int \frac{\mathbf{s}}{s} u'_{\alpha\beta}(s)[n^{(2)}_{\alpha\beta}(\mathbf{r}-\mathbf{s}, \mathbf{r}) - n^{(2)}_{\alpha\beta}(\mathbf{r}, \mathbf{r}-\mathbf{s})]d^3s. \quad (8.3.29)$$

The so-called asymmetric force $\mathbf{B}_\alpha$ on species α results because the distribution $n^{(2)}_{\alpha\beta}(\mathbf{r}, \mathbf{r}')$ is not equal to $n^{(2)}_{\alpha\beta}(\mathbf{r}', \mathbf{r})$ when species α and β are different.

We anticipated a species hydrostatic equation such as eq. (8.3.29) in Section 8.2, but continuum mechanics is not able to provide a definition for the asymmetric force beyond admitting that it can exist. As is implied by eq. (8.3.30) the force arises from the fact that species α and β react differently to an inhomogeneous environment. The sum of $\mathbf{B}_\alpha$ over all species vanishes and so the sum of the species hydrostatic eq. (8.3.29) over all α reduces as expected to

$$\nabla \cdot \mathbf{P} = -\sum_\alpha n_\alpha \nabla u^e_\alpha \equiv -\rho \mathbf{F}^e, \quad (8.3.30)$$

the ordinary equation of hydrostatics. The proof that $\sum_\beta \mathbf{B}_\beta = 0$ is based on the properties $u_{\alpha\beta}(s) = u_{\beta\alpha}(s)$ and $n^{(2)}_{\alpha\beta}(\mathbf{r}, \mathbf{r}') = n^{(2)}_{\beta\alpha}(\mathbf{r}', \mathbf{r})$ and makes use of the interchange of symbols for the dummy indices $\alpha\beta$ in the double sum over α and β.

To summarize this section, we have deduced the pressure tensor for the species of a multicomponent classical fluid of particles interacting via pairwise additive, centrally symmetric forces. We have shown that the YBG equation for the density distribution function of species α is equivalent to the equation of hydrostatic equilibrium for this species and the sum of the YBG equation over all species is equivalent to the ordinary equation of hydrostatic equilibrium.

8.4 Interfacial Tension of Planar Interfaces

In Chapter 7 it was shown that the interfacial tension between two fluids separated by a planar interface is equal to $\int_{-\infty}^{\infty}(P_N - P_T)dz$, where P_N and P_T are components of pressure normal and transverse to the interface. The convention chosen is that the z axis is normal to the interface and the x and y axes are transverse to the interface, that is, the xy plane is parallel to the plane of the interface. Thus, $P_N = P_{zz}$ and $P_T = P_{xx} = P_{yy} = (P_{xx} + P_{yy})/2$. In terms of the normal and transverse components of the formula for $\mathbf{P}$ given by eqs. (8.3.25) and (8.3.26), the interfacial tension reads as follows:

$$\gamma = \sum_{\alpha,\beta} \frac{1}{4} \int_{-\infty}^{\infty} \int \int_0^1 \frac{s_x^2 + s_y^2 - 2s_z^2}{s} u'_{\alpha\beta}(s) n^{(2)}_{\alpha\beta}(\mathbf{r} - \eta\mathbf{s}, \mathbf{r} - \eta\mathbf{s} + \mathbf{s}) d\eta d^3s dz. \quad (8.4.1)$$

Note that the kinetic part of the pressure tensor does not contribute to the interfacial tension because $\mathbf{P}^K$ is isotropic.

Because of symmetry the density distribution $n_{\alpha\beta}^{(2)}(\mathbf{r}, \mathbf{r}')$ in a planar system depends only on the distances z and z' of the two particles from the interface and on the distance $|\mathbf{r}-\mathbf{r}'|$ of separation of the pair. Thus, $n_{\alpha\beta}^{(2)} = n_{\alpha\beta}^{(2)}(z-\eta s_z, z-\eta s_z - s_z, s)$. Consider the coordinate transformation $z' = z - \eta s_z$ in eq. (8.4.1). With this transformation $dz' = dz$, the limits of integration of z' are $-\infty$ and ∞, and the integrand of eq. (8.4.1) becomes independent of η. Thus, eq. (8.4.1) reduces to

$$\gamma = \sum_{\alpha,\beta} \frac{1}{4} \int_{-\infty}^{\infty} \int \frac{s_x^2 + s_y^2 - 2s_z^2}{s} u'_{\alpha\beta}(s) n_{\alpha\beta}^{(2)}(z', z' + s_z, s) d^3s dz', \qquad (8.4.2)$$

a formula first derived by Kirkwood and Buff (1949). Because the origin of the coordinate system for z' is arbitrary, we can drop the prime on z' and rewrite eq. (8.4.2) in the simpler form

$$\gamma = \sum_{\alpha,\beta} \frac{1}{4} \int_{-\infty}^{\infty} \int \frac{s^2 - 3s_z^2}{s} u'_{\alpha\beta}(s) n_{\alpha\beta}^{(2)}(\mathbf{r}, \mathbf{r}+\mathbf{s}) d^3s dz, \qquad (8.4.3)$$

where use is made of the relation $s_x^2 + s_y^2 = s^2 - s_z^2$.

The pair correlation function, $g_{\alpha\beta}^{(2)}$, is defined by

$$g_{\alpha\beta}^{(2)}(\mathbf{r}, \mathbf{r}') = \frac{n_{\alpha\beta}^{(2)}(\mathbf{r}, \mathbf{r}')}{n_\alpha(\mathbf{r}) n_\beta(\mathbf{r}')}. \qquad (8.4.4)$$

In a planar system, the density depends only on z, that is, $n_\alpha = n_\alpha(z)$. Equation (8.4.4) can be substituted into eq. (8.4.3) to get still another expression for tension, namely,

$$\gamma = \sum_{\alpha,\beta} \frac{1}{4} \int \int \frac{s^2 - 3s_z^2}{s} u'_{\alpha\beta}(s) n_\alpha(z) n_\beta(z + s_z) g_{\alpha\beta}^{(2)}(\mathbf{r}, \mathbf{r}+\mathbf{s}) d^3s dz. \qquad (8.4.5)$$

A more general version of eq. (8.4.5) can be derived using the formula $\gamma = (\partial F/\partial A)_{T,V,N}$. We suppose again that the partition function factors into a term depending on T and $\mathbf{N}$ and Z_N, where Z_N has the classical mechanical form

$$Z_N = \int \cdots \int e^{-\beta u^N(\mathbf{r}_1,\ldots,\mathbf{r}_N)} d^3r_1 \cdots d^3r_N. \qquad (8.4.6)$$

However, unlike the assumption leading to eq. (8.4.5), we do not assume that the intermolecular interactions are pair additive. Consider a one-component fluid in a rectangular box whose sides are of length L_x, L_y, and L_z, respectively. The volume of the box is $V = L_x L_y L_z$. Assume there are two phases separated by a planar interface of area $A = L_x L_y$. We will imagine a change in A in which L_x is constant, that is, $dA = L_x dL_y$. The constraint of constant V yields the relation $dL_z/dL_y = -L_x/L_y$ between expansion dL_z of the box normal to the interface and the stretch dL_y of the interface.

The interfacial tension is to be computed from

$$\gamma = -kT\left(\frac{\partial \ln Z_N}{\partial A}\right)_{T,V,N} = -\frac{kT}{L_x}\left(\frac{\partial \ln Z_N}{\partial L_y}\right)_{T,V,N}. \tag{8.4.7}$$

Introducing the coordinates $\xi_i = x_i/L_x$, $\eta_i = y_i/L_y$, and $\zeta_i = z_i/L_z$, we obtain

$$Z_N = V^N \int_0^1 \cdots \int_0^1 e^{-\beta u^N(\{L_x\xi_i, L_y\eta_i, L_z\zeta_i\})} d\xi_1 d\eta_1 d\zeta_1 \cdots d\xi_N d\eta_N d\zeta_N. \tag{8.4.8}$$

The derivative of the function u^N in the integrand of Z_N with respect to L_y is

$$\begin{aligned}\left(\frac{\partial u^N}{\partial L_y}\right)_V &= \sum_i \left[\frac{\partial(L_y\eta_i)}{\partial L_y}\frac{\partial u^N}{\partial(L_y\eta_i)} + \frac{\partial(L_z\zeta_i)}{\partial L_y}\frac{\partial u^N}{\partial(L_z\zeta)}\right] \\ &= \sum_i \frac{1}{L_y}\left[u_i\frac{\partial u^N}{\partial y_i} - z_i\frac{\partial u^N}{\partial z_i}\right],\end{aligned} \tag{8.4.9}$$

where the final form is obtained by using the relationship $\partial L_z/\partial L_y = -L_z/L_y$ and by transforming the variables ξ_i, η_i, ζ_i back to x_i, y_i, z_i.

Using eq. (8.4.9) to evaluate the rhs of eq. (8.4.7), we obtain

$$\gamma = \frac{1}{A}\sum_{i=1}^{N}\langle y_i F_{iy} - z_i F_{iz}\rangle = \frac{N}{A}\langle y_1 F_{1y} - z_1 F_{1z}\rangle, \tag{8.4.10}$$

where F_{iy} and F_{iz} are the x and z components of the force on molecule $i(F_{iy} = -\partial u^N/\partial y_i,\ F_{iz} = -\partial u^N/\partial z_i)$. This result is analogous to the virial equation of state,

$$P = \frac{NkT}{V} + \frac{1}{3V}\sum_i \langle \mathbf{r}_i \cdot \mathbf{F}_i\rangle, \tag{8.4.11}$$

of a homogeneous fluid. This can be derived from $P = kT(\partial \ln Z_N/\partial V)_{T,N}$ by taking the system to be a cube of volume V and evaluating the derivative of Z_N by changing the cube uniformly in all directions. Equation (8.4.11) is also valid for arbitrary u^N.

By the same procedure as that used to find eq. (8.4.10), it is easy to show that for a multicomponent system

$$\begin{aligned}\gamma &= \sum_\alpha \sum_i \frac{1}{A}\langle y_i F_{iy}^\alpha - z_i F_{iz}^\alpha\rangle \\ &= \sum_\alpha \frac{N_\alpha}{A}\langle y_1 F_{1y}^\alpha - z_1 F_{1z}^\alpha\rangle,\end{aligned} \tag{8.4.12}$$

where F_{iy}^α and F_{iz}^α are the forces on particle i of component α.

If the intermolecular interactions of the particles of the system are centrally symmetric and pair additive, then the force on particle i is $\mathbf{F}_i = -\sum_j \hat{\mathbf{r}}_{ij} u'_{ij}(r_{ij})$, where $\hat{\mathbf{r}}_{ij} = (\mathbf{r}_i - \mathbf{r}_j)/r_{ij}$. With this form for $\mathbf{F}_i$, it is straightforward to derive eq. (8.4.5) from eq. (8.4.12).

Let us close this section by considering several approximations enabling one to estimate interfacial tensions from the Kirkwood–Buff formula (1949).

With the approximation that an interface is a mathematical surface dividing two homogeneous bulk phases, eq. (8.4.5) provides a simple means of estimating the surface tension of a liquid at low vapor pressure. Because in this case the liquid density if much larger than the vapor density, we assume

$$\begin{aligned} n_\alpha(z) &= n^l_\alpha, \qquad z < 0, \\ &= 0, \qquad z > 0, \end{aligned} \tag{8.4.13}$$

where n^l_α is the density of component α in the bulk liquid phase. Moreover, in the liquid phase $g^{(2)}_{\alpha\beta} = g^l_{\alpha\beta}$ that depends only on the magnitude of s, temperature, and the bulk densities $\mathbf{n}^l(=\{n^l, \ldots, n^l\})$. With this approximation eq. (8.4.5) becomes

$$\gamma = \sum_{\alpha,\beta} \frac{n^l_\alpha n^l_\beta}{4} \int d^3s \int_0^L dz \frac{s^2 - 3s_z^2}{s} u'_{\alpha\beta}(s)\eta(z + s_x) g^l_{\alpha\beta}(s), \tag{8.4.14}$$

where η is the Heaviside function $[\eta(z) = 0,\ z < 0$ and $\eta(z) = 1,\ z > 0]$. The upper limit placed on z is L, which is the height of the vapor above the liquid surface.

To evaluate eq. (8.4.7) further, note that

$$\int s_z^2 f(s) d^3s = \int s_x^2 f(s) d^3s = \int s_y^2 f(s) d^3s = \frac{1}{3} \int s^2 f(s) d^3s, \tag{8.4.15}$$

where $f(s)$ depends only on the magnitude s, that in a Cartesian coordinate system is $s = (s_x^2 + s_y^2 + s_z^2)^{1/2}$. From this property it follows that

$$\int \frac{s^2 - 3s_z^2}{s} u'_{\alpha\beta}(s) g^l_{\alpha\beta}(s) d^3s = 0. \tag{8.4.16}$$

Thus, the integrals in eq. (8.4.7) can be manipulated as follows:

$$\begin{aligned} I &\equiv \int_{-\infty}^{\infty} ds_x \int_{-\infty}^{\infty} ds_y \int_{-\infty}^{\infty} ds_z \int_0^L dz \eta(z + s_z) \frac{s^2 - 3s_z^2}{s} u'_{\alpha\beta}(s) g^l_{\alpha\beta}(s) d^3s \\ &= \int_{-\infty}^{\infty} ds_x \int_{-\infty}^{\infty} ds_y \int_{-\infty}^{0} ds_z \int_{-s_z}^{L} dz \frac{s^2 - 3s_z^2}{s} u'_{\alpha\beta}(s) g^l_{\alpha\beta}(s) \end{aligned}$$

$$+\int_{-\infty}^{\infty} ds_x \int_{-\infty}^{\infty} ds_y \int_{0}^{\infty} ds_z \int_{0}^{L} dz \frac{s^2 - 3s_z^2}{s} u'_{\alpha\beta}(s) g^l_{\alpha\beta}(s)$$

$$= \int_{-\infty}^{\infty} ds_x \int_{-\infty}^{\infty} ds_y \int_{-\infty}^{0} ds_z \int_{-s_z}^{0} dz \frac{s^2 - 3s_z^2}{s} u'_{\alpha\beta}(s) g^l_{\alpha\beta}(s)$$

$$+ \int_{-\infty}^{\infty} ds_x \int_{-\infty}^{\infty} ds_y \int_{-\infty}^{\infty} ds_z \int_{0}^{L} dz \frac{s^2 - 3s_z^2}{s} u'_{\alpha\beta}(s) g^l_{\alpha\beta}(s). \tag{8.4.17}$$

The last term in eq. (8.4.17) vanishes because of the property shown at eq. (8.4.9). The other term (with the transformation $s_z \to -s_z$) becomes

$$I = \int_{-\infty}^{\infty} ds_x \int_{-\infty}^{\infty} ds_y \int_{0}^{\infty} ds_z s_z \frac{s^2 - 3s_z^2}{s} u'_{\alpha\beta}(s) g^l_{\alpha\beta}(s). \tag{8.4.18}$$

It is next convenient to introduce a polar coordinate system in which $\mathbf{s} = s(\sin\theta\cos\phi\hat{\imath} + \sin\theta\sin\phi\hat{\jmath} + \cos\theta\hat{k})$. Then $d^3s = \sin\theta d\theta d\phi s^2 ds$, $s_z = s\cos\theta$, and the constraint that $s_z > 0$ in eq. (8.4.17) is satisfied by the condition $0 < \theta < \pi/2$. Thus, eq. (8.4.17) becomes

$$I = \int_0^{\infty}\int_0^{2\pi}\int_0^{\pi/2} s^4 \cos\theta(1 - 3\cos^2\theta) u'_{\alpha\beta}(s) g^l_{\alpha\beta}(s) \sin\theta d\theta d\phi ds \tag{8.4.19}$$

$$= \frac{\pi}{2} \int_0^{\infty} s^4 u'_{\alpha\beta}(s) g^l_{\alpha\beta}(s) ds.$$

This result when inserted into eq. (8.4.14) finally gives the formula,

$$\gamma = \sum_{\alpha,\beta} \frac{\pi}{8} n^l_\alpha n^l_\beta \int_0^{\infty} s^4 u'_{\alpha\beta}(s) g^l_{\alpha\beta}(s) ds, \tag{8.4.20}$$

whose one-component version was derived by Fowler and later by Kirkwood and Buff.

At low vapor pressures, in which case the interface is rather sharp, the Fowler–Kirkwood–Buff (FKB) formula appears to be a reasonable approximation for simple fluids. For example, for liquid argon at $T = 84$ K, the experimental surface tension is 13.3 dyn/cm. Using the superposition approximation for $g^l(s)$ and the 6-12 Lennard–Jones potential with $\varepsilon/k = 116.4$ K and $\sigma = 3.37$ Å, Salter and Davis (1975) predicted with the FKB formula the value 13.3 dyn/cm, agreement of which is fortuitously close. As temperature increases, the discontinuous interface approximation worsens and, as expected, whereas the formula predicts 10.1 dyn/cm for argon at 115 K, the experimental value is 5.1 dyn/cm.

Exercise

For a very rough approximation of thermodynamic quantities one sometimes neglects the density dependence of $g^\ell(s)$. If we assume this is so in eq. (8.4.20), then we can set $g^\ell(s) = exp(-\beta u)$ and integrate the right hand side to obtain the approximation

$$\gamma = \sum_{\alpha,\beta} \frac{\pi}{2} n^\ell_\alpha n^\ell_\beta \int_0^{\infty} s^3 e^{-\beta u_{\alpha\beta}} ds.$$

Use the 6-12 Lennard-Jones potential for argon and pentane and the experimental saturation density to predict from this expression γ versus T for the pure liquids. Compare with experiment. Do the same for the square-well potential of argon ($\epsilon/k = 69.4K, d_1 = 3.16Å$, and $d_2 = 1.85d_1$).

An extension of the discontinuous FKB model of the interface between places 1 and 2 is

$$n_\alpha(z) = n_\alpha^{(1)}, \quad z < 0, \\ = n_\alpha^{(2)}, \quad z > 0, \tag{8.4.21}$$

where $n_\alpha^{(i)}$ is the bulk density of component α in phase i. When both particles α and β lie in phase (i), we assume $g_{\alpha\beta}^{(i,i)} = g_{\alpha\beta}^{(i)}(s; \mathbf{n}^{(i)})$, the pair correlation function in bulk phase i; when one particle is in phase (1) and the other is in phase (2), we assume

$$g_{\alpha\beta}^{(1,2)} \equiv \frac{1}{2}\left[g_{\alpha\beta}^{(1)}(s; \mathbf{n}^{(1)}) + g_{\alpha\beta}^{(2)}(s; \mathbf{n}^{(2)})\right]. \tag{8.4.22}$$

With these approximations, eq. (8.4.5) yields

$$\gamma = \sum_{\alpha,\beta}\left[n_\alpha^{(1)} n_\beta^{(1)} B_{\alpha\beta}^{(1,1)} + n_\alpha^{(2)} n_\beta^{(2)} B_{\alpha\beta}^{(2,2)} - 2n_\alpha^{(1)} n_\alpha^{(2)} B_{\alpha\beta}^{(1,2)}\right], \tag{8.4.23}$$

where

$$B_{\alpha\beta}^{(i,j)} = \frac{\pi}{8}\int_0^\infty s^4 u'_{\alpha\beta}(s) g_{\alpha\beta}^{(i,j)}(s) ds. \tag{8.4.24}$$

When phase 1 is a liquid and phase 2 is a vapor of negligible density compared to liquid density, then eq. (8.4.16) reduces to the FKB result. When phases 1 and 2 are a binary pair immiscible liquids (e.g., water and oil or mercury and a nonmetallic liquid), then $n_1^{(2)} \simeq n_2^{(1)} \simeq 0$ and eq. (8.4.16) becomes

$$\gamma \simeq (n_1^{(1)})^2 B_{11}^{(1,1)} + \left(n_2^{(2)}\right)^2 B_{22}^{(2)} - 2n_1^{(1)} n_2^{(2)} B_{12}^{(1,2)}. \tag{8.4.25}$$

Identifying $(n_i^{(i)})^2 B_{ii}^{(i,i)}$ with the surface tension γ_i^0 of pure liquid i against its vapor and introducing "mixing rule" approximation

$$B_{12}^{(1,2)} = \left[B_{11}^{(1,1)} B_{22}^{(2,2)}\right]^{1/2} \Phi, \tag{8.4.26}$$

where $\Phi = d_1 d_2/(d_1 + d_2)^2$, $d_i^{-3} \equiv n_i^{(i)}$. Girifalco and Good (1957) obtained the formula

$$\gamma \simeq \gamma_1^0 + \gamma_2^0 - 2\Phi\sqrt{\gamma_1^0 \gamma_2^0}. \tag{8.4.27}$$

This approximation enables one to estimate the interfacial tension between immiscible liquids from the surface tensions and densities of pure liquids. The estimate is accurate to within several percent for quite a few fluids. Fowkes (1963) introduced an empirical variation on the theme of Girifalco and Good and increased the accuracy of the approximation for polar and hydrogen bonded fluids.

Along similar lines, if in the multicomponent FKB formula ($n_\alpha^{(1)} = n_\alpha^l$ and $n_\alpha^{(2)} \approx 0$) one assumes $B_{\alpha\alpha}^{(ll)} = \gamma_\alpha^0/(n_\alpha^0)^2$ and $B_{\alpha\beta}^{(ll)} =$

$\left[B_{\alpha\alpha}^{(ll)} B_{\beta\beta}^{(ll)}\right]^{1/2}$, then for the surface tension of a multicomponent, low vapor pressure liquid one obtains the approximating formula

$$\gamma = \sum_{\alpha\beta} \frac{n_\alpha^{(l)} n_\beta^{(l)}}{n_\alpha^0 n_\beta^0} \gamma_\alpha^0 \gamma_\beta^0, \tag{8.4.28}$$

where n_α^0 is the number density of pure liquid and γ_α^0 its surface tension, and $n_\alpha^{(l)}$ is the number density of component α in the liquid mixture. Estimating the mixture component density from Amagat's law, $n^{-1} = \sum x_\alpha / n_\alpha^0$, x_α the mole fraction of component α, Winterfeld et al. (1978) applied eq. (8.4.21) to predict the composition dependence of 32 binary mixtures involving polar and nonpolar hydrocarbons, alcohols and water. Excluding the water–alcohol mixtures, the average error of the predictions was about 1% (never more than 3%). For methanol–water and ethanol–water solutions average errors were 18 and 32%, respectively.

Using a lattice model in conjunction with the discontinuous density profile, Davis and Scriven (1976) derived the approximate formula

$$\gamma = \frac{1}{8(n^o)^{1/3}} \left(\frac{\partial U}{\partial V}\right)_{T,N} \tag{8.4.29}$$

for the surface tension of a low vapor pressure liquid. U is the thermodynamic energy of the liquid phase. When applied to benzene, n-heptane, diethyl ether, argon, nitrogen, methane and carbon tetrachloride at various temperatures the formula was in error by no more than 17%. The error was usually significantly less than this. The largest errors were for methane and oxygen.

As illustrated above, the exact Kirkwood–Buff result for tension when exploited with the discontinuous interface model produces approximations useful for rough estimates of interfacial tensions in some cases. An interface is, however, not continuous and thus the model is limited and does not provide insight into the actual structure and stresses in the interfacial zone. Thus, in the next section, we pursue a model that retains some of the simplicity of the discontinuous interface model but gives a more realistic description of a fluid–fluid interface.

8.5 Gradient Theory of Inhomogeneous Fluids and Applications to Fluid–Fluid Interface

In this section we want to investigate inhomogeneous fluids in which the density varies slowly enough that if $\mathbf{s}$ is the separation of a pair of interacting particles, then the Taylor series

$$n(\mathbf{r} + \mathbf{s}) = \sum_{i=0}^{\infty} \frac{(\mathbf{s} \cdot \nabla)^i}{i!} n(\mathbf{r}) \tag{8.5.1}$$

can be truncated after terms of order $\nabla_{\mathbf{r}}\nabla_{\mathbf{r}}n$. In this case, it is consistent to expand the density quantities in **P** and keep terms of order $\nabla_{\mathbf{r}}\nabla_{\mathbf{r}}n$ and $(\nabla_{\mathbf{r}}n)(\nabla_{\mathbf{r}}n)$.

With the aid of the Taylor expansion

$$n^{(2)}(\mathbf{r}-\eta\mathbf{s},\ \mathbf{r}-\eta\mathbf{s}+\mathbf{s})=\sum_{i=0}^{\infty}\frac{(-\eta\mathbf{s}\cdot\nabla)^i}{i!}n^{(2)}(\mathbf{r},\ \mathbf{r}+\mathbf{s}), \tag{8.5.2}$$

the integration over $d\eta$ in eq. (8.3.12) can be carried out explicitly to obtain the expression

$$\mathbf{P}=n(\mathbf{r})kT\mathbf{I}-\frac{1}{2}\int\frac{\mathbf{ss}}{s}u'(s)\sum_{i=0}^{\infty}\frac{(-\mathbf{s}\cdot\nabla)^i}{(i+1)!}n^{(2)}(\mathbf{r},\ \mathbf{r}+\mathbf{s})d^3s \tag{8.5.3}$$

for the pressure tensor of a one-component fluid. The doublet density function can be expressed as

$$n^{(2)}(\mathbf{r},\ \mathbf{r}+\mathbf{s})=n(\mathbf{r})n(\mathbf{r}+\mathbf{s})g^{(2)}(\mathbf{r},\ \mathbf{r}+\mathbf{s}). \tag{8.5.4}$$

In addition to assuming gradients are small enough to truncate eq. (8.5.1) we shall assume that $g^{(2)}$ depends only on the magnitude of the separation between the particles. With this approximation, the stress tensor to order second order in density gradients becomes

$$\begin{aligned}\mathbf{P}=n(\mathbf{r})kT\mathbf{I}&-\frac{1}{2}\int\frac{\mathbf{ss}}{s}u'(s)g^{(2)}(s)n(\mathbf{r})\left[n(\mathbf{r})+\mathbf{s}\cdot\nabla n(\mathbf{r})+\frac{1}{2}\mathbf{ss}:\nabla\nabla n(\mathbf{r})\right]d^3s\\&+\frac{1}{4}\int\frac{\mathbf{ss}}{s}u'(s)g^{(2)}(s)\mathbf{s}\cdot\nabla\{n(\mathbf{r})[n(\mathbf{r})+\mathbf{s}\cdot\nabla_{\mathbf{r}}n(\mathbf{r})]\}d^3s\\&-\frac{1}{12}\int\frac{\mathbf{ss}}{s}u'(s)g^{(2)}(s)\mathbf{ss}:\nabla\nabla[n(\mathbf{r})]^2d^3s.\end{aligned} \tag{8.5.5}$$

The integrals over d^3s in eq. (8.5.5) are of the form

$$\mathbf{I}_m=\int\mathbf{s}\cdots\mathbf{s}f(s)d^3s, \tag{8.5.6}$$

where $f(s)$ is an isotropic function of s, that is, $f(s)$ depends only on the magnitude of s. Thus, because the integral over d^3s is over all space, $\mathbf{I}_m$ is an mth rank isotropic tensor. If m is odd, the coordinate transformation $\mathbf{s}\to-\mathbf{s}'$ proves that $\mathbf{I}_m=-\mathbf{I}_m$, and so the isotropic tensors of odd rank are zero, that is,

$$\int\mathbf{s}f(s)d^3s=\int\mathbf{sss}f(s)d^3s=0. \tag{8.5.7}$$

Expressing s in Cartesian coordinates $\hat{e}_1$, $\hat{e}_2$, and $\hat{e}_3$, we find

$$\begin{aligned}\mathbf{I}_2&=\sum_{i,j=1}^{3}\hat{e}_i\hat{e}_j\int s_is_jf(s)d^3s=\sum_{i=1}^{3}\hat{e}_i\hat{e}_i\int s_i^2f(s)d^3s\\&=\sum_{i+1}^{3}\hat{e}_i\hat{e}_i\int\frac{s^2}{3}f(s)d^3s=\left[\frac{1}{3}\int s^2f(s)d^3s\right]\mathbf{I},\end{aligned} \tag{8.5.8}$$

where **I** is the unit tensor. In terms of the same Cartesian coordinate system

$$\mathbf{I}_4 = \sum_{i,j,k,l} \hat{e}_i \hat{e}_j \hat{e}_k \hat{e}_l \int s_i s_j s_k s_l f(s) d^3 s \qquad (8.5.9)$$

There are two types of nonzero terms in eq. (8.5.9). These are

$$J_1 \equiv \int s_j^4 f(s) d^3 s, \qquad (8.5.10)$$

and, for $i \neq j \neq k$,

$$\begin{aligned} J_2 &\equiv \int s_i^2 s_j^2 f(s) d^3 s = \frac{1}{2} \int (s_i^2 + s_k^2) s_j^2 f(s) d^3 s \\ &= \frac{1}{2} \int (s^2 - s_j^2) s_j^2 f(s) d^3 s = \frac{1}{6} \int s^4 f(s) d^3 s - \frac{1}{2} J_1. \end{aligned} \qquad (8.5.11)$$

The rearrangements shown in eq. (8.5.11) take advantage of the equivalence of different pairs of Cartesian coordinates. To simplify J_1 further, we note that in spherical coordinates $\mathbf{s} = [\sin\phi \cos\theta \hat{e}_1 + \sin\phi \sin\theta \hat{e}_2 + \cos\theta \hat{e}_3]$ and $d^3 s = s^2 \sin\theta d\theta d\phi$. Choose $j = 3$ in eq. (8.5.10), and so $s_3 = s\cos\theta$ and

$$\begin{aligned} J_1 &= \int_0^\infty \int_0^{2\pi} \int_0^\pi s^6 f(s) \cos^4\theta \sin\theta d\theta d\phi ds \\ &= 2\pi \int_0^\infty s^6 f(s) \left[\frac{-\cos^5\theta}{5} \right]_0^\pi ds = \frac{4\pi}{5} \int_0^\infty s^6 f(s) ds \\ &= \frac{1}{5} \int s^4 f(s) d^3 s. \end{aligned} \qquad (8.5.12)$$

This result combined with eq. (8.5.11) yields

$$J_2 = \frac{1}{15} \int s^4 f(s) d^3 s. \qquad (8.5.13)$$

The above results lead to

$$\mathbf{I}_4 = \left[\frac{1}{15} \int s^4 f(s) d^3 s \right] \sum_{i,j} \left[\hat{e}_i \hat{e}_j \hat{e}_j \hat{e}_i + \hat{e}_i \hat{e}_j \hat{e}_i \hat{e}_j + \hat{e}_i \hat{e}_i \hat{e}_j \hat{e}_j \right] \qquad (8.5.14)$$

It is now easy to show that

$$\mathbf{I}_4 \mathbf{:AB} = \frac{1}{15} \int s^4 f(s) d^3 s [\mathbf{AB} + \mathbf{BA} + (\mathbf{A} \cdot \mathbf{B})\mathbf{I}], \qquad (8.5.15)$$

where **A** and **B** are arbitrary vectors.

The properties established in eqs. (8.5.7)–(8.5.15) can be used to reduce eq. (8.5.5) to

$$\mathbf{P} = P_o(n(\mathbf{r}))\mathbf{I} + 2c \left[\frac{1}{6} \nabla n \nabla n + \frac{1}{12} (\nabla n)^2 \mathbf{I} - \frac{1}{3} n \nabla\nabla n - \frac{1}{6} n \nabla^2 n \mathbf{I} \right], \qquad (8.5.16)$$

where $P_o(n)$ is the pressure of homogeneous fluid at local density n, that is,

$$P_o(n) = nkT - \frac{n^2}{6}\int su'(s)g^{(2)}(s)d^3s, \qquad (8.5.17)$$

and c is a parameter measuring the influence of density gradients on the local pressure tensor. The formula for c is

$$c = \frac{1}{30}\int s^3u'(s)g^{(2)}(s)d^3s. \qquad (8.5.18)$$

From this it follows that for an interaction potential of a given strength (say, given $u_{\min}$) the longer the range of the potential the larger the influence of parameter c.

Evaluation of the gradient approximation for a multicomponent fluid poses no special problems. The result for **P** is

$$\mathbf{P} = P_o(\mathbf{n}(\mathbf{r}))\mathbf{I} + \sum_{\alpha,\beta} 2c_{\alpha\beta}\left[\frac{1}{6}\nabla n_\alpha \nabla n_\beta + \frac{1}{12}\nabla n_\alpha \cdot \nabla n_\beta \mathbf{I} - \frac{1}{3}n_\alpha \nabla\nabla n_\beta - \frac{1}{6}n_\alpha \nabla^2 n_\beta \mathbf{I}\right], \qquad (8.5.19)$$

and for $\mathbf{P}_\alpha$ is

$$\begin{aligned}\mathbf{P}_\alpha = {} & P_{o,\alpha}(\mathbf{n}(\mathbf{r}))\mathbf{I} + \sum_\beta 2c_{\alpha\beta}\left[\frac{1}{12}\nabla_\mathbf{r} n_\alpha \nabla n_\beta\right. \\ & + \frac{1}{12}\nabla n_\beta \nabla n_\alpha + \frac{1}{12}\nabla n_\alpha \cdot \nabla n_\beta \mathbf{I} \\ & \left. - \frac{1}{6}n_\alpha \nabla\nabla n_\beta - \frac{1}{6}n_\beta \nabla\nabla n_\alpha - \frac{1}{12}n_\alpha \nabla^2 n_\beta \mathbf{I} - \frac{1}{12}n_\beta \nabla^2 n_\alpha \mathbf{I}\right],\end{aligned} \qquad (8.5.20)$$

where

$$c_{\alpha\beta} = \frac{1}{30}\int s^3 u'_{\alpha\beta}(s)g^{(2)}_{\alpha\beta}(s)d^3s, \qquad (8.5.21)$$

and

$$P^{(\mathbf{n})}_{o,\alpha} = n_\alpha kT - \sum_\beta \frac{1}{6}n_\alpha n_\beta \int su'_{\alpha\beta}(s)g_{\alpha\beta}(s)d^3s. \qquad (8.5.22)$$

Similarly, to order $\nabla^2_\mathbf{r}$, the gradient approximation to the asymmetric force $\mathbf{B}_\alpha$ is

$$\mathbf{B}_\alpha = \sum_\beta \left[-\frac{1}{6}\int su'_{\alpha\beta}(s)g_{\alpha\beta^{(2)}}(s)d^3s\right]\left[n_\beta \nabla n_\alpha - n_\alpha \nabla n_\beta\right] \qquad (8.5.23)$$

Substitution of eqs. (8.5.20) and (8.5.23) into eq. (8.3.29), the equation of hydrostatic equilibrium for species α, results in a set of differential equations that with appropriate boundary conditions can be solved to find the density profiles $n_\alpha(\mathbf{r})$.

Let us now consider a planar fluid–fluid interface. For this case $n_\alpha = n_\alpha(z)$ and external forces are negligible so that $P_N \equiv$

P_{zz} = the bulk pressure of the phases and $P_T = P_{xx} = P_{yy}$. From gradient theory, if follows that

$$P_N = P_o(\mathbf{n}(z)) + \sum_{\alpha,\beta} c_{\alpha\beta}\left[\frac{1}{2}\frac{dn_\alpha}{dz}\frac{dn_\beta}{dz} - n_\alpha\frac{d^2 n_\beta}{dz^2}\right] \quad (8.5.24)$$

and

$$P_T = P_o(\mathbf{n}(z)) + \sum_{\alpha,\beta} c_{\alpha\beta}\left[\frac{1}{6}\frac{dn_\alpha}{dz}\frac{dn_\beta}{dz} - \frac{n_\alpha}{3}\frac{d^2 n_\beta}{dz^2}\right] \quad (8.5.25)$$

The boundary conditions for a planar interface are that $n_\alpha(z) \to n_\alpha^b$ as $z \to \pm\infty$, where b = bulk phase 1 as $z \to -\infty$ and b = bulk phase 2 as $z \to \infty$. Thus, all derivatives of n_α vanish far from the interface and the component densities become those of the coexisting bulk phases.

Two interesting results can be obtained immediately. First, because $\gamma = \int(P_N - P_T)dz$, an integration by parts of the difference between eqs. (8.5.24) and (8.5.25) leads to the expression

$$\gamma = \sum_{\alpha,\beta}\int_{-\infty}^{\infty} c_{\alpha\beta}\frac{dn_\alpha}{dz}\frac{dn_\beta}{dz}dz. \quad (8.5.26)$$

Second, multiplying eq. (8.5.25) by 3 and subtracting it from eq. (8.5.24), we obtain by rearranging the result

$$P_T(z) = \frac{2}{3}P_o(\mathbf{n}(z)) + \frac{1}{3}P_N. \quad (8.5.27)$$

Thus, the influence parameters and density gradients determine directly the tension of the planar interface. Even more interestingly, according to eq. (8.5.27), the variation of the transverse stress across a planar interface is controlled by the pressure of homogeneous fluid at the local compositions occurring in the interfacial zone.

To obtain more insight into the structure and stress in the interfacial zone, let us consider the one-component liquid–vapor interface. The normal pressure for this case is

$$P_N = P_o(\mathbf{n}(z)) + c\left[\frac{1}{2}\left(\frac{dn}{dz}\right)^2 - n\frac{d^2 n}{dz^2}\right]. \quad (8.5.28)$$

Differentiating eq. (8.5.28) with respect to and using the Gibbs–Duhem equation, $dP_0 = n d\mu_0$, and the equilibrium condition $dP_N/dz = 0$, we obtain

$$0 = n\frac{d\mu_o}{dz} - nc\frac{d^3 n}{dz^3} \quad (8.5.29)$$

The factor of n can be canceled, leaving an equation that can be integrated to yield

$$c\frac{d^2 n}{dz^2} = \mu_o(n) - \mu, \quad (8.5.30)$$

where the constant of integration μ is the chemical potential because far from the interface $d^2n/dz^2 = 0$, and so

$$0 = \mu_o(n_l) - \mu = \mu_o(n_g) - \mu \quad \text{or} \quad \mu_o(n_l) = \mu_o(n_g) = \mu. \tag{8.5.31}$$

n_l and n_g are the coexisting bulk liquid and vapor densities and eq. (8.5.31) is the usual condition of chemical equilibrium. Equation (8.5.30) can be integrated by quadrature. To do this it is useful to define a kind of local "grand potential" density $\omega(n)$:

$$\omega(n) \equiv f_o(n) - n\mu \tag{8.5.32}$$

$f_o(n)$ is the Helmholtz free energy density of homogeneous fluid at density n. If μ were replaced by the chemical potential $\mu_o(n)$ of homogeneous fluid at density n, $\omega(n)$ would be in fact the grand potential density of homogeneous fluid. However, $\mu_o(n)$ is not constant in the interfacial zone and is only equal to the chemical potential μ of the system only in the bulk phases. Because $f_o(n) = F_o/V$ and $\mu_o = (\partial F_o/\partial N)_{T,V} = [\partial(F_o/V)/\partial(N/V)]_{T,V}$, it follows that

$$\mu_0(n) = \frac{\partial f_o}{\partial n} \quad \textit{and} \quad \mu_o(n) - \mu = \frac{\partial \omega}{\partial n}, \tag{8.5.33}$$

and so eq. (8.5.30) can be expressed as

$$c\frac{d^2n}{dz^2} = \frac{\partial \omega}{\partial n} \tag{8.5.34}$$

Now, multiplication of each side of eq. (8.5.34) by dn/dz leads to

$$\frac{d}{dz}\left[\frac{c}{2}\left(\frac{dn}{dz}\right)^2 - \omega(n)\right] = 0$$

that integrates to

$$\frac{c}{2}\left(\frac{dn}{dz}\right)^2 = \omega(n) + K. \tag{8.5.35}$$

Again the boundary conditions that $n \to$ bulk densities as $z \to \pm\infty$ enable evaluation of the constant K of integration:

$$K = -\omega(n_l) = -\omega(n_g). \tag{8.5.36}$$

But because $\omega(n_b) = f_o(n_b) - n_b\mu_o(n_b)$ for $b = l$ or g, it follows that $-\omega(n_b) = P_o(n_b)$ and so eq. (8.5.36) becomes

$$K = P_o(n_l) = P_o(n_g) = P_N, \tag{8.5.37}$$

the condition of mechanical equilibrium.

Equation (8.5.35) can now be solved for dn/dz and rearranged to yield

$$\sqrt{\frac{c}{2}}\frac{dn}{\sqrt{\Delta\omega(n)}} = \pm dz, \tag{8.5.38}$$

where $\Delta\omega(n) \equiv \omega(n) - \omega_B$, with $\omega_B = \omega(n_l) = \omega(n_g)$ = the negative of the normal or bulk phase pressure. The sign in eq. (8.5.38) will be negative if the liquid phase is located in the direction of $-\infty$. The origin of the coordinate of the coordinate system for z is arbitrary. We can pick $n_0 = n\ (z = 0)$ to be somewhere between n_g and n_l, say $n_0 = (n_g + n_l)/2$, and find the interfacial density profile from

$$\int_{n_0}^{n(z)} \frac{\sqrt{c/2}dn}{\sqrt{\Delta\omega(n)}} = -z. \tag{8.5.39}$$

One sets $n(z)$ and calculates the position z of this density. Because $\Delta\omega(n) \rightarrow 0$ as $n \rightarrow n_l$ and as $n \rightarrow n_g$, the range of z will be from $-\infty$ to $+\infty$. However, as we shall see shortly, $n(z)$ approaches its asymptotic values over the range of several molecular diameters, and so the interfacial zone of appreciable density variation is very narrow on a macroscopic scale.

Before examining density profiles predicted with the aid of a particular equation of state, a most interesting implication of gradient theory can now be obtained. Using eq. (8.5.35) to eliminate $c(dn/dz)^2$ from the tension formula eq. (8.5.26) and eq. (8.5.38) to transform from the variable dz to dn, we find

$$\gamma = \int_{n_g}^{n_l} \sqrt{2c\Delta\omega(n)}dn. \tag{8.5.40}$$

The tension can be calculated directly from the influence parameter c and a thermodynamic function, $\Delta\omega(n)$, of homogeneous fluid! One need not compute the density profile if only surface tension is desired.

Geometrically $\Delta\omega(n)$ can be represented as the vertical distance between the curve of $f_o(n)$ versus n and a straight line touching $f_o(n)$ at the vapor and liquid densities, n_g and n_l. The construction of $\Delta\omega(n)$ from $f_o(n)$ is illustrated in Figure 8.5: $\Delta\omega(n)$ is the Helmholtz free energy density of homogeneous fluid measured relative to the Helmholtz free energy of bulk liquid and vapor at the overall density n.

As the critical point T_c is approached, $\Delta\omega(n)$ flattens out rapidly between n_g and n_l (Fig. 8.5). Thus, the surface tension goes to zero as the critical point is approached not only because the vapor and liquid densities approach each other, but also because $[\Delta\omega(n)]^{1/2}$ becomes smaller as the free energy curve flattens out between n_g and n_l. Interfaces near the critical point become broad because $\Delta\omega(n) \rightarrow 0$ and $dz \propto \Delta\omega^{-1/2}dn$ according to eq. (8.5.38).

To illustrate the behavior of the density and transverse pressure in the interfacial zone, we have calculated the density profile from eq. (8.5.39) and the transverse pressure profile from eq.

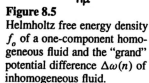

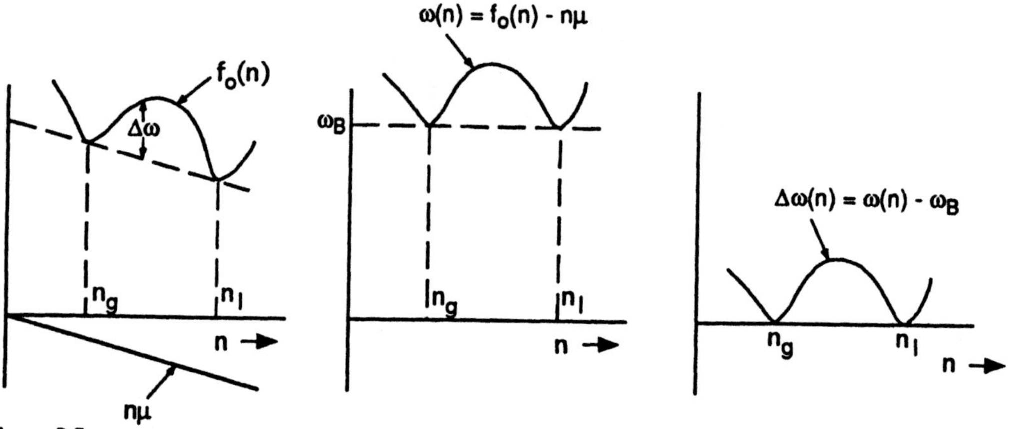

Figure 8.5
Helmholtz free energy density f_o of a one-component homogeneous fluid and the "grand" potential difference $\Delta\omega(n)$ of inhomogeneous fluid.

(8.5.27) for a van der Waals fluid whose hard sphere pressure is approximated by the Carnahan–Starling formula. In this case (see Section 5.6 of Chapter 5),

$$f_o(n) = nkT \ln n\sigma^3 + nkT(1 + 2y - 2y^2)/(1 - y)^2 - n^2 a, \qquad (8.5.41)$$

and

$$P_o(n) = nkT(1 + y + y^2 - y^3)/(1 - y)^3 - n^2 a, \qquad (8.5.42)$$

where

$$y = \pi\sigma^3 n/6 \qquad (8.5.43)$$

Results are shown in Figures 8.6–8.8. The following dimensional variables are introduced to display the results in these figures:

$$n^* = n\sigma^3, \quad P^* = \left(\frac{8\pi}{3}\right)\frac{p\sigma^6}{a}, \quad \gamma^* = \left(\frac{8\pi}{3}\right)\frac{\gamma\sigma^5}{a}, \quad z^* = \frac{z}{\sigma}.$$

The value of c was chosen so that $c/a\sigma^2 = 1$. For this fluid the critical point is $T_c = 1.50925(3a/8\pi k\sigma^3)$, $n_c = 0.249/\sigma^3$, and $P_c = 0.1349(3a/8\pi^6)$.

In Figure 8.6 the trend is that as the temperature is lowered the pressure minimum between n_g and n_l becomes increasingly negative. From this and eq. (8.5.27) it can be anticipated that the surface tension will increase with decreasing temperature. As seen in Figure 8.7 the density profile sharpens and the width of the interfacial zone narrows with decreasing temperature. The transverse pressure profiles and the corresponding surface tensions are shown in Figure 8.8. The interface has regions of tension, $P_N - P_T > 0$, and of compression, $P_N - P_T < 0$, in the interfacial zone. Overall, however, the tension dominates and results in a positive surface tension in a stable system.

Near the critical point, n_g and n_l are close to each other and $\Delta\omega(n)$ is small throughout the interface. With the aid of eq. (8.5.40), we can predict how the surface tension approaches zero as

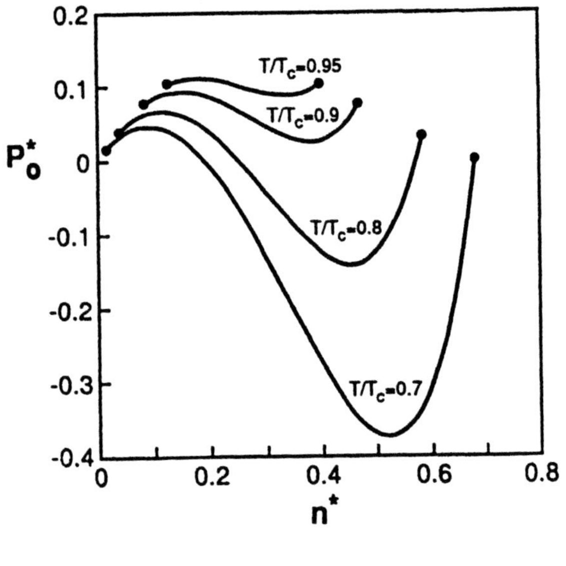

Figure 8.6
Homogeneous fluid pressure isotherms between coexisting vapor and liquid densities (filled circles).

the critical point is approach. Consider the Taylor series of $\Delta\omega(n)$ about n_l:

$$\begin{aligned}\Delta\omega(n) &= \Delta\omega(n_l) + (n - n_l)(\mu_o(n_l) - \mu) \\ &\quad + \frac{1}{2}(n - n_l)^2 \left[\frac{\partial^2 \Delta\omega(n)}{\partial n^2}\right]_{n'} + \cdots \\ &= \frac{1}{2}(n - n_l)^2 \left[\frac{\partial^2 f_o(n)}{\partial n^2}\right]_{n'} + \cdots .\end{aligned} \tag{8.5.44}$$

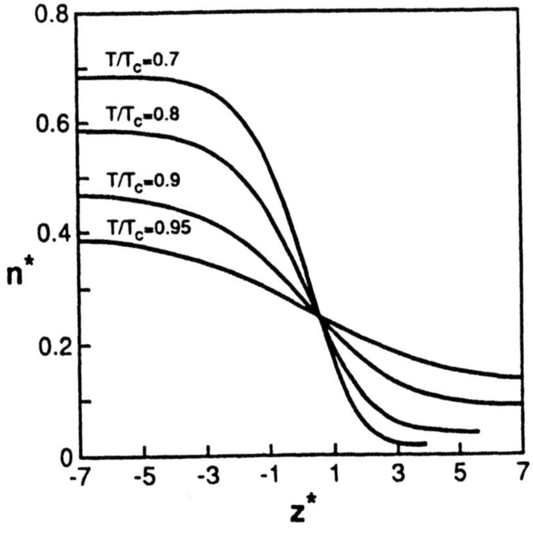

Figure 8.7
Density profile of the liquid-vapor interface at several temperatures.

Figure 8.8
Transverse pressure profile of the liquid–vapor interface at several temperatures.

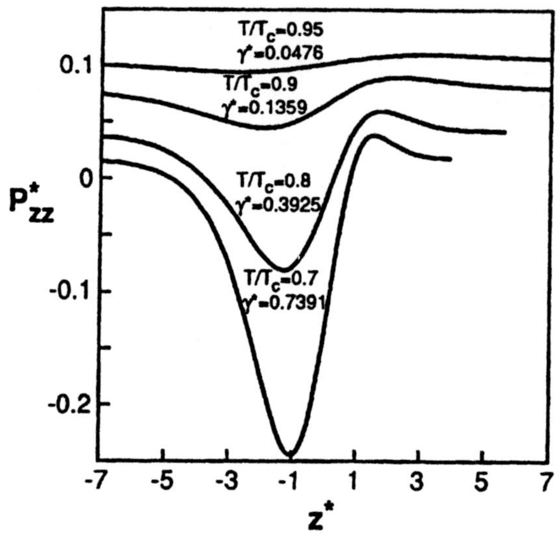

The first two terms in the series vanish because of the equilibrium conditions $\Delta\omega(n_l)0 = \mu_o(n_l) - \mu = 0$. With the thermodynamic relations,

$$\frac{\partial f_o}{\partial n} = \mu_o, \qquad \frac{\partial^2 f_0}{\partial n^2} = \frac{1}{n}\frac{\partial P_o}{\partial n} = \frac{1}{n^2 K_T}, \tag{8.5.45}$$

where K_T is the isothermal compressibility $[K_T = n^{-1}(\partial n/\partial P_o)]$, we obtain

$$\Delta\omega(n) = \frac{(n_l - n)^2}{n_l^2 K_T(n_l)} + O(n_l - n)^3. \tag{8.5.46}$$

Sufficiently near the critical point, $(n_l - n)$ is very small for values of n between n_g and n_l and so the terms of order $(n_l - n)^3$ and higher can be neglected in eq. (8.5.46). With this approximation eq. (8.5.40) yields

$$\gamma = \frac{\sqrt{2c}}{4n_l} K_T^{-1/2}(n_l - n_g)^2. \tag{8.5.47}$$

According to modern statistical mechanical theory (in particular the renormalization group theory), quantities such as γ, K_T^{-1}, and $(n_l - n_g)$ approach zero by obeying universal scaling laws as the critical point is approached. These laws take the form

$$\gamma \propto \left(1 - \frac{T}{T_c}\right)^{\zeta}, \quad K_T \propto \left(1 - \frac{T}{T_c}\right)^{-\nu}, \quad n_l - n_g \propto \left(1 - \frac{T}{T_c}\right)^{\beta}, \tag{8.5.48}$$

as $T \to T_c$, where ζ, ν, and β are universal exponents, that is, they are the same for any fluid. (They do depend on the dimensionality of space, a fact important to theory but not to our three-dimensional world of experimentation.) The theory is even more general than is indicated by eq. (8.5.48). In multicomponent fluid–fluid equilibria, the scaling laws are

$$\gamma \propto \left(1 - \frac{h}{h_c}\right)^{\zeta}, \quad K_T \propto \left(1 - \frac{h}{h_c}\right)^{-\nu}, \quad n_i^1 - n_i^2 \propto \left(1 - \frac{h}{h_c}\right)^{\beta}, \tag{8.5.49}$$

where h is a thermodynamic field variable and n_i^j is the density of component i in phase j. A field variable is one that is the same in all coexisting phases. Temperature, pressure, and chemical potentials are common field variables. The implication of the theory is that γ, K_T^{-1}, and $n_i^1 - n_i^2$ approach zero with the same scaling exponents as the critical point is approached along any field variable.

The scaling theory has been verified in numerous experiments on liquid–vapor and liquid–liquid systems as well as on magnetic systems, which obey the same scaling laws. Theory and experiment give $\zeta = 1.3$ (e.g., experiment gives 1.253 for CO_2 and 1.302 for Xe), $\nu = 1.2$ (e.g., 1.20 ± 0.02 for CO_2) and $\beta = 0.34$ (e.g., $\beta = 0.3447 \pm 0.0007$ for CO_2).

Returning now to the gradient theory, we note that if we choose $K_T \propto (1 - T/T_c)^{-1.2}$ and $(n_l - n_g) \propto (1 - T/T_c)^{0.34}$, we predict with eq. (8.5.47) that $\zeta = 1.28$, in agreement with experiment. The general prediction of eq. (8.5.47) is that the scaling coefficients β, ζ, and ν are not independent. In particular, gradient theory predicts

$$\zeta = \frac{\nu}{2} + 2\beta, \tag{8.5.50}$$

a relationship implied by more rigorous theory and verified experimentally.

If mean field equations of state, such as eqs. (8.5.41) and (8.5.42), as used to describe phase equilibria, one finds $\nu = 1$ and $\beta = 1/2$ and so predicts from eq. (8.5.50) that $\zeta = 1.5$. The problem is that sufficiently near a critical point there are enormous fluctuations in density and composition (e.g., because K_T is approaching ∞, tiny pressure fluctuations can result in large density fluctuation). Mean field theory, in which one usually assumes that $\langle e^{-\beta u^{N,s}} \rangle \simeq e^{-\langle \beta u^{N,s} \rangle}$, does not correctly account for these fluctuations. It was this failure of mean field models very near the critical point that fueled the development of renormalization group theory. Because this theory is rather advanced and is usually important only for systems very near the critical point, it will not be explored further in this text.

Although the density profile of a liquid–vapor interface cannot be measured presently, one can compute it by molecular dynamics. In Figure 8.9, gradient theory is compared with a computer simulation for a 6–12 Lennard–Jones fluid. The ripples in the computer generated curve are now known to be computational noise, and so the agreement between gradient theory and the simulated data is quite good. Surface tension is predicted by gradient theory

Figure 8.9
Liquid–vapor density profile of a 6–12 Lennard–Jones fluid. Comparison of a— computer simulation with (- - -) gradient theory. (Adapted from Bongiorno and Davis (1975). (– - –) Prediction of gradient theory for van der waals approximation of the Lennard–Jones fluid.

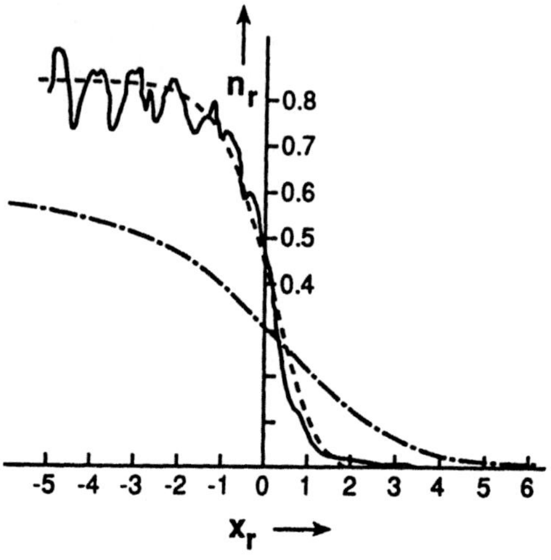

to be 19.6 dyn/cm, where the computer simulation gave 16.5 ± 2.3 dyn/cm.

A possible improvement of gradient theory is to approximate the inhomogeneous pair correlation function by the equilibrium correlation function at the local density midway between the particles, that is,

$$g^{(2)}(\mathbf{r}_1, \mathbf{r}_2) = g_o^{(2)}\left(|\mathbf{r}_1 - \mathbf{r}_2|; n\frac{(\mathbf{r}_1 + \mathbf{r}_2)}{2}\right) \tag{8.5.51}$$

Inserting this into the YBG equation, we obtain

$$kT\nabla n(\mathbf{r}) = -\nabla u^e(\mathbf{r})n(\mathbf{r}) - \int \nabla u(|\mathbf{r} - \mathbf{r}'|)n(\mathbf{r})n(\mathbf{r}')g_o^{(2)}(|\mathbf{r} - \mathbf{r}'|; n\left(\frac{(\mathbf{r} + \mathbf{r}')}{2}\right)d^3r' \tag{8.5.52}$$

Given the homogeneous fluid pair correlation function as a function of density, eq. (8.5.52) can be integrated to yield an integral equation from which the density distribution $n(\mathbf{r})$ can be computed Toxvaerd (1976).

Toxvaerd applied eq. (8.5.52) to the planar liquid–vapor interface of a square-well fluid, for which the pair potential is

$$\begin{aligned} u(s) &= \infty, \quad s < d, \\ &= -\varepsilon, \quad \sigma < s < Rd \\ &= 0, \quad s > Rd. \end{aligned} \tag{8.5.53}$$

His results are compared to gradient theory in Figure 8.10. The comparison is typical of extensive comparisons of gradient theory with integral equation models (e.g., see Bongiorno, et al., (1976). The profiles are in good agreement, which is good news for the gradient theory of fluid–fluid interfaces. The gradient theoretical problem is computationally much less costly than the integral equation derived from eq. (8.5.52).

Consider next a fluid in contact with a planar solid. Suppose the solid can be described by an external potential $u^e(z)$ that is a

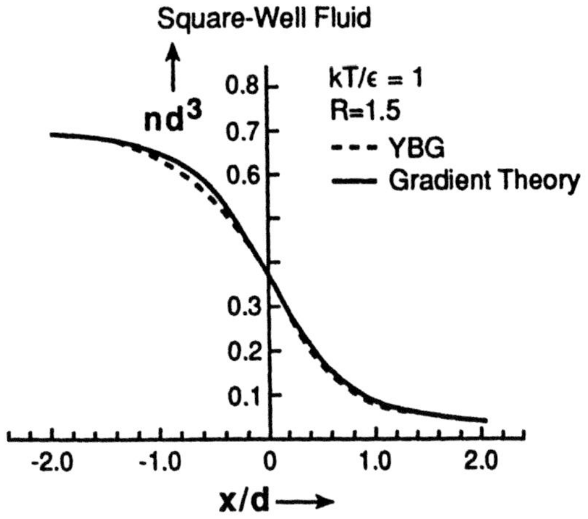

Figure 8.10
Density profile of the liquid–vapor interface of a square-well fluid. Comparison of results of Toxvaerd's YBG equation and gradient theory. (Adapted from Davis and Scriven, 1982).

function only of the distance z from the solid surface. For a one-component fluid the gradient approximation to P_N, when inserted into the equation of hydrostatic equilibrium, leads to the equation

$$c\frac{d^2n}{dx^2} = \mu_o(n) + u^e(x) - \mu. \tag{8.5.54}$$

Given the success of gradient theory with the liquid–vapor interface, one might expect eq. (8.5.54) to yield a good qualitative description of the fluid–solid interface. Unfortunately, this is not the case. The density profile of fluid near a solid surface is much more structured than what is predicted by gradient theory. For example, for a hard sphere fluid confined by a hard wall [$u^e(z) = \infty$, $z \leq 0$, $u^e(z) = 0$, $z > 0$], one predicts from eq. (8.5.54) that $n(z) = n_b$ for $z > 0$, where n_b is the bulk density determined by the chemical potential μ [i.e., $\mu_o(n_b) = \mu$]. Computer simulation on the other hand gives a density profile $n(z)$ that oscillates rapidly as a function of z (see Fig. 8.11).

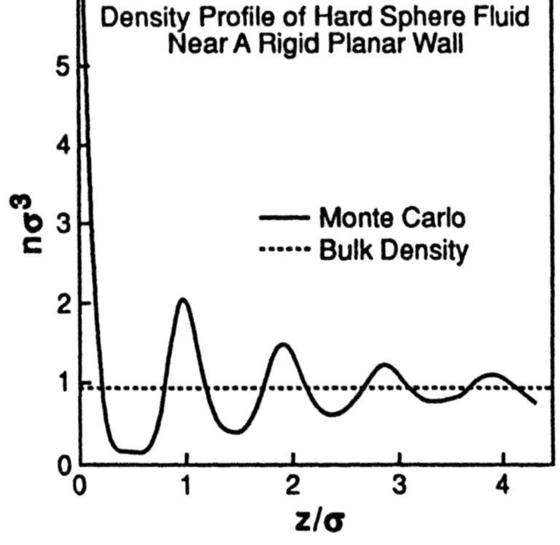

Figure 8.11
Density profile of a hard sphere fluid at a hard wall. Results from a Monte Carlo simulation. (Snook and Henderson, 1978).

The YBG equation combined with the local density approximation given by eq. (8.5.51) is not a significant improvement over the gradient theory of the fluid–solid interface. Thus, a better closure approximation must be found for the YBG equation if one wishes to describes fluid–solid interfaces. Such an improvement is the topic of the next section.

8.6 Fluids at Solid Surfaces or in Porous Media

In the last section, we saw gradient theory [and the local density approximation for $g^{(2)}(\mathbf{r}, \mathbf{r}')$ in the YBG equation] fails to account for the rapid oscillations in the density profile of a fluid near a solid wall. As we learn in a later chapter, the problem is that the pair correlation function of strongly inhomogeneous fluid is poorly approximated by the local homogeneous fluid models such as $g_0^{(2)}\left(|\mathbf{r}-\mathbf{r}'|, n\left(\frac{\mathbf{r}+\mathbf{r}'}{2}\right)\right)$. The correlations between neighboring particles depend on the density profile in their vicinity not just at their midpoint $(\mathbf{r}+\mathbf{r}')/2$.

A more accurate approximation of the pair correlation function of strongly inhomogeneous fluid is thus needed for fluids confined by solid surfaces. One model is to retain the approximation that $g^{(2)}$ can be estimated from a homogeneous fluid correlation function, but to evaluate it at a local average density $\overline{n}(\mathbf{r})$ instead of the local density $n(\mathbf{r})$. Fischer and Methfessel (1980) introduced such an approximation to the YBG equation and others have done so in the context of the density functional free energy theory described in a later chapter. The local average density is defined by

$$\overline{n}(\mathbf{r}) = \int w(\mathbf{r}, \mathbf{r}')n(\mathbf{r}')d^3r', \tag{8.6.1}$$

where $w(\mathbf{r}, \mathbf{r}')$ is a weighting function with the property

$$\int w(\mathbf{r}, \mathbf{r}')d^3\mathbf{r}' = 1, \tag{8.6.2}$$

required so that in homogeneous fluid $\overline{n}(\mathbf{r}) = n$. Fischer and Methfessel chose

$$\overline{n}(\mathbf{r}) = \left[\frac{4\pi}{3}\left(\frac{\sigma}{2}\right)^3\right]^{-1} \int \eta\left(\frac{\sigma}{2} - |\mathbf{r}-\mathbf{r}'|\right) n(\mathbf{r}')d^3r', \tag{8.6.3}$$

where σ is the particle diameter and η is the Heaviside step function. As given by eq. (8.6.3), $\overline{n}(\mathbf{r})$ is the average of the density in a sphere of radius $\frac{\sigma}{2}$ centered at point $\mathbf{r}$. This is of course an arbitrary choice for $w(\mathbf{r}, \mathbf{r}')$. However, as will be seen in a later chapter, it has the attractive feature that it gives the exact result for the YBG equation of an inhomogeneous fluid of one-dimensional hard rods.

With this "local average density" approximation, the YBG equation for a one-component fluid becomes

$$kT\nabla \ln n(\mathbf{r}) = -\int \nabla u(|\mathbf{r}-\mathbf{r}'|)n(\mathbf{r}')g_0^{(2)}\left(|\mathbf{r}-\mathbf{r}'|, \bar{n}\left(\frac{\mathbf{r}+\mathbf{r}'}{2}\right)\right)d^3r' - \nabla u^e(\mathbf{r}). \tag{8.6.4}$$

With a suitable theory of $g_0^{(2)}$ for homogeneous fluids, eq. (8.6.4) is now an integro-differential equation from which the density profile can be computed for a given external field $u^e(\mathbf{r})$.

Instead of applying eq. (8.6.4), Fischer and Methfessel introduced the further simplifying approximations that (1) the pair potential is of the form

$$u(|\mathbf{r}-\mathbf{r}'|) = u^H(|\mathbf{r}-\mathbf{r}'|) + u^S(|\mathbf{r}-\mathbf{r}'|), \tag{8.6.5}$$

where u^H is the hard sphere potential and u^S is the continuous longer ranged "soft" part of the potential lying outside the hard core and (2) the structureless fluid approximation $[g^{(2)}(\mathbf{r}, \mathbf{r}') = 1]$ is sufficient for evaluation ofthe contribution of u^S to eq. (8.6.4). Because $e^{-\beta u^H(s)} = \eta(s-\sigma)$, if follows that

$$\delta(s-\sigma) = \frac{d}{ds}e^{-\beta u^H(s)} = -\beta\frac{du^H}{ds}(s)e^{-\beta u^H(s)}, \tag{8.6.6}$$

or

$$\frac{du^H}{ds}(s) = -kT\delta(s-\sigma)e^{\beta u^H(s)}. \tag{8.6.7}$$

Thus,

$$\frac{du^H}{ds}(s)g^{(2)}(\mathbf{r}, \mathbf{r}+\mathbf{s}) = -kT\delta(s-\sigma)\left[e^{\beta u^H(s)}g^{(2)}(\mathbf{r}, \mathbf{r}+\mathbf{s})\right]. \tag{8.6.8}$$

From its definition, $g^{(2)}$ has the property that $g^{(2)}(\mathbf{r}, \mathbf{r}+\mathbf{s}) = e^{-\beta u^H(s)}\tilde{g}^{(2)}(\mathbf{r}, \mathbf{r}+\mathbf{s})$, where $\tilde{g}^{(2)}(\mathbf{r}, \mathbf{r}+\mathbf{s})$ is a continuous function at $s=\sigma$. Consequently, eq. (8.6.8) becomes

$$\frac{du^H}{ds}(s)g^{(2)}(\mathbf{r}, \mathbf{r}+\mathbf{s}) = -kT\delta(s-\sigma)\tilde{g}^{(2)}(\mathbf{r}, \mathbf{r}+\sigma\hat{s}), \tag{8.6.9}$$

where $\hat{s}$ is a unit vector in the direction of $\mathbf{s}$. Because $u^H(s) = 0$ for $s > \sigma$, it follows that $\tilde{g}^{(2)}(\mathbf{r}, \mathbf{r}+\sigma\hat{s}) = \lim_{s\to\sigma+} g^{(2)}(\mathbf{r}, \mathbf{r}+s\hat{s})$, where $s \to \sigma+$ means the limit $s=\sigma$ is approached from values of s greater than σ. Thus, $\tilde{g}^{(2)}(\mathbf{r}, \mathbf{r}+\sigma\hat{s})$ is simply the value of the correlation function of a pair of particles at hard core contact.

With the aid of eq. (8.6.9), Fischer and Methfessel's approximations transform eq. (8.6.4) to

$$\begin{aligned} kT\nabla \ln n(\mathbf{r}) &= -\nabla \int n(\mathbf{r}')u^s(|\mathbf{r}-\mathbf{r}'|)d^3r' \\ &+ kT\int n(\mathbf{r}')g_0^{(2)}\left(\sigma; \bar{n}\left(\frac{\mathbf{r}+\mathbf{r}'}{2}\right)\right)\frac{\mathbf{r}-\mathbf{r}'}{|\mathbf{r}-\mathbf{r}'|}\delta(|\mathbf{r}'-\mathbf{r}|-\sigma)d^3r' - \nabla u^e(\mathbf{r}). \end{aligned} \tag{8.6.10}$$

We have dropped the tilde from $\tilde{g}$ with the understanding that $g_0^{(2)}(\sigma)$ is the contact value of the correlation function.

For a planar system, $n = n(z)$ and $u^e = u^e(z)$, and so eq. (8.6.10) can be integrated to yield

$$K = u^e(z) + kT \ln n(z) + \int n(z')\Phi^S(|z - z'|)dz' + 2\pi kT \int_0^z dz'' \int_{-\sigma}^{\sigma} dz' z' n(z' + z'') g_0^{(2)}\left(\sigma; \bar{n}\left(z'' + \frac{1}{2}z'\right)\right), \tag{8.6.11}$$

where K is a constant of integration,

$$\Phi^S(|z - z'|) = \iint_{-\infty}^{\infty} u^S\left(\sqrt{s_x^2 + s_y^2 + (z - z')^2}\right) ds_x ds_y \tag{8.6.12}$$

and

$$\bar{n}(z) = \frac{6}{\sigma^2} \int_{z-\frac{\sigma}{2}}^{z+\frac{\sigma}{2}} [(\frac{\sigma}{2})^2 - (z - z')^2] n(z') dz'. \tag{8.6.13}$$

The hard core contact value of the pair correlation function must be specified to render eqs. (8.6.10) or (8.6.11) solvable. For this purpose, Fischer and Methfessel chose the Carnahan–Starling hard sphere formula

$$g_0^{(2)}(\sigma, n) = \frac{1 - \pi n \sigma^3/12}{(1 - \pi n \sigma^3/6)^3}, \tag{8.6.14}$$

which is known to agree well with computer simulations of hard sphere fluids.

Vanderlick et al. (1989) used eq. (8.6.11) to predict the density profile of a hard sphere fluid near a rigid wall. The value of K is determined by the temperature and density n_b of bulk phase as follows. First note that for this system $\Phi^s \equiv 0$, $u^e(z) = 0, z > 0$, and so $K = kT \ln n(0)$. The contact theorem for hard spheres is that at a planar hard wall $P_N = n(0)kT$ (the wall pressure is the same as that of a dilute gas because the only force exerted on the wall is from collisional transfer of momentum). But because P_N is constant, $P_N = P_o(n_b) = n(0)kT$. With the Carnahan–Starling formula for pressure,

$$P_o(n) = nkT \frac{1 + y + y^2 - y^3}{(1 - y)^3}, \tag{8.6.15}$$

where $y = \pi n \sigma^3/6$, we can compute $n(0)$ from n_b and hence K from $K = kT \ln n(0)$.

In Figure 8.12 the predicted density profile is compared to the profile computed by Monte Carlo simulations. Although the predicted profile does not agree quantitatively with the computer simulations, it does predict density oscillations, a feature missed entirely by gradient theory and the local density approximation $g_o^{(2)}(s; n(\frac{\mathbf{r}+\mathbf{r}'}{2}))$.

Figure 8.12
Density profile of a hard sphere fluid at a planar rigid wall. Dashed cure is Monte Carlo results from Snook and Henderson (1978). The origin of the coordinate system has been chosen arbitrarily. The solid curve is the Fischer–Methfessel YBG model results. (Vanderlick et al., 1989).

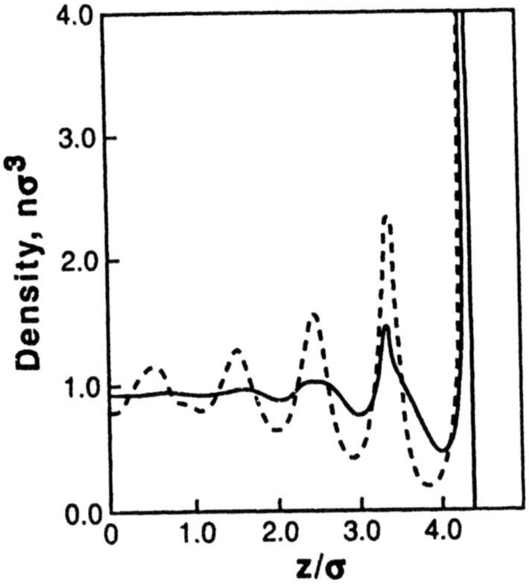

Another application of the Fischer–Methfessel model was carried out by Vanderlick et al. This was a comparison between predictions and molecular dynamics results for a 6–12 Lennard–Jones fluid confined between flat parallel Lennard–Jones walls separated by a distance L. The fluid–wall potential energy function used in the simulations and calculations was

$$u^e(z) = v^e(z) + v^e(L - z), \tag{8.6.16}$$

where

$$v^e(z) = 2\pi\varepsilon_w \left\{ \frac{2}{5}\left(\frac{\sigma_w}{z}\right)^{10} - \left(\frac{\sigma_w}{z}\right)^4 - \frac{\sqrt{2}\sigma_w^3}{3[z + 0.61\sigma_w/\sqrt{2}]^3} \right\}. \tag{8.6.17}$$

The fluid–fluid pair potential used in the model calculations was $u^S(s) = 0$, $s < \sigma$, and $u^s(s) = -4\varepsilon(\sigma/s)^6$, $s > \sigma$, that yields

$$\begin{aligned} \Phi^S(|z - z'|) &= -2\pi\varepsilon\sigma^2, \quad |z - z'| < \sigma \\ &= -2\pi\varepsilon\sigma^2[\sigma/(z - z')]^4, \quad |z - z'| > \sigma. \end{aligned} \tag{8.6.18}$$

The integration constant K in eq. (8.6.11) was determined by requiring that the number of confined particles in the model application equals that of the simulation, that is,

$$n_{\text{pore}} = \frac{1}{L}\int_0^L n_c(z)dz = \frac{1}{L}\int_0^L n_s(z)dz \tag{8.6.19}$$

where $n_c(z)$ and $n_s(z)$ are the calculated and simulated density profiles.

A comparison of theory and simulation are shown in Figure 8.13. The wall and fluid parameters were $\varepsilon/k = \varepsilon_w/k = 1.21$K $\sigma_w = \sigma$, $n_{\text{pore}} = 0.5297\sigma^{-3}$ for $L = 3\sigma$ and $n_{\text{pore}} = 0.4811\sigma^{-3}$ for

Figure 8.13
Comparison of density profiles predicted by Fischer–Methfessel YBG model with molecular dynamics for a Lennard–Jones system. (Adapted from Vanderlick et al. (1989). Solid curves are molecular dynamics results and dashed curves are YBG predictions.

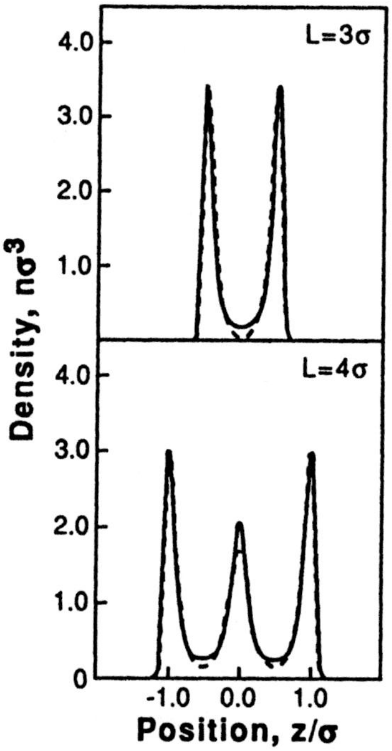

$L = 4\sigma$. The agreement is quite good and indicates that the wall–fluid interactions and the repulsive intermolecular forces cause strong layering of the fluid that is sensitive to wall separation. It was remarked in Chapter 7 that experimentally the pressure of confined fluid films at constant chemical potential is an oscillating function of distance of separation of the confining solid surfaces. Assuming that constant K corresponds to constant chemical potential (the assumption that constant K is equivalent to constant μ is only approximately true, but mean pore densities predicted by constant K are roughly equal to those found in the simulations), one can

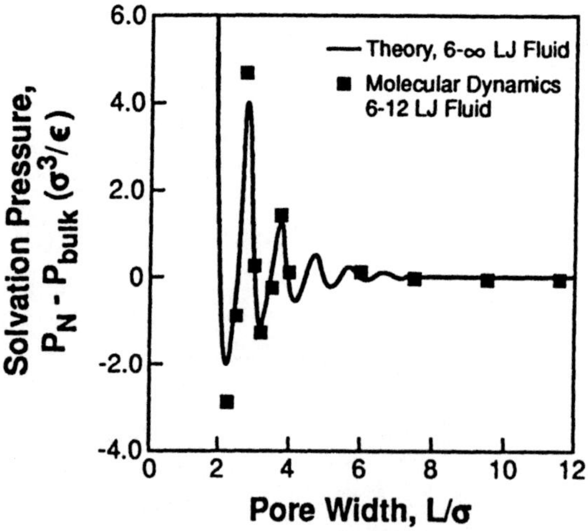

Figure 8.14
Comparison of predictions of the Fischer–Methfessel YBG model with molecular dynamics. Model calculations are for 6–∞ LJ fluid, molecular dynamics are for 6–12 LJ fluid, and 10–4–3 LJ walls are used in both cases. Adapted from Davis et al. (1987)

Figure 8.15
Density profiles of a fluid confined between planar walls of separation the profiles illustrate the effect of wall separation (i.e., porewidth) on the layering structure. YBG theory with $6-\infty$ LJ fluid and 10-4-3 LJ walls. (From Davis et al. as in Fig. 8.14).

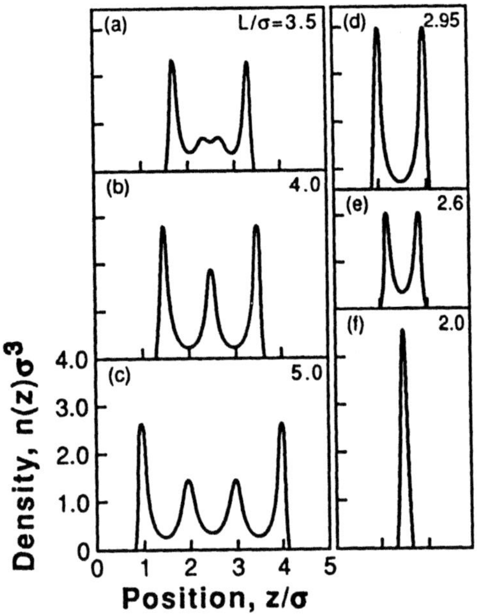

compute from eq. (8.6.11) the density profiles and from $P_N = (1/2)\int_0^L n(z)(\partial u^e/\partial z)dz$ the normal pressure as a function of L. A comparison of theory and molecular dynamics simulation is give in Figure 8.14. Again the agreement is quite good. The result as solvation pressure, $P_N - P_{\text{bulk}}$, versus planar wall separation or "porewidth" L where P_{bulk} is the pressure of a homogeneous fluid having the same K value. Predicted density profiles corresponding to Figure 8.14 are shown in Figure 8.15. From these one sees that the minima in the normal pressure occur when the pore width L favors an integral number of fluid layers and the maxima occur at pore widths least favorable to an integral number of layers.

The fact that the constant K is not the chemical potential poses a problem for applications of the YBG equation to fluids in pores. One either has to take mean pore density, $L^{-1}\int_0^L n(z)dz$, as the thermodynamic variable controlled, or develop a free energy theory in which the chemical potential occurs as the natural control parameter. This is what will be done in subsequent chapters.

Supplementary Reading

Davis, H. T. and Scriven, L. E. 1981. Adv. Chem. Phys., **49**, 357

Henderson, D., Ed. 1992. *Fundamentals of Inhomogeneous Fluids*, Dekker, New York.

Hill, T. L. 1956. *Statistical Mechanics*, McGraw-Hill, New York.

Rice, S. A. and Gray, P. 1965. *The Statistical Mechanics of Simple Liquids*, Interscience, New York.

Exercises

1 Using eq. (8.4.28) compute as a function of mole fraction the surface tension of a benzene–decane solution at room temperature.

2 Assume that eq. (8.4.29) provides a good estimate for the surface tension γ of a low vapor pressure liquid. Derive a formula for γ for a Peng–Robinson liquid. Calculate from this formula the surface tensions of benzene, n-heptane, and n-octane at 20°C. Experimentally $\gamma = 28.9, 20.1$, and 21.8 dyn/cm^2 for these liquids.

3 Using eq. (8.4.29) and the equation

$$U = \frac{3}{2}NkT + \frac{1}{2}\frac{N^2}{V}\int u(r)g^{(2)}(r)d^3r,$$

with

$$g^{(2)}(r) = 0, \quad r < \sigma$$
$$= 1, \quad r > \sigma$$

and

$$u(r) = 4\varepsilon\left[\left(\frac{\sigma}{r}\right)^{12} - \left(\frac{\sigma}{r}\right)^{6}\right],$$

find a formula for the surface tension. Predict the surface tension of liquid argon as a function of temperature using the experimental coexistence liquid density for argon. $\sigma = 3.4$ Åand $\varepsilon/k = 120$ K for argon.

4 Using gradient theory and the Peng–Robinson equation of state, predict the density profile transverse pressure profile and the surface tension for decane at $T = 0.5T_c, 0.75T_c$ and $0.9T_c$. Also, calculate the moment $\int_{-\infty}^{\infty}(P_T(z) - P_N)z\,dz$ for the three temperatures. This moment measures the tendency of an interface to curve toward the vapor or liquid side.

5 For a one-component fluid, the Kirkwood–Buff formula for tension becomes

$$\gamma = \frac{\pi}{8}n_l^2\int_0^\infty s^4u'(s)g^{(2)}ds.$$

Assume that the pair correlation function can be estimated from

$$g^{(2)}(s) = e^{-\beta u(s)}.$$

Show that this expression can be rearranged to

$$\gamma = \frac{\pi}{4}kTn_l^2\int_0^\infty s^3\left[e^{-\beta u(s)} - 1\right]ds.$$

Compute $\gamma/n_l^2\sigma^4kT$ versus kT/ε for the 6–12 Lennard–Jones potential. Use experimental liquid density data for argon to predict γ versus T. $\sigma = 3.4$ Å, $\varepsilon/k = 120$ K for argon.

6 With the aid of bipolar coordinates show that the Kirkwood–Born–Green–Yvon (BGYK) equation, eq. (8.1.36), can be expressed in the form

$$\ln g^{(2)}(r) + \beta u(r) = \pi n \int_0^\infty dt\, t \left[g^{(2)}(t) - 1\right] \int_{|r-t|}^{r+t} ds g(s) u'(s) \frac{s^2 - (r-t)^2}{r}$$

$$+ \pi n \int_0^\infty dt 4t^2 \left[g^{(2)}(t) - 1\right] \int_{r+t}^\infty ds g(s) u'(s).$$

This is now a nonlinear integral equation for the pair correlation function of homogeneous fluid. Its solutions for simple model potentials are discussed in Chapter 9.

7 To lowest order in density, the pair correlation function is $g^{(2)}(r) = e^{-\beta u(r)}$. Use this in the rhs of the equation in Problem 8.6 and compute the BGYK pair correlation function to first order in density for a 6–12 Lennard–Jones fluid at $kT/\varepsilon = 1.5$ and $n\sigma^3 = 0.1$ and 0.3.

References

Bongiorno, V. and Davis, H. T. 1975. *Phys. Rev. A.* 1976. **12**, 2213.

Bongiorno, V., Scriven, L. E., and Davis, H. T. 1976. *J. Colloid Interface Sci.*, **57**, 462.

Davis, H. T., Bitsanis, I., Vanderlick, T. K., and Tirrel, M. in 1987. *Supercomputer Research in Chemistry and Chemical Engineering*, K. S. Jensen and D. G. Truhlar, Eds. ACS Symposium Series 352, ACS, Washington, D.C.

Davis, H. T. and Scriven, L. E. 1976. *J. Chem. Phys.*, **80**, 2805; 1982. *J. Chem. Phys.*, **59**, 357.

Girifalco, L. A. and Good, R. J. 1957. *J. Chem. Phys.*, **61**, 904.

Eisenreich, A. and Gingrich, N. S. 1942. *J. Chem. Phys.*, **62**, 261.

Fischer, J. and Methfessel, M. 1980. *Phys. Rev. A*, **22**, 2836.

Fowkes, F. M. 1963. *J. Chem. Phys.*, **67**, 2538.

Irving, J. H. and Kirkwood, J. G. 1950. *J. Chem. Phys.*, **18**, 817.

Kerins, J., Scriven, L. E., and Davis, H. T. 1986. *Adv. Chem. Phys.*, **65**, 215.

Kirkwood, J. G. and Buff, F. 1949. *J. Chem. Phys.*, **17**, 817.

Kirkwood, J. G., Lewinson, V. A., and Alder, B. J. 1952. *J. Chem. Phys.*, **20**, 929.

Salter, S. J. and Davis. H. T. 1975. *J. Chem. Phys.*, **63**, 3295.

Snook, I. K. and Henderson, D. 1978. *J. Chem. Phys.*, **68**, 2134.

Toxvaerd, S. 1976. *J. Chem. Phys.*, **64**, 2863.

Vanderlick, T. K., Scriven, L. E., and Davis, H. T. 1989. *J. Chem. Phys.*, **90**, 2422.

Winterfeld, P. H., Scriven, L. E., and Davis, H. T., 1978. American Institute of Chemical Engineers J., **24**, 1010.

9

DENSITY FUNCTIONAL THEORY OF STRUCTURE AND THERMODYNAMICS

9.1 Calculus of Variations and Functional Derivatives

9.1.1 Functional Differentiation

We are all familiar with the concept of a function, the simplest being a function $y(x)$ of the variable x, but common examples also include a function $y(\mathbf{x})$ of the D-dimensional vector $\mathbf{x}$ and the n-dimensional vector function $\mathbf{y}(\mathbf{x})$ of the D-dimensional vector $\mathbf{x}$. We think of a function as the set of all of its values for every value of its argument. On the other hand, there are many problems in science in which a number $\mathcal{F}(\{y\})$ is related to each function $y(x)$. A function $\mathcal{F}$ that relates a definite number $\mathcal{F}(\{y\})$ to each function y (or $\mathbf{y}$) is called a *functional.*

The simplest example of a functional is the value of a function y or some derivative of the function $d^k y/dx^k$ at some particular point x. Another example of a functional is

$$\mathcal{F}(\{y\}) = \int_0^1 \{\frac{1}{2}[y(x)]^2 - y(x)e^{-x}\}dx. \qquad (9.1.1)$$

Still another example is the configuration partition function Z_N of classical fluids. The partition function is a functional of the potential energy u^N that itself is a function of the $3N$-dimensional vector $\{\mathbf{r}_1, \ldots, \mathbf{r}_N\}$.

In the calculus of functions one often explores the extremal points of $y(\mathbf{x})$, that is, points $\mathbf{x}$ at which y is a maximum, a minimum, or a point of inflection. At these points $\nabla_{\mathbf{x}} y(\mathbf{x}) = 0$ and the eigenvalues of the matrix $\mathbf{H} = [\partial^2 y/\partial x_i \partial x_j]$ at an extremal point distinguish whether it is a maximum, minimum, or inflection. Analogously, in the calculus of variations one explores the functions for which the functional $\mathcal{F}(\{y\})$ is a maximum, a minimum, or a point of inflection.

For example, consider eq. (9.1.1) and suppose we are seeking the function y_s for which $\mathcal{F}$ is a minimum. If $\mathcal{F}(\{y_s\})$ is a minimum, then it follows that the derivative of $\mathcal{F}(\{y_s + \varepsilon v\})$ with respect to ε is zero at $\varepsilon = 0$ for arbitrary square integrable functions $v(x)$. Thus, from

$$\frac{d}{d\varepsilon}\mathcal{F}(\{y_s + \varepsilon v\}) = \int_0^1 \{y_s(x)v(x) + \varepsilon[v(x)]^2 - v(x)e^{-x}\}dx, \qquad (9.1.2)$$

it follows that

$$\left[\frac{d}{d\varepsilon}\mathcal{F}(\{y_s + \varepsilon v\})\right]_{\varepsilon=0} = \int_0^1 [y_s(x) - e^{-x}]v(x)dx = 0 \qquad (9.1.3)$$

for arbitrary $v(x)$ if $\mathcal{F}(\{y_s\})$ is a minimum. Because eq. (9.1.3) must hold for arbitrary $v(x)$, we can choose $v(x) = y_s(x) - e^{-x}$, from which it follows that $\int_0^1 [y_s(x) - e^{-x}]^2 dx = 0$, or

$$y_s(x) = e^{-x}. \qquad (9.1.4)$$

We have established that $\mathcal{F}(\{y_s\})$ is an extremum. However, because

$$\frac{d^2}{d\varepsilon^2}\mathcal{F}(\{y_s + \varepsilon v\}) = \int_0^1 [v(x)]^2 dx > 0 \qquad (9.1.5)$$

for arbitrary $v(x)$ (we are considering only real functions), it follows that $y_s(x)$ is indeed the function that minimizes $\mathcal{F}(\{y\})$.

In this case, the functional is a quadratic function of εv, namely,

$$\mathcal{F}(\{y_s + \varepsilon v\}) = \frac{\varepsilon^2}{2}\int_0^1 [v(x)]^2 dx - \frac{1}{4}(1 - e^{-2}). \qquad (9.1.6)$$

The integral of the square of $v(x)$ determines the curvature of the parabola defined by eq. (9.1.6). A few special cases are presented in Figure 9.1.

In the simple example just analyzed, the extremum condition for $\mathcal{F}(\{y\})$ yielded an algebraic equation for $y_s(x)$. This is not the case in general. For example, consider the functional

$$\begin{aligned}\mathcal{F}(\{y\}) &= \frac{1}{2}\int_0^1 \int_0^1 e^{-(x+x')}y(x)y(x')dxdx' \\ &\quad - \frac{1}{2}\int_0^1 \{[y(x)]^2 + 2y(x)\}dx.\end{aligned} \qquad (9.1.7)$$

Figure 9.1
The variation of the functional defined by eq. (9.1.1) about its minimum.

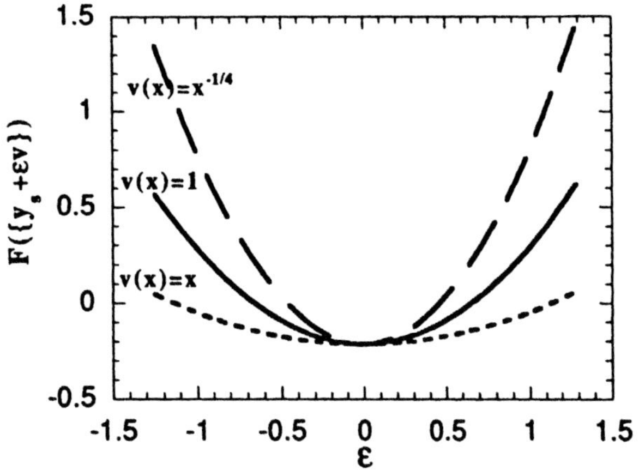

For this functional

$$\left[\frac{d}{d\varepsilon}\mathcal{F}(\{y_s+\varepsilon v\})\right]_{\varepsilon=0} = \frac{1}{2}\int_0^1\int_0^1 e^{-(x+x')}[v(x)y_s(x')+v(x')y_s(x)]dxdx'$$
$$-\int_0^1 [v(x)y_s(x)+v(x)]dx \tag{9.1.8}$$
$$=\int_0^1\left\{\int_0^1 e^{-(x+x')}y_s(x')dx' - y_s(x) - 1\right\}v(x)dx.$$

The second form of the right-hand side (rhs) of eq. (9.1.8) is obtained by interchange of the dummy variable x', x in the term involving $v(x')y_s(x)$. The condition that $d\mathcal{F}/d\varepsilon = 0$ at $\varepsilon = 0$ for arbitrary $v(x)$ requires that terms between braces is zero, that is, $y_s(x)$ satisfies the integral equation

$$\int_0^1 e^{-(x+x')}y_s(x')dx' - y_s(x) = 1. \tag{9.1.9}$$

Thus, the function for which $\mathcal{F}(\{y\})$ is an extremum must be found by solving an integral equation. For this contrived case, the solution is simple:

$$y_s(x) = -\left[1 + 2\left(\frac{1-e^{-1}}{1+e^{-2}}\right)e^{-x}\right]. \tag{9.1.10}$$

Functionals of vector functions of a multidimensional space pose no special problems. If $\mathbf{y}$ is an n-dimensional vector function of the D-dimensional vector $\mathbf{x}$, a generalization of the functional at eq. (9.1.7) is

$$\mathcal{F}(\{\mathbf{y}\}) = \frac{1}{2}\int\int_{V_D} e^{-\mathbf{a}^T(\mathbf{x}+\mathbf{x}')}\mathbf{y}^T(\mathbf{x})\mathbf{y}(\mathbf{x}')d^Dxd^Dx'$$
$$-\frac{1}{2}\int_{V_D}[\mathbf{y}^T(\mathbf{x})\mathbf{y}(\mathbf{x}) + 2\mathbf{b}^T\mathbf{y}(\mathbf{x})]d^Dx, \tag{9.1.11}$$

where $\mathbf{y}^T\mathbf{z} = \sum_{i=1}^n y_i z_i$ is the inner product of the vectors $\mathbf{y}$ and $\mathbf{z}$, V_D the volume of the D-dimensional space of interest, $d^D x$ an infinitesimal in the space, $\mathbf{a}$ a vector in the D-dimensional space of $\mathbf{x}$, and $\mathbf{b}$ a vector in the n-dimensional space of $\mathbf{y}$. Again we are only considering real functions and variables. The equation for the extremum of this functional is easily found to be

$$\int_{V_D} e^{-\mathbf{a}^T(\mathbf{x}+\mathbf{x}')}\mathbf{y}_s(\mathbf{x}')d^D\mathbf{x}' - \mathbf{y}_s(\mathbf{x}) = \mathbf{b} \tag{9.1.12}$$

and the solution is

$$\mathbf{y}_s(\mathbf{x}) = -\mathbf{b}\left[1 - e^{-\mathbf{a}^T\mathbf{x}}\int_{V_D} e^{-\mathbf{a}^T\mathbf{x}}d^D x/\left(\int_{V_D} e^{-2\mathbf{a}^T\mathbf{x}}d^D x - 1\right)\right]. \tag{9.1.13}$$

We will return to functionals of multicomponent functions in what follows.

A functional $\mathcal{F}(\{y\})$ can involve functions and integrals of y but it also can involve derivatives of y. If y represents the position of a particle of mass m in a conservative potential $V(y)$, then the Lagrangian for the particle is

$$\mathcal{F}(\{y\}) = \int_{t_1}^{t_2}\left[\frac{m}{2}\left(\frac{dy}{dt}\right)^2 - V(y)\right]dt, \tag{9.1.14}$$

where t is time. Consider the path of a particle $y(t)$ that begins at $y(t_1)$ at time t_1 and ends at $y(t_2)$ at time t_2. Let us explore the path $y_s(t)$ for which the Lagrangian is a minimum. We need to examine all paths $y(t) = y_s(t) + \varepsilon v(t)$, where $v(t)$ is arbitrary except that $v(t_1) = v(t_2) = 0$, that is, we want every path $y(t)$ beginning and ending at the same points $y(t_1)$ and $y(t_2)$. Differentiation of $\mathcal{F}(\{y + \varepsilon v\})$ with respect to ε yields

$$\begin{aligned}\left[\frac{d}{d\varepsilon}\mathcal{F}(\{y_s + \varepsilon v\})\right]_{\varepsilon=0} &= \int_{t_1}^{t_2}\left[m\frac{dy_s}{dt}\frac{dv}{dt} - \frac{\partial V}{\partial y_s}v\right]dt \\ &= \int_{t_1}^{t_2}\left[-m\frac{d^2y_s}{dt^2} - \frac{\partial V}{\partial y_s}\right]v(t)dt,\end{aligned} \tag{9.1.15}$$

where we have used the chain rule, $\partial V(y)/\partial\varepsilon = [\partial V(y)/\partial y]\partial y/\partial\varepsilon = [\partial V(y)/\partial y]v$, and have integrated the term $m(dy_s/dt)(dv/dt)$ by parts with the aid of the boundary conditions $v(t_1) = v(t_2) = 0$. The extremum condition that eq. (9.1.15) is zero for arbitrary $v(t)$ requires that y_s satisfy

$$m\frac{d^2y_s}{dt^2} = -\frac{\partial V(y_s)}{\partial y_s}. \tag{9.1.16}$$

This is Newton's equation of motion for a particle in a conservative force field, that is, the particle path that minimizes Lagrange's functional is in fact the path determined by Newton's law of motion. For N classical particles in a D-dimensional space, the Lagrangian is the time integral from t_1 to t_2 of the difference between the total kinetic energy and the total potential energy of the system. Similarly to the one-dimensional single particle case just described, one can show that the particle trajectories that minimize Lagrange's functional are those trajectories obeying Newton's equations of motion. Thus is established the equivalence of Lagrangian mechanics, couched in the calculus of variations, and Newtonian mechanics.

The calculus of variations, for which simple examples are given above, eventually led to the concept of *functional differentiation*, that in turn led to new insights in functional behavior in general and to new approaches in quantum and statistical mechanics in particular. We note that for all the examples, considered above the quantity $d\mathcal{F}(\{y+\varepsilon v\})/d\varepsilon$ at $\varepsilon = 0$, can be expressed in the form

$$\left[\frac{d\mathcal{F}}{d\varepsilon}(\{y+\varepsilon v\})\right]_{\varepsilon=0} \equiv \int \frac{\delta\mathcal{F}(\{y\})}{\delta y(\mathbf{x})} v(\mathbf{x}) d^D x, \tag{9.1.17}$$

where the object $\delta\mathcal{F}(\{y\})/\delta y(\mathbf{x})$ depends on y and $\mathbf{x}$ but not on ε or $v(\mathbf{x})$. Thus, $\delta\mathcal{F}(\{y\})/\delta y(\mathbf{x})$ is a functional itself and is totally determined by the original functional $\mathcal{F}(\{y\})$. We call $\delta\mathcal{F}/\delta y$ the *functional derivative* of $\mathcal{F}$ and *define* it by eq. (9.1.17). Similarly, the second functional derivative can be defined by

$$\left[\frac{d^2\mathcal{F}}{d\varepsilon^2}(\{y+\varepsilon v\})\right]_{\varepsilon=0} \equiv \int\int \frac{\delta^2\mathcal{F}(\{y\})}{\delta y(\mathbf{x})\delta y(\mathbf{x}')} v(\mathbf{x})v(\mathbf{x}') d^D x d^D x', \tag{9.1.18}$$

or, equivalently, by

$$\left[\frac{d}{d\varepsilon}\frac{\delta\mathcal{F}}{\delta y(\mathbf{x})}(\{y+\varepsilon v\})\right]_{\varepsilon=0} \equiv \int \frac{\delta^2\mathcal{F}(\{y\})}{\delta y(\mathbf{x})\delta y(\mathbf{x}')} v(\mathbf{x}') d^D x'. \tag{9.1.19}$$

Functional derivatives for several of the above examples are

$$\frac{\delta\mathcal{F}}{\delta y(x)}(\{y\}) = y(x) - e^{-x} \quad \text{[from eq. (9.1.3)]}$$

$$= \int_0^1 e^{-(x+x')} y(x')dx' - y(x) - 1 \quad \text{[from eq. (9.1.8)]}$$

$$= -m\frac{d^2y}{dx^2} - \frac{\partial V}{\partial y} \quad \text{[from eq. (9.1.15) with } x \equiv t].$$

The functional derivative of a function generates a function times the Dirac delta function. Take $\mathcal{F}(\{y\}) = f(y(\mathbf{x}), \mathbf{x})$ where $f(y, \mathbf{x})$ is a differentiable function of y. We can write

$$\mathcal{F}(\{y\}) = \int f(y(\mathbf{x}'), \mathbf{x}')\delta(\mathbf{x}' - \mathbf{x}) d^D x'. \tag{9.1.20}$$

It follows then that

$$\left[\frac{d}{d\varepsilon}\mathcal{F}(\{y+\varepsilon v\})\right]_{\varepsilon=0} = \int \frac{\partial f(y(\mathbf{x}'), \mathbf{x}')}{\partial y(\mathbf{x}')} \delta(\mathbf{x}' - \mathbf{x}) v(\mathbf{x}') d^D x'. \tag{9.1.21}$$

Thus, by comparison with eq. (9.1.17) it follows that

$$\frac{\delta f(y(\mathbf{x}),(\mathbf{x}))}{\delta y(\mathbf{x}')} = \frac{\partial f(y(\mathbf{x}'),\mathbf{x}')}{\partial y(\mathbf{x}')}\delta(\mathbf{x}'-\mathbf{x}) = \frac{\partial f(y(\mathbf{x}),\mathbf{x})}{\partial y(\mathbf{x})}\delta(\mathbf{x}'-\mathbf{x}), \tag{9.1.22}$$

where $\mathbf{x}'$ in $\partial f/\partial y$ can be replaced by $\mathbf{x}$ because of the delta function in the expression.

With eq. (9.1.22) all the simple rules of differential calculus can be used to generate functional derivatives. For instance if $f = [y(\mathbf{x})]^k$, then $\partial f/\partial y = k[y(\mathbf{x})]^{k-1}$ and $\delta f(\mathbf{x})/\delta y(\mathbf{x}') = k[y(\mathbf{x})]^{k-1}$ $\delta(\mathbf{x}'-\mathbf{x})$. Or, if $f = \exp[-\alpha y(\mathbf{x})]$, then $\partial f/\partial y = -\alpha \exp[-\alpha y]$ and $\delta f(\mathbf{x})/\delta y(\mathbf{x}') = -\alpha \exp[-\alpha y(\mathbf{x})]\delta(\mathbf{x}'-\mathbf{x})$. As another example, let

$$\mathcal{F}(\{y\}) = \int \prod_{i=1}^{N} e^{-\beta y(\mathbf{x}_i)} d^D x_1 \cdots d^D x_N. \tag{9.1.23}$$

Then, because $\delta e^{-\beta y(\mathbf{x}_i)}/\delta y(\mathbf{x}) = -\beta e^{-\beta y(\mathbf{x}_i)}\delta(\mathbf{x}-\mathbf{x}_i)$, it is easy to show that

$$\begin{aligned}\frac{\delta \mathcal{F}(\{y\})}{\delta y(\mathbf{x})} &= -\beta \sum_{j=1}^{N} \int \cdots \int \prod_{i=1}^{N} e^{-\beta y(\mathbf{x}_i)} \delta(\mathbf{x}_j - \mathbf{x}) d^D x_1 \cdots d^D x_N \\ &= -N\beta e^{-\beta y(\mathbf{x})} \int \cdots \int \prod_{i=2}^{N} e^{-\beta y(\mathbf{x}_i)} d^D x_2 \cdots d^D x_N,\end{aligned} \tag{9.1.24}$$

where we have used the fact that the value of the integral is the same for all j.

The functional derivative of

$$\mathcal{F}(\{y\}) = \frac{1}{2}\int [\nabla_{\mathbf{x}'} y(\mathbf{x}')]^2 d^D x' \tag{9.1.25}$$

is

$$\begin{aligned}\frac{\delta F}{\delta y(\mathbf{x})} &= \int \left[\nabla_{\mathbf{x}'} \frac{\delta y(\mathbf{x}')}{\delta y(\mathbf{x})}\right] \cdot [\nabla_{\mathbf{x}'} y(\mathbf{x}')]\, d^D x' \\ &= \int [\nabla_{\mathbf{x}'} \delta(\mathbf{x}'-\mathbf{x})] \cdot [\nabla_{\mathbf{x}'} y(\mathbf{x}')]\, d^D x' \\ &= -\nabla_{\mathbf{x}}^2 y(\mathbf{x}),\end{aligned} \tag{9.1.26}$$

where the property, $\int f(\mathbf{x}')\nabla_{\mathbf{x}'}\delta(\mathbf{x}'-\mathbf{x})d^D x' = -\nabla_{\mathbf{x}} f(\mathbf{x})$, of a Dirac delta function has been used.

In fact, the general property of derivatives of a Dirac delta function,

$$\int f(\mathbf{x}') \underbrace{\nabla_{\mathbf{x}'} \cdots \nabla_{\mathbf{x}'}}_{k \text{ factors}} \delta(\mathbf{x}'-\mathbf{x}) d^D x = (-1)^k \underbrace{\nabla_{\mathbf{x}} \cdots \nabla_{\mathbf{x}}}_{k \text{ factors}} f(\mathbf{x}), \tag{9.1.27}$$

enables one to show from the first functional derivative of

$$\mathcal{F} = \frac{1}{2}\int \left[\underbrace{\nabla_{\mathbf{x}''} \cdots \nabla_{\mathbf{x}''}}_{k/2 \text{ factors}} y(\mathbf{x}'')\right]\left[\underbrace{\nabla_{\mathbf{x}''} \cdots \nabla_{\mathbf{x}''}}_{k/2 \text{ factors}} y(\mathbf{x}'')\right] d^D x'', \tag{9.1.28}$$

$$\frac{\delta \mathcal{F}}{\delta y(\mathbf{x})} = \int \left[\nabla_{\mathbf{x}''} \cdots \nabla_{\mathbf{x}''}\delta(\mathbf{x}'' - \mathbf{x})\right]\left[\nabla_{\mathbf{x}''} \cdots \nabla_{\mathbf{x}''} y(\mathbf{x}'')\right] d^D x'', \tag{9.1.29}$$

that the second functional derivative is

$$\begin{aligned} \frac{\delta^2 \mathcal{F}}{\delta y(\mathbf{x})\delta y(\mathbf{x}')} &= \int \left[\nabla_{\mathbf{x}''} \cdots \nabla_{\mathbf{x}''}\delta(\mathbf{x}'' - \mathbf{x})\right]\left[\nabla_{\mathbf{x}''} \cdots \nabla_{\mathbf{x}''}\delta(\mathbf{x}'' - \mathbf{x}')\right] d^D x'' \\ &= (-1)^{k/2} \underbrace{\nabla_{\mathbf{x}} \cdots \nabla_{\mathbf{x}}}_{k \text{ factors}} \delta(\mathbf{x} - \mathbf{x}'), \end{aligned} \tag{9.1.30}$$

where k is an even integer, including $k = 0$. Thus, the second functional derivative of $\mathcal{F} = (1/2) \int [\nabla_{\mathbf{x}'} y(\mathbf{x}')][\nabla_{\mathbf{x}'} y(\mathbf{x}')] d^D x'$ is

$$\frac{\delta^2 \mathcal{F}}{\delta y(\mathbf{x})\delta y(\mathbf{x}')} = -\nabla_{\mathbf{x}}\nabla_{\mathbf{x}}\delta(\mathbf{x} - \mathbf{x}'). \tag{9.1.31}$$

If $\mathcal{F} = 1/2 \int [\nabla_{\mathbf{x}'} y(\mathbf{x}')]^2 d^D x'$, then its second functional derivative is simply the trace of eq. (9.1.31), that is, it is the double dot product of eq. (9.1.31) with the unit tensor $\mathbf{I}$, that is,

$$\frac{\delta^2 \mathcal{F}}{\delta y(\mathbf{x})\delta y(\mathbf{x}')} = -\nabla_{\mathbf{x}}^2\delta(\mathbf{x} - \mathbf{x}'). \tag{9.1.32}$$

Similarly, if $\mathcal{F}$ is the trace of eq. (9.1.28), that is the result of $k/2$ dot products between the quantities in each pair of brackets in eq. (9.1.28), then

$$\frac{\delta^2 \mathcal{F}}{\delta y(\mathbf{x})\delta y(\mathbf{x}')} = (-1)^{k/2}(\nabla_{\mathbf{x}}^2)^k\delta(\mathbf{x} - \mathbf{x}'). \tag{9.1.33}$$

Next consider a vector function $\mathbf{y}$ whose components are y_α and the vector function $\boldsymbol{\varepsilon} \cdot \mathbf{v}$ whose components are $\varepsilon_\alpha v_\alpha$. The partial derivatives of $\mathcal{F}(\{\mathbf{y} + \boldsymbol{\varepsilon} \cdot \mathbf{v}\})$ with respect to ε_α enable one to define the partial functional derivatives of $\mathcal{F}(\{\mathbf{y}\})$, namely,

$$\left[\frac{\partial \mathcal{F}(\{\mathbf{y} + \boldsymbol{\varepsilon} \cdot \mathbf{v}\})}{\partial \varepsilon_\alpha}\right]_{\boldsymbol{\varepsilon}=0} \equiv \int \frac{\delta \mathcal{F}(\{\mathbf{y}\})}{\delta y_\alpha(\mathbf{x})} v_\alpha(\mathbf{x}) d^D x. \tag{9.1.34}$$

The condition of an extremum in $\mathcal{F}$ is then

$$\frac{\delta \mathcal{F}(\{\mathbf{y}\})}{\delta y_\alpha(\mathbf{x})} = 0, \quad \text{for all components } \alpha. \tag{9.1.35}$$

Higher order partial functional derivatives are obtained by taking appropriate successive functional derivatives of functional derivatives.

9.1.2 Functional Taylor's Series

We have now developed sufficient functional calculus to expand $\mathcal{F}$ in a functional Taylor series. We simply expand $\mathcal{F}(\{\mathbf{y} + \boldsymbol{\varepsilon} \cdot \mathbf{v}\})$

about $\mathbf{y}$ and invoke the definition of functional derivatives. The result is

$$\begin{aligned}\mathcal{F}(\{\mathbf{y}+\varepsilon\cdot\mathbf{v}\}) &= \mathcal{F}(\{\mathbf{y}\}) + \sum_\alpha \varepsilon_\alpha \left[\frac{\partial \mathcal{F}}{\partial \varepsilon_\alpha}\right]_{\varepsilon=0} + \sum_{\alpha,\beta} \frac{1}{2!}\varepsilon_\alpha \varepsilon_\beta \left[\frac{\partial^2 \mathcal{F}}{\partial \varepsilon_\alpha \partial \varepsilon_\beta}\right]_{\varepsilon=0} + \dots \\ &= \mathcal{F}(\{\mathbf{y}\}) + \sum_\alpha \int \frac{\delta \mathcal{F}(\{\mathbf{y}\})}{\delta y_\alpha(\mathbf{x})} \varepsilon_\alpha v_\alpha(\mathbf{x}) d^D\mathbf{x} \\ &+ \sum_{\alpha,\beta} \frac{1}{2!} \int\int \frac{\delta^2 \mathcal{F}(\{\mathbf{y}\})}{\delta y_\alpha(\mathbf{x})\delta y_\beta(\mathbf{x}')} \varepsilon_\alpha v_\alpha(\mathbf{x}) \varepsilon_\beta v_\beta(\mathbf{x}') d^D x d^D x' + \dots .\end{aligned} \tag{9.1.36}$$

Because $\mathbf{v}$ is arbitrary, the quantities ε_α can now be set equal to one without loss of generality to obtain the multivariable functional Taylor expansion

$$\begin{aligned}\mathcal{F}(\{\mathbf{y}+\mathbf{v}\}) &= \sum_{k=0}^{\infty} \frac{1}{k!} \sum_{\alpha_1} \cdots \sum_{\alpha_k} \\ &\int \cdots \int \frac{\delta^k \mathcal{F}(\{\mathbf{y}\})}{\delta y_{\alpha_1}(\mathbf{x}_1)\cdots \delta y_{\alpha_k}(\mathbf{x}_k)} v_{\alpha_1}(\mathbf{x}_1)\cdots v_{\alpha_k}(\mathbf{x}_k) d^D x_1 \cdots d^D x_k.\end{aligned} \tag{9.1.37}$$

9.1.3 Chain Rule and Inverse Functional Derivative

The chain rule of differential calculus is also valid for functional differentiation. To show this assume $\mathcal{F}(\{y\})$ is a functional of y that in turn is a functional of η, that is, $y = y(\{\eta\})$. The functional Taylor expansion of $\mathcal{F}(\{y+u\})$ about $y(\{\eta\})$ is

$$\mathcal{F}(\{y+u\}) = \sum_{k=o}^{\infty} \frac{1}{k!} \int \cdots \int \frac{\delta^k \mathcal{F}(\{y\})}{\delta y(\mathbf{x}_1)\cdots \delta y(\mathbf{x}_k)} u(\mathbf{x}_1)\cdots u(\mathbf{x}_k) d^D x_1 \dots d^D x_k. \tag{9.1.38}$$

Suppose next that $u = y(\{\eta + \varepsilon v\}) - y(\{\eta\})$. Then differentiation of each side of eq. (9.1.38) yields

$$\left[\frac{d\mathcal{F}}{d\varepsilon}\right]_{\varepsilon=0} = \int \frac{\delta \mathcal{F}}{\delta \eta(\mathbf{x})} v(\mathbf{x}) d^D x = \int \frac{\delta \mathcal{F}}{\delta y(\mathbf{x}_1)} \frac{\delta y(\mathbf{x}_1)}{\delta \eta(\mathbf{x})} v(\mathbf{x}) d^D x_1 d^D x \tag{9.1.39}$$

for arbitrary $v(\mathbf{x})$, that implies

$$\frac{\delta \mathcal{F}}{\delta \eta(\mathbf{x})} = \int \frac{\delta \mathcal{F}(\{y\})}{\delta y(\mathbf{x}_1)} \frac{\delta y(\mathbf{x}_1)}{\delta \eta(\mathbf{x})} d^D x_1, \tag{9.1.40}$$

the chain rule for functional differentiation for $\mathcal{F}(\{y(\{\eta\})\})$. If, in particular, $\mathcal{F} = \eta(\mathbf{x}')$, eq. (9.1.40) implies

$$\delta(\mathbf{x}' - \mathbf{x}) = \int \frac{\delta \eta(\mathbf{x}')}{\delta y(\mathbf{x}_1)} \frac{\delta y(\mathbf{x}_1)}{\delta \eta(\mathbf{x})} d^D x_1. \tag{9.1.41}$$

The meaning of eq. (9.1.41) is that $\delta\eta(\mathbf{x}')/\delta y(\mathbf{x}_1)$ and $\delta y(\mathbf{x}_1)/\delta\eta(\mathbf{x})$ are functional reciprocals of one another, that is, if $\delta y(\mathbf{x}_1)/\delta\eta(\mathbf{x})$ is the kernel of the integral operator K, then $\delta\eta(\mathbf{x}')/\delta y(\mathbf{x}_1)$ is the kernel of the inverse operator K^{-1}. To see this multiply each side of eq. (9.1.41) by $\psi(\mathbf{x}')d^D x'$ and integrate to obtain $\psi = K^{-1}K\psi$, the property required of inverse operators. Equation (9.1.41) is also the functional derivative analogue of $1 = (d\eta/dy)(dy/d\eta)$.

The multicomponent versions of eqs. (9.1.40) and (9.1.41) are

$$\frac{\delta \mathcal{F}}{\delta \eta_\alpha(\mathbf{x})} = \sum_\beta \int \frac{\delta \mathcal{F}}{\delta y_\beta(\mathbf{x}_1)} \frac{\delta y_\beta(\mathbf{x}_1)}{\delta \eta_\alpha(\mathbf{x})} d^D x_1 \tag{9.1.42}$$

and

$$\delta_{\alpha\gamma} \delta(\mathbf{x}' - \mathbf{x}) = \sum_\beta \int \frac{\delta \eta_\gamma(\mathbf{x}')}{\delta y_\beta(\mathbf{x}_1)} \frac{\delta y_\beta(\mathbf{x}_1)}{\delta \eta_\alpha(\mathbf{x})} d^D x_1, \tag{9.1.43}$$

where $\delta_{\alpha\gamma}$ is the Kronecker delta.

9.1.4 Implicit Functional Theorem

There is also a functional equivalent of the implicit function theorem. If $\mathcal{F}$ is a functional of $u(\mathbf{x})$ and $v(\mathbf{x})$, that is, $\mathcal{F} = \mathcal{F}(\{u\}, \{v\})$, it can be shown from the property of the functional derivative,

$$\begin{aligned} \delta F &\equiv \mathcal{F}(\{u + \delta u\}, \{v + \delta v\}) - \mathcal{F}(\{u\}, \{v\}) \\ &= \int \frac{\delta \mathcal{F}}{\delta u(\mathbf{x})} \delta u(\mathbf{x}) d^D x + \int \frac{\delta \mathcal{F}}{\delta v(\mathbf{x})} \delta v(\mathbf{x}) d^D x \\ &\quad + \text{higher order terms in } \delta u \text{ and } \delta v, \end{aligned} \tag{9.1.44}$$

that

$$\int \frac{\delta \mathcal{F}}{\delta u(\mathbf{x}')} \frac{\delta u(\mathbf{x}')}{\delta v(\mathbf{x})} d^D x' + \frac{\delta \mathcal{F}}{\delta v(\mathbf{x})} = 0, \tag{9.1.45}$$

where the functional derivative $\delta u(\mathbf{x}')/\delta v(\mathbf{x})$ is taken with constant $\mathcal{F}$. Equation (9.1.45) is the analogue of the implicit function theorem.

9.1.5 Functional Differential

Equation (9.1.44) suggests a "functional differential." If we imagine that the $\delta y_\alpha(\mathbf{x})$ are infinitesimal functions, then it follows that

$$\begin{aligned} \delta \mathcal{F} &= \sum_\alpha \int \frac{\delta \mathcal{F}}{\delta y_\alpha(\mathbf{x})} \delta y_\alpha(\mathbf{x}) d^D x, \\ \delta^2 \mathcal{F} &= \frac{1}{2!} \sum_{\alpha\beta} \int \int \frac{\delta^2 \mathcal{F}}{\delta y_\alpha(\mathbf{x}) \delta y_\beta(\mathbf{x}')} \delta y_\alpha(\mathbf{x}) \delta y_\beta(\mathbf{x}') d^D x d^D x', \quad \text{etc.} \end{aligned} \tag{9.1.46}$$

In fact, eq. (9.1.46) can be taken as the definition of functional differentiation if it is assumed that $\delta \mathcal{F}/\delta y_\alpha(\mathbf{x})$ is dependent on $\mathbf{y}$ but not on δy_α.

9.1.6 Conditions for Extremum for a Functional

This completes our primer on functional differentiation. In terms of the extremal properties of $\mathcal{F}(\{y\})$, it can now be stated that if $y(\mathbf{x})$ satisfies the conditions

$$\frac{\delta\mathcal{F}(\{y\})}{\delta y(\mathbf{x})} = 0 \tag{9.1.47}$$

then the function y of the vector $\mathbf{x}$ is an extremum of $\mathcal{F}$. The second functional derivative of $\mathcal{F}$ determines what kind of extremum $y(\mathbf{x})$ is. In particular, if all of the eigenvalues of the equation

$$K\psi = \lambda\psi, \tag{9.1.48}$$

where

$$K\psi \equiv \int \frac{\delta^2\mathcal{F}}{\delta y(\mathbf{x})\delta y(\mathbf{x}')}\psi(\mathbf{x}')d^Dx', \tag{9.1.49}$$

are positive the extremum is a minimum and if they are all negative it is a maximum. If any eigenvalues are zero then the extremum may be an inflection point or higher order functional derivatives may be needed to decide the nature of the extremum (as in the calculus of functions where if at x_0 the conditions $\partial f/\partial x = \partial^2 f/\partial x^2 = 0$ and $\partial^3 f/\partial x^3 \neq 0$ hold, x_0 is an inflection point but if $\partial f/\partial x = \partial^2 f/\partial x^2 = \partial^3 f/\partial x^3 = 0$ and $\partial^4 f/\partial x^4 > 0$ or < 0 the point is a minimum or a maximum).

For the multicomponent (multifunction) problem the extremal conditions are

$$\frac{\delta\mathcal{F}(\{\mathbf{y}\})}{\delta y_\alpha(\mathbf{x})} = 0, \quad \alpha = 1, \ldots, n \tag{9.1.50}$$

and the stability eigenproblem is

$$\sum_\beta \int \frac{\delta^2\mathcal{F}}{\delta y_\alpha(\mathbf{x})\delta y_\beta(\mathbf{x}')}\psi_\beta(\mathbf{x}')d^Dx' = \lambda\psi_\alpha(\mathbf{x}), \quad \alpha = 1, \ldots, n \tag{9.1.51}$$

or

$$\mathbf{K}\boldsymbol{\psi} = \lambda\boldsymbol{\psi}, \tag{9.1.52}$$

where $\boldsymbol{\psi}(\mathbf{x})$ is a vector function and $\mathbf{K}$ is an integral operator whose kernel is of a matrix form. The elements of the kernel matrix are the second-order partial functional derivatives of $\mathcal{F}$ evaluated with the solution $\mathbf{y}(\mathbf{x})$ to the extremal problem.

9.1.7 Functional Integration

Let us close this section by showing how to integrate a functional derivative. First note that

$$\left[\frac{d\mathcal{F}}{d\zeta}(\{\varepsilon y + \zeta y\})\right]_{\zeta=0} = \int \frac{\delta\mathcal{F}(\{\varepsilon y\})}{\delta(\varepsilon y(\mathbf{x}))}y(\mathbf{x})d^Dx. \tag{9.1.53}$$

But by the chain rule of ordinary calculus $d\mathcal{F}(\{\varepsilon y+\zeta y\})/d\zeta = d\mathcal{F}(\{\varepsilon y+\zeta y)/d(\varepsilon+\zeta)$, and so eq. (9.1.53) can be expressed as

$$\frac{d\mathcal{F}(\{\varepsilon y\})}{d\varepsilon} = \int \frac{\delta\mathcal{F}(\{\varepsilon y\})}{\delta(\varepsilon y(\mathbf{x}))} y(\mathbf{x}) d^D x, \tag{9.1.54}$$

which can be integrated with respect to ε to obtain

$$\mathcal{F}(\{y\}) = \mathcal{F}(\{0\}) + \int_0^1 \int \frac{\delta\mathcal{F}(\{\varepsilon y\})}{\delta(\varepsilon y(\mathbf{x}))} y(\mathbf{x}) d^D x\, d\varepsilon. \tag{9.1.55}$$

The multicomponent version of eq. (9.1.48) is

$$\mathcal{F}(\{\mathbf{y}\}) = \mathcal{F}(\{\mathbf{0}\}) + \sum_\alpha \int_0^1 \int \frac{\delta\mathcal{F}(\{\varepsilon \mathbf{y}\})}{\delta(\varepsilon y_\alpha(\mathbf{x}))} y_\alpha(\mathbf{x}) d^D x\, d\varepsilon. \tag{9.1.56}$$

9.2 Density Distributions and Correlation Functions

An important point to bear in mind is that the results derived in this section require only that the particles behave classically with respect to their center of mass motion. It is *not* assumed that the intermolecular interactions are pair additive or centrally symmetric. The s-body density distribution function for a one-component system of molecules whose center of mass motion behaves classically is of the form

$$n^{(s)}(\mathbf{r}_1, \ldots, \mathbf{r}_s) = \sum_{N=s}^{\infty} \frac{q_I^N}{(N-s)!\Xi} \int \cdots \int e^{-\beta u^N - \beta \sum_{i=1}^N \phi(\mathbf{r}_i)} d^3 r_{s+1} \cdots d^3 r_N \tag{9.2.1}$$

in a grand canonical ensemble. $q_I(T)$ denotes the contribution of kinetic and internal energy to the single particle partition function. The quantity $\phi(\mathbf{r})$ is

$$\phi(\mathbf{r}) \equiv v(\mathbf{r}) - \mu, \tag{9.2.2}$$

where $v(\mathbf{r})$ is the external potential energy on a particle at position $\mathbf{r}$ and μ is the chemical potential of fluid. Ξ is the grand canonical ensemble partition function, namely,

$$\Xi = \sum_{N=0}^{\infty} \frac{q_I^N}{N!} \int \cdots \int e^{-\beta u^N - \beta \sum_{i=1}^N \phi(\mathbf{r}_i)} d^3 r_1 \cdots d^3 r_N. \tag{9.2.3}$$

It turns out that the density distribution functions can be computed as functional derivatives of Ξ. Consider the ordinary density

$n(\mathbf{r})(n^{(1)}(\mathbf{r}_1)$ in the notation of eq. (9.2.1). Using the property $\delta[\exp(-\beta\phi(\mathbf{r}_j))]/\delta\phi(\mathbf{r}) = -\beta\exp[-\beta\phi(\mathbf{r}_j)]\delta(\mathbf{r}_j - \mathbf{r})$, we find

$$
\begin{aligned}
-kT\frac{\delta \ln \Xi}{\delta\phi(\mathbf{r})} &= \frac{1}{\Xi}\sum_{N=1}^{\infty}\frac{q_I^N}{N!}\int\cdots\int e^{-\beta u^N-\beta\sum_i\phi(\mathbf{r}_i)}\sum_j\delta(\mathbf{r}_j-\mathbf{r})d^3r_1\cdots d^3r_N \\
&= \frac{1}{\Xi}\sum_{N=1}^{\infty}\frac{q_I^N}{(N-1)!}\int\cdots\int e^{-\beta u^N-\beta\sum_i\phi(\mathbf{r}_i)}\delta(\mathbf{r}_1-\mathbf{r})d^3r_1\cdots d^3r_N,
\end{aligned}
\tag{9.2.4}
$$

and so

$$
n(\mathbf{r}) = -kT\frac{\delta\ln\Xi}{\delta\phi(\mathbf{r})}. \tag{9.2.5}
$$

Because the particles are identical, the average of $\sum_j\delta(\mathbf{r}_j-\mathbf{r})$ in eq. (9.2.4) was replaced by $N\delta(\mathbf{r}_1-\mathbf{r})$. Similarly, we can carry out a second functional differentiation and obtain

$$
(kT)^2\frac{\delta^2\ln\Xi}{\delta\phi(\mathbf{r}')\delta\phi(\mathbf{r})} = -kT\frac{\delta n(\mathbf{r})}{\delta\phi(\mathbf{r}')} = n^{(2)}(\mathbf{r},\mathbf{r}') - n(\mathbf{r})n(\mathbf{r}') + n(\mathbf{r})\delta(\mathbf{r}'-\mathbf{r}). \tag{9.2.6}
$$

From eq. (9.2.6) we see that the functional derivative of $n(\mathbf{r})$ with respect to external potential $\phi(\mathbf{r}')$ is related to the pair correlations between particles. In Section 9.1 we derived the inverse relationship

$$
\int\frac{\delta\phi(\mathbf{r})}{\delta n(\mathbf{r}'')}\frac{\delta n(\mathbf{r}'')}{\delta\phi(\mathbf{r}')}d^3r'' = \delta(\mathbf{r}-\mathbf{r}') \tag{9.2.7}
$$

between the functional derivative of n with respect to ϕ and the functional derivative of ϕ with respect to n. In the special limit of no correlations, $n^{(2)}(\mathbf{r},\mathbf{r}') = n(\mathbf{r})n(\mathbf{r}')$, and so

$$
\frac{\delta n(\mathbf{r})}{\delta\phi(\mathbf{r}')} = -\beta n(\mathbf{r})\delta(\mathbf{r}'-\mathbf{r}), \tag{9.2.8}
$$

that when combined with eq. (9.2.7) yields

$$
\frac{\delta\phi(\mathbf{r})}{\delta n(\mathbf{r}')} = -[\beta n(\mathbf{r})]^{-1}\delta(\mathbf{r}'-\mathbf{r}). \tag{9.2.9}
$$

Correlations will, of course, affect the functional derivative $\delta\phi(\mathbf{r}')/\delta n(\mathbf{r})$. The part of this derivative due to correlations is given the symbol $\beta^{-1}C^{(2)}(\mathbf{r}',\mathbf{r})$ and $C^{(2)}(\mathbf{r}',\mathbf{r})$ is called the "direct correlation function." Thus, in general

$$
\frac{\delta\phi(\mathbf{r})}{\delta n(\mathbf{r}')} = -[\beta n(\mathbf{r})]^{-1}\delta(\mathbf{r}'-\mathbf{r}) + \beta^{-1}C^{(2)}(\mathbf{r},\mathbf{r}'). \tag{9.2.10}
$$

Insertion of eqs. (9.2.6) and (9.2.10) into eq. (9.2.7) yields the equation

$$
g^{(2)}(\mathbf{r},\mathbf{r}') - 1 = C^{(2)}(\mathbf{r},\mathbf{r}') + \int C^{(2)}(\mathbf{r},\mathbf{r}'')n(\mathbf{r}'')[g^{(2)}(\mathbf{r}'',\mathbf{r}') - 1]d^3r'', \tag{9.2.11}
$$

where the pair correlation function

$$
g^{(2)}(\mathbf{r},\mathbf{r}') = n^{(2)}(\mathbf{r},\mathbf{r}')/n(\mathbf{r})n(\mathbf{r}') \tag{9.2.12}
$$

was used in obtaining the final form.

Equation (9.2.11) is known as the Ornstein–Zernike equation. It is an exact relationship and, as noted at the beginning of this section, it is valid for classical systems independently of the detailed nature of the interaction potential. In particular, we have not assumed pair additive forces. Unfortunately, because $g^{(2)}$, $C^{(2)}$ and n are unknown in general, the Ornstein–Zernike equation suffers the same problem as Yvon–Born–Green (YBG) equations—some closure approximation must be introduced. Happily, it turns out some useful approximations have been found in connection with the Ornstein–Zernike equation. For example, in homogeneous fluid the Percus–Yevick approximation, namely, $C^{(2)} = [1 - \exp(\beta u)]g^{(2)}$, gives the nonlinear equation

$$g^{(2)}(R) = e^{-\beta u(R)} + ne^{-\beta u(R)} \int [1 - e^{\beta u(R')}]g^{(2)}(R')[g^{(2)}(|\mathbf{R} - \mathbf{R}'|) - 1]d^3R', \qquad (9.2.13)$$

whose solution gives a pair correlation function and thermodynamic functions that are qualitatively correct over a wide range of temperatures and densities.

If the system were an ideal gas, the chemical potential would be given by

$$\mu = \mu^+(T) + v(\mathbf{r}) + kT \ln n(\mathbf{r}). \qquad (9.2.14)$$

In general the system is not ideal and consequently one introduces the singlet direct correlation function $C^{(1)}(\mathbf{r})$, defined by

$$\mu = \mu^+(T) + v(\mathbf{r}) + kT \ln n(\mathbf{r}) - kTC^{(1)}(\mathbf{r}), \qquad (9.2.15)$$

to account for the contribution of intermolecular interactions to the chemical potential. Rearrangement of eq. (9.2.15) yields

$$C^{(1)}(\mathbf{r}) = \ln\{n(\mathbf{r})e^{\beta[\phi(\mathbf{r})+\mu^+]}\}. \qquad (9.2.16)$$

The functional derivative of $C^{(1)}$ with respect to density is

$$\frac{\delta C^{(1)}(\mathbf{r})}{\delta n(\mathbf{r}')} = [n(\mathbf{r})]^{-1}\delta(\mathbf{r} - \mathbf{r}') + \beta\frac{\delta\phi(\mathbf{r})}{\delta n(\mathbf{r}')}, \qquad (9.2.17)$$

which when combined with eq. (9.2.10) yields

$$C^{(2)}(\mathbf{r}, \mathbf{r}') = \frac{\delta C^{(1)}(\mathbf{r})}{\delta n(\mathbf{r}')} \qquad (9.2.18)$$

This formula is suggestive of a systematic way to define s-body correlation functions, namely,

$$C^{(s)}(\mathbf{r}_1, \ldots, \mathbf{r}_s) = \frac{\delta C^{(s-1)}}{\delta n(\mathbf{r}_s)}(\mathbf{r}_1, \ldots, \mathbf{r}_{s-1}) = \frac{\delta^{s-1}C^{(1)}(\mathbf{r}_1)}{\delta n(\mathbf{r}_2)\cdots\delta n(\mathbf{r}_s)}. \qquad (9.2.19)$$

There turns out to be an intimate relationship between the direct correlation functions and the thermodynamic potentials. At equilibrium in an open system the grand potential, $\Omega(\{n\}) =$

$F(\{n\}) - \mu \int n(\mathbf{r})d^3r$, is a minimum. This means that the equilibrium density distribution is the one for which

$$\frac{\delta \Omega}{\delta n(\mathbf{r})} = 0 \quad \text{or} \quad \mu = \frac{\delta F(\{n\})}{\delta n(\mathbf{r})}. \tag{9.2.20}$$

Next we note that $n(\mathbf{r}) = \delta\Omega/\delta\phi(\mathbf{r})$. This follows from the relationship

$$F_N - N\mu = -kT \ln\left[\frac{q_I^N}{N!}\int \cdots \int e^{-\beta u^N - \beta\sum_i \phi(\mathbf{r}_i)} d^3r_1 \cdots d^3r_N\right], \tag{9.2.21}$$

in which F_N is the canonical ensemble Helmholtz free energy. The functional derivative of this equation with respect to $\phi(\mathbf{r})$ yields

$$\frac{\delta[F_N - N\mu]}{\delta\phi(\mathbf{r})} = n_N^{(1)}(\mathbf{r}) \tag{9.2.22}$$

and the grand canonical ensemble average of eq. (9.2.22) gives the desired result

$$n(\mathbf{r}) = \frac{\delta\Omega(\{n\})}{\delta\phi(\mathbf{r})}, \tag{9.2.23}$$

which by the way implies that $\delta^s\Omega = -kT\delta^s \ln \Xi$ for $s \geq 1$.

We can now introduce a generating potential for the direct correlation functions. Define

$$\Phi(\{n\}) = \int \{\mu^+(T) + \phi(\mathbf{r}) + kT[\ln n(\mathbf{r}) - 1]\} n(\mathbf{r})d^3r - \Omega(\{n\}). \tag{9.2.24}$$

Then

$$\frac{\delta\Phi(\{n\})}{\delta n(\mathbf{r}_1)} = \mu^+(T) + \phi(\mathbf{r}) + kT\ln n(\mathbf{r}) + \left[\int n(\mathbf{r})\frac{\delta\phi(\mathbf{r})}{\delta n(\mathbf{r}_1)}d^3r - \frac{\delta\Omega(\{n\})}{\delta n(\mathbf{r}_1)}\right]. \tag{9.2.25}$$

With the aid of the chain rule,

$$\frac{\delta\Omega}{\delta n(\mathbf{r}_1)} = \int \frac{\delta\Omega}{\delta\phi(\mathbf{r})}\frac{\delta\phi(\mathbf{r})}{\delta n(\mathbf{r}_1)}d^3r = \int n(\mathbf{r})\frac{\delta\phi(\mathbf{r})}{\delta n(\mathbf{r}_1)}d^3r, \tag{9.2.26}$$

we see that the term in brackets in eq. (9.2.25) vanishes. Combination of eqs. (9.2.25) and (9.2.15) yields the desired results:

$$C^{(1)}(\mathbf{r}_1) = \beta\frac{\delta\Phi(\{n\})}{\delta n(\mathbf{r}_1)} \tag{9.2.27}$$

and so

$$C^{(s)}(\mathbf{r}_1, \ldots, \mathbf{r}_s) = \beta\frac{\delta^s\Phi(\{n\})}{\delta n(\mathbf{r}_1)\cdots\delta n(\mathbf{r}_s)}. \tag{9.2.28}$$

Another interesting relationship can be derived from the expression

$$F(\{n\}) = \int \{\mu^+(T) + v(\mathbf{r}) + kT[\ln n(\mathbf{r}) - 1]\} n(\mathbf{r})d^3r - \Phi(\{n\}), \tag{9.2.29}$$

obtained by rearranging eq. (9.2.24). Bcause $n(\mathbf{r})$ is a functional of $\phi(\mathbf{r}) = v(\mathbf{r}) - \mu$, we can generate a density variation $\delta n(\mathbf{r})$ by a variation in either $v(\mathbf{r})$ or μ. Thus, we can take the functional derivatives of $F(\{n\})$ for a fixed $v(\mathbf{r})$. The first such functional derivative, $\mu = \delta F/\delta n(\mathbf{r})$, generates eq. (9.2.15) again and the second functional derivative is

$$\frac{\delta^2 F}{\delta n(\mathbf{r})\delta n(\mathbf{r}')} = [\beta n(\mathbf{r})]^{-1}\delta(\mathbf{r}-\mathbf{r}') - \beta^{-1}C^{(2)}(\mathbf{r},\mathbf{r}'). \quad (9.2.30)$$

Because the stability of an equilibrium density distribution is determined by $\delta^2 F$, it follows that the direct correlation function $C^{(2)}(\mathbf{r},\mathbf{r}')$ determines stability. To see this denote the equilibrium density distribution by $n^e(\mathbf{r})$ and expand $\Omega(\{n^e + v\})$ in a Taylor series about n_s to obtain

$$\begin{aligned}\Omega(\{n^e + v\}) = \Omega(\{n^e\}) &+ \int \left[\frac{\delta F(\{n^e\})}{\delta n^e(\mathbf{r})} - \mu\right] v(\mathbf{r})d^D r \\ &+ \frac{1}{2}\int\int \frac{\delta^2 F(\{n^e\})}{\delta n^e(\mathbf{r})\delta n^e(\mathbf{r}')} v(\mathbf{r})v(\mathbf{r}')d^D r d^D r' + \cdots.\end{aligned} \quad (9.2.31)$$

The linear term in $v(\mathbf{r})$ is zero and so $\Omega(\{n^e\})$ is a minimum if

$$\int \frac{\delta^2 F(\{n^e\})}{\delta n^e(\mathbf{r})\delta n^e(\mathbf{r}')} v(\mathbf{r})v(\mathbf{r}')d^D r d^D r' > 0 \quad (9.2.32)$$

for all $v(\mathbf{r})$. In other words, if the integral operator K whose kernel is $\delta^2 F(\{n^e\})/\delta n^e(\mathbf{r})\delta n^e(\mathbf{r}')$ has only positive eigenvalues, then the density distribution $n^e(\mathbf{r})$ represents a state of stable, or at least locally stable (metastable), thermodynamic equilibrium. Thus, according to eq. (9.2.30), the stability eigenproblem, $K\psi = \lambda\psi$, can be expressed as

$$[\beta n^e(\mathbf{r})]^{-1}\psi(\mathbf{r}) - \beta^{-1}\int C^{(2)}(\mathbf{r},\mathbf{r}',\{n^e\})\psi(\mathbf{r}')d^D r' = \lambda\psi(\mathbf{r}). \quad (9.2.33)$$

Moreover, because $C^{(2)}$ and the pair correlation function are connected by the integral equation, eq. (9.2.11), it follows that stability is also directly related to the pair correlations among particles.

By functional integration, the chemical potential and Helmholtz free energy density can be related to the two-body direct correlation function. From its definition $C^{(1)}(\mathbf{r},\{n\})$ is zero at zero density. Thus, the expression $C^{(2)}(\mathbf{r},\mathbf{r}',\{n\}) = \delta C^{(1)}(\mathbf{r},\{n\})/\delta n(\mathbf{r}')$ can be integrated using the formula at eq. (9.1.55) to obtain

$$C^{(1)}(\mathbf{r},\{n\}) = \int_0^1\int C^{(2)}(\mathbf{r},\mathbf{r}',\{\varepsilon n\})n(\mathbf{r}')d^3r'd\varepsilon. \quad (9.2.34)$$

The chemical potential is then

$$\mu = \mu^+(T) + kT\ln n + u^e(\mathbf{r}) - kT\int_0^1\int C^{(2)}(\mathbf{r},\mathbf{r}',\{\varepsilon n\})n(\mathbf{r}')d^3r'd\varepsilon. \quad (9.2.35)$$

If we define the Helmholtz free energy density $f(\mathbf{r}, \{n\})$ by

$$F(\{n\}) = \int f(\mathbf{r}, \{n\}) d^3 r \tag{9.2.36}$$

and integrate the functional integral of $\delta F/\delta n(\mathbf{r}) = \mu$, where μ is given by eq. (9.2.35), we obtain

$$\begin{aligned} f(\mathbf{r}) = {} & n(\mathbf{r})\{\mu^+(T) + kT[\ln n(\mathbf{r}) - 1] + u^e(\mathbf{r})\} \\ & - n(\mathbf{r})kT \int_0^1 \int (1-\varepsilon) C^{(2)}(\mathbf{r}, \mathbf{r}', \{\varepsilon n\}) n(\mathbf{r}') d^3 r' d\varepsilon. \end{aligned} \tag{9.2.37}$$

The generalization of the one-component results given above to a multicomponent system is straightforward. Let us simply tabulate the multicomponent results here.

$$C_\alpha^{(1)}(\mathbf{r}) = \ln\{n_\alpha(\mathbf{r}) e^{\beta[\phi_\alpha(\mathbf{r}) + \mu_\alpha^+(T)]}\} = \beta \frac{\delta \Phi(\{\mathbf{n}\})}{\delta n_\alpha(\mathbf{r})} \tag{9.2.38}$$

$$C_{\alpha\beta\cdots\omega}^{(s)}(\mathbf{r}_1, \ldots, \mathbf{r}_s) = \frac{\delta^{s-1} C_\alpha^{(1)}(\mathbf{r}_1)}{\delta n_\beta(\mathbf{r}_2) \cdots \delta n_\omega(\mathbf{r}_s)} = \beta \frac{\delta^s \Phi(\{\mathbf{n}\})}{\delta n_\alpha(\mathbf{r}_1) \cdots \delta n_\omega(\mathbf{r}_s)} \tag{9.2.39}$$

$$\Phi(\{\mathbf{n}\}) = \sum_\alpha \int \{\mu_\alpha^+(T) + \phi_\alpha(\mathbf{r}) + kT[\ln n_\alpha(\mathbf{r}) - 1]\} n_\alpha(\mathbf{r}) d^3 r - \Omega(\{\mathbf{n}\}) \tag{9.2.40}$$

$$\Omega(\{\mathbf{n}\}) = F(\{\mathbf{n}\}) - \sum_\alpha \mu_\alpha \int n_\alpha(\mathbf{r}) d^3 r \tag{9.2.41}$$

$$g_{\alpha\beta}^{(2)}(\mathbf{r}, \mathbf{r}') - 1 = C_{\alpha\beta}^{(2)}(\mathbf{r}, \mathbf{r}') + \sum_\gamma \int C_{\alpha\gamma}^{(2)}(\mathbf{r}, \mathbf{r}'') n_\gamma(\mathbf{r}'')[g_{\gamma\beta}^{(2)}(\mathbf{r}'', \mathbf{r}') - 1] d^3 r'' \tag{9.2.42}$$

$$\begin{aligned} \mu_\alpha = {} & \mu_\alpha^+(T) + kT \ln n_\alpha + v_\alpha(\mathbf{r}) \\ & - \sum_\beta kT \int_0^1 \int C_{\alpha\beta}^{(2)}(\mathbf{r}, \mathbf{r}'; \{\varepsilon n\}) n_\beta(\mathbf{r}') d^3 r' d\varepsilon \end{aligned} \tag{9.2.43}$$

$$\begin{aligned} f(\mathbf{r}) = {} & \sum_\alpha n_\alpha(\mathbf{r})\{\mu^+(T) + [\ln n_\alpha(\mathbf{r}) - 1] + v_\alpha(\mathbf{r})\} \\ & - \sum_{\alpha,\beta} n_\alpha(\mathbf{r}) kT \int_0^1 \int (1-\varepsilon) C_{\alpha\beta}^{(2)}(\mathbf{r}, \mathbf{r}'; \{\varepsilon n\}) n_\beta(\mathbf{r}') d^3 r' d\varepsilon \end{aligned} \tag{9.2.44}$$

$$\frac{\delta^2 F}{\delta n_\alpha(\mathbf{r}) \delta n_\beta(\mathbf{r}')} = \frac{kT}{n_\alpha(\mathbf{r})} \delta_{\alpha\beta} \delta(\mathbf{r}' - \mathbf{r}) - kT C_{\alpha\beta}^{(2)}(\mathbf{r}, \mathbf{r}'). \tag{9.2.45}$$

Up to this point we have assumed that the center-of-mass interactions of the molecules are independent of the internal degrees of freedom. This is of course not always the case. For example, the interaction between a pair of spherical molecules with point dipoles

$$u(\mathbf{r}_i - \mathbf{r}_j, \mathbf{D}_i, \mathbf{D}_j) = u_o(|\mathbf{r}_i - \mathbf{r}_j|) + \frac{D_i D_j \hat{e}_i \hat{e}_j}{|\mathbf{r}_i - \mathbf{r}_j|^3} : \left[\mathbf{I} - \frac{3(\mathbf{r}_i - \mathbf{r}_j)(\bar{\mathbf{r}}_i - \mathbf{r}_j)}{(\mathbf{r}_i - \mathbf{r}_j)^2} \right], \tag{9.2.46}$$

where $\hat{e}_i$ is a unit vector denoting the orientation of the dipole of molecule i and D_i is the magnitude of the dipole.

If the translational and rotational degrees of freedom of a dipolar system behave classically and are independent of any other internal degrees of freedom, the canonical ensemble configuration partition function is of the form

$$Z_N = \int \cdots \int e^{-\beta u^N - \beta \sum_{i=1}^N v(\mathbf{r}_i, \mathbf{e}_i)} d^3 r_1 d^2 e_1 \cdots d^3 r_N d^2 e_N$$
$$= \int \cdots \int e^{-\beta u^N - \beta \sum_{i=1}^N v(\tau_i)} d^5 \tau_1 \cdots d^5 \tau_N, \tag{9.2.47}$$

where τ_i denote $\mathbf{r}_i, \hat{e}_i$, the position and orientation of the dipoles and $d^5\tau_i \equiv d^3 r_i d^2 e_i$ is the corresponding volume element in configuration space.

Equation (9.2.47) suggests the generalization of everything in this section to systems in which τ denotes the configuration vectors of all degrees of freedom (e.g., translational, rotational, and some vibrational or hindered rotational degrees of freedom) that behave classically and are independent of the remaining degrees of freedom (e.g., some vibrational, electronic, and nuclear degrees of freedom). The s-body density distribution function for a one-component system then becomes

$$n^{(s)}(\tau_1, \ldots, \tau_s) = \sum_{N=s}^{\infty} \frac{q_I^N}{(N-s)!\Xi} \int \cdots \int e^{-\beta u^N - \beta \sum_{i=1}^N \phi(\tau_i)} d^D \tau_{s+1} \cdots d^D \tau_N \tag{9.2.48}$$

in a grand canonical ensemble. The quantity D denotes the number of independent coordinates possessed by the configuration vectors of the classical degrees of freedom. For this system,

$$\Xi = \sum_{N=0}^{\infty} \frac{q_I^N}{N!} \int \cdots \int e^{-\beta u^N - \beta \sum_{i=1}^N \phi(\tau_i)} d^D \tau_1 \cdots d^D \tau_N \tag{9.2.49}$$

The quantity ϕ is as defined before by $\phi(\tau) = u_e(\tau) - \mu$.

Analogously to what was done above, the density and doublet density distributions can be computed from appropriate functional derivatives:

$$n(\tau) = -kT \frac{\delta \ln \Xi}{\delta \phi(\tau)} \tag{9.2.50}$$

$$n^{(2)}(\tau, \tau') = -\frac{\delta n(\tau)}{\delta \phi(\tau')} + n(\tau) n(\tau') - n(\tau)\delta(\tau - \tau'). \tag{9.2.51}$$

In fact, all of the previous results for a multicomponent system can be obtained by analogy. A summary of these are the following

$$C_\alpha^{(1)}(\tau) = \ln\{n_\alpha(\tau) e^{\beta[\phi_\alpha(\tau) + \mu_\alpha^+(T)]}\} = \beta \frac{\delta \Phi(\{\mathbf{n}\})}{\delta n_\alpha(\tau)} \tag{9.2.52}$$

$$C_{\alpha\beta\ldots\omega}^{(s)}(\tau_1, \ldots, \tau_s) = \frac{\delta^{s-1} C_\alpha^{(1)}(\tau_1)}{\delta n_\beta(\tau_2) \cdots \delta n_\omega(\tau_s)} = \frac{\delta^s \Phi(\{\mathbf{n}\})}{\delta n_\alpha(\tau_1) \cdots \delta n_\omega(\tau_s)} \tag{9.2.53}$$

$$\Phi(\{\mathbf{n}\}) = \sum_\alpha \int \{\mu_\alpha^+(T) + \phi_\alpha(\tau) + kT[\ln n_\alpha(\tau) - 1]\} n_\alpha(\tau) d^D\tau - \Omega(\{\mathbf{n}\}) \tag{9.2.54}$$

$$\Omega(\{\mathbf{n}\}) = F(\{\mathbf{n}\}) - \sum_\alpha \mu_\alpha \int n_\alpha(\tau) d^D\tau \tag{9.2.55}$$

$$g_{\alpha\beta}^{(2)}(\tau, \tau') - 1 = C_{\alpha\beta}^{(2)}(\tau, \tau') + \sum_\gamma \int C_{\alpha\gamma}^{(2)}(\tau, \tau'') n_\gamma(\tau'') [g_{\gamma\beta}^{(2)}(\tau'', \tau') - 1] d^D\tau'' \tag{9.2.56}$$

$$\begin{aligned} \mu_\alpha &= \mu_\alpha^+(T) + kT \ln n_\alpha + v_\alpha(\mathbf{r}) \\ &\quad - \sum_\beta kT \int_0^1 \int C_{\alpha\beta}^{(2)}(\tau, \tau'; \{\varepsilon n\}) n_\beta(\tau') d^D\tau d\varepsilon \end{aligned} \tag{9.2.57}$$

$$\begin{aligned} f(\tau) &= \sum_\alpha n_\alpha(\tau) \{\mu^+(T) + [\ln n_\alpha(\tau) - 1] + v_\alpha(\tau)\} \\ &\quad - \sum_{\alpha,\beta} n_\alpha(\tau) \int_0^1 \int (1 - \varepsilon) C_{\alpha\beta}^{(2)}(\tau, \tau'; \{\varepsilon n\}) n_\beta(\tau') d^D\tau d\varepsilon \end{aligned} \tag{9.2.58}$$

$$\frac{\delta^2 F}{\delta n_\alpha(\tau) \delta n_\beta(\tau')} = \frac{kT}{n_\alpha(\tau)} \delta_{\alpha\beta} \delta(\tau' - \tau) - kT C_{\alpha\beta}^{(2)}(\tau, \tau'). \tag{9.2.59}$$

The relationships established in this section have provided the starting point for most of the last several decades of theory of the structure and thermodynamics of classical systems. We expand on this work in the following sections and in subsequent chapters.

9.3 Homogeneous Fluids: Some Exact Results

In a one-component fluid the mean number $\overline{N}$ of particles in a volume V of an open system (grand canonical ensemble) at temperature T and chemical potential μ is given by

$$\overline{N} = \frac{\partial \ln \Xi}{\partial \beta\mu} = \sum_{N=0}^{\infty} \frac{N e^{\beta N \mu} Q_N}{\Xi}. \tag{9.3.1}$$

Because $n d\mu = dP$, $n = \overline{N}/V$, the isothermal compressibility

$$\kappa_T = \frac{1}{n} \left(\frac{\partial n}{\partial P} \right)_T, \tag{9.3.2}$$

can be computed from the formula $\kappa_T = (\beta V / \overline{N}^2) \partial \overline{N} / \partial \beta\mu$.

But from eq. (9.3.1), it follows that

$$\begin{aligned} \frac{\partial \overline{N}}{\partial \beta\mu} &= \sum_{N=0}^{\infty} \frac{N^2 e^{\beta N \mu} Q_N}{\Xi} - \left(\sum_{N=0}^{\infty} \frac{N e^{\beta N \mu} Q_N}{\Xi} \right)^2 \\ &= \overline{N^2} - (\overline{N})^2, \end{aligned} \tag{9.3.3}$$

a result that is true in general (i.e., not restricted to a homogeneous fluid). Recall that

$$\overline{N(N-1)} = \int n^{(2)}(\mathbf{r}, \mathbf{r}')d^3r d^3r' \quad \text{and} \quad \overline{N} = \int n(\mathbf{r})d^3r, \tag{9.3.4}$$

and so

$$\frac{\partial \overline{N}}{\partial \beta\mu} = \overline{N^2} - (\overline{N})^2 = \overline{N} + \int n(\mathbf{r})n(\mathbf{r}')[g^{(2)}(\mathbf{r}, \mathbf{r}') - 1]d^3r d^3r'. \tag{9.3.5}$$

In this section we consider systems whose interactions depend only on center-of-mass positions. As seen at the end of the previous section, generalization to molecules with more complicated classical interactions is straightforward.

For a homogeneous fluid $n(\mathbf{r})$ is a constant $(\overline{N}/V)$ and $g^{(2)}(\mathbf{r}, \mathbf{r}')$ depends only on $|\mathbf{r} - \mathbf{r}'|$. Thus, changing coordinates from $\mathbf{r}, \mathbf{r}'$ to $\mathbf{r}, \mathbf{R} = \mathbf{r}' - \mathbf{r}$, we can integrate eq. (9.3.5) over d^3r and use the results to obtain

$$kT\left(\frac{\partial n}{\partial P}\right)_T = nkT\kappa_T = 1 + n\int [g^{(2)}(R) - 1]d^3R. \tag{9.3.6}$$

By integration of eq. (9.3.6) one can in principle determine the equation of state of a homogeneous fluid solely from the pair correlation function of the fluid, namely

$$P(n) = kT\int_0^n \frac{dn'}{1 + n'\int[g^{(2)}(R, n') - 1]d^3R}. \tag{9.3.7}$$

This result is valid for any one-component classical fluid and in particular does not require the assumption of pair additive forces.

The fact that the isothermal compressibility diverges at the critical point implies that the pair correlation function will become infinitely long ranged at the critical point. For instance if we assume $g^{(2)}(R) - 1 = 0$ for $R > R_c$, R_c is some finite number, then according to eq. (9.3.6) the isothermal compressibility will remain finite at the critical point in contradiction to experimental observation. What is required for consistency with the definition of $g^{(2)}(R)$ and the observed critical point behavior is that at $g^{(2)}(R) - 1 \rightarrow 0$ as $R \rightarrow \infty$ whereas the integral $\int[g^{(2)}(R) - 1]d^3R \rightarrow \infty$ as $T \rightarrow T_c$.

The isothermal compressibility can also be expressed in terms of the direct correlation function. The Ornstein–Zernike equation for a homogeneous fluid is

$$g^{(2)}(R) - 1 = C^{(2)}(R) + n\int C^{(2)}(|\mathbf{R} - \mathbf{R}'|)[g^{(2)}(\mathbf{R}') - 1]d^3R'. \tag{9.3.8}$$

Multiplying this equation by d^3R, integrating [noting that $\int C^{(2)}(|\mathbf{R} - \mathbf{R}|)d^3R = \int C^{(2)}(R)d^3R$], and rearranging the results, we find

$$\int [g^{(2)}(R) - 1]d^3R = \frac{\int C^{(2)}(R)d^3R}{1 - n\int C^{(2)}(R)d^3R}, \tag{9.3.9}$$

that when combined with eq. (9.3.6) yields

$$(nkT\kappa_T)^{-1} = 1 - n \int C^{(2)}(R)d^3R. \tag{9.3.10}$$

From eq. (9.3.10) it follows that, unlike the pair correlation function, the direct correlation function remains short ranged even at the critical point. In particular, $n \int C^{(2)}(R)d^3R = 1$ at the critical point because the left-hand side (lhs) of eq. (9.3.10) must be zero at this point.

The Ornstein–Zernike equation can be simplified by the method of Fourier transforms. The Fourier transform of a function $f(R)$ is defined by

$$\tilde{f}(\mathbf{k}) \equiv \frac{1}{(2\pi)^{3/2}} \int e^{-i\mathbf{k}\cdot\mathbf{R}} f(\mathbf{R})d^3R. \tag{9.3.11}$$

The inverse is

$$f(\mathbf{R}) = \frac{1}{(2\pi)^{3/2}} \int e^{i\mathbf{k}\cdot\mathbf{R}} \tilde{f}(\mathbf{k})d^3\mathbf{k}. \tag{9.3.12}$$

A very useful property of the Fourier transform is

$$\int f(\mathbf{R}')h(\mathbf{R} - \mathbf{R}')d^3R' = \int e^{i\mathbf{k}\cdot\mathbf{R}} \tilde{f}(\mathbf{k})\tilde{h}(\mathbf{k})d^3k, \tag{9.3.13}$$

which can be found in standard texts on Fourier transforms. One can establish the relationship by substituting expressions of the form of eq. (9.3.11) into the lhs of eq. (9.3.13) and using the property

$$\frac{1}{(2\pi)^3} \int e^{i(\mathbf{k}-\mathbf{k}')\cdot\mathbf{R}} d^3R = \delta(\mathbf{k} - \mathbf{k}') \quad \text{or} \quad \frac{1}{(2\pi)^3} \int e^{i(\mathbf{R}-\mathbf{R}')\cdot\mathbf{k}} d^3k = \delta(\mathbf{R} - \mathbf{R}'). \tag{9.3.14}$$

Equation (9.3.14) can also be used to verify that eq. (9.3.11) is the inverse transform.

Defining $h(R) \equiv g^{(2)}(R) - 1$ and taking the Fourier transform of eq. (9.3.8), we obtain

$$\tilde{C}^{(2)}(\mathbf{k}) = \frac{\tilde{h}(\mathbf{k})}{1 + n(2\pi)^{3/2}\tilde{h}(\mathbf{k})}, \tag{9.3.15}$$

or

$$\tilde{h}(\mathbf{k}) = \frac{\tilde{C}^{(2)}(\mathbf{k})}{1 - n(2\pi)^{3/2}\tilde{C}^{(2)}(\mathbf{k})}. \tag{9.3.16}$$

Because $h(R)$ and $C^{(2)}(R)$ are real and depend only on the magnitude of R, the Fourier transforms $\tilde{h}(k)$ and $\tilde{C}^{(2)}(k)$ are also real and depend only on the magnitude of $\mathbf{k}$. It turns out that the long- range behavior of a function $f(R)$ is determined primarily by the short-range behavior of its Fourier transform $\tilde{f}(k)$. We will use this property to investigate the behavior of the pair correlation function near the critical point. Thus, to investigate the long-range behavior of $h(R)$, we will expand the rhs of eq. (9.3.16) about $k = 0$, keep the first two terms in k, and invert the Fourier transform. The first two nonzero terms of $\tilde{C}^{(2)}(k)$ are

$$\tilde{C}^{(2)}(k) \approx \tilde{C}_0^{(2)}(0) - \tilde{C}_2^{(2)}(0)k^2, \tag{9.3.17}$$

where

$$\tilde{C}_0^{(2)}(0) = \frac{1}{(2\pi)^{3/2}} \int C^{(2)}(R) d^3R \tag{9.3.18}$$

and

$$\tilde{C}_2^{(2)}(0) = \frac{1}{6(2\pi)^{3/2}} \int R^2 C^{(2)}(R) d^3R. \tag{9.3.19}$$

Note that according to eq. (9.3.10)

$$(nkT\kappa)^{-1} = 1 - n(2\pi)^{3/2}\tilde{C}_0^{(2)}(0), \tag{9.3.20}$$

and so eq. (9.3.16) becomes

$$\tilde{h}(k) = \frac{\tilde{C}_0^{(2)}(0) + \mathcal{O}(k^2)}{(nkT\kappa)^{-1} - n(2\pi)^{3/2}\tilde{C}_2^{(2)}(0)k^2 + \mathcal{O}(k^4)}. \tag{9.3.21}$$

As $T \to T_c$, the quantity $(nkT\kappa)^{-1} \to 0$. Thus, if we define

$$\xi = [nkT\kappa n(2\pi)^{3/2} C_2^{(2)}(0)]^{1/2}, \quad A = \tilde{C}_0^{(2)}/n(2\pi)^{3/2}\tilde{C}_2^{(2)}(0), \tag{9.3.22}$$

the small k behavior of $\tilde{h}(k)$ near T_c is determined by

$$\tilde{h}(k) = \frac{A}{\xi^{-2} + k^2}. \tag{9.3.23}$$

The inverse Fourier transform of eq. (9.3.23) yields

$$h(R) \equiv g^{(2)}(R) - 1 = \frac{\sqrt{2\pi}A}{2} \frac{e^{-R/\xi}}{R}. \tag{9.3.24}$$

The quantity ξ is thus seen to be the range of pair correlations, that is, the characteristic correlation length, when the fluid is near its critical point. Because we know that near the critical point $\kappa_T \propto |1 - T/T_c|^{-\gamma}$, it follows from eq. (9.3.22) that near the critical point the correlation length goes as

$$\xi = \xi_0|1 - T/T_c|^{-\nu}, \quad \nu = \gamma/2, \tag{9.3.25}$$

if $T \to T_c$ from above $\gamma = \gamma^+ \simeq 1.1$ and if $T \to T_c$ from below $\gamma = \gamma^- \simeq 1.2$. Correspondingly, eq. (9.3.22) implies that $\nu^+ \simeq 0.55$ and $\nu^- \simeq 0.6$. Light scattering near the critical point can be used to determine ξ. Values deduced for ν are about 0.6, consistent with the analysis just presented. The link between ξ and κ_T, eq. (9.3.22), reveals the origin of the long-range correlations that arise near the critical point: they are the result of thermally driven density fluctuations which are of large scale near the critical point because the compressibility diverges at the critical point.

Example
A ferromagnetic system near the Curie point obeys the following critical point scaling laws when no magnetic field is present:

$$\begin{aligned} M(T) &\propto (-\epsilon)^{\beta} \\ C(T) &\propto (-\epsilon)^{-\alpha'}, T < T_c \\ &\propto \epsilon^{\alpha}, T > T_c \\ &\textit{and} \\ X_T &\propto (-\epsilon)^{-\gamma'}, T < T_c \\ &\propto \epsilon^{\gamma}, T > T_c, \end{aligned}$$

where $\epsilon = (T_c - T)/T_c$ and M, C and X_T are the magnetization, the heat capacity and the magnetic susceptibility. Experimentally it is known that the critical exponents are are $\beta \simeq 1/3$, α, $\alpha' \simeq 0$ and $\gamma \simeq 1.3$. M and X_T are analogous to density and compressibility of a one-component fluid. According to renormalization theory the analogous critical exponents are the same.

Insertion of eq. (9.3.23) into eq. (9.3.15) and inversion of the Fourier transform leads to

$$C^{(2)}(R) = \frac{\sqrt{2\pi}A}{2}\frac{e^{-R/\delta}}{R}, \tag{9.3.26}$$

where

$$\delta^{-2} = \xi^{-2} + (2\pi)^{3/2} nA. \tag{9.3.27}$$

As $T \to T_c$, $\xi^{-2} \to 0$ and $\delta \to ((2\pi)^{3/2} nA)^{-1/2}$, and so as expected the direct correlation function remains short ranged at the critical point.

Not too near the critical point of a fluid whose intermolecular potential is centrally symmetric, $C^{(2)}(R)$ is usually expected [March, 1968] to have the asymptotic form

$$C^{(2)}(R) \to -\beta u(R) \quad \text{as } R \to \infty, \tag{9.3.28}$$

where $u(R)$ is the pair potential between fluid particles (again it is not assumed that intermolecular interactions are pair additive).

In a stable fluid the isothermal compressibility must be positive. According to eq. (9.3.10) this places the following condition on the direct pair correlation function

$$n \int C^{(2)}(R) d^3R \le 1, \tag{9.3.29}$$

where the equality is reached only at the critical point. Actually, the condition of thermodynamic stability places even greater constraints on $C^{(2)}(R)$. From eq. (9.2.34) it follows that a homogeneous fluid of density n is stable (or at least metastable) if the eigenvalues λ of the equation

$$(\beta n)^{-1}\psi(\mathbf{r}) - \beta^{-1} \int C^{(2)}(|\mathbf{r}-\mathbf{r}'|)\psi(\mathbf{r}') d^3r' = \lambda\psi(\mathbf{r}) \tag{9.3.30}$$

are positive. Taking the Fourier transform of eq. (9.3.30), using eq. (9.3.13), we obtain

$$[(\beta n)^{-1} - \beta^{-1}(2\pi)^{3/2}\tilde{C}^{(2)}(k) - \lambda]\tilde{\psi}(\mathbf{k}) = 0, \tag{9.3.31}$$

and so the eigenvalues $\lambda(k)$ are given by

$$\beta n \lambda(k) = -(2\pi)^{3/2} n\tilde{C}^{(2)}(k) + 1 \tag{9.3.32}$$

Thus, the thermodynamic stability condition, $\lambda > 0$, becomes

$$(2\pi)^{3/2} n\tilde{C}^{(2)}(k) < 1 \quad \text{or} \quad n \int e^{-i\mathbf{k}\cdot\mathbf{R}} C^{(2)}(R) d^3R < 1, \tag{9.3.33}$$

for all k, which condition is considerably more demanding than the condition $(2\pi)^{3/2} n\tilde{C}^{(2)}(0) < 1$ imposed to assure positive compressibility.

Exercise

For a one-component homogeneous fluid of hard rods the stability condition is

$$n \int_{-\infty}^{\infty} e^{-ikx} C^{(2)}(x) dx < 1,$$

where the direct correlation function $C^{(2)}(x)$ is given on p. 491. Taking the Fourier transform we find that the stability condition is that

$$\frac{-2na}{1-na}\left[\frac{\sin ka}{ka} + na\frac{(1-\cos ka)}{(1-na)(ka)^2}\right]$$

is less than 1. Demonstrate that this inequality holds for all k and allowable na $(0 < na < 1)$, which proves that a homogeneous, one-component hard rod fluid is stable at all densities. Thus, as implied by its equation of state, a hard rod fluid undergoes no phase transition.

Let us close this section by giving results for a multicomponent system that are relevant to the osmotic compressibility components. The relationship

$$\frac{\partial \overline{N}_\alpha}{\partial \beta \mu_\beta} = \overline{N_\alpha N_\beta} - \overline{N}_\alpha \overline{N}_\beta$$
$$= \int n_\alpha(\mathbf{r}) n_\beta(\mathbf{r}') h_{\alpha\beta}(\mathbf{r}, \mathbf{r}') d^3r d^3r' + \overline{N}_\alpha \delta_{\alpha\beta}, \tag{9.3.34}$$

becomes

$$\frac{\partial \overline{N}_\alpha}{\partial \beta \mu_\beta} = V n_\alpha n_\beta \int h_{\alpha\beta}(R) d^3R + \overline{N}_\alpha \delta_{\alpha\beta} \tag{9.3.35}$$

for a homogeneous fluid. Here

$$h_{\alpha\beta}(\mathbf{r}, \mathbf{r}') \equiv g^{(2)}_{\alpha,\beta}(\mathbf{r}, \mathbf{r}') - 1. \tag{9.3.36}$$

The Ornstein–Zernike equation, eq. (9.2.42), for the special case of a homogeneous fluid yields, when integrated over d^3R, the matrix equation

$$\mathbf{H} = \mathbf{C} + \mathbf{CDH}, \tag{9.3.37}$$

where

$$H_{\alpha\beta} = \int [g^{(2)}_{\alpha\beta}(R) - 1] d^3R, \quad C_{\alpha\beta} = \int C^{(2)}_{\alpha\beta}(R) d^3R, \quad \text{and} \quad D_{\alpha\beta} = n_\alpha \delta_{\alpha\beta}. \tag{9.3.38}$$

Solving eq. (9.3.37) for $\mathbf{H}$, we find

$$\mathbf{H} = [\mathbf{I} - \mathbf{CD}]^{-1} \mathbf{C}. \tag{9.3.39}$$

In matrix form, eq. (9.3.35) becomes

$$\left[\frac{\partial n_\alpha}{\partial \beta \mu_\beta} \right] = \mathbf{DHD} + \mathbf{D}, \tag{9.3.40}$$

and so, with the aid of eq. (9.3.39),

$$\left[\frac{\partial n_\alpha}{\partial \beta \mu_\beta} \right] = \mathbf{D}(\mathbf{I} - \mathbf{CD})^{-1}, \tag{9.3.41}$$

or

$$\left[\frac{\partial \beta \mu_\beta}{\partial n_\alpha} \right] = \mathbf{D}^{-1} - \mathbf{C}, \tag{9.3.42}$$

or in components

$$\frac{\partial \beta \mu_\beta}{\partial n_\alpha} = n_\alpha^{-1} \delta_{\alpha\beta} - \int C^{(2)}_{\alpha\beta}(R) d^3R, \tag{9.3.43}$$

the multicomponent version of eq. (9.3.10).

In the dilute gas limit, for centrally symmetric forces, $C^{(2)}_{\alpha\beta}(R) = e^{-\beta u_{\alpha\beta}(R)} - 1$ (a result that follows from the Ornstein–Zernike equation because $g^{(2)}(R) = e^{-\beta u(R)}$ in the dilute gas limit). Because μ_β is a function of $n_1, \ldots, n_c$, it follows that

$$d\beta\mu_\beta = \sum_\alpha \frac{\partial \beta\mu_\beta}{\partial n_\alpha} dn_\alpha. \tag{9.3.44}$$

With the aid of eq. (9.3.43), we can integrate eq. (9.3.44) to obtain

$$\mu_\beta = \mu^+_\beta(T) + kT \ln n_\beta - \sum_\alpha n_\alpha kT \int [e^{-\beta u_{\alpha\beta}(R)} - 1] d^3R. \tag{9.3.45}$$

This is the chemical potential of a nonideal gas sufficiently dilute to be described by a density expansion through terms linear in density.

Comparison of eqs. (9.2.34) and (9.3.45) yields

$$C^{(1)}_\beta = \sum_\alpha n_\alpha \int [e^{-\beta u_{\alpha\beta}(R)} - 1] d^3R, \tag{9.3.46}$$

the value of the singlet direct distribution function for a homogeneous fluid to first order in density. For a system of hard spheres

$$C^{(1)}_\beta = -\sum_\alpha n_\alpha \frac{4}{3}\pi d^3_{\alpha\beta}, \tag{9.3.47}$$

where $d_{\alpha\beta} = (d_\alpha + d_\beta)/2$ and d_α is the diameter of a molecule of species α. Clearly, for hard spheres the term $C^{(1)}_\beta$ represents an excluded volume effect.

The Gibbs–Duhem equation at constant temperature, $dP = \sum_\alpha n_\alpha d\mu_\alpha$, can be integrated using eq. (9.3.45) to obtain

$$P = nkT - \frac{1}{2} \sum_{\alpha,\beta} n_\alpha n_\beta kT \int [e^{-\beta u_{\alpha\beta}(R)} - 1] d^3R, \tag{9.3.48}$$

where $n = \sum_\alpha n_\alpha$ is the total number density of the gas. Empirically it is sometimes convenient to express the pressure of a gas in terms of the virial expansion, namely,

$$P = nkT \left[1 + \sum_{j=2}^{\infty} n^{j-1} B_j(T, x_1, \ldots, x_c) \right], \tag{9.3.49}$$

where B_j is the jth virial coefficient and $x_\alpha = n_\alpha / n$ is the mole fraction of species α. Equation (9.3.48) enables us to find the second virial coefficient. It is

$$B_2(T, x_1, \ldots, x_c) = \sum_{\alpha,\beta} x_\alpha x_\beta B^{\alpha\beta}_2(T), \tag{9.3.50}$$

where

$$B^{\alpha\beta}_2 = -\frac{1}{2} \int [e^{-\beta\mu_{\alpha\beta}(R)} - 1] d^3R. \tag{9.3.51}$$

For hard spheres, $B_2^{\alpha\beta}$ is simply

$$B_2^{\alpha\beta} = \frac{2\pi}{3} d_{\alpha\beta}^3, \tag{9.3.52}$$

the pair excluded volume described in Chapter 5.

9.4 Homogeneous Fluids: Approximate Theories

9.4.1 Mean Spherical Approximation

As was pointed out previously, for the Ornstein–Zernike equation to be useful for investigating fluid structure, some kind of closure approximation must be introduced. The simplest and intuitively most appealing such approximation was introduced by Lebowitz and Percus (1966) as the Mean Spherical Approximation (MSA). The model is logically suitable for fluids whose molecules interact via pair potentials of the form

$$\begin{aligned} u(R) &= \infty, \quad R < d \\ &= u^c(R), \quad R > d, \end{aligned} \tag{9.4.1}$$

where $u^c(R)$ is a continuous function of intermolecular separation R. Because of the hard sphere cutoff, the pair correlation function obeys the condition $g^{(2)}(R) = 0,\ R < d$. Furthermore, the asymptotic behavior of the direct correlation function is $C^{(2)}(R) \simeq -\beta u^c(R)$ for large R. The MSA model is based on the assumption that $C^{(2)}(R) = -\beta u^c(R)$ for $R > d$. Thus, the approximation is

$$\begin{aligned} g^{(2)}(R) &= 0, \quad R < d \\ C^{(2)}(R) &= -\beta u^c(R), \quad R > d. \end{aligned} \tag{9.4.2}$$

For a multicomponent system the MSA model is defined by

$$\begin{aligned} g_{\alpha\beta}^{(2)}(R) &= 0, \quad R < d_{\alpha\beta} \\ C_{\alpha\beta}^{(2)}(R) &= -\beta u_{\alpha\beta}^c(R), \quad R > d_{\alpha\beta}. \end{aligned} \tag{9.4.3}$$

A special attraction of MSA is that it renders the Ornstein–Zernike equation exactly solvable for the correlation functions of hard sphere fluids and for the primitive electrolyte solution for which $u_{\alpha\beta}^c(R) = z_\alpha z_\beta e^2/\varepsilon R$, where z_α and z_β are the valences of ionic species α and β and ε is the dielectric constant of the solvent (usually water). In fact, if $d_{\alpha\beta} = 0$, the MSA model becomes the Debye–Hückel theory of dilute electrolytes (in which case the short range behavior of $g_{\alpha\beta}^{(2)}$ is not important).

To obtain the Debye–Hückel limiting law for dilute electrolytes let us recall that the multicomponent Ornstein–Zernike equation for a homogeneous fluid is

$$h_{\alpha\beta}(R) = C^{(2)}_{\alpha\beta}(R) + \sum_{\gamma} n_{\gamma} \int C^{(2)}_{\alpha\gamma}(|\mathbf{R}' - \mathbf{R}|) h_{\gamma\beta}(\mathbf{R}') d^3R', \qquad (9.4.4)$$

where $h_{\alpha\beta} \equiv g^{(2)}_{\alpha\beta} - 1$. The Fourier transform of this equation yields

$$\tilde{h}_{\alpha\beta}(k) = \tilde{C}^{(2)}_{\alpha\beta}(k) + (2\pi)^{3/2} \sum_{\gamma} n_{\gamma} \tilde{C}^{(2)}_{\alpha\gamma}(k) \tilde{h}_{\gamma\beta}(k). \qquad (9.4.5)$$

If we set $d_{\alpha\beta} = 0$ in eq. (9.4.3), then for $u^C_{\alpha\beta}(R) = z_\alpha z_\beta e^2/\varepsilon R$ the Fourier transform of $C^{(2)}_{\alpha\beta}(R)$ is

$$\tilde{C}^{(2)}_{\alpha\beta}(k) = -\frac{2\beta z_\alpha z_\beta e^2}{\sqrt{2\pi}\varepsilon k^2}. \qquad (9.4.6)$$

Substituting this expression into eq. (9.4.5), we find $\tilde{h}_{\alpha\beta}(k) = -2\beta z_\alpha z_\beta e^2/[\sqrt{2\pi}\varepsilon(k^2 + \kappa^2)]$, the inverse of which yields

$$h_{\alpha\beta}(R) = -\frac{z_\alpha z_\beta e^2}{kT} \frac{e^{-\kappa R}}{\varepsilon R}, \qquad (9.4.7)$$

where the Debye kappa is defined by

$$\kappa = \sqrt{\frac{4\pi}{kT\varepsilon} \sum_{\alpha} n_\alpha z_\alpha^2 e^2} \qquad (9.4.8)$$

[the k in eqs. (9.4.7) and (9.4.8) is the Boltzmann constant, not the Fourier variable]. The implications of eq. (9.4.7) is that although the Coulombic potential is long ranged (falling off as R^{-1}), the pair correlations decay exponentially because of ionic screening. The characteristics screening length, κ^{-1}, goes to infinity as $n^{-1/2}$ as the electrolyte density n goes to zero.

Equation (9.4.7) can be used in the equations

$$U_{\text{ion}} = \frac{3}{2} NkT + \frac{1}{2} \sum_{\alpha,\beta} n_\alpha n_\beta \int u_{\alpha\beta}(s) g^{(2)}_{\alpha\beta}(s) d^3s \qquad (9.4.9)$$

$$P_{\text{ion}} = nkT - \frac{1}{6} \sum_{\alpha,\beta} n_\alpha n_\beta \int s \frac{du_{\alpha\beta}(s)}{ds} g^{(2)}_{\alpha\beta}(s) d^3s \qquad (9.4.10)$$

and

$$F_{\text{ion}} = \sum_{\alpha} N_\alpha [\mu^+_\alpha(T) + \ln n_\alpha - 1] + \int_V^\infty [P_{\text{ion}} - nkT] dV \qquad (9.4.11)$$

Example
The Debye-Hückel approximation to the electrolyte activity coefficient $\gamma_\pm$ of a dilute aqueous solution of a strong electrolyte is

$$\ell n \gamma_\pm = \frac{-z_+ z_- e^z \kappa}{z\epsilon kT},$$

where κ is the Debye kappa, z_α the valence of ion α, ϵ the solvent dielectric constant, e the electronic charge, k Boltzmann's constant and T the temperature. At 0°C, $\kappa = 0.3244 \times 10^8 \sqrt{s}$ and

$$\ell n \gamma_\pm = -0.4896 z_+ z_- \sqrt{s}.$$

where the ionic strength is $s = \frac{1}{2} \sum_\alpha m_\alpha z_\alpha^2$. m_α is the concentration of ions α in moles per liter. $\ell n \gamma_\pm$ is dimensionless and κ is in units of cm^{-1}. For potassium chloride at 0.01 molar, $s = 0.01$, $\kappa = 3.244 \times 10^6 cm^{-1}$ and $\ell n \gamma_\pm = 0.04896$. The electromotive force between two electrolyte solutions at concentrations c_1 and c_2 is proportional to $\ell n[\gamma_\pm(c_2)c_2/\gamma_\pm(c_1)c_1)]$, and so provides a convenient way to measure the ionic activity coefficient when c_1 is chosen to be so small that $\gamma_\pm(c_1) = 1$.

and

$$\mu_\alpha^{\text{ion}} = \frac{\partial F_{\text{ion}}}{\partial n_\alpha} \tag{9.4.12}$$

to obtain the contribution of the ions to the thermodynamic properties of a dilute electrolyte solution. Because of electroneutrality, $\sum_\alpha n_\alpha z_\alpha = 0$, $g_{\alpha\beta}^{(2)}$ can be replaced by $h_{\alpha\beta}$ in eqs. (9.4.9) and (9.4.10). Evaluation of the above expressions gives

$$\frac{U_{\text{ion}}}{VkT} = \frac{3n}{2} - \frac{\kappa^3}{8\pi}, \tag{9.4.13}$$

$$\frac{P_{\text{ion}}}{kT} = n - \frac{\kappa^3}{24\pi}, \tag{9.4.14}$$

$$\frac{F_{\text{ion}}}{VkT} = \frac{1}{kT}\sum_\alpha n_\alpha\left[\mu_\alpha^+(T) + \ln n_\alpha - 1\right] - \frac{\kappa^3}{12\pi}, \tag{9.4.15}$$

and

$$\mu_\alpha^{\text{ion}} = \mu_\alpha^+(T) + kT\ln n_\alpha - \frac{z_\alpha^2 e^2}{2\varepsilon}\kappa. \tag{9.4.16}$$

These results have been verified experimentally for electrolytic solutions sufficiently dilute that contributions to μ_α^{ion} of the order of n_α or κ^2 are negligible; terms of this order arise from ion–ion short-range repulsive interactions and from ion–solvent interactions.

Let us next consider a neutral hard sphere fluid. For a one-component fluid of hard spheres the MSA model can be expressed as

$$\begin{aligned} g^{(2)}(R) &= 0, \quad R < d \\ g^{(2)}(R) &= \tau(R), \quad R > d \\ C^{(2)}(R) &= -\tau(R), \quad R < d \\ C^{(2)}(R) &= 0, \quad R > d. \end{aligned} \tag{9.4.17}$$

With this notation the Ornstein–Zernike equation becomes

$$\tau(R) = 1 + n\int_{R<d} \tau(R')d^3R' - n\int_{\substack{R'<d \\ |\mathbf{R}'-\mathbf{R}|>d}} \tau(R')\tau(|\mathbf{R}'-\mathbf{R}|)d^3R'. \tag{9.4.18}$$

For the special case of hard spheres the mean spherical approximation (MSA) model turns out to be the same as the earlier and more accurate Percus–Yevick (PY) approximation (the PY approximation will be discussed in the next subsection). Because the PY approximation was introduced earlier, we will refer to the hard sphere case as the PY/MSA model.

By the methods of Laplace transforms Wertheim (1963) solved eq. (9.4.18). The result is

$$\begin{aligned} C^{(2)}(R) &= -(1-\eta)^{-4}\left[(1+2\eta)^2 - 6\eta(1+\frac{1}{2}\eta)^2(R/d) + \frac{1}{2}\eta(1+2\eta)^2(R/d)^3\right], \quad R < d, \\ &= 0, \quad R > d \end{aligned} \tag{9.4.19}$$

where

$$\eta = \frac{\pi}{6}nd^3. \tag{9.4.20}$$

The pair correlation function can be computed from the formula

$$g^{(2)}(R) = (2\pi i R/d)^{-1} \int_{\delta-i\infty}^{\delta+i\infty} \frac{tL(t)e^{tR/d}}{12\eta[L(t) + S(t)e^t]}dt, \tag{9.4.21}$$

where $\delta > 0$ and $S(t) = (1-\eta)^2t^3 + 6\eta(1-\eta)t^2 + 18\eta^2 t - 12\eta(1+2\eta)$ and $L(t) = 12\eta[(1+\eta/2)t + (1+2\eta)]$. Using the fact that the denominator of the above expression has no zeros in the right half of the complex plane, Wertheim was able to close the contour in the left-half plane and use the method of residues to find $g^{(2)}(R)$. The results are somewhat tedious and will not be given here. However, the result for the contact value of the pair correlation function is simply

$$g^{(2)}(d) \equiv \lim_{\varepsilon\to 0} g^{(2)}(R = d + \varepsilon) = \frac{1+\eta/2}{(1-\eta)^2}. \tag{9.4.22}$$

Using the property $-kT(du(R)/dR)e^{-\beta u(R)} = de^{-\beta u(R)}/dR = \delta(R - d+)$ when $u(R)$ is the hard sphere potential, from the virial equation of state, eq. (8.3.14), we can show that pressure of a hard sphere fluid is given by

$$P = nkT\left[1 + \frac{2\pi}{3}nd^3 g^{(2)}(d)\right]. \tag{9.4.23}$$

Insertion of eq. (9.4.22) into eq. (9.4.23) yields the PY/MSA "virial pressure" for a hard sphere fluid

$$P^v = nkT\left[\frac{1+2\eta+3\eta^2}{(1-\eta)^2}\right]. \tag{9.4.24}$$

Alternatively, the pressure can be computed from the compressibility formula, eq. (9.3.10), with the aid of eq. (9.4.19) and the result integrated to obtain the PY/MSA "compressibility pressure"

$$P^c = nkT\left[\frac{1+\eta+\eta^2}{(1-\eta)^3}\right]. \tag{9.4.25}$$

If the PY/MSA model were exact for hard spheres, P^v and P^c would be identical. That they differ attests to the approximate nature of the model.

The quality of the approximation can be measured by comparing the exact virial coefficients of the density expansion of the hard sphere equation of state,

$$\frac{P}{nkT} = 1 + Bn + Cn^2 + Dn^3 + En^4 + \cdots, \tag{9.4.26}$$

with those predicted by eqs. (9.4.24) and (9.4.25). The results are shown in Table 9.1. Both P^v and P^c yield exact second and third virial coefficients. The compressibility equation of state appears somewhat better. Its fourth and fifth virial coefficients are only 3 and 10% high, respectively.

Another comparison is made in Figure 9.2 where predictions of eqs. (9.4.24) and (9.4.25) are compared with computer simulations for hard spheres. We see that both approximations are quite accurate up to a density of $0.5d^{-3}$ and that the compressibility equation of state is better than the virial equation of state at densities greater than $0.5d^{-3}$.

Given that the model is so simple, the PY/MSA predicted equations of state of a hard sphere fluid are surprisingly good. It also provides analytic expressions for the pair correlation functions $g^{(2)}$ and $C^{(2)}$. These have proven useful in perturbation theories (e.g., Weeks, et al., (1971) for treating more realistic fluid models than the hard sphere model.

Waisman and Lebowitz (1972) solved the MSA equations for the primitive electrolyte and Wertheim (1973) solved the MSA equation for a fluid of hard spheres with point dipoles at their

TABLE 9.1 Virial Coefficients for a Hard Sphere Fluid Compared With Predictions of the PY/MSA Theory ($b \equiv 2\pi d^3/3$)

		PY/MSA Predictions	
	Exact	*Virial Pressure*	*Compressibility*
B/b	1	1	1
C/b^2	$\frac{5}{8}$	$\frac{5}{8}$	$\frac{5}{8}$
D/b^3	0.2869	0.2500	0.2969
E/b^4	0.1103	0.0859	0.1214

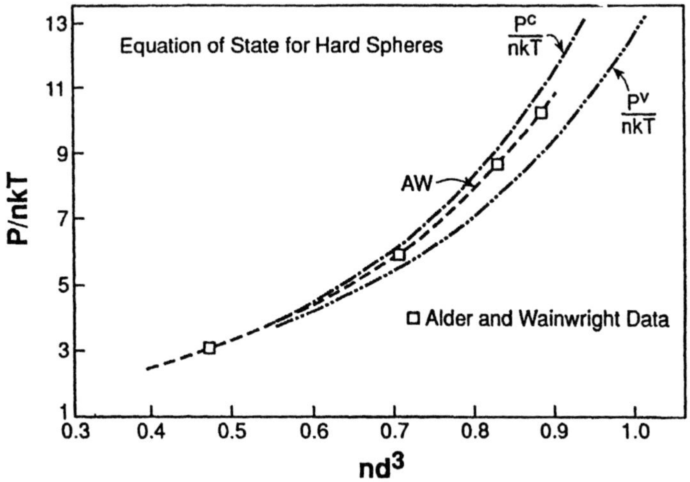

Figure 9.2
Comparison of the PY/MSA predicted pressure with those determined by molecular dynamics for a hard sphere fluid. Adapted from Rice and Gray, (1965).

centers. The results are far superior to the previous Debye–Hückel theory of dilute electrolytes and in the case of dipolar hard spheres the model furnishes analytic results where few theoretical results exist.

9.4.2 PY Approximation

For continuous interaction potentials, the MSA does not perform as well as its predecessor the approximation (Percus and Yevick, (1958). According to this approximation the Ornstein–Zernike equation is truncated by the assumption

$$C^{(2)}(R) = g^{(2)}(R)[1 - e^{\beta u(R)}], \tag{9.4.27}$$

which yields the following expression for the pair correlation function

$$e^{\beta u(R)} g^{(2)}(R) = 1 + n \int [g^{(2)}(|\mathbf{R}' - \mathbf{R}|) - 1][1 - e^{\beta u(R')}] g^{(2)}(R') d^3 R'. \tag{9.4.28}$$

The multicomponent version of the PY equation is

$$e^{\beta u_{\alpha\beta}(R)} g_{\alpha\beta}^{(2)}(R) = 1 + \sum_{\gamma} n_{\gamma} \int [g_{\alpha\gamma}^{(2)}(|\mathbf{R} - \mathbf{R}'|) - 1][1 - e^{-\beta u_{\alpha\gamma}(R')}] g_{\gamma\beta}^{(2)}(R') d^3 R'. \tag{9.4.29}$$

There are several derivations of the PY equations. The best and most suggestive one is that given by Percus (1964) using density functional expansions. Because it admits an analytic solution for hard sphere fluids, the PY approximation has been the most frequently used approximation to the pair correlation function over the last few decades. The solution of the PY equations for hard sphere mixtures will be discussed in Section 9.4.5.

9.4.3 Hypernetted Chain Approximation

Another popular approximation to the pair correlation function is the hypernetted chain approximation (HNC),

$$g^{(2)}(R) = 1 + C^{(2)}(R) + \ln g^{(2)}(R) + \beta u(R), \tag{9.4.30}$$

or for a multicomponent system.

$$g_{\alpha\beta}^{(2)}(R) = 1 + C_{\alpha\beta}^{(2)}(R) + \ln g_{\alpha\beta}^{(2)}(R) + \beta u_{\alpha\beta}(R). \tag{9.4.31}$$

This approximation can be derived by a diagram theory (E. Meeron, J. Math. Phys., **1**, 192 (1960)) which amounts to trying to identify and sum the dominant terms in an infinite series in density. Percus has also used density functional expansions to obtain the approximation.

Insertion of eq. (9.4.30) into the Orstein-Zernike equation yields the HNC equation for a one-component fluid:

$$\ln\left[g^{(2)}_{\alpha\beta}(R)e^{\beta u_{\alpha\beta}(R)}\right] = n\int\left[g^{(2)}(|\mathbf{R}'-\mathbf{R}|)-1\right]$$
$$\left[g^{(2)}(R')-1-\ln g^{(2)}(R')-\beta u(R')\right]d^3R'. \tag{9.4.32}$$

For a multicomponent system

$$\ln\left[g^{(2)}_{\alpha\beta}(R)e^{\beta u_{\alpha\beta}(R)}\right] = \sum_{\gamma} n_\gamma \int\left[g^{(2)}_{\alpha\beta}(|\mathbf{R}'-\mathbf{R}|)-1\right]$$
$$\left[g^{(2)}_{\alpha\beta}(R')-1-\ln g^{(2)}_{\alpha\beta}(R')-\beta u_{\alpha\beta}(R')\right]d^3R'. \tag{9.4.33}$$

The HNC approximation (or variations of it) has been the approximation of choice in the theory of model electrolyte solutions. It has been quite successful (i.e., compares well with simulations) in applications to the model solution in which the solvent is given the dielectric constant of water, is assumed to be composed of neutral spheres and the ions are assumed to be charged spheres.

9.4.4 Comparison of HNC, PY, and the BGYK approximations

By the BGYK approximation, we mean eq. (8.1.36), which is Kirkwood's superposition approximation to the Yvon–Born–Green hierarchy of equations. This is the oldest and most simply motivated approximation available.

For homogeneous fluids of molecules with pair additive forces, the pair and direct correlation functions, the pressure and the thermodynamic energy are the usual objects of interest. The pressure can be computed either from the virial equation of state

$$P = nkT - \frac{n^2}{6}\int s\frac{du(s)}{ds}g^{(2)}(s)d^3s \tag{9.4.34}$$

or the compressibility equation of state

$$P = kT\int_0^n \frac{dn'}{1+n'\int[g^{(2)}(s,n')-1]d^3s}. \tag{9.4.35}$$

The thermodynamic energy can be computed from

$$U = \frac{3}{2}NkT + \frac{1}{2}Nn\int g^{(2)}(s)u(s)d^3s. \tag{9.4.36}$$

Equations (9.4.34) and (9.4.36) are valid only for continuous potentials $u(s)$. In the special case of hard spheres [$u(s) = \infty, s < d$ and $u(s) = 0,\ s > d$] the properties, found by taking the hard core limit of a continuous potential,

$$\delta(s-d) = \frac{d}{ds}[e^{-\beta u(s)}] = -\beta e^{-\beta u(s)} \frac{du(s)}{ds} \tag{9.4.37}$$

$$e^{-\beta u(s)} g^{(2)}(s) \Big|_{s=d} = g^{(2)}(s = d^+) \tag{9.4.38}$$

$$u(s) g^{(2)}(s) = 0 \tag{9.4.39}$$

yield

$$P = nkT + \frac{2\pi}{3} kTn^2 d^3 g^{(2)}(d^+) \tag{9.4.40}$$

and

$$U = \frac{3}{2} NkT \tag{9.4.41}$$

Equations (9.4.38) and (9.4.39) follow because the function

$$y(s) \equiv e^{\beta u(s)} g^{(2)}(s) \tag{9.4.42}$$

is continuous at $s = d$ in the limit of a hard sphere interaction.

Because the PY approximation for hard spheres is the same as MSA, the PY virial and compressibility equations of state are given by eqs. (9.4.24) and (9.4.25). As indicated in Table 9.1, the PY compressibility equation of state is better than the virial equation of state.

The first four virial coefficients, defined in the density series

$$\frac{P}{nkT} = 1 + Bn + Cn^2 + Dn^3 + En^4 + \cdots, \tag{9.4.43}$$

computed from the three approximations for hard spheres are compared in Table 9.2 with the exact virial coefficients. The superscripts "v" and "c" indicate the values obtained from the virial equation of state and the compressiblity equation of state, respectively. The PY compressibility equation of state is superior to the other two approximations.

Consider next a Lennard-Jones fluid whose particles have the pair potential

$$u(s) = 4\varepsilon \left[(\frac{\sigma}{s})^{12} - (\frac{\sigma}{s})^{6} \right]. \tag{9.4.44}$$

The pressure and energy predicted by the three approximations for the temperature $T = 2.74\varepsilon/k$ are shown in Table 9.3. The system corresponds to a high pressure gas. Also shown in the table are results of Monte Carlo simulations of the fluid. The PY

TABLE 9.2 Virial Coefficients for Hard Sphere Fluid from Three Approximate Theories

	$B^v(b)$	$B^c(b)$	$C^v(b^2)$	$C^c(b^2)$	$D^v(b^3)$	$D^c(b^3)$	$E^v(b^4)$	$E^c(b^4)$
Exact:	1	1	$\frac{5}{8}$	$\frac{5}{8}$	0.2869	0.2869	0.1103	0.1103
YBG	1	1	$\frac{5}{8}$	$\frac{5}{8}$	0.2252	0.3424	0.0475	0.1335
HNC	1	1	$\frac{5}{8}$	$\frac{5}{8}$	0.4453	0.2092	0.1447	0.0493
PY	1	1	$\frac{5}{8}$	$\frac{5}{8}$	0.2500	0.2969	0.0859	0.121

The unit $b = 2\pi d^3/3$, where d is the hard core diameter. Adapted from Rice and Gray (1965).

approximation compares the most favorably with simulations and the BGYK approximation compares the least favorably.

The pair and direct correlation functions predicted at $T = 1.65\varepsilon/k$ by the three approximations are shown for various densities in Figure 9.3.

At the thermodynamic conditions considered in the figure, all three approximations perform rather well and are in good agreement with each other. At higher densities, the approximations begin to differ appreciably, and for a Lennard–Jones fluid the PY approximation seems to fare best.

For those readers interested in methods of solution of the nonlinear equations for the pair correlation functions, the paper by Kerins, et al. (1986) is recommended.

9.4.5 Hard Sphere Mixtures

For a hard sphere mixture the PY (or PY/MSA) approximation is $C^{(2)}_{\alpha\beta}(R) = 0$, $R > d_{\alpha\beta}$, and $g^{(2)}_{\alpha\beta}(R) = 0$, $R < d_{\alpha\beta}$, where $d_{\alpha\beta} = (d_\alpha + d_\beta)/2$. d_α and d_β are the diameters of spheres of species α and β. With this approximation, Lebowitz (1964) solved the multicomponent Ornstein–Zernike equation, eq. (9.4.4), for hard

TABLE 9.3 Comparison of Thermodynamic Functions from Calculations

$\rho\sigma^3$	*MC*	*PY*	*HNC*	*BGYK*
		$(p/\rho kT)^v$		
0.400	1.2–1.5	1.24	1.28	1.26
0.833	4.01	4.01	5.11	2.3
1.000	7.0	6.8	9.1	3.1
1.111	7.8	9.2	13.2	3.8
		$U(NkT)$		
0.400	−0.86	−0.865	−0.859	−0.85
0.833	−1.58	−1.615	−1.409	−1.8
1.000	−1.60	−1.67	−1.19	−2.2
1.111	−1.90	−1.59	−0.78	−2.6

Adapted from Rice and Gray (1965)

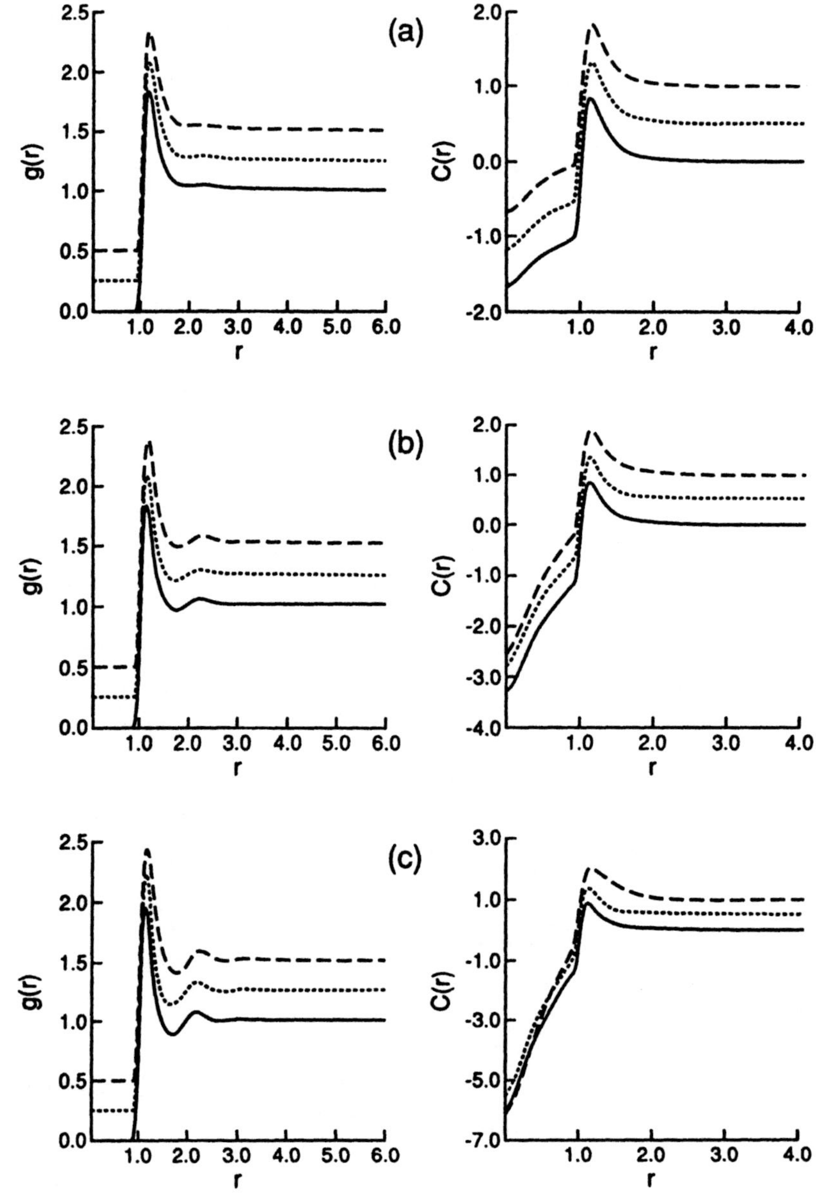

Figure 9.3
Pair and direct correlation functions, $g(r)$ and $C(r)$, under the (—) HNC, (- - -) PY, and (– –) BGYK approximations at $kT/\varepsilon = 1.65$ and density n equal to (a) 0.10, (b) 0.30, and (c) 0.50. PY and BGYK curves have been shifted up by 0.25 and 0.50 for $g(r)$ and by 0.50 and 1.0 for $C(r)$, respectively.

sphere mixtures. The contact values of the pair correlation functions he found are

$$g_{\alpha\beta}^{(2)}(d_\alpha) = \left[(1-\eta_3) + \frac{3}{2} d_\alpha \eta_2\right](1-\eta_3)^{-2} \tag{9.4.45}$$

and

$$g_{\alpha\beta}^{(2)}(d_{\alpha\beta}) = \left[d_\beta g_{\alpha\beta}^{(2)}(d_\alpha) + d_\alpha g_{\beta\beta}^{(2)}(d_\beta)\right] / 2d_{\alpha\beta}, \tag{9.4.46}$$

where

$$\eta_i = \sum_{\alpha=1}^{c} \frac{\pi}{6} n_\alpha d_\alpha^i. \tag{9.4.47}$$

Carnahan and Starling's approximation (based on a rational polynomial approximation to an expansion motivated by the virial expansion and computer simulations) was generalized to a multicomponent fluid by Mansoori, et al. (1971). Their result is

$$g_{\alpha\beta}^{(2)} = \frac{1}{1-\eta_3} + \frac{3d_\alpha d_\beta}{d_{\alpha\beta}} \frac{\eta_2}{(1-\eta_3)^2} + 2\left(\frac{d_\alpha d_\beta}{d_{\alpha\beta}}\right)^2 \frac{\eta_2^2}{(1-\eta_3)^3}. \tag{9.4.48}$$

The virial equation,

$$P^v = \sum_\alpha n_\alpha kT + kT \sum_{\alpha,\beta} \frac{2\pi}{3} n_\alpha n_\beta d_{\alpha\beta}^3 g_{\alpha\beta}^{(2)}(d_{\alpha\beta}), \tag{9.4.49}$$

enables one to compute the pressure from the contact value of the pair correlation functions for hard spheres.

From the Gibbs–Duhem equation, $dP = \sum n_\beta d\mu_\beta$ and eq. (9.3.43), it follows that

$$\frac{\partial P^c}{\partial n_\alpha} = 1 - \sum_\beta n_\beta \int C_{\alpha\beta}^{(2)}(R) d^3R, \tag{9.4.50}$$

where the superscript on P^c indicates that eq. (9.4.50) yields the "compressibility" pressure. Lebowitz showed that for the PY theory of hard spheres

$$P^c = P^v + \frac{18}{\pi} kT \eta_3 \eta_2^3 (1-\eta_3)^{-3}. \tag{9.4.51}$$

Of course, if the theory were exact P^c and P^v would be the same, and so the second term on the rhs of eq. (9.4.51) is a measure of the inaccuracy of the PY theory.

For a one-component hard sphere fluid, the PY and Carnahan–Starling (CS) equations of state are

$$\frac{P^v}{nkT} = \frac{1+2\eta+3\eta^2}{(1-\eta)^2}, \quad \frac{P^c}{nkT} = \frac{1+\eta+\eta^2}{(1-\eta)^3}, \quad \frac{P^{CS}}{nkT} = \frac{1+\eta+\eta^2-\eta^3}{(1-\eta)^3} \tag{9.4.52}$$

where $\eta = \eta_3 = (\pi/6)nd^3$. We saw in Section 9.4.1 that the PY compressibility pressure P^c is more accurate than the PY virial pressure P^v. In Figure 9.4, we see that the CS pressure P^{CS} is more accurate than either of the PY estimates of pressure. This is not surprising because the CS formula was arrived at by trying to get a rational polynomial that fit the known virial coefficients very accurately.

Because the simplest and most practicable approximate theories of real fluids treat the repulsive interactions as hard sphere interactions, the analytic equations of the PY theory and the CS

Figure 9.4
Comparision of PY and CS pressures with molecular dynamics results for the one-component hard sphere fluid.

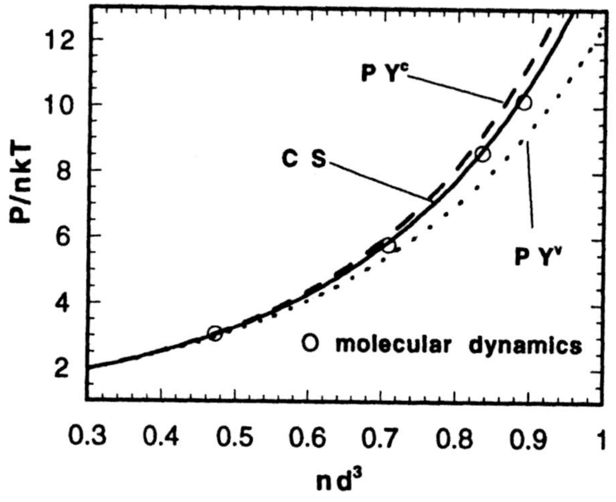

equation are especially useful. In the next section a few perturbation theories requiring hard sphere equations will be discussed.

The molar volume of dense fluid mixtures is often estimated from Amagat's law of ideal mixing,

$$\tilde{V} = \sum_\alpha x_\alpha \tilde{V}^o_\alpha \quad \text{or} \quad n^{-1} = \sum_\alpha x_\alpha (n^o_\alpha)^{-1}, \tag{9.4.53}$$

where $\tilde{V}^o_\alpha$ is the molar volume of pure fluid α and x_α is the mole fraction of α in the mixture. It is argued that the reason eq. (9.4.53) is a good approximation is that repulsive forces dominate the packing arrangements of molecules in dense fluids. In the PY or CS equations, the density is a complicated, cubic function of mole fraction. Nevertheless, in the high density region, these equations yield molar volumes obeying eq. (9.4.53) quite well. For example, at $P^c d_1^3/kT = 13$, the PV compressibility equation predicts molar volumes obeying eq. (9.4.52) remarkably well. The straight line in Figure 9.5 is eq. (9.4.53) and the points are predictions of the PY model.

The Helmholtz free energy and the chemical potentials can be computed from the equation of state for pressure. For the PY compressibility pressure, we find

$$\begin{aligned} \frac{\mu^c_\alpha}{kT} &= \frac{\mu^+_\alpha(T)}{kT} + \ln\left(\frac{n_\alpha}{1-\eta_3}\right) + \frac{\pi P^c d^3_\alpha}{6kT} \\ &\quad + \frac{3\eta_2 d_\alpha}{1-\eta_3} + \frac{3\eta_1 d^2_\alpha}{1-\eta_3} + \frac{9\eta_2^2 d^2_\alpha}{2(1-\eta_3)^2}, \end{aligned} \tag{9.4.54}$$

where $\mu^+_\alpha(T) = kT\ln(\Lambda^3_\alpha/q^I_\alpha)$. Λ_α is the thermal deBroglie wave length of species α and $q^I_\alpha(T)$ is the internal energy partition function of α. The chemical potential from the PY virial pressure is

Figure 9.5
Molecular density of a binary hard sphere mixture predicted by the PY compressibility pressure and by the ideal mixing law, eq. (9.4.53). The ratio d_2/d_1 is set equal to 1.25 in the calculations.

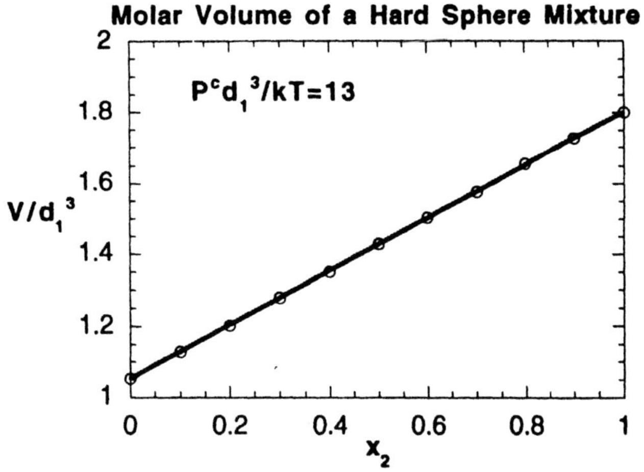

given by

$$\frac{\mu_\alpha^v}{kT} = \frac{\mu_\alpha^c}{kT} + \frac{\partial}{\partial n_\alpha}\left\{\frac{18}{\pi}\eta_3\eta_2^3\left[\frac{1}{\eta_3^3}\ln(1-\eta_3) + \frac{2-3\eta_3}{2\eta_3^2(1-\eta_3)^2}\right]\right\}. \tag{9.4.55}$$

The CS chemical potential is given by

$$\begin{aligned}\frac{\mu_\alpha^{CS}}{kT} = {} & \frac{\mu_\alpha^+(T)}{kT} + \ln\left(\frac{n_\alpha}{1-\eta_3}\right) + \frac{\pi P^{CS} d_\alpha^3}{6kT} \\ & + \frac{3\eta_2 d_\alpha + 3\eta_1 d_\alpha^2}{1-\eta_3} + \frac{9\eta_2^2 d_\alpha^2}{2(1-\eta_3)^2} \\ & + 3\left(\frac{\eta_2 d_\alpha}{\eta_3}\right)^2\left[\ln(1-\eta_3) + \frac{\eta_3}{1-\eta_3} - \frac{\eta_3^2}{2(1-\eta_3)}\right] \\ & - \left(\frac{\eta_2 d_\alpha}{\eta_3}\right)^3\left[2\ln(1-\eta_3) + \frac{\eta_3(2-\eta_3)}{1-\eta_3}\right].\end{aligned} \tag{9.4.56}$$

For binary mixtures, Lebowitz found for the direct correlation functions that (with the convention that $d_2 \geq d_1$)

$$C_{\alpha\alpha}^{(2)}(R) = -a_\alpha - b_\alpha R - dR^3, \quad R < d_\alpha, \quad \alpha = 1, 2 \tag{9.4.57}$$

and

$$\begin{aligned}C_{12}(R) = C_{21}(R) &= a_1, \quad R < \Delta_{21} \equiv (d_2 - d_2), \\ &= a_1 + \left[b(R-\Delta_{21})^2 + 4\Delta_{21} d(R-\Delta_{21})^3\right. \\ &\quad \left. + d(R-\Delta_{21})^4\right]/R, \quad \Delta_{21} < R < d_{12}\end{aligned} \tag{9.4.58}$$

where

$$a_\alpha = \beta\frac{\partial P^c}{\partial n_\alpha}$$

$$
\begin{aligned}
b_1 &= -\pi \left[n_1 (d_1 g_{11}^{(2)}(d_1))^2 + n_2 (d_{12} g_{12}^{(2)}(d_{12}))^2 \right] \\
b &= -\pi \left[n_1 d_1 g_{11}^{(2)}(d_1) + n_2 d_2 g_{22}^{(2)}(d_2) \right] d_{12} g_{12}^{(2)}(d_{12}) \\
d &= \frac{\pi}{12}(n_1 a_1 + n_2 a_2).
\end{aligned}
\tag{9.4.59}
$$

b_2 is obtained by interchanging indices 1 and 2 in b_1.

9.4.6 Perturbation Approximations

Several perturbation theories based on the van der Waals approximation

$$
F = F^{HS} + \Delta F^A \tag{9.4.60}
$$

have been developed in which F^{HS} is a hard sphere fluid with a temperature dependent diameter $d(T)$ determined by their repulsive forces of the fluid and ΔF^A is a perturbation free energy accounting for the attractive interactions of the fluid.

Formally, we can motivate eq. (9.4.60) from an exact result. Consider a λ system in which the potential energy can be written as $u_N = u_N^R + \lambda u_N^A$, where u_N^R is the reference (often repulsive) part of the potential energy of the system of interest and u_N^A is the perturbation (often attractive) part. If $\lambda = 0$, the λ system is the reference system and if $\lambda = 1$ the λ system is the system of interest. The configuration Helmholtz free energy ΔF of the λ system in a canonical ensemble is given by

$$
e^{-\beta \Delta F} = \int \cdots \int e^{-\beta u_N^R - \lambda \beta u_N^A} d^3 r_1 \cdots d^3 r_N. \tag{9.4.61}
$$

Differentiating eq. (9.4.57) with respect to λ yields

$$
\frac{d\Delta F}{d\lambda} = \frac{\int \cdots \int e^{-\beta u_N^R - \lambda \beta u_N^A} u_N^A d^3 r_1 \cdots d^3 r_N}{Z_N}, \tag{9.4.62}
$$

that, when integrated from $\lambda = 0$ to $\lambda = 1$, gives

$$
\Delta F = \Delta F^R + \Delta F^A, \quad \text{or} \quad F = F^R + \Delta F^A \tag{9.4.63}
$$

where F^R is the configuration free energy of the reference system and

$$
\Delta F^A = \int_0^1 \int \cdots \int p_N(\mathbf{r}_1, \ldots, \mathbf{r}_N; \lambda) u_N^A d^3 r_1 \cdots d^3 r_N d\lambda. \tag{9.4.64}
$$

For a one-component fluid, particles interacting via central, pair potentials

$$
\Delta F^A = \frac{1}{2} N n \int_0^1 \int g^{(2)}(r_{12}, \lambda) u^A(r_{12}) d^3 r_{12} d\lambda. \tag{9.4.65}
$$

In the multicomponent case

$$\Delta F^A = \frac{1}{2}\sum_{\alpha,\beta} n_\alpha n_\beta \int_0 \int g_{\alpha\beta}^{(2)}(r_{12};\lambda)u_{\alpha\beta}^A(r_{12})d^3r_{12}d\lambda. \quad (9.4.66)$$

If the reference system is chosen to be a hard sphere system and the pair correlation function is approximated by the step function $g_{\alpha\beta}^{(2)}(r_{12};\lambda) = \eta(r_{12} - d_{\alpha\beta})$, the van der Waals free energy is obtained.

In a dense fluid, structure is determined primarily by the repulsive forces, and so the pair correlation function $g_{\alpha\beta}^{(2)}(r_{12};\lambda)$ is approximated by the pair correlation function $g_{\alpha\beta}^{(2),R}(r_{12})$ of the reference system. In a dilute fluid $g_{\alpha\beta}^{(2)}(r_{12};\lambda) = e^{-\lambda\beta u_{\alpha\beta}^A(r_{12})}$. These two limits suggest the approximation

$$g_{\alpha\beta}^{(2)}(r_{12};\lambda) = g_{\alpha\beta}^{(2),R}(r_{12})\left[1 - \alpha_{\alpha\beta}(1 - e^{-\lambda\beta u_{\alpha\beta}^A(r_{12})})\right], \quad (9.4.67)$$

where we chose $\alpha_{\alpha\beta}$ so that $\alpha_{\alpha\beta} \to 1$ as $n \to 0$ and $\alpha_{\alpha\beta} \to 0$ with increasing density. A function with these properties is

$$\alpha_{\alpha\beta} = \frac{n_\alpha}{kT}\frac{\partial \mu_\beta^R}{\partial n_\alpha}, \quad (9.4.68)$$

where μ_β^R is the chemical potential of the reference fluid. The motivation for choosing eq. (9.4.64) is a systematic perturbation theory developed for a one-component fluid by Barker and Henderson (1976). They obtained the result

$$g^{(2)}(r_{12};\lambda) = g^{(2),R}(r_{12})\left[1 - \alpha(n)\lambda\beta u^A(r_{12})\right], \quad (9.4.69)$$

where

$$\alpha(n) = \frac{1}{kT}\frac{\partial P^R}{\partial n}. \quad (9.4.70)$$

Taking the reference system to be a fluid of hard spheres of diameter

$$d = \int_0^\infty (1 - e^{-\beta u^R(r)})dr, \quad (9.4.71)$$

they found that the free energy $F = F^{HS} + \Delta F^A$, with $g^{(2)}(r_{12}, \lambda)$ approximated by eq. (9.4.68) gave good agreement with computer simulations for the 6–12 Lennard–Jones fluid.

Keeping the first two terms in the expansion $e^{-\lambda\beta u^A} = 1 - \lambda\beta u^A + (\lambda\beta u^A)^2/2 + \cdots$ and using the one-component Gibbs–Duhem equation, $dP^R = n d\mu^R$, it is easy to show that eq. (9.4.66) reduces to the Barker–Henderson approximation. Because eq. (9.4.66) has the right low and high density limits and agrees with the Barker–Henderson result for a high temperature, one-component fluid, we submit it as a logical extension of the Barker–Henderson model to multicomponents. Choosing the reference system as a hard sphere fluid in which

$$d_\alpha = \int_0^\infty (1 - e^{-\beta u^R_{\alpha\alpha}(r)})dr \tag{9.4.72}$$

and $d_{\alpha\beta} = (d_\alpha + d_\beta)/2$, one can use the PY theory for the reference pair correlation function $g^{(2),R}_{\alpha\beta}$ and either the PY theory or the CS approximation for F^R and $\alpha_{\alpha\beta}$. The result is a fairly simple molecular theory with which to explore the phase behavior of multicomponent fluids.

The liquid–vapor coexistence curve for a one-component 6–12 Lennard–Jones is plotted in Figure 9.6 using perturbation theory for the pair correlation function approximated by

$$g^{(2)}(r, \lambda) = g^{(2),HS}(r), \tag{9.4.73}$$

and by eqs. (9.4.67 and (9.4.69), respectively. In computing the coexistence curve F^{HS} and $\alpha(n)$ were calculated from the CS hard sphere equation of state using eq. (9.4.71) to determine the hard sphere diameter d. The pair potential was taken to be $u(r) = 4\varepsilon[(\sigma/r)^{12} - (\sigma/r)^6]$ and $u^R(r) = u(r),\ 0 < r < \sigma$ and $u^A(r) = u(r),\ \sigma < r < \infty$. u^R and u^A are zero outside the specified ranges of r. The hard sphere pair correlation function $g^{(2),HS}$ was estimated from the PY theory (eq. 9.4.21).

Also shown in Figure 9.6 is the liquid–vapor coexistence curve computed from the empirical equation of state of Nicolas *et al.* (1979). Nicolas *et al.* fitted a 33 parameter equation of state to extensive computer simulation data for a 6–12 Lennard–Jones fluid. It accurately reproduces the known computer results for the 6–12 LJ fluid. As seen in Figure 9.6, the perturbation theory with the pair correlation function estimated from eq. (9.4.67) or (9.4.69)

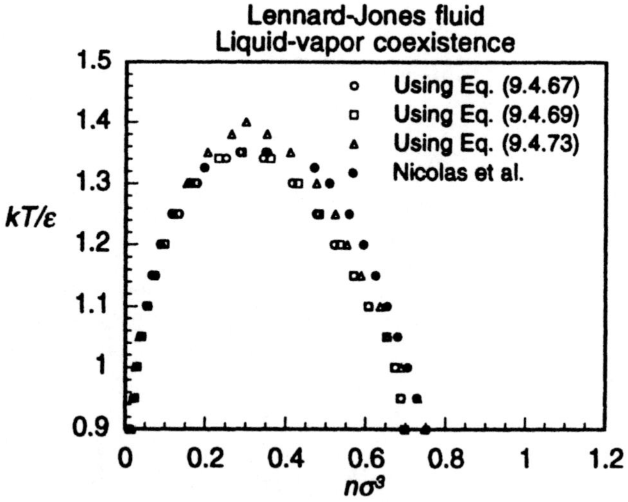

Figure 9.6
Liquid–vapor coexistence curve of a 6–12 Lennard–Jones fluid. Perturbation theory, using three approximations to the pair correlation function, is compared to the empirical equation of state of Nicolas *et al.* (1979).

is a pretty good approximation along the vapor branch of the coexistence curve all the way to the critical point. Along the liquid branch of the coexistence curve the theory performs less well. With eq. (9.4.73), the perturbation theory has significant error around the critical point, but, surprisingly, at the highest liquid densities it is more accurate than the other two model pair correlation functions.

A comparison of the predicitions of the critical point is shown in Table 9.4. The computer simulated critical point for a 6–12 LJ fluid is $kT_c/k = 1.36$ and $n_c\sigma^3 = 0.36$, and so one sees that the equation of state of Nicolas *et al.* (1979) is quite good. The perturbation theory provides a good estimate of T_c but gives a critical density that is about 15% too low.

Perturbation theory using the Barker–Henderson estimate of the hard core diameter is a reasonably easy model to use with any of the pair correlation functions considered. Given its simplicity, the agreement illustrated in Figure 9.6 and Table 9.4 is quite satisfactory.

Weeks, Chandler, and Andersen (WCA; 1971) developed a perturbation theory for dense one-component fluids. In the WCA theory $g^{(2)}(r, \lambda) \approx \tilde{g}^{(2)}_{\text{hs}}(r)e^{-\beta u^R(r)}$, where $\tilde{g}^{(2)}_{\text{hs}}(r) = e^{-\beta u^{HS}(r)}g^{(2)}_{\text{hs}}$ and $g^{(2)}_{\text{hs}}(r)$ is the pair correlation function of a fluid of hard spheres whose diameter is determined by

$$0 = \int \tilde{g}^{(2)}_{\text{hs}}(r)\left[\eta(r-d) - e^{-\beta u^R(r)}\right]d^3r. \qquad (9.4.74)$$

The repulsive reference potential in the WCA theory is

$$\begin{aligned} u^R(r) &= u(r) - u(r_m), \quad r < r_m \\ &= 0, \quad r > r_m, \end{aligned} \qquad (9.4.75)$$

and the attractive potential is

$$\begin{aligned} u^A(r) &= u(r_m), \quad r < r_m \\ &= u(r), \quad r > r_m \end{aligned} \qquad (9.4.76)$$

where r_m is the dividing point between repulsive and attractive pair forces, that is, $\partial u(r)/\partial r = 0$ at r_m. The WCA theory is based on choosing the hard sphere diameter to minimize the difference between F^{HS} and F^R. At densities above $nd^3 = 0.6$, the WCA theory is a good approximation. A measure of the success of the theory is seen in Figure 15.4 where the predicted pair correlation

TABLE 9.4 Predicted Critical Points for a 6–12 Lennard-Jones Fluid

	Nicolas et al. *(1979)*	*Perturbation Theory* Eq. (9.4.67)	Eq. (9.4.69)	Eq. (9.4.73)
T_c (ε/k)	1.35	1.35	1.35	1.4
n_c (σ^{-3})	0.35	0.29	0.28	0.3

function agrees well with computer simulation of a 6–12 Lennard–Jones fluid.

9.5 Inhomogeneous Fluids: Some Exact Results

In Chapter 8, we presented the Kirkwood–Buff formula for the tension of a planar interface between fluids composed of classical particles interacting via pair additive, centrally symmetric forces. The formula calls for the density profile, the pair potential, and the pair correlation function. With the aid of density functional theory, we can derive an alternative formula, that calls only for the density profile and the pair direct correlation function and the result is not restricted to fluids of molecules interacting via pair, central forces.

Imagine a large droplet of one phase present in another phase. Assume the droplet radius R is very large compared to the width of the interfacial zone between the phases. Recall the equation of hydrostatics,

$$\nabla \cdot \mathbf{P} = -n\nabla\phi, \tag{9.5.1}$$

where $\phi(\mathbf{r}) = v^e(\mathbf{r}) - \mu$ ($\nabla\phi = \nabla v^e$ becausee μ is a constant). If the external field ϕ is spherically symmetrical about the center of the drop, then the principal components of the pressure tensor is of the form $\mathbf{P} = P_N(r)\hat{r}\hat{r} + P_T(r)\mathbf{I}^{(2)}$, where $\hat{r}$ is a unit tensor along $\mathbf{r}$ and $\mathbf{I}^{(2)} = \mathbf{I} - \hat{r}\hat{r}$, where $\mathbf{I}$ is the unit tensor. The radial component of eq. (9.5.1) is

$$\frac{dP_N}{dr} + 2\frac{P_N - P_T}{r} = -n\frac{d\phi}{dr}. \tag{9.5.2}$$

The difference $P_N - P_T$ is nonzero only in the interfacial zone where $r \simeq R$. Thus, for sufficiently large R, the integral of eq. (9.5.2) along the x axis ($y = z = 0$) yields

$$\Delta P_N + \frac{2\gamma}{R} = -\int_{-\infty}^{\infty} n\frac{\partial\phi}{\partial x}dx + \text{terms of order } R^{-2}. \tag{9.5.3}$$

The origin of the x axis is chosen to be centered on R in the interfacial zone and Φ is chosen so that it vanishes outside the interfacial zone.

Next we will use a trick introduced by Lovett et al. (1973). They note that because ϕ is arbitrary, it can be chosen such that $\Delta P_N = 0$, and so eq. (9.5.3) reads

$$\gamma \simeq -\frac{R}{2}\int_{-\infty}^{\infty} n\frac{\partial\phi}{\partial x}dx = \frac{R}{2}\int_{-\infty}^{\infty} \frac{\partial n}{\partial x}\phi dx, \tag{9.5.4}$$

where the neglected terms in eq. (9.5.4) are of order R^{-1}. But from density functional theory we know that $\phi(r)$ is a functional of the density distribution $n^s(r)$ of the spherical drop. Let us denote by $n(x)$ the density distribution of a planar interface and expand $\phi(\mathbf{r}, \{n^s\})$ in a functional Taylor's series about $n(x)$ to obtain

$$\phi(\mathbf{r}, \{n^s\}), \quad \{n^s\} \simeq \phi(\mathbf{r}, \{n\}) + \sum_{k=1}^{\infty} \frac{1}{k!} \int \left[\frac{\delta^k \phi}{\delta n(\mathbf{r}_1) \cdots \delta n(\mathbf{r}_k)} \right]_{n^s(r)=n(x)} \prod_{i=1}^{k} [n^s(\mathbf{r}_i) - n(x_i)] d^3 r_i. \quad (9.5.5)$$

Because $\Delta P_N \equiv 0$ for a planar interface, it follows that $\phi(\mathbf{r}, \{n\}) = 0$ when $n = n(x)$, and because for large R the quantity $n^s(r)$ differs little from the density distribution of a planar interface, eq. (9.5.5) yields

$$\begin{aligned} \phi(\mathbf{r}) &\simeq \int \left[\frac{\delta \phi}{\delta n(\mathbf{r}')} \right]_{n^s(\mathbf{r}')=n(x')} [n^s(\mathbf{r}') - n(x')] d^3 r' \\ &\simeq \int \left[-\frac{kT}{n(x')} \delta(\mathbf{r}' - \mathbf{r}) + kTC^{(2)}(\mathbf{r}, \mathbf{r}', \{n\}) \right] [n^s(\mathbf{r}') - n(x')] d^3 r', \end{aligned} \quad (9.5.6)$$

where the neglected terms are of order R^{-2}. At sufficiently large R, the density distribution n^s along $\hat{r}$ becomes the same as that of a planar interface, that is, the coordinated system used in eq. (9.5.3),

$$n^s(\mathbf{r}') \simeq n([(x' + R)^2 + y'^2 + z'^2]^{1/2} - R). \quad (9.5.7)$$

Because we want $\phi(\mathbf{r})|_{y=z=0}$ for eq. (9.5.4) and because $C^{(2)}$ is short ranged in $\mathbf{r} - \mathbf{r}'$, the values of y' and z' of interest in eq. (9.5.6) will be small compared to R. Thus, because

$$\begin{aligned} [(x' + R)^2 + y'^2 + z'^2]^{1/2} - R &= (x' + R) \left[1 + \frac{y'^2 + z'^2}{(x' + R)^2} \right]^{1/2} - R \\ &= (x' + R) \left[1 + \frac{(y'^2 + z'^2)}{2(x' + R)^2} + \mathcal{O}\left(\frac{y'^2 + z'^2}{(x' + R)^2} \right)^2 \right] - R \\ &= x' + \frac{y'^2 + z'^2}{2R} + \mathcal{O}R^{-2}, \end{aligned} \quad (9.5.8)$$

we find

$$n^s(\mathbf{r}') \simeq n(x') + \frac{y'^2 + z'^2}{2R} \frac{dn(x')}{dx'}. \quad (9.5.9)$$

Inserting eq. (9.5.9) into eq. (9.5.6), substituting the result into eq. (9.5.4), and passing to the limit $R = \infty$, we obtain

$$\gamma = \frac{kT}{4} \int \int (s_y^2 + s_z^2) C^{(2)}(\mathbf{r}, \mathbf{r} + \mathbf{s}, \{n\}) \frac{dn}{dx}(x) \frac{dn}{dx}(x + s_x) dx d^3 s, \quad (9.5.10)$$

where the coordinate transformation $\mathbf{s} = \mathbf{r}' - \mathbf{r}$ has been used. For a planar interface $C^{(2)}(\mathbf{r}, \mathbf{r} + \mathbf{s})$ depends only on s, x and s_x. This formula is rigorous because all of the approximation relations leading to it become equalities in the limit $R = \infty$. The result was derived independently by Triezenberg and Zwanzig (1972) and by Lovett et al.

For a multicomponent fluid a similar derivation leads to the formula

$$\gamma = \frac{kT}{4}\sum_{\alpha,\beta}\int\int (s_y^2 + s_z^2)C_{\alpha\beta}^{(2)}(\mathbf{r}, \mathbf{r}+\mathbf{s}, \{\mathbf{n}\})\frac{dn_\alpha(x)}{dx}\frac{dn_\beta(x+s_x)}{dx}dx d^3s. \tag{9.5.11}$$

Let us close this section by establishing an analogue of the Yvon-Born-Green (YBG) equation. In Chapter 8, it was shown that the equilibrium density distribution obeys the YBG equation, which is an integrodifferential equation relating the density distribution $n(\mathbf{r})$ to the pair correlation function $g^{(2)}(\mathbf{r}, \mathbf{r}')$. Density function theory yields an equation similar to the YBG equation but in which $C^{(2)}(\mathbf{r}, \mathbf{r}')$ appears instead of $g^{(2)}(\mathbf{r}, \mathbf{r}')$. To derive this equation, recall that the potential $\phi(\mathbf{r})$ is a functional of $n(\mathbf{r})$. Thus, given $n(\mathbf{r})$, then $\phi(\mathbf{r})$ in principle can be computed from eq. (9.2.15). Because $\phi(\mathbf{r})$ is generated by a functional of $n(\mathbf{r})$, the potential $\phi(\mathbf{r}+\mathbf{a})$ would be generated by a functional of $\tilde{n}(\mathbf{r}) \equiv n(\mathbf{r}+\mathbf{a})$. $\phi(\mathbf{r}+\mathbf{a})$ can be expanded in a functional Taylor expansion about $\tilde{n} = n(\mathbf{r})$ to obtain

$$\phi(\mathbf{r}+\mathbf{a}) = \phi(\mathbf{r}) + \int \frac{\delta\phi(\mathbf{r})}{\delta n(\mathbf{r}')}[n(\mathbf{r}+\mathbf{a}) - n(\mathbf{r}')]d^3r'$$
$$\sum_{k=2}^{\infty}\frac{1}{k!}\int\frac{\delta^k\phi(\mathbf{r})}{\delta n(\mathbf{r}_1)\cdots\delta n(r_k)}\sum_{j=1}^{k}[n(\mathbf{r}_j+\mathbf{a}) - n(\mathbf{r}_j)]d^3r_j. \tag{9.5.12}$$

If the vector $\mathbf{a}$ is reduced to an arbitrary infinitesimal, then $\phi(\mathbf{r}+\mathbf{a}) - \phi(\mathbf{r}) = \mathbf{a}\cdot\nabla_r\phi(\mathbf{r}) + \phi(a^2)$ and $n(\mathbf{r}'+a) - n(\mathbf{r}') = \mathbf{a}\cdot\nabla_{\mathbf{r}'}n(\mathbf{r}') + \mathcal{O}(a^2)$, and so eq. (9.5.12) yields

$$\nabla_{\mathbf{r}}\phi(\mathbf{r}) = \int\frac{\delta\phi(\mathbf{r})}{\delta n(\mathbf{r}')}\nabla_{\mathbf{r}'}n(\mathbf{r}')d^3r'. \tag{9.5.13}$$

$\nabla_{\mathbf{r}}\phi(\mathbf{r}) = \nabla_{\mathbf{r}}u^3(\mathbf{r})$ because μ is constant at equilibrium. Using this and eq. (9.2.10) for $\delta\phi(\mathbf{r})/\delta n(\mathbf{r}')$, we derive

$$kT\nabla_{\mathbf{r}}\ln n(\mathbf{r}) + \nabla_{\mathbf{r}}v^e(\mathbf{r}) = kT\int C^{(2)}(\mathbf{r}, \mathbf{r}')\nabla_{\mathbf{r}'}n(\mathbf{r}')d^3r', \tag{9.5.14}$$

an equation analogous to the YBG equation. It has the same problem as the YBG equation, namely, that to use eq. (9.5.14) to find the density distribution $n(\mathbf{r})$ some sort of closure approximation must be introduced. However, eq. (9.5.14) does not require forces to be pair additive and does not involve the potential energy function explicitly.

In the case of a multicomponent system, eq. (9.5.14) becomes

$$kT\nabla_{\mathbf{r}}\ln n_\alpha(\mathbf{r}) + \nabla_{\mathbf{r}}v_\alpha^e(\mathbf{r}) = kT\sum_\beta\int C_{\alpha\beta}^{(2)}(\mathbf{r}, \mathbf{r}')\nabla_{\mathbf{r}'}n_\alpha(\mathbf{r}')d^3r', \tag{9.5.15}$$

for $\alpha = 1, \ldots c$.

Supplementary Reading

Bogoliubov, N. N. (1965). *The Dynamical Theory in Statistical Physics*, Hindustan Publishing Corporation, India.

Frisch, H. L. and Lebowitz, J. L., eds. 1965. *Equilibrium Theory of Classical Fluids*, Benjamin, New York.

Henderson, D., ed. 1992. *Fundamentals of Homogeneous Fluids*, Dekker, New York.

Münster, A. 1974. *Statistical Thermodynamics*, Academic Press, New York.

Reed, T. M. and Gubbins, K. E. 1973. *Applied Statistical Mechanics*, McGraw–Hill, New York.

Rice, S. A. and Gray, P. 1965. *The Statistical Mechanics of Simple Liquids*, Interscience, New York.

Volterra, V. 1959. *Theory of Functionals and of Integral and Intro-Differential Equations*, Dover, New York.

Exercises

1 Consider a functional $H(\{n\})$ of the function $n(\mathbf{r})$. Define the functional derivative $\delta H/\delta n(\mathbf{r})$ by

$$\frac{d}{d\varepsilon}H(\{n+\varepsilon v\})|_{\varepsilon=0} \equiv \int \frac{\delta H}{\delta n(\mathbf{r})} v(\mathbf{r}) d^D r,$$

where D is a dimensionality of space. With this definition derive the following formulas:

a) if $H = \int f(n(\mathbf{r}_1)) d^D r_1$,

$$\frac{\delta H}{\delta n(\mathbf{r})} = \frac{df(n(\mathbf{r}))}{dn}$$

b) if $H = f(n(\mathbf{r}))$

$$\frac{\delta H}{\delta n(\mathbf{r}')} = \frac{df(n(\mathbf{r}))}{dn}\delta(\mathbf{r}-\mathbf{r}')$$

c) if $H = \int f(n(\mathbf{r}_1)) d^D r_1$

$$\frac{\delta^2 H}{\delta n(\mathbf{r})\delta n(\mathbf{r}')} = \frac{d^2 f(n(\mathbf{r}))}{dn^2}\delta(\mathbf{r}-\mathbf{r}')$$

d) if $H = \int\int\int e^{-\beta n(\mathbf{r}_1)-\beta n(\mathbf{r}_2)-\beta n(\mathbf{r}_3)} d^D r_1 d^D r_2 d^D r_3$,

$$\frac{\delta H}{\delta n(\mathbf{r})} = -3\beta e^{-\beta n(\mathbf{r})} \int\int e^{-\beta n(\mathbf{r}_2)-\beta n(\mathbf{r}_3)} d^D r_2 d^D r_3.$$

2 Find the first and second functional derivatives of

$$H = \int u(\mathbf{r}) e^{-u(\mathbf{r})} d^D r$$

with respect to $u(\mathbf{r})$. D is the dimensionality of space.

3 Find the functional derivative of

$$H = \int_a^b \frac{[n(x)]^2}{1 + \int_a^b k(x, y)n(y)dy} dx$$

with respect to $n(x)$.

4 Find the functional derivative of

$$H = \frac{1}{2} \int_a^b \left[\frac{du(x)}{dx} \right]^2 dx$$

with respect to $u(x)$. In defining the functional derivative as in Problem 1, assume $v(a) = v(b) = 0$.

a) Find the relationship between n, ϕ and μ.

What is the functional derivative with respect to $n(x)$ of

b) $F = \int_{-\infty}^{\infty} n(x)dx$

c) $F = \int_{-\infty}^{\infty} K(y, x)n(x)dx$

d) $F = \int_{-\infty}^{\infty} K(y, x)\frac{dn(x)}{dx}n(x)dx$

5 Under certain conditions, the internal energy and entropy of a system of charged particles in a charged static continuum are given by

$$U = \frac{3}{2}NkT + \frac{1}{2}q^2 \int\int \frac{n(\mathbf{r})n(\mathbf{r}')}{D_o|\mathbf{r} - \mathbf{r}'|} d^3r d^3r' + q \int n(\mathbf{r})\phi_{ext} d^3r$$

and

$$S = \frac{3}{2}Nk \ln\left(\frac{2\pi mkTe}{h^2}\right) - k \int n(\mathbf{r})[\ln n(\mathbf{r}) - 1]d^3r,$$

where q is the charge on the particles, ϕ_{ext} is the external electric potential, e is the base of the natural logarithm, and D_o is the dielectric constant of the medium.

At equilibrium the density distribution $n(\mathbf{r})$ is the function that minimizes the grand potential, $\Omega = F - N\mu$, where F is the Helmholtz free energy, μ the chemical potential, and $N = \int n(\mathbf{r})d^3r$.

a) Find the equation satisfied by $n(\mathbf{r})$ at equilibrium.

b) The electrostatic potential is given by

$$\phi = \psi_{ext} + q \int \frac{n(\mathbf{r}')}{D_o|\mathbf{r} - \mathbf{r}'|} d^3r'.$$

c) Find the relationship n, ϕ, and μ.

6 The excess free energy of a two-phase fluid system with meniscus at position $h(x, y)$, above the xy plane given by

$$\Delta F(h) = \gamma \int \sqrt{1 + h_x^2 + h_y^2} dxdy - \Delta P \int \eta(h(x, y) - z)dxdydz,$$

where $h_x = \partial h/\partial x$, $h_y = \partial h/\partial y$ and $\eta(\zeta)$ is a step function that is zero when $\zeta < 0$ and one when $\xi > 0$. Gravity is assumed to be negligible.

The meniscus takes the equilibrium shape $h(x, y)$ that minimizes ΔF. Thus,

$$\frac{\delta \Delta F}{\delta h(x, y)} = 0.(A)$$

Show that condition (A) yields the Young–Laplace equation.

$$\Delta P = 2H\gamma,$$

where

$$2H = \text{eq.}(7.3.30).$$

7 Consider Problem 9.5. Find the density $n(r)$ of counterions surrounding an isolated sphere of radius $R = 1\ \mu$ m. Assume that the charge density on the sphere is $e/50$ Å^2, that the temperature is 298 K, $D_o = 80$, and $n(r) = 3 \times 10^{19}$ counterions/cm^3 from the charged sphere.

8 Compare CS and the PY compressibility and virial pressures versus density for a binary mixture of hard spheres in which $d_2 = 1.5d_1$. Calculate the pressures for the mole fractions $x_1 = 0,\ 0.25,\ 0.5,\ 0.75$, and 1.

9 From the PY and CS equations of state, determine the second and third virial coefficients for hard sphere mixtures.

10 Using eq. (9.4.19) for $C^{(2)}(R)$, discretize (using the trapezoidal rule) and solve the Ornstein–Zernike equation for $g^{(2)}(R)$ for a one-component hard core fluid at the densities $nd^3 = 0.5$ and $nd^3 = 1$.

11 Consider the free energy model of a one component fluid,

$$F = F^{HS} + \frac{Nn}{2} \int g_R^{(2)}(R) u^A(R) d^3R,$$

where F^{HS} is the free energy of a fluid to hard spheres of diameter

$$d = \int_o^\infty (1 - e^{-\beta u^R(R)}) dR$$

and $g_R^{(2)}(R)$ is the pair correlation function of the hard sphere fluid. Assume

$$u(R) = u^R(R) + u^A(R),$$
$$u^R(R) = 4\varepsilon \left(\frac{\sigma}{R}\right)^{12}; \quad u^A(R) = -4\varepsilon \left(\frac{\sigma}{R}\right)^6.$$

Use the method of finding the common tangent of an $f(n)$ versus n curve (see Figure 8.5) to find the temperature–density liquid–vapor binodal for the model. Use the PY model for the pair correlation function and take argon parameters for ε and σ. Compare the results with experimental data for argon.

12 Repeat Problem 9.11 using the WCA theory.

13 Explore the validity of Amagat's law by comparing it with the PY theory of hard spheres for values of d_2/d_1 larger than the value chosen in Figure 9.5.

References

Barker, J. A. and Henderson, D. 1976. *Rev. Mod. Phys.* **48,** 587–671.

Kerins, J., Scriven, L. EE., and Davis, H. T. 1986. *Adv. Chem. Phys.,* **65,** 215.

Lebowitz, J. L. 1964. *Phys. Rev.,* **133,** A895.

Lebowitz, J. L. and Percus, J. K. 1966. *Phys. Rev.,* **144,** 251.

Lovett, R., DeHaven, P. W., Vieceli, J. J. Jr., and Buff, F. P. 1973. *J. Chem. Phys.,* **58,** 1880.

Mansoori, A., Carnahan, N. F., Starling, K. E., and Leland, T. W. 1971. *J. Chem Phys.,* **54,** 1523.

March, N. H. 1968. Liquid Metals, Pergamon Press, New York.

Meeron, E. 1960. *J. Math. Phys.,* **1,** 192.

Nicolas, J. J., Gubbins, K. E., Street, W. B., and Tildesley, D. J. 1979. *Moleecular Physics,* **37,** 1429.

Percus, J. K., H. C. Frisch and J. L. Lebowitz, eds. 1964. *The Equilibrium Theory of Classical Fluids,* W. A. Benjamin, Inc., New York.

Percus, J. K. and Yevick, G. Y. 1958. *Phys. Rev.,* **110,** 1.

Rice, S. A. and Gray, P. 1965. *Statistical Mechanics of Simple Liquids,* Wiley, New York.

Triezenberg, D. G. and Zwanzig, R. 1972. *Phys. Rev. Lett.,* **28,** 1183.

Waisman, E. and Lebowitz, J. L. 1972. *J. Chem. Phys.,* **56,** 3086; 1972. *J. Chem. Phys.,* **56,** 3093.

Weeks, J. D., Chandler, D., and Anderseen, H. C. 1971. *J. Chem. Phys.,* **55,** 5522.

Wertheim, M. S. 1963. *Phys. Rev. Lett.,* **10,** 321; 1973. *J. Chem. Phys.,* **55,** 4291.

10

CONFINED ONE-DIMENSIONAL FLUIDS

10.1 Thermodynamic Properties of Hard-Rod Fluids Between Rigid Walls

10.1.1 Evaluation of Partition Functions and Thermodynamic Functions

Consider a multicomponent hard-rod fluid. The pair potential between molecules of species α and β separated by a distance x is

$$\begin{aligned} u_{\alpha\beta}(x) &= \infty, \quad |x| < a_{\alpha\beta} \\ &= 0, \quad |x| > a_{\alpha\beta}, \end{aligned} \tag{10.1.1}$$

where $a_{\alpha\beta} = (a_\alpha + a_\beta)/2$. a_α and a_β denote the lengths of rods of species α and β, respectively. Such a fluid confined by hard walls to a box of length L is illustrated in Figure 10.1.

The canonical ensemble partition function of this system is

$$Q_{\mathrm{N}} = \prod_{\alpha=1}^{c} \frac{1}{N_\alpha! \Lambda_\alpha^{N_\alpha}} Z_{\mathrm{N}}, \tag{10.1.2}$$

where N_α is the number of particles of species α, Λ_α the de Broglie thermal wavelength of species α,that is,

$$\Lambda_\alpha = \left(\frac{h^2}{2\pi m_\alpha kT} \right)^{1/2}, \tag{10.1.3}$$

and Z_{N} the configuration partition function,

$$Z_{\mathrm{N}} = \int_0^L \cdots \int_0^L e^{-\beta u^{\mathrm{N}}(x_1,\ldots,x_N)} dx_1 \ldots dx_N, \quad \beta \equiv 1/kT. \tag{10.1.4}$$

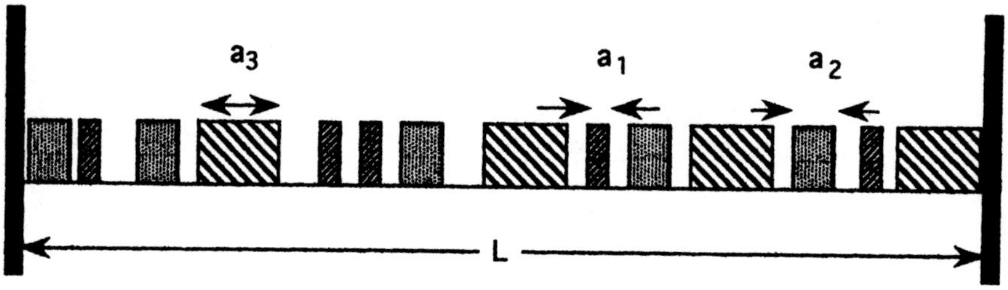

Figure 10.1
Illustration of a hard-rod fluid on a line of length L.

As a first step in evaluating the partition function Z_N, suppose the particles $\alpha_1, \cdots, \alpha_N$ are arranged in the order $x_1 \le x_2 \le \cdots \le x_N$. The contribution to Z_N is

$$Z_{\{\alpha_i\}} = \int\limits_{a_{\alpha_N\alpha_{N-1}}+\cdots+a_{\alpha_2\alpha_1}+a_{\alpha_1}/2}^{L-a_{\alpha_N}/2} dx_N \cdots \int\limits_{a_{\alpha_2\alpha_1}+a_{\alpha_1}/2}^{x_3-a_{\alpha_3\alpha_2}} dx_2 \int\limits_{a_{\alpha_1}/2}^{x_2-a_{\alpha_2\alpha_1}} dx_1$$

where the ranges of integration account for the hard-rod interactions. With the coordinate transformation $\xi_1 = x_1 - a_{\alpha_1}/2$ and $\xi_j = x_j - \sum_{i=1}^{j-1} a_{\alpha_{i+1}\alpha_i} - a_{\alpha_1}/2$ for $j > 1$, $Z_{\{\alpha_i\}}$ can be easily integrated to yield

$$Z_{\{\alpha_i\}} = \frac{(L - \sum_{\alpha=1}^{c} N_\alpha a_\alpha)^N}{N!}, \quad L > \sum_\alpha N_\alpha a_\alpha. \tag{10.1.5}$$

Note that $Z_{\{\alpha_i\}}$ depends only on the number of particles of each species, not on the particle configuration chosen. Thus, because there are $N!$ ways the particles $\alpha_1, \ldots, \alpha_N$ can be ordered and because each configuration gives the same contribution to Z_N, we find

$$Z_N = (L - \sum_{\alpha=1}^{c} N_\alpha a_\alpha)^N, \quad L > \sum_{\alpha=1}^{c} N_\alpha a_\alpha. \tag{10.1.6}$$

Z_N is zero if $L < \sum_\alpha N_\alpha a_\alpha$ because $L = \sum_\alpha N_\alpha a_\alpha$ is the smallest box that the hard-rod fluid can be put into (i.e., the fluid cannot exceed the close-packed density).

Combining eqs. (10.1.6) and (10.1.2) we can easily compute the thermodynamic properties of the hard-rod mixture in the canonical ensemble. From $U_N = kT^2(\partial \ln Q_N/\partial T)_{L,N}$, we find

$$U_N = \frac{1}{2}NkT, \quad N = \sum_\alpha N_\alpha. \tag{10.1.7}$$

The pressure, $P_N = kT(\partial \ln Q_N/\partial L)_{T,N}$, is

$$P_N = \frac{NkT}{L - \sum_\alpha N_\alpha a_\alpha}, \tag{10.1.8}$$

and the chemical potential of species α, $\mu_\alpha = kT(\partial \ln Q_{\mathbf{N}}/\partial N_\alpha)_{T,L,N_{\gamma\neq\alpha}}$, is

$$\mu_\alpha(N) = \frac{Na_\alpha kT}{L - \sum_\gamma N_\gamma a_\gamma} + kT \ln\left(\frac{\Lambda_\alpha N_\alpha}{L - \sum_\gamma N_\gamma a_\gamma}\right)$$
$$= P_{\mathbf{N}} a_\alpha + kT \ln\left(\frac{x_\alpha P_{\mathbf{N}} \Lambda_\alpha}{kT}\right), \qquad (10.1.9)$$

where $x_\alpha = N_\alpha/N$ is the overall mole fraction of species α in the system. Because L is finite the system is not homogeneous and so the local mole fraction of species α may vary. In the limit that N_α, $L \to \infty$ while $N_\alpha/L \to n_\alpha =$ constant, eqs. (10.1.8) and (10.1.9) become equations of state of homogeneous fluid.

The grand canonical ensemble partition function of an open isothermal mixture of hard rods is of the form

$$\Xi = \sum_{N_1=0}^{\infty} \cdots \sum_{N_c=0}^{\infty} \prod_{\alpha=1}^{c} e^{\beta N_\alpha \mu_\alpha} Q_{\mathbf{N}}$$
$$= \sum_{N_1=0}^{\infty} \cdots \sum_{N_c=0}^{\infty} \prod_{\alpha=1}^{c} \left(\frac{\zeta_\alpha^{N_\alpha}}{N_\alpha!}\right) Z_{\mathbf{N}} \qquad (10.1.10)$$
$$= \sum_{N_1=0}^{\infty} \cdots \sum_{N_c=0}^{\infty} \prod_{\alpha=1}^{c} \left(\frac{\zeta_\alpha^{N_\alpha}}{N_\alpha!}\right) \left(L - \sum_\alpha N_\alpha a_\alpha\right)^N \eta\left(L - \sum_\alpha N_\alpha a_\alpha\right),$$

where

$$\zeta_\alpha \equiv \frac{e^{\beta\mu_\alpha}}{\Lambda_\alpha} \qquad (10.1.11)$$

and $\eta(x)$ is the step function with the property $\eta(x) = 0$, $x < 0$, and $\eta(x) = 1$, $x > 0$.

We can now derive the grand canonical ensemble averages of a thermodynamic quantity X from the formula

$$\langle X \rangle_{GC} = \sum_{N_1=0}^{\infty} \cdots \sum_{N_c=0}^{\infty} X_{\mathbf{N}} \mathcal{P}_{\mathbf{N}}, \qquad (10.1.12)$$

where $X_{\mathbf{N}}$ is the canonical ensemble value of the quantity and $\mathcal{P}_{\mathbf{N}}$ is the probability of finding the occupancy $\mathbf{N} = \{N_1, \cdots, N_c\}$ in a grand canonical ensemble. From eq. (10.1.10) it follows that

$$\mathcal{P}_{\mathbf{N}} = \frac{1}{\Xi} \prod_{\alpha=1}^{c} \left(\frac{\zeta_\alpha^{N_\alpha}}{N_\alpha!}\right) \left(L - \sum_{\beta=1}^{c} N_\beta a_\beta\right)^N \eta\left(L - \sum_{\beta=1}^{c} N_\beta a_\beta\right). \qquad (10.1.13)$$

The pressure P or average occupancy by species α can be computed by substituting $P_{\mathbf{N}}$ or N_α into eq. (10.1.12) for $X_{\mathbf{N}}$, or from the formulas

$$P = kT\left(\frac{\partial \ln \Xi}{\partial L}\right)_{T,\mu_1,\cdots,\mu_c} \qquad (10.1.14)$$

and

$$\overline{N}_\alpha = kT\left(\frac{\partial \ln \Xi}{\partial \mu_\alpha}\right)_{T,\mu_{\beta\neq\alpha},L}, \tag{10.1.15}$$

which can be derived from eq. (10.1.12).

For finite L the pressure P and density $\overline{N}_\alpha/L$ of an open isothermal system will differ from the pressure P_N and density N_α/L of a closed isothermal system. However, in the limit that $L \to \infty$ while $N_\alpha/L = n_\alpha =$ constant, the thermodynamic properties are the same in either case. Thus, for a homogeneous fluid, the chemical potential and pressure obey the relationships

$$\mu_\alpha = Pa_\alpha + kT\ln(x_\alpha P\Lambda_\alpha/kT) \tag{10.1.16}$$

and

$$P = \frac{nkT}{1-\sum_\beta n_\beta a_\beta}, \tag{10.1.17}$$

for any ensemble. The mole fraction x_α, densities n_α, and $n = \sum_\alpha n_\alpha$ are of course constant in a homogeneous fluid.

10.1.2 Pore Occupancy and Disjoining Pressure of a Pure Fluid

The probability $\mathcal{P}_N$ that N particles of a one-component hard-rod fluid will occupy a pore of width L (i.e., a box with a wall separation L) is

$$\mathcal{P}_N = \frac{e^{\beta N\mu}}{\Lambda^N}\frac{(L-Na)^N\eta(L-Na)}{N!\,\Xi}, \tag{10.1.18}$$

$$\Xi = \sum_{N=0}^{\infty}\frac{e^{\beta N\mu}}{\Lambda^N}\frac{(L-Na)^N\eta(L-Na)}{N!}, \tag{10.1.19}$$

where μ is the chemical potential of the hard-rod fluid. If the confined fluid is at equilibrium with a bulk fluid at bulk pressure P_B, then the chemical potential is given by

$$\mu = P_B a + kT\ln(P_B\Lambda/kT). \tag{10.1.20}$$

For highly confined systems,that is, L/a is not too large, the partition function is easily evaluated from eq. (10.1.19) and the average pore density and pore pressure can be economically computed from

$$n_p = \sum_{N=0}^{\infty}(N/L)\mathcal{P}_N \tag{10.1.21}$$

and

$$P = \sum_{N=0}^{\infty} P_N\mathcal{P}_N, \tag{10.1.22}$$

with

$$P_N = \frac{NkT}{L - Na}. \qquad (10.1.23)$$

Because the thermodynamic properties of the confined fluid are controlled by $\mathcal{P}_N$, it is interesting to study this quantity as a function of pore width L. To do this we choose a chemical potential corresponding to a bulk fluid in which the bulk density is $n_B = 0.75a^{-1}$. In this case $\beta Pa = 3$ and $\beta\mu = 4.099 + \ln(\Lambda/a)$, or $e^{\beta\mu}/\Lambda = 60.26a^{-1}$. For a confined hard-rod fluid $\mathcal{P}_N$, βPa and $n_p a$ are universal functions of $e^{\beta\mu}a/\Lambda$ and L/a. The probabilities $\mathcal{P}_1$, $\mathcal{P}_2$, and $\mathcal{P}_3$ of one, two and three particle occupancy are plotted versus pore width in Figure 10.2. The probability $\mathcal{P}_N$ is of course zero if Na exceeds the close-packing density (i.e., if $L < Na$). As L increases beyond its close-packed value, $\mathcal{P}_N$ increases rapidly until the pore is wide enough to admit $N + 1$ particles. Then the rate of increase of $\mathcal{P}_N$ slows down because of the competition with occupancy states of $N + 1$. Ultimately, $\mathcal{P}_N$ begins decreasing and approaches zero with increasing L as the states of occupancy greater than N begin to win the competition. The maximum in $\mathcal{P}_N$ occurs at a width that is large enough to easily admit N particles but is not large enough to easily admit $N + 1$ or more particles.

As a result of the competitive occupancy seen in Figure 10.2, the pressure and mean pore density are oscillating functions of porewidth. The pressure and mean pore density are plotted versus pore width L in Figure 10.3. Also shown in Figure 10.3 are plots of $P_N\mathcal{P}_N$ versus L. The total pressure is the envelope of sum of these quantities. The peaks in the pressure and mean pore density are dominated by the maxima of $P_N\mathcal{P}_N$: the peak at $L = a$ comes entirely from $P_1\mathcal{P}_1$, that at $L \simeq 2.3a$ coming mostly from $P_2\mathcal{P}_2$, that at $L \simeq 3.3a$ comes largely from $P_3\mathcal{P}_3$, etc. The larger N, the broader the distribution $\mathcal{P}_N$, and so with increasing L the overlap between neighboring distributions smooths out the maxima and

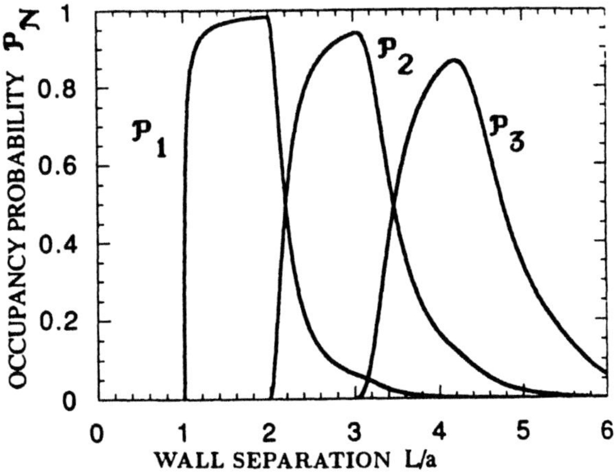

Figure 10.2
The probability $\mathcal{P}_N$ that N hard rods will occupy a pore of width (i.e., wall separation) L. The confined fluid is in equilibrium with a bulk fluid at reduced pressure $\beta P_B a = 3$.

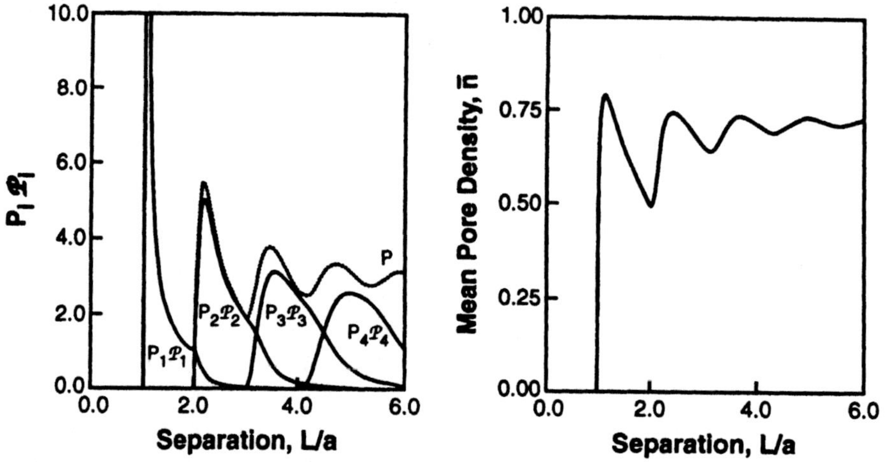

Figure 10.3
(a) Pressure and (b) mean pore density of a confined hard-rod fluid versus pore width L. $\beta P_B a = 3$. Adapted from Vanderlick (1988).

minima and the pressure tends to a constant (bulk) value. The minima in the pressure tend to occur near $L = Na$, where there is ample room for $N - 1$ or fewer particles but not enough for N or more. Thus, at these widths, the mean pore density is a minimum and the pore pressure is a maximum.

Another way to display the pressure of a confined fluid is as the disjoining pressure,

$$\Pi = P - P_B, \tag{10.1.24}$$

that is simply the difference between the pore pressure and the pressure of the bulk fluid in thermodynamic equilibrium with the confined fluid. For the case examined here, the disjoining pressure is shown in Figure 10.4.

The disjoining force F/R between crossed mica cylinders of radius R confining thin films of fluids can be measured by the

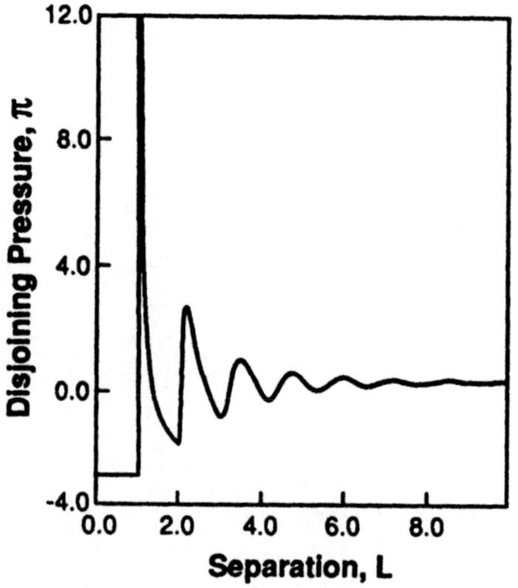

Figure 10.4
Disjoining pressure versus pore width for a hard-rod fluid. $\beta P_B a = 3$.

Figure 10.5
Disjoining force versus separation of crossed mica cylinders confining cyclohexane versus separation L of the cylinders. Adapted from Vanderlick (1988).

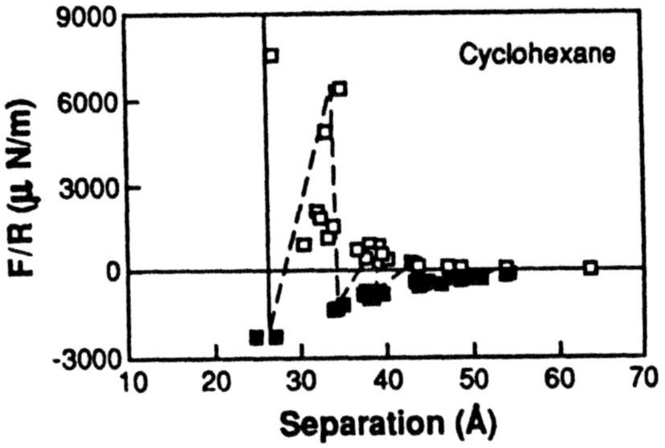

surface forces apparatus. A typical result is shown in Figure 10.5. The dashed part of the curve is not accessible to the surface forces apparatus.

The disjoining pressure is actually the derivative with respect to L of the quantity F/R plotted in Figure 10.4. The quantity F/R is actually equal to $\int_{\infty}^{L} \Pi(L')dL'$, not $\Pi(L)$. Moreover, one certainly does not expect quantitative agreement between a real, three-dimensional fluid and hard-rod fluid. It is nevertheless interesting that as is the case of hard rods, the period of oscillation of F/R is roughly equal to the van der Waals diameter (about 5.6 Å) of cyclohexane. We conclude from this that, although the magnitude of the disjoining pressure of a highly confined real fluid will be sensitive to the attractive interactions, the oscillations in the disjoining pressure result primarily from the short-range repulsive forces between fluid molecules.

Next let us consider a binary hard-rod fluid. In this case, the average pore occupancy of component α is computed from

$$\overline{N}_\alpha = \sum_{N_1=0}^{\infty} \sum_{N_2=0}^{\infty} N_\alpha \mathcal{P}_{N_1,N_2} \tag{10.1.25}$$

and the pressure from

$$P = \sum_{N_1=0}^{\infty} \sum_{N_2=0}^{\infty} \frac{(N_1 + N_2)kT}{L - N_1 a_1 - N_2 a_2} \mathcal{P}_{N_1,N_2} \tag{10.1.26}$$

Occupancy probabilities $\mathcal{P}_{N_1,N_2}$ are given in Table 10.1 for $\beta P_B a_1 = 3$ and $a_2 = 2.5a_1$. As for a one-component fluid, $\mathcal{P}_{N_1,N_2} = 0$ if $L < \sum_\alpha N_\alpha a_\alpha$, the close-packing width, and for fixed N_1, N_2 the occupancy probability tends to zero as L increases because the larger L the larger the most probable favored occupancy numbers. For the same excluded volume, $\sum_\alpha N_\alpha a_\alpha$, and for a given pore width, L, occupancies of fewer total particles, $\sum_\alpha N_\alpha$, tend to be favored. For example, with $L = 6a_1$ and $\sum_\alpha N_\alpha a_\alpha = 5a_1$, $\mathcal{P}_{0,2} = 0.37$ whereas $\mathcal{P}_{5,0} = 0.02$.

TABLE 10.1 Probabilities of Permissible States, $\mathcal{P}_{N_1,N_2}$, for a Binary Fluid ($a_2 = 2.5a_1$) Confined Between Hard Walls of Separation L (Given by Top Row of Numbers in Units of a_1)

		$\mathcal{P}_{N_1,N_2}$ (*Length*)											
(N_1, N_2)	$\sum N_i d_i$	*2.0*	*2.5*	*3.0*	*3.5*	*4.0*	*4.5*	*5.0*	*5.5*	*6.0*	*6.5*	*7.0*	*7.5*
(0, 0)	0.0	0.03	0.01	0.00	0.00	0.00	0.00	0.00	0.00	0.00	0.00	0.00	0.00
(1, 0)	1.0	0.97	0.28	0.03	0.00	0.00	0.00	0.00	0.00	0.00	0.00	0.00	0.00
(2, 0)	2.0	–	0.71	0.24	0.23	0.06	0.03	0.01	0.00	0.00	0.00	0.00	0.00
(0, 1)	2.5	–	–	0.73	0.61	0.13	0.05	0.02	0.00	0.00	0.00	0.00	0.00
(3, 0)	3.0	–	–	–	0.13	0.15	0.14	0.09	0.03	0.01	0.00	0.00	0.00
(1, 1)	3.5	–	–	–	–	0.66	0.76	0.44	0.12	0.05	0.02	0.00	0.00
(4, 0)	4.0	–	–	–	–	–	0.02	0.08	0.06	0.06	0.03	0.01	0.01
(2, 1)	4.5	–	–	–	–	–	–	0.37	0.45	0.42	0.20	0.09	0.04
(0, 2)	5.0	–	–	–	–	–	–	–	0.34	0.37	0.17	0.07	0.03
(5, 0)	5.0	–	–	–	–	–	–	–	0.00	0.02	0.03	0.03	0.02
(3, 1)	5.5	–	–	–	–	–	–	–	–	0.08	0.26	0.28	0.23
(1, 2)	6.0	–	–	–	–	–	–	–	–	–	0.29	0.50	0.43
(6, 0)	6.0	–	–	–	–	–	–	–	–	–	0.00	0.00	0.01
(4, 1)	6.5	–	–	–	–	–	–	–	–	–	–	0.01	0.11
(2, 2)	7.0	–	–	–	–	–	–	–	–	–	–	–	0.12
(7, 0)	7.0	–	–	–	–	–	–	–	–	–	–	–	0.00

(–) Nonpermissible Configurations.

Of special importance for multicomponent fluid separations is the adsorptive selectivity of the pores to each species. If x_α^b is the mole fraction of species α in the bulk phase, then the adsorptive selectivity of a pore of width L is measured by the difference between x_α^b and the pore average mole fraction $\overline{x}_\alpha$,

$$\overline{x}_\alpha = \frac{\overline{N}_\alpha}{\sum_\beta \overline{N}_\beta}. \tag{10.1.27}$$

The pore average mole fractions of component 1 of a binary hard-rod mixture are given in Figure 10.6 as functions of pore width for several mole fractions of the bulk fluid in equilibrium with the confined fluid. The bulk pressure is $\beta P a_1 = 3$ in all cases and the two cases, $a_2 = 1.5a$ and $a_2 = 2.5a$, are considered. The chemical potentials μ_1 and μ_2 at each concentration are computed from eq. (10.1.16) using $\beta P a_1 = 3$ and the mole fractions of the bulk phase.

Figure 10.6 illustrates that, depending on pore width, confinement in ultranarrow pores can enrich or deplete a particular component. The high specificity of zeolites, periodic nanoporous solids used in separations, may have qualitative origins similar to the excluded volume effects captured by the one-dimensional hard-rod model.

The disjoining pressure Π versus pore width L is shown in Figure 10.7 for hard-rod mixtures in equilibrium with bulk phases at pressure $\beta P_B a_1 = 3$ and at various bulk concentration ($a_2 = 1.5a$,

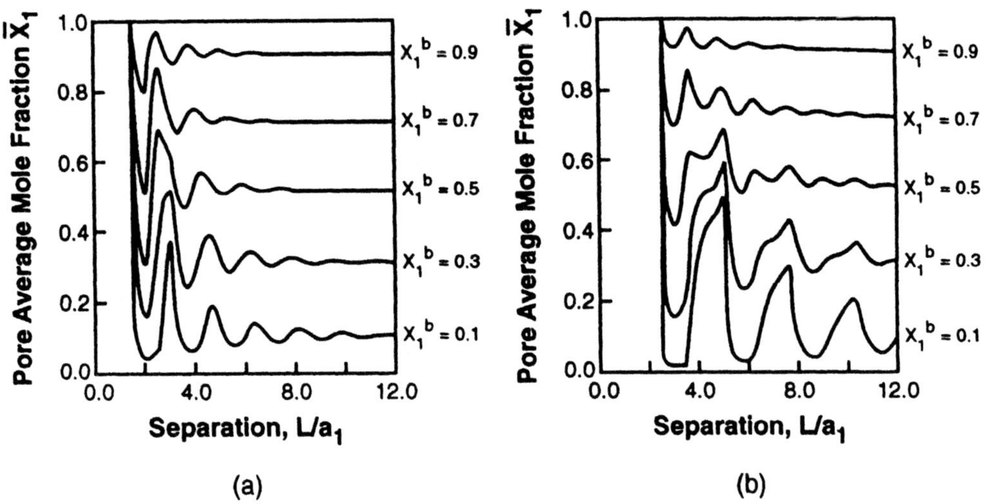

Figure 10.6
Pore average mole fraction $\bar{x}_1$ of component 1 is a function of pore width L for several bulk phase mole fractions in (a) a binary mixture with $a_2 = 1.5a$ and (b) a binary mixture with $a_2 = 2.5a_1$. $\beta P_B a_1 = 3$. (Vanderlick et al., 1989).

in the case presented). The oscillations in the disjoining pressure are more complex in the binary mixture than in the pure fluid. Furthermore, certain minima and maxima in the disjoining pressure curve for the mixture would be missed by the surface forces apparatus (see Vanderlick, 1988). The apparatus detects successive maxima only if the magnitudes of the successive maxima increase with decreasing wall separation (pore width) L. Thus, for the $x_1^b = 0.25$ case in Figure 10.7, the third maximum from the left would be missed by the apparatus. Correspondingly, if, after the mica cylinders approach to a separation lying on the second (from the left) pressure peak, the mica cylinders are slowly moved away from each other they will spring far apart upon reaching the minimum to the left of the maximum missed by the surface forces apparatus. Thus, the minimum to the right of the missed maximum is also inaccessible to the surface forces apparatus. Several maxima and minima inaccessible to the surface forces apparatus appear in the various panels of Figure 10.7.

The thermodynamic functions of hard-rod fluids confined between hard walls are relatively easily computed from the grand partition function or the occupancy distribution $\mathcal{P}_N$ derived in this section. The local density distribution functions of this system can also be computed from the grand partition function. However, to obtain tractable formulas for the density distribution functions we need to consider hard-rod fluids in the presence of external fields. Thus, in the next section, the theory of hard-rod mixtures in the presence of arbitrary confining external potentials will be presented. Except for the special case considered in this section in which the external potential is represented by a pair of hard walls, the general theory results in a nonlinear set of integral equations for the particle density distributions $n_\alpha(x)$ that must be solved to compute thermodynamic properties such as pressure, free energy, and mean pore densities.

Figure 10.7
Disjoining pressure versus pore width for a binary hard-rod fluid. The bulk pressure is $\beta P_B a_1 = 3, a_2 = 1.5a_1$, and the bulk mole fractions are given in the graphs (Vanderlick et al., 1989).

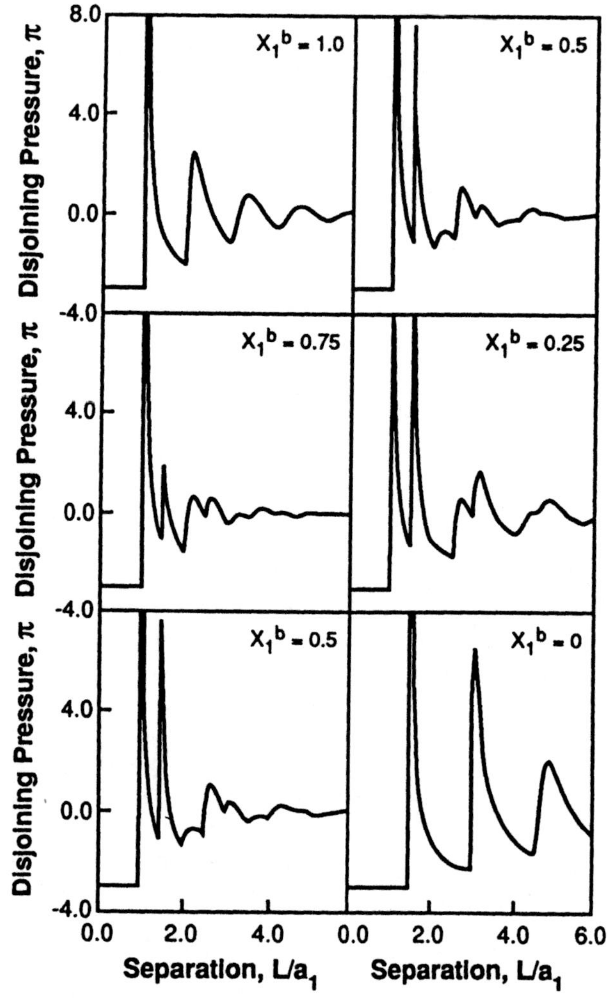

10.2 Density Distribution Functions for Hard-Rod Fluids in Arbitrary External Fields

The theory presented in this section is based on methods introduced by Percus (1976) and Vanderlick, et al. (1989). To derive formulas for the density distribution functions, we need to consider the partition function of hard rods in the presence of external potentials $v_\alpha(x)$. For this purpose we consider a hard-rod mixture of **N** particles confined between a pair of hard walls located at y and z $(z > y)$, respectively. The external potentials $v_\alpha(x_i)$ are otherwise arbitrary.

To obtain closed expressions for the density distributions it is convenient to transform the grand canonical partition function. First, for a given **N** we note that the canonical partition function can be expressed as

$$Z_N(y,z) = \int_y^z \cdots \int_y^z e^{-\sum_{i>j=1}^N \beta u_{\alpha_i\alpha_j}(x_i - x_j) - \sum_{i=1}^N \beta v_{\alpha_i}(x_i)} dx_1 \cdots dx_N, \tag{10.2.1}$$

where $u_{\alpha_i\alpha_j}$ is the pair potential between particles species α_i and α_j, respectively, and v_{α_i} is the external potential on a particle of species α_i. The grand partition function of the system is

$$\Xi(y,z) = \sum_{N_1=0}^{\infty} \cdots \sum_{N_C=0}^{\infty} \prod_{i=1}^{N} \frac{\zeta_{\alpha_i}}{\prod_{\alpha=1} N_\alpha!} \int_y^z \cdots \int_y^z e^{-\sum_{i>j=1}^N \beta u_{(x_i - x_j)} - \sum_{i=1}^N \beta v_{\alpha_i}(x_i)} dx_1 \cdots dx_N. \tag{10.2.2}$$

Next, it is useful to introduce in eq. (10.2.1) the summation transformation

$$\sum_{N_1=0}^{\infty} \cdots \sum_{N_c=0}^{\infty} (\cdots) \to \sum_{N=0}^{\infty} \sum_{\alpha_1=1}^{c} \cdots \sum_{\alpha_N=1}^{c} (\cdots). \tag{10.2.3}$$

A typical term in the left-hand side (lhs) summation has N_1 particles of species 1, N_2 of species 2, etc. If on the right-hand side (rhs) N is set, then there are $N!/\prod_{\alpha=1}^{c} N_\alpha!$ ways that the values of $\alpha_1, \ldots, \alpha_N$ can be chosen to coincide with the occupancy distribution $\{N_1, N_2, \ldots, N_c\}$. Thus, the transformation at eq. (10.2.3) is

$$\sum_{N_1=0}^{\infty} \cdots \sum_{N_c=0}^{\infty} \frac{N!}{\prod_\alpha N_\alpha!} (\cdots) = \sum_{N=0}^{\infty} \sum_{\alpha_1=1}^{c} \cdots \sum_{\alpha_N=1}^{c} (\cdots), \tag{10.2.4}$$

which leads to the result

$$\Xi(y,z) = \sum_{N=0}^{\infty} \sum_{\alpha_1=1}^{c} \cdots \sum_{\alpha_N=1}^{c} \frac{\prod_{i=1}^{N} \zeta_{\alpha_i}}{N!} \int_y^z \cdots \int_y^z e^{-\beta u^N} dx_1 \cdots dx_N. \tag{10.2.5}$$

Next we note that for a given N, the quantity

$$\sum_{\alpha_1=1}^{c} \cdots \sum_{\alpha_N=1}^{c} \prod_{i=1}^{N} \zeta_{\alpha_i} e^{-\beta u^N} \tag{10.2.6}$$

is invariant to interchange of x_i and x_j, and so by the coordinate transformation used in Chapter 5, the unconstrained integration variables $x_1, \ldots, x_N$ can be replaced by the ordered variables $x_N \geq x_{N-1} \geq \cdots \geq x_1$ to obtain the expression

$$\Xi(y,z) = \sum_{N=0}^{\infty} \sum_{\alpha_1=1}^{c} \cdots \sum_{\alpha_N=1}^{c} \prod_{i=1}^{N} \zeta_{\alpha_i} \int_y^z dx_N \int_y^{x_N} dx_{N-1} \cdots \int_y^{x_2} dx_1 e^{-\beta u^N}, \tag{10.2.7}$$

or, noting that the center of the left-most particle can approach the wall only as close as $y + a_{\alpha_1}/2$ and the center of the right-most particle can approach the wall only as close as $z - a_{\alpha_N}/2$, we find

$$\Xi(y, z) = \sum_{N=0}^{\infty} \sum_{\alpha_1=1}^{c} \cdots \sum_{\alpha_N=1}^{c} \prod_{i=1}^{N} \zeta_{\alpha_i} \int_{y+a_{\alpha_1}/2}^{z-a_{\alpha_N}/2} dx_N \cdots \int_{y+a_{\alpha_1}/2}^{x_2} dx_1 e^{-\sum_{i>j=1}^{N} \beta u_{\alpha_i \alpha_j} - \sum_{i=1}^{N} \beta v_{\alpha_i}}$$
$$= \sum_{N=0}^{\infty} X_N(y, z). \tag{10.2.8}$$

The density distribution $n_\alpha(x)$ can be computed from the functional derivative

$$n_\alpha(x) = \frac{\delta \ln \Xi(y, z)}{\delta(-\beta v_\alpha(x))}. \tag{10.2.9}$$

Using eq. (10.2.8) and the relation $\delta e^{-\beta v_{\alpha_k}(x_k)}/\delta(-\beta v_\alpha(x)) = e^{-\beta v_\alpha(x)} \delta_{\alpha\alpha_k} \delta(x_k - x)$, we obtain

$$\frac{\delta \Xi(y, z)}{\delta(-\beta v_\alpha(x))} = \zeta_\alpha e^{-\beta v_\alpha(x)} \sum_{N=1}^{\infty} \sum_{k=1}^{N} K_{N-k} L_{k-1}, \tag{10.2.10}$$

where

$$K_{N-k} \equiv \sum_{\alpha_{k+1}=1}^{c} \cdots \sum_{\alpha_N=1}^{c} \prod_{j=k+1}^{N} \zeta_{\alpha_j} \int_{x+a_{\alpha_1}/2}^{z-a_{\alpha_N}/2} dx_N \cdots \int_{x+a_{\alpha_1}/2}^{x_{k+2}} dx_{k+1} e^{-\beta u_{\alpha\alpha_{k+1}}(x_{k+1}-x)}$$
$$\times e^{-\sum_{i>j=k+1}^{N} \beta u_{\alpha_i \alpha_j}(x_i - x_j) - \sum_{i=k+1}^{N} \beta v_{\alpha_i}(x_i)} \tag{10.2.11}$$

and

$$L_{k-1} \equiv \sum_{\alpha_1}^{c} \cdots \sum_{\alpha_{k-1}}^{c} \prod_{j=1}^{k-1} \zeta_{\alpha_j} \int_{y+a_{\alpha_1}/2}^{x} dx_{k-1} \cdots \int_{y+a_{\alpha_1}/2}^{x_2} dx_1 e^{-\beta u_{\alpha\alpha_{k-1}}(x - x_{k-1})}$$
$$\times e^{-\sum_{i>j=1}^{k-1} \beta u_{\alpha_i \alpha_j}(x_i - x_j) - \sum_{i=1}^{k-1} \beta v_\alpha(x_i)}. \tag{10.2.12}$$

The factor $e^{-\beta u_{\alpha\alpha_{k+1}}(x_{k+1}-x)}$ in K_{N-k} is zero unless $x_{k+1} > x + a_{\alpha\alpha_{k+1}}$. Thus, the lower limit of integration in K_{N-k} is $x + a_\alpha/2 + a_{\alpha_{k+1}}/2$ instead of $a_{\alpha_1}/2$. This means that K_{N-k} belongs to a system with the left-hand wall at $x + a_\alpha/2$. Furthermore, the dummy variables of integration in K_{N-k} can be transformed so that K_{N-k} reads

$$K_{N-k} = \sum_{\alpha_1=1}^{c} \cdots \sum_{\alpha_{N-k}=1}^{c} \prod_{j=1}^{N-k} \zeta_{\alpha_j} \int_{x+a_\alpha/2+a_{\alpha_1}/2}^{z-a_{\alpha_{N-k}}/2} dx_{N-k} \cdots \int_{x+a_\alpha/2+a_{\alpha_1}/2}^{x_2} dx_1$$
$$\times e^{-\sum_{i>j=1}^{N-k} \beta u_{\alpha_i \alpha_j}(x_i - x_j) - \sum_{i=1}^{N-k} \beta v_{\alpha_i}(x_i)}$$
$$= X_{N-k}(x + a_\alpha/2, z). \tag{10.2.13}$$

The factor $e^{-\beta u_{\alpha\alpha_{k-1}}(x-x_{k-1})}$ in L_{k-1} is zero unless $x - x_{k-1} > a_{\alpha\alpha_{k-1}}$, that is, the upper limit on x_{k-1} is $x - a_\alpha/2 - a_{\alpha_{k-1}}/2$. This means that L_{k-1} belongs to a system with the right-hand wall at $x - a_\alpha/2$. Thus it follows that

$$L_{k-1} = X_{k-1}(y, x - a_\alpha/2). \tag{10.2.14}$$

Combining eqs. (10.2.8), (10.2.10), (10.2.13), and (10.2.14), we find

$$\begin{aligned}\frac{\delta \Xi(y,z)}{\delta(-\beta v_\alpha(x))} &= \zeta_\alpha e^{-\beta v_\alpha(x)} \sum_{N=1}^{\infty}\sum_{k=1}^{N} X_{k-1}(y, x - a_\alpha/2) X_{N-k}(x + a_\alpha/2, z) \\ &= \zeta_\alpha e^{-\beta v_\alpha(x)} \sum_{N'=0}^{\infty}\sum_{k'=0}^{\infty} X_{k'}(y, x - a_\alpha/2) X_{N'}(x + a_\alpha/2, z) \qquad (10.2.15) \\ &= \zeta_\alpha e^{-\beta v_\alpha(x)} \Xi(y, x - a_\alpha/2)\Xi(x + a_\alpha/2, z).\end{aligned}$$

This result is obtained by transforming the order of summation as follows

$$\sum_{N=1}^{\infty}\sum_{k=1}^{N} f_{N-k,k-1} = \sum_{k=1}^{\infty}\sum_{N=k}^{\infty} f_{N-k,k-1} = \sum_{k'=0}^{\infty}\sum_{N'=0}^{\infty} f_{N',k'}. \tag{10.2.16}$$

The final result for the density distribution of species α is

$$n_\alpha(x) = \zeta_\alpha e^{-\beta v_\alpha(x)} \frac{\Xi(y, x - a_\alpha/2)\Xi(x + a_\alpha/2, z)}{\Xi(y,z)}, \qquad y < x < z. \tag{10.2.17}$$

The derivation of eq. (10.2.17) was tedious but the result is simple! Moreover, higher order density distribution functions can now be derived easily with the aid of eq. (10.2.15).

For example, because

$$n_{\alpha\beta}^{(2)}(x_1, x_2) = \frac{\delta n_\alpha(x_1)}{\delta(-\beta v_\beta(x_2))} + n_\alpha(x_1) n_\beta(x_2), \qquad x_2 \neq x_1, \tag{10.2.18}$$

the result for $y < x_1, x_2 < z$,

$$\begin{aligned} n_{\alpha\beta}^{(2)}(x_1, x_2) &= \zeta_\alpha \zeta_\beta e^{-\beta v_\alpha(x_1) - \beta v_\beta(x_2)} \\ &\quad \frac{\Xi(y, x_1 - a_\alpha/2)\Xi(x_1 + a_\alpha/2, x_2 - a_\beta/2)\Xi(x_2 + a_\beta/2, z)}{\Xi(y,z)}, \qquad x_2 > x_1, \\ &= \zeta_\alpha \zeta_\beta e^{-\beta v_\alpha(x_1) - \beta v_\beta(x_2)} \\ &\quad \frac{\Xi(y, x_2 - a_\beta/2)\Xi(x_2 + a_\beta/2, x_1 - a_\alpha/2)\Xi(x_1 + a_\alpha/2, z)}{\Xi(y,z)}, \qquad x_2 < x_1,\end{aligned} \tag{10.2.19}$$

is easily obtained by functional differentiation of eq. (10.2.17) with the aid of eq. (10.2.15). The general result for the s-body distribution is

$$n_{\alpha}^{(s)}(x_1, \ldots, x_s) = \prod_{i=1}^{s} \zeta_{\alpha_i} e^{-\beta v_{\alpha_i}(x_i)} \Xi(y, x_1 - a_{\alpha_1}/2) \times \frac{\prod_{i=2}^{s} \Xi(x_{i-1} + a_{\alpha_{i-1}}/2, x_i - a_{\alpha_i}/2) \Xi(x_s + a_{\alpha_s}/2, z)}{\Xi(y, z)} \tag{10.2.20}$$

where $x_1 < x_2 \cdots < x_s$ and α denotes the set $\{\alpha_1, \alpha_2, \ldots, \alpha_s\}$.

Although eqs. (10.2.17), (10.2.19), and (10.2.20) are rigorous formulas for the density distribution functions for a multicomponent hard-rod fluid confined between a pair of walls located at y and z ($y < z$) and further subjected to an arbitrary external field, there still remains the task of evaluating the partition function $\Xi(y, z)$. In the case of hard rods between hard walls without other external forces, we have already evaluated $\Xi(y, z)$. By simply shifting the origin of the coordinate system $\Xi(y, z) = \Xi(0, z - y)$ when no external fields are present, the expression for $\Xi(y, z)$ can be obtained from eq. (10.1.10) by substituting $z - y$ for L, namely

$$\begin{aligned}\Xi(y, z) &= \Xi_{z-y} \\ &= \sum_{N_1=0}^{\infty} \cdots \sum_{N_c=0}^{\infty} \prod_{\alpha=1}^{c} \left(\frac{\zeta_{\alpha}^{N_{\alpha}}}{N_{\alpha}!} \right) (z - y - \sum_{\alpha} N_{\alpha} a_{\alpha})^{N} \eta \left(z - y - \sum_{\alpha} N_{\alpha} a_{\alpha} \right).\end{aligned} \tag{10.2.21}$$

In this case all of the Ξ's appearing in the density distribution function can be computed directly from this formula using appropriate values of y and z.

Let us now consider further the case of arbitrary external fields. Without loss of generality we can let $y = -\infty$ and $z = +\infty$. We will still require confinement of the system in the sense that $v_{\alpha}(x) \to \infty$ as $x \to \pm\infty$. In this case the grand canonical partition function Ξ is equal to $\Xi(-\infty, \infty)$ and eq. (10.2.17) becomes

$$n_{\alpha}(x) = \zeta_{\alpha} e^{-\beta v_{\alpha}(x)} \frac{\Xi(-\infty, x - a_{\alpha}/2) \Xi(x + a_{\alpha}/2, \infty)}{\Xi}. \tag{10.2.22}$$

From the definition of $\Xi(y, z)$ it follows

$$\Xi(-\infty, x) = \sum_{N=0}^{\infty} \sum_{\alpha_1=1}^{c} \cdots \sum_{\alpha_N=1}^{c} \prod_{i=1}^{N} \zeta_{\alpha_i} \int_{-\infty}^{x-a_{\alpha_N}/2} dx_N \int_{-\infty}^{x_N - a_{\alpha_N \alpha_{N-1}}} dx_{N-1} \cdots \int_{-\infty}^{x_2 - a_{\alpha_2 \alpha_1}} dx_1 \times e^{-\sum_{i=1}^{N} \beta v_{\alpha_i}(x_i)}, \tag{10.2.23}$$

where the upper limits of the integrals reflect the fact that the centers of neighboring hard rods i and $i + 1$ cannot approach more closely than $a_{\alpha_{i+1}\alpha_i}$. With these nearest neighbor constraints taken account of through the limits of integration, all of the hard-rod Boltzmann factors, $e^{-\beta u_{\alpha_i \alpha_j}}$, are equal to one.

By summing over all species but the Nth one, α_N, which we denote as γ, we can rewrite eq. (10.2.23) as

$$\Xi(-\infty, x) = 1 + \sum_{\gamma=1}^{c} \zeta_\gamma \int_{-\infty}^{x-a_\gamma/2} dx' e^{-\beta v_\gamma(x')} \sum_{N'=0}^{\infty} \sum_{\alpha_1=1}^{c} \cdots \sum_{\alpha_{N'}=1}^{c} \prod_{i=1}^{N'} \zeta_{\alpha_i} \int_{-\infty}^{x'-a_{\gamma\alpha_{N'}}} dx_{N'} \cdots \int_{-\infty}^{x_2-a_{\alpha_2\alpha_1}} dx_1 e^{-\sum_{i=1}^{N'} \beta v_{\alpha_i}(x_i)} \tag{10.2.24}$$

$$= 1 + \sum_{\gamma=1}^{c} \zeta_\gamma \int_{-\infty}^{x-a_\gamma/2} dx' e^{-\beta v_\gamma(x')} \Xi(-\infty, x' - a_\gamma/2).$$

Equation (10.2.24) is a recursion formula for $\Xi(-\infty, x)$. Using eq. (10.2.22) to eliminate $\zeta_\gamma e^{-\beta v_\gamma}$ in eq. (10.2.24), we obtain the relation

$$\Xi(-\infty, x) = 1 + \Xi \sum_{\gamma=1}^{c} \int_{-\infty}^{x-a_\gamma/2} dx' \frac{n_\gamma(x')}{\Xi(x' + a_\gamma/2, \infty)}$$
$$= 1 + \Xi \int_{-\infty}^{x} dx'' \frac{n_-(x'')}{\Xi(x'', \infty)}, \tag{10.2.25}$$

where

$$n_-(x) \equiv \sum_{\gamma=1}^{c} n_\gamma(x - a_\gamma/2). \tag{10.2.26}$$

An expression conjugate to eq. (10.2.25) can be derived for $\Xi(x, \infty)$. Determination of the recursion relation is, however, a bit more subtle. Recall that $\Xi(x, \infty)$ is the grand canonical ensemble partition function for a system of hard rods subjected to arbitrary external forces and confined on the left by a hard wall at x. Going back to the configuration partition function

$$Z_N = \int_x^\infty \cdots \int_x^\infty e^{-\beta u^N(x_1,\cdots,x_N)} dx_1 \cdots dx_N, \tag{10.2.27}$$

we note that the variables of integration can be reordered such that $x_1 \geq x_2 \cdots \geq x_N$ (instead of $x_1 \leq x_2 \cdots \leq x_N$ as done before) with the result

$$Z_N = N! \int_x^\infty dx_1 \int_{x_1}^\infty dx_2 \cdots \int_{x_{N-1}}^\infty e^{-\beta u^N(x_1,\cdots,x_N)}. \tag{10.2.28}$$

With this set of variables, $\Xi(x, \infty)$ becomes

$$\Xi(x,\infty) = \sum_{N=0}^{\infty}\sum_{\alpha_1=1}^{c}\cdots\sum_{\alpha_N=1}^{c}\prod_{i=1}^{N}\zeta_{\alpha_i}\int_{x+a_{\alpha_1}/2}^{\infty}dx_1\int_{x_1+a_{\alpha_2\alpha_1}}^{\infty}dx_2\cdots\int_{x_{N-1}+a_{\alpha_N\alpha_{N-1}}}^{\infty}dx_N e^{-\sum_{i=1}^{N}\beta v_{\alpha_i}(x_i)} \quad (10.2.29)$$

where again the hard-rod interactions have been accounted for through the limits of integration.

Summation over all species except α_1, which we denote as γ, leads to the following recursion formula

$$\Xi(x,\infty) = 1 + \sum_{\gamma=1}^{c}\zeta_\gamma\int_{x+a_\gamma/2}^{\infty}dx' e^{-\beta v_\gamma(x')}\sum_{N'=0}^{\infty}\sum_{\alpha_1=1}^{c}\cdots\sum_{\alpha_{N'}=1}^{c}\prod_{i=1}^{N'}\zeta_{\alpha_i}$$
$$\times\int_{x'+a_{\alpha_1\gamma}}^{\infty}dx_1\int_{x_1+a_{\alpha_2\alpha_1}}^{\infty}dx_2\cdots\int_{x_{N-1}+a_{\alpha_N\alpha_{N'-1}}}dx_{N'}e^{-\sum_{i=1}^{N'}\beta v_{\alpha_i}(x_i)} \quad (10.2.30)$$
$$= 1 + \sum_{\gamma=1}^{c}\zeta_\gamma\int_{x+a_\gamma/2}^{\infty}dx' e^{-\beta v_\gamma(x')}\Xi(x'+a_\gamma/2,\infty).$$

Use of eq. (10.2.22) to eliminate $\zeta_\gamma e^{-\beta v_\gamma}$ and the change of variable $x'' = x' - a_\gamma/2$ leads to the relation

$$\Xi(x,\infty) = 1 + \Xi\int_{x}^{\infty}dx''\frac{n_+(x'')}{\Xi(-\infty,x'')}, \quad (10.2.31)$$

where

$$n_+(x) \equiv \sum_{\gamma=1}^{c}n_\gamma(x+a_\gamma/2). \quad (10.2.32)$$

Equation (10.2.31) is the conjugate of eq. (10.2.25).

Equations (10.2.25) and (10.2.31) provide a closed set from which to determine $\Xi(-\infty, x)$ and $\Xi(x, \infty)$ as functionals of the density distributions. The derivatives of these equations are

$$\frac{d\Xi(-\infty,x)}{dx} = \Xi\frac{n_-(x)}{\Xi(x,\infty)} \quad (10.2.33)$$

and

$$\frac{d\Xi(x,\infty)}{dx} = -\Xi\frac{n_+(x)}{\Xi(-\infty,x)}, \quad (10.2.34)$$

that combine to give

$$\frac{d}{dx}[\Xi(-\infty,x)\Xi(x,\infty)] = \Xi[n_-(x) - n_+(x)]. \quad (10.2.35)$$

Because $\lim_{x\to-\infty} \Xi(-\infty, x) = 1$, $\lim_{x\to-\infty} \Xi(x, \infty) = \Xi$, and $\lim_{x\to-\infty} n_\pm(x) = 0$, eq. (10.2.35) integrates to

$$\Xi(-\infty, x)\Xi(x, \infty) = \Xi \left\{ 1 + \int_{-\infty}^{x} [n_-(y) - n_+(y)]dy \right\}. \tag{10.2.36}$$

Substitution of $\Xi(x, \infty)$ from this expression into eq. (10.2.33) and integration yields

$$\Xi(-\infty, x) = \exp \int_{-\infty}^{x} \frac{n_-(y)}{1 + \int_{-\infty}^{y}[n_-(z) - n_+(z)]dz} dy. \tag{10.2.37}$$

Similarly the combination of eqs. (10.2.34) and (10.2.36) [with $\lim_{x\to-\infty} \Xi(x, \infty) = \Xi$] leads to

$$\Xi(x, \infty) = \Xi \exp \int_{-\infty}^{x} \frac{-n_+(y)}{1 + \int_{-\infty}^{y} [n_-(z) - n_+(z)]dz} dy. \tag{10.2.38}$$

The partition function is then given by

$$\Xi = \exp \int_{-\infty}^{\infty} \frac{n_-(y)}{1 + \int_{-\infty}^{y}[n_-(z) - n_+(z)]dz}. \tag{10.2.39}$$

Combination of eqs. (10.2.22), (10.2.37), and (10.2.38) leads to

$$\begin{aligned} \beta[\mu_\alpha - v_\alpha(x)] - \ln \Lambda_\alpha n_\alpha(x) = &- \int_{-\infty}^{x-a_\alpha/2} \frac{n_-(y)}{1 + \int_{-\infty}^{y}[n_-(z) - n_+(z)]dz} dy \\ &+ \int_{-\infty}^{x+a_\alpha/2} \frac{n_+(y)}{1 + \int_{-\infty}^{y}[n_-(z) - n_+(z)]dz} dy. \end{aligned} \tag{10.2.40}$$

The rhs of eq. (10.2.40) can be rearranged into a local form. First it can be rearranged as follows:

$$\begin{aligned} \text{rhs} &= - \int_{y=-\infty}^{y=x+a_\alpha/2} d \ln \left\{ 1 + \int_{-\infty}^{y} [n_-(z) - n_+(z)]dz \right\} \\ &\quad + \int_{x-a_\alpha/2}^{x+a_\alpha/2} \frac{n_-(y)}{1 + \int_{-\infty}^{y}[n_-(z) - n_+(z)]} dy \\ &= - \ln \left\{ 1 + \int_{-\infty}^{x+a_\alpha/2} [n_-(z) - n_+(z)]dz \right\} \end{aligned} \tag{10.2.41}$$

$$+\int_{x-a_\alpha/2}^{x+a_\alpha/2} \frac{n_-(y)}{1+\int_{-\infty}^{y}[n_-(z)-n_+(z)]dz}dy.$$

By transformation of dummy variables, we find

$$\int_{-\infty}^{y}[n_-(z)-n_+(z)]dz = \sum_\beta \left[\int_{-\infty}^{y} n_\beta\left(z-\frac{a_\beta}{2}\right)dz - \int_{-\infty}^{y} n_\beta(z+\frac{a_\beta}{2}a)dz\right]$$
$$= -\sum_\beta \int_{y-a_\beta/2}^{y+a_\beta/2} n_\beta(z)dz \qquad (10.2.42)$$

and

$$\int_{x-a_\alpha/2}^{x+a_\alpha/2} \frac{n_-(y)}{1+\int_{-\infty}^{y}[n_-(z)-n_+(z)}dy = \sum_\gamma \int_{x-a_\alpha/2}^{x+a_\alpha/2} \frac{n_\gamma(y-\frac{a_\gamma}{2})dy}{1+\int_{-\infty}^{y}[n_-(z)-n_+(z)]dz}$$
$$= \sum_\gamma \int_{x-a_{\alpha\gamma}}^{x+\Delta_{\alpha\gamma}} \frac{n_\gamma(y)}{1-\sum_\beta \int_{y+\Delta_{\gamma\beta}}^{y+a_{\gamma\beta}} n_\beta(z)dz}dy, \qquad (10.2.43)$$

where

$$\Delta_{\alpha\gamma} \equiv (a_\alpha - a_\gamma)/2 \quad \text{and} \quad a_{\alpha\gamma} \equiv (a_\alpha + a_\gamma)/2. \qquad (10.2.44)$$

The combination of eqs. (10.2.40)–(10.2.44) yields the equation

$$\beta\mu_\alpha = \beta v_\alpha(x) + \ln\left[\frac{n_\alpha(x)\Lambda_\alpha}{1-\sum_\beta \int_{x+\Delta_{\alpha\beta}}^{x+a_{\alpha\beta}} n_\beta(z)dz}\right]$$
$$+\sum_\gamma \int_{x-a_{\alpha\gamma}}^{x+\Delta_{\alpha\gamma}} \frac{n_\gamma(y)}{1-\sum_\beta \int_{y+\Delta_{\gamma\beta}}^{y+a_{\gamma\beta}} n_\beta(z)dz}dy, \qquad (10.2.45)$$

Comment
Equation (10.2.45) is the key result in this section. It provides a closed solution to the problem of finding density profiles and all correlation functions of a hard rod mixture in the presence of an arbitrary external field.

which is the desired local density functional equation determining the density distribution of a hard-rod mixture in an arbitrary external field.

Because $\mu_\alpha = \delta F/\delta n_\alpha(x)$, one can construct the Helmholtz energy as a functional of densities by inspection. The result is

$$\beta F = \sum_\alpha \int n_\alpha(x)[\beta v_\alpha(x) + \ln(\Lambda_\alpha n_\alpha(x)) - 1]dx$$
$$-\int dx \sum_\alpha \frac{1}{2}\left[n_\alpha\left(x+\frac{a_\alpha}{2}\right) + n_\alpha\left(x-\frac{a_\alpha}{2}\right)\right]\ln\left[1-\sum_\beta \int_{x-a_\beta/2}^{x+a_\beta/2} n_\beta(y)dy\right]. \qquad (10.2.46)$$

The pair direct correlation function, $C_{\alpha\beta}^{(2)}(x, y)$, can be obtained from the functional derivative

$$C_{\alpha\beta}^{(2)}(x, y) = \frac{\delta\{\beta[v_\alpha(x) - \mu_\alpha] + \ln n_\alpha(x)\}}{\delta n_\beta(y)}. \tag{10.2.47}$$

The result is

$$\begin{aligned} C_{\alpha\beta}^{(2)}(x, y) = &- \frac{\eta(x + a_{\alpha\beta} - y)\eta(y - x - \Delta_{\alpha\beta})}{1 - \sum_\gamma \int_{x+\Delta_{\alpha\gamma}}^{x+a_{\alpha\gamma}} n_\gamma(z)dz} \\ &- \frac{\eta(x + \Delta_{\alpha\beta} - y)\eta(y - x + a_{\alpha\beta})}{1 - \sum_\gamma \int_{x+\Delta_{\beta\gamma}}^{x+a_{\beta\gamma}} n_\gamma(z)dz} \\ &- \sum_\gamma \int_{x-a_{\alpha\gamma}}^{x+\Delta_{\alpha\gamma}} n_\gamma(z) \frac{\eta(z + a_{\alpha\beta} - y)\eta(y - z - \Delta_{\gamma\beta})}{\left[1 - \sum_\lambda \int_{z+\Delta_{\gamma\lambda}}^{z+a_{\gamma\lambda}} n_\lambda(w)dw\right]^2} dz. \end{aligned} \tag{10.2.48}$$

If follows from the step functions in eq. (10.2.48) that

$$C_{\alpha\beta}^{(2)}(x, y) = 0, \quad |y - x| > a_{\alpha\beta}. \tag{10.2.49}$$

Exercise
The direct correlation function $C^{(2)}(x - y)$ of a homogeneous, one-component hard rod fluid is

$$-\frac{\eta(a-|x-y|)}{1-na}\left[1 - \frac{(|x-y|-a)n}{1-na}\right]$$

which is negative for all physically allowed densities, i.e., $na < 1$.

Thus, the Percus–Yevick or mean spherical approximation that $C_{\alpha\beta}^{(2)}$ vanishes outside the range of the hard-rod pair interaction is rigorously valid for hard-rod mixtures.

Given the direct pair correlation functions $C_{\alpha\beta}^{(2)}$ and the density distribution functions n_α, the pair correlation functions can be computed from the Ornstein–Zernike equation. However, this route would still require solving a linear integral equation for $g_{\alpha\beta}^{(2)}$. The alternative is to compute $g_{\alpha\beta}^{(2)}$ directly from $n_{\alpha\beta}^{(2)}/n_\alpha n_\beta$, where $n_{\alpha\beta}^{(2)}$ is given by eq. (10.2.19). The combination of eqs. (10.2.17) and (10.2.19) yields

$$\begin{aligned} g_{\alpha\beta}^{(2)}(x_1, x_2) &= \frac{\Xi(y, z)\Xi(x_1 + a_\alpha/2, x_2 - a_\beta/2)}{\Xi(x_1 + a_\alpha/2, z)\Xi(y, x_2 - a_\beta/2)}, \quad x_2 > x_1 \\ &= \frac{\Xi(y, z)\Xi(x_2 + a_\beta/2, x_1 - a_\alpha/2)}{\Xi(x_2 + a_\beta/2, z)\Xi(y, x_1 - a_\alpha/2)}, \quad x_2 < x_1. \end{aligned} \tag{10.2.50}$$

For a hard-rod fluid confined by hard walls, we can set $y = 0$ and $z = L$ and calculate from eq. (10.2.21) the various Ξ's in eq. (10.2.50). In the general case [with $y = -\infty$, $z = +\infty$, $\Xi = \Xi(-\infty, \infty)$], eq. (10.2.50) combines with eqs. (10.2.37)–(10.2.39) to yield

$$\begin{aligned} g_{\alpha\beta}^{(2)}(x_1, x_2) &= \frac{\Xi(x_1 + a_\alpha/2, x_2 - a_\beta/2)\exp\left\{-\int_{x_1+a_\alpha/2}^{x_2-a_\beta/2} \frac{n_-(y)}{1+\int_{-\infty}^{y}[n_-(z)-n_+(z)]dz} dy\right\}}{1 + \int_{-\infty}^{x_1+a_\alpha/2}[n_-(z) - n_+(z)]dz}, \quad x_2 > x_1, \\ &= \frac{\Xi(x_2 + a_\beta/2, x_1 - a_\alpha/2)\exp\left\{-\int_{x_2+a_\beta/2}^{x_1-a_\alpha/2} \frac{n_-(y)}{1+\int_{-\infty}^{y}[n_-(z)-n_+(z)]dz} dy\right\}}{1 + \int_{-\infty}^{x_2+a_\beta/2}[n_-(z) - n_+(z)]dz}, \quad x_2 < x_1. \end{aligned} \tag{10.2.51}$$

Because

$$\int_{-\infty}^{y} [n_-(z) - n_+(z)]dz = \sum_{\gamma} \int_{y+a_\gamma/2}^{y-a_\gamma/2} n_\gamma(z)dz, \tag{10.2.52}$$

eq. (10.2.51) provides a local formula for the pair correlation function. However, evaluation of $\Xi(x_1 + a_\alpha/2, x_2 - a_\beta/2)$ or $\Xi(x_2 + a_\beta/2, x_1 - a_\alpha/2)$ is required to compute $g_{\alpha\beta}^{(2)}$ from eq. (10.2.51).

Because $\Xi(x_1 + a_\alpha/2, x_2 - a_\beta/2) = 1$ when $x_2 = x_1 + a_{\alpha\beta}$ and $\Xi(x_2 + a_\beta/2, x_1 - a_\alpha/2) = 1$ when $x_2 = x_1 - a_{\alpha\beta}$, it follows from eq. (10.2.51) that the contact value of the pair correlation function is

$$g_{\alpha\beta}^{(2)}(x_1, x_2 = x_1 + a_{\alpha\beta}) = \frac{1}{1 + \sum_\gamma \int_{x_1+a_{\alpha\gamma}}^{x_1+\Delta_{\alpha\gamma}} n_\gamma(z)dz} \tag{10.2.53}$$

and

$$g_{\alpha\beta}^{(2)}(x_1, x_2 = x_1 - a_{\alpha\beta}) = \frac{1}{1 + \sum_\gamma \int_{x_1-\Delta_{\alpha\gamma}}^{x_1-a_{\alpha\gamma}} n_\gamma(z)dz} \tag{10.2.54}$$

Let us close this section by considering a couple of special cases. First consider a semi-confined fluid, that is, a fluid with a hard wall at $y = 0$ and in contact with bulk fluid at large distance z from the wall. Moreover, the external potential will be assumed to be zero except to impose the hard wall at $y = 0$. In this case, the density distribution at eq. (10.2.17) becomes

$$\begin{aligned} n_\alpha(x) &= \lim_{z\to\infty} \zeta_\alpha \Xi(0, x - a_\alpha/2) \frac{\Xi(x + a_\alpha/2, z)}{\Xi(0, z)} \\ &= \zeta_\alpha \Xi_{x-a_\alpha/2} e^{-\beta P_B(x+a_\alpha/2)}, \end{aligned} \tag{10.2.55}$$

where Ξ_y is given by eq. (10.2.21) and the relationship [see eq. (2.6.25) in Chapter 3]

$$\Xi(x + a_\alpha/2, z)/\Xi(0, z) \to \frac{e^{\beta P_B(z-x-a_\alpha/2)}}{e^{\beta P_B z}} \to e^{-\beta P_B(x+a_\alpha/2)}, \quad z \to \infty \tag{10.2.56}$$

has been used. P_B is the pressure of the bulk phase far from the confining wall.

The other special case is the pair correlation function $g_{\alpha\beta}^{(2)}(x_1, x_2)$ for a homogeneous or bulk phase hard-rod fluid. For x_1 and x_2 far from confining hard walls, n_γ equals is bulk value n_γ^B. Thus, eq. (10.2.51) becomes

$$\begin{aligned} g_{\alpha\beta}^{(2)}(x_1, x_2) &= \Xi_{x_2-x_1-a_{\alpha\beta}} \frac{e^{-\beta P_B(x_2-x_1-a_{\alpha\beta})}}{1 - \sum_\gamma n_\gamma^B a_\gamma}, \quad x_2 > x_1 \\ &= \Xi_{x_1-x_2-a_{\alpha\beta}} \frac{e^{-\beta P_B(x_1-x_2-a_{\alpha\beta})}}{1 - \sum_\gamma n_\gamma^B a_\gamma}, \quad x_1 > x_2. \end{aligned} \tag{10.2.57}$$

where P_B is the bulk fluid pressure and n_γ^B is the bulk phase density of species γ. With the aid of eqs. (10.1.16) and (10.2.55), eq. (10.2.57) can be rearranged to

$$g_{\alpha\beta}^{(2)}(x_1, x_2) = \frac{n_\alpha(x_2 - x_1 - a_\beta/2)}{n_\alpha^B}, \quad x_2 > x_1$$
$$= \frac{n_\beta(x_1 - x_2 - a_\alpha/2)}{n_\beta^B}, \quad x_1 > x_2. \tag{10.2.58}$$

The pair correlation functions of a homogeneous hard-rod fluid can thus be calculated from the normalized density distributions of a semiconfined hard-rod fluid. Higher order correlation functions in homogeneous hard-rod fluids can similarly be computed from the density distributions of semiconfined hard-rod fluids.

10.3 Computation of Density Distributions of Inhomogeneous Hard-Rod Fluids

10.3.1 Reformulation of Integral Equations for Density Distributions

In the previous section we found that the density distributions $n_\alpha(x)$ for a given set $v_\alpha(x)$ of external potentials obey eq. (10.2.45), namely,

$$\beta\mu_\alpha = \beta v_\alpha(x) + \ln\left[\frac{n_\alpha(x)\Lambda_\alpha}{1 - \sum_\beta \int_{x+\Delta_{\alpha\beta}}^{x+a_{\alpha\beta}} n_\beta(z)dz}\right]$$
$$+ \sum_\gamma \int_{x-a_{\alpha\gamma}}^{x+\Delta_{\alpha\gamma}} \frac{n_\gamma(y)}{1 - \sum_\beta \int_{y+\Delta_{\gamma\beta}}^{y+a_{\gamma\beta}} n_\beta(z)dz} dy, \tag{10.3.1}$$

where $\Delta_{\alpha\beta} \equiv (a_\alpha - a_\beta)/2$ and $a_{\alpha\beta} \equiv (a_\alpha + a_\beta)/2$. This set of nonlinear integral equations must in general be solved by some numerical method. In seeking such a solution it turns out to be convenient to reformulate the equations.

One approach of reformulation is to define the function $h_\alpha(x)$:

$$h_\alpha(x) \equiv \frac{n_\alpha(x)}{1 - \sum_\beta \int_{x+\Delta_{\alpha\beta}}^{x+a_{\alpha\beta}} n_\beta(x)dx}. \tag{10.3.2}$$

In terms of the h functions, eq. (10.3.1) becomes

$$h_\alpha(x) = w_\alpha(x) e^{-\sum_\gamma \int_{x-a_{\alpha\gamma}}^{x+\Delta_{\alpha\gamma}} h_\gamma(y)dy}, \tag{10.3.3}$$

where $w_\alpha(x)$ is defined by the expression

$$w_\alpha(x) = \frac{e^{\beta(\mu_\alpha - v_\alpha(x))}}{\Lambda_\alpha}. \tag{10.3.4}$$

Equation (10.3.3) is simpler and more robust to an iterative solution than is eq. (10.3.1).

For a c-component system, finding the density distribution involves solving c coupled nonlinear integral equations whether eq. (10.3.1) or eq. (10.3.2) is used. It turns out, however, there is another reformulation (Tang et al., (1992)) that reduces the problem to that of solving two uncoupled nonlinear integral equations no matter how many components there are in the system. The starting point for this approach is eq. (10.2.40), which is a nonlocal set of equations for the density distributions.

If one defines

$$p_\pm(x) = \frac{n_\pm(x)}{1 - \int_{-\infty}^{x} [n_+(y) - n_-(y)]dy}, \tag{10.3.5}$$

where $n_-(x)$ and $n_+(x)$ are defined at eqs. (10.2.26) and (10.2.32), one can rewrite eq. (10.2.40) in the form

$$n_\alpha(x) = w_\alpha(x) \exp\left\{ -\int_{-\infty}^{x} \left[p_+\left(y + \frac{1}{2}a_\alpha\right) - p_-\left(y - \frac{1}{2}a_\alpha\right)\right] dy \right\}. \tag{10.3.6}$$

Using the identity

$$\frac{d}{dx} \ln \left\{ 1 - \int_{-\infty}^{x} [n_+(y) - n_-(y)]dy \right\} = -p_+(x) + p_-(x), \tag{10.3.7}$$

we can transform eq. (10.3.6) into either

$$\frac{n_\alpha(x)}{1-\int_{-\infty}^{x-a_\alpha/2}[n_+(y)-n_-(y)]dy} = w_\alpha(x)\exp\left[-\int_{x-a_\alpha/2}^{x+a_\alpha/2} p_+(y)dy\right] \tag{10.3.8}$$

or

$$\frac{n_\alpha(x)}{1-\int_{-\infty}^{x+a_\alpha/2}[n_+(y)-n_-(y)]dy} = w_\alpha(x)\exp\left[-\int_{x-a_\alpha/2}^{x+a_\alpha/2} p_-(y)dy\right]. \tag{10.3.9}$$

Shifting the argument x in eq. (10.3.8) to $x + a_\alpha/2$ and that in eq. (10.3.9) to $x - a_\alpha/2$ and summing each expression over α we obtain

$$p_+(x) = \sum_\alpha w_\alpha(x + a_\alpha/2)\exp\left[-\int_x^{x+a_\alpha} p_+(y)dy\right] \tag{10.3.10}$$

and

$$p_-(x) = \sum_\alpha w_\alpha(x - a_\alpha/2)\exp\left[-\int_{x-a_\alpha}^{x} p_-(y)dy\right], \tag{10.3.11}$$

or in summary

$$p_\pm(x) = \sum_\alpha w_\alpha(x \pm a_\alpha/2)\exp\left[\mp\int_x^{x\pm a_\alpha} p_\pm(y)dy\right]. \tag{10.3.12}$$

Thus, solution of these two uncoupled equations for $p_+(x)$ and $p_-(x)$ enables direct computation of $n_\alpha(x)$ from eq. (10.3.6) without solving any other equation. This p-function approach reduces to solving two uncoupled nonlinear equations, a problem that requires solution of c nonlinear equations and c linear integral equations in the h-function approach. The h-function approach is desirable over solution of the original density distribution set of equations, eq. (10.3.1), because the h-function set of equations is better behaved and simpler when using the iterative methods to be discussed in what follows. However, it appears that the p-function formulation, which shares the numerical robustness of the earlier developed h-function formulation, but which requires less (increasingly less as the number of components c increases) computation, is the preferred formulation.

Another interesting result in the p-function formulation is that the direct correlation function $C^{(2)}_{\alpha\beta}$ can be expressed as a functional of the p_+. In terms of the quantities

$$\mathcal{T}(x) = -\exp\left\{\int_{-\infty}^{x}[p_\pm(z) - p_-(z)]dz\right\} \tag{10.3.13}$$

and

$$\mathcal{U}(x, y) = -\int_{x}^{y} p_-(z) \exp\left\{\int_{-\infty}^{z} [p_+(w) - p_-(w)]dw\right\} dz, \tag{10.3.14}$$

the direct correlation function becomes

$$\begin{aligned} C_{\alpha\beta}^{(2)}(x, y) &= \mathcal{T}(x + a_\alpha/2) + \mathcal{U}(y - a_\beta/2, x + a_\alpha/2), \ \Delta_{\alpha\beta} < y - x < a_{\alpha\beta} \\ &= \mathcal{T}(y + a_\beta/2) + \mathcal{U}(y - a_\beta/2, y + a_\beta/2), \ -\Delta_{\alpha\beta} < y - x < \Delta_{\alpha\beta} \\ &= \mathcal{T}(y + a_\beta/2) + \mathcal{U}(x - a_\alpha/2, y + a_\beta/2), \ -a_{\alpha\beta} < y - x < -\Delta_{\alpha\beta}. \end{aligned} \tag{10.3.15}$$

$C_{\alpha\beta}^{(2)} = 0$ when $|y - x| > a_{\alpha\beta}$. Thus, the two functions $\mathcal{T}(x)$ and $\mathcal{U}(x, y)$, which depend only on $p_+(x)$ and $p_-(x)$, contain all of the direct correlation functions of the multicomponent hard-rod system. The diameters of a particular molecular pair $\alpha\beta$ enter only in determining the ranges of x and y over which $\mathcal{T}$ and $\mathcal{U}$ contribute to $C_{\alpha\beta}^{(2)}$.

Finally, let us note that when the system is symmetric about its midplane, i.e., $v_\alpha(x) = v_\alpha(L - x)$, where L is the domain in which $v_\alpha(x) < +\infty$, the p functions obey the symmetry condition

$$p_\pm(x) = p_\mp(L - x). \tag{10.3.16}$$

Thus, in this case solution of one nonlinear integral equation provides all the input needed to calculate the density distributions and $C_{\alpha\beta}^{(2)}$ for a system with an arbitrary number of components.

10.3.2 Numerical Methods

A major advantage of the p- and h-function formulations is that their nonlinear integral equations can be solved easily by Picard iteration. If $p_+^{(k)}(x)$, $p_-^{(k)}(x)$ and $h_\alpha^{k)}(x)$ are the kth estimates of p_+, p_- and h_α, then the $k + 1$st estimates can be computed from

$$p_\pm^{(k+1)}(x) = \sum_\alpha \omega_\alpha(x \pm a_\alpha/2)e^{\mp\int_x^{x\pm a_\alpha} p_\pm^{(k)}(y)dy} \tag{10.3.17}$$

and

$$h_\alpha^{(k+1)}(x) = \omega_\alpha(x)e^{-\sum_\gamma \int_{x-a_{\alpha\gamma}}^{x+\Delta_{\alpha\gamma}} h_\gamma^{(k)}(y)dy}. \tag{10.3.18}$$

Thus, a simple procedure for computing p_+ or p_- or h_α is to begin with a first guess say 0, and integrate eq. (10.3.17) or eq. (10.3.18) until the solution converges to within some predetermined accuracy, for example, until

$$\int_{-\infty}^{\infty} |p_+^{(k+1)}(x) - p_+^{(k)}(x)|dx < \varepsilon, \tag{10.3.19}$$

where ε is some small preselected number. Although Picard's interation is not guaranteed to converge it has been the author's experience that for many choices of external potentials (including all of those examined in the next section), the Picard method does converge to a solution.

A more complicated, but often more efficient method of solving nonlinear integral equations is the Newton–Raphson method. Because the method requires solution of an associated integral equation, the Newton–Raphson method has the disadvantage that it requires more computer storage than does the Picard method. The Newton–Raphson method for a functional $f(\{u\})$ of the function $u(x)$ is based on the functional Taylor's series

$$f(\{u\}) = f(\{u^\circ\}) + \int \frac{\delta f(\{u\})}{\delta u^o(y)}[u(y) - u^o(y)]dy + O(u - u^o)^2. \quad (10.3.20)$$

If the equation to be solved is

$$f(\{u\}) = 0, \quad (10.3.21)$$

then from the estimate u^o one obtains the next estimate u from the linear integral equation

$$0 = f(\{u^o\}) + \int \frac{\delta f\{u^o\})}{\delta u^o(y)}[u(y) - u^o(y)]dy. \quad (10.3.22)$$

The Newton–Raphson method is to iterate eq. (10.3.22) until the new estimate is as close as desired to the previous estimate.

For the quantity $p_+(x)$ the functional is

$$f(\{p_+\}) = p_+(x) - \sum_\alpha \omega_\alpha(x + a_\alpha/2)e^{-\int_x^{x+a_\alpha} p_+(z)dz},$$

and

$$\frac{\delta f(\{p_+\})}{\delta p_+(y)} = \delta(x - y) + \sum_\alpha \omega_\alpha(x + a_\alpha/2)e^{-\int_x^{x+a_\alpha} p_+(z)dz}\eta(x + a_\alpha - y)\eta(y - x) \quad (10.3.23)$$

A similar result can be obtained for $p_-(x)$. Thus, the Newton–Raphson iteration equations for the functions $p_\pm$ are

$$\begin{aligned} p_\pm^{(k+1)}(x) - p_\pm^{(k)}(x) \mp \sum_\alpha w_\alpha(x \pm a_\alpha/2)e^{\mp \int_x^{x\pm a_\alpha} p_\pm^{(k)}(z)dz} \int_x^{x\pm a_\alpha} [p_\pm^{(k+1)}(y) - p_\pm^{(k)}(y)]dy \\ = -p_\pm^{(k)}(x) + \sum_\alpha w_\alpha(x \pm a_\alpha/2)e^{\mp \int_x^{x\pm a_\alpha} p_\pm^{(k)}(z)dz}. \end{aligned} \quad (10.3.24)$$

Similarly, the Newton–Raphson set of equations for the h_α can be shown to be

$$\begin{aligned} h_\alpha^{(k+1)}(x) - h_\alpha^{(k)}(x) + w_\alpha(x)e^{-\sum_\gamma \int_{x-a_{\alpha\gamma}}^{x+\Delta_{\alpha\gamma}} h_\gamma^{(k)}(z)dz} \sum_\beta \int_{x-a_{\alpha\beta}}^{x+\Delta_{\alpha\beta}} [h_\beta^{(k+1)}(z) - h_\beta^{(k)}(z)]dz \\ = -h_\alpha^{(k)}(x) + w_\alpha(x)e^{-\sum_\gamma \int_{x-a_{\alpha\gamma}}^{x+\Delta_{\alpha\gamma}} h_\gamma^{(k)}(z)dz}, \quad \alpha = 1, \cdots, c. \end{aligned} \quad (10.3.25)$$

Equations (10.3.24) and (10.3.25) are linear integral equations that must be solved to obtain the $k+1$st estimate of p_+, p_- or h_α from the kth estimate. This solution must be iterated until some convergence criterion, such as eq. (10.3.19), is attained. When the Newton–Raphson method converges it does so quadratically, and consequently when computer storage is not a problem the method is much more efficient than the Picard method. By quadratic convergence we mean that if at step $k = \nu$ the error, say as measured by the lhs of eq. (10.3.19), is 10^{-1}, then, if the scheme is converging, the error will be of the order of 10^{-2}, 10^{-4} and 10^{-8} at steps $k = \nu + 1$, $\nu + 2$ and $\nu + 3$.

The basic problem with the Newton–Raphson method is to get a good first guess. A good way to do this for a given T and set μ_α is to fix T and first seek a solution at small μ_α (say $\beta\mu_\alpha = -2$). In this case a good first guess is

$$p_\pm^{(1)}(x) = \sum_\alpha w_\alpha(x \pm a_\alpha/2) \quad \text{or} \quad h_\alpha^{(1)}(x) = w_\alpha(x). \tag{10.3.26}$$

Iterate eq. (10.3.24) or eq. (10.3.25) until the criterion at eq. (10.3.19) is satisfied. Then advance the values of the μ_α by increments $\Delta\mu_\alpha$, solving eq. (10.3.24) or eq. (10.3.25) at each increment and using the solution as the first guess for the Newton–Raphson solution at the next increment, until the desired chemical potential values have been reached. The maximum size of the increments $\Delta\mu_\alpha$ that can be used without losing convergence will depend on the value of T and the form of the external potentials $v_\alpha(x)$, $\alpha = 1, \ldots, c$.

In using either the Picard method or the Newton–Raphson method, some technique for evaluating integrals numerically must be introduced. The trapezoidal rule provides a simple way to evaluate integrals. According to this rule,

$$\int_{x'}^{x''} g(x)dx \approx \sum_{i=1}^{M-1} \frac{1}{2}[g(x_{i+1}) + g(x_i)]\Delta x_i, \tag{10.3.27}$$

where the interval $x'' - x'$ has been divided into M subintervals in which $x_1 = x'$, $x_M = x''$ and $x_{i+1} - x_i = \Delta x_i$, $i = 1, \ldots, M-1$. If the Δx_i are small enough, then the rhs of eq. (10.3.27) will be a good approximation to the integral. Because x is a continuous variable either a variable mesh size must be used or, if $\Delta x_i = \Delta x$ for all i, the mesh must be chosen fine enough that one of the grid points $x_1, \ldots, x_M$ spanning the space to which the fluid is confined is close to any limit of the integrals occurring in the iteration equations. For example, in the case of a system bounded by $x = 0$ and $x = L$ we could pick a large value of M, let $x_1 = 0$ and $x_M = L$, choose each interval as $\Delta x_i = \Delta x \equiv L/M$, and set $x_i + \Delta_{\alpha\gamma}$ or $x_i - a_{\alpha\gamma}$ equal to the grid point x_j closest to its value. In this case the Picard equation for $h_\alpha^{(k+1)}$ becomes

$$h_\alpha^{(k+1)}(x_i) = w_\alpha(x_i)e^{-\sum_\gamma \sum_{j=l}^{m-1} \frac{1}{2}[h_\gamma^{(k)}(x_{j+1}) + h_\gamma^{(k)}(x_j)]\Delta x}, \tag{10.3.28}$$

where l and m are chosen from the conditions $|x_m - x_i - \Delta_{\alpha\gamma}| < \Delta x/2$ and $|x_l - x_i + a_{\alpha\gamma}| < \Delta x/2$. This procedure will convert the Newton–Raphson equations, eqs. (10.3.24) and (10.3.25), into linear algebraic equations for $p_{\pm}^{(k+1)}(x_i)$ and $h_{\alpha}^{(k+1)}(x_i)$.

More complicated discretization procedures and/or integration methods may be used or more complicated numerical schemes, such as finite element analysis, can be used to solve the Newton–Raphson equation. However, experience has shown that use of the trapezoidal rule as outlined above suffices as long as Δx is chosen to be sufficiently small.

10.3.3 Applications

Let us first consider the density profile of one- and two-component hard-rod fluids confined by a pair of hard walls a distance L apart. Aside from the walls there is no other external field [i.e., $v_\alpha(x) = 0$ for $a_\alpha/2 < x < L - a_\alpha/2$ (or $-(L - a_\alpha)/2 < x < (L - a_\alpha)/2$ if the coordinate system is put at the midpoint between the walls]. For this situation there is no need to solve the integral equations discussed above. Instead the density profiles can be computed directly from

$$n_\alpha(x) = \zeta_\alpha \Xi_{x-a_\alpha/2} \Xi_{L-x-a_\alpha/2} / \Xi_L, \qquad (10.3.29)$$

where Ξ_y is given by eq. (10.2.21). All the results presented in Figure 10.8 were computed from eq. (10.3.29). However, as a test of the robustness of the Picard solution of the h-function equations presented above, it was established that Picard iteration with the trapezoidal rule for integration converges for all the systems described in this figure. Given this, it is very likely that Picard iteration will converge equally well for the p-function equations although it was not tried for these systems. The density profiles

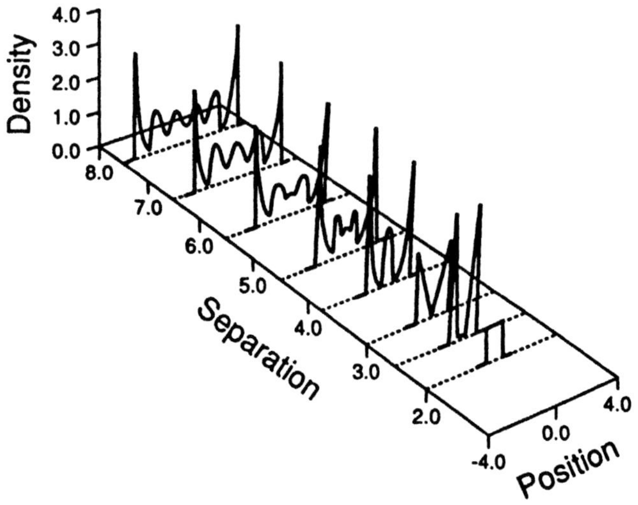

Figure 10.8
Density profiles for a one-component hard-rod fluid confined between hard walls at various separations. The confined fluid is in equilibrium with a bulk fluid at $\beta P_B a = 3$ (Vanderlick, 1988).

of a one-component hard-rod fluid are shown in Figure 10.8 for various pore widths (wall separations) L. Position in the pore is chosen so that $x = 0$ is at the center of the system. The systems are in equilibrium with a bulk hard-rod fluid at a reduced pressure of $\beta P_B a = 3$. The parameter $\zeta(\equiv e^{\beta\mu}/\Lambda)$ is determined by the equation of state of bulk phase hard rods,

$$\beta\mu = \beta P_B a + \ln(\beta P_B a) + \ln(\Lambda/a), \tag{10.3.30}$$

that is

$$\zeta a = \beta P_B a \exp(\beta P_B a).$$

The density profiles show that ordering (or layering) is induced by the confining walls. When the pore width L lies between a and $2a$ the pore is wide enough to admit only one particle at a time. Thus, the density distribution is a single peak of uniform height, the uniform height indicating that an occupying particle has equal probability of lying anywhere in the allowed pore space between $-(L-a)/2$ and $(L-a)/2$.

At $L = 2.5a$, the pore space is wide enough to hold one or two particles (or zero), whose occupancy probability $\mathcal{P}_N$ is given in Table 10.2. At the high chemical potential considered here, two-particle occupancy dominates, $\mathcal{P}_2 = 0.832$, and hence the density profile has two sharp, symmetrical peaks adjacent to the left and right walls, respectively. At $L = 3a$, two-particle occupancy still dominates, but the peaks are not as sharp the two particles have more free volume to occupy.

For $L > 3a$, one, two, and three particle occupancy can occur. Indeed, for the $L = 3.75a$ profile shown in Figure 10.8 the density profile has three well-developed peaks indicating that three particle occupancy is favored at the chemical potential of the system.

At $L = 4.5a$, in which case one, two, three, and four particle occupancy can occur, the density profile has five peaks. This of course does not mean five particle occupancy. Rather, it is the result of the grand canonical ensemble average of the density profiles of zero, one, two, three, and four particle systems. In this case,

TABLE 10.2 Pore Occupancy Probability $\mathcal{P}_N$ for a One-Component Hard-Rod Fluid Confined Between Hard Walls of Separation L

Length	$N=0$	$N=1$	$N=2$	$N=3$	$N=4$	$N=5$	$N=6$	$N=7$
2.0	0.016	0.984	–	–	–	–	–	–
2.5	0.002	0.166	0.832	–	–	–	–	–
3.0	0.001	0.062	0.937	–	–	–	–	–
3.75	0.000	0.008	0.236	0.729	–	–	–	–
4.5	0.000	0.001	0.067	0.728	0.203	–	–	–
5.5	0.000	0.000	0.006	0.159	0.777	0.058	–	–
6.5	0.000	0.000	0.000	0.021	0.289	0.676	0.014	–
7.5	0.000	0.000	0.000	0.002	0.055	0.433	0.507	0.003

The confined fluid is in equilibrium with a bulk fluid at $\beta P_B a = 3$.

the occupancy probabilities $\mathcal{P}_N$ are 0.000, 0.001, 0.067, 0.728, and 0.203 for $N = 0, 1, 2, 3$, and 4. These values show that the peaks in the density profile for $L = 4.5a$ come primarily from an average of three and four particle occupancy states. Reference to the occupancy probabilities given in Table 10.2 provides a similar explanation for the symmetry and number of peaks observed in the density profiles for the larger pore widths $L = 5.5a, 6.5a$, and $7.5a$ in Figure 10.8.

Let us next study a two-component hard-rod fluid confined between hard walls of separation L. Occupancy probabilities of the two-component case, $a_2 = 2.5a_1$ were presented in Table 10.1 for the conditions $\beta P_B a_1 = 3$ and $x_1^b = 0.5$. The chemical potentials are computed from the bulk phase equation of state

$$\beta\mu_\alpha = \beta P_B a_\alpha + \ln(x_\alpha^b \beta P_B a_\alpha) + \ln(\Lambda_\alpha / a_\alpha), \qquad (10.3.31)$$

where x_α^b is the mole fraction of species α in the bulk phase in equilibrium with the confined fluid.

Density profiles for various binary mixtures in pores of different widths L are presented in Figure 10.9. Results for two different mixtures, $a_2 = 1.5a_1$ and $a_2 = 2.5a_1$, are presented. For $a_2 = 2.5a_1$, only the case of an equimolar coexisting phase is considered (i.e., in the bulk phase $x_1^b = 0.5$). Three different bulk phase concentrations are examined for the $a_2 = 1.5a_1$ mixture. The peaks or layering structure of the density profiles $n_1(x)$ and $n_2(x)$ are more complicated than those for a one-component fluid. However, the pore occupancy probabilities reveal the basic patterns. For example, at $L = 4a_1$ when $a_2 = 2.5a_1$ and $x_1^b = 0.5$, the profile $n_1(x)$ has three peaks and the $n_2(x)$ profile has two peaks. Referring to Table 10.1, we see that the three peaks of n_1 result primarily from the average of the occupancy states $N_1 = 1$, $N_2 = 1$, and $N_1 = 3$, $N_2 = 0$, and that the two peaks of n_2 result from an average of the occupancy states $N_1 = 0, N_2 = 1$ ($\mathcal{P}_N = 0.13$) and $N_1 = 1, N_2 = 1$ ($\mathcal{P}_N = 0.66$). The symmetry of the two peaks in $n_2(x)$ results from the fact that particle 2 will be on the left or the right of particle 1 with equal probablility. The fact that the peaks of $n_2(x)$ are closer together than a_2 bears witness that the peaks do not arise from pair occupancy by particles of species 2. In the case $L = 5.5a_1$, however, the two large peaks in $n_2(x)$ do arise from primarily pair occupancy of species 2 ($\mathcal{P}_N = 0.34$ when $N_1 = 0$, $N_2 = 2$); the small peak in the middle comes from the state $N_1 = 2$, $N_2 = 1$ ($\mathcal{P}_N = 0.45$). We leave it as an exercise for the reader to rationalize the rest of the density profiles shown in Figure 10.9.

The next thing we wish to explore is the effect of an external potential on the properties of a confined hard-rod fluid. We consider a binary mixture confined by walls whose external potential is

$$\begin{aligned} v_\alpha(x) &= \infty, \quad x < a_\alpha/2, \quad x > L - a_\alpha/2, \\ &= -\varepsilon_\alpha \left\{ \left(\frac{a_\alpha}{x + a_\alpha/2} \right)^3 + \left(\frac{a_\alpha}{L + a_\alpha/2 - x} \right)^3 \right\}, \quad a_\alpha/2 < x < L - a_\alpha/2. \end{aligned} \qquad (10.3.32)$$

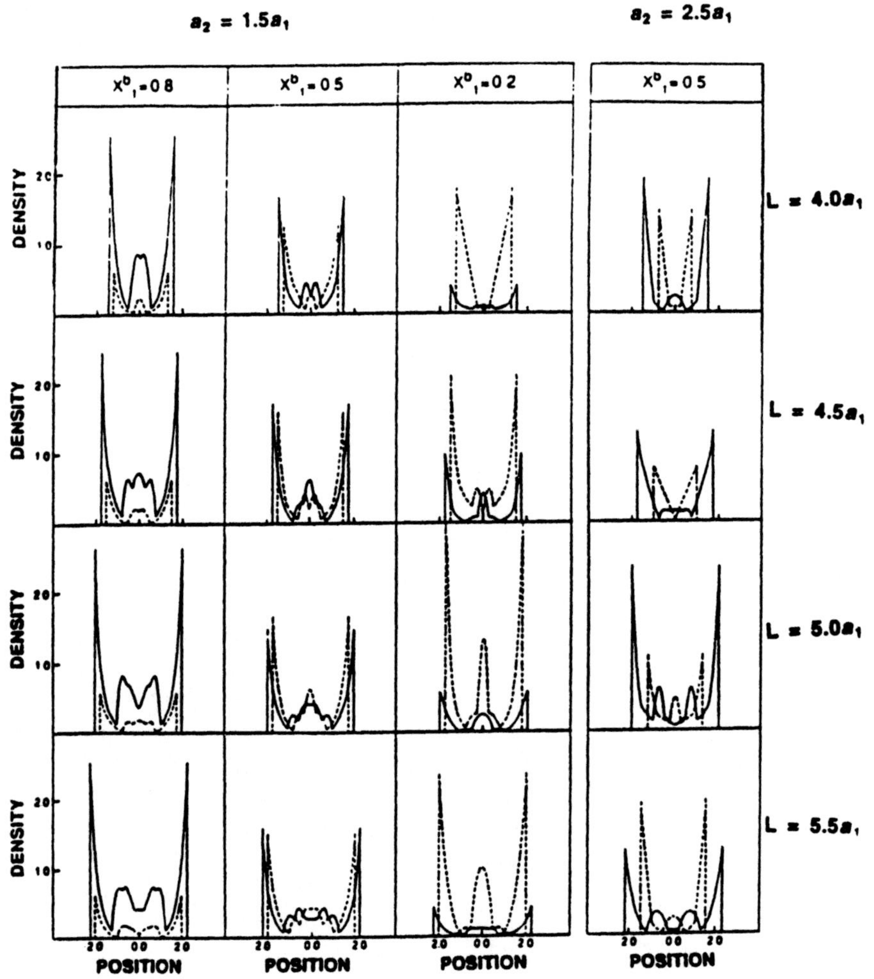

Figure 10.9a
Density profiles of two-component hard-rod fluids confined between hard walls of separation L. The coexisting bulk phases are at the reduced pressure $\beta P_B a_1 = 3$ and concentration x_α^b (Vanderlick et al., 1989).

We compare two cases, $\beta\varepsilon_\alpha = 1$ and $\varepsilon_\alpha = 0$, the latter case being a hard-rod mixture confined by hard walls. A wall-fluid particle inverse third power attraction corresponds to an inverse sixth power attraction between a molecule and an infinite slab of uniformly distributed molecules. The pressure of the system can be computed from

$$\beta P = \sum_\alpha n_\alpha(a_\alpha/2) - \sum_\alpha \frac{\beta}{2} \int_{a_\alpha/2}^{L-a_\alpha/2} n_\alpha(x) \frac{\partial v_\alpha(x)}{\partial x}\, dx. \quad (10.3.33)$$

The equation of state of the coexisting bulk phase is again eq. (10.3.31).

Density profiles are presented in Figure 10.10 for the two cases $\beta\varepsilon_\alpha = 1$ and 0. The density profiles were computed from the h-function formulation using Picard's method with the trapezoidal rule.

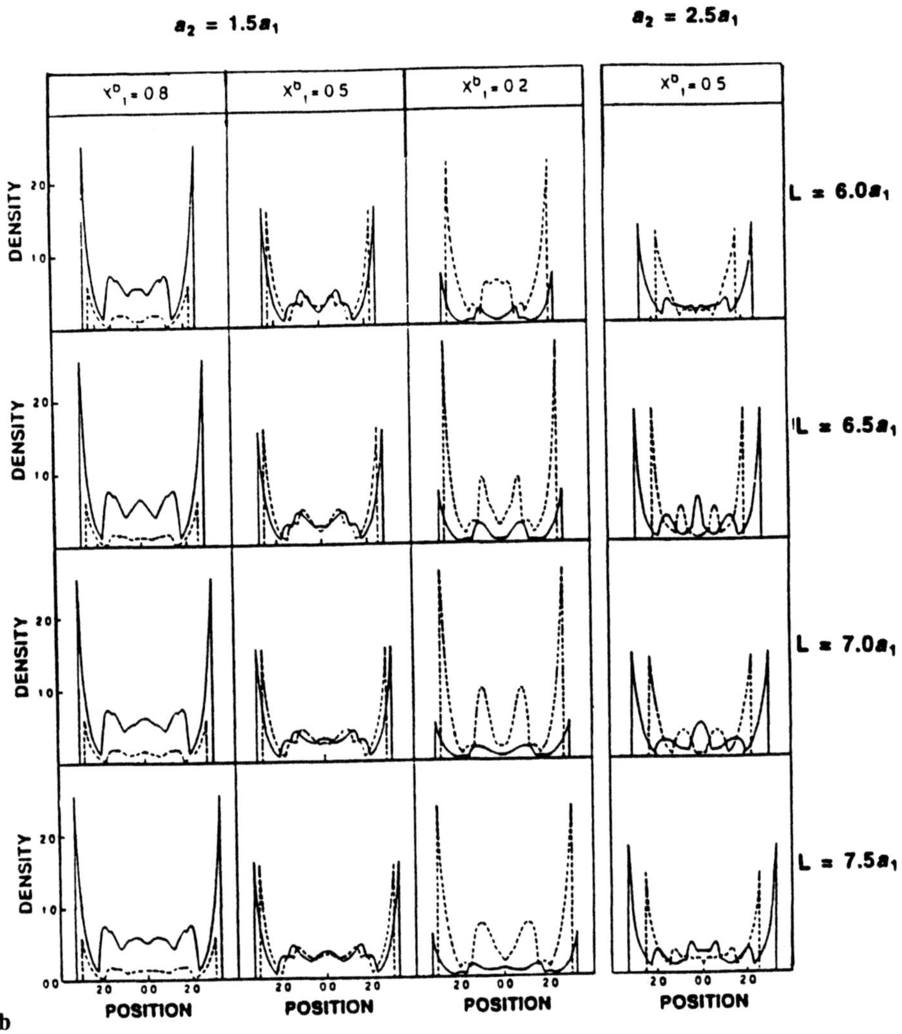

Figure 10.9b
Density profiles of two-component hard-rod fluids confined between hard walls of separation L. The coexisting bulk phases are at the reduced pressure $\beta P_B a_1 = 3$ and concentration x_α^b (Vanderlick et al., 1989).

As illustrated in Figure 10.10, the main effect of the attractive interaction between fluid particles and the confining walls is to sharpen the peaks in the density profiles. The number and spacing of peaks remains the same. Thus, at the high chemical potentials chosen in this example the hard-rod interaction (i.e., the excluded volume effect) controls the periods of the maxima and minima in the density profiles while the wall attractions simply sharpen the structure induced by the hard-rod interactions.

The disjoining pressure Π $(\equiv P - P_B)$ and the pore average mole fraction $\overline{x}_1^p$ are plotted versus pore width in Figures 10.11 and 10.12. Again we see that the periods of oscillation of these quantities are not changed appreciably by the attractive wall–particle interaction whereas the amplitudes of the maxima and minima are increased.

Let us close this section with an application of the p-function formulation to a polydisperse hard-rod fluid, that is, to a situation

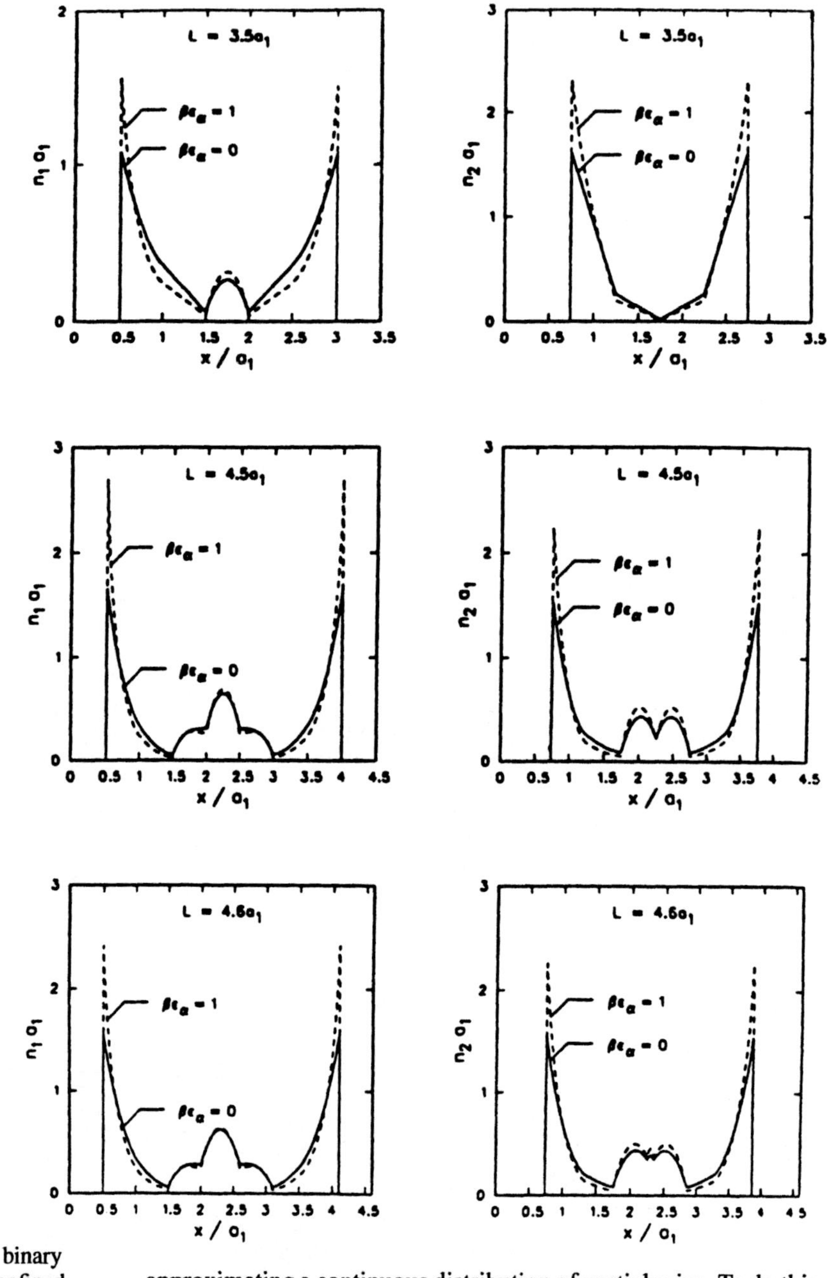

Figure 10.10
Density profiles for binary hard-rod mixtures confined between hard walls ($\varepsilon_\alpha = 0$) and attractive walls ($\beta\varepsilon_\alpha = 1$) for various wall separations L. $a_2 = 1.5a_1$, $\beta P_B a_1 = 3$, and $x_1^b = 0.5$.

approximating a continuous distribution of particle size. To do this we assume a bulk phase at pressure P_B in which the mole fraction of rods with diameter between a and $a + da$ is

$$f_B(a)da. \tag{10.3.34}$$

For a discrete set of components α whose mole fraction are x_α^b, $\alpha = 1, \ldots, c$, the distribution function is

$$f_B(a) = \sum_\alpha x_\alpha^b \delta(a - a_\alpha). \tag{10.3.35}$$

Figure 10.11
(a) Disjoining pressure versus wall separation L. (b) Pore average mole fraction of component 1. $a_2 = 1.5a_1$, $\beta P_B a_1 = 3$, and $x_1^b = 0.5$.

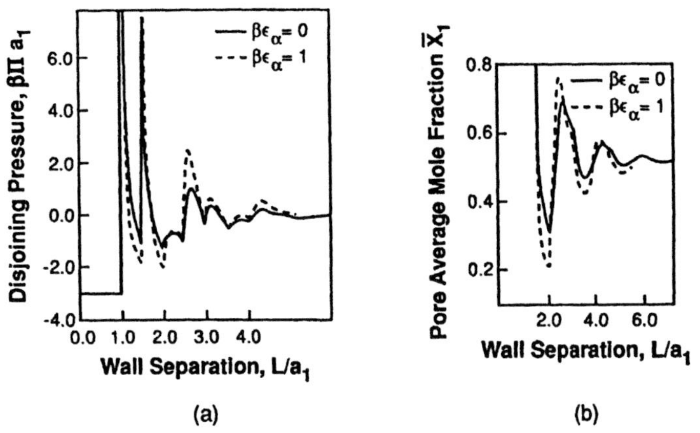

When distribution function is continuous, we can define pseudocomponents α as those hard rods whose diameter lies between a_α and $a_\alpha + \Delta a$. Then the mole fraction of the pseudocomponent in the bulk phase is

$$x_\alpha^b = \int_{a_\alpha}^{a_\alpha + \Delta a} f_B(a)da \qquad (10.3.36)$$

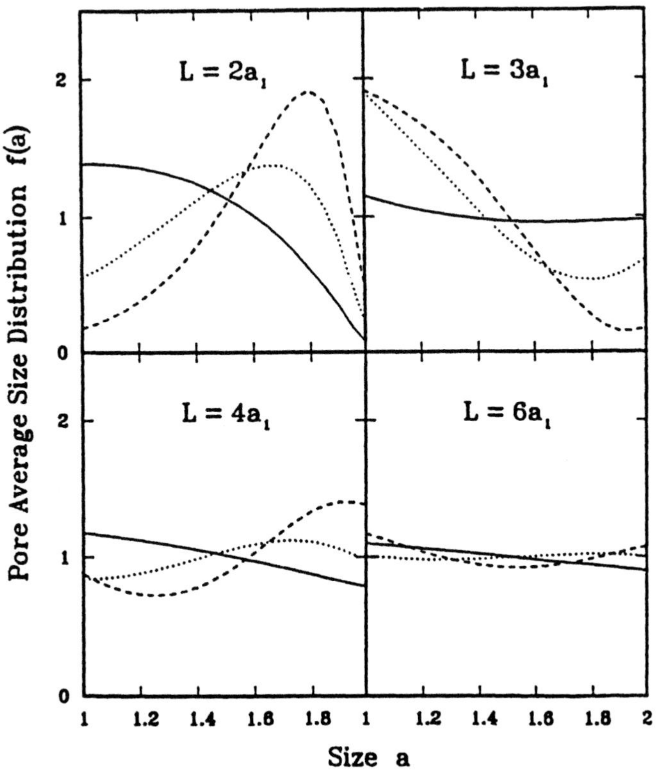

Figure 10.12
Pore average size distribution $f(a)$ for a polydisperse hard-rod fluid confined by hard walls at various separations L. The solid, dotted, and dashed curves correspond to bulk pressure $\beta P_B a_1 = 1$, 3 and 5, (solid, dotted and dashed curves) respectively. Adapted from Tang et al., (1992).

and the chemical potential of this pseudocomponent is given by eq. (10.3.31).

To illustrate the effect of size polydispersity on the behavior of confined fluids, we compared the case of a uniform size distribution,

$$f_B(a) = \frac{1}{a_2 - a_1}, \quad a_1 < a < a_2 = 2a_1$$
$$= 0, \quad \text{otherwise,} \tag{10.3.37}$$

with a binary mixture consisting of equal numbers of particles of size a_1 and of size $a_2 = 2a_1$.

Pseudocomponents were defined by eq. (10.3.36), their density distribution functions were computed by solving the p_+-function equation, eq. (10.3.24), using the Newton–Raphson method with the trapezoidal rule, the pore average mole fraction was computed from

$$\bar{x}_\alpha^p = \int_{a_\alpha}^{a_\alpha+\Delta a} f(a)da = \int_0^L n_\alpha(x)dx \Big/ \sum_\beta \int_0^L n_\beta(x)dx, \tag{10.3.38}$$

where $a_\alpha = a_1 + \alpha\Delta a$, $\alpha = 0, 1, \ldots, a_1/\Delta a$, and the pressure from the equation

$$\beta P = \sum_\alpha n_\alpha(x = a_\alpha/2) = p_+(x = 0). \tag{10.3.39}$$

The pore average size distribution $f(a)$ was estimated from $\bar{x}_\alpha^p/\Delta a$. Δa was chosen small enough that the curve $f(a)$ versus a changed little with further refinement.

Pore average size distributions $f(a)$ are given in Figure 10.12 for three bulk pressures $\beta P_B a_1 = 1, 3$, and 5 for several pore widths L. For $L = 2a_1$ increasing pressure skews the distribution from favoring small particle occupancy to favoring larger particle occupancy. At $L = 3a_1$ the trend with increasing pressure is from an almost uniform distribution to a distribution favoring smaller particle occupancy. At larger separations, for example, $L = 6a_1$, the pore occupancy distribution remains close to the uniform distribution $f_B(a)$ of the bulk phase.

The disjoining pressure Π of the polydisperse system is compared in Figure 10.13 with the disjoining pressure of a binary hard-rod system with $a_2 = 2a_1$ that is in equilibrium with an equimolar bulk phase. Because the diameter a_2 of large rods in the binary mixture is exactly twice the diameter a_1 of the small rods, the neighboring maxima and minima of the disjoining pressure lie almost exactly the distance a_1 apart. Results given in Section 10.1 for $a_2 = 1.5a_1$ and $2.5a_2$ illustrate that a much more complicated disjoining pressure is predicted for binary mixtures when particle diameters are not multiples of one another. The oscillations of the disjoining pressure of the polydisperse fluid are much attenuated but not eliminated. At the lower pressure ($\beta P_B a_1 = 1$) the main

Figure 10.13
Disjoining pressure Π as a function of pore width L at three bulk pressures. Comparison of polydisperse systems (solid curves) with a binary mixture (dashed curves, $a_2 = 2a_1$) which is in equilibrium with an equimolar bulk phase. (Tang et al., 1992).

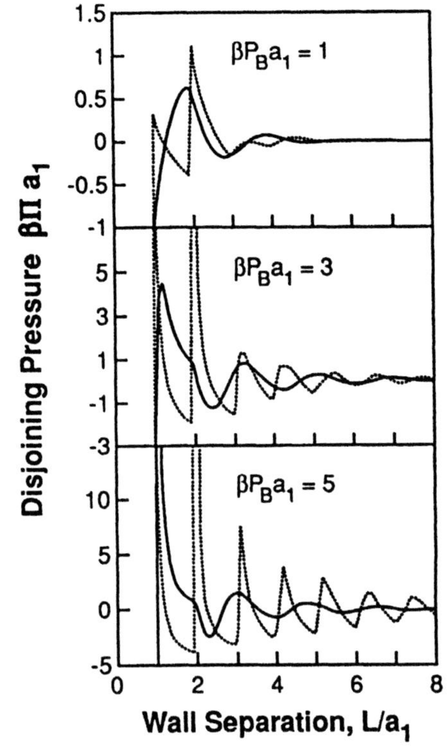

peak coincides with the large particle diameter, in spite of the fact that the distribution is skewed toward small particles at this pressure when $L = 2a_1$. Equally curiously, the main peak at $\beta P_B a_1 = 5$ arises from the small particles whereas the particle distribution is skewed toward the large particle occupancy at this bulk pressure.

In Section 10.1 and in this section we have examined selective adsorption (occupancy) of components of a mixture in pores of a given width. In connection with separations processes it is also of interest to study selective adsorption of fluid mixtures in porous media with a distribution of pore sizes. The p-function version of the hard-rod theory should provide an economical vehicle for studying this problem, although such an investigation has not been made at this point.

10.4 Confined Tonks–Takahashi Fluids

10.4.1 Fluids Confined and in the Presence of an Arbitrary External Field

The Tonks–Takahashi fluid obeys the nearest neighbor pair potential

$$u(x_{ij}) = \infty, \quad |x_{ij}| < a \tag{10.4.1}$$

$$= \psi(x_{ij}), \quad a < |x_{ij}| < \nu a \tag{10.4.2}$$

$$= 0, \quad |x_{ij}| > \nu a, \tag{10.4.3}$$

where $1 < \nu < 2$.

Consider a Tonks–Takahashi fluid confined by fixed Tonks–Takahashi particles as shown in Figure 10.14. The confining walls consist of Tonks–Takahashi particles whose centers are fixed at $y - a/2$ and $z + a/2$. Suppose the particles are also subject to an external field whose potential is $v(x_i)$. The canonical ensemble partition function for this system is

$$Z_N(y, z) = N! \int_y^z dx_N \int_y^{x_N} dx_{N-1} \cdots \int_y^{x_2} dx_1 e^{-\beta u(z+a/2-x_N)} \\ \times e^{-\sum_{i>j=1}^N \beta u(x_i - x_j) - \sum_{i=1}^N \beta v(x_i) - \beta u(x_1 - y + a/2)} \tag{10.4.4}$$

and the grand canonical ensemble partition function is

$$\Xi(y, z) = \sum_{N=0}^{\infty} X_N(y, z), \tag{10.4.5}$$

where

$$X_N(y, z) = \zeta^N \int_y^z dx_N \int_y^{x_N} dx_{N-1} \cdots \int_y^{x_2} dx_1 e^{-\beta u(z+a/2-x_N)} \\ \times e^{-\sum_{i>j=1}^N \beta u(x_i - x_j) - \sum_{i=1}^N \beta v(x_i) - \beta u(x_1 - y + a/2)}, \tag{10.4.6}$$

and $\zeta = e^{\beta u}/\Lambda$.

The density distribution $n(x)$ for the fluid can be computed from the formula $n(x) = \delta \ln \Xi(y, z)/\delta(-\beta v(x))$. From the property $\delta(\beta v(x_i))/\delta(\beta v(x)) = \delta(x_i - x)$ it follows that

$$\frac{\delta X_N(y, z)}{\delta(-\beta v(x))} = \zeta e^{-\beta v(x)} \sum_{k=1}^N \zeta^{N-k} \int_y^z dx_N \cdots \int_y^{x_{k+2}} dx_{k+1} e^{-\beta u(z+a/2-x_N)} \\ \times e^{-\sum_{i>j=k+1}^N \beta u(x_i - x_j) - \sum_{i=k+1}^N \beta v(x_i) - \beta u(x_{k+1} - x)} \\ \times \zeta^{k-1} \int_y^x dx_{k-1} \cdots \int_y^{x_2} dx_1 e^{-\beta u(x - x_{k-1})} e^{-\sum_{i>j=1}^{k-1} \beta u(x_i - x_j)} e^{-\beta u(x_1 - y + a/2) - \sum_{i=1}^{k-1} \beta v(x_i)}. \tag{10.4.7}$$

The factor $e^{-\beta u(x_{k+1} - x)}$ in this expression enforces the constraint $x_{k+1} > x + a$ and the factor $e^{-\beta u(x - x_{k-1})}$ enforces the constraint $x_{k-1} < x - a$. Thus, because $x_N > x_{N-1} \cdots > x_{k+1}$ and $x_1 < x_2 \cdots < x_{k-1}$, eq. (10.4.5) factors into a partition function of a fluid confined by Tonks–Takahashi particles with centers fixed at x and $z + a/2$ and that of a fluid confined by Tonks–Takahashi particles fixed at $y - a/2$ and x, that is,

$$\frac{\delta X_N(y, z)}{\delta(-\beta v(x))} = \zeta e^{-\beta v(x)} \sum_{k=1}^N X_{N-k}(x + a/2, z) X_{k-1}(y, x - a/2). \tag{10.4.8}$$

With this result it follows that

$$\frac{\delta \Xi(y,z)}{\delta(-\beta v(x))} = \zeta e^{-\beta v(x)} \sum_{N=1}^{\infty} \sum_{k=1}^{\infty} X_{k-1}(y, x-a/2) X_{N-k}(x+a/2, z)$$
$$= \zeta e^{-\beta v(x)} \Xi(y, x-a/2) \Xi(x+a/2, z), \tag{10.4.9}$$

where the summation transformation introduced at eq. (10.2.14) has been used.

The density distribution is thus found to be given by the formula

$$n(x) = \zeta e^{-\beta v(x)} \frac{\Xi(y, x-a/2)\Xi(x+a/2, z)}{\Xi(y,z)}, \quad y < x < z, \tag{10.4.10}$$

which is formally identical to that given by eq. (10.2.17) for a hard-rod fluid. The partition function of the hard-rod fluid is of course different from that of the general Tonks–Takahashi fluid.

The s-body distribution function for a Tonks–Takahashi fluid can similarly be shown to be given by

$$n^{(s)}(x_1, \ldots, x_s) = \zeta^s \frac{e^{-\sum_{i=1}^{s} \beta v(x_i)}}{\Xi(y,z)} \Xi(y, x_1 - a/2)$$
$$\prod_{i=2}^{s} \Xi(x_{i-1} + a/2, x_i - a/2) \Xi(x_s + a/2, z), \tag{10.4.11}$$

with $x_1 < x_2 < \cdots < x_N$. This is a result that is formally identical to the hard-rod expression for $n^{(s)}$.

In a multicomponent Tonks–Takahashi fluid the pair potential is of the form

$$u_{\alpha_i\alpha_j}(x_{ij}) = \infty, \quad |x_{ij}| < a_{\alpha_i\alpha_j}$$
$$= \psi_{\alpha_i\alpha_j}(x_{ij}), \quad a_{\alpha_i\alpha_j} < |x_{ij}| < \tilde{a}_{\alpha_i\alpha_j} \tag{10.4.12}$$
$$= 0, \quad |x_{ij}| > \tilde{a}_{\alpha_i\alpha_j},$$

where $\tilde{a}_{\alpha_i\alpha_j}$ must be small enough that no next nearest neighbors can interact with α_i or α_j. For example, if in a two-component system $a_2 = 1.5a_1$ ($a_2 \equiv a_{22}, a_1 \equiv a_{11}$), then $\tilde{a}_{11} \leq 2a_1$, $\tilde{a}_{12} \leq$

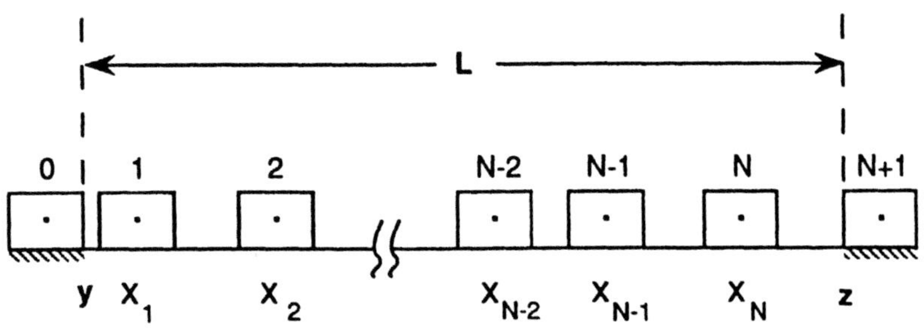

Figure 10.14
Tonks–Takahashi fluid of N particles confined to a subregion $L = z - y$ by a pair of the same kind of particles whose centers are fixed at $y - a/2$ and $z + a/2$.

$(3a_1 + a_2)/2$, and $\tilde{a}_{22} \le (3a_2 + a_1)/2$. The grand canonical ensemble partition function for the confined multicomponent fluid is

$$\Xi^{w_1 w_2}(y,z) = \sum_{N_1=0}^{\infty} \cdots \sum_{N_c=0}^{\infty} \prod_{\alpha=1}^{c} \frac{\zeta_\alpha^{N_\alpha}}{N_\alpha!} Z_{\mathbf{N}}^{w_1 w_2}(y,z)$$
$$= \sum_{N=0}^{\infty} \sum_{\alpha_1=1}^{c} \cdots \sum_{\alpha_N=1}^{c} \prod_{i=1}^{N} \zeta_{\alpha_i} \int_y^z dx_N \int_y^{x_N} dx_{N-1} \cdots \int_y^{x_2} dx_1 e^{-\beta u_{w_2 \alpha_N}(z + a_{w_2}/2 - x_N)} \tag{10.4.13}$$
$$\times e^{-\sum_{i>j=1}^{N} \beta u_{\alpha_i \alpha_j}(x_i - x_j) - \sum_{i=1}^{N} \beta v_{\alpha_i}(x_i) - \beta u_{\alpha_1 w_1}(x_1 - y + a_{w_1}/2)}.$$

The superscripts $w_1 w_2$ on $\Xi^{w_1 w_2}(y,z)$ mean that a Tonks–Takahashi particle of species w_1 is the confining particle whose center is fixed at $y - a_{w_1}/2$ and one of species w_2 is the confining particle whose center is fixed at $z + a_{w_2}/2$.

The density distribution of species α can be derived from the general relation $n_\alpha(x) = \delta \ln \Xi^{w_1 w_2}/\delta(-\beta v_\alpha(x))$. The details are similar to those presented for a hard-rod mixture in Section 10.2 and again lead to the equation

$$n_\alpha(x) = \zeta_\alpha e^{-\beta v_\alpha(x)} \frac{\Xi^{w_1 \alpha}(y, x - a_\alpha/2)\Xi^{\alpha w_2}(x + a_\alpha/2, z)}{\Xi^{w_1 w_2}(y,z)}, \quad y < x < z. \tag{10.4.14}$$

The s-body density distribution function is also of the same form as eq. (10.2.18), that is,

$$n_{\boldsymbol{\alpha}}^{(s)}(x_1, \ldots, x_s) = \prod_{i=1}^{s} \zeta_{\alpha_i} e^{-\beta v_{\alpha_i}(x_i)} \Xi^{w_1 \alpha_1}(y, x_1 - a_{\alpha_1}/2)$$
$$\times \frac{\prod_{i=2}^{s} \Xi^{\alpha_{i-1}\alpha_i}(x_{i-1} + a_{\alpha_{i-1}}/2, x_i - a_{\alpha_i}/2)\Xi^{\alpha_s w_2}(x_s + a_{\alpha_s}/2, z)}{\Xi^{w_1 w_2}(y,z)}. \tag{10.4.15}$$

$\Xi^{\alpha_{i-1}\alpha_i}(r,s)$ denotes the grand canonical partition function for a system in which confining Tonks–Takahashi particles of types α_{i-1} and α_i have their centers fixed at positions $r - a_{\alpha_{i-1}}/2$ and $s + a_{\alpha_i}/2$, respectively. $\boldsymbol{\alpha}$ denotes the set $\{\alpha_1, \ldots, \alpha_s\}$.

For the semiconfined system with $v_\alpha(x_i) \equiv 0$, $y = 0$, and $z \to \infty$, the property $\lim_{z\to\infty} \Xi^{\alpha w_2}(x + a_\alpha/2, z)/\Xi^{w_1 w_2}(0,z) = e^{-\beta P_B(x + a_\alpha/2)}$ can be used to derive from eq. (10.4.14) the following formula for the density distribution function:

$$n_\alpha^{w_1}(x) = \zeta_\alpha \Xi^{w_1 \alpha}(0, x - a_\alpha/2) e^{-\beta P_B(x + a_\alpha/2)}, \tag{10.4.16}$$

where P_B is the bulk pressure of the Tonks–Takahashi fluid far from the confining wall. The superscript on $n_\alpha^{w_1}$ indicates that the confining wall is a Tonks–Takahashi particle of species w_1.

Finally, if $v_\alpha(x_i) = 0$ and $z - y$ is very large, the fluid will be homogeneous far from the walls, that is, where $x_1, x_2 \ll z$ and $x_1, x_2 \gg y$. In this case

$$\begin{aligned} n_{\alpha\beta}^{(2)}(x_1, x_2) &= n_{\alpha,B}\zeta_\beta \frac{\Xi^{\beta w_2}(x_2 + a_\beta/2, z)}{\Xi^{\alpha w_2}(x_1 + a_\alpha/2, z)} \Xi^{\alpha\beta}(x_1 + a_\alpha/2, x_2 - a_\beta/2) \\ &= n_{\alpha,B}\zeta_\beta e^{-\beta P_B(x_2 - x_1 + \Delta_{\beta\alpha})} \Xi^{\alpha\beta}(0, x_2 - x_1 - a_{\alpha\beta}), \end{aligned} \quad (10.4.17)$$

where $n_{\alpha,B}$ is the bulk phase density of species α. Thus, the pair correlation function of homogeneous phase ($g_{\alpha\beta}^{(2)} \equiv n_{\alpha\beta}^{(2)}/n_{\alpha,B}n_{\beta,B}$) is given by

$$g_{\alpha\beta}^{(2)}(x_1, x_2) = \frac{\zeta_\beta}{n_{\beta,B}} e^{-\beta P_B(x_2 - x_1 + \Delta_{\beta\alpha})} \Xi^{\alpha\beta}(0, x_2 - x_1 - a_{\alpha\beta}), \quad x_2 > x_1. \quad (10.4.18)$$

With the aid of eq. (10.4.14), we can rewrite eq. (10.4.16) as

$$g_{\alpha\beta}^{(2)}(x_1, x_2) = \frac{n_\beta^\alpha(x_2 - x_1 - a_\alpha/2)}{n_{\beta,B}}, \quad x_2 > x_1, \quad (10.4.19)$$

where $n_\beta^\alpha(x)$ is the density of species β semiconfined by a Tonks–Takahashi particle of species α whose center is fixed at $x - a_\alpha/2$. For $x_1 > x_2$, $g_{\alpha\beta}^{(2)}(x_1, x_2)$ is computed by interchanging x_1 and x_2 and α and β on the rhs of eq. (10.4.19).

10.4.2 One-Component Fluid: No External Field ($v(x) = 0$)

At this writing no one has found a closed set of integral equations for the density distributions of Tonks–Takahashi fluids in the presence of external forces (except for hard rods). Thus, the rest of this section will be devoted to the special case in which $v_\alpha(x) \equiv 0$. In this subsection we examine one-component Tonks–Takahashi fluids whose confining particles are identical to the fluid particles.

The basic problem is to evaluate the quantity $X_N(y, z)$ given by eq. (10.4.6) for the special case $v(x_i) = 0$. By transforming variables $x_j \to x_j - y$ and noting that only nearest neighbor particles interact, we obtain

$$\begin{aligned} X_N(y, z) = X_N(0, z - y) &= \zeta^N \int_0^{z-y} dx_N \int_0^{x_n} dx_{N-1} \cdots \int_0^{x_2} dx_1 e^{-\beta u(z - y + a/2 - x_N)} \\ &\times e^{-\sum_{i=1}^{N-1} \beta u(x_{i+1} - x_i) - \beta u(x_1 + a/2)}. \end{aligned} \quad (10.4.20)$$

This expression has the form of a convolution integral, that is,

$$X_N(0, \xi) = \zeta^N \int_0^\xi dx_N \int_0^{x_N} dx_{N-1} \cdots \int_0^{x_2} dx_1 g(\xi - x_N) \prod_{i=1}^{N-1} k(x_{i+1} - x_i) g(x_1), \quad (10.4.21)$$

where

$$g(x) = e^{-\beta u(x+a/2)} \quad \text{and} \quad k(x) = e^{-\beta u(x)}. \tag{10.4.22}$$

The Laplace transform,

$$\tilde{X}_N(s) = \int_0^\infty X_N(0, \xi) e^{-s\xi} d\xi, \tag{10.4.23}$$

of a convolution integral is simply a product of the Laplace transform of the convolved functions, that is,

$$\tilde{X}_N(s) = \zeta^N [\tilde{g}(s)]^2 [\tilde{k}(s)]^{N-1}. \tag{10.4.24}$$

Note that this expression is valid only if $N \geq 1$.

Because the inverse Laplace transform is

$$X_N(0, \xi) = \frac{1}{2\pi i} \int_{-i\infty+\tau_0}^{i\infty+\tau_0} e^{\xi s} \tilde{X}_N(s) ds, \tag{10.4.25}$$

where τ_0 lies to the right of the poles of $\tilde{X}_N(s)$ in the complex plane, the grand canonical partition function for this case ($\Xi(y, z) = \Xi(0, z-y) \equiv \Xi_{z-y}$) becomes

$$\Xi_\xi = e^{-\beta u(\xi+a)} + \sum_{N=1}^{\infty} \frac{\zeta^N}{2\pi i} \int_{-i\infty+\tau_0}^{i\infty+\tau_0} e^{\xi s} [\tilde{g}(s)]^2 [\tilde{k}(s)]^{N-1} ds. \tag{10.4.26}$$

The first term on the rhs of eq. (10.4.24) results from interaction between the fixed Tonks–Takahashi particles (the pore walls) when their separation is too small to allow occupancy by a fluid particle.

In Section 4.9 we introduced the following three model Tonks–Takahashi fluids:

1 hard-rod fluid

$$\psi(x) = 0, \quad x > a, \tag{10.4.27}$$

2 square-well fluid

$$\begin{aligned} \psi(x) &= -\varepsilon, \quad a < x < \nu a \\ &= 0, \quad x > \nu a, \end{aligned} \tag{10.4.28}$$

3 triangle-well fluid

$$\begin{aligned} \psi(x) &= \varepsilon[(x - \nu a)]/(\nu - 1)a, \quad a < x < \nu a \\ &= 0. \end{aligned} \tag{10.4.29}$$

For the Tonks–Takahashi fluids it is easy to show that

$$\tilde{g}(s) = e^{sa/2} \tilde{k}(s). \tag{10.4.30}$$

The Laplace transforms of $k(x)$ were given in Section 4.9. They are

1 hard-rod fluid

$$\tilde{k}(s) = \frac{e^{-sa}}{s} \tag{10.4.31}$$

2 square-well fluid

$$\tilde{k}(s) = \frac{e^{-sa+\beta\varepsilon}}{s} + (1 - e^{\beta\varepsilon})\frac{e^{-sav}}{s} \tag{10.4.32}$$

3 triangle-well fluid

$$\tilde{k}(s) = \frac{e^{-sa+\beta\varepsilon} - e^{-sva}}{s + \beta\varepsilon/(v-1)a} + \frac{e^{-sva}}{s}. \tag{10.4.33}$$

Evaluation of the inverse Laplace transforms in eq. (10.4.24) is tedious and the details will not be given here. However, the following properties of Laplace transforms are all that are required to carry out the inversion:

$$I_1 \equiv \frac{1}{2\pi i}\int_{-i\infty+\tau_0}^{i\infty+\tau_0} \frac{e^{ys}}{s - s_1}ds = e^{ys_1}\eta(y), \tag{10.4.34}$$

$$I_n \equiv \frac{1}{2\pi i}\int_{-i\infty+\tau_0}^{i\infty+\tau_0} \frac{e^{ys}}{(s - s_1)^{n+1}}ds = \frac{1}{n!}\frac{d^n I_1}{ds_1^n} = \frac{y^n e^{ys_1}}{n!}\eta(y), \tag{10.4.35}$$

$$J_{11} \equiv \frac{1}{2\pi i}\int_{-i\infty+\tau_0}^{i\infty+\tau_0} \frac{e^{ys}}{(s - s_1)(s - s_2)}ds = \left[\frac{e^{ys_1}}{s_1 - s_2} + \frac{e^{ys_2}}{s_2 - s_1}\right]\eta(y), \quad s_1 \neq s_2 \tag{10.4.36}$$

$$J_{nm} \equiv \frac{1}{2\pi i}\int_{-i\infty+\tau_0}^{i\infty+\tau_0} \frac{e^{ys}}{(s - s_1)^{n+1}(s - s_2)^{m+1}}ds = \frac{1}{n!m!}\frac{d^n}{ds_1^n}\frac{d^m}{ds_2^m}J_{11}, \quad s_1 \neq s_2, \tag{10.4.37}$$

where $\eta(y)$ is the Heaviside step function.

Application of eqs. (10.4.34)–(10.4.37) to eq. (10.4.26) for the various models yields

1 hard-rod fluid

$$\Xi_\xi = \sum_{N=0}^{\infty} \frac{\zeta^N}{N!}(\xi - Na)^N \eta(\xi - Na), \tag{10.4.38}$$

2 square-well fluid

$$\Xi_\xi = \sum_{N=0}^{\infty}\sum_{k=0}^{N+1} \frac{\zeta^N}{N!}\frac{(N+1)!}{(N+1-k)!k!}e^{(N+1-k)\beta\varepsilon} \times (1 - e^{\beta\varepsilon})^k(\xi - Na - (v-1)ka)^N \eta(\xi - Na - (v-1)ka) \tag{10.4.39}$$

3 triangle-well fluid

$$\begin{aligned}\Xi_\xi &= \sum_{N=0}^{\infty}\zeta^N \sum_{k=1}^{N+1}\frac{(N+1)!\alpha^k}{(N+1-k)!k!}\eta(\xi - Na - k(\nu-1)a)e^{(N+1-k)\beta\varepsilon}\\ &\times\left\{\frac{(-1)^k}{N!}\sum_{j=0}^{N}(\xi - Na - k(\nu-1)a)^{N-j}\frac{N!}{(N-j)!j!}\frac{(k+j-1)!}{(k-1)!}\frac{e^{-\alpha(\xi-Na-k(\nu-1)a)}}{\alpha^{k+j}}\right.\\ &\left.+\sum_{j=0}^{k-1}(-1)^j(\xi - Na - k(\nu-1)a)^{k-1-j}\frac{1}{(k-1-j)!j!}\frac{(N+j)!}{N!}\frac{1}{\alpha^{N+1+j}}\right\}\\ &+\sum_{N=0}^{\infty}\zeta^N\frac{e^{(N+1)\beta\varepsilon}}{N!}\eta(\xi - Na)(\xi - Na)^N e^{-\alpha(\xi-Na)},\end{aligned} \tag{10.4.40}$$

where $\alpha \equiv \beta\varepsilon/(\nu-1)a$.

In Section 4.9 we saw that the equations of state for the chemical potential and density of a one-component bulk Tonks–Takahashi fluid are

$$\mu = kT\ln\Lambda - kT\ln\tilde{k}(\beta P_B) \tag{10.4.41}$$

$$n_B^{-1} = -\tilde{k}'(\beta P_B)/\tilde{k}(\beta P_B). \tag{10.4.42}$$

For the three Tonks–Takahashi models considered here these equations become

1 hard-rod fluid

$$\beta\mu = \beta P_B a + \ln(\beta P_B a) + \ln(\Lambda/a) \tag{10.4.43}$$

$$n_B^{-1} = a + 1/\beta P \tag{10.4.44}$$

2 square-well fluid

$$\beta\mu = \beta P_B a + \ln\beta P_B a - \ln\{e^{\beta\varepsilon} + (1-e^{\beta\varepsilon})e^{-\beta P_B(\nu-1)a}\} + \ln(\Lambda/a) \tag{10.4.45}$$

$$n_B^{-1} = a + 1/\beta P + \frac{(\nu-1)a(1-e^{\beta\varepsilon})e^{-\beta P_B(\nu-1)a}}{e^{\beta\varepsilon} + (1-e^{\beta\varepsilon})e^{-\beta P_B(\nu-1)a}} \tag{10.4.46}$$

3 triangle-well fluid

$$\beta\mu = \beta P_B a + \ln[\beta P_B a(\beta P_B a + \alpha a)] - \ln[\beta P_B a e^{\beta\varepsilon} + \alpha a e^{-\beta P_B(\nu-1)a}] + \ln(\Lambda/a) \tag{10.4.47}$$

$$n_B^{-1} = a + \frac{2\beta P_B a + \beta\alpha a}{(\beta P a)^2 + (\alpha a)(\beta P a)} - a\frac{e^{\beta\varepsilon} - \alpha(\nu-1)a e^{-\beta P_B(\nu-1)a}}{\beta P_B a e^{\beta\varepsilon} + \alpha a e^{-\beta P_B(\nu-1)a}}, \tag{10.4.48}$$

with

$$\alpha \equiv \beta\varepsilon/(\nu-1)a. \tag{10.4.49}$$

To illustrate the effect of the attractive interactions on the properties of confined fluids, we calculated the pressure and density profiles versus pore width (wall separation) L. The fluids compared are in equilibrium with homogeneous phases at nearly the same bulk density, namely, $n_B = (3/4)a$. The well width is chosen to be $2a$ (that is, $\nu = 2$) for the square- and triangle-well fluids and well depth is set at $\beta\varepsilon = 2$. For these conditions, the bulk properties of the fluids are

1 hard-rod fluid

$$\beta P_B a = 3, \quad \beta\mu = 4.0986, \quad na = 3/4 \tag{10.4.50}$$

2 square-well fluid

$$\beta P_B a = 2.4, \quad \beta\mu = 1.3572, \quad n_B a = 0.7510 \qquad (10.4.51)$$

3 triangle-well fluid

$$\beta P_B a = 1.6, \quad \beta\mu = 0.8474, \quad n_B a = 0.7511 \qquad (10.4.52)$$

The datum of the chemical potential is $kT \ln(\Lambda/a)$, that is, the numbers given above for $\beta\mu$ are actually the values of $\beta\mu - \ln(\Lambda/a)$.

The pressure, computed from

$$P = kT\left(\frac{\partial \ln \Xi_L}{\partial L}\right)_{T,\mu}, \qquad (10.4.53)$$

is plotted versus pore width L in Figure 10.15 for the three model fluids.

The pressure versus pore width behaves qualitatively similarly in the three fluids. The peaks in the pressure of the triangle-well fluids are shifted to lower L compared to the hard-rod fluid, whereas those in the square-well fluids are shifted to higher L. Also, the differences between neighboring maxima and minima in the pressure are greatest for the triangle-well fluid and least for the square-well fluid. Thus, the triangle-well attraction seems to sharpen the layering whereas the square-well attraction attenuates it relative to the layering induced in hard-rod fluids by confinement.

The density profiles of the three fluids are shown in Figure 10.16 for three different pore widths. As expected from the pressure behavior versus L, the peaks in the density profiles of the triangle-well fluids are greater than those of the hard-rod fluid, but those of the square-well fluids are smaller. Thus, the triangle-well attraction tends to reinforce the layering induced by the hard-rod excluded volume effect and the square-well attraction decreases the layering tendency.

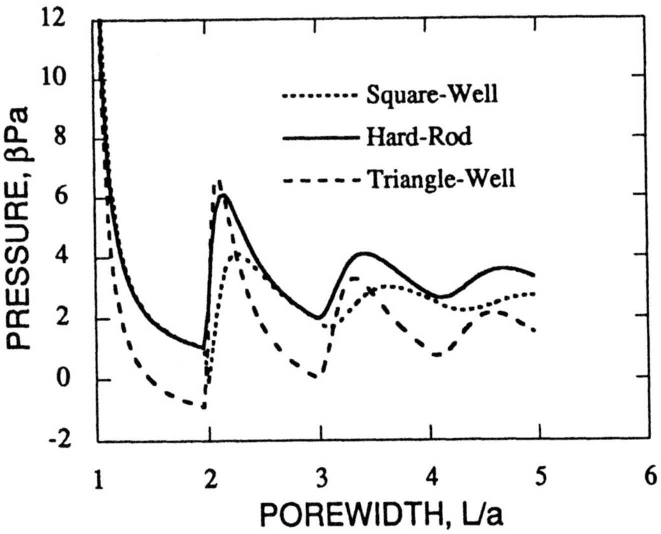

Figure 10.15
Pressure of confined fluids versus confining wall separation L. $\nu = 2$ and $\beta\varepsilon = 2$. Adapted from Davis (1990).

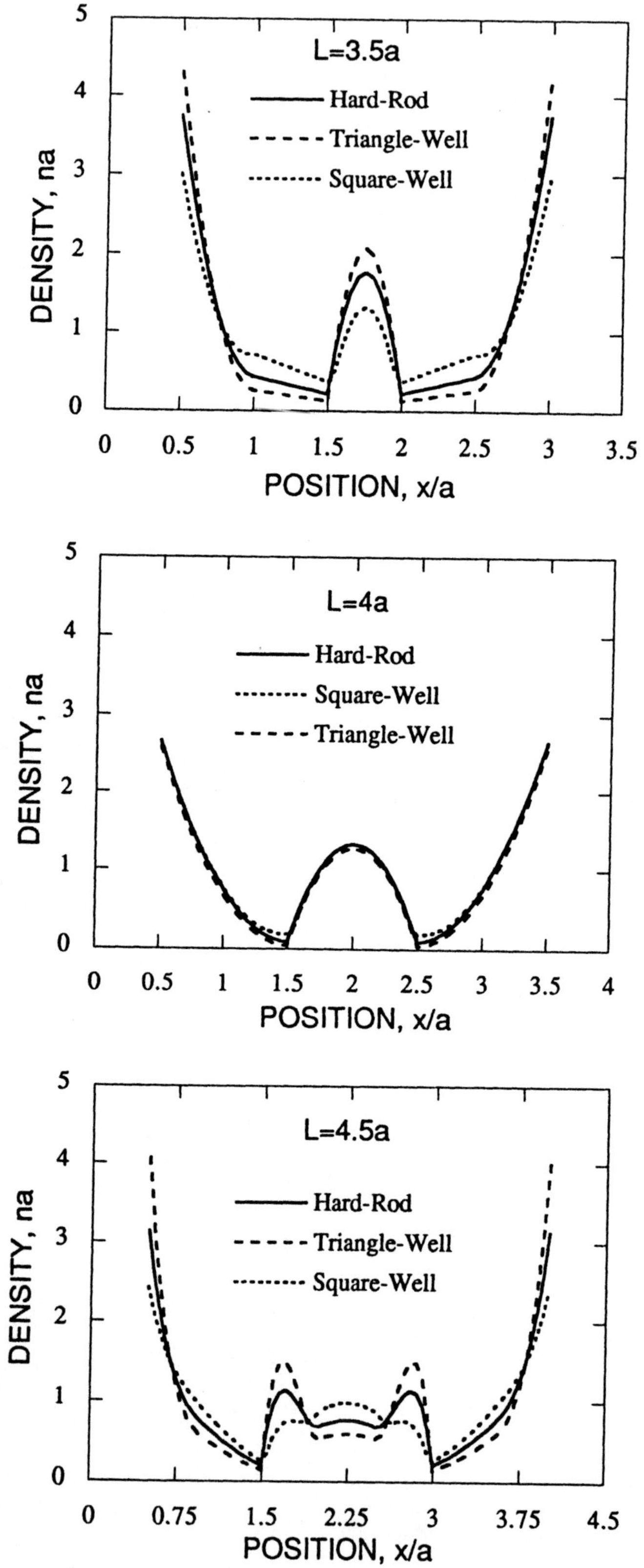

Figure 10.16
Density of confined fluids versus distance from pore wall for wall separations $L = 3.5a$, $4a$, and 4.5a. $\nu = 2$, $\beta\varepsilon = 2$, and $n_B a \simeq 3/4$. Adapted from Davis (1990).

In all three fluids the pressure minima occur near integral values of L/a. This qualitative pattern can be understood from the density profiles shown in Figure 10.16. At a width of $L = 3.5a$, as many as three particles can occupy the pore, but the particles have only half a particle diameter of free volume (length) to distribute among themselves. As a consequence, the density peaks are narrow and high resulting in a relatively high pressure. At a width of $L = 4a$ there is still only room for three particles (the probability of four particle occupancy is zero when $L \leq 4a$), but now the three particles share an entire particle diameter of free volume (length). The result is that the density peaks are lower and broader and the pressure is lower than when the pore width is $L = 3.5a$. At $L = 4.5a$ four particle occupancy can occur, but because the particles share only a half a particle diameter of free volume, the peaks are sharper and the pressure is higher than for $L = 4a$ (the pattern is shifted in L and muted in the case of the square-well fluid).

For a semiconfined Tonks–Takahashi fluid

$$n(x) = \frac{e^{\beta\mu}}{\Lambda} e^{-\beta P_B(x+a/2)} \Xi_{x-a/2}. \tag{10.4.54}$$

The density profiles for the three model fluids at the bulk conditions given at eqs. (10.4.50)–(10.4.52) are shown in Figure 10.17.

In the homogeneous bulk phase, the pair correlation function is given by

$$g^{(2)}(x_1, x_2) = n(x_2 - x_1 - a/2)/n_B, \quad x_2 > x_1, \tag{10.4.55}$$

where $n(x)$ is the density profile of semiconfined fluid. Thus, the pair correlation function in bulk fluid as a function of x_{21} can be deduced directly from Figure 10.16 by shifting the profiles to the right by the amount 1/2 and dividing the abscissa by the bulk density (3/4 in this case).

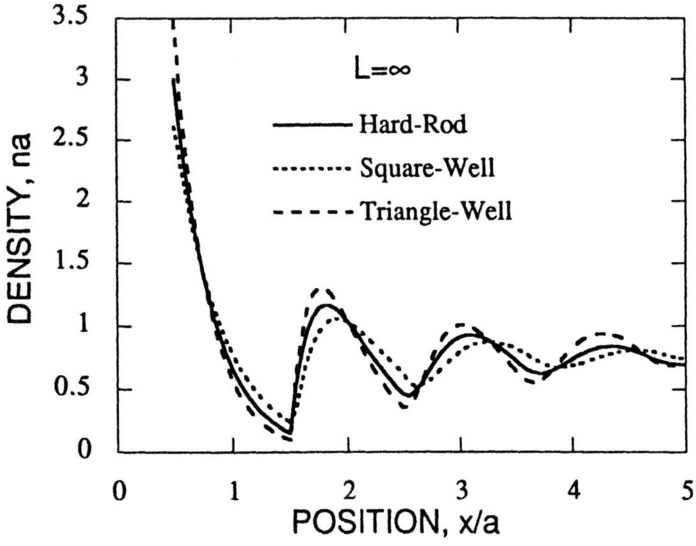

Figure 10.17
Density of semiconfined fluids versus distance from pore wall. $\nu = 2$, $\beta\varepsilon = 2$, $n_B a \simeq 3/4$. Adapted from Davis (1990).

Supplementary Reading

Henderson, D., Ed. 1991. *Fundamentals of Inhomogeneous Fluids,* Dekker, New York.

Leib, E. H. and Mattis, D. C. 1966. *Mathematical Physics in One Dimension,* Collection of Reprints, Academic Press, New York.

Exercises

1 The definition of the singlet direct correlation function for a one-dimensional fluid is

$$C^{(1)}(x) = \beta\Big[-\mu + kT\ln\Lambda + v(x)\Big] + \ln n(x)$$

From $C^{(1)}$, the doublet direct correlation function can be determined from

$$C^{(2)}(x,x') = \delta C^{(1)}(x)/\delta n(x').$$

The density $n(x)$ of hard rods obeys the equation

$$\beta\Big[v(x)-\mu\Big] + \ln n(x) + \ln\Lambda = \ln\left[1 - \int_x^{x+a} n(w)dw\right] - \int_{x-a}^{x} \frac{n(z)dz}{1-\int_z^{z+a} n(w)dw}.$$

a) Obtain an expression for $C^{(1)}(x)$ as a functional of density.

b) Derive a density functional expression for $C^{(2)}(x,x')$.

2 Adsorption in a pore may be modeled as a 1-dimensional gas. Suppose the adsorbates obey the following potential in a pore of length L

$$\phi(x) = -|a|x, \quad 0 \le x \le L.$$

Assume that the adsorbate particles do not interact with each other.

a) Calculate the partition function, Ξ, of the adsorbed gas. (Remember to include kinetic energy.)
Note: $\Xi = \sum_{N=0}^{\infty} e^{\beta N\mu} Q_N/N!\Lambda^N$.

b) Determine the adsorption isotherm (i.e., obtain $\overline{N}$) by taking an appropriate derivative of Ξ.

c) How does $\overline{N}$ vary with pressure? Assume $\mu = kT\ln\left(\frac{P}{P_o}\right)$. What effect would an increase in $|a|$ have?

d) Is the isotherm realistic at high pressures? What model might be a qualitative improvement?

3 Calculate the pressure versus L for hard rods confined by an external potential of the form

$$\begin{aligned} v(x) &= \infty, \quad x < a/2, \quad x > L - a/2 \\ &= -\varepsilon e^{-(x-a/2)^2/a^2} - \varepsilon e^{-(L-a/2-x)^2/a^2}, \; (a/2) < x < L - (a/2) \end{aligned}$$

where the bulk pressure is $P_B a/kT = 3$, where P_B is the pressure of bulk phase in equilibrium with the confined fluid. Assume $kT/\varepsilon = 1$ and 2, respectively.

4 Calculate the disjoining pressure versus pore width for a hard wall confined hard-rod fluid for $P_B a/kT = 1, 2,$ and 4. Compare the results with Figure 10.4.

5 Repeat the calculations in Figures 10.6 and 10.7 for $a_2 = 2a_1$ and $P_B a_1/kT = 1$ and 4.

6 Repeat the calculations in Figure 10.12 for $f(a) = e^{-(a-a_1)^2/2\sigma^2} / \int_{a_1}^{a_2} e^{-(a-a_1)^2/2\sigma^2} da, \sigma = a_1, a_2 = 4a_1$.

References

Davis, H. T. 1990. *J. Chem. Phys.*, **93,** 4343.

Percus, J. K. 1976. *J. Stat. Phys.*, **15,** 505.

Tang, Z., Scriven, L. E., and Davis, H. T. 1992. *J. Chem. Phys.*, **97,** 5732.

Vanderlick, T. K. 1988. Ph.D. Thesis, University of Minnesota at Minneapolis, MN.

Vanderlick, T. K., Davis, H. T., and Percus, J. K. 1989. *J. Chem. Phys.*, **91,** 7136.

11

DENSITY FUNCTIONAL THEORY OF FLUID INTERFACES

11.1 Local Density Functional Free Energy Models

For strongly inhomogeneous systems such as fluids at solid walls, the fluids are so highly structured that nonlocal density functional theories must be used to describe the systems. An exactly solvable example of such a system is provided by the one-dimensional (ID) hard-rod theory in Chapter 10. For fluid–fluid interfaces, however, density gradients are smaller and so local density functional theory is sufficient to capture the physics of these interfaces.

11.1.1 The van der Waals Model

The oldest approximation to the free energy of strongly inhomogeneous fluid is the model of van der Waals (VDW). Some predictions of the model are still good qualitative approximations (sometimes even quantitative estimates, e.g., of the tension of fluid interfaces), and the model forms the motivational starting point for several more quantitative theories that have been developed. Furthermore, it remains the simplest attempt to understand the molecular nature of fluid interfaces.

As in the case of homogeneous fluids, the starting point of VD theory is the "mean field" approximation to the partition function, namely,

$$Q_N \simeq Q_N^H e^{-\beta U^A}, \tag{11.1.1}$$

where Q_N^H is the partition function of a fluid of hard spheres and $U^A \equiv \langle u^A \rangle^H$ is the contribution of the long-range attractive interactions between fluid particles. $\langle u^A \rangle^H$ indicates an average of the interaction potential of u^A over an ensemble of hard spheres. The density distributions, $n_\alpha(\mathbf{r})$, of the hard sphere fluid are chosen to be the same as those in the actual fluid. From eq. (11.1.1) it follows that the Helmholtz free energy of the fluid is

$$F = F^H + U^A, \tag{11.1.2}$$

where F^H is the free energy of the hard sphere fluid.

According to the VDW approximation the attractive interactions are pair additive and are sufficiently long ranged that the pair correlation function can be approximated by unity in $\langle u^A \rangle^H$. This is to say that the fluid is assumed to be structureless in computing U^A, and so

$$U^A = \sum_{\alpha,\beta} \frac{1}{2} \int n_\alpha(\mathbf{r}) n_\beta(\mathbf{r}') u^A_{\alpha\beta}(|\mathbf{r}' - \mathbf{r}|) d^3r d^3r', \tag{11.1.3}$$

where $u^A_{\alpha\beta}(s)$ is the attractive part of the pair potential between particles of species α and β. For example, for a 6–12 Lennard–Jones fluid

$$\begin{aligned} u^A_{\alpha\beta}(s) &= 0, \quad s < d_{\alpha\beta} \\ &= 4\varepsilon_{\alpha\beta} \left[\left(\frac{\sigma_{\alpha\beta}}{s}\right)^{12} - \left(\frac{\sigma_{\alpha\beta}}{s}\right)^{6} \right], \quad s > d_{\alpha\beta} \end{aligned} \tag{11.1.4}$$

where $d_{\alpha\beta}$ is the hard sphere cutoff for the interaction between species α and β. The simplest choice of $d_{\alpha\beta}$ is $\sigma_{\alpha\beta}$, the characteristic length of the Lennard–Jones potential, but one could make other choices.

The next assumption of VDW theory is that the local Helmholtz free energy density, $f^H(\mathbf{r})$, of hard spheres depends only on the local densities $n_\alpha(\mathbf{r})$ at position $\mathbf{r}$. The argument is that because the hard sphere interactions are so short ranged the Helmholtz free energy density of an inhomogeneous hard sphere fluid is a function of the local density only. This is the sense in which the VDW model is a local density functional approximation. With this approximation,

$$F^H = \int f^H(\mathbf{n}(\mathbf{r})) d^3r, \tag{11.1.5}$$

where $f^H(\mathbf{n})$ is the Helmholtz free energy density of a homogeneous fluid of hard spheres at the composition $\mathbf{n} = \{n_1, \ldots, n_c\}$. Combining eqs. (11.1.3) and (11.1.5) and using the fact that the dummy indices in the summation $\sum_{\alpha\beta}$ can be interchanged, we obtain ofter some rearrangement

$$F = \int f^\circ(\mathbf{n}(\mathbf{r})) - \sum_{\alpha,\beta} \frac{1}{4} \int [n_\alpha(\mathbf{n}') - n_\alpha(\mathbf{r})][n_\beta(\mathbf{r}') - n_\beta(\mathbf{r})] u^A_{\alpha\beta}(|\mathbf{r}' - \mathbf{r}|) d^3r d^3r', \tag{11.1.6}$$

where

$$f^\circ(\mathbf{n}(\mathbf{r})) \equiv f^H(\mathbf{n}(\mathbf{r})) - \sum_{\alpha,\beta} \frac{1}{2} n_\alpha(\mathbf{r}) n_\beta(\mathbf{r}) a_{\alpha\beta} \tag{11.1.7}$$

with

$$a_{\alpha\beta} \equiv -\int u^A_{\alpha\beta}(s) d^3 s. \tag{11.1.8}$$

$f^\circ(\mathbf{n})$ is the Helmholtz free energy density of a homogeneous VD fluid at composition $\mathbf{n}$.

If an external field is present, the VDW free energy functional becomes

$$F = \int f^\circ(\mathbf{n}(\mathbf{r})) - \sum_{\alpha,\beta} \frac{1}{4} \int [n_\alpha(\mathbf{n}') - n_\alpha(\mathbf{r})][n_\beta(\mathbf{r}') - n_\beta(\mathbf{r})] u^A_{\alpha\beta}(|\mathbf{r}' - \mathbf{r}|) d^3 r d^3 r' + \sum_\alpha \int n_\alpha(\mathbf{r}) v_\alpha(\mathbf{r}) d^3 r, \tag{11.1.9}$$

where $v_\alpha(\mathbf{r})$ is the potential of the external force on a particle of species α at position $\mathbf{r}$.

The chemical potential μ_α of species α is equal to the density functional derivative of F, that is, $\mu_\alpha = \delta F / \delta n_\alpha(\mathbf{r})$, and so for the VDW model

$$\mu_\alpha = \mu^\circ_\alpha(\mathbf{n}(\mathbf{r})) + \sum_\beta \int [n_\alpha(\mathbf{r}') - n_\beta(\mathbf{r})] u^A_{\alpha\beta}(|\mathbf{r}' - \mathbf{r}|) d^3 r' + v_\alpha(\mathbf{r}). \tag{11.1.10}$$

where $\mu^\circ_\alpha(\mathbf{r})(\equiv \partial f^\circ(\mathbf{n})/\partial n_\alpha)$ is the chemical potential of α in a homogeneous fluid at composition $\mathbf{n}$. At equilibrium the μ_α are constant. Thus, eq. (11.1.9) provides a set of nonlinear integral equations that determine the density distributions for a given system.

Physically the VDW model divides the free energy F into a first part, $\int f^\circ d^3 r$, characteristic of homogeneous fluid, a second part that is nonzero only because of fluid density inhomogeneities, and a third part, $\sum_\alpha \int n_\alpha v_\alpha d^3 r$, arising from external forces. In the limit of a homogeneous phases in the absence of external fields only the first term, $\int f^\circ(n) d^3 r = f^\circ(n) V = F^\circ$, survives.

11.1.2 A Modified VDW Model

One can improve the VDW theory by replacing the hard sphere partition function $Q^H_\mathbf{N}$ by the partition function $Q^R_\mathbf{N}$ of a fluid whose particles interact with the purely repulsive forces of the particles of the subject fluid. Also, in taking the average $\langle u^A \rangle^R$ the pair correlation function of the reference repulsive fluid can be used, instead of unity, to improve the quantitative estimate of the average.

Thus, as a quantitative improvement, we replace the VDW free energy functional by

$$F = \sum_{\alpha} \int n_\alpha(\mathbf{r}) v_\alpha(\mathbf{r}) d^3r + \int f^\circ(\mathbf{n}(\mathbf{r})) d^3r$$
$$- \sum_{\alpha,\beta} \frac{1}{4} \int [n_\alpha(\mathbf{r}') - n_\alpha(\mathbf{r})][n_\beta(\mathbf{r}') - n_\beta(\mathbf{r})] \tag{11.1.11}$$
$$\times g^R_{\alpha\beta}(|\mathbf{r}' - \mathbf{r}|, \bar{\mathbf{n}}) u^A_{\alpha\beta}(|\mathbf{r}' - \mathbf{r}|) d^3r d^3r'$$

where

$$\bar{\mathbf{n}} \equiv [\mathbf{n}(\mathbf{r}) + \mathbf{n}(\mathbf{r}')]/2 \tag{11.1.12}$$

and

$$f^\circ(\mathbf{n}(\mathbf{r})) = f^R(\mathbf{n}(\mathbf{r})) + \sum_{\alpha,\beta} \frac{1}{2} n_\alpha(\mathbf{r}) n_\beta(\mathbf{r}) \int g^R_{\alpha\beta}(|\mathbf{r}' - \mathbf{r}|, \mathbf{n}(\mathbf{r})) u^A_{\alpha\beta}(|\mathbf{r}' - \mathbf{r}|) d^3r d^3r'. \tag{11.1.13}$$

The quantity $g^R_{\alpha\beta}(|\mathbf{r}' - \mathbf{r}|, \mathbf{n}(\mathbf{r}))$ denotes the pair correlation function of a homogeneous fluid of the repulsive particles at a composition $\mathbf{n}(\mathbf{r})$. Because these particles have purely repulsive forces, $g^R_{\alpha\beta}(|\mathbf{r}' - \mathbf{r}, \mathbf{n})$ exists at any composition, i.e., such a fluid does not have spinodal regions where the homogeneous phase and their correlation functions do not exist.

In the second term in F, we have introduced the correlation function of homogeneous fluid at the mean composition of the correlated particles, that is, at $\bar{\mathbf{n}} = [\mathbf{n}(\mathbf{r}) + \mathbf{n}(\mathbf{r}')]/2$. It would have been equally logical to have used $g^R_{\alpha\beta}(|\mathbf{r}' - \mathbf{r}|, \mathbf{n}(\mathbf{r} + \mathbf{r}')/2)$ or $[g^R_{\alpha\beta}(|\mathbf{r}' - \mathbf{r}|, \mathbf{n}(\mathbf{r})) + g^R_{\alpha\beta}(|\mathbf{r}' - \mathbf{r}|, \mathbf{n}(\mathbf{r}')]/2$ or some other local composition for $\bar{\mathbf{n}}$, but there is evidence that the different choices make little quantitative difference (McCoy and Davis, for fluid–fluid interfaces.

The chemical potential μ_α for the modified VDW fluid is obtained from the functional derivative of eq. (11.1.11) with respect to $n_\alpha(\mathbf{r})$. The result is

$$\mu_\alpha = v_\alpha(\mathbf{r}) + \mu^\circ_\alpha(\mathbf{n}(\mathbf{r})) + \sum_\beta \int [n_\beta(\mathbf{r}) - n_\beta(\mathbf{r})] g^R_{\alpha\beta}(|\mathbf{r}' - \mathbf{r}|, \bar{\mathbf{n}}) u^A_{\alpha\beta}(|\mathbf{r}' - \mathbf{r}|) d^3r'$$
$$- \sum_{\beta\gamma} \frac{1}{4} \int [n_\beta(\mathbf{r}') - n_\beta(\mathbf{r})][n_\gamma(\mathbf{r}') - n_\gamma(\mathbf{r})] \frac{\partial g^R_{\beta\gamma}(|\mathbf{r}' - \mathbf{r}|, \bar{\mathbf{n}})}{\partial \bar{n}_\alpha} u^A_{\beta\gamma}(|\mathbf{r}' - \mathbf{r}|) d^3r'. \tag{11.1.14}$$

Given $g^R_{\beta\gamma}(|\mathbf{r} - \mathbf{r}|, \mathbf{n})$ as a function of composition, one takes the partial derivative $\partial g_{\beta\gamma}/\partial n_\alpha$ and sets $\mathbf{n}$ equal to $[n(\mathbf{r}) + n(\mathbf{r}')]/2$ to obtain the function called for in the integrand of eq. (11.1.14). $\mu^\circ_\alpha(\mathbf{n})(\equiv \partial f^\circ(\mathbf{n})/\partial n_\alpha)$ is the chemical potential of homogeneous fluid at composition $\mathbf{n}$.

11.1.3 An Approximate Density Functional (ADF) Model

The exact function form of F for a classical fluid is [see eq. (9.2.44)]

$$F = \sum_\alpha \int n_\alpha(\mathbf{r}) v_\alpha(\mathbf{r}) d^3r + \sum_\alpha \int n_\alpha(\mathbf{r})[\mu_\alpha^+(T) + \ln n_\alpha(\mathbf{r}) - 1] d^3r - \sum_{\alpha,\beta} kT \int \int_0^1 (1-\varepsilon) C_{\alpha\beta}^{(2)}(\mathbf{r}, \mathbf{r}'; \{\varepsilon \mathbf{n}\}) n_\alpha(\mathbf{r}) n_\beta(\mathbf{r}') d^3r d^3r' d\varepsilon. \tag{11.1.15}$$

In the limit of low density, $C_{\alpha\beta}^{(2)}(\mathbf{r}, \mathbf{r}')$ is independent of density and depends only on $|\mathbf{r}' - \mathbf{r}|$. Also in this low density limit, its asymptotic value is $-u_{\alpha\beta}/kT$, which is the exact asymptotic value of $C_{\alpha\beta}^{(2)}$ for any density distribution. With the low density formula for $C_{\alpha\beta}^{(2)}$, eq. (11.1.14) can be rearranged to

$$F = \int f^\circ(\mathbf{n}(\mathbf{r})) d^3r + \sum_{\alpha,\beta} \frac{kT}{4} \int C_{\alpha\beta}^{(2)}(|\mathbf{r}' - \mathbf{r}|)[n_\alpha(\mathbf{r}') - n_\alpha(\mathbf{r})][n_\beta(\mathbf{r}') - n_\beta(\mathbf{r})] d^3r d^3r' + \sum_\alpha \int n_\alpha(\mathbf{r}) v_\alpha(\mathbf{r}) d^3r \tag{11.1.16}$$

where $f^\circ(\mathbf{n})$ is the Helmholtz free energy density of homogeneous fluid (with $C_{\alpha\beta}^{(2)}$ approximated by its low density limit) at local composition $\mathbf{n}$. This free energy functional is quite similar to VDW because $C_{\alpha\beta}^{(2)} \approx -\beta u_{\alpha\beta}^A$ for large interparticle separations.

Ebner (1976) introduced an approximate extension of eq. (11.1.16) to higher densities by replacing $C_{\alpha\beta}^{(2)}$ by the direct correlation function of homogeneous fluid at composition $\bar{\mathbf{n}} = [\mathbf{n}(\mathbf{r}) + \mathbf{n}(\mathbf{r}')]/2$ and by choosing a model for $f^\circ(\mathbf{n})$ valid at any density. Thus, their model is

$$F = \int f^\circ(\mathbf{n}(\mathbf{r})) d^3r + \sum_{\alpha,\beta} \frac{kT}{4} \int C_{\alpha\beta}^{(2)}(|\mathbf{r}' - \mathbf{r}|, \bar{\mathbf{n}})[n_\alpha(\mathbf{r}') - n_\alpha(\mathbf{r})][n_\beta(\mathbf{r}') - n_\beta(\mathbf{r})] d^3r d^3r' + \sum_\alpha \int n_\alpha(\mathbf{r}) v_\alpha(\mathbf{r}) d^3r. \tag{11.1.17}$$

In keeping with past usage, we will refer to this as the approximate density functional (ADF) model, although all the models discussed in this and the next two chapters are approximate density functional models. With this model Ebner and Saam (1977) predicted a first-order, prewetting thin-film transition similar to the one predicted by Cahn (1977). The theory of wetting transitions will be discussed in detail in Chapter 13. In implementing eq. (11.1.17) one has to choose functional forms for the $C_{\alpha\beta}^{(2)}$ that exist at all the compositions of interest. The chemical potential μ_α corresponding to eq. (11.1.17) is

$$\mu_\alpha = v_\alpha(\mathbf{r}) + \mu_\alpha^\circ(\mathbf{n}(\mathbf{r})) - \sum_\beta kT \int [n_\beta(\mathbf{r}') - n_\beta(\mathbf{r})] C_{\alpha\beta}^{(2)}(|\mathbf{r}' - \mathbf{r}|, \bar{\mathbf{n}}) d^3r' + \sum_{\beta,\gamma} \frac{kT}{4} \int \frac{\partial C_{\alpha\beta}^{(2)}(|\mathbf{r}' - \mathbf{r}|, \bar{\mathbf{n}})}{\partial \bar{n}_\alpha} [n_\beta(\mathbf{r}') - n_\beta(\mathbf{r})][n_\gamma(\mathbf{r}') - n_\gamma(\mathbf{r})] d^3r'. \tag{11.1.18}$$

As is the case of the VDW models, the equilibrium condition of constant chemical potentials leads to a nonlinear set of integral equations for the density distributions of fluids in equilibrium situations.

11.1.4 Density Gradient Theory

When the density variations are not too strong, as when two coexisting fluids approach a critical point, density differences in the free energy functionals can be expanded in a Taylor's series and truncated after a few terms. The result is that the equations controlling the equilibrium distributions become nonlinear differential equations instead of nonlinear integral equations. Because the density variations are always strong sufficiently near an impenetrable solid, the theory developed in this section is at best valid only for fluid far from such confining solids and in cases for which the external potential $v_\alpha(\mathbf{r})$ is slowly varying, as for example the gravitational potential energy.

With such restrictions on theory in mind, consider the Taylor series

$$n_\alpha(\mathbf{r}') = n_\alpha(\mathbf{r}) + (\mathbf{r}' - \mathbf{r}) \cdot \nabla n_\alpha(\mathbf{r}) + \frac{1}{2}(\mathbf{r}' - \mathbf{r})(\mathbf{r}' - \mathbf{r}) : \nabla\nabla n_\alpha(\mathbf{r}) + O(\nabla^3 n_\alpha). \tag{11.1.19}$$

Inserting this equation into the VDW free energy functional, we obtain

$$\begin{aligned} F = &\int f^\circ(\mathbf{n}(\mathbf{r}))d^3r + \sum_\alpha \int n_\alpha(\mathbf{r})v_\alpha(\mathbf{r})d^3r \\ &+ \sum_{\alpha,\beta} \frac{kT}{4} \int \{(\mathbf{r}' - \mathbf{r})(\mathbf{r}' - \mathbf{r}) : \nabla n_\alpha(\mathbf{r})\nabla n_\alpha(\mathbf{r}) \\ &+ \frac{1}{2}(\mathbf{r}' - \mathbf{r})(\mathbf{r}' - \mathbf{r})(\mathbf{r}' - \mathbf{r}) : \left[\nabla n_\alpha(\mathbf{r})\nabla\nabla n_\beta(\mathbf{r}) + \nabla\nabla n_\alpha(\mathbf{r})\nabla n_\beta(\mathbf{r})\right] \\ &+ O(\nabla^4)\} u^A_{\alpha\beta}(|\mathbf{r}' - \mathbf{r}|)d^3r' . \end{aligned} \tag{11.1.20}$$

Transformation of variables from $\mathbf{r}'$ to $\mathbf{s} = \mathbf{r}' - \mathbf{r}$ casts the integrals in eq. (11.1.20) into the forms

$$\int \mathbf{ss}u^A_{\alpha\beta}(s)d^3s \quad \text{and} \quad \int \mathbf{sss}u^A_{\alpha\beta}(s)d^3s.$$

Because $u^A_{\alpha\beta}$ is an even function of s, the second of these expressions vanishes and the first is isotropic, that is, it is equal to a constant times the unit tensor $\mathbf{I}$. Thus,

$$\int \mathbf{ss}u^A_{\alpha\beta}(s)d^3s = \frac{1}{3}\mathbf{I}\int s^2u^A_{\alpha\beta}(s)d^3s. \tag{11.1.21}$$

This result can be obtained by introducing Cartesian coordinates, $\mathbf{s} = s_x\hat{\imath} + s_y\hat{\jmath} + s_z\hat{k}$, and noting that

$$\int s_\nu s_\mu u^A_{\alpha\beta}(s)d^3s = 0, \quad \nu \neq \mu, \quad \textit{and} \quad \int s_\nu^2 u^A_{\alpha\beta}(s)d^3s = \frac{1}{3}\int s^2u^A_{\alpha\beta}(s)d^3s. \tag{11.1.22}$$

Thus, to order $\nabla^4 n$, the VDW free energy functional becomes

$$F = \int \left\{ f^{\circ}(\mathbf{n}(\mathbf{r})) + \sum_{\alpha,\beta} \frac{1}{2} c_{\alpha\beta} \nabla n_{\alpha}(\mathbf{r}) \cdot \nabla n_{\beta}(\mathbf{r}) + \sum_{\alpha} n_{\alpha}(\mathbf{n}) v_{\alpha}(\mathbf{r}) \right\} d^3 r, \qquad (11.1.23)$$

where the "influence parameters" $c_{\alpha\beta}$, measuring the effect of density gradients on the free energy of inhomogeneous fluid, are defined by

$$c_{\alpha\beta} = -\frac{1}{6} \int s^2 u^A_{\alpha\beta}(s) d^3 s. \qquad (11.1.24)$$

The parameters $c_{\alpha\beta}$ figure importantly in the gradient theory of inhomogeneous fluids. In the context of the VDW theory, $c_{\alpha\beta}$ is proportional to the pair potential weighted mean square range of intermolecular interactions.

To the same order $\nabla^4 n$ in density gradients, the modified VDW model [eq. (11.1.11)] and the direct correlation function model [eq. (11.1.16)] have the same form as eq. (11.1.23) except that the influence parameters are composition dependent and are given by

$$c_{\alpha\beta}(\mathbf{n}) = -\frac{1}{6} \int s^2 u^A_{\alpha\beta}(s) g^R_{\alpha\beta}(s, \mathbf{n}) d^3 s \qquad (11.1.25)$$

and

$$c_{\alpha\beta}(\mathbf{n}) = \frac{kT}{6} \int s^2 C^{(2)}_{\alpha\beta}(s, \mathbf{n}) d^3 s, \qquad (11.1.26)$$

respectively. The influence parameters of the modified VDW model are similar but not identical to those obtained [eq. (8.5.21)] in the gradient approximation to the Yvon–Born–Green (YBG) equations. With the approximations $g^A_{\alpha\beta} = 1$ and $C^{(2)}_{\alpha\beta} = -\beta u^A_{\alpha\beta}(s)$, all three free energy models give the same influence parameters. Also, if $u'_{\alpha\beta}(s) g_{\alpha\beta}(s, \mathbf{n})$ is approximated as 0 when $s < d_{\alpha\beta}$ and $du^A_{\alpha\beta}(s)/ds$ when $s > d_{\alpha\beta}$, integration of eq. (8.5.21) by parts also results in the VDW value of $c_{\alpha\beta}$. In each model, the symmetry condition $c_{\alpha\beta} = c_{\beta\alpha}$ obtains.

From a continuum point of view, if we postulate that the quantity $\Phi(\mathbf{r}) \equiv f(\mathbf{r}) - \sum n_{\alpha}(\mathbf{r}) v_{\alpha}(\mathbf{r})$ depends on the density distributions through the local densities and their derivatives, that is, $\Phi = \Phi(\mathbf{n}(\mathbf{r}), \nabla \mathbf{n}(\mathbf{r}), \nabla\nabla \mathbf{n}(\mathbf{r}), \ldots)$, then, because Φ is a scalar, a Taylor expansion about $\nabla \mathbf{n} = \nabla\nabla \mathbf{n} = \cdots = 0$ yields to order ∇^4

$$\Phi = \Phi^{(0)}(\mathbf{n}(\mathbf{r})) + \sum_{\alpha\beta} \{ \Phi^{(1)}_{\alpha\beta}(\mathbf{n}(\mathbf{r})) \nabla n_{\alpha} \cdot \nabla n_{\beta} + \Phi^{(2)}_{\alpha\beta}(\mathbf{n}(\mathbf{r}))[n_{\alpha} \nabla^2 n_{\beta} + n_{\beta} \nabla^2 n_{\alpha}] \}. \qquad (11.1.27)$$

Insertion of eq. (11.1.27) for Φ into $F = \int [\Phi + \sum n_{\alpha} v_{\alpha}] d^3 r$ and integration by parts with the condition that n_{α} vanishes on the boundaries of the system, yields the general result

$$F = \int \left[f^{\circ}(\mathbf{n}(\mathbf{r})) + \sum_{\alpha\beta} \frac{1}{2} c_{\alpha\beta}(\mathbf{n}(\mathbf{r})) \nabla n_{\alpha} \cdot \nabla n_{\beta} + \sum_{\alpha} n_{\alpha}(\mathbf{r}) v_{\alpha}(\mathbf{r}) \right] d^3 r, \qquad (11.1.28)$$

where $f^\circ(\mathbf{n}) \equiv \Phi^{(0)}(\mathbf{n})$ and $c_{\alpha\beta}$ is an appropriate combination of the phenomenological quantities $\Phi^{(1)}_{\alpha\beta}$ and $\Phi^{(2)}_{\alpha\beta}$. It follows that from the fact that $\delta^2 F/\delta n_\alpha(\mathbf{r})\delta n_\beta(\mathbf{r}')$ is invariant to the transformation $\alpha, \beta; \mathbf{r}, \mathbf{r}' \to \beta, \alpha; \mathbf{r}', \mathbf{r}$ that $c_{\alpha\beta} = c_{\beta\alpha}$. Thus, the density gradient free energy results derived here for the three models described above can be viewed as special approximations to a general gradient theory. The virtue of the molecular approximations is of course that they lead to explicit formulas for the influence parameters $c_{\alpha\beta}$. In Section 11.2 we will see that for a planar interface an exact molecular theoretical formula, eq.(11.2.24), for the influence parameters can be derived. However, the exact result still contains the pair direct correlation function of inhomogeneous fluid and so cannot be used without some sort of closure approximation.

Recalling from Chapter 9 the properties

$$\frac{\delta n_\beta(\mathbf{r}')}{\delta n_\alpha(\mathbf{r})} = \delta(\mathbf{r}-\mathbf{r}')\delta_{\alpha\beta}, \quad \frac{\delta h(\mathbf{n}(\mathbf{r}'))}{\delta n_\alpha(\mathbf{r})} = \frac{\partial h(\mathbf{n}(\mathbf{r}))}{\partial n_\alpha}\delta(\mathbf{r}-\mathbf{r}'), \tag{11.1.29}$$

we find from eq. (11.1.23) that the chemical potential μ_α ($= \delta F/\delta n_\alpha(\mathbf{r})$) of density gradient theory is of the general form

$$\mu_\alpha = \mu^\circ_\alpha(\mathbf{n}(\mathbf{r})) + v_\alpha(\mathbf{r}) - \sum_\beta \nabla\cdot(c_{\alpha\beta}\nabla n_\beta) + \sum_{\beta,\gamma}\frac{1}{2}\frac{\partial c_{\beta\gamma}}{\partial n_\alpha}\nabla n_\beta\cdot\nabla n_\gamma. \tag{11.1.30}$$

In the special case that the $c_{\alpha\beta}$ are constant, as in the VDW approximation, eq. (11.1.20) becomes

$$\mu_\alpha = \mu^\circ_\alpha(\mathbf{n}) + v_\alpha - \sum_\beta c_{\alpha\beta}\nabla^2 n_\beta. \tag{11.1.31}$$

In a later section of this chapter we see that gradient theory provides a relatively simple, qualitatively (sometimes quantitatively) correct description of fluid–fluid interfaces. As a matter of fact, Cahn and Hilliard's successful theory (1958) of spinodal decomposition of alloys (in which composition gradients develop continuously from a homogeneous state) is based on a gradient free energy theory. van der Waals himself was the first to derive a density gradient free energy functional. In recent decades a similar gradient approximation, called the Landau–Ginzberg model, has been used frequently in the theory of magnetism and of microstructural materials such as surfactant bilayers and lyotropic liquid crystals and bicontinuous phases.

11.2 Local Density Functional Theory of Planar Fluid–Fluid Interfaces

In the case of a planar interface the density $n_\alpha(x)$ is a function only of the potential x normal to the plane of the interface. Thus if A is the area of the interface and in the absence of external forces, the integral local density functional free energy models introduced in the previous section reduce to the form

$$F = A\int f^\circ(\mathbf{n}(x))dx - \sum_{\alpha,\beta}\frac{A}{4}\int [n_\alpha(x') - n_\alpha(x)][n_\beta(x') - n_\beta(x)]u_{\alpha\beta}^{eff}(x,x')dxdx', \qquad (11.2.1)$$

where

$$u_{\alpha\beta}^{eff}(x,x') = \int_0^\infty u_{\alpha\beta}^A\left(\sqrt{(x'-x)^2+\rho^2}\right)2\pi\rho d\rho \qquad (11.2.2)$$

for the VDW model,

$$u_{\alpha\beta}^{eff}(x,x') = \int_0^\infty u_{\alpha\beta}^A\left(\sqrt{(x'-x)^2+\rho^2}\right)g^R\left(\sqrt{(x'-x)^2+\rho^2},\bar{n}\right)2\pi\rho d\rho \qquad (11.2.3)$$

for the modified VDW (MVDW) model, and

$$u_{\alpha\beta}^{eff}(x,x') = -kT\int_0^\infty C_{\alpha\beta}^{(2)}\left(\sqrt{(x'-x)^2+\rho^2},\bar{n}\right)2\pi\rho d\rho, \qquad (11.2.4)$$

for the ADF model. In these expressions

$$\bar{\mathbf{n}} \equiv \frac{1}{2}[\mathbf{n}(x) + \mathbf{n}(x')]. \qquad (11.2.5)$$

We are free to introduce a hypothetical system with the density profile

$$\begin{aligned} n_\alpha(x) &= n_\alpha^{(1)}, \quad x < 0 \\ &= n_\alpha^{(2)}, \quad x > 0 \end{aligned} \qquad (11.2.6)$$

where $\alpha = 1, \ldots c$ and $n_\alpha^{(1)}$ and $n_\alpha^{(2)}$ are the densities of component α in the coexisting phases. The coexistence densities are determined by the equilibrium conditions

$$P_B = P^\circ(\mathbf{n}^{(1)}) = P^\circ(\mathbf{n}^{(2)}) \quad and \quad \mu_\alpha = \mu_\alpha^\circ(\mathbf{n}^{(1)}) = \mu_\alpha^\circ(\mathbf{n}^{(2)}), \quad \alpha = 1, \ldots, c, \qquad (11.2.7)$$

where $P^\circ(\mathbf{n})$ is the pressure of homogeneous phase at composition $\mathbf{n}$ and P_B is the pressure of the coexisting bulk phases. Of course, according to the Gibbs phase rule some of the intensive variables must be set to completely specify the phases in equilibrium.

Choosing the volumes of the actual and hypothetical systems to be large and to be the same, we obtain for the excess free energy associated with the interface

$$\begin{aligned} \Delta F = A\int_{-\infty}^{\infty} [f^\circ(\mathbf{n}(x)) - (1-\eta(x))f^\circ(\mathbf{n}^{(1)}) - \eta(x)f^\circ(\mathbf{n}^{(2)})]dx \\ - A\sum_{\alpha,\beta}\frac{1}{4}\int_{-\infty}^{\infty}\int_{-\infty}^{\infty}[n_\alpha(x') - n_\alpha(x)][n_\beta(x') - n_\beta(x)]u_{\alpha\beta}^{eff}(x,x')dxdx' \end{aligned} \qquad (11.2.8)$$

where

$$\eta(x) = 0, \quad x < 0$$
$$= 1, \quad x > 0. \tag{11.2.9}$$

Because the integrands of ΔF vanish outside a small region near the interface, the limits of integration in eq. (11.2.8) have been extended to plus and minus infinity without loss of generality. The formula of the interfacial tension, obtained from $\gamma = \partial \Delta F / \partial A$, is thus

$$\gamma = \int_{-\infty}^{\infty} [f^{\circ}(\mathbf{n}(x)) - (1 - \eta(x)) f^{\circ}(\mathbf{n})^{(1)} - \eta(x) f^{o}(\mathbf{n}^{(2)})] dx$$
$$- \sum_{\alpha,\beta} \frac{1}{4} \int_{-\infty}^{\infty} \int_{-\infty}^{\infty} [n_\alpha(x') - n_\alpha(x)][n_\beta(x') - n_\beta(x)] u_{\alpha\beta}^{eff}(x, x') dx dx'. \tag{11.2.10}$$

The density profiles needed to compute γ from eq. (11.2.10) must be determined from the set of two equations

$$\mu_\alpha = \mu_\alpha^{\circ}(\mathbf{n}(x)) + \sum_\beta \int_{-\infty}^{\infty} [n_\beta(x') - n_\beta(x)] u_{\alpha\beta}^{eff}(x, x') dx'$$
$$- \sum_{\beta,\gamma} \frac{1}{4} \int_{-\infty}^{\infty} [n_\beta(x') - n_\beta(x)][n_\gamma(x') - n_\gamma(x)] \frac{\partial u_{\beta\gamma}^{eff}}{\partial \bar{n}_\alpha}(x, x') dx'. \tag{11.2.11}$$

Of course, $\partial u_{\alpha\beta}^{eff} / \partial \bar{n}_\alpha = 0$ in the VDW model. The value of μ_α is set by the chemical equilibrium condition ($\mu_\alpha = \mu_\alpha^{\circ}(\mathbf{n}^{(1)}) = \mu_\alpha^{\circ}(\mathbf{n}^{(2)})$). The nonlinear system in eq. (11.2.11) must be solved for the $n_\alpha(x)$ and the results inserted into eq. (11.2.10) to calculate γ.

The gradient theory of fluid interfaces is considerably simpler than the integral theories. The free energy and chemical potentials in this case are

$$F = A \int \left[f^{\circ}(\mathbf{n}(x)) + \sum_{\alpha,\beta} \frac{1}{2} c_{\alpha\beta}(n(x)) \frac{dn_\alpha}{dx} \frac{dn_\beta}{dx} \right] dx \tag{11.2.12}$$

and

$$\mu_\alpha = \mu_\alpha^{o}(\mathbf{n}(x)) - \sum_\beta \frac{d}{dx} \left(c_{\alpha\beta} \frac{dn_\beta}{dx} \right) + \sum_{\beta\gamma} \frac{1}{2} \frac{\partial c_{\beta\gamma}}{\partial n_\alpha} \frac{dn_\gamma}{dx} \frac{dn_\gamma}{dx}. \tag{11.2.13}$$

The density profiles are found by solving the set of equations (11.2.13) with the boundary conditions

$$n_\alpha(x) \to n_\alpha^{(1)}, \quad x \to -\infty$$
$$\to n_\alpha^{(2)}, \quad x \to +\infty, \tag{11.2.14}$$

where $\alpha = 1, \ldots, c$. The values of $n_\alpha^{(1)}, n_\alpha^{(2)}$, and μ_α are again determined by the equilibrium conditions at eq. (11.2.7).

Equation (11.2.13) can be put into an integrable form. First define the function

$$\omega(\mathbf{n}) = f^{\circ}(\mathbf{n}) - \sum_{\alpha} n_{\alpha}\mu_{\alpha}, \tag{11.2.15}$$

whose partial derivative with respect to n_{α} is

$$\frac{\partial \omega}{\partial n_{\alpha}} = \mu^{\circ}(\mathbf{n}_{\alpha}) - \mu_{\alpha}, \tag{11.2.16}$$

so that

$$\frac{d\omega}{dx} = \sum_{\alpha} \frac{dn_{\alpha}}{dx}\frac{\partial \omega}{\partial n_{\alpha}}. \tag{11.2.17}$$

Then multiply eq. (11.2.13) by dn_{α}/dx, sum over α, and rearrange the results to obtain

$$\frac{d}{dx}\left[\sum_{\alpha\beta} \frac{1}{2} c_{\alpha\beta} \frac{dn_{\alpha}}{dx}\frac{dn_{\beta}}{dx}\right] = \frac{d\omega}{dx}. \tag{11.2.18}$$

Integration of this expression yields

$$\sum_{\alpha,\beta} \frac{1}{2}\frac{dn_{\alpha}}{dx}\frac{d\mu_{\beta}}{dx} = \omega(\mathbf{n}) + K. \tag{11.2.19}$$

The integration constant K is determined by the boundary conditions to be

$$K = -\omega(\mathbf{n}^{(1)}) = -\omega(\mathbf{n}^{(2)}). \tag{11.2.20}$$

But at a coexisting bulk phase density, according to eq. (11.2.7), $\omega(\mathbf{n}^B) = f^{\circ}(\mathbf{n}^B) - \sum_{\alpha} n_{\alpha}^B \mu_{\alpha}^{\circ}(\mathbf{n}^B)$. Because $\sum_{\alpha} n_{\alpha}^B \mu_{\alpha}^{\circ}(\mathbf{n}^B)$ is the Gibbs free energy density, we find $\omega(\mathbf{n}^B) = -P^{\circ}(\mathbf{n}^g) = -P^{\circ}(\mathbf{n}^l) = -P_B$. Thus, eq. (11.2.19) becomes

$$\sum_{\alpha,\beta} \frac{1}{2} c_{\alpha\beta}\frac{dn_{\alpha}}{dx}\frac{dn_{\beta}}{dx} = \Delta\omega(\mathbf{n}) \equiv \omega(\mathbf{n}) - \omega(\mathbf{n}^B) = f^{\circ}(\mathbf{n}) - \sum_{\alpha} n_{\alpha}\mu_{\alpha} + P_B, \tag{11.2.21}$$

where $\omega(\mathbf{n}^B)$ can be evaluated at $\mathbf{n}^B = \mathbf{n}^{(1)}$ or $\mathbf{n}^{(2)}$ or can be set equal to the negative of the bulk pressure of coexisting phases.

Equation (11.2.21) can be used to eliminate f^o from eq. (11.2.12) to obtain

$$F = \sum_{\alpha} N_{\alpha}\mu_{\alpha} - P_B V + A \int_{-\infty}^{\infty} \sum_{\alpha,\beta} c_{\alpha\beta} \frac{dn_{\alpha}}{dx}\frac{dn_{\beta}}{dx} dx, \tag{11.2.22}$$

where $N_{\alpha} = \int n_{\alpha} d^3r$ and $P_B V = \int P_B d^3r$. In the last term of F we have extended the upper and lower limits to $\pm\infty$ because the derivatives dn_{α}/dx go to zero outside a small region near the interface. With the relation $\gamma = (\partial F/\partial A)_{T,\mathbf{N},V}$, it follows from eq. (11.2.22) that

$$\gamma = \int_{-\infty}^{\infty} \sum_{\alpha,\beta} c_{\alpha\beta} \frac{du_\alpha}{dx} \frac{du_\beta}{dx} dx. \tag{11.2.23}$$

This is the same formula as the one obtained from the gradient approximation to the stress tensor [eq. (8.5.26). The only difference is that different approximations give different expressions for $c_{\alpha\beta}$. If an exact expression could be obtained by the different approaches it would of course be the same in all cases. In fact, to second order in density gradients $dn_\beta(x + s_x)dx$ can be replaced by $dn_\beta(x)/dx$ in the exact formula, eq. (9.5.10), to give eq. (11.2.23) with the exact expression,

$$c_{\alpha\beta} = \frac{kT}{4} \int (s_y^2 + s_z^2) C_{\alpha\beta}^{(2)}(\mathbf{s}, x, x + s_x, \{\mathbf{n}\}) d^3 s, \tag{11.2.24}$$

of gradient theory. The problem with eq. (11.2.24) is that the direct correlation function, $C_{\alpha\beta}^{(2)}(\mathbf{s}, x, x', \{\mathbf{n}\})$, is still an unknown functional of the density distributions. This is of course the reason for introducing approximations such as the VDW, MVDW and ADF models. However, at the level of density gradient theory one can take the set of equations at eq. (11.2.13) as the starting point and introduce at eq. (11.2.24) an approximation for $C_{\alpha\beta}^{(2)}$ to bring closure to the theory. If one approximates $C_{\alpha\beta}^{(2)}(\mathbf{s}, x, x', \{\mathbf{n}\})$ by the homogeneous fluid function $C_{\alpha\beta}^{(2)}(s, (\mathbf{n}(\mathbf{r}) + n(\mathbf{r}'))/2)$, one obtains the ADF model approximation [eq. (11.1.21)] for $c_{\alpha\beta}$.

11.3 Liquid–Vapor Interfaces: One-Component Fluids

We saw in Chapter 8 that the gradient theory of one-component liquid–vapor interfaces agrees well with computer simulation of a simple fluid (Fig. 8.9). Also, the gradient approximation to the YBG equation was shown to agree well (Fig. 8.10) with an approximation of the YBG equation similar to the approximation underlying the MVDW model. The density profile $n(x)$ and the surface tension of a 6–12 Lennard–Jones fluid predicted by the MVDW model and the ADF model are compared in Figure 11.1. The predicted profiles are quite similar and the predicted tensions agree to within about 12%.

In Figure 11.2, density profile predictions of the MVDW model are compared with predictions of gradient theory for a 6–12 Lennard–Jones fluid at various temperatures. The influence parameter $c(n)$ is computed in these calculations from the MVDW formula, eq. (11.2.23). The results of the two models are quite similar.

Figure 11.1
Density profile of the vapor–liquid interface of a 6–12 Lennard–Jones fluid. Comparison of the MVDW and ADF models (Adapted from McCoy and Davis, 1979).

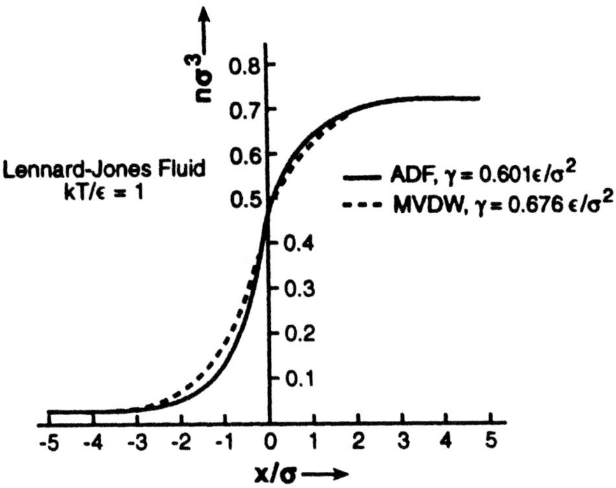

On the basis of the comparisons given above, we conclude that for simple liquid–vapor interfaces the local density functional approximations are reasonably accurate and that density gradient theory is as accurate as the integral approximations. Thus, becausee gradient theory is simpler, we will use it in the description of fluid–fluid interfaces in this and the next section. Furthermore, because it has been shown (McCoy and Davis, (1979) that $c(n)$ is not a strong function of density (because the long-range part of $u_{\alpha\beta}$ dominates), we will also simplify the gradient theory further by assuming that the $c_{\alpha\beta}$ are constant.

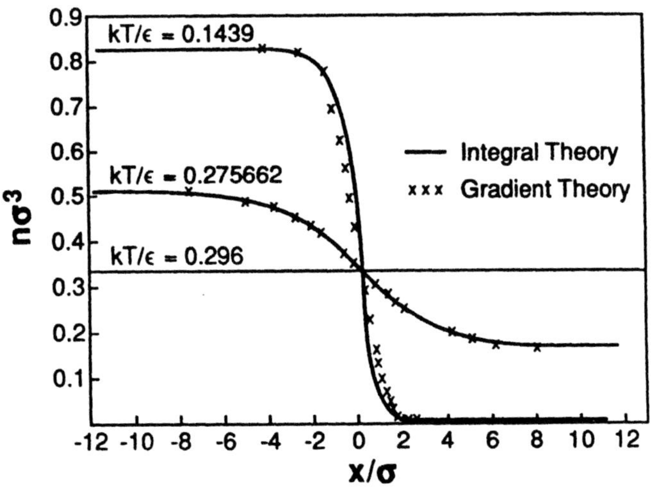

Figure 11.2
Density profile of the vapor–liquid interface of a 6–12 Lennard–Jones fluid. Comparison of the MVDW model (integral equation) with its gradient theoretical version (differential equation). (Adapted from Bongiorno, et al., 1976).

We saw in Section 8.5 that for a one-component fluid eq. (11.2.21) can be integrated to give

$$\int_{n_o}^{n(x)} \sqrt{\frac{c}{\Delta\omega}}\, dn = -x, \qquad (11.3.1)$$

where $n_o = n(x = 0)$ can be chosen as any value lying between n^g and n^l. Using eq. (11.3.1) in eq. (11.2.23), we get

$$\gamma = \int_{n^g}^{n^l} \sqrt{2c\Delta\omega}\, dn. \qquad (11.3.2)$$

We also saw in Section 8.5 that near the critical point eq. (11.3.2) implies that the scaling law $\gamma \propto \kappa_T^{1/2}(n^l - n^g)^2$, where κ_T is the isothermal compressiblity. The known scaling laws for κ_T and $n^l - n^g$, namely that

$$\kappa_T \propto \left(1 - \frac{T}{T_c}\right)^{-\nu} \quad \text{and} \quad n^l - n^g \propto \left(1 - \frac{T}{T_c}\right)^{\beta}, \qquad (11.3.3)$$

imply that

$$\gamma \propto \left(\frac{1 - T}{T_c}\right)^{\zeta}, \qquad (11.3.4)$$

where

$$\zeta = \frac{\nu}{2} + 2\beta. \qquad (11.3.5)$$

From experiment and theory it is known that $\nu \simeq 1.2$ and $\beta \simeq 0.34$. from which gradient theory predicts $\zeta \simeq 1.28$ in good agreement with the experiment on xenon ($\zeta = 1.302$) and carbon dioxide (1.253). Thus, gradient theory, which is expected to become increasingly accurate near the critical point (where density and composition differences between phases become increasingly small), predicts the correct scaling law for γ.

Gradient theory also predicts a law of corresponding states for simple fluids. According to the law of corresponding states for homogeneous fluids the quantity $\Delta\omega(n)$, having the units of pressure, when divided by the critical pressure P_c becomes a universal function of $T_r \equiv T/T_c$ and $n_r \equiv n/n_c$. Thus, for constant c, gradient theory obeys the law of corresponding states

$$\gamma_r = \gamma_r(T_r) = \int_{n_r^g}^{n_r^l} \sqrt{\Delta\omega_r(n_r, T_r)}\, dn_r, \qquad (11.3.6)$$

where $\Delta\omega_r \equiv \Delta\omega/P_c$ and

$$\gamma_r \equiv \gamma/n_c\sqrt{cP_c}. \qquad (11.3.7)$$

n_r^l and n_r^g are functions only of T_r according to the law of corresponding states. As seen earlier, the law of corresponding states is obeyed fairly well for nonpolar fluids. As is the case for other thermodynamic properties, hydrogen bonded fluids and liquid metals do not belong to the same corresponding states class as the nonpolar fluids.

Incidently, if it is assumed that all one-component fluids belong to a class whose interaction potentials are of the form

$$u(s) = \varepsilon \tilde{u}\left(\frac{s}{\sigma}\right), \tag{11.3.8}$$

$\tilde{u}$ the same function for all fluids, the exact formula for the surface tension, eq. (9.5.10), implies the law of corresponding states

$$\gamma^* = \gamma^*(T^*), \tag{11.3.9}$$

where $T^* \equiv kT/\varepsilon$, $n^* \equiv n\sigma^3$, and $\gamma^* \equiv \gamma\sigma^2/\varepsilon$. This microscopic law of corresponding states in turn implies the macroscopic law of corresponding states $\gamma_r = \gamma_r(T_r)$.

In applying eq. (11.3.2) to the estimate of surface tension of fluids one needs an equation of state of homogeneous fluid and the value of c. For nonpolar fluids the Peng–Robinson equation, eq. (4.5.24), is a simple one giving fairly good vapor pressure predictions. In the theory a/b goes as the average energy and b goes as the molecular volume. Because dimensionally c is proportional to a times the interaction length squared, we expect c to go as $ab^{2/3}$. For the 6–12 Lennard–Jones model, eq. (11.1.4), with $d = \sigma$, the VDW estimates of a, b, and c are $a = 16\varepsilon\pi d^3/9$, $b = 2\pi d^3/3$, and $c = 8\varepsilon\pi d^5/7$. In this case $c = (9/14)a(3b/2\pi)^{2/3} = 0.39ab^{2/3}$.

In comparing experimental and gradient theory using the Peng–Robinson equation of state for the normal alkane homologous series Carey et al. (1978) found that $c = 0.33ab^{2/3}$ gives good predictions of surface tension. Their predictions are shown in Figure 11.3. The agreement is quite good.

Thus, in applying the Peng–Robinson equation to estimate surface tension of a given temperature, we recommend taking $c = 0.33ab^{2/3}$ for normal and not too branched alkanes. If no other data are available then this value of c should be viewed as the best guess for any fluid. If tension is available at a temperature T', then we recommend that one assume $c = (\text{constant})ab^{2/3}$, use the experimental value of γ at T' to fix the constant, and compute tension at other temperatures with the formula. Where equations of state other than the Peng–Robinson are used, other choices of c will of course be necessary. Nevertheless, theory suggests that a scaling relationship of the form $c/\varepsilon d^5 =$ universal function of T should be sought, where ε and d are the characteristic energy and length in the other models.

Figure 11.3
Surface tensions predicted by gradient theory with Peng–Robinson equation of state and $c = 0.33ab^{2/3}$ for several normal alkanes.

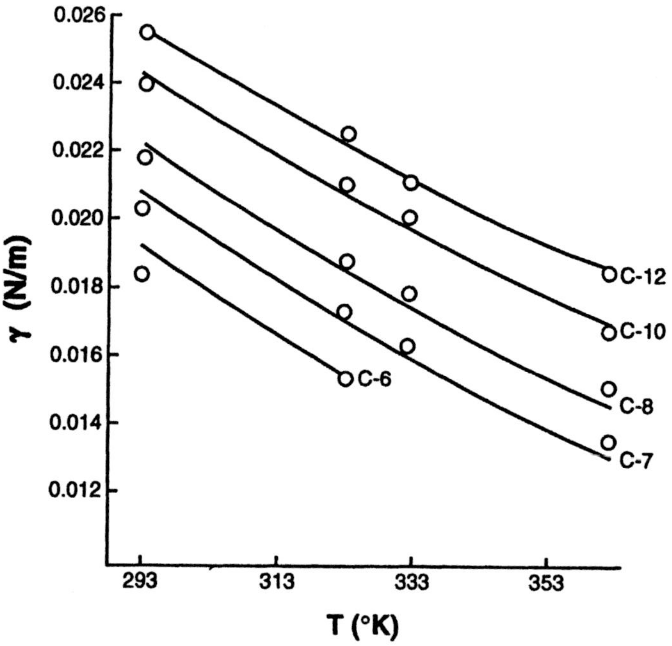

Poser and Sanchez (1979) employed gradient theory with the lattice model equation of state (Sanchez and Lacombe, (1976))

$$P^\circ(n) = -\frac{kT}{\upsilon}\left[\ln(1 - nz\upsilon) + \left(1 - \frac{1}{z}\right)nz\upsilon\right] - n^2\varepsilon z^2\upsilon, \qquad (11.3.10)$$

where υ, z, and ε are constants of the model. They obtained accurate tension predictions for a large variety of fluids. The parameters υ, z, and ε were fit previously with pressure-volumne-temperature data for each fluid. Then they used the formula $c = 1.24\varepsilon\upsilon^{2/3}$. Over a wide range of temperatures their estimates of surface tension agreed with the experiment to within 5% for many normal and branched alkanes, benzene, the xylenes, diethyl ether and carbon tetrachloride. Also, for polymers ($z = \infty$), with $c = 1.1\varepsilon\upsilon^{2/3}$, the theory is quite accurate as illustrated by Figure 11.4.

Given the successes of gradient theory for the one-component liquid–vapor interface and the demonstrated similarity between the results of computer simulations, integral theory and gradient theory, we use gradient theory to describe fluid–fluid interfaces in multicomponent fluids.

11.4 Liquid–Vapor Interfaces: Multicomponent Fluids

For bulk phase densities, fixed by the equilibrium conditions at eq. (11.2.7) and by setting additional intensive variables as required by

Figure 11.4
Gradient theory of surface tension of polymer melts. (Adapted from Poser and Sanchez, 1979).

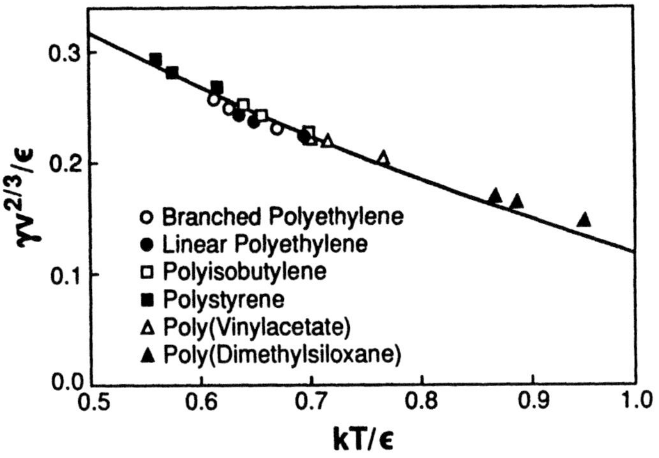

the phase rule, the nonlinear system of equations at eq. (11.2.13) can be discretized by some convenient technique such as the finite difference or the finite element method. With c components there are c equations to solve and the domain of the problem (i.e., the range of x) is infinite. However, the system of equations can be integrated once to obtain eq. (11.2.21) and, if one of the component densities n_1 varies monotonically across an interface, this equation enables one to eliminate one equation and to exchange the unbounded independent variable x for the bounded independent variable n_1. Unfortunately, one does not know a priori whether or not a density variable is monotonic across an interface.

Consider a two-component fluid and assume that the influence parameters $c_{\alpha\beta}$ are independent of composition. The operative equations are then

$$c_{11}\frac{d^2n_1}{dx^2} + c_{12}\frac{d^2n_2}{dx^2} = \mu_1^\circ(\mathbf{n}) - \mu_1 \equiv \Delta\mu_1(\mathbf{n}) \quad (11.4.1)$$

$$c_{21}\frac{d^2n_1}{dx^2} + c_{22}\frac{d^2n_2}{dx^2} = \mu_2^\circ(\mathbf{n}) - \mu_2 \equiv \Delta\mu_2(\mathbf{n}) \quad (11.4.2)$$

and

$$\sum_{\alpha,\beta=1}^{2} \frac{1}{2} c_{\alpha\beta}\frac{dn_\alpha}{dx}\frac{dn_\beta}{dx} = \Delta\omega(\mathbf{n}), \quad (11.4.3)$$

where $c_{12} = c_{21}$.

It is required for thermodynamic stability (see Davis and Scriven, (1982) that for constant $c_{\alpha\beta}$ the determinant of the matrix $[c_{\alpha\beta}]$ must be greater than or equal to zero. For the two-component case this means that the condition $c_{11}c_{22} - c_{12}c_{21} \geq 0$ must hold.

Because one does not know whether or not one of the densities varies monotonically across the interface, we will describe a direct method for reducing eqs. (11.4.1) and (11.4.2) to a problem on a finite domain. We will assume that to the left of the interval [0,

L] the density has nearly the bulk liquid values and to the right it has nearly the bulk vapor value $\mathbf{n}^{(2)}$. A more accurate method, but one that takes more analysis will now be described. We choose L large enough that $\mathbf{n}$ is slowly varying outside the inverval [0, L]. The simplest way to solve eqs. (11.4.1) and (11.4.2) is to approximate the boundary conditions, $\mathbf{n} \to \mathbf{n}^{(1)}$ as $x \to -\infty$ and $\mathbf{n} \to \mathbf{n}^{(2)}$ as $x \to -\infty$, by the conditions

$$\mathbf{n} = \mathbf{n}^{(1)} \quad \text{at} \quad x = 0,$$
$$\mathbf{n} \equiv \mathbf{n}^{(2)} \quad at \quad x = L,$$

known as Dirichlet boundary conditions. The procedure is then to solve the equations for increasing values of L and accept the solution when further increase in L does not give appreciable change in the density distributions.

Because of its simplicity, solution of the profile equations using the Dirichlet boundary conditions is the method of choice if computer time is not a problem. An alternative is to carry out more complicated asymptotic analysis and cast the boundary conditions as Robin or radiation boundary conditions. The trade-off is faster computation times for the price of more complicated analysis. For achieving Robin boundary conditions, we suppose $x = 0$ and $x = L$ lie far enough from the interface for $\mathbf{n}$ to be slowly varying. Where $\mathbf{n}$ is slowly varying, the right-hand side of eqs. (11.4.1) and (11.4.2) can be linearized,

$$\Delta\mu_\alpha(\mathbf{n}) \approx \sum_\beta \frac{\partial \mu_\alpha(\mathbf{n}^B)}{\partial n_\beta}(n_\beta - n_\beta^B) = \sum_{\alpha,\beta} \frac{\partial^2 f^\circ}{\partial n_\alpha \partial n_\beta}(\mathbf{n}^B)(n_B - n_\beta^B) \tag{11.4.4}$$

where $\mathbf{n}^B = \mathbf{n}^{(1)}$ or $\mathbf{n}^{(2)}$ depending on whether we are referring to the bulk phase to the left or the right of the interface. In the regions where linearization is valid, eqs. (11.4.1) and (11.4.2) obey the vector equation

$$\mathbf{C}\frac{d^2\mathbf{y}}{dx^2} = \mathbf{F}\mathbf{y}, \tag{11.4.5}$$

where the matrices $\mathbf{C}$ and $\mathbf{F}$ and the vector $\mathbf{y}$ are given by

$$\mathbf{C} = [c_{\alpha\beta}], \quad \mathbf{F} = \left[\frac{\partial^2 f^\circ(\mathbf{n}^B)}{\partial n_\alpha \partial n_\beta}\right], \quad y = \begin{bmatrix} n_1 - n_1^B \\ n_2 - n_2^B \end{bmatrix}. \tag{11.4.6}$$

$\mathbf{C}$ and $\mathbf{F}$ are self-adjoint matrices (because they are real and symmetric). They must be positive definite for thermodynamic stability. This requires that their determinants $|\mathbf{F}|$ and $|\mathbf{C}|$ be positive (the limiting stability case, $|\mathbf{C}| = 0$, will be analyzed later). Because $|\mathbf{C}| \neq 0$, the inverse $\mathbf{C}^{-1}$ exists and so eq. (11.4.5) can be rearranged to

$$\frac{d^2\mathbf{y}}{dx^2} = \mathbf{C}^{-1}\mathbf{F}\mathbf{y} = \mathbf{K}\mathbf{y} \tag{11.4.7}$$

where

$$\mathbf{C}^{-1} = \frac{1}{|\mathbf{C}|}\begin{bmatrix} c_{22} & -c_{12} \\ -c_{12} & c_{11} \end{bmatrix}, \quad \mathbf{K} = \begin{bmatrix} c_{22}f_{11} - c_{12}f_{12} & c_{22}f_{12} - f_{22}c_{12} \\ c_{11}f_{12} - c_{12}f_{11} & c_{11}f_{22} - c_{12}f_{12} \end{bmatrix}. \tag{11.4.8}$$

The matrix **K** is a product of positive definite matrices and so although it is not itself self-adjoint, it has several useful properties (see Davis, (1986). Its eigenvalues λ_i are positive, its eigenvectors $\mathbf{x}_i$ are a basis set, and so there exists a nonsingular matrix

$$\mathbf{S} = [\mathbf{x}_1, \mathbf{x}_2] = \begin{bmatrix} x_{11} & x_{12} \\ x_{21} & x_{22} \end{bmatrix} \tag{11.4.9}$$

with the properties

$$\mathbf{S}^{-1}\mathbf{S} = \mathbf{T} \equiv \begin{bmatrix} 1 & 0 \\ 0 & 1 \end{bmatrix} \tag{11.4.10}$$

and

$$\mathbf{S}^{-1}\mathbf{K}\mathbf{S} = \Lambda \equiv \begin{bmatrix} \lambda_1 & 0 \\ 0 & \lambda_1 \end{bmatrix}, \tag{11.4.11}$$

where the volumns of **S** are the eigenvectors of **K**. The eigenvalues and eigenvectors are determined from the eigenvalue equation

$$\mathbf{K}\mathbf{x} = \lambda\mathbf{x}, \quad \text{or} \quad (\mathbf{K} - \lambda\mathbf{I})\mathbf{x} = 0 \tag{11.4.12}$$

A homogeneous equation of the form of eq. (11.4.12) has a solution if and only if the characteristic determinant is zero, that is, if

$$|\mathbf{K} - \lambda\mathbf{I}| = 0 \tag{11.4.13}$$

or

$$(k_{11} - \lambda)(k_{22} - \lambda) - k_{12}k_{21} = 0 \tag{11.4.14}$$

Solving this quadratic equation we find the roots

$$\lambda_1, \lambda_2 = \frac{1}{2}\left[k_{11} + k_{22} \pm \sqrt{(k_{11} + k_{22})^2 - 4(k_{11}k_{22} - k_{12}k_{21})}\right] \tag{11.4.15}$$

The corresponding eigenvectors $\mathbf{x}_i$ follow from eq. (11.4.12), which yields

$$x_{1i} = -\frac{k_{12}}{k_{11} - \lambda_i}x_{2i}, \tag{11.4.16}$$

and the condition of orthonormality,

$$x_{1i}^2 + x_{2i}^2 = 1. \tag{11.4.17}$$

Returning to eq. (11.4.17), we can see why these properties of **K** are useful. If we introduce the following linear transformation

$$\zeta = \mathbf{S}^{-1}\mathbf{y}, \tag{11.4.18}$$

multiply eq. (11.4.17) by $\mathbf{S}^{-1}$, and use the property $\mathbf{S}^{-1}\mathbf{S} = \mathbf{I}$, we obtain

$$\frac{d^2\mathbf{S}^{-1}\mathbf{y}}{dx^2} = \mathbf{S}^{-1}\mathbf{K}\mathbf{S}\mathbf{S}^{-1}\mathbf{y}$$

or

$$\frac{d^2\zeta}{dx^2} = \Lambda\zeta \quad \text{or} \quad \frac{d^2\zeta_\alpha}{dx^2} = \lambda_\alpha\zeta_\alpha, \quad \alpha = 1, 2. \tag{11.4.19}$$

For the liquid side of the interface $\zeta \to 0$ as $x \to -\infty$, and so the solution to eq. (11.4.19) is

$$\zeta_1 = \zeta_1^0 e^{\sqrt{\lambda_1^l}x}, \quad \zeta_2 = \zeta_2^0 e^{\sqrt{\lambda_2^l}x}. \tag{11.4.20}$$

The eigenvalues are computed from eq. (11.4.19) using bulk liquid densities to evaluate the quantities $f_{\alpha\beta}$.

The equations at (11.4.20) enable us to derive boundary conditions at $x = 0$. Taking the derivatives of these equations with respect to x and evaluating the results at $x = 0$, we find

$$\left.\frac{d\zeta_\alpha}{dx}\right|_{x=0} = \left.\sqrt{\lambda_\alpha^l}\zeta_\alpha\right|_{x=0}, \quad \alpha = 1, 2, \tag{11.4.21}$$

or because $\zeta = \mathbf{S}^{-1}\mathbf{y}$, we obtain

$$\sum_\beta \chi_{\beta\alpha}^l \left.\frac{dn_\beta}{dx}\right|_{x=0} = \left.\sum_\beta \sqrt{\lambda_\alpha^l}\chi_{\beta\alpha}^l(n_\beta - n_\beta^l)\right|_{x=0}, \quad \alpha = 1, 2, \tag{11.4.22}$$

that are the boundary conditions to use on the liquid side of the interface in solving eqs. (11.4.1) and (11.4.2). $\chi_{\alpha\beta}$ denotes the $\alpha\beta$ element of $\mathbf{S}^{-1}$. The superscript on λ_β^l and $\chi_{\alpha\beta}^l$ is a reminder that the $f_{\alpha\beta}$ are to be evaluated at bulk liquid density in computing the eigenvalues from eq. (11.4.15).

Similarly on the vapor side of the interface $(x > L)$, the solution to eq. (11.4.19) is

$$\zeta_1 = \zeta_1^0 e^{-\sqrt{\lambda_1^g}(x-L)}, \quad \zeta_2 = \zeta_2^0 e^{-\sqrt{\lambda_2^g}(x-L)}. \tag{11.4.23}$$

This is the vapor side asymptotic solution, where $\zeta \to 0$ as $x - L \to \infty$. The boundary conditions at $x = L$ are thus

$$\sum_\beta \chi_{\beta\alpha}^g \left.\frac{dn_\beta}{dx}\right|_{x=L} = \left.\sum_\beta \sqrt{\lambda_\alpha^g}\chi_{\beta\alpha}^g(n_\beta - n_\beta^g)\right|_{x=L}, \quad \alpha = 1, 2. \tag{11.4.24}$$

The superscript on λ_α^g and $\chi_{\beta\alpha}^g$ indicates that bulk vapor densities are to be used in computing the $f_{\alpha\beta}$ as inputs for calculating the eigenvalues and eigenvectors.

We have now transformed the problem to that of solving the differential eqs. (11.4.1) and (11.4.2) as a boundary value problem on the domain $0 < x < L$ with the boundary conditions at eqs. (11.4.22) and (11.4.24). As mentioned above, these boundary conditions are known in the literature as radiation boundary conditions or Robin boundary conditions. The best value of L is not known a priori. Thus, in practice one guesses the value of L, say 5 times the diameter of the largest molecule, solves the problem, and then uses a slightly larger value, say $1.1L$, to see if the density profiles change appreciably. If they do, one continues increasing L until no appreciable change occurs at successive values. "No appreciable change" is a measure that must be set by the accuracy of the solution sought.

If it turns out that the density of one of the components, say n_1, increases monotonically across the interface, yet another method can be used to solve the profile equation. As mentioned this is the situation in which the independent variable x can be replaced by n_1 and the two equations at eqs. (11.4.1) and (11.4.2) can be reduced to a single equation defined on the finite domain $n_1^g < n_1 < n_1^l$. The connection between x and n_1 is furnished by eq. (11.4.3), namely,

$$x = \int_{n_1^0}^{n_1(x)} \sqrt{\frac{\sum_{\alpha,\beta} c_{\alpha\beta} \frac{dn_\alpha}{dn_1} \frac{dn_\beta}{dn_1}}{2\Delta\omega(\mathbf{n})}}\, dn_1. \tag{11.4.25}$$

Because $dn_1/dx > 0$, it follows that $n_2(x) = n_2(n_1)$ and so

$$\frac{dn_2}{dx} = \frac{dn_1}{dx}\frac{dn_2}{dn_1} \quad \text{and} \quad \frac{d^2n_2}{dx^2} = \frac{d^2n_1}{dx^2}\frac{dn_2}{dn_1} + \frac{dn_1}{dx}\frac{d^2n_2}{dn_1^2}. \tag{11.4.26}$$

The quantities dn_1/dx and d^2n_1/dx^2 can be determined from eq. (11.4.25) as functions of $\mathbf{n}$, dn_2/dn_1 and d^2n_2/dn_1^2. These results can be used in eqs. (11.4.1) and (11.4.2) to obtain (see Carey et al. (1980).

$$\begin{aligned}
&[c_{11} + 2c_{12}\frac{dn_2}{dn_1} + c_{22}\left(\frac{dn_2}{dn_1}\right)^2]\left[(c_{22}\Delta\mu_1 - c_{12}\Delta\mu_2)\frac{dn_2}{dn_1} - (c_{11}\Delta\mu_2 - c_{12}\Delta\mu_1)\right] \\
&+ 2\Delta\omega(\mathbf{n})|\mathbf{C}|\frac{d^2n_2}{dn_1^2} = 0.
\end{aligned} \tag{11.4.27}$$

This differential equation is to be solved with the Dirichlet boundary conditions

$$\begin{aligned}
n_2(n_1) &= n_2^g \quad \text{at} \quad n_1 = n_1^g \\
&= n_2^l \quad \text{at} \quad n_1 = n_1^l.
\end{aligned} \tag{11.4.28}$$

known in the literature as Dirichlet boundary conditions. In principle, this single equation is cheaper to solve than the set, eqs. (11.4.1) and (11.4.2). However, because the coefficient $\Delta\omega(\mathbf{n})$ of the highest derivative (d^2n_2/dn_1^2) in eq. (11.4.28) has the properties that $\Delta\omega = d\Delta\omega/dn_1 = 0$ at the boundaries, care must be taken to avoid numerical difficulties in solving eq. (11.4.27). The value of dn_2/dn_1 near $n_1 = n_1^g$ or n_1^l must be estimated from

$$\frac{dn_2}{dn_1} = \lim_{n_1 \to n_1^B} \frac{c_{11}\Delta\mu_2 - c_{12}\Delta\mu_1}{c_{22}\Delta\mu_1 - c_{12}\Delta\mu_2} = \frac{\sum_\beta \left[\frac{\partial}{\partial n_\beta}(c_{11}\Delta\mu_2 - c_{12}\Delta\mu_1)\right]_{\mathbf{n}^B} \left(\frac{dn_\beta}{dn_1}\right)_{\mathbf{n}^B}}{\sum_\beta \left[\frac{\partial}{\partial n_\beta}(c_{22}\Delta\mu_1 - c_{12}\Delta\mu_2)\right]_{\mathbf{n}^B} \left(\frac{dn_\beta}{dn_1}\right)_{\mathbf{n}^B}}. \qquad (11.4.29)$$

Because it is not known a priori whether or not a component density is monotonic across the interface and because of the numerical problems associated with the vanishing of $\Delta\omega$ and its derivative at the boundaries, the author prefers the formulation of the problem in x space with asymptotic boundary conditions. The x space approach also generalizes quite naturally to a c-component system.

Consider next the special case for which $|\mathbf{C}| = 0$. In this case, which corresponds incidentally to the limit of thermodynamic stability, the problem of finding the component density profiles of gradient theory turns out to be relatively simple. For a two-component fluid with constant $c_{\alpha\beta}$ and $|\mathbf{C}| = 0$, or $c_{12} = \sqrt{c_{11}c_{22}}$, one can multiply eq. (11.4.1) by $\sqrt{c_{22}}$ and eq. (11.4.2) by $\sqrt{c_{11}}$ and take the difference between the results to obtain

$$\sqrt{c_{22}}\Delta\mu_1(\mathbf{n}) = \sqrt{c_{11}}\Delta\mu_2(\mathbf{n}). \qquad (11.4.30)$$

Thus, in this special case n_2 as a function of n_1 or n_1 as a function of n_2 can be computed as roots of an algebraic equation. Possiblities of composition paths between n_1^g, n_2^g and n_1^l, n_2^l are shown in Figure 11.5.

In Figure 11.5a, n_1 and n_2 are monotonically increasing functions from the vapor to the liquid side of an interface. Thus one can compute n_1 as a function of n_2 or vice versa from eq. (11.4.30). Figure 11.5b, however, points up a problem that can occur in the search for roots of eq. (11.4.30). If one begins at $n_1 = n_1^g$ and computes $n_2(n_1)$ as a function of n_1 by increasing n_1 in small increments, no problem would occur until the turning point at A is reached. At this point, a small increase in n_1 requires the root n_2 to be far removed from its value at point A. A Newton–Raphson root finding scheme would fail at this point. An easy check to see that a turning point is being approached is to check dn_1/dn_2 at each step increase of n_1. If dn_1/dn_2 begins to approach zero (or dn_2/dn_1 begins to diverge), one knows a turning point may be near. One then merely changes independent variables to n_2 and calculates successive values of n_1 as a function of increasing n_2. With the independent variable n_2, the turning point at B poses no problem. On the other hand, at C there is turning point at which $dn_2/dn_1 = 0$. Near this turning point (recognized by keeping track of the values of dn_2/dn_1 at successive steps in n_2), one switches independent variables from n_2 to n_1. The turning point at D causes no problem because now n_1 is the independent variable that is being increased.

If eqs. (11.4.1) and (11.4.2) have been solved directly, then the density profiles are known and tension can be computed from eq. (11.2.23). If n_1 and n_2 are found by solving eq. (11.4.27) or

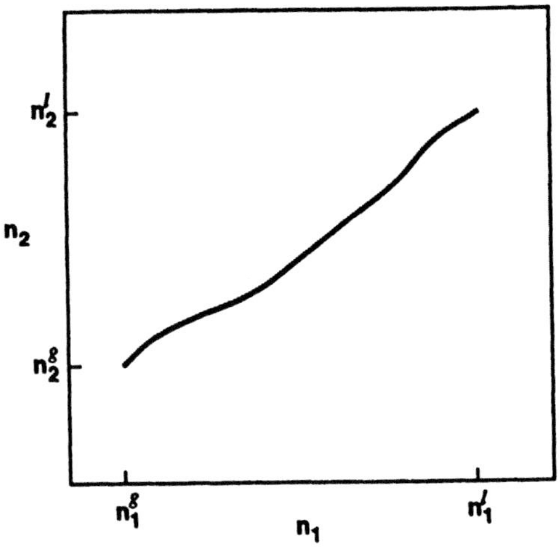

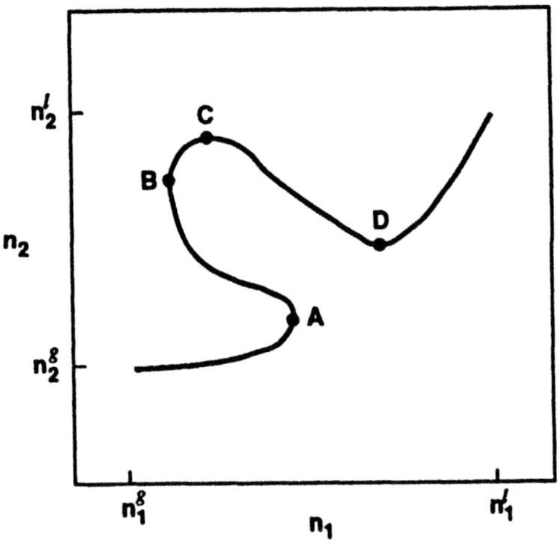

Figure 11.5
Possible component density values across a liquid–vapor interface of a two-component fluid. (a) n_1 and n_2 increase monotonically across the interface. (b) n_2 increases monotonically across the interface but n_1 does not.

(11.4.30), then the density profiles and interfacial tension can be computed eq. (11.4.3). When n_1 and n_2 are not monotonic across the interface, it is convenient to introduce a monotonic composition variable for integrating eq. (11.4.3) and evaluating eq. (11.2.23). The arc length ρ, defined by

$$(d\rho)^2 = (dn_1)^2 + (dn_2)^2, \qquad (11.4.31)$$

can serve as such a quantity. In terms of ρ,

$$x = \int_{\rho(0)}^{\rho(x)} \sqrt{\frac{\sum_{\alpha,\beta} c_{\alpha\beta} \frac{dn_\alpha}{d\rho} \frac{dn_\beta}{d\rho}}{2\Delta w(\mathbf{n})}} d\rho \qquad (11.4.32)$$

and

$$\gamma = \int_{\rho^g}^{\rho^l} \sqrt{2\Delta w(\mathbf{n}) \sum_{\alpha\beta} c_{\alpha\beta} \frac{dn_\alpha}{d\rho} \frac{dn_\beta}{d\rho}} d\rho. \qquad (11.4.33)$$

The steps in the analysis are to first determine the pairs of n_1, n_2 for a set of points across the interface whose spacing is adequate for the fineness desired for a given accuracy of evaluation of eqs. (11.4.31)–(11.4.33). Suppose there are M successive points and denote by dn_α^i the density difference of component α between point i and point $i+1$. ρ^g can be set to zero and its value at the kth point can be computed from

$$\rho^k = \sum_{i=1}^{k} \sqrt{(dn_1^i)^2 + (dn_2^i)^2}. \qquad (11.4.34)$$

Then $\rho^M = \rho^l$. We are free to choose the origin of the coordinate system so that $\rho(0) = \rho^{M/2}$.

At the kth point the densities n_1^k and n_2^k are known and $d\rho^i = \sqrt{(dn_1^i)^2 + (dn_2^i)^2}$. Thus, eqs. (11.4.32) and (11.4.33) for the ρ profile and the surface tension become

$$x(k) = \sum_{i=M/2}^{k} \sqrt{\frac{\sum_{\alpha\beta} c_{\alpha\beta} \frac{dn^i_\alpha}{d\rho^i} \frac{dn^i_\beta}{d\rho^i}}{2\Delta w(\mathbf{n}^i)}} d\rho^i \tag{11.4.35}$$

and

$$\gamma = \sum_{i=1}^{M} \sqrt{2\Delta\omega(\mathbf{n}^i) \sum_{\alpha\beta} c_{\alpha\beta} \frac{dn^i_\alpha}{d\rho^i} \frac{dn^i_\beta}{d\rho^i}} d\rho^i. \tag{11.4.36}$$

Equation (11.4.35) gives x versus ρ. This plus the known connection between n_1, n_2 pairs and ρ enables one to compute the density profiles. Tension can be computed directly from eq. (11.4.36) from the known values of the density pairs as a function of ρ. The density profiles are not needed for computation of the tension.

The Peng–Robinson equation of state has been used to study the structure and tension of the liquid–vapor interface of two- and three-component fluids (Carey, 1978); Sahimi, 1982). The free energy density $f^\circ(\mathbf{n})$, chemical potentials $\mu^\circ_\alpha(\mathbf{n})$, and pressure $P^\circ(\mathbf{n})$ of homogeneous fluid are given by eqs. (6.1.72), (6.3.4), and (6.3.1). The parameters of the model are

$$b = \sum_{\alpha\alpha} b_\alpha \quad \text{and} \quad a = \sum_{\alpha,\beta} x_\alpha x_\beta c_{\alpha\beta} \tag{11.4.37}$$

b_α and $a_{\alpha\alpha}$ are determined as described in Section 4.5 of Chapter 4. The $a_{\alpha\beta}$, $\alpha \neq \beta$, are given by the mixing rule,

$$a_{\alpha\beta} = (1 - k_{\alpha\beta})\sqrt{a_{\alpha\alpha} a_{\beta\beta}}, \quad \alpha \neq \beta, \tag{11.4.38}$$

where the composition independent parameters $k_{\alpha\beta}$ are determined from vapor pressure data on binary systems. Some typical values of $k_{\alpha\beta}$ are given in Table 6.1. For nonpolar hydrocarbon mixtures the $k_{\alpha\beta}$ are often quite small and the ideal or geometric mixing law $a_{\alpha\beta} = \sqrt{a_{\alpha\alpha} a_{\beta\beta}}$ is a good approximation.

Because of the similarity between $a_{\alpha\beta}$ and $c_{\alpha\beta}$, one expects $c_{\alpha\beta}$ to obey a mixing rule similar to eq. (11.4.38). For the VDW model with the Lennard–Jones potential [eq. (11.1.4)] and $d_{\alpha\beta} = \sigma_{\alpha\beta}$, $c_{\alpha\beta} \propto \sigma^5_{\alpha\beta}\varepsilon_{\alpha\beta}$. With the conventional mixing rules of the Lennard–Jones potential, $\sigma_{\alpha\beta} = (\sigma_\alpha + \sigma_\beta)/2$, $\varepsilon_{\alpha\beta} = \sqrt{\varepsilon_{\alpha\alpha}\varepsilon_{\beta\beta}}$, the mixing rule for $c_{\alpha\beta}$ becomes

$$c_{\alpha\beta} = (1 - h_{\alpha\beta})\sqrt{c_{\alpha\alpha} c_{\beta\beta}}, \tag{11.4.39}$$

where

$$h_{\alpha\beta} = 1 - \frac{1}{32}\left[\frac{\sigma_\alpha + \sigma_\beta}{\sqrt{\sigma_\alpha \sigma_\beta}}\right]^5. \tag{11.4.40}$$

If the molecular volumes of α and β are in the ratio 1, 2, and 4, the values of $h_{\alpha\beta}$ are 0, -0.033 and -0.141 according to eq. (11.4.40).

Because many molecular pairs have volume ratios in the range 0.25 to 4 and because the inequality $h_{\alpha\beta} \geq 0$ is required for thermodynamic stability, one might expect from the estimate at eq.(11.4.40) that the geometric mixing rule $c_{\alpha\beta} = \sqrt{c_{\alpha\alpha}c_{\beta\beta}}$ would be the best approximation for many nonpolar hydrocarbons. This is indeed what was found in a study carried out by Carey et al. (1980). Carey et al. used the Peng–Robinson equation of state, fit the pure fluid influence parameters to experimental surface tension data at a given temperature T, and predicted the surface tension of the binary fluid as a function of gas phase mole fraction y_α for various mixing rules. Generally as $h_{\alpha\beta}$ was increased from zero, agreement between experiment and theory worsened. Figure 11.6 presents some of their results for the geometric mixture rule. The predictions agree quite well with the experiment. The results of Carey et al. for several fluid pairs is summarized in Table 11.1.

For a given data point $\gamma^i_{\rm exp}$ the error is defined as $\Delta_i = (\gamma^i_{\rm exp} - \gamma^i_{\rm pred})/\gamma^i_{\rm exp}$. The average error is

$$\Delta = \frac{1}{M}\sum_{i=1}^{M}\Delta_i, \qquad (11.4.41)$$

where M is the number of data points available. The maximum error $\Delta_{\rm max}$ is the value of Δ_i with the largest magnitude.

The density profiles predicted for an iso-octane and cyclohexane mixture are shown in Figure 11.7. The density of iso-octane is not monotonically varying across the interface, thus illustrating the need for keeping track of turning points in solving eq. (11.4.27)

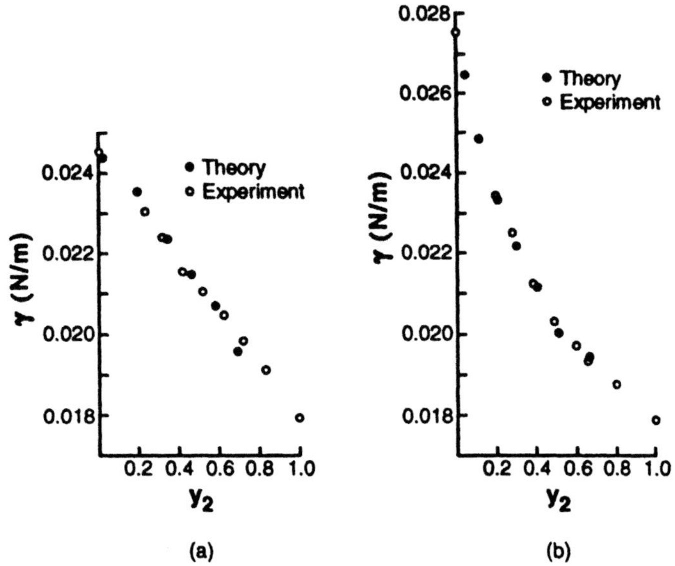

Figure 11.6
Surface tensions of two-component mixtures versus vapor phase composition. Comparison of prediction gradient theory predictions and experiment for the geometric mixing rule $c_{12} = \sqrt{c_{11}c_{22}}$. (a) Benzene (1) and iso-octane (2); (b) n-dodecane (1) and n-hexane (2). (Carey et al., 1980, figures 8 and 10).

TABLE 11.1 Average and Maximum Percent Error in Surface Tensions for Binary Solutions With Geometric Mixing Rule $c_{12} = \sqrt{c_{11}c_{22}}$

	T (K)	γ_1^0 (N/m)	γ_2^0 (N/m)	$\overline{\Delta}$	Δ_{max}
Cyclopentane-carbon tetrachloride	298	0.02185	0.02613	0.00	0.00
Cyclopentane-tetrachlorethylene	298	0.02185	0.03130	0.01	0.01
Cyclopentane-benzene	298	0.02185	0.02820	0.02	0.02
Cyclopentane-toluene	298	0.02185	0.02794	0.01	0.01
Cyclohexane-benzene	293	0.02438	0.02886	0.01	0.02
Cyclohexane-(*cis*)decalin	298	0.02438	0.03224	0.00	0.01
Cyclohexane-(*trans*)decalin	298	0.02438	0.02997	0.01	0.01
Cyclohexane-toluene	298	0.02438	0.02794	0.02	0.02
Cyclohexane-(*n*)hexadecane	298	0.02438	0.01980	0.00	0.01
Iso-cyclohexane	303	0.01789	0.02377	0.00	0.00
Iso-octane-dodecane	303	0.01789	0.02447	0.01	0.07
Iso-octane-benzene	303	0.01789	0.02753	0.01	0.02
(*n*)Hexane-(*n*)dodecane	298	0.01794	0.02469	0.00	0.00
(*n*)Hexane-(*n*)dodecane	313	0.01638	0.02342	0.01	0.01
Methanol-(*n*)butanol	298	0.02210	0.02418	0.04	0.05
Methanol-(*t*)butanol	298	0.02210	0.02011	0.07	0.12

Adapted from Carey et al. (1980).

or (11.4.30) for the values of n_1 and n_2 across the interface. The iso-octane density goes through a small maximum in the interface indicating that iso-octane is slightly surface active at the liquid–vapor interface of the iso-octane/cyclohexane system. One or more components will generally be somewhat surface active at a fluid–fluid interface even if all components are nonpolar molecules.

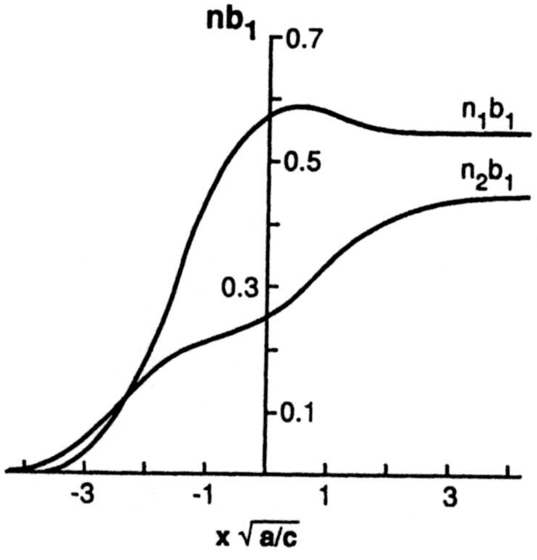

Figure 11.7
Density profiles of an interface of an iso-octane (1) and cyclohexane (2) solution at 303 K. Predicted by gradient theory with the geometric mixing rule $c_{12} = \sqrt{c_{11}c_{12}}$. (Adapted from Carey et al., 1980).

For a multicomponent fluid ($c \geq 2$) most of the results given above for a two-component fluid generalize easily. Again we consider only the case of constant $c_{\alpha\beta}$. The differential equations for the density profiles are then

$$\sum_{\beta=1}^{c} c_{\alpha\beta}\frac{d^2 n_\beta}{dx^2} = \Delta\mu_\alpha(\mathbf{n}), \quad \alpha = 1, \ldots, c. \tag{11.4.42}$$

These integrate to

$$\sum_{\alpha,\beta=1}^{c} \frac{1}{2} c_{\alpha\beta}\frac{dn_\alpha}{dx}\frac{dn_\beta}{dx} = \Delta\omega(\mathbf{n}), \tag{11.4.43}$$

where $c_{\alpha\beta} = c_{\beta\alpha}$, $|\mathbf{C}| \geq 0$ for thermodynamic stability, and the boundary conditions of the differential equations are

$$\begin{aligned} n_\alpha &\to n_\alpha^l, \quad x \to -\infty \\ n_\alpha &\to n_\alpha^g, \quad x \to \infty, \ \alpha = 1, \ldots, c \end{aligned} \tag{11.4.44}$$

where the bulk values of $\mathbf{n}$ are given by the equilibrium conditions

$$P_B = P^\circ(\mathbf{n}^l) = P^\circ(\mathbf{n}^g), \quad \mu_\alpha^\circ(\mathbf{n}^l) = \mu_\alpha^\circ(\mathbf{n}^g), \quad \alpha = 1, \ldots, c. \tag{11.4.45}$$

Of course, the temperature and $c - 1$ bulk phase concentrations (or temperature, pressure, and $c - 2$ concentrations) must be set to fully specify the system.

The asymptotic equations of the Robin boundary conditions are

$$\begin{aligned} \sum_{\beta=1}^{c} \chi_{\beta\alpha}^l \frac{dn_\beta}{dx}\bigg|_{x=0} &= \sum_{\beta=1}^{c} \sqrt{\lambda_\alpha^l}\chi_{\beta\alpha}^l (n_\beta - n_\beta^l)\bigg|_{x=0}, \\ \sum_{\beta=}^{c} \chi_{\beta\alpha}^g \frac{dn_\beta}{dx}\bigg|_{x=L} &= \sum_{\beta=1}^{c} \sqrt{\lambda_\alpha^g}\chi_{\beta\alpha}^g (n_\beta - n_\beta^g)\bigg|_{x=L}, \quad \alpha = 1, \ldots, c, \end{aligned} \tag{11.4.46}$$

where $\chi_{\alpha\beta}$ are components of the matrix, $\mathbf{S} = [\mathbf{x}, \ldots, \mathbf{x}_c]$, in which the quantities $\mathbf{x}_i$ are the eigenvectors of $\mathbf{K}$,

$$\mathbf{K} = \mathbf{C}^{-1}\mathbf{F}, \quad \mathbf{C} = \left[c_{\alpha\beta}\right], \ \mathbf{F} = \left[\frac{\partial^2 f^\circ(\mathbf{n})}{\partial n_\alpha \partial n_\beta}\right], \tag{11.4.47}$$

and λ_α are the eigenvalues of $\mathbf{K}$. The superscripts l and g indicate that $\mathbf{x}_i$ and λ_i are computed at liquid and vapor compositions, respectively.

Because eigenvalue analysis becomes increasingly more complicated with an increasing number of components c, it becomes even more attractive to use the Dirichlet boundary conditions,

$$\begin{aligned} n_\alpha|_{x=0} &= n_\alpha^l \\ n_\alpha|_{x=L} &= n_\alpha^g, \quad \alpha = 1, \ldots, c \end{aligned} \tag{11.4.48}$$

if computer time is not a problem. Again L has to be chosen arbitrarily and then increased until successive solutions change sufficiently little. Because the Robin boundary conditions better capture the actual density profiles, the value of L in eq. (11.4.46) will be smaller than that in eq. (11.4.48).

Once the $n_\alpha(x)$ are found by solving the differential equations, the surface tension can be computed from

$$\gamma = \sum_{\alpha,\beta=1}^{c} \int_{-\infty}^{\infty} c_{\alpha\beta} \frac{dn_\alpha}{dx} \frac{dn_\beta}{dx} dx. \qquad (11.4.49)$$

For the geometric mixing rule $c_{\alpha\beta} = \sqrt{c_{\alpha\alpha} c_{\beta\beta}}$, the determinant of $\mathbf{C}$ is zero and the differential equations at eq. (11.4.42) can be combined to obtain the algebraic system

$$\frac{\Delta\mu_\alpha(\mathbf{n})}{\sqrt{c_{\alpha\alpha}}} = \frac{\Delta\mu_\beta(\mathbf{n})}{\sqrt{c_{\beta\beta}}} \qquad (11.4.50)$$

that can be solved for densities $\mathbf{n}$ across the interface similarly to the method described above for the two-component case. Turning points must be kept track of and at each step in the root-finding computation the component density chosen as the independent variable must be one that is monotonically increasing toward the liquid side of the interface. Once the sequence of compositions $\mathbf{n}^i$ are known across the interface, the arc-length ρ can be computed from

$$\rho^k = \sum_{h=1}^{k} \sqrt{\sum_{\beta=1}^{c} (dn_\beta^i)^2} \qquad (11.4.51)$$

and the density profile and tension can be computed from

$$x(k) = \sum_{i=M/2}^{k} \sqrt{\frac{\sum_{\alpha,\beta=1}^{c} c_{\alpha\beta} \frac{dn_\alpha^i}{d\rho^i} \frac{dn_\beta^i}{d\rho^i}}{2\Delta\omega(\mathbf{n}^i)}} d\rho^i \qquad (11.4.52)$$

and

$$\gamma = \sum_{i=1}^{M} \sqrt{2\Delta\omega(\mathbf{n}^i) \sum_{\alpha,\beta=1}^{c} c_{\alpha\beta} \frac{dn_\alpha^i}{d\rho^i} \frac{dn_\beta^i}{d\rho^i}} d\rho^i. \qquad (11.4.53)$$

An example of the surface tension predicted for a ternary system obeying the Peng–Robinson equation of state with the geometric mixing rule is shown in Figure 11.8. Some experimental data are also shown in the figure. For this mixture of normal alkanes, experiment and prediction tend to be closer than 0.5 dyn/cm. Other results for ternary systems can be found in Carey's and Sahimi's theses (1978; 1982). The density profiles predicted for one of the mixtures of Figure 11.8 is shown in Figure 11.9. Both normal octane and normal hexane are surface active in this solution. Thus, if

Figure 11.8
Experimental (filled circles) versus predicted surface tension of a ternary mixture. Gradient theory, Peng–Robinson equation of state, geometric mixing rule $c_{\alpha\beta} = \sqrt{c_{\alpha\alpha} c_{\beta\beta}}$. (Adapted from Carey, 1978).

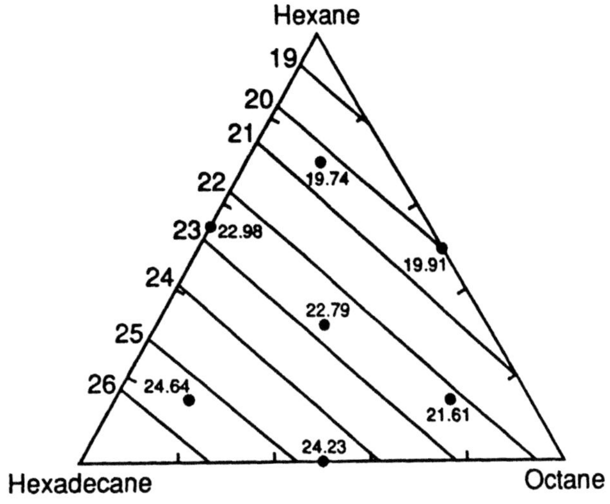

one had wisely, or luckily, chosen the density of normal hexadecane as the independent variable to advance in solving the equations at eq. (11.4.50), no turning point would be encountered and the entire set of interfacial compositions between the vapor and liquid phases would result without changing the independent variable. With either octane or hexane as the initial independent variable, a turning point would be reached and a change of independent variable would be required by the solution scheme we described earlier. If all three- component densities passed through maxima at nearly the same place in the interface, a more complicated analysis than that described earlier would have to be used. This latter case is not common.

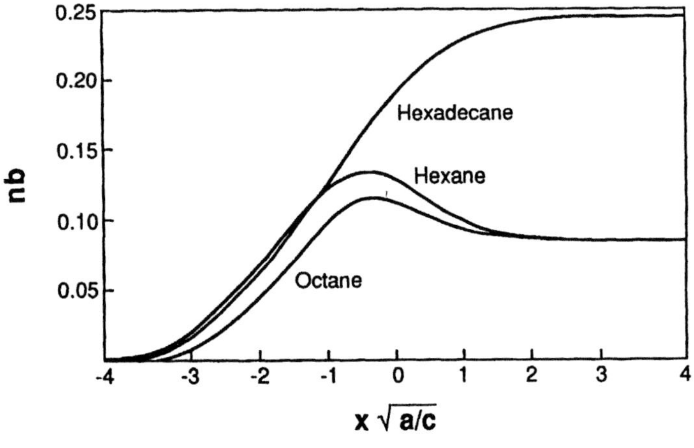

Figure 11.9
Density profiles for one of the ternary solutions shown in Figure 11.8. (Adapted from Carey, 1978).

11.5 Liquid–Liquid Interfaces

The gradient theory of liquid–liquid interfaces for constant influence parameters and an equation of state, such as the VDW or Peng–Robinson equation differs from that of a liquid–vapor interface only in the bulk phase densities, $\mathbf{n}^l$ and $\mathbf{n}^g$ are replaced by the bulk phase densities, $\mathbf{n}^{(1)}$ and $\mathbf{n}^{(2)}$ of coexisting liquids 1 and 2. Thus, the operative equations are eqs. (11.4.42)–(11.4.53), but liquid–liquid solutions are sought at eq. (11.4.45) instead of liquid–vapor solutions. The strategy for solving the equilibrium equations was described in Chapter 6.

The density profiles predicted for the liquid–liquid interface of an oil (component 1), water (component 2), and alcohol interface (component 3) are shown in Figure 11.10. The oil and water phases are quite immiscible and the alcohol is strongly surface active at the interface. Interestingly, the oil is slightly surface active as witnessed by a small density maximum at the right of the vertical axis in Figure 11.10. Because of the high concentration of alcohol in the interfacial region excess oil is "dissolved" in the interfacial region.

As discussed in Chapter 6, to describe liquid–liquid equilibrium one often introduces a lattice theory in which the fluid densities obey the lattice constraint

$$\sum_{\alpha=1}^{c} n_\alpha v_\alpha = 1, \tag{11.5.1}$$

where v_α is the volume of the lattice occupied by a molecule of species α. The advantage of a lattice approximation is that the number of independent concentration variables is reduced to $c-1$. Thus, the theory of binary and ternary equilibria described by a lattice model is computationally equivalent to the theory of one- and two-component equilibria described by a full equation of state.

With the aid of the condition at eq. (11.5.1), the density of component c can be eliminated from the equations at eq. (11.4.42). If one subtracts from the equation for each component α the product of v_α/v_c and the equation for component c, the following set of equations results:

$$\sum_{\beta=1}^{c-1} \tilde{c}_{\alpha\beta}\frac{d^2 n_\beta}{dx^2} = \Delta\tilde{\mu}_\alpha, \quad \alpha = 1, \ldots, c-1, \tag{11.5.2}$$

where

$$\tilde{c}_{\alpha\beta} \equiv c_{\alpha\beta} + \frac{v_\alpha v_\beta c_{cc}}{v_c^2} - \frac{v_\alpha c_{\alpha c}}{v_c} - \frac{v_\beta c_{\beta c}}{v_c} \quad \text{and} \quad \Delta\tilde{\mu}_\alpha \equiv \Delta\mu_\alpha - \frac{v_\alpha}{v_c}\Delta\mu_c. \tag{11.5.3}$$

Similarly, it can be shown that

$$\sum_{\alpha,\beta=1}^{c-1} \frac{1}{2}\tilde{c}_{\alpha\beta}\frac{dn_\alpha}{dx}\frac{dn_\beta}{dx} = \Delta\omega(\mathbf{n}), \tag{11.5.4}$$

Figure 11.10
Density profiles at the liquid–liquid interface of an oil, water, and alcohol system. Predicted by gradient theory with VDW equation of state, and geometric mixing rules $a_{\alpha\beta} = \sqrt{a_{\alpha\alpha}a_{\beta\beta}}$ and $c_{\alpha\beta} = \sqrt{c_{\alpha\alpha}c_{\beta\beta}}$. (Adapted from Carey, 1978).

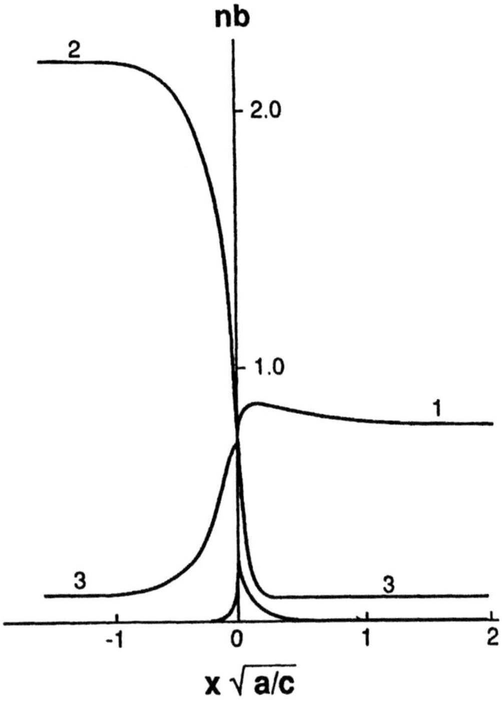

where $\Delta\omega(\mathbf{n}) = \omega(\mathbf{n}) - \omega(\mathbf{n}^B)$ and $\omega(\mathbf{n}) = f^\circ(\mathbf{n}) - \sum_{\alpha=1}^{c} n_\alpha \mu_\alpha$. To see this, note that with eq. (11.5.1),

$$d\omega(\mathbf{n}) = \sum_{\alpha=1}^{c} \Delta\mu_\alpha(\mathbf{n}) dn_\alpha = \sum_{\alpha=1}^{c-1} \left[\Delta\mu_\alpha(\mathbf{n}) - \frac{v_c}{v_\alpha}\Delta\mu_c(\mathbf{n})\right] dn_\alpha, \qquad (11.5.5)$$

and so it follows that

$$\frac{\partial\omega(\mathbf{n})}{\partial n_\alpha} = \Delta\tilde{\mu}_\alpha, \quad \alpha = 1, \ldots, c-1, \qquad (11.5.6)$$

where it is understood that in $\omega(\mathbf{n})$ the variable n_c has been replaced by $-\sum_{\beta=1}^{c-1} (v_\beta/v_c) n_\beta$ before the partial derivative is taken with respect n_α. The thermodynamic conditions of equilibrium in the lattice model are

$$\Delta\omega(\mathbf{n}^{(1)}) = \Delta\omega(\mathbf{n}^{(2)}) = 0 \quad \text{and} \quad \Delta\tilde{\mu}_\alpha(\mathbf{n}^{(1)}) = \Delta\tilde{\mu}_\alpha(\mathbf{n}^{(2)}), \quad \alpha = 1, \ldots, c-1. \qquad (11.5.7)$$

If the model lattice is pressure independent, as is frequently the case, the condition $\Delta\omega(\mathbf{n}^B) = 0$ is automatically satisfied by the conditions of chemical equilibrium, $\Delta\tilde{\mu}_\alpha(\mathbf{n}^{(1)}) = \Delta\tilde{\mu}_\alpha(\mathbf{n}^{(2)})$.

For gradient theory with constant $c_{\alpha\beta}$, the interface tension for lattice models is given by

$$\gamma = \int_{-\infty}^{\infty} \sum_{\alpha,\beta=1}^{c-1} \tilde{c}_{\alpha\beta} \frac{dn_\alpha}{dx}\frac{dn_\beta}{dx} dx. \qquad (11.5.8)$$

In the simplest case of a lattice $v_\alpha = v$ for all α and so the total density n $(=1/v)$ is constant and $n_\alpha = nx_\alpha$, where x_α is the mole fraction. An example is the two-component regular solution. In this case, the density profile obeys the equation

$$\frac{\tilde{c}_{11}}{2v^2}\left(\frac{dx_1}{dx}\right)^2 = \Delta\omega, \quad \tilde{c}_{11} = c_{11} + c_{22} - 2c_{12}, \tag{11.5.9}$$

whose solution is

$$x = \int_{1/2}^{x_1(x)} \sqrt{\frac{\tilde{c}_{11}}{2v^2\Delta\omega(x_1)}}dx_1, \tag{11.5.10}$$

where we have chosen the position $x = 0$ to be where the mole fraction $x_1 = 1/2$. As shown in Section 6.3 the bulk concentrations of the coexisting liquid phases are given by $x_1^{(1)} = x_2^{(2)}$, where $x_1^{(1)}$ is the root of

$$T = (a/k)(1 - 2x_1^{(1)})/\ln[(1 - x_1^{(1)})/x_1^{(1)}]. \tag{11.5.11}$$

In the regular solution model the parameter a has the units of energy. The interaction parameter a can be computed from liquid–liquid critical point T_c with the relationship $a = kT_c/2$. Because [see eqs. (6.4.5) and (6.4.8)]

$$f^\circ(x_1) = \frac{1}{v}\left[-\sum_{\alpha=1}^{2} x_\alpha\mu_\alpha^+(T) + kT\sum_{\alpha=1}^{2} x_\alpha \ln x_\alpha + x_1x_2a\right] \tag{11.5.12}$$

$$\mu_\alpha = \mu_\alpha^+ + kT\ln x_\alpha + (1 - x_\alpha)^2 a \tag{11.5.13}$$

for the regular solution, it follows that

$$\begin{aligned}\omega(x_1) = {} & \frac{kT}{v}\left\{x_1 \ln\left[x_1/x_1^{(1)}\right] + (1 + x_1)\ln\left[(1 - x_1)/(1 - x_1^{(1)})\right]\right\} \\ & + \frac{a}{v}\left\{x_1(1 - x_1) - x_1(1 - x_1^{(1)})^2 - (1 - x_1)x_1^{(1)^2}\right\}.\end{aligned} \tag{11.5.14}$$

It turns out that $\omega(x_1^{(1)}) = \omega(x_1^{(2)}) = 0$ for this model and so $\Delta\omega(x_1)) = \omega(x_1)$.

The interfacial tension of the binary regular solution is

$$\gamma = \int_{-\infty}^{\infty} \tilde{c}_{11}\frac{d^2n_1}{dx^2}dx = \int_{-\infty}^{\infty} \frac{\tilde{c}_{11}}{v^2}\left(\frac{dx_1}{dx}\right)^2 dx = \sqrt{\frac{2\tilde{c}_{11}}{v^3}}\int_{x_1^{(1)}}^{x_1^{(2)}} \sqrt{v\Delta\omega(x_1)}dx_1. \tag{11.5.15}$$

In terms of the dimensionless variables,

$$\begin{aligned}&\Delta\omega(x_1)^* \equiv v\Delta\omega(x_1)/kT_c, \qquad T^* = T/T_c, \\ &\gamma^* \equiv \gamma\sqrt{v^3/2kT_c\tilde{c}_{11}}, \qquad x^* = x\sqrt{2v^2kT_c/\tilde{c}_{11}},\end{aligned} \tag{11.5.16}$$

Figure 11.11
Reduced interfacial tension γ^* versus reduced temperature T^* [see eq. (11.5.16) for units] for a regular solution.

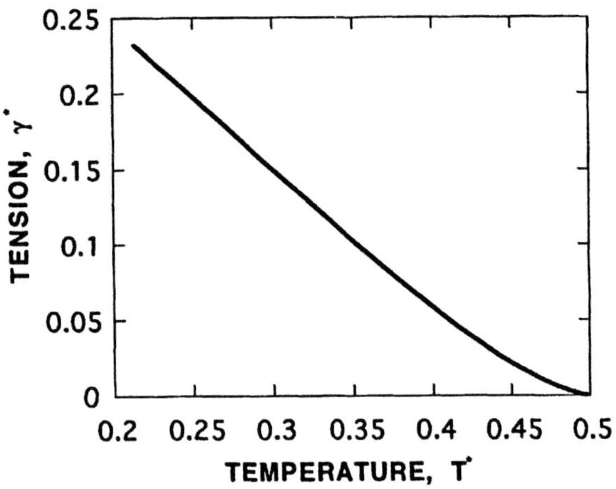

the interfacial tension, γ^*, and the mole fraction profiles, x_1 versus x^*, are universal functions of T^*. The profiles of x_1 are similar in form to those of the profile of n/n_c at the one-component liquid–vapor interface. The plot of γ^* versus T^* is given in Figure 11.11.

As a rough extimate of the tension of a liquid–liquid system at a given T from its value at some other temperature T', one can take

$$\gamma(T) = \gamma(T')\frac{\gamma^*(T/T_c)}{\gamma^*(T'/T_c)}, \qquad (11.5.17)$$

where the γ^* are read from Figure 11.11 at the appropriate T^*'s.

All the binary lattice models introduced in Chapter 6 can be reduced to pseudo one-component models and analogues of eqs. (11.5.10) and (11.5.15) can be used for calculating the composition profiles and interfacial tensions. c-Component lattice models must be analyzed by the methods introduced for the $c - 1$ component VDWaals-like models. The quality of the results from any of the models will of course be limited first by the accuracy of these models in predicting phase diagrams and next by the approximations inherent in gradient theory. The former limitation is probably more serious than the latter for fluid–fluid interfaces. However, it follows from what we learned in studying one-dimensional hard rods that the detailed structure of the density at the solid–liquid interface is not well predicted by local density functional theories. In the next chapter we present nonlocal density functional models that do capture the qualitative physics of liquids at solid surfaces.

Supplementary Reading

Davis, H. T. and Scriven, L. E., 1981. Adv. Chem. Phys., **49**.

Henderson, D., Ed. 1992. Dekker, New York.

Rowlinson, J. S. and Widom, B. 1982. *Molecular Theory of Capillarity*, Clarendon Press, Oxford, UK.

Exercises

1 Find the density profile at a planar liquid–vapor interface using VDW gradient theory. Assume the fluid obeys the Carnahan–Starlinĝ equation of state given by

$$P_o = P_{HS} + \frac{n^2}{2}\int \bar{u}^A(s)d^3s$$

where

$$P_{HS} = nkT\frac{(1+y+y^2-y^2)}{(1-y)^3}; \quad y = \frac{n\pi\sigma^3}{6}$$

$$\bar{u}^A(s) = \begin{cases} 0, & s < \sigma \\ -4\varepsilon\left(\frac{\sigma}{s}\right)^6, & s > \sigma. \end{cases}$$

Plot the profile for $T/T_c = 0.95, 0.9, 0.8, 0.7$ where T_c is the critical temperature. Compute the reduced tension γ_r [Eq. (11.3.7)] as a function of T/T_c. Recall that

$$c = \frac{1}{30}\int_{s>\sigma} s^3 \frac{d\bar{u}^A}{ds} d^3s$$

for the VDW approximation. Compare theory with experiment for liquid argon. Set $\sigma = 3.4$ Å and fit ε at one temperature, and then predict γ versus T.

2 Calculate the surface tension of n-decane and n-dodecane as a function of temperature using gradient theory with the Sanchez–Lacombe equation of state [Eq. (11.3.10)]. Compare with experimental data and predictions using the Peng–Robinson equatin of state (Figure 11.3).

3 Using the mixing rule $c_{12} = \sqrt{c_{11}c_{22}}$ with gradient theory and the VDW equation of state, calculate the surface tension of an n-hexane and n-dodecane mixture at room temperature and for a range of concentrations. Compare with the results shown in Figure 11.6.

4 Consider a one-component spherical drop in its vapor. According to VDW gradient theory,

$$c\frac{1}{r^2}\frac{d}{dr}\left(r^2\frac{dn}{dr}\right) = \mu^\circ(n) - \mu.$$

The boundary conditions are $dn/dr = 0$ at $r = 0$ and $n(r) \to n_b$, where $\mu = \mu^\circ(n_b)$. n_b is the density of bulk vapor. Solve this equation for the droplet density profiles at $T = 0.8T_c$ for several bulk vapor densities ranging from $1.5n_s^v$ to $1.05n_s^v$, where n_s^v is the coexistence vapor density at T. Use the VDW equation for $\mu^\circ(n)$ and the relationship $c = 0.39ab^{2/3}$ and do the calculation for decane.
An estimate of the tension of the droplet can be computed from the formula

$$\gamma = \int_0^\infty (P_N - P_T)dr,$$

where P_N is the component of the pressure tensor normal to the drop center and P_T is the component transverse to the drop. Use the gradient approximation to the stress tensor (Chapter 8) and the computed density profiles to compute the surface tension of the drop as a function of n_b. Plot the results as γ versus R, where R is the drop radius computed as the distance r at which $n(r) = [n((0) + n(\infty)]/s$.

What are the implications of the results found here for homogeneous nucleation?

5 Use the regular solution model [Eq. (11.5.2)] to estimate the interfacial tension of the benzene–water interface at 50°C given the tension at room temperature.

References

Bongiorno, V., Scriven, L. E., and Davis, H. T. 1976. *J. Colloid Interface Sci.*, **57**, 462.

Cahn, J. W. 1977. *J. Chem. Phys.*, **66**, 3667.

Cahn, J. W. and Hilliard, J. E. 1958. *J. Chem. Phys.*, **28**, 258.

Carey, B. S., Scriven, L. E., and Davis, H. T. 1978. *AIChE J.*, **24**, 1076.

Carey, B. S., Scriven, L. E., and Davis, H. T. 1980. *AIChE J.*, **26**, 705.

Carey, B. S. 1978. Ph.D. Thesis, University of Minnesota at City, Minneapolis, MN.

Davis, H. T. and Scriven, L. E. 1982. *Adv. Chem. Phys.*, **49**, 357.

Davis, H. T. 1986. *Chem. Eng. Commun.*, **41**, 267.

Ebner, C., Saam, W. F., and Stroud, D. 1976. *Phys. Rev. A*, **14**, 2264.

Ebner, C. and Saam, W. F. 1977. *Phys. Rev. Lett.*, **38**, 1486.

McCoy, B. F. and Davis, H. T. 1979. *Phys. Rev. A*,**20**, 1201.

Poser, C. I. and Sanchez, I. C. 1979. *J. Colloid Sci.*, **69**, 539.

Sahimi, M. 1982. Ph.D. Thesis, University of Minnesota at City, Minnesota, MN.

Sanchez, I. C. and Lacombe, R. H. 1976. *J. Phys. Chem.*, **80**, 2568.

12

DENSITY FUNCTIONAL THEORY OF CONFINED FLUIDS

12.1 Nonlocal Density Functional Free Energy Models

The simplest case one can imagine for a fluid at a solid surface is that of a one-component hard-rod fluid at a hard wall. The wall potential on the fluid particles is then

$$\begin{aligned} v(x) &= \infty, \quad x < d/2 \\ &= 0, \quad x > d/2. \end{aligned} \tag{12.1.1}$$

For all the local density functional models described in Chapter 11, the chemical potential of a hard sphere fluid is $\mu = \mu^\circ(n) + v(x)$. Thus, if the density far from the wall is n^B, the density profile predicted by local density functional theory for a hard sphere fluid at a hard wall is

$$\begin{aligned} n(x) &= 0, \quad x < d/2 \\ &= n^B, \quad x > d/2. \end{aligned} \tag{12.1.2}$$

However, we know from Chapter 10 (see for example Fig. 10.8) that the density profile of a hard-rod fluid at a hard wall is highly structured. Thus, local density functional theories fail to capture the detailed structure of a fluid confined by an impenetrable solid surface.

In this section we present *nonlocal* density functional approximations that can account for the detailed structure of confined fluids. The models presented derive motivation and intuition from the exact results for one-dimensional hard-rod fluids given in Chapter 10.

We restrict our attention to systems whose particles interact classically with pair additive, central forces. It is convenient to decompose the potential as

$$u = u^R + u^A, \tag{12.1.3}$$

where u^R is the pair potential of a reference system and u^A is the part of u not contained in u^R. In most modern theories of liquids, u^R is taken to be the strongly repulsive short-ranged pair interaction. However, at this point we leave the division arbitrary.

Consider a system in which the pair potential is $u^R + \lambda u^A$, where λ is an arbitrary parameter. By varying λ from 0 to 1, this system can be made to range from the reference system to the subject system. The configuration Helmholtz free energy, ΔF, of the λ-system in a canonical ensemble is given by

$$e^{-\beta \Delta F_N} = \int \cdots \int e^{-\beta u_N^R - \lambda \beta u_N^A} d^3 r_1 \cdots d^3 r_N. \tag{12.1.4}$$

u^R contains the reference system pair interactions plus contributions of any external forces present in the subject system. When $\lambda = 0$, ΔF_N is the free energy of the reference system, and when $\lambda = 1$, ΔF_N is the free energy of the subject system.

Differentiating eq. (12.1.4) with respect to λ, we obtain

$$\frac{d\Delta F_N}{d\lambda} = \sum_{i>j=1}^{N} \frac{\int e^{-\beta u_N} u^A(|\mathbf{r}_i - \mathbf{r}_j'|) d^3 r_1 \cdots d^3 r_N}{Z_N}. \tag{12.1.5}$$

Integrating eq. (12.1.5), taking grand canonical ensemble average of the results, and setting $n^{(2)}(\mathbf{r}, \mathbf{r}') = n(\mathbf{r}) n(\mathbf{r}') g^{(2)}(\mathbf{r}, \mathbf{r}')$, we find

$$\Delta F = \Delta F^R + \Delta F^A \tag{12.1.6}$$

where ΔF^R is the configuration free energy of the reference system and

$$\Delta F^A = \frac{1}{2} \int_0^1 \int n(\mathbf{r}) n(\mathbf{r}') g^{(2)}(\mathbf{r}, \mathbf{r}', \lambda) u^A(|\mathbf{r} - \mathbf{r}'|) d^3 r d^3 r' d\lambda. \tag{12.1.7}$$

This is an exact result. $g^{(2)}(\mathbf{r}, \mathbf{r}', \lambda)$ is the pair correlation function of a system whose pair potential is $u^R + \lambda u^A$.

Thus, the total Helmholtz free energy of the subject system can be decomposed as

$$F = F^R + \Delta F^A. \tag{12.1.8}$$

The free energy F^R of the reference system can be further decomposed as

$$F^R = F^e + F^I + F^{\text{ex}} \tag{12.1.9}$$

where (for a one-component system) F^e is the external field contribution,

$$F^e = \int n(\mathbf{r})v(\mathbf{r})d^3r, \tag{12.1.10}$$

F^I is the "ideal gas" part,

$$F^I = \int n(\mathbf{r})\{\mu^+(T) + kT[\ln n(\mathbf{r}) - 1]\}d^3r, \tag{12.1.11}$$

and F^{ex} is the excess free energy arising from the interactions among the particles of the reference system.

In analogy with the exact hard-rod results, Percus (1981) defined a generic free energy functional for F^{ex}, namely,

$$F^{\text{ex}} = \int \bar{n}^\sigma(\mathbf{r})\mathcal{F}(\bar{n}^\tau(\mathbf{r}))d^3r, \tag{12.1.12}$$

where $\mathcal{F}(n)$ is the excess free energy per particle of a homogeneous hard sphere fluid of density n. The quantities $\bar{n}^\sigma(\mathbf{r})$ and $n^\tau(\mathbf{r})$ are coarse-grained densities that are linear functionals of density and are related to the local density by weighting functions, that is,

$$\begin{aligned}\bar{n}^\sigma(\mathbf{r}) &\equiv \int \sigma(\mathbf{r}-\mathbf{r}'; \{n\})n(\mathbf{r}')d^3r' \\ \bar{n}^\tau(\mathbf{r}) &\equiv \int \tau(\mathbf{r}-\mathbf{r}'; \{n\})n(\mathbf{r}')d^3r'.\end{aligned} \tag{12.1.13}$$

The requirement that $\bar{n}^\sigma$ and $\bar{n}^\tau$ be the constant density n in the limit of a homogeneous fluid requires that

$$\int \sigma(\mathbf{r}-\mathbf{r}; \{n\})d^3r' = \int \tau(\mathbf{r}-\mathbf{r}'; \{n\})d^3r' = 1, \quad n \text{ an arbitrary constant.} \tag{12.1.14}$$

Equation (12.1.12) will be the starting point of all but one of the nonlocal density functional models presented in this section. The basic problem will then be to find reasonable approximations to the weighting functionals σ and τ, the function $\mathcal{F}$, and the pair correlation function $g^{(2)}(\mathbf{r}, \mathbf{r}', \lambda)$.

As an example of σ, τ, and $\mathcal{F}$, consider the one- dimensional hard-rod fluid. From eq. (10.2.44) it follows that for this fluid

$$\bar{n}^\sigma(x) = 2^{-1}\left[n\left(x+\frac{a}{2}\right) + n\left(x-\frac{a}{2}\right)\right] = 2^{-1}\int \delta\left(|x-x'| - \frac{a}{2}\right)n(x')dx' \tag{12.1.15}$$

and

$$\bar{n}^\tau(x) = a^{-1}\int_{x-\frac{a}{2}}^{x+\frac{a}{2}} n(x')dx' = a^{-1}\int \eta\left(\frac{a}{2} - |x-x'|\right)n(x')dx' \tag{12.1.16}$$

and so the density weighting functionals are

$$\sigma(x-x') = \delta\left(|x'-x| - \frac{a}{2}\right)/2 \tag{12.1.17}$$

and

$$\tau(x - x') = \eta\left(\frac{a}{2} - |x'|\right)/a, \qquad (12.1.18)$$

where $\delta(x)$ and $\eta(x)$ are the Dirac delta functions and the step function. The corresponding formula for $\mathcal{F}$ is

$$\mathcal{F}(n) = -kT\ln(1 - na). \qquad (12.1.19)$$

These exact results form the basis of one of the approximations presented later in this section.

Van der Waals' (VDW) local density functional approximation is obtained by choosing

$$\sigma(\mathbf{r}) = \tau(\mathbf{r}) = \delta(\mathbf{r}) \qquad (12.1.20)$$

and using the Clausius approximation, $P^R(n) = nkT/(1 - nb)$, for which

$$\mathcal{F}(n) = -kT\ln(1 - nb). \qquad (12.1.21)$$

As discussed in Chapter 5, eq. (12.1.21) is an approximation in which one assumes that multiparticle packing is pair additive. The more accurate Carnahan–Starling equation of state for hard spheres, $P^R(n) = nkT(1 + y + y^2 - y^3)/(1 - y)^3$, $y \equiv \pi nd^3/6$, gives the formula

$$\mathcal{F}(n) = kT\frac{y(4 - 3y)}{(1 - y)^2}. \qquad (12.1.22)$$

The simplest approximation to ΔF^A is the VDW approximation in which it is assumed that the pair interaction u^A is sufficiently long ranged to approximate the pair correlation function by unity. In this case

$$\Delta F^A = \frac{1}{2}\int n(\mathbf{r})n(\mathbf{r}')u^A(|r - \mathbf{r}'|)d^3rd^3r', \qquad (12.1.23)$$

The contribution of F^A to the chemical potential in the VDW approximation is

$$\Delta\mu^A = \frac{\delta\Delta F^A}{\delta n(\mathbf{r})} = \int n(\mathbf{r}')u^A(|\mathbf{r} - \mathbf{r}'|)d^3r'. \qquad (12.1.24)$$

A significant quantitative improvement on the VDW approximation was introduced by Tang et al., (1991). They used the approximation

$$g^{(2)}(\mathbf{r}, \mathbf{r}', \lambda) \approx g^R(|\mathbf{r} - \mathbf{r}'|, \bar{n}^\tau)[1 - \alpha(\bar{n}^\tau)\lambda u^A(|\mathbf{r} - \mathbf{r}'|)/kT], \qquad (12.1.25)$$

where $\bar{n}^\tau$ is the coarse grained density evaluated at $(\mathbf{r} + \mathbf{r}')/2$, $g^R(|\mathbf{r}|, n)$ is the pair correlation function of a homogeneous reference fluid at density n and

$$\frac{1}{\alpha(n)} = \frac{1}{kT}\frac{\partial P^R(n)}{\partial n}, \qquad (12.1.26)$$

where $P^R(n)$ is the pressure of the homogeneous hard sphere fluid at the coarse grained density n. The Tang–Scriven–Davis (TSD) formula for F^A is then

$$\Delta F^A = \frac{1}{2}\int n(\mathbf{r})n(\mathbf{r}')u^A(|\mathbf{r}-\mathbf{r}'|)g^R(|\mathbf{r}-\mathbf{r}'|,\bar{n}^\tau)\left[1-\frac{1}{2}\beta\alpha(\bar{n}^\tau)u^A(|\mathbf{r}-\mathbf{r}'|)\right]d^3r d^3r'. \quad (12.1.27)$$

The formula for $g^{(2)}(\mathbf{r},\mathbf{r}',\lambda)$ is a heuristic extension to inhomogeneous fluids of the approximation introduced by Barker and Henderson (see Section 9.4.6) for homogeneous fluids. The contribution of F^A to the chemical potential is

$$\begin{aligned}\Delta\mu^A &= \int n(\mathbf{r}')u^A(|\mathbf{r}-\mathbf{r}'|)g^{(2)}(\mathbf{r},\mathbf{r}',\lambda)d^3r \\ &+ \frac{1}{2}\iint n(\mathbf{r}')n(\mathbf{r}'')u^A(|\mathbf{r}'-\mathbf{r}''|)\frac{\partial g^{(2)}(\mathbf{r}',\mathbf{r}'')}{\partial\bar{n}^\tau}\frac{\delta\bar{n}^\tau((\mathbf{r}'+\mathbf{r}'')/2)}{\delta n(\mathbf{r})}d^3r'd^3r''.\end{aligned} \quad (12.1.28)$$

Once $\bar{n}^\tau$ is specified μ^A can be determined from eq. (12.1.28).

We will now discuss several nonlocal density functional models that have proven useful in the theory of strongly inhomogeneous systems.

12.1.1 The Generalized VDW Model

Nordholm and coworkers (1980) and Johnson and Nordholm (1981) introduced the generalized VDW (GVDW) model. The model is defined by the weighting functions

$$\sigma(\mathbf{r}) = \delta(\mathbf{r}), \quad \tau(\mathbf{r}) = \eta(d-|\mathbf{r}|)\Big/\left(\frac{4\pi d^3}{3}\right). \quad (12.1.29)$$

With this choice the $\bar{n}^\sigma(\mathbf{r}) = n(\mathbf{r})$, the local density and $\bar{n}^\tau(\mathbf{r})$ is the density locally averaged over a volume equal to the volume excluded to a neighboring particle by a particle centered on $\mathbf{r}$, that is,

$$\bar{n}^\tau(\mathbf{r}) = \frac{1}{(4\pi d^3/3)}\int \eta(d-|\mathbf{r}-\mathbf{r}'|)n(\mathbf{r}')d^3r' = \frac{1}{(4\pi d^3/3)}\int_{|\mathbf{r}-\mathbf{r}'|<d} n(\mathbf{r}')d^3\mathbf{r}'. \quad (12.1.30)$$

Using the functional derivative properties

$$\frac{\delta n(\mathbf{r})}{\delta n(\mathbf{r}')} = \delta(\mathbf{r}-\mathbf{r}'), \quad \frac{\delta\mathcal{F}(\bar{n}^\tau(\mathbf{r}))}{\delta n(\mathbf{r}')} = \mathcal{F}'(\bar{n}^\tau(\mathbf{r}))\frac{\delta\bar{n}^\tau(\mathbf{r})}{\delta n(\mathbf{r}')}, \quad \frac{\delta\bar{n}^\tau(\mathbf{r})}{\delta n(\mathbf{r}')} = \frac{\eta(d-|\mathbf{r}'-\mathbf{r}|)}{(4\pi d^3/3)}, \quad (12.1.31)$$

where $\mathcal{F}'(n) = \partial\mathcal{F}(n)/\partial n$, we find for the chemical potential of the GVDW model

$$\begin{aligned}\mu = \frac{\delta F}{\delta n(\mathbf{r})} &= \mu^+(T) + kT\ln n(\mathbf{r}) + \mathcal{F}(\bar{n}^\tau(\mathbf{r})) + \frac{1}{(4\pi d^3/3)}\int \eta(d-|\mathbf{r}'-\mathbf{r}|)n(\mathbf{r})\mathcal{F}'(\bar{n}^\tau(\mathbf{r}'))d^3r' \\ &+\Delta\mu^A + v(\mathbf{r}).\end{aligned} \quad (12.1.32)$$

Nordholm used the VDW formula, eq. (12.1.24) for μ^A. Equation (12.1.28) could, of course, be used for quantitative improvement.

12.1.2 The Generalized Hard-Rod Model

In analogy with the exact one-dimensional hard-rod weighting functions, Fischer and Heinbuch (1988) introduced the generalized hard-rod (GHR) model whose weighting functions are

$$\sigma(\mathbf{r}) = \delta\left(\frac{d}{2} - |\mathbf{r}|\right) \Big/ \left[4\pi\left(\frac{d}{2}\right)^2\right] \tag{12.1.33}$$

and

$$\tau(\mathbf{r}) = \eta\left(\frac{d}{2} - |\mathbf{r}|\right) \Big/ \left[\frac{4\pi}{3}\left(\frac{d}{2}\right)^3\right]. \tag{12.1.34}$$

With these weighting functions, $\bar{n}^\sigma(\mathbf{r})$ is the local density averaged over the surface of a hard sphere and $\bar{n}^\tau(\mathbf{r})$ is the local density averaged over the volume of a hard sphere.

With F^A given by eq. (12.1.17) and with the functional derivatives given by

$$\frac{\delta \bar{n}^\sigma(\mathbf{r})}{\delta n(\mathbf{r}')} = \frac{\delta(\frac{d}{2} - |\mathbf{r} - \mathbf{r}|)}{[4\pi(d/2)^2]}, \quad \frac{\delta \bar{n}^\tau(\mathbf{r})}{\delta n(\mathbf{r}')} = \frac{\eta(\frac{d}{2} - |\mathbf{r} - \mathbf{r}'|)}{[(4\pi/3)(d/2)^3]}, \tag{12.1.35}$$

the chemical potential of the GHR model is

$$\mu = \mu^+(T) + kT \ln n(\mathbf{r}) + \frac{1}{4\pi(d/2)^2}\int \delta\left(|\mathbf{r} - r'| - \frac{d}{2}\right)\mathcal{F}(\bar{n}^\tau(\mathbf{r}'))d^3r'$$
$$\frac{1}{[(4\pi/3)(d/2)^3]}\int \eta(\frac{d}{2} - |\mathbf{r} - \mathbf{r}'|)\bar{n}^\sigma(\mathbf{r}')\mathcal{F}'(\bar{n}^\tau(\mathbf{r}'))d^3r' + \Delta\mu^A + v(\mathbf{r}). \tag{12.1.36}$$

12.1.3 The Tarazona Model

The two models just presented are based on heuristically obtained weighting functions that depend on relative position alone. Tarazona (1985) presented a somewhat less heuristic way to determine τ. In his approach he took $\sigma(\mathbf{r}) = \delta(\mathbf{r})$, or $\bar{n}^\sigma(\mathbf{r}) = n(\mathbf{r})$, and defined for $\bar{n}^\tau(\mathbf{r})$ a density weighted spatial average of the local density of the form

$$\bar{n}^\tau(\mathbf{r}) = \int \tau(\mathbf{r} - \mathbf{r}', \bar{n}^\tau(\mathbf{r}))n(\mathbf{r}')d^3r'. \tag{12.1.37}$$

He further expressed the function τ as a quadratic function of $\bar{n}^\tau$:

$$\begin{aligned}\tau(\mathbf{r} - \mathbf{r}', \bar{n}^\tau(\mathbf{r}) &= \omega_o(|\mathbf{r} - \mathbf{r}'|) + \omega_1(|\mathbf{r} - \mathbf{r}'|)\bar{n}^\tau(\mathbf{r}) \\ &\quad + \omega_2(|\mathbf{r} - \mathbf{r}'|)[\bar{n}^\tau(\mathbf{r})]^2,\end{aligned} \tag{12.1.38}$$

so that eq. (12.1.37) becomes

$$\bar{n}^{\tau}(\mathbf{r}) = \sum_{i=0}^{2} \bar{n}_i(\mathbf{r})[\bar{n}^{\tau}(\mathbf{r})]^i, \qquad (12.1.39)$$

where

$$\bar{n}_i(\mathbf{r}) \equiv \int \omega_i(|\mathbf{r}-\mathbf{r}'|)n(\mathbf{r}')d^3r', \quad i = 0, 1, 2. \qquad (12.1.40)$$

The homogeneous fluid limit requires that

$$\int \omega_o(|\mathbf{r}-\mathbf{r}'|)d^3r = 1 \quad \text{and} \quad \int \omega_\iota(|\mathbf{r}-\mathbf{r}'|)d^3r' = 0, \quad i = 1, 2. \qquad (12.1.41)$$

The criterion used to fix the functional forms of the weighting functions $\omega_\iota(\mathbf{r})$ was to require that the direct correlation function calculated from the exact relation

$$\delta^2 F^R/\delta n(\mathbf{r})\delta n(\mathbf{r}') = -kTC^R(\mathbf{r}, \mathbf{r}'), \quad \mathbf{r} \neq \mathbf{r}',$$

for a homogeneous hard sphere fluid agree over a range of densities with the direct correlation function predicted by the Percus-Yevick approximation for a homogeneous hard sphere fluid. The results of Tarazona's fit (in the version not containing typographical errors) are

$$\begin{aligned} \omega_o(\mathbf{r}) &= \frac{3}{4\pi d^3}, \quad |\mathbf{r}| < d \\ &= 0, \quad |\mathbf{r}| > d \end{aligned} \qquad (12.1.42a)$$

$$\begin{aligned} \omega_1(\mathbf{r}) &= 0.475 - 0.648(|\mathbf{r}|/d) + 0.113(|\mathbf{r}|/d)^2, \quad |\mathbf{r}| < d \\ &= 0.288(d/|\mathbf{r}|) - 0.924 + 0.764(|\mathbf{r}|/d) - 0.187(|\mathbf{r}|/d)^2, \quad d < r < 2d \\ &= 0, \quad |\mathbf{r}| > 2d, \end{aligned} \qquad (12.1.42b)$$

$$\begin{aligned} \omega_2(\mathbf{r}) &= \frac{5\pi d^3}{144}\left[6 - 12(|\mathbf{r}|/d) + 5(|\mathbf{r}|/d)^2\right], \quad |\mathbf{r}| < d \\ &= 0, \quad |\mathbf{r}| > d. \end{aligned} \qquad (12.1.42c)$$

With the omission of ω_1 and ω_2, Tarazona's model reduces to the GVDW model. Thus, in a sense, Tarazona's model can be viewed as a systematic extension of the GVDW model.

The chemical potential of the Tarazona model is

$$\begin{aligned} \mu = \mu^+ + kT\ln n(\mathbf{r}) + \mathcal{F}(\bar{n}^{\tau}(\mathbf{r})) + \int \frac{\delta \bar{n}^{\tau}(\mathbf{r}')}{\delta n(\mathbf{r})} n(\mathbf{r}')\mathcal{F}'(\bar{n}^{\tau}(\mathbf{r}'))d^3r' \\ + \Delta\mu^A + v(\mathbf{r}), \end{aligned} \qquad (12.1.43)$$

where

$$\frac{\delta \bar{n}^{\tau}(\mathbf{r}')}{\delta n(\mathbf{r})} = \frac{\tau(\mathbf{r}'-\mathbf{r}; \bar{n}^{\tau}(\mathbf{r}))}{1 - \bar{n}_1(\mathbf{r}') - 2\bar{n}_2(\mathbf{r}')\bar{n}^{\tau}(\mathbf{r}')}, \qquad (12.1.44)$$

with τ given by eq. (12.1.38) with $\bar{n}^\tau(\mathbf{r})$ replaced by the physically appropriate root of eq. (12.1.39), namely,

$$\bar{n}^\tau(\mathbf{r}) = \frac{1 - \bar{n}_1(\mathbf{r}) - \{[1 - \bar{n}_1(\mathbf{r})]^2 - 4\bar{n}_0(\mathbf{r})\bar{n}_2(\mathbf{r})\}^{1/2}}{2\bar{n}_2(\mathbf{r})}. \quad (12.1.45)$$

The "weighted density functional theory" of Curtin and Ashcroft (1985) and Denton and Ashcroft (1989) represents a refinement of the Tarazona approach.

12.1.4 The Curtin–Ashcroft Model

The model of Curtin and Ashcroft (1985) is similar to that of Tarazona. They do not break up the pair potential into two parts, that is, F^{ex} includes the entire interaction contribution. As did Tarazona, they assume $\sigma(\mathbf{r}, \mathbf{r}') = \delta(\mathbf{r} - \mathbf{r}')$, and relate $\bar{n}^\tau$ and $\tau(\mathbf{r} - \mathbf{r}', \bar{n})^\tau$ by eq. (12.1.26). Then, with F^{ex} given by eq. (12.1.12), they use the exact relationship

$$C^{(2)}(\mathbf{r}, \mathbf{r}') = -\beta \frac{\delta^2 F^{ex}}{\delta n(\mathbf{r})\delta n(\mathbf{r}')} \quad (12.1.46)$$

and its homogeneous fluid limit to obtain the equation

$$\begin{aligned} -\beta^{-1} C_o^{(2)}(|\mathbf{r} - \mathbf{r}'|, n) &= 2\mathcal{F}'(n)\tau(\mathbf{r} - \mathbf{r}', n) + n\mathcal{F}''(n) \int \tau(\mathbf{r} - \mathbf{r}'', n)\tau(\mathbf{r}' - \mathbf{r}'', n) d^3r'' \\ &+ n\mathcal{F}'(n) \int \frac{\partial}{\partial n}[\tau(\mathbf{r} - \mathbf{r}'', n)\tau(\mathbf{r}' - \mathbf{r}'', n)] d^3r'', \end{aligned} \quad (12.1.47)$$

where n and $C_o^{(2)}$ are the density and direct correlation function of homogeneous fluid and $\mathcal{F}'$ and $\mathcal{F}''$ are the first and second derivatives of $\mathcal{F}(n)$ with respect to n. This equation is an integrodifferential equation that determines the functional τ in terms of the homogeneous fluid properties $C_o^{(2)}$ and $\mathcal{F}$.

The Fourier transform of a function is defined as

$$h(\mathbf{k}) = \frac{1}{(2\pi)^{3/2}} \int e^{i\mathbf{k}\cdot\mathbf{r}} h(\mathbf{r}) d^3r. \quad (12.1.48)$$

If $h(\mathbf{r})$ depends only on r then $h(\mathbf{k})$ depends only on the magnitude of $\mathbf{k}$.

The Fourier transform of eq. (12.1.47) yields

$$\begin{aligned} -\beta^{-1} C_o^{(2)}(k, n) &= 2\mathcal{F}'(n)\tau(k, n) + n\mathcal{F}''(n)[\tau(k, n)]^2 \\ &+ n\mathcal{F}'(n)\frac{\partial}{\partial n}[\tau(k, n)]^2. \end{aligned} \quad (12.1.49)$$

This is now a first-order differential equation in density for $\tau(k, n)$. Given $C_o^{(2)}(k, n)$ and $\mathcal{F}(n)$, τ must be computed by numerical solution of eq. (12.1.49) for each k. Then $\tau(r, n)$ must be computed by numerical inversion of the Fourier transform. Plainly, this is a more expensive procedure than the quadratic fit used by Tarazona, but of course it is a more rigorous way to find τ than is Tarazona's.

12.1.5 The Meister–Kroll Model

The model developed by Meister and Kroll (1985) takes advantage of exact density functional theory in a way that can be systematically improved if homogeneous fluid theory improves. They too decompose the free energy into the sum $F^H + F^A$ where F^A is the VDW mean field contribution from attractive forces and F^H the free energy of an inhomogeneous reference fluid of hard spheres (a reference fluid of repulsive particles of any sort could be chosen without loss of generality). F^H is again decomposed as

$$F^H = F^I + F^{\text{ex}} + F^e$$
$$= \int n(\mathbf{r})[\mu^+(\tau) + \ln n(\mathbf{r}) - 1]d^3r + F^{\text{ex}} + \int n(\mathbf{r})v(\mathbf{r})d^3r. \tag{12.1.50}$$

F^{ex} is the excess free energy from hard sphere interactions. The chemical potential corresponding to F^{ex} is

$$\mu^{\text{ex}}(\mathbf{r}, \{n\}) = \frac{\delta F^{\text{ex}}}{\delta n(\mathbf{r})}, \tag{12.1.51}$$

and so, with the condition $F^{\text{ex}}(\{0\}) = 0$, eq. (12.1.51) can be integrated [see eq. (9.1.55)] to obtain

$$F^{\text{ex}} = \int_0^1 \int \mu^{\text{ex}}(\mathbf{r}, \{\varepsilon n\})n(\mathbf{r})d^3r d\varepsilon \tag{12.1.52}$$

Meister and Kroll (1985) introduced the coarse grained density $n_o(\mathbf{r})$ [which is equivalent to the $\bar{n}^\tau(\mathbf{r})$ in the notation of the previous models] that they assumed to be smoothly varying compared to the local density $n(\mathbf{r})$. They then expanded $\mu^{\text{ex}}(r, \{n\})$ in a functional Taylor's series about a homogeneous state at density $n_o(\mathbf{r})$. Because $\mu^{\text{ex}}(\mathbf{r}\{n\}) = -\delta\Phi^H(\{n\})/\delta n(\mathbf{r})$, where $\Phi(\{n\})$ is defined by eq. (9.2.40), and because the kth functional derivative of $\beta\Phi^H$ equals the kth direct correlation function [see eq. (9.3.39)], it follows that

$$\mu^{\text{ex}}(\mathbf{r}, \{n\}) = \mu^{\text{ex}}(\mathbf{r}, n_o(\mathbf{r})) - \sum_{k=1}^{\infty} \frac{1}{k!} kT \int C^{(k+1),H}(\mathbf{r}, \mathbf{r}_1, \cdots, \mathbf{r}_k, n_o(\mathbf{r}))$$
$$\times \prod_{\iota=1}^{k} [n(\mathbf{r}_\iota) - n_o(\mathbf{r})]d^3r_1 \cdots d^3r_k, \tag{12.1.53}$$

where $\mu^{\text{ex}}(\mathbf{r}, n_o(\mathbf{r}))$ and $C^{(k+1),H}(\mathbf{r}, \mathbf{r}_1, \ldots, \mathbf{r}_k, n_o(\mathbf{r}))$ are the excess chemical potential and direct correlation function of a homogeneous hard sphere fluid at density $n_o(\mathbf{r})$.

Combining eqs. (12.1.50)–(12.1.53) and keeping terms through the first order in $n - n_o$ in eq. (12.1.53), we find for the grand potential of the hard sphere system, $\Omega^H = F^H - N\mu^H$, the Meister–Kroll result

$$
\begin{aligned}
\Omega = kT \int & n(\mathbf{r})\{\beta\mu^+(T) + \ln n(\mathbf{r}) - 1 + v(\mathbf{r}) - \mu \\
& + [\beta f^H(n_o(\mathbf{r})) - \beta n_o(\mathbf{r})\mu^+ - n_o(\mathbf{r})(\ln n_o(\mathbf{r}) - 1)]/n_o(\mathbf{r})\}d^3r \\
& - kT \int n(\mathbf{r}) \left\{ \int_0^1 \int C^H(|\mathbf{r} - \mathbf{r}'|, \varepsilon n_o(\mathbf{r}))\varepsilon[n(\mathbf{r}') - n_o(\mathbf{r})]d^3r'd\varepsilon \right\} d^3r \\
& + \frac{1}{2}\int n(\mathbf{r})n(\mathbf{r}')u^A(|r - \mathbf{r}'|)d^3rd^3r',
\end{aligned}
\tag{12.1.54}
$$

where $C^H(|\mathbf{r} - \mathbf{r}'|, \varepsilon n_o)$ denotes the pair direct correlation function of a homogeneous fluid of hard spheres at density εn_o.

The coarse grained density $n_o(\mathbf{r})$ has not been specified in deriving eq. (12.1.39). Meister and Kroll chose $n(\mathbf{r})$ and $n_o(\mathbf{r})$ to be independent variables and impose the conditions that Ω^H be a minimum with respect to both variables, that is,

$$
\frac{\delta\Omega^H}{\delta n(\mathbf{r})} = \frac{\delta\Omega^H}{\delta n_o(\mathbf{r})} = 0. \tag{12.1.55}
$$

These conditions provide a self-consistent way of computing $n_0(\mathbf{r})$ as a functional of $n(\mathbf{r})$ from which to calculate $n(\mathbf{r})$ for a given hard sphere chemical potential and external potential.

With the aid of the following properties of homogeneous fluid

$$
f^H(n_o) - f^I(n_o) = n_o[\mu^H(n_o) - \mu^I(n_o)] - [P^H(n_o) - P^I(n_o)] \tag{12.1.56}
$$

$$
\mu^H(n_o) - \mu^I(n_o) = \frac{\partial}{\partial n_o}[f^H(n_o) - f^I(n_o)] \tag{12.1.57}
$$

and

$$
\lambda n_o \int C^H(|\mathbf{r} - \mathbf{r}'|, \lambda n_o)d^3r' = 1 - \frac{1}{kT}\frac{\partial P^H(\lambda n_o)}{\partial(\lambda n_o)}, \quad \text{[given at eq. (9.3.6)]}, \tag{12.1.58}
$$

it is straightforward to show that the condition $\delta\Omega^H/\delta n_o(\mathbf{r})$ yields

$$
n_o(\mathbf{r}) = \int \tau(|\mathbf{r} - \mathbf{r}'|, \quad n_o(\mathbf{r}'))n(\mathbf{r}')d^3r', \tag{12.1.59}
$$

where

$$
\tau(|\mathbf{r} - \mathbf{r}'|, n_o(\mathbf{r})) = \frac{\chi(|(\mathbf{r} - \mathbf{r}'|, n_o(\mathbf{r}))}{\int \chi(|\mathbf{r} - \mathbf{r}''|, n_o(\mathbf{r}))d^3r''}, \tag{12.1.60}
$$

with

$$
\chi(|\mathbf{r} - \mathbf{r}'|, n_o(\mathbf{r})) \equiv \int_o^1 \varepsilon^2 \frac{\partial C^H}{\partial(\varepsilon n_o(\mathbf{r}))}(|\mathbf{r} - \mathbf{r}'|, \varepsilon n_o(\mathbf{r}))d\varepsilon. \tag{12.1.61}
$$

If the three-body or higher direct correlation functions of homogeneous fluid were available, higher order terms in the expansion at eq. (12.1.53) could be kept with a resultant systematic improvement in the theory of $n_o(\mathbf{r})$ and the estimate of the density profile $n(\mathbf{r})$.

The chemical potential of the Meister–Kroll model is then

$$\mu^H = \mu^+(T) + kT \ln n(\mathbf{r}) + [f^H(n_o(\mathbf{r})) - f^I(n_o(\mathbf{r}))]/n_o(\mathbf{r})$$
$$+ \int n(\mathbf{r}')u^A(|\mathbf{r}-\mathbf{r}'|)d^3r' - \int_o^1 \int \varepsilon C^H(|\mathbf{r}-\mathbf{r}'|, \varepsilon n_o(r))[n(\mathbf{r}') - n_o(\mathbf{r})]d^3r'd\varepsilon \quad (12.1.62)$$
$$- \int_o^1 \int \varepsilon C^H(|\mathbf{r}-\mathbf{r}'|, \varepsilon n_o(\mathbf{r}'))n(\mathbf{r}')d^3r'd\varepsilon.$$

Completion of the model requires a way to calculate $f^H(n)$ and $C^H(|\mathbf{r}-\mathbf{r}'|, n)$. Meister and Kroll made the self-consistent choice of the Percus–Yevick result for C^H, that is, eq. (9.4.19), and the Percus–Yevick compressibility pressure, $P^H(n) = nkT(1 + y + y^2)/(1-y)^3$, $y \equiv \pi n d^3/6$, for which $f^H(n)$ is given by

$$f^H(n) - f^I(n) = n \int_0^n \frac{P^H(n) - P^I(n)}{n^2} dn$$
$$= nkT\left[\frac{5/2 - y^2}{(1-y)^2} - \frac{2}{1-y} - \ln(1-y) - \frac{1}{2}\right], \quad (12.1.63)$$

and, of course,

$$P^I(n) = nkT, \quad f^I(n) = n\mu^+(T) + nkT(\ln n - 1), \quad \mu^I(n) = \mu^+ + kT \ln n, \quad (12.1.64)$$

and

$$\mu^H(n) = \mu^I(n) + nkT\left[\frac{1+y+y^2}{(1-y)^3} + \frac{5/2-y^2}{(1-y)^2} - \frac{2}{1+y} - \ln(1-\eta) - \frac{1}{2}\right]. \quad (12.1.65)$$

Finally, to apply the Meister–Kroll model to fluids with attractive as well as repulsive forces we set $\mu = \mu^H + \Delta\mu^A$, where μ^H is taken from eq. (12.1.62) and $\Delta\mu^A$ is taken from some convenient approximation [e.g., the VDW formula at eq. (12.1.24) or the TSD formula at eq. (12.1.28)].

12.1.6 Multicomponent Generalizations of the Models

The quantity F^H has the same general form in the GVDW, GHR, Tarazona, and Curtin–Ashcroft models. In the mixture version of these models, the parts of F^H are

$$F^e = \sum_\alpha \int n_\alpha(\mathbf{r})v_\alpha(\mathbf{r})d^3r \quad (12.1.66)$$

$$F^I = \sum_\alpha n_\alpha \int n_\alpha(\mathbf{r})\left\{\mu_\alpha^+(T)kT[\ln n_\alpha(\mathbf{r}) - 1\right\}d^3r, \quad (12.1.67)$$

and

$$F^{\text{ex}} = \sum_{\alpha} \int \bar{n}_{\alpha}^{\sigma}(\mathbf{r}) \mathcal{F}(\bar{\mathbf{n}}^{\tau}(\mathbf{r})) d^3 r, \tag{12.1.68}$$

where $\mathcal{F}(\bar{\mathbf{n}}^{\tau})$ is the free energy per molecule of a homogeneous hard sphere mixture at composition $\bar{\mathbf{n}}^{\tau} = \{\bar{n}_1^{\tau_1}, \ldots, \bar{n}_c^{\tau_c}\}$. Selection of the coarse grained densities $\bar{n}_{\alpha}^{\sigma_\alpha}$ and $\bar{n}_{\alpha}^{\tau_\alpha}$ is needed to specify the particular model.

For the GVDW model we choose $\sigma_{\alpha}(\mathbf{r}) = \delta(\mathbf{r})$ and $\tau(\mathbf{r})$ and $\tau_{\alpha}(\mathbf{r}) = \eta(d_{\alpha} - |\mathbf{r}|)/(4\pi d_{\alpha}^3/3)$, where d_{α} is the diameter of a hard sphere of species α. For the GHR model, eq. (11.2.44) suggests the choices

$$\sigma_{\alpha}(\mathbf{r}) = \delta\left(\frac{d_{\alpha}}{2} - |\mathbf{r}|\right) \tag{12.1.69}$$

and

$$\tau_{\alpha}(\mathbf{r}) = \eta\left(\frac{d_{\alpha}}{2} - |\mathbf{r}|\right) \Big/ \left[\frac{4\pi}{3}\left(\frac{d_{\alpha}}{2}\right)^3\right]. \tag{12.1.70}$$

For the Tarazona model, $\sigma_{\alpha}(\mathbf{r}) = \sigma(\mathbf{r})$ as before. The multicomponent expression for τ_{α} has not been worked out except for the special case that all the components of the mixture have the same hard sphere diameter d. Then τ_{α} is the same as that of the pure fluid for each of the species. The exact formula for the attractive contribution to F for a multicomponent system is

$$\Delta F^A = \sum_{\alpha,\beta=1}^{c} \int_0^1 \int n_{\alpha}(\mathbf{r}) n_{\beta}(\mathbf{r}') g^{(2)}(\mathbf{r}, \mathbf{r}', \lambda) u_{\alpha\beta}^A(|\mathbf{r} - \mathbf{r}'|) d^3 r d^3 r' d\lambda, \tag{12.1.71}$$

the VDW approximation is

$$\Delta F^A = \sum_{\alpha,\beta=1}^{c} \frac{1}{2} \int n_{\alpha}(\mathbf{r}) n_{\beta}(\mathbf{r}') u_{\alpha\beta}^A(|\mathbf{r} - \mathbf{r}') d^3 r d^3 r', \tag{12.1.72}$$

and the TSD approximation is

$$\Delta F^A = \sum_{\alpha,\beta=1}^{c} \frac{1}{2} \int n_{\alpha}(\mathbf{r} n_{\beta}(\mathbf{r}') g_{\alpha\beta}^H(|\mathbf{r} - \mathbf{r}'|, \bar{\mathbf{n}}) u_{\alpha\beta}^A(|\mathbf{r} - \mathbf{r}'|) \left[1 - \frac{1}{2}\beta\alpha_{\alpha\beta}(\bar{\mathbf{n}}) u_{\alpha\beta}^A(|\mathbf{r} - \mathbf{r}'|)\right] d^3 r d^3 r', \tag{12.1.73}$$

where

$$\frac{1}{\alpha_{\alpha\beta}(\bar{\mathbf{n}})} = \frac{1}{kT} \frac{\partial \mu_{\alpha}^H(\bar{\mathbf{n}})}{\partial \bar{n}_{\beta}}, \tag{12.1.74}$$

with $\mu_{\alpha}^H(\bar{\mathbf{n}})$ the chemical potential of α in a homogeneous hard sphere fluid at composition $\bar{\mathbf{n}}(\frac{1}{2}(\mathbf{r} + \mathbf{r}'))$ and $\bar{\mathbf{n}} = \{\bar{n}_1^{\tau_1}, \ldots, \bar{n}_c^{\tau_c}\}$. The course grained densities, $\bar{n}_{\alpha}^{\tau_\alpha}$ are chosen to be the same as those that enter the formula chosen for F^H.

In the Meister–Kroll model F is again decomposed into the sum $F^H + F^A$, where F^A is given by eq. (12.1.64), and F^H is the sum $F^e + F^I + F^{ex}$, with F^e and F^I given by eqs. (12.1.66) and (12.1.67) and

$$F^{ex} = \sum_\alpha \int_o^1 \int \mu_\alpha^{ex}(\mathbf{r}; \{\varepsilon\mathbf{n}\}) n_\alpha(\mathbf{r}) d^3r d\varepsilon. \tag{12.1.75}$$

Following the procedure of Meister and Kroll for a one-component fluid, we introduce the coarse grained densities $n_{\alpha o}(\mathbf{r})$ and expand the μ_α^{ex} in a functional Taylor series about a homogeneous hard sphere fluid at composition $\mathbf{n}_o(\mathbf{r})$. The result, which makes use of eq. (9.2.39), is

$$\begin{aligned}\mu_\alpha^{ex}(\mathbf{r}, \{\mathbf{n}\}) = \mu_\alpha^{ex}(\mathbf{n}_o(\mathbf{r})) - \sum_{k=1}^{\infty} \frac{1}{k!} \sum_{\beta_1} \cdots, \sum_{\beta_k} C^H_{\alpha,\beta_1,\ldots,\beta_k}(\mathbf{r}, \mathbf{r}_1, \ldots, \mathbf{r}_k, \mathbf{n}_o(\mathbf{r})) \\ \times \prod_{i=1}^{k} [n_{\beta_i}(\mathbf{r}_i) - n_{\beta_i o}(\mathbf{r})] d^3r_1 \cdots d^3v_k,\end{aligned} \tag{12.1.76}$$

$\mu_\alpha^{ex}(\mathbf{n}_o(\mathbf{r}))$ is the excess chemical potential of species α in a homogeneous hard sphere fluid at composition $\mathbf{n}_o(\mathbf{r})$ and $C^H_{\alpha,\beta_1,\ldots,\beta_k}$ is the $k+1$st direct correlation function of a homogeneous hard sphere fluid at a composition $\mathbf{n}_o(\mathbf{r})$. Keeping terms through first order in $n_\beta - n_{\beta o}$ in eq. (12.1.75), we obtain for the multicomponent version of the Meister–Kroll model

$$\begin{aligned}\Omega = \sum_\alpha \int n_\alpha(\mathbf{r}) \Bigg\{ \mu_\alpha^+(T) + \ln n_\alpha(\mathbf{r}) - 1 + \int_o^1 \mu_\alpha^{ex}(\varepsilon\mathbf{n}_o(\mathbf{r})) d\varepsilon d^3r \\ - \sum_\beta kT \int_o^1 \int \varepsilon C^H_{\alpha\beta}(|\,\mathbf{r} - \mathbf{r}'\,|, \varepsilon\mathbf{n}_o(\mathbf{r}))[n_\beta(\mathbf{r}') - n_{\beta o}(\mathbf{r})] d^3r' d\varepsilon \\ + \sum_\beta \frac{1}{2} \int n_\beta(\mathbf{r}') u^A_{\alpha\beta}(|\,(\mathbf{r} - \mathbf{r}'\,|)) d^3r' + v_\alpha(\mathbf{r}) - \mu_\alpha \Bigg\} d^3r.\end{aligned} \tag{12.1.77}$$

To obtain $\mathbf{n}_o$ we impose the conditions that $\Omega^H(\mathbf{n}, \mathbf{n}_o)$ be a minimum with respect to independent variations of $\mathbf{n}(\mathbf{r})$ and $\mathbf{n}_o(\mathbf{r})$, that is,

$$\frac{\delta\Omega^H}{\delta n_\beta(\mathbf{r})} = \frac{\delta\Omega^H}{\delta n_{\beta o}(\mathbf{r})} = 0, \quad \beta = 1, \ldots, c. \tag{12.1.78}$$

Using the definition $\mu_\alpha^{ex}(\mathbf{n}_o) \equiv \mu_\alpha^H(\mathbf{n}_o) - \mu_\alpha^I(\mathbf{n}_o)$ and the homogeneous fluid property [eq. (9.3.43)]

$$\frac{\partial \mu_\alpha^{ex}(\varepsilon\bar{\mathbf{n}}_o(\mathbf{r}))}{\partial n_{\beta o(\mathbf{r})}} = -kT \int \varepsilon C^H_{\alpha\beta}(|\,\mathbf{r} - \mathbf{r}'|, \varepsilon\mathbf{n}_o(\mathbf{r})) d^3r', \tag{12.1.79}$$

we find from the condition $\delta\Omega^H/\delta n_{\gamma 0} = 0$, the relations that determine coarse grained densities $n_{\beta 0}(\mathbf{r})$:

$$\sum_{\alpha,\beta} n_\alpha(\mathbf{r}) n_{\beta o}(\mathbf{r}) \int \chi_{\alpha\beta\gamma}(|\mathbf{r}-\mathbf{r}'|, \mathbf{n}_o(\mathbf{r})) d^3r' = \sum_{\alpha,\beta} n_\alpha(\mathbf{r}) \int \chi_{\alpha\beta\gamma}(|\mathbf{r}-\mathbf{r}'|, \mathbf{n}_o(\mathbf{r})) n_{\beta o}(\mathbf{r}') d^3r', \tag{12.1.80}$$

for $\gamma = 1, \ldots, c$, where

$$\chi_{\alpha\beta\gamma}(|\mathbf{r}-\mathbf{r}'|, \mathbf{n}(\mathbf{r})) \equiv \sum_0^1 \varepsilon^2 \frac{\partial C^H_{\alpha\beta}}{\partial(\varepsilon n_{\gamma o}(\mathbf{r}))}(|\mathbf{r}-\mathbf{r}'|, \varepsilon \mathbf{n}_o(\mathbf{r})) d\varepsilon. \tag{12.1.81}$$

The chemical potential of the Meister–Kroll model is

$$\begin{aligned}\mu_\alpha &= \mu^I_\alpha(n_\alpha(\mathbf{r})) + \int_0^1 \mu^{\mathrm{ex}}_\alpha(\varepsilon \mathbf{n}_o(\mathbf{r})) d\varepsilon + \sum_\beta kT \int_0^1 \int \varepsilon C^H_{\alpha\beta}(|\mathbf{r}-\mathbf{r}'|, \varepsilon \mathbf{n}_o(\mathbf{r})) n_{\beta o}(\mathbf{r}) d^3r' d\varepsilon \\ &- \sum_\beta kT \int \varepsilon \left[C^H_{\alpha\beta}(|\mathbf{r}-\mathbf{r}', \varepsilon \mathbf{n}_o(\mathbf{r})) + C^H_{\alpha\beta}(|\mathbf{r}-\mathbf{r}'|, \varepsilon n_o(\mathbf{r}'))\right] n_{\beta\delta}(\mathbf{r}) d^3r' d\varepsilon \\ &+ \sum_\beta \int u^A_{\alpha\beta}(|\mathbf{r}-\mathbf{r}'|) n_\beta(\mathbf{r}') d^3r' + v_\alpha(\mathbf{r}).\end{aligned} \tag{12.1.82}$$

The Percus–Yevick direct correlation functions $C^H_{\alpha\beta}$ are available for homogeneous multicomponent hard sphere fluids. The quantities $\mu^{Ex}_\alpha(\mathbf{n}_o)$ can be computed from these correlation functions and so, in principle, the Meister–Kroll model can be used for multicomponent systems. The Meister–Kroll model is much more computationally intensive than either the MVDW model or the GHR model. On the other hand, τ_α has not been determined for the Tarazona model and the Curtin–Ashcroft approach is computationally more costly than all the others.

12.2 Simple Fluids Confined to Slit Pores

We learned in Chapters 8 and 10 that fluids confined by parallel planar walls, that is, slit pores, are highly layered. In a comparison of computer simulations and the GVDW, GHR, and Tarazona models (Vanderlick et al., (1989), it has been demonstrated that the Tarazona model is the most accurate of the three for fluids confined by smooth slit pores.

The interaction potential $v(x)$ between smooth slit pores and confined fluid depends only on the distance x from the pore walls, and so the fluid density $n(x)$ also depends only on the position x. In this case the chemical potential of the confined fluid according to the three models becomes

$$\begin{aligned}\mu &= v(x) + \mu^+ + kT \ln n(x) + \int \frac{\delta \bar{n}^\sigma(x')}{\delta n(x)} \mathcal{F}(\bar{n}^\tau(x')) dx' \\ &+ \int \frac{\delta \bar{n}^\tau(x')}{\delta n(x)} \bar{n}^\sigma(x') \mathcal{F}'(\bar{n}^\tau(x')) dx' + \int n(x) \bar{u}^A(x'-x) dx',\end{aligned} \tag{12.2.1}$$

if the VDW approximation is used for $\mathcal{F}^A$. Because of planar symmetry the quantities in μ have become functions only of the x coordinate, namely,

$$\begin{aligned}\bar{u}^A(x-x') &= \int u^A(|\mathbf{r}-\mathbf{r}'|)dy'dz'\\ \bar{n}^w(x) &= \int w_x(x-x';\{n\})n(x')dx', \qquad w=\tau \text{ or } \sigma, \qquad (12.2.2)\\ w_x(x-x';\{n\}) &= \int w(\mathbf{r}-\mathbf{r}',\{n\})dy'dz'.\end{aligned}$$

In particular, for the GVDW model

$$\begin{aligned}\sigma_x(x) &= \sigma(x)\\ \tau_x(x) &= \frac{3}{4d^3}\eta(d-|x|)(d^2-x^2), \qquad (12.2.3)\end{aligned}$$

for the GHR model

$$\begin{aligned}\sigma_x(x) &= \frac{1}{d}\eta\left(\frac{d}{2}-|x|\right)\\ \tau_x(x) &= \frac{6}{d^3}\eta\left(\frac{d}{2}-|x|\right)\left[\left(\frac{d}{2}\right)^2-x^2\right], \qquad (12.2.4)\end{aligned}$$

and for the Tarazona model

$$\begin{aligned}\sigma_x(x-x') &= \sigma(x-x')\\ \tau_x(x-x') &= \omega_{ox}(x-x')+\omega_{1x}(x-x')\bar{n}^\tau(x)+\omega_{2x}(x-x')[\bar{n}^\tau(x)]^2, \qquad (12.2.5)\end{aligned}$$

where

$$\begin{aligned}\omega_{0x}(x) &= \frac{3}{4d^3}\eta(d-|x|)(d^2-x^2),\\ \omega_{1x}(x) &= 2\pi\eta(d-|x|)\left[\frac{0.475}{2}(d^2-|x|^2)-\frac{0.648}{3d}(d^3-|x|^3)\right.\\ &\quad\left.+\frac{0.113}{4d^2}(d^4-|x|^4)-0.01895d^2\right] \qquad (12.2.6)\\ &\quad+2\pi\eta(|x|-d)\eta(2d-|x|)\left[0.288d(2d-|x|)-0.463(4d^2-|x|^2)\right.\\ &\quad\left.+\frac{0.764}{3d}(8d^3-|x|^3)-\frac{0.187}{4d^2}(16d^4-|x|^4)+0.00181662d^2\right],\end{aligned}$$

and

$$\begin{aligned}\omega_{2x}(x) &= \frac{5\pi^2d^3}{72}\eta(d-|x|)[3(d^2-|x|^2)-\frac{4}{d}(d^3-|x|^3)\\ &\quad+\frac{5}{4d^2}(d^4-|x|^4)]. \qquad (12.2.7)\end{aligned}$$

$\eta(x)$ is the Heaviside step function. The numbers as displayed in eqs. (12.2.7) and (12.2.8) preserve the conditions $\int\omega_{1x}dx = \int\omega_{2x}dx = 0$ to sufficient accuracy for the calculation of density profiles of confined and semiconfined ($0 < x < \infty$) fluids. For the Carnahan–Starling hard sphere equation of state

$$\mathcal{F}(n) = kT\frac{y(4-3y)}{(1-y)^2}, \qquad y=\frac{\pi n d^3}{6}, \qquad (12.2.8)$$

whereas for the Clausius hard sphere equation of state

$$\mathcal{F}(n) = kT \ln(1 - nb)^{-1}, \quad b = \frac{2}{3}\pi d^3. \tag{12.2.9}$$

The simplest example of fluids confined to slit pores is a hard sphere fluid between hard walls. Then $\bar{u}^A \equiv 0$ and

$$\begin{aligned} v(x) &= \infty, \quad x < \frac{d}{2} \quad \text{and} \quad x > L - \frac{d}{2} \\ &= 0, \quad \frac{d}{2} < x < L - \frac{d}{2}. \end{aligned} \tag{12.2.10}$$

Predictions of the density profile $n(x)$ of these three models for a pore of width $L = 9.74d$ is compared with Monte Carlo simulations in Figure 12.1. The Tarazona model is the most accurate of the three. This is not too surprising because the weighting functional τ was chosen so that the free energy gave a direct correlation function in agreement with the Percus–Yevick theory of a homogeneous hard sphere fluid.

The normal pressure obeys the contact theorem,

$$P_N = n\left(\frac{d}{2}\right) kT, \tag{12.2.11}$$

for this system. The derivation of this theorem will be given in the next section. The mean pore density in the Monte Carlo simulation was $0.897d^{-3}$. Predicting normal pressures for the same mean pore density, using the Carnahan–Starling equation of state, Vanderlick et al. (1989) found

$$P_N d^3\, kT = 7.32, 11.29, \text{and } 11.11 \tag{12.2.12}$$

for the GVDW, GHR, and Tarazona models, respectively. The Monte Carlo result was

$$P_N d^3\, kT = 10.3. \tag{12.2.13}$$

With the Clausius equation of state they found $P_N d^3\, kT = 7.73$, 16.54, and 17.28. Clearly the Carnahan–Starling model treats hard sphere interactions better than does the Clausius model. Vanderlick et al. also compared the three models with molecular dynamics simulations (Magda et al., 1985), of 6–12 Lennard–Jones fluids confined planar walls obeying the Lennard–Jones wall potential

$$v(x) = 2\pi\varepsilon_w \left\{ \frac{2}{5}\left(\frac{\sigma_w}{x}\right)^{10} - \left(\frac{\sigma_w}{x}\right)^4 - \frac{\sqrt{2}\sigma_w^3}{3[|x| + (0.61/\sqrt{2})\sigma_w]^3} \right\}. \tag{12.2.14}$$

In their calculations Vanderlick et al. chose

$$\begin{aligned} u^A(r) &= -4\pi\varepsilon\left(\frac{d}{r}\right)^2, \quad |\mathbf{r}| \geq d \\ &= 0, \quad |\mathbf{r}| < d, \end{aligned} \tag{12.2.15}$$

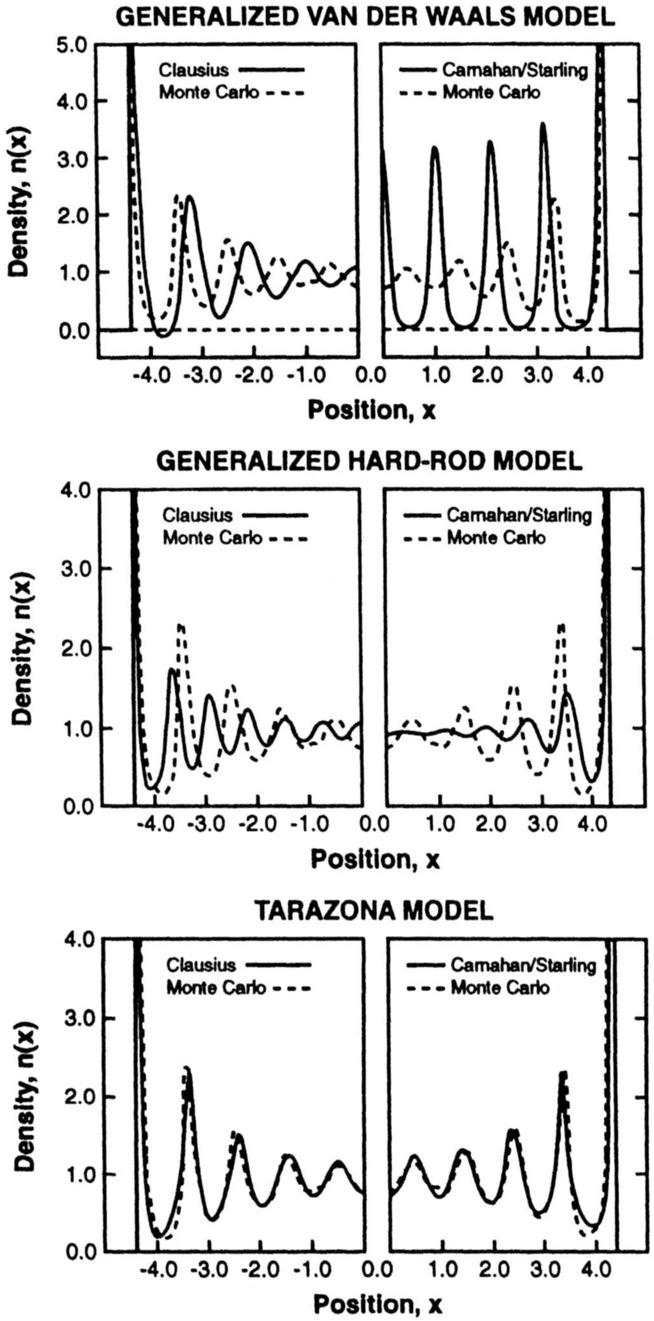

Figure 12.1
Density profile of a hard sphere fluid confined to a hard slit pore. Comparison of density functional theories and Monte Carlo simulations. Adapted from Vanderlick et al. (1989).

from which it follows from eq. (12.2.2) that

$$\begin{aligned}\overline{u}^A(x) &= -2\pi\varepsilon d^2\left(\frac{d}{x}\right)^4, \quad |x| \geq d \\ &= -2\pi\varepsilon d^2, \quad |x| < d. \end{aligned} \tag{12.2.16}$$

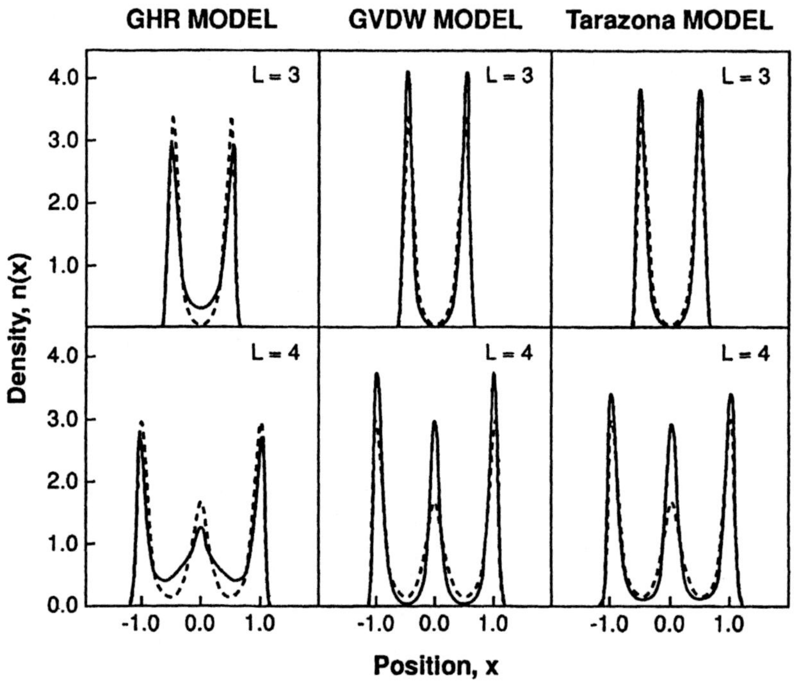

Figure 12.2
Density profiles of a Lennard–Jones fluid confined in Lennard–Jones slit pores. Comparison of density functional theories and molecular dynamics simulations. Units of pore width L and position x are d and of density $n(x)$ are d^{-3}. See caption of Figure 12.1.

It is likely that the results of their calculations would be improved by making a better choice of $u^A(r)$ and by taking the TSD formula of F^A instead of the VDW formula. Such an improved calculation has, however, not been pursued.

The simulated and predicted density profiles are compared in Figure 12.2. The temperature of the fluid was $T = \varepsilon/1.21k$. The wall–fluid parameters were set equal to the fluid–fluid parameters, that is, $\varepsilon_w = \varepsilon$ and $\sigma_w = \sigma = d$. Two pore widths, $L = 3d$ and $4d$, were studied. The mean pore densities in the simulation were, respectively, $0.5297d^{-3}$ and $0.4811d^{-3}$. The predictions were compared at chemical potentials yielding the same pore widths as the simulations.

Qualitatively all three models predict the significant layering that occurs in confined fluids of spherical molecules. The length scale of the layering is d. At $L = 3d$, only two layers can form, but at $L = 4d$ a third layer appears. The Tarazona model is somewhat better than the other two, but not as superior as it was in predicting the layering of hard spheres in a hard slit pore. As mentioned earlier, a better approximation to F^A might improve all three models.

The adsorption of a Lennard–Jones mixture chosen to correspond to an argon–krypton mixture in a graphite slit pore was studied by Sokolowski and Fischer (1990). They carried out molecular dynamics simulations and compared these with predictions of a version of the Meister–Kroll theory. For the intermolecular potential they assumed

$$u_{\alpha\beta}(r) = 4\varepsilon_{\alpha\beta}\left[\left(\frac{\sigma_{\alpha\beta}}{r}\right)^{12} - \left(\frac{\sigma_{\alpha\beta}}{r}\right)^{6}\right], \tag{12.2.17}$$

and used the Lorentz–Berthelet mixing rules

$$\varepsilon_{\alpha\beta} = (\varepsilon_{\alpha\alpha}\varepsilon_{\beta\beta})^{1/2}, \quad \sigma_{\alpha\beta} = \frac{1}{2}(\sigma_{\alpha\alpha} + \sigma_{\beta\beta}). \tag{12.2.18}$$

The Ar and Kr values of the parameters were chosen to be

	σ Å	ε/k (K)
Ar	3.405	119.8
Kr	3.630	163.1

They used the Barker–Henderson hard sphere approximation for the repulsive fluid, that is, in F^H they used the effective hard sphere diameter

$$d_{\alpha\alpha} = \int [1 - e^{-\beta u^R_{\alpha\alpha}(r)}]dr,$$

where

$$\begin{aligned} u^R_{\alpha\alpha}(r) &= u_{\alpha\alpha}(r), \quad r < 2^{1/6}\sigma_{\alpha\alpha} \\ &= 0, \quad r > 2^{1/6}\sigma_{\alpha\alpha}, \end{aligned} \tag{12.2.19}$$

and the mixing rule

$$d_{\alpha\beta} = \frac{1}{2}[d_{\alpha\alpha} + d_{\alpha\beta}]. \tag{12.2.20}$$

They computed F^A from the VDW approximation with the attractive potential

$$\begin{aligned} u^A_{\alpha\beta}(r) &= -\varepsilon_{\alpha\beta}, \quad r \leq 2^{1/6}\sigma_{\alpha\beta} \\ &= u_{\alpha\beta}(r), \quad r > 2^{1/6}\sigma_{\alpha\beta}. \end{aligned} \tag{12.2.21}$$

The fluid–wall potential was modeled as

$$v_\alpha(x) = \frac{3^{3/2}\varepsilon_{\alpha w}}{2}\left[\left(\frac{\sigma_{\alpha w}}{x}\right)^9 - \left(\frac{\sigma_{\alpha w}}{x}\right)^3\right], \tag{12.2.22}$$

with the parameters set to

	Ar	*Kr*
$\sigma_{\alpha w}/\sigma_{\alpha\alpha}$	0.5621	0.588
$\varepsilon_{\alpha w}/\varepsilon_{\alpha\alpha}$	9.2367	12.1744

They consider these values appropriate for a graphite wall.

The simulated and predicted density profiles for adsorption from two different bulk solutions are shown in Figure 12.3. The agreement is good and is typical of results at other temperatures and compositions. Krypton is adsorbed more strongly than argon.

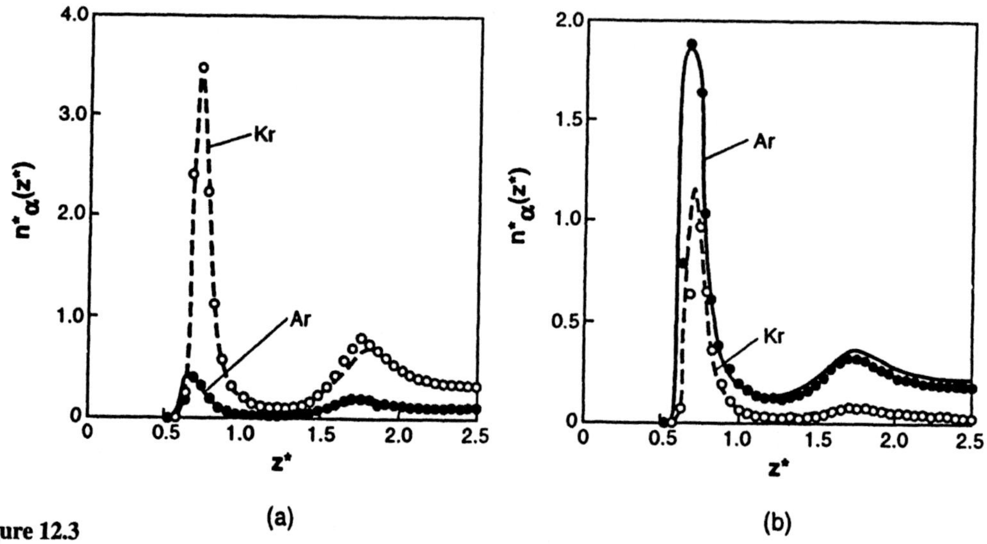

Figure 12.3
Argon and krypton density profiles for adsorption by a graphite slit pore at $T = \varepsilon_{Ar}/k$ and of width $L = 5\sigma_{Ar}$. Comparison of simulation and density of functional theory. Bulk density and mole fraction are $n^B\sigma_{Ar}^3$ and x_{Ar}^B (a) = 0.444 and 0.262 and (b) = 0.103 and 0.891. Adapted from Sokolowski and Fischer (1990).

TABLE 12.1 Excess Densities of Adsorption of Argon and Krypton in Graphite Slit Pore at $kT/\varepsilon_{Ar} = 1$ and Pore Width $L = 5\sigma_{Ar}$.

No.	$n^B\sigma_{Ar}^3$	x_{Ar}^B	Γ_{Ar}^*	Γ_{Kr}^*
1	0.0001	0	0	1.36
2	0.00011	0	0	1.40
3	0.00018	0	0	2.85
4	0.00029	0.48	0.123	2.73
5	0.00043	0.646	0.211	2.64
6	0.000314	0.672	0.264	1.32
7	0.0001	0.690	0.152	0.880
8	0.00076	0.74	0.327	2.46
9	0.00004	0.81	0.088	0.352
10	0.00094	0.836	0.563	2.29
11	0.00052	0.850	0.215	1.20
12	0.000044	0.855	0.098	0.350
13	0.00151	0.864	0.740	2.11
14	0.00059	0.878	0.264	1.14
15	0.00122	0.878	0.431	1.31
16	0.00168	0.884	1.00	1.85
17	0.00059	0.924	0.352	1.06
18	0.000752	0.944	0.528	0.880
19	0.00071	0.972	0.730	0.710
20	0.00077	0.980	0.752	0.700
21	0.00335	1.0	1.81	0
22	0.00433	1.0	2.85	0

n^B and x_{Ar}^B denote density and mole fraction of argon in bulk phase . $\Gamma_\alpha^* = \Gamma_\alpha\sigma_{Ar}^2$. Molecular dynamics from Sokolowski and Fischer (1990).

The excess density of adsorption of component α is defined by

$$\Gamma_\alpha = \int_0^L [n_\alpha(x) - n_\alpha^B(x)]dx. \qquad (12.2.23)$$

Capillary condensation can occur in slit pores and so an isotherm of Γ_α can be discontinuous. This was observed in the molecular simulations and predicted by the density functional theory for $T = \varepsilon_{Ar}/k$. The simulation results are given in Table 12.1 and compared to theory in Figure 12.4.

At $kT/\varepsilon_{Ar} = 1$, pure Ar and pure Kr are seen in Figure 12.4 to undergo capillary condensation. At $kT/\varepsilon_{Ar} = 1.5$ neither theory nor simulation indicate capillary condensation for Ar, but the theory predicts capillary condensation in the Kr-rich regime. No capillary condensation was found in the simulations at this temperature. The reason is undoubtedly because the equation of state used by Sokolowski and Fischer predicts a bulk fluid critical point that is too high. Thus, the critical point predicted for capillary condensation in a pore of width $L = 5\sigma_{Ar}$ is also too high. For $kT/\varepsilon_{Ar} = 2$, Sokolowski and Fischer found that no capillary condensation occurs in an Ar–Kr mixture for the equation of state they used.

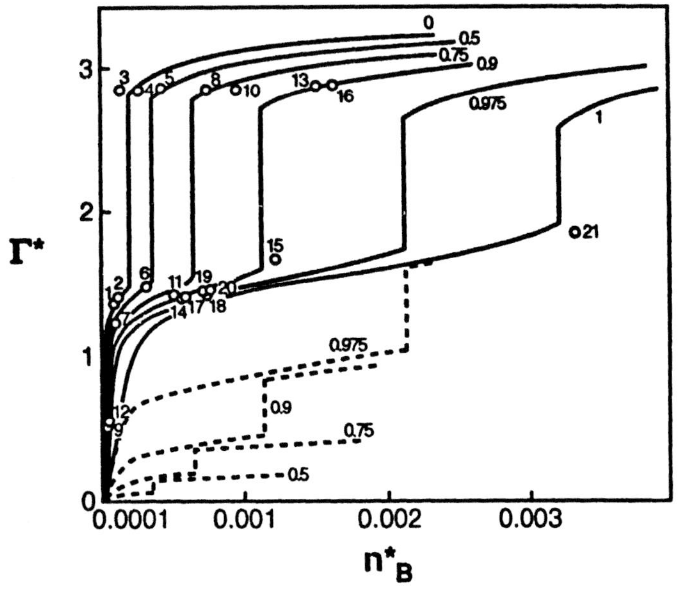

Figure 12.4
Total adsorption density isotherms, $\Gamma = \Gamma_{Ar} + \Gamma_{Kr}$ versus n_B^* for different bulk mole fractions x_{Ar}^B at $kT/\varepsilon_{Ar} = 1$. $\Gamma^* = \Gamma\sigma_{Ar}^2$ and $n_B^* = n_B\sigma_{Ar}^3$. Points are simulations (Table 12.1) and solid curves are from density functional theory. (Sokolowski and Fischer, 1990).

12.3 Interactions Between Electrically Charged Confining Surfaces

12.3.1 The Contact Theorem

If two planar, parallel charged plates are brought close together in the presence of a dielectric constant ε, they will exert a force on each other. If the walls are uniformly charged with changes of densities σ_l and σ_r, respectively, and if the left wall is at $x = 0$ and the right wall is at $x = L$, then the Coulombic potential between the walls is

$$v^{\rm ww} = \int \frac{\sigma_l \delta(x) \sigma_r \delta(L - x')}{|\mathbf{r} - \mathbf{r}'|} d^3r d^3r'. \qquad (12.3.1)$$

We assume the walls are a pair of disks whose radius R is much larger than their separation L. Then in cylindrical coordinates eq. (12.3.1) becomes

$$\begin{aligned} v_{\rm ww} &= A\sigma_l\sigma_r \int_0^R \frac{1}{\varepsilon[L^2 + \rho^2]^{1/2}} 2\pi p dp \\ &= \frac{2\pi}{\varepsilon} A\sigma_l\sigma_r[(R^2 + L^2)^{1/2} - L] \\ &= \frac{2\pi}{\varepsilon} A\sigma_l\sigma_r[R - L + O(L^2/R)], \end{aligned} \qquad (12.3.2)$$

where A is the area of the plates.

Similarly, we can compute the Coulombic potential of the walls on an ion of species α.

$$\begin{aligned} q_\alpha v^c(\mathbf{v}) &= \sum_{i=l}^{r} q_\alpha \int \frac{\sigma_i \delta(x - x_i)}{\varepsilon|\mathbf{r} - \mathbf{r}'|} d^3r' \\ &= \sum_{i=l}^{r} \frac{\pi\sigma_i q_\alpha}{\varepsilon} \int_o^R \frac{2\rho}{[(x - x_i)^2 + \rho^2]^{1/2}} d\rho \\ &= \sum_{i=l}^{r} \frac{2\pi\sigma_i q_\alpha}{\varepsilon} \{[R^2 + (x - x_i)^2]^{1/2} - |x - x_i|\} \\ &= \frac{2\pi q_\alpha}{\varepsilon} \left[(\sigma_l - \sigma_r) x + \sigma_r L + (\sigma_l + \sigma_r) R\right] + O\left(\frac{|x - x_i|}{R}\right). \end{aligned} \qquad (12.3.3)$$

Thus, for large charged walls (large R), $v(\mathbf{r})$ varies linearly with x when $\sigma_l \neq \sigma_r$ and is a constant when the wall charge densities are the same.

We assume that the non-Coulombic interaction between the molecules of the fluid and the wall is a hard wall interaction, that is,

$$v_\alpha^{\rm hw}(x) = \infty, \quad x \geq d_\alpha/2, \quad x \leq L - d_\alpha/2, \qquad (12.3.4)$$

and so

$$e^{-\beta v_\alpha^{\text{hw}}(x)} = \eta\left(L - \frac{d_\alpha}{2} - x\right)\eta\left(x - \frac{d_\alpha}{2}\right). \qquad (12.3.5)$$

The distance $d_\alpha/2$ is the distance of closest approach of the center of molecules α to the confining walls. The pressure on the wall at L can be computed from

$$P = -\frac{kT}{A}\left(\frac{\partial \ln Z_{\text{N}}}{\partial L}\right)_{T,\text{N}} \qquad (12.3.6)$$

where

$$Z_{\text{N}} = \int \cdots \int e^{-\sum_\alpha \beta v_\alpha^{\text{hw}}(x_\alpha) - \sum_\alpha \beta q_\alpha v^c(x_\alpha) - \sum_v \beta w^{\text{ww}}(L)} e^{-\beta u^w} (d^3 r)^N. \qquad (12.3.7)$$

u^N is the intermolecular potential energy. With the conditions

$$\frac{\partial e^{-\beta v_\alpha^{\text{hw}}(x_\alpha)}}{\partial L} = \delta\left(L - \frac{d_\alpha}{2} - x_\alpha\right), \quad \frac{\partial e^{-\beta q_\alpha v^c(x_\alpha)}}{\partial L} = -\beta \frac{2\pi q_\alpha \sigma_r}{\varepsilon} e^{-\beta q_\alpha v^c(x_\alpha)},$$

and

$$\frac{\partial}{\partial L} e^{-\beta v^{\text{ww}}(L)} = \frac{2\pi \sigma_l \sigma_r A\beta}{\varepsilon} e^{-\beta v^{\text{ww}}(L)}, \qquad (12.3.8)$$

we find from eq. (12.3.6)

$$P = kT \sum_\alpha n_\alpha \left(L - \frac{d_\alpha}{2}\right) - \sum_\alpha \frac{2\pi N_\alpha q_\alpha \sigma_r}{\varepsilon A} - \frac{2\pi \sigma_l \sigma_r}{\varepsilon}. \qquad (12.3.9)$$

But from overall electroneutrality of the confined system it follows that

$$\sum_\alpha N_\alpha q_\alpha = -A(\sigma_r + \sigma_l), \qquad (12.3.10)$$

and so eq. (12.3.9) becomes

$$P = kT \sum_\alpha n_\alpha \left(L - \frac{d_\alpha}{2}\right) + \frac{2\pi \sigma_r^2}{\varepsilon}. \qquad (12.3.11)$$

Equation (12.3.11) is known as the contact theorem.

For a neutral fluid between neutral hard walls the contact theorem is

$$P = kT \sum_\alpha n_\alpha \left(L - \frac{d_\alpha}{2}\right). \qquad (12.3.12)$$

By symmetry arguments, it follows that

$$P = kT \sum_\alpha n_\alpha \left(\frac{d_\alpha}{2}\right).$$

In obtaining eq. (12.3.11), the assumptions are that (1) the fluid behaves as a dielectric continuum in transmitting Coulombic interactions between the walls, (2) the fluid is classical, and (3) its molecules behave as hard spheres of diameter d_α, $\alpha = 1, \ldots, c$, in interaction with the walls. Otherwise, the intermolecular interactions are arbitrary. The generality of the result makes it useful for testing approximate theories and, in some cases, for carrying out calculations.

12.3.2 Disjoining Pressure of Electrical Double Layer: DLVO Theory

We consider here a dilute electrolyte in a continuum solvent of dielectric constant ε confined between two identical, uniformly charged, parallel flat plates. The electric potential $\psi(\mathbf{r})$ at a position $\mathbf{r}$ between the plates is

$$\psi(\mathbf{r}) = v(\mathbf{r}) + \sum_{\beta} q_\beta \int \frac{n_\beta(\mathbf{r}')}{\varepsilon|\mathbf{r} - \mathbf{r}'|} \tag{12.3.13}$$

where $v(\mathbf{r})$ is the wall contribution

$$\begin{aligned} v(\mathbf{r}) = &\int \frac{1}{\varepsilon|\mathbf{r} - \mathbf{r}'|}\sigma\delta\left(x' + \frac{L}{2}\right) d^3r' \\ &+ \int \frac{1}{\varepsilon|\mathbf{r} - \mathbf{r}'|}\sigma\delta\left(x' - \frac{L}{2}\right) d^3r' \end{aligned} \tag{12.3.14}$$

We assume that walls are located at $-L/2$ and $L/2$, respectively. The charge densities on the walls are $\sigma_l = \sigma_r = \sigma$. Because $\nabla^2|\mathbf{r} - \mathbf{r}'| = -4\pi\delta(\mathbf{r} - \mathbf{r}')$, it follows from eq. (12.3.13) that ψ obeys the Poisson equation

$$\nabla^2\psi = \frac{-4\pi}{\varepsilon}\sum_{\alpha} q_\alpha n_\alpha(\mathbf{r}). \tag{12.3.15}$$

The simplest theory of double layer forces is based on the Poisson–Boltzmann equation for point ions. This theory is known as the DLVO theory after the Russian scientists, Derjaguin and Landau and the Dutch scientists, Verway and Overbeek, who independently developed the theory (see Derjaguin and Landau 1941; and Verwey and Overbeek 1948).

When the concentration of the electrolyte is low enough for the solution to be ideal, the chemical potential of species α is

$$\mu_\alpha = kT \ln \Lambda_\alpha n_\alpha(\mathbf{r}) + q_\alpha \psi(\mathbf{r}) \tag{12.3.16}$$

and eq. (12.3.15) becomes the Poisson–Boltzmann equation

$$\nabla^2\psi = \frac{-4\pi}{\varepsilon}\sum_{\alpha} q_\alpha n_\alpha^B e^{-\beta q_\alpha \psi(\mathbf{r})}, \tag{12.3.17}$$

where n_α^B is the concentration of species α in a bulk electrolyte in equilibrium with the confined electrolyte (for which $\mu_\alpha =$

$kT \ln \Lambda_\alpha n_\alpha^B$). In the bulk electrolyte the condition of electroneutrality must hold:

$$\sum_\alpha q_\alpha n_\alpha^B = 0. \qquad (12.3.18)$$

For the planar geometry considered here, $\psi = \psi(x)$ and x ranges from $-L/2$ to $L/2$. Thus, the equation to solve is

$$\frac{d^2\psi}{dx^2} = -\frac{4\pi}{\varepsilon} \sum_\alpha q_\alpha n_\alpha^B e^{-\beta q_\alpha \psi(x)} \qquad (12.3.19)$$

with appropriate boundary conditions. Because the walls are identical, one of the boundary conditions is the symmetry condition

$$\frac{d\psi}{dx} = 0 \quad \text{at} \quad x = 0. \qquad (12.3.20)$$

The other can be *fixed surface potential*,

$$\psi = \psi_o \quad \text{at} \quad x = L/2, \qquad (12.3.21)$$

or *fixed charge density*,

$$\frac{d\psi}{dx} = \frac{4\pi\sigma}{\varepsilon} \quad \text{at} \quad x = L/2, \qquad (12.3.22)$$

or *surface charge regulation*

$$K(\sigma, \{n_\alpha(L/2)\}) = 0. \qquad (12.3.23)$$

Equation (12.3.23) represents an equilibrium between surface charge on the wall and the electrolyte concentration in the fluid next to the wall. For example, if the adsorption obeys a Langmuir isotherm the boundary condition at eq. (12.3.23) becomes

$$\sigma = \sum_\alpha \frac{k_\alpha n_\alpha(L/2)}{1 + h_\alpha n_\alpha(L/2)}, \qquad (12.3.24)$$

where k_α and h_α are adsorption constants. An example with more explicit chemistry is that of a surface weak-acid dissociation for a 1–1 electrolyte (Ninham and Parsegian (1971):

$$\frac{\sigma 10^{-\text{pH}}}{qN_s - \sigma} = K_a e^{-\beta q \psi_m}, \qquad (12.3.25)$$

where at the given solution pH the charge is regulated by the surface density N_s of acid groups and the acid dissociation constant K_a. $\psi_m = \psi(0)$, the potential midway between the plates.

Multiplying eq. (12.3.19) by $d\psi/dx$ we get

$$\frac{d}{dx}\left[\frac{1}{2}\left(\frac{d\psi}{dx}\right)^2\right] = \frac{d}{dx}\left[\sum_\alpha \frac{4\pi kT}{\varepsilon} n_\alpha^B e^{-\beta q_\alpha \psi}\right] \qquad (12.3.26)$$

and integrating once with the condition at eq. (12.3.20), we obtain

$$\left(\frac{d\psi}{dx}\right)^2 = \sum_\alpha \frac{4\pi kT}{\varepsilon} n_\alpha^B \left[e^{-\beta q_\alpha \psi} - e^{-\beta q_\alpha \psi_m}\right]. \tag{12.3.27}$$

The potential ψ_m midway between the walls is of course still unknown. It is fixed by the second boundary condition.

Equation (12.3.27) yields

$$dx = \pm\frac{1}{\sqrt{2}}\left[\sum_\alpha \frac{4\pi kT}{\varepsilon} n_\alpha^B \left(e^{-\beta q_\alpha \psi} - e^{-\beta q_\alpha \psi_m}\right)\right]^{1/2} d\psi. \tag{12.3.28}$$

Choice of the sign in eq. (12.3.28) depends on the boundary condition. The sign is the same as the sign of ψ_o for the fixed surface potential boundary condition and is the same as σ for the fixed or regulated charge boundary condition. In any case, the sign of ψ_m will be the same as the sign of the appropriate sign in eq. (12.3.28).

In the case of the fixed potential boundary condition, ψ_m is determined from

$$\frac{L}{2} = \operatorname{sgn}(\psi_m)\frac{1}{\sqrt{2}}\int_{\psi_m}^{\psi_o}\left[\sum_\alpha \frac{4\pi kT}{\varepsilon} n_\alpha^B \left(e^{-\beta q_\alpha \psi} - e^{-\beta q_\alpha \psi_m}\right)\right]^{-1/2} d\psi, \tag{12.3.29}$$

where $\operatorname{sgn}(\psi_m) = \pm 1$ according to whether the sign of ψ_m is + or −. For fixed surface charge, ψ_m is determined from eq. (12.3.29) and

$$\left(\frac{4\pi\sigma}{\varepsilon}\right)^2 = \sum_\alpha \frac{4\pi kT}{\varepsilon} n_\alpha^B \left[e^{-\beta q_\alpha \psi_o} - e^{-\beta q_\alpha \psi_m}\right]. \tag{12.3.30}$$

In this case $\psi_o = \psi\,(x = L/2)$ and $\psi_m = \psi\,(x = 0)$ are unknowns. In the charge regulation case, the unknowns σ, ψ_o, and ψ_m are determined from eqs. (12.3.23), (12.3.29), and (12.3.30).

Because the electrolyte is an ideal solution, the local pressure is given by the sum of the osmotic pressure $\sum_\alpha n_\alpha(x)kT$ and the electrostatic contribution $(\varepsilon/4\pi)E^2$, where $E = -d\psi/dx$. At $x = 0,\ E = 0$, and so, evaluating the disjoining pressure there, we find

$$\begin{aligned}\Pi(L) &= kT\sum_\alpha \left[n_\alpha\,(x=0, L) - n_\alpha\,(x = 0, L = \infty)\right] \\ &= \sum_\alpha kTn_\alpha^B\left[e^{-\beta q_\alpha \psi_m} - 1\right]\end{aligned} \tag{12.3.31}$$

Thus, we see that in the Poisson–Boltzmann theory the disjoining pressure is determined by the potential ψ_m midway between the confining walls and the bulk electrolyte concentration n_α^B.

Of course, to compute $\Pi(L)$ one must obtain ψ_m numerically from the combination of eqs. (12.3.23), (12.3.29), and (12.3.30) that corresponds to the given boundary conditions.

For a one-component, charge symmetric electrolyte, $q_\alpha = z_\alpha e$, $z_+ = -z_- = z$, $n_+^B = n_-^B = zn^B$, the problem of finding ψ simplifies considerably (Chan et al., 1989). With the definitions

$$Y = \beta z e \psi, \quad X = \kappa x, \quad \kappa^2 = \frac{8\pi z^2 e^2 n^B}{\varepsilon k T}, \tag{12.3.32}$$

the pertinent equations become

$$dX = \left[\left(\frac{Q^2}{2} + \cosh Y_m\right)^2 - 1\right]^{-1/2} dQ \tag{12.3.33}$$

$$\Pi(L) = 2zn^B kT(\cosh Y_m - 1), \tag{12.3.34}$$

where

$$Q \equiv \sqrt{2}(\cosh Y - \cosh Y_m)^{1/2}, \tag{12.3.35}$$

The boundary condition at $x = 0$ becomes

$$X = 0 \quad \text{at} \quad Q = 0. \tag{12.3.36}$$

The fixed potential boundary condition becomes

$$Q_o = \sqrt{2}(\cosh Y_o - \cosh Y_m)^{1/2} \tag{12.3.37}$$

and the fixed surface charge density one becomes

$$Q_o = \frac{4\pi e}{\varepsilon k T}|\sigma|. \tag{12.3.38}$$

The procedure to solve these equations is to set Y_m, calculate Q_o and integrate eq. (12.3.33) from $Q = 0$ to Q_o to obtain $X_o = \kappa L/2$ and thus the wall separation L and disjoining pressure Π corresponding to Y_m. By varying Y_m and repeating the process we find L as a function of Y_m and Π as a function of L for given electrolyte and boundary conditions.

For large L and far enough from a wall for the condition $|\beta q_\alpha \psi| \ll 1$ to hold, eq. (12.3.19) can be linearized to obtain

$$\frac{d^2\psi}{dx^2} = \kappa^2 \psi, \tag{12.3.39}$$

where

$$\kappa^2 = \sum_\alpha \frac{4\pi q_\alpha^2 n_\alpha^B}{\varepsilon k T}. \tag{12.3.40}$$

Let us ignore the fact that the approximation may break down too near a wall (when the wall charge or potential is sufficiently large), then solution of eq. (12.3.39) with the fixed surface potential and boundary conditions gives

$$\psi = \psi_o \frac{e^{\kappa x} + e^{-\kappa x}}{e^{(\kappa L/2)} + e^{-(\kappa L/2)}}. \tag{12.3.41}$$

and for the fixed surface charge case gives

$$\psi = \frac{4\pi\sigma}{\varepsilon\kappa}\frac{e^{\kappa x} + e^{-\kappa x}}{e^{\kappa L/2} - e^{-(\kappa L/2)}}. \tag{12.3.42}$$

The disjoining pressure, also in the limit of small $\beta q_\alpha \psi$ to be consistent, is

$$\Pi(L) = \frac{\varepsilon\kappa^2\psi_o^2}{2\pi}\frac{e^{-\kappa L}}{(1 + e^{-\kappa L})^2} \tag{12.3.43}$$

for fixed surface potential and is

$$\Pi(L) = \frac{8\pi\sigma^2}{\varepsilon}\frac{e^{-\kappa L}}{(1 - e^{-\kappa L})^2} \tag{12.3.44}$$

for fixed surface charge.

Although eqs. (12.3.43) and (12.3.44) are valid for sufficiently large L, they fail at sufficiently small L for large values of ψ_o or for any value of fixed charge density. In this small L domain, one term, $e^{-\beta q_c \psi}$, dominates all the other exponentials. Here q_c is the charge of the ion of largest valence whose charge is opposite the sign of ψ. Equation (12.3.29) then can be approximated as

$$\frac{L}{2} \simeq \left[\frac{\varepsilon}{8\pi kTn_c^B}\right]^{1/2}\frac{kTe^{\beta q_c \psi_m/2}}{|q_c|}\int_0^\infty \frac{1}{[e^s - 1]^{1/2}}ds = \left[\frac{\varepsilon}{4\pi kT}\right]^{1/2}\frac{\pi kT}{|q_c|}e^{\beta q_c \psi_m/2}. \tag{12.3.45}$$

Correspondingly,

$$\Pi(L) \simeq kTn_c^B e^{-\beta q_o \psi_m/2} \approx \frac{\pi}{2}\varepsilon\left(\frac{kT}{q_c}\right)^2\frac{1}{L^2}. \tag{12.3.46}$$

This rather general result is somewhat surprising. It was first achieved by Derjaguin and Landau for the case of fixed ψ_o. It says that at very small L and any wall charge σ or if $e|\psi_o|kT \gg 1$, the disjoining pressure is independent of the wall potential or charge and of the electrolyte concentration. It depends only on the charge of the ion of largest valence whose sign is the opposite of the sign of ψ.

Double layer repulsion is one of the chief mechanisms for stabilizing colloidal dispersions. In addition to Coulombic forces in confined systems, there are also dispersion forces. Asymptotically the contribution of dispersion or VDW forces to the disjoining pressure is of the form

$$\Pi^d = -\frac{A_H}{6\pi L^\nu}, \tag{12.3.47}$$

where $\nu = 3$ when L is 100 nm and $\nu = 4$ when L is $> \mu$m. A_H is called the Hamaker constant and depends on the composition of the confining walls and of the confined fluid. The Hamaker constant can be negative, but it is usually positive. In cases for which $\kappa L/2$ is large compared to unity, $\psi_o \simeq 4\pi\sigma/\kappa\varepsilon$

and $\Pi(L) \simeq (8\pi\sigma^2/\varepsilon)e^{-\kappa L}$ for either constant surface potential or surface charge. For this case, combination of eqs. (12.3.44) and (12.3.47) gives the asymptotic disjoining pressure:

$$\begin{aligned}\Pi &= \frac{8\pi\sigma^2}{\varepsilon}e^{-\kappa L} - \frac{A_H}{6\pi L^\nu} \\ &= \frac{\varepsilon\kappa^2}{2\pi}\psi_o^2 e^{-\kappa L} - \frac{A_H}{6\pi L^\nu}.\end{aligned} \tag{12.3.48}$$

In the case that $e|\psi_o|/kT$ is not small compared to unity, but $e^{-\kappa L} \ll 1$, an approximation introduced by Reerink and Overbeek (1954), yields the double-layer disjoining pressure

$$\Pi(L) = \frac{2\varepsilon\kappa^2}{\pi}\left(\frac{kT}{q_c}\right)^2 \gamma^2 e^{-\kappa L} - \frac{A_H}{6\pi L^\nu} \tag{12.3.49}$$

where q_c is the charge of the counterion of the electrolyte and

$$\gamma = \frac{\exp\left[|q_c\psi_o|/2kT\right] - 1}{\exp\left[|q_c\psi_o|/2kT\right] + 1}. \tag{12.3.50}$$

Equation (12.3.49) is superior to eq. (12.3.48). Of course, when $|q_c\psi_o| \ll kT$, eq. (12.3.49) reduces to eq. (12.3.48).

Up to this point, we have considered only flat charged surfaces. In the case of interactions between colloidal particles the surfaces are curved, and so one has to consider this curvature in predicting colloidal interactions. It is frequently the case in dealing with colloidal dispersions that the curvature of the charged surfaces are large compared to their separation. For this case Derjaguin (1931) showed that the mean force Φ between the surfaces can be calculated from the disjoining pressure of flat surfaces and a geometric factor. In particular he found that the mean force between two convex bodies whose closest distance of separation is L is small compared to the radii of curvature is

$$\Phi(L) = 2\pi C_F \int_L^\infty \Pi(L')dL'. \tag{12.3.51}$$

C_F is a curvature factor and $\Pi(L)$ is the disjoining pressure between the flat surfaces of bodies of the same material as the convex bodies.

In the case of two spheres of radii R_1 and R_2,

$$C_F = \frac{R_1 R_2}{R_1 + R_2}. \tag{12.3.52}$$

For long, crossed cylinders (axes orthogonally oriented) of radii R_1 and R_2,

$$C_F = \sqrt{R_1 R_2}. \tag{12.3.53}$$

For a sphere of radius R interacting with the flat surface of a semi-infinite body, $C_F = R$, the same factor as that of identical crossed cylinders of radius R.

The disjoining pressure at eq. (12.3.49) gives for the mean force

$$\Phi = C_F \left[4\varepsilon\kappa \left(\frac{kT}{q_c} \right)^2 \gamma^2 e^{-\kappa L} - \frac{A_H}{3(\nu - 1)L^{\nu-1}} \right] \qquad (12.3.54)$$

between charged bodies having the curvature factor C_F. In Figure 12.5, we compare the force predicted by eq. (12.3.54) with the DLVO theory, according to which the double layer part of the disjoining pressure is computed from eq. (12.3.34). The Reerink–Overbeek approximation is in good qualitative agreement with the full DLVO theory, but the former overestimates the peak in the force between the bodies. The peak is the result of competition between double layer repulsion and VDW attraction. As we will discuss in the following text, the interplay of this competition is an important factor in colloidal stability.

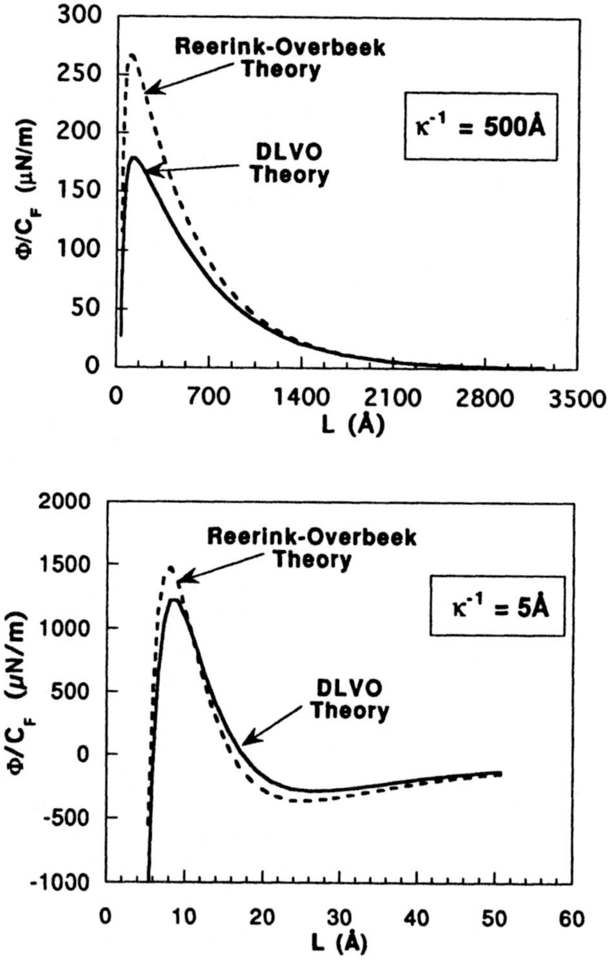

Figure 12.5
Force between large colloidal bodies versus separation when double layer repulsion and VDW attraction are present. The Reerink–Overbeek approximation [eq. (12.3.54)] is compared to the DLVO theory. Parameters used in the predictions are surface potential $\psi_o = 50$ mV, $T = 298$ K, $\varepsilon/\varepsilon_o = 78.5$, $\varepsilon_o = 8.854 \times 10^{-12}$ F/m, $\nu = 3, q_+ = -q_- = e$, and Hamaker constant $A_H = 2 \times 10^{-19}$ J. According to Derjaguin's approximation the curvature factor C_F is the only thing distinguishing the mean force between convex bodies.

The surface forces apparatus measures the force Φ/R between opposed, orthogonal mica cylinders of radius R. The radius of the cylinders is usually chosen to be about 1 cm and so Derjaguin's formula, eq. (12.3.51), with $C_F = R$ is valid for the separations of interest to colloidal phenomena. An example of the measured force is compared in Figure 12.6 with the prediction of the DLVO theory. The theory accounts accurately for the long-range part of the force. The distance of adhesive contact is not so accurately predicted as might be expected because complicated solvation forces are operative at short distances of separation.

In closing this section, let us consider the problem of colloidal stability. Because the long-ranged double layer forces are repulsive, they tend to keep colloidal particles from aggregating. The probability that particle–particle encounter will result in aggregation goes as $e^{-W/kT}$, where W is a colloidal pair interaction energy defined by

$$W = \int_L^\infty \Phi(L')dL'. \tag{12.3.55}$$

When the particles are large enough for Derjaguin's approximation to be valid,

$$W = 2\pi C_F \int_L^\infty dL' \int_{L'}^\infty dL''\Pi(L''). \tag{12.3.56}$$

The Reerink–Overbeck approximation at eq. (12.3.54) yields

$$W = 2\pi C_F \left[4\varepsilon \left(\frac{kT}{q_c} \right)^2 \gamma^2 e^{-\kappa L} - \frac{A_{\mathrm{H}}}{3(\nu-1)(\nu-2)L^{\nu-3}} \right]. \tag{12.3.57}$$

In Figure 12.7 values of W predicted by eq. (12.3.57) for a pair of charged spheres of radius $R = 1000$ Åare given for several electrolyte concentrations. However, the DLVO theory ignores effects associated with finite ionic diameters. That these can be important at very small separations L will be shown in the next section.

For a 1:1 electrolyte at the concentration 9.6×10^{-4} mol/L, the Debye length is $\kappa^{-1} = 100$ Å. The double-layer repulsion provides a barrier of over 100 kT and so the probability of the particles aggregating is vanishingly small. Thus, a dispersion of these colloidal particles would be stable at this electrolyte concentration. Increasing the electrolyte concentration to 2.4×10^{-2} mol/L (so $\kappa^{-1} = 20$ Å) lowers the repulsive maximum in W considerably, but it remains high enough to make irreversible particle aggregation unlikely. There is, however, an attracting minimum in the energy at about 100 Å. The minimum is a few kT in magnitude and so pairs of particles can be sometimes trapped in the minimum and will occasionally escape one another. This phenomenon is known as flocculation. At any instant many of the colloidal particles will

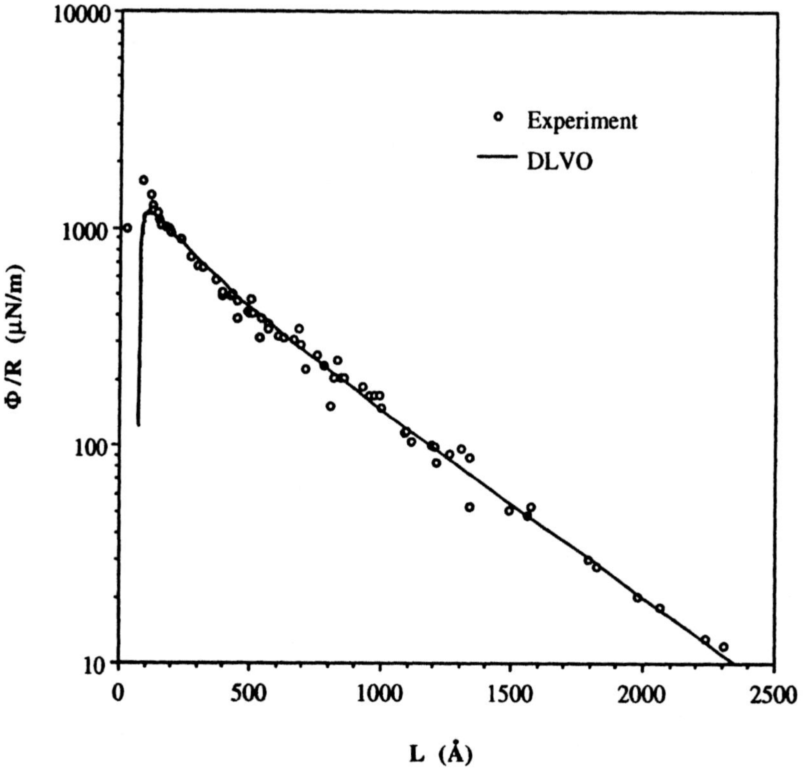

Figure 12.6
Force, Φ/R, versus separation of crossed mica cylinders. The mica has a charged monolayer on it. The net charge is positive. The solid curve is a fit to DLVO theory, assuming constant surface potential, $\psi_o = 106$ mV, and a Hamaker constant $A_H = 1.0 \times 10^{-20}$ J. The double layer contribution to Φ is computed from eq. (12.3.34). The Debye κ is set to 489 Å, which corresponds to the ion density 3.89×10^{-5} mol/l, the value expected for CO_2 dissolved in water in equilibrium with ambient air. Adapted from Mao et al. (1993).

be found in loosely held aggregates but they will not fuse or coalesce. The phenomenon is reversible: if the dispersion is diluted by distilled water back to $\kappa^{-1} = 100$ Å, the colloid interaction energy will recover its original form and the flocs will dissociate. On the other hand, at an electrolyte concentration of 9.6×10^{-2} mol/L ($\kappa^{-1} = 10$ Å), the energy W is purely attractive and so the colloidal particles are irreversibly aggregated by the attractive VDW forces.

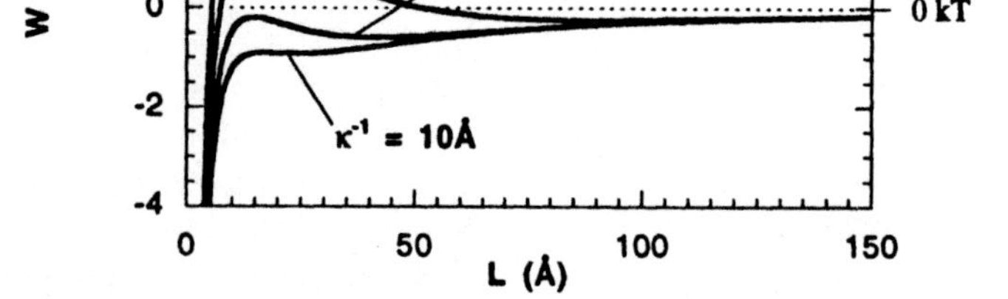

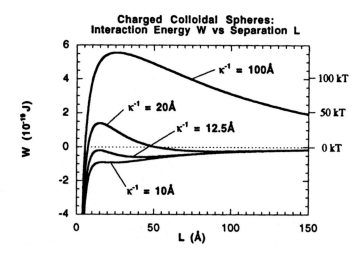

Figure 12.7
The effect of electrolyte concentration on the colloidal interaction energy of two spherical particles. $R = 1000$ Å, $T = 298$ K, $q_+ = -q_- = e$, $\nu = 3$, $A_H = 6.1 \times 10^{-20}$ J, $\varepsilon/\varepsilon_o = 78.5$, $\varepsilon_o = 8.854 \times 10^{-12}$ Fm^{-1}, and $\psi_o = 50$ mV. Calculated from eq. (12.3.57).

If the system is an emulsion, the particles are liquid and upon aggregation will quickly coalesce. These coalesced aggregates will coalesce with other particles or aggregates and very quickly the entire emulsion will separate into the two phases comprising the emulsion. Addition of salt is in fact a common way to break an emulsion. Raising the temperature is another means because the probability of aggregation goes as $e^{-W/kT}$. If the colloidal particles are solid, aggregation and breakage of the dispersion is also rather fast. The liquid phase will separate out and, depending on relative densities, the aggregated particle mass will sink to the bottom (e.g., clay particles in water) or rise to the top (e.g., latex particles in water) of the container. The aggregated solid particles will ultimately fuse, but can be an extremely slow process and so in a practical time span little fusion may take place unless processing is carried out (e.g., compaction, elevation of temperature, drying, etc.)

Although the DLVO theory adequately explains colloidal stability, the double-layer theory ignores effects associated with finite ionic diameters. Ultimately, aggregating colloidal particles will approach closely enough that the molecularity of the ions might contribute importantly to the disjoining forces. That this is the case at very small separations L is shown in the next section.

12.3.3 Disjoining Pressure of Electrical Double Layer: Density Functional Theory

As in the preceding section, we consider a primitive electrolyte model in which the solvent appears only through a dielectric constant ε. We assume all ions have the same diameter d. Again the walls are assumed to be uniformly charged parallel plates. The charge density of each wall is σ and is positive. The results for positive σ can be converted to those for negative σ by interchanging the roles of anions and cations.

The Poisson equation for planar symmetry is again

$$\frac{d^2\psi}{dx^2} = -\frac{4\pi}{\varepsilon}\sum_\alpha q_\alpha n_\alpha(x) \tag{12.3.58}$$

and the boundary conditions for fixed charge are

$$\frac{d\psi}{dx} = -\frac{4\pi\sigma}{\varepsilon}, \quad x = 0; \quad \frac{d\psi}{dx} = \frac{4\pi\sigma}{\varepsilon}, \quad x = L. \tag{12.3.59}$$

Integration of the Poisson equation with the aid of eq. (12.3.59) gives

$$\int_0^L \sum_\alpha q_\alpha n_\alpha(x')dx' = -2\sigma, \tag{12.3.60}$$

which is the condition of overall electroneutrality of the system.

Because $d\psi/dx = 0$ at $x = L/2$, the Poisson equation can be integrated twice to yield

$$\psi(x) = \frac{4\pi}{\varepsilon} \int_x^{L/2} (x - x') \sum_\alpha q_\alpha n_\alpha(x')dx + \psi\left(\frac{L}{2}\right). \quad (12.3.61)$$

$\psi(L/2)$ appears in eq. (12.3.61) as an extra unknown. However, the electroneutrality condition at eq. (12.3.60) is an extra equation, and so the number of unknowns balances.

To proceed further we need a free energy model. We choose the one developed by Tang et al. (1992) namely, an extension of the Tarazona model:

$$\begin{aligned} F = &\sum_\alpha \int_{d/2<x<L-d/2} n_\alpha(\mathbf{r})\{kT[\ln \Lambda_\alpha^3 n_\alpha(\mathbf{r}) - 1] + q_\alpha v(\mathbf{r})\}d^3r \\ &+ \sum_{\alpha,\beta} q_\alpha q_\beta \int \int_{d/2<x,x'<L-d/2} \frac{n_\alpha(\mathbf{r})n_\beta(\mathbf{r}')}{\varepsilon|\mathbf{r} - \mathbf{r}'|} d^3r d^3r' + \sum_\alpha \int_{d/2<x<L-d/2} n_\alpha(\mathbf{r})\mathcal{F}(\bar{\mathbf{n}}^\tau(\mathbf{r}))d^3r \quad (12.3.62) \\ &- kT \sum_{\alpha,\beta} \int \int_{d/2<x,x'<L-d/2} n_\alpha(\mathbf{r})n_\beta(\mathbf{r}') \int_0^1 \Delta c_{\alpha\beta}(|\mathbf{r} - \mathbf{r}'|, \lambda)(1 - \lambda)d\lambda d^3r d^3r' \end{aligned}$$

where $\mathcal{F}(n)$ is the excess free energy per molecule of a homogeneous hard sphere fluid at density $n\ (= \sum_\alpha n_\alpha)$ and $\Delta c_{\alpha\beta}$ is the contribution to the direct correlation function from the Coulomb–hard sphere interactions. $\Delta c_{\alpha\beta}$ is computed from the mean spherical approximation for the bulk electrolyte. The coarse grained densities $\bar{n}^\tau(\mathbf{r})$ are computed from Tarazona's formulas, that are the same as the one-component formula because we are assuming that all ionic species have the same diameter. Finally, Tang et al. use the Carnahan–Starling formula for $\mathcal{F}(\mathbf{n})$, namely,

$$\mathcal{F}(n) = \frac{kTy(4 - 3y)}{(1 - y)^2}, \quad (12.3.63)$$

where $y = \pi n d^3/6$.

Tang et al. proved that the contact theorem,

$$P(L) = kT \sum_\alpha n_\alpha\left(L - \frac{d}{2}\right) + \frac{2\pi\sigma^2}{\varepsilon}, \quad (12.3.64)$$

holds for the free energy model given at eq. (12.3.62) if the course grained density for planar symmetry obeys the relation

$$\bar{n}^\tau(x) \equiv \sum_\alpha \bar{n}_\alpha^\tau(x) = \sum_\alpha \int_0^L n_\alpha(x')\tau(|x - x'|; \bar{n}^\tau(x)). \quad (12.3.65)$$

The Tarazona model obeys eq. (12.3.65) for a mixture of hard spheres having the same diameter.

The chemical potential corresponding to the above free energy model is

$$\mu_\alpha = kT \ln[\lambda_\alpha^3 n_\alpha(x)] + q_\alpha \psi(x) + \mathcal{F}(\bar{n}^\tau(x)) + \sum_\beta \int_0^L n_\beta(x')\mathcal{F}(\bar{n}^\tau(x')) \frac{\delta \bar{n}^\tau(x')}{\delta n_\alpha(x)} dx' - kT \sum_\beta \int_O^L n_\beta(x') \int_0^1 \Delta c_{\alpha\beta}(|x - x'|; \lambda)(1-\lambda) d\lambda dx', \quad (12.3.66)$$

where

$$\Delta c_{\alpha\beta}(x) \equiv \int\int_{-\infty}^{\infty} \Delta c_{\alpha\beta}(\mathbf{r}) dy dz. \quad (12.3.67)$$

Tang et al. used finite element analysis and solved eqs. (12.3.60), (12.3.61), and (12.3.66) simultaneously for various electrolyte concentrations (various values of μ_α). The disjoining pressure was then computed from the formula

$$\Pi = kT \sum_\alpha \left[n_\alpha \left(L - \frac{d}{2} \right) - n_\alpha(\infty) \right]. \quad (12.3.68)$$

Predicted density profiles of a 1:1 electrolyte ($q_+ = -q_- = e$) at 0.1 molar bulk concentration and for a charge density of $\sigma = 0.1e/d^2$ are shown in Figure 12.8. Although not shown the density profiles predicted by the Poisson–Boltzmann equation are almost identical. The coions are almost shut out of the confined region until separations of $7d$ are reached. Disjoining pressures predicted by the density functional theory and the Poisson–Boltzmann equation are compared in Figure 12.9. The results are qualitatively quite similar: both are positive, monotonically decreasing functions, except that the disjoining pressures of the density functional theory are displaced to the left of the Poisson–Boltzmann equation.

At higher ionic strength, larger κ, the disjoining pressure predicted by density functional theory becomes nonmonotonic, in disagreement with the Poisson–Boltzmann or DLVO theory. Density profiles predicted by density function theory for a 2:2 electrolyte ($q_+ = -q_- = 2e$) are shown in Figures 12.10 and 12.11. At a bulk electrolyte concentration of 0.1655 M, the ion density profiles are monotonic. At the higher bulk electrolyte concentration the ion density profiles become nonmonotonic, reflecting charge inversion. The coion density actually exceeds the counterion density in the vicinity of $x = 1.5d$. This arises because the balance between Coulombic attraction and hard sphere repulsion favors layering of alternate charges near the charged wall. Agreement between density functional theory and computer simulations is good.

The competition between Coulombic interactions and hard sphere repulsion is even more evident in the disjoining pressures shown in Figure 12.12.

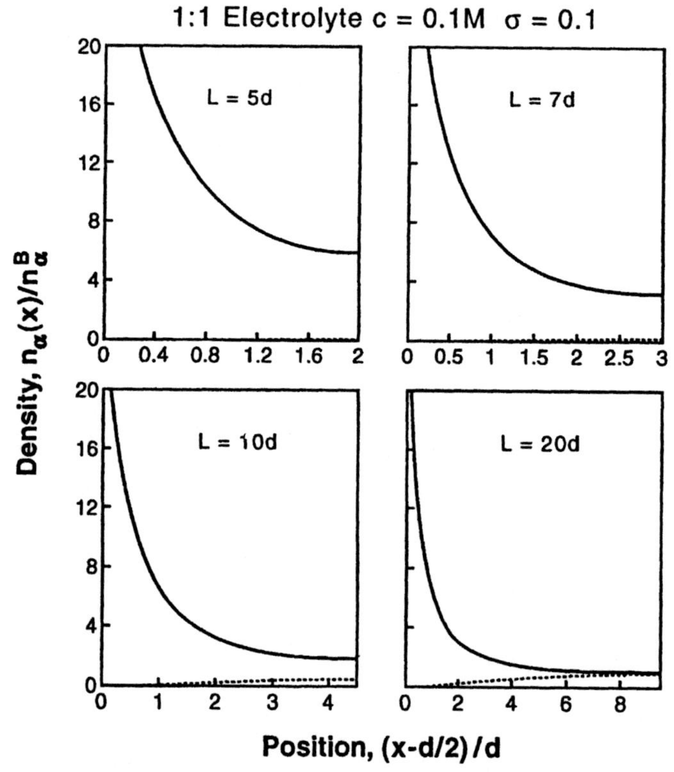

Figure 12.8
Ion density profiles of a 1:1 electrolyte confined between charged walls of separation L. Adapted from Tang et al. (1992).

The minima in the disjoining pressure curves occur for wall separations near $L = 2d$, where there is just room for one layer of ions. This layer is almost entirely counterions that attract both walls and result in a net negative disjoining pressure. At the higher bulk electrolyte concentration there is a maximum in the disjoining pressure at a pore separation near $L = 3d$, where two layers of

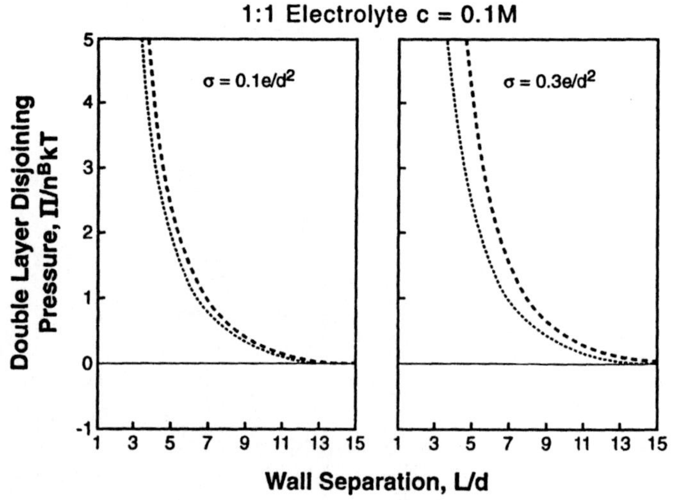

Figure 12.9
Disjoining pressures of a 1:1 electrolyte confined between charged walls versus separation L. See caption of Figure 12.7.

Figure 12.10
Ion density profiles predicted by density functional theory for a 2:2 electrolyte in equilibrium with a bulk electrolyte of concentration 0.1655 M. The charge density σ and wall separation L are indicated in the figure. The open circles with error bars are results of Monte Carlo simulations (Valleau et al. 1984). See caption of Figure 12.8.

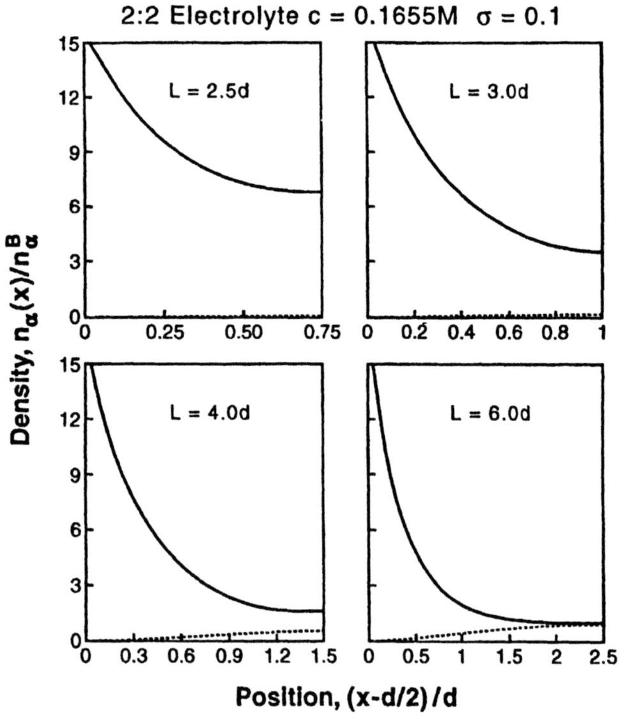

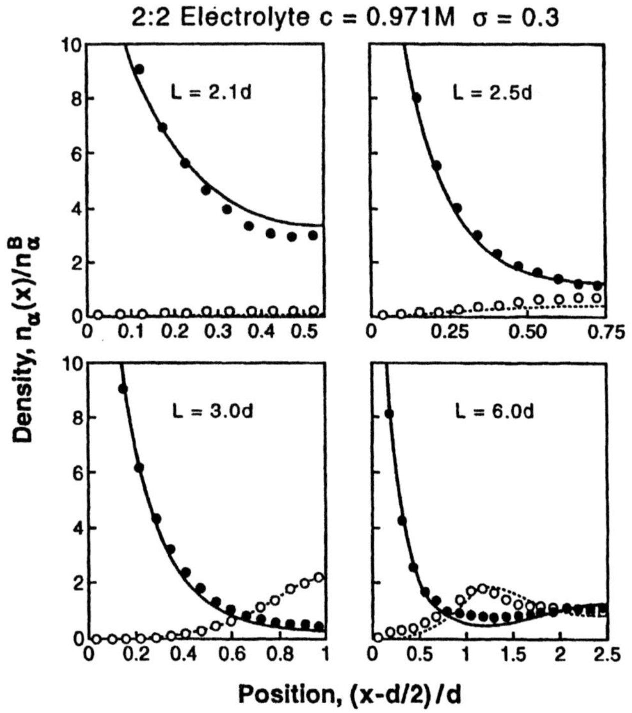

Figure 12.11
Ion density profiles predicted by density functional theory for a 2:2 electrolyte in equilibrium with a bulk electrolyte in equilibrium with a bulk electrolyte at 0.971 M. See caption of Figure 12.8.

Figure 12.12
The electrical double layer force in a 2:2 electrolyte. Bulk concentrations and surface charge densities σ are indicated in the figure. The dotted and dashed lines denote predictions of the density function theory and the DLVO theory, respectively. The density functional theory agrees quite well with the computer experiment, whereas the DLVO theory misses the nonmonotonic features in the disjoining theory. See caption of Figure 12.8.

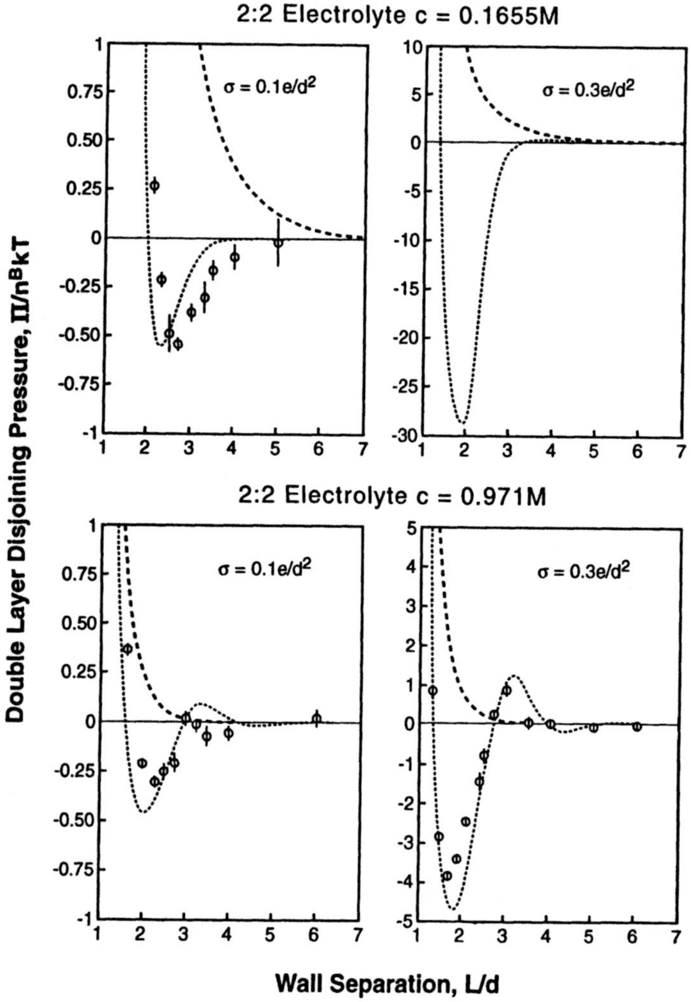

counterions form at the wall and one layer of coions form in the center.

The minimum in the double layer contribution to the disjoining pressure might be important in the final stages of colloidal particle coalescense or flocculation. For example, for mica surfaces in aqueous solution the dispersion or VDW disjoining pressure is

$$\Pi^{\text{VDW}} = -\frac{2.2 \times 10^{-20}\text{J}}{6\pi L^3}. \qquad (12.3.69)$$

In Table 12.2 we compare Π^{VDW} and the double layer Π^d predicted by the density functional theory at the wall separation L_0 where the double layer disjoining pressure has a minimum.

For all the cases considered in Table 12.2, the electrical double layer minimum value of Π is comparable to or significantly greater than the corresponding VDW attraction.

The message in this section is that although the simple DLVO theory can be qualitatively reliable at sufficiently low electrolyte concentrations, the more complicated density functional theory is required at higher electrolyte concentrations or in the case of greater than univalent ions.

TABLE 12.2 Comparisons Between Electrical Double Layer Disjoining Pressure Π^d and VDW Disjoining Pressure Π^{VDW}

$c\,(M)$	σ	L_0	$\Pi^d\,(L_0)$	$\Pi^{VDW}\,(L_0)$
		1:1 Electrolytes		
1.0	0.3	3.5	−0.214	−0.367
1.0	0.7	4.8	−0.091	−0.142
		2:2 Electrolytes		
0.1655	0.1	2.3	−0.455	−1.295
0.1655	0.3	2.0	−22.85	−1.969
0.971	0.1	2.1	−2.232	−1.701
0.971	0.3	1.8	−22.49	−2.701
		2:1 Electrolytes		
0.1	−0.1	2.3	−0.496	−1.295
1.0	−0.3	3.0	−2.147	−0.583
1.0	−0.3	1.8	−25.70	−2.701

The pressures are evaluated at the wall separation of L_0 where the double layer attraction is maximum. The Hamaker constant is $A = 2.2 \times 10^{-20}$ J. The units of pressure, charge density, and wall separation are $10^6\,\mathrm{Nm}^{-2}$, e/d, and d. $d = 4.25$ Å. See caption to Figure 12.9.

Supplementary Reading

Derjaguin, B. V., Churaev, N. V., and Maller, V. M., V. I. Kisin, Trans., J. A. Kitchner, ed. 1987. *Surface Forces,* Consultants Bureau, New York.

Evans, D. F. and Wennerström, H. 1994. *The Colloidal Domain. Where Physics, Chemistry, Biology and Technology Meet,* VCH Publishers, New York.

Henderson, D., Ed. 1992. *Fundamentals of Inhomogeneous Fluids,* Dekker, New York.

Israelachvili, J. N. 1992. *Intermolecular and Surface Forces,* Academic Press, New York.

Mahanty, J. and Ninham, B. W. 1976. *Dispersion Forces,* Academic Press, New York.

Exercises

1 This problem illustrates what is involved in computing density profiles and disjoining pressure. It also shows why nonlocal functional theories are needed to treat confined fluids. Consider a

fluid confined between two planar solids that interact with the fluid via the wall potential

$$\phi_w(x) = 1.5\sqrt{3\varepsilon_w}\left[\left(\frac{\sigma_w}{z}\right)^9 - \left(\frac{\sigma_w}{z}\right)^3\right],$$

where ε_w and σ_w are energy and length parameters. There are two walls separated by a distance L and so the total external potential exerted by the solids is

$$u^e(x) = \phi_w(z) + \phi_w(L - z).$$

According to gradient theory

$$c\frac{d^2n}{dz^2} = \mu_0(n) - \mu + u^e(z).$$

The walls are identical and so with the boundary condition

$$\frac{dn}{dx} = 0 \quad \text{at} \quad z = L/2.$$

You need solve (*) only in the range $0 < z < L/2$. Close to the wall the density becomes very small because u^e is strongly repulsive there. Thus, a second boundary condition is

$$n(z) = e^{\beta[\mu - u^e(z)]} \quad \text{at} \quad z = \sigma.$$

A reasonable choice for σ might be $\sigma = 0.7\sigma_w$. You should try other values and pick the larges σ below which the solution is independent of σ.

Use the equation of state that adds a Carnahan–Starling pressure of hard spheres to a VDW pressure of attraction, that is,

$$f_0(n) = nkT\left[\ln\sigma^3 - 1\right] + nkTy(4 - 3y)/(1 - y)^2 - n^2a,$$

where $y = \pi\sigma^3 n/6$.

a) Solve (*) for the density profile for $L/\sigma = 1.5, 2, 2.5, 3, 3.5$, and 4. Assume $\sigma = \sigma_w$ and use the dimensionless variables z/σ, L/σ, $\mu_o(n)\sigma^3/a$, $kT\sigma^3/a$, $n\sigma^3$, $\varepsilon_w\sigma^3/a$, and $c/\sigma^2 a$.

Choose

$$\frac{\mu\sigma^3}{a} = -4.8997; \quad \frac{kT\sigma^3}{a} = 0.12611; \quad \frac{c}{\sigma^2 a} = 1; \quad \frac{\varepsilon_w\sigma^3}{a} = \frac{5}{\sqrt{3\pi}}.$$

Plot the density profiles.

b) Calculate the pressure exerted by the fluid on a solid as a function of L. The pressure is given by

$$P_w = -\int_0^L n(z)\frac{d\phi_w(z)}{dz}dz.$$

Plot P_w versus L.

2 Compare the Reerink–Overbeek theory and the DLVO theory for the conditions in Figure 12.5 except set $\psi_o = 100$ mV. Explain the dependence on ψ_o.

3 Suppose the probability of coalescence of a pair of colloids goes as $p_{co} = \exp(-W_{max}/kT)$, where W_{max} is the maximum in the colloidal pair interaction energy. Use the Reerink–Overbeek

approximation and the conditions given in Figure 13.11, except let $R = 500$ Å, to compute p_{co} versus the Debye κ.

4 Use the generalized hard-rod model with the Carnahan–Starling equation of state for the hard sphere fluid and the VDW attractive energy to compute the density profile of the fluid confined by a planar wall whose wall–molecule interaction is

$$v_w(x) = \infty, \quad x < 0, \quad x > L$$
$$= -\varepsilon_w \exp(-x/d_w) - \varepsilon_w \exp(-(L-x)/d_w), \quad 0 < x < L.$$

Assume the molecule–molecule attraction is

$$\bar{u}^A(x) = -2\pi\varepsilon d^2, \quad |x| < d$$
$$= -2\pi\varepsilon d^2, \quad \left(\frac{d}{x}\right)^4, \quad |x| > d.$$

Calculate the disjoining pressures and density profiles at $T/T_c = 0.8$ and $n/n_c = 0.5$. Consider the cases $\varepsilon_w = \varepsilon$, $d_w = d$; $\varepsilon_w = 3\varepsilon$, $d_w = d$; $\varepsilon_w = 3\varepsilon$, $d_w = 3d$; $\varepsilon_w = \varepsilon$, $d_w = 3d$. Note that you can solve the problem for the dimensionless variables $n_r = n\sigma^3$, $T_r = kT/\varepsilon$, $x_r = x/d$, $L_r = L/d$, and $P_r = P\sigma^3/\varepsilon$.

5 Consider the wall–fluid potential

$$v(x) = \infty, \quad x < 0, \quad x > L$$
$$= -\varepsilon_w e^{-x/d} - \varepsilon_w e^{-(L-x)/d}, \quad 0 < x < L$$

Use the Tarazona model with the Carnahan–Starling equation of state to compute the density profile and the disjoining pressure of a confined hard sphere fluid as a function of L/d. Choose $\varepsilon_w = 1.5$ kT. d is the diameter of the hard spheres. Choose the chemical potential to correspond to a bulk pressure $Pd^3 = 10$ kT. Choose appropriate dimensionless variables for solving the problem and plotting the results.

6 Consider aqueous solutions of sodium chloride and sodium sulfate, respectively. Using the parameters in Figure 12.5 and the Reerink–Overbeek approximation, calculate the force between colloidal bodies versus separation for 0.001, 0.01, and 0.1 molar solutions of each of the salts. Plot the curves and explain the differences between the results for the different salts.

References

Chan, D. Y. C., Pashley, R. M., and White, L. R. 1989. *J. Colloid Interface Sci.*, **127,** 283.

Curtin, W. A. and Ashcroft, N. W. 1985. *Phys. Rev. A,* **32,** 2909.

Denton, A. R. and Ashcroft, N. W. 1989. *Phys. Rev. A,* **39,** 4701.

Derjaguin, B. V. 1931. *Kolloidn. Z.,* **69,** 155.

Derjaguin, B. V. and Landau, L. 1941. *Act. Phys.-Chim. USSR,* **14,** 633.

Fischer, J. and Heinbuch, U. 1988. *J. Chem. Phys.,* **88,** 1909.

Johnson, M. and Nordholm, S. 1981. *J. Chem. Phys.* **75,** 1953.

Magda, J. J., Tirrell, M. V., and Davis, H. T. 1985. *J. Chem. Phys.,* **83,** 1888.

Mao, G., Tsao, Y., Tirrell, M., Davis, H. T., Hessel, V., and Ringsdorf, H. 1993. *Langmuir,* **9,** 3461.

Meister, T. F. and Kroll, D. M. 1985. *Phys. Rev. A,* **31,** 4055.

Ninham, B. W., and Parsegian, V. A. 1971. *J. Theor. Biol.,* **31,** 405.

Nordholm, S., Johnson, J., and Freasier, B. C. 1980. *Aust. J. Chem.,* **33,** 2139.

Percus, J. K. 1981. **75,** 1316.

Reerink, H. and Overbeek, J. Th. G. 1954. *Disc. Faraday Soc.,* **18,** 74.

Sokolowski, S. and Fischer, J. 1990. *Mole. Phys.,* **71,** 393.

Tang, Z., Scriven, L. E., and Davis, H. T. 1992. *J. Chem. Phys.,* **97,** 9258.

Tarazona, P. 1985. *Phys. Rev. A,* **31,** 2672; **32,** 4055.

Vanderlick, T. K., Scriven, L. E., and Davis, H. T. 1989. *J. Chem. Phys.,* **90,** 2422.

Valleau, J. P., Ivokov, R., and Torrie, G. M. 1984. *J. Cheem. Phys.,* **80,** 2221.

Verwey, E. J. W. and Overbeek, J. Th. G. 1948. *Theory of Stability of Lyophobic Colloids,* Elsevier, Amsterdam.

13

THIN FILMS AND WETTING TRANSITIONS

13.1 Introduction

The adsorption and wetting behavior of fluids on solid surfaces and at fluid–fluid interfaces is important in countless natural and technological situations. The way water wets their surfaces can determine the survival of land animals, birds, and water-walking insects. The effectiveness of pesticides sprayed onto plants, of flotation agents, and of capillary delivery devices is controlled by wetting characteristics of the contacting phases. The quality of paints and inks is strongly dependent on wetting properties.

As discussed in Chapter 7, if a drop of fluid phase β is placed at the interface between fluid phase α and the γ phase (fluid or solid), it will be rearranged into a droplet at capillary equilibrium or it will spread into a thin film between the phases to find a thickness determined by the volume of the droplet β. The first case, illustrated in Figure 13.1, represents a nonwetting situation. Through the balance of forces the three interfacial tensions between the phases determine the dihedral contact angles θ_α, θ_β, and θ_γ of a lens of phase β at the $\alpha\gamma$ fluid–fluid interface (Figure 13.1a) and the contact angle θ of a drop of β at the $\alpha\gamma$ fluid–solid interface (Figure 13.1b).

In the perfect wetting situation, where β is the wetting phase, the interfacial tensions obey the equation

$$\gamma_{\alpha\gamma} = \gamma_{\alpha\beta} + \gamma_{\beta\gamma}, \tag{13.1.1}$$

corresponding to the limits $\theta_\gamma, \theta_\alpha \to 180°$ and $\theta_\beta \to 0°$ in the case of the lens of β at the fluid–fluid interface and corresponding to the limit $\theta \to 0°$ in the case of a drop of β at the fluid–solid interface. Illustrated in Figure 13.2 is what would happen if, in the two situations, phases α and β mutually filled a container of solid γ?

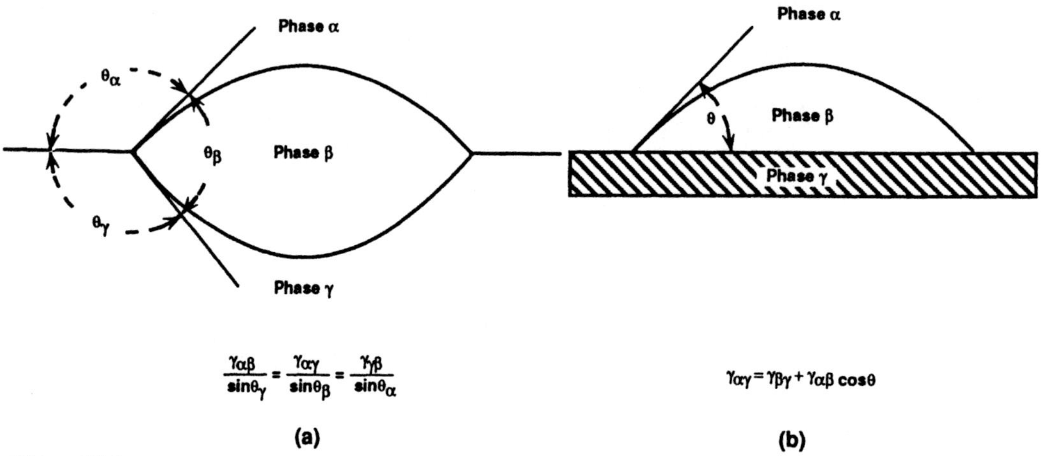

Figure 13.1
A nonwetting drop β at the interface between phases α and γ.

One of the purposes of this chapter is to explore theoretically what controls how a droplet of phase undergoes a transition from a nonwetting phase to a perfect wetting phase.

The reason for referring to the wetting case as "perfect" wetting or "complete" wetting is to distinguish from a common practice in the literature of referring to β as a nonwetting phase on solid γ when the contact angle θ is $> 90°$ and as a wetting phase when θ is $< 90°$. The old usage is unfortunate, but confusion is likely if it is not followed.

Figure 13.2
Conditions for perfect wetting by either phase α or phase β. Adapted from Davis et al. (1986).

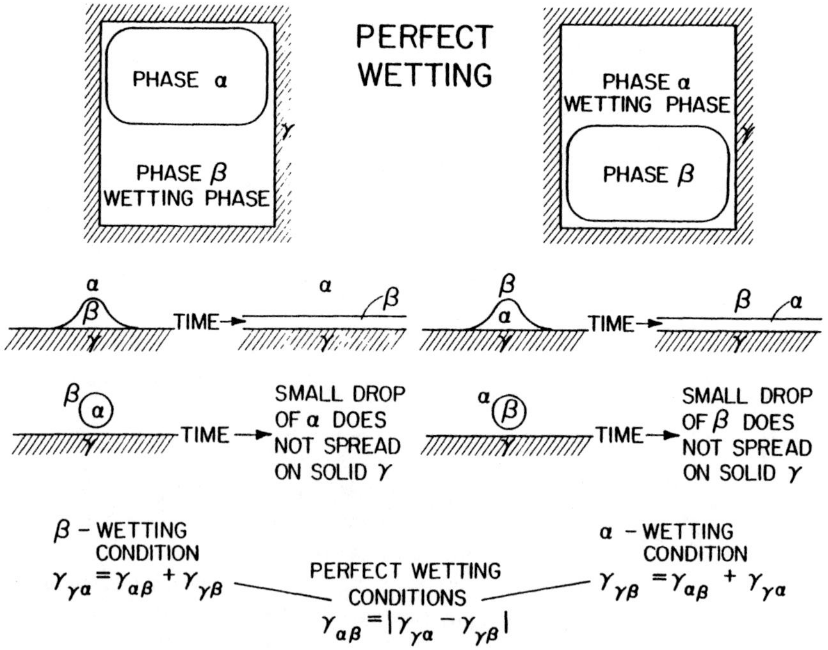

In the case of confined fluid films, one commonly changes the thickness of a film by externally controlling the distance of separation of opposed solid surfaces while the film is in thermodynamic contact with bulk fluid that has fixed temperature, pressure, and composition. Consequently, the film is at fixed temperature and all its components are at fixed chemical potentials. Because the normal pressure P_f in the confined film is different from the pressure P_B of the bulk phase, it is convenient to describe the film in terms of its disjoining pressure $\Pi \equiv P_f - P_B$. In the wetting situation envisioned in this chapter, the film will always be assumed to be planar and to be bordered at least on one side by an unconstrained fluid phase. Thus, the normal pressure in the film and bordering bulk fluid phase (or phases) will always be equal and so the disjoining pressure cannot be defined as the force of confinement of the film. For this reason it is frequently more useful to introduce the concept of disjoining potential.

Consider an open system consisting of a planar interface between phases α and γ. Suppose the chemical potentials of all components except i are equal to their values when phases α, β, and γ are in three-phase equilibrium. Suppose, however, that μ_i differs from its value at three-phase coexistence. Under these conditions there is no phase β at the $\alpha\gamma$ interface. As μ_i approaches its three-phase coexistence value μ_i^e, the $\alpha\gamma$ interface will change in composition and when μ_i^e is reached phase β will appear at the $\alpha\gamma$ interface, either as a thin layer of phase if β is a wetting phase or as a droplet of phase. Thus, we define the disjoining potential $\Delta\mu_i$,

$$\Delta\mu_i = \mu_i - \mu_i^e, \tag{13.1.2}$$

to characterize the adsorption or "thin film" state at the $\alpha\gamma$ interface as μ_i approaches the three-phase coexistence value. With respect to some arbitrary dividing surface at $x = 0$ in the $\alpha\gamma$ interfacial zone, we can characterize the adsorption state by the excess surface density

$$\Gamma_i = \int_{-\infty}^{\infty} [n_i(x) - n_i^\alpha \eta(x) - n_i^\gamma (1 - \eta(x))]dx, \tag{13.1.3}$$

where $n_i(x)$ is the density of component i at x, n_i^α and n_i^γ are the densities of component i in bulk phases α and γ at the chemical potential μ_i and $\eta(x)$ is the step function. A "film thickness" h is defined by

$$h = \Gamma_i / n_i^\beta, \tag{13.1.4}$$

where n_i^β is the density of i in phase β at the three-phase coexistence potential μ_i^e.

A simple example of the general situation just described is a one-component fluid near its liquid–vapor coexistence condition and in contact with an immiscible solid. Then

$$\Delta\mu \equiv \mu - \mu(n_{\text{sat}}^\beta, T) \tag{13.1.5}$$

and

$$h = \frac{1}{n^{\beta}_{\text{sat}}} \int_0^{\infty} [n(x) - n^{\alpha}]dx, \tag{13.1.6}$$

where μ is the chemical potential of the fluid and n^{α} its bulk phase density. If the bulk fluid is a vapor, then n^{β}_{sat} is the saturated liquid density at temperature T. If the bulk fluid is a liquid, then n^{β}_{sat} is the saturated vapor density at temperature T. Of course, at liquid–vapor coexistence $\mu(n^{v}_{\text{sat}}, T) = \mu(n^{l}_{\text{sat}}, T)$. Because the value of h is determined by μ, we can denote μ as a function of h, i.e., $\mu = \mu(h, T)$ is the chemical potential of the film and $\mu(h = \infty, T) = \mu(n^{\beta}_{\text{sat}}, T)$, $\beta = v$ or l.

As was done in Chapter 7, we can define a disjoining pressure in terms of the disjoining potential. In isothermal bulk phase α at pressure P (and density n^{α}) the Gibbs–Duhem equation is $d\mu(P, T) = dP/n^{\alpha}$. We define a disjoining pressure as $\Pi = P - P^{\text{sat}}$, P^{sat} being the liquid–vapor coexistence pressure at T. Then the relationship between $\Delta\mu$ and Π is obtained by integration of the Gibbs–Duhem equation, that is,

$$\Delta\mu = -\int_P^{P+\Pi} \frac{1}{n^{\alpha}} dP. \tag{13.1.7}$$

If phase α is incompressible the simpler result,

$$\Delta\mu(h) = -\frac{1}{n^{\alpha}} \Pi(h), \tag{13.1.8}$$

is obtained.

The multicomponent generalizations of eqs. (13.1.7) and (13.1.8) are

$$\Delta\mu_i = -\int_P^{P+\Pi} \frac{1}{n^{\alpha}_i} dP \tag{13.1.9}$$

and

$$\Delta\mu_i(h) = -\frac{1}{n^{\alpha}_i} \Pi(h), \tag{13.1.10}$$

where in the integration of eq. (13.1.9) the temperature and the chemical potentials μ_j for all species except i are held constant.

As we see in the following section, there is a close connection between the dependence of $\Delta\mu$ on h and the wetting behavior.

13.2 Gradient Theory of Wetting Transitions

Although local density functional theories do a poor job of predicting the molecular details of adsorption at a solid surface, we will present evidence in the next section that the qualitative patterns of wetting transitions are correctly predicted by local density functional theories. Consequently, in this section the simplest model, density gradient theory, is used to explore the wetting behavior of a one-component fluid on a solid.

The Helmholtz free energy of density gradient theory is

$$F = \int \left[f_o(n) + \frac{c}{2}(\nabla n)^2 + n v_w \right] d^3r, \qquad (13.2.1)$$

where $f_o(n)$ is the free energy density of homogeneous fluid at density n, c the influence parameter (assumed in what follows to be constant) and, v_w the wall–fluid interaction potential. For a planar system $n = n(x)$ and $(\nabla n)^2 = (dn/dx)^2$. At equilibrium the density distribution $n(x)$ is the one that minimizes the grand potential $\Omega = F - \mu A \int n dx$. The minimization condition, $\delta\Omega/\delta n(x) = 0$, yields the equation

$$c \frac{d^2 n}{dx^2} = \mu_o(n) + v_w(x) - \mu. \qquad (13.2.2)$$

This equation has been used (Teletzke, et al., 1982) for a system with the fluid–solid interaction potential

$$v_w = W \left[\frac{1}{45} \left(\frac{\sigma_w}{x} \right)^9 - \frac{1}{6} \left(\frac{\sigma_w}{x} \right)^3 \right]. \qquad (13.2.3)$$

This is the potential energy for solid molecules uniformly distributed in the region $x < 0$ and interacting with a fluid molecule by a 6–12 Lennard-Jones pair potential. W is the energy parameter and σ_w is related to the intermolecular collision diameter.

The appropriate boundary conditions for the problem are that far from the wall, density has a constant bulk value n_B, and at the wall, the density approaches zero as the repulsive region of the wall potential is reached. Mathematically stated these conditions are

$$n(x) \to e^{-(v_w - \mu)/kT} \quad \text{as} \quad x \to 0; \quad n(x) \to n_B \quad \text{as} \quad x \to \infty. \qquad (13.2.4)$$

Teletzke et al. (1982) used the Peng–Robinson equation of state (see Chapter 6) and found from gradient theory the wetting behavior summarized in the phase diagram shown in Figure 13.3. In addition to the liquid–vapor binodal shown in the figure is a line along which two thin films coexist. The quantities in this figure and others to follow are given in the units in Table 13.1.

According to Figure 13.3, there is a perfect wetting temperature T_w below which a drop of saturated liquid placed at the solid–vapor interface will have a contact angle such that $0° < \theta < 180°$. Above T_w, a drop of saturated liquid placed at the solid–vapor interface will spread into a thin film. It will completely wet the interface. Consider a temperature $T < T_w$ and a vapor density less than the saturated vapor density. There is an adsorbed layer of fluid at the solid–vapor interface. If density is increased by compression at constant temperature, the adsorbed layer changes only slightly

Figure 13.3
Phase and film diagram predicted by gradient theory of a Peng–Robinson fluid and a 9–3 Lennard–Jones wall. $W = 6.4$, $d = 1$. Adapted from Davis et al. (1986).

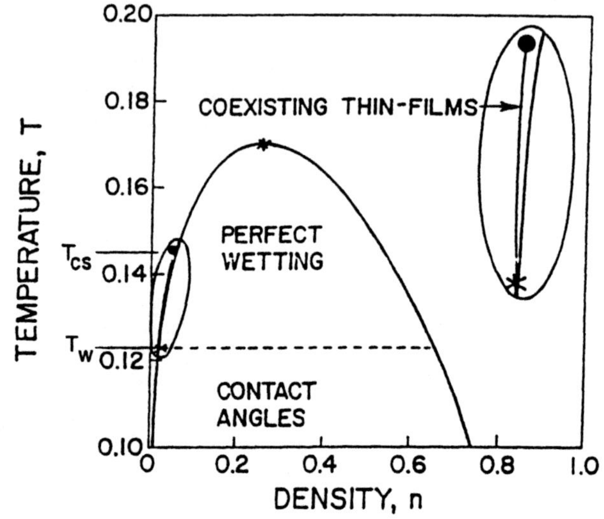

as the saturated vapor density is approached. In the limit of the saturated density, further compression of the vapor will cause small liquid droplets to appear at the solid–vapor interface. The menisci of the droplets will take on an equilibrium contact angle as they are formed.

A quite different thing happens at a temperature T lying between T_w and T_{cs}. In this case, as the density is compressed isothermally from a very low value, the adsorbed layer of fluid at the solid–vapor interface changes only slightly until the density reaches the curve labeled "coexisting thin-films" in Figure 13.3. At this density a thicker adsorbed layer appears that is in thermodynamic equilibrium with the thinner adsorbed layer. We refer to these two states of adsorption as coexisting thin films. With continued compression the thicker film completely covers the solid and thickens continuously with increasing density until at saturation it thickens into a perfect wetting layer of liquid.

As the temperature is increased toward T_{cs} the coexisting thin films become more and more similar until at T_{cs} they are identical. Above T_{cs} only one thin-film appears at the interface. Thus, T_{cs}

TABLE 13.1 Units of Quantities Used in Section 13.2

Quantity	*Units*
W	a/b
n	b^{-1}
T	a/kb
d	$\sigma_w/(c/a)^{1/2}$
u	a/b
x	$(c/a)^{1/2}$
h	$(c/a)^{1/2}$

is a critical point and is called the *surface critical temperature.* Above T_{cs} and below the liquid–vapor critical point T_c, isothermal compression of subsaturated vapor causes a continuous thickening of the thin film at the solid–vapor interface until at saturation it grows into a perfect wetting layer of liquid.

The origin of the predicted wetting behavior can be understood in terms of the dependence of the disjoining potential on film thickness. A qualitative plot of the potential calculated by Teletzke et al. is shown in Figure 13.4.

At temperatures about T_{cs} the disjoining chemical potential is a monotonic function of film thickness. Thus, as the negative disjoining potential approaches zero (and bulk vapor density approaches the coexistence density), the film thickens continuously and results in a layer of wetting liquid at zero disjoining potential. At temperatures below T_{cs} the disjoining pressure has a van der Waals loop and so two-film coexistence is possible. At coexistence the chemical potential of the two films must be equal and their film tension must be equal. The solid–fluid tension obeys the Gibbs isotherm

$$d\gamma = -\Gamma d\mu = -n_l^{\text{sat}} h d(\Delta\mu), \qquad (13.2.5)$$

and so the film coexistence condition becomes

$$d\gamma = 0 = -n_l^{\text{sat}} \int_{h_1}^{h_2} h d(\Delta\mu). \qquad (13.2.6)$$

Thus, the coexisting film thickness h_1 and h_2 at a given T are determined by placing an intersecting line (constant $\Delta\mu$) so that it intersects the isotherm at, say, A, B, and C such that the area between the isotherm and the line segment $\overrightarrow{AB}$ equals the area between the isotherm and the line segment $\overrightarrow{BC}$. Then h_1 and h_2

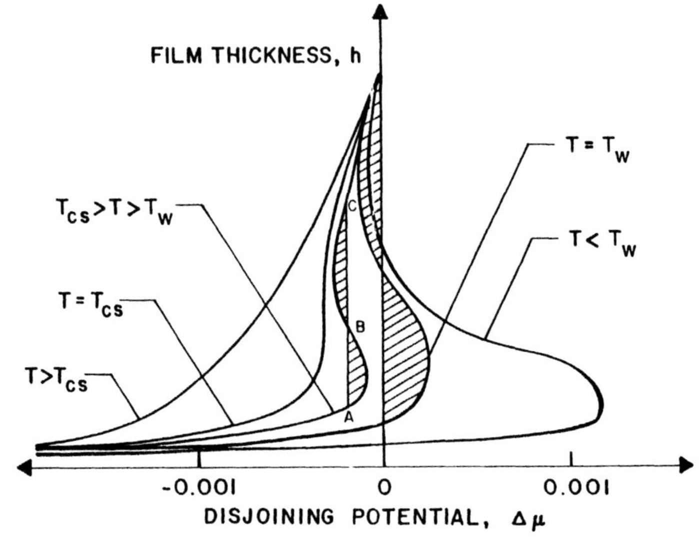

Figure 13.4
Disjoining potential isotherms (schematic). Adapted from Davis et al. (1986).

are the coexisting film thickness. The situation is analogous to the Maxwell tie-line construction of liquid–vapor phase equilibria.

With decreasing temperature the minimum of the disjoining pressure isotherm decreases faster than the maximum increases. The consequence is that at a temperature below T_w the conditions of two-film coexistence cannot be satisfied and so at a given negative disjoining potential only one thin film exists. The tension of the thinner film is smaller than that of the thicker film and so the equilibrium film is the thinner one (referred to in some works as the adsorbed layer). Thus, if along the isotherm $T < T_w$ in Figure 13.4 the negative disjoining potential approaches zero (liquid–vapor coexistence), the thin film changes relatively little. At liquid–vapor coexistence this thin film will coexist with droplets of liquid whose contact angle will depend on T.

The transition from thin to thick thin films across the film coexistence curve is a first-order transition because the film thickness changes discontinuously at a coexistence point {correspondingly, there is a discontinuity in the derivative $\partial\gamma/\partial\mu$ because $\partial\gamma/\partial\mu = \int_0^\infty [n(x) - n_B]dx$.} The thin film phase diagram corresponding to the coexistence curve in Figure 13.3 is shown in Figure 13.5. The first-order character of the transition between T_w and T_{cs} is apparent.

Because the thin film coexistence curve intersects the vapor side of the vapor–liquid coexistence curve, it follows that the transition from nonwetting ($0° < \theta < 180°$) to perfect wetting occurring at T_w is also a first-order transition (because there is a discontinuity in $\partial\gamma/\partial\mu$ at T_w) for the model presented here. Experimentally the wetting transition appears to be first-order as predicted here, but, as we shall see in a later section, some models predict that the transition is continuous, or second order.

The wetting transition temperature depends on the strength W of the fluid–wall attraction; the larger W the lower the temperature T_w of the transition. A plot of the perfect wetting transition temperature versus W is shown in Figure 13.6. Above $W = 5$, T_w is a linearly decreasing function of W. The liquid–vapor surface tension of the fluid at the wetting temperature is also shown in Figure 13.6. Below the wetting transition point, a droplet of liquid does not spread because the solid–vapor surface tension γ_{SV} is less than the sum of the solid–vapor and liquid–vapor tensions, that is,

$$\gamma_{SV} < \gamma_{SL} + \gamma_{LV}. \tag{13.2.7}$$

The perfect wetting equality, $\gamma_{SV} = \gamma_{SL} + \gamma_{LV}$, can be reached by increasing temperature, thereby lowering γ_{LV} faster than $\gamma_{SV} - \gamma_{SL}$, or by increasing the energy W of the solid, thereby increasing $\gamma_{SV} - \gamma_{SL}$ without affecting γ_{LV}. We see in Figure 13.6, that as W becomes smaller, the perfect wetting transition temperature approaches the liquid–vapor critical temperature. In fact, as we shall see in Section 13.4, for sufficiently small W, vapor can become the wetting phase and the prewetting line will move onto the liquid side of the binodal.

Figure 13.5
Thin film phase diagram. Gradient theory of a Peng–Robinson fluid at a 9–3 Lennard–Jones wall. $W = 6.4$, $d = 1$. Adapted from Davis et al. (1986).

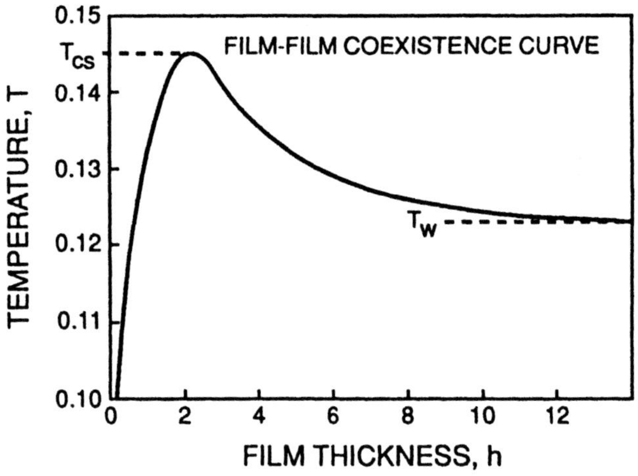

Teletzke et al. (1982) also calculated the contact angle of a liquid droplet on the solid as a function of temperature for various interaction parameters. The results are displayed in Figure 13.7 as a plot of $\cos\theta$ versus $\gamma_{LV}(T)/\gamma_w$, where γ_w is the liquid–vapor surface tension at the perfect wetting temperature T_w. As discussed in Chapter 7, the theory of fluids near a critical fluid point implies that the contact angle will obey the critical point scaling law

$$\cos\theta = (\gamma_w/\gamma_{LV})^{1-\beta_1/\phi} \tag{13.2.8}$$

where β_1 and ϕ are the critical exponents of the tension difference, $\gamma_{SV} - \gamma_{SL}$, and of the surface tension, γ_{LV}, that is, $|\gamma_{SV} - \gamma_{SL}| \propto (1 - T/T_c)^{\beta_1}$ and $\gamma_{LV} \propto (1 - T/T_c)^{\phi}$ near the critical point T_c. According to Cahn's hypothesis, β_1 is equal to the density exponent β, where $n_l - n_g \propto (1 - T/T_c)^{\beta}$ for T near T_c. Thus, for a mean field model, as is the Peng–Robinson model, Cahn predicts

$$\cos\theta = (\gamma_w/\gamma_{LV})^{2/3}. \tag{13.2.9}$$

For a low energy solid with interaction parameter $W \simeq 3$, T_w is close enough for critical point scaling also to hold, and indeed

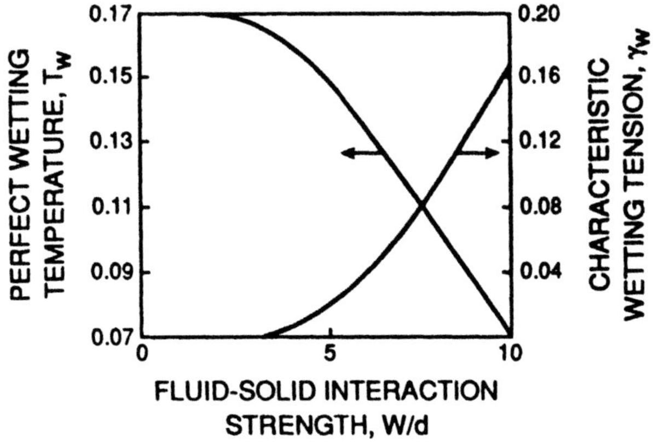

Figure 13.6
Perfect wetting temperature and characteristic wetting tension. Adapted from Davis et al. (1986).

Figure 13.7
Variation of contact angle with liquid–vapor tension for several values of the fluid–solid interaction parameter. $d = 1$. Adapted from Davis et al. (1986).

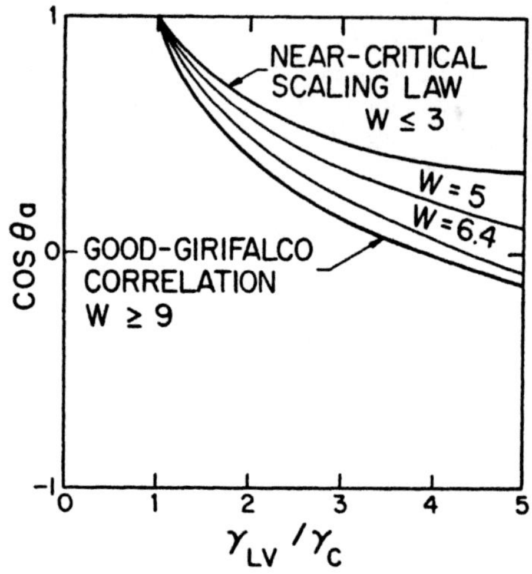

Cahn's theoretical scaling holds. For a high energy solid, T_w is not close enough for critical point scaling to hold. For example, for $W \geq 9$ the scaling law

$$\cos\theta = 2\sqrt{\frac{\gamma_w}{\gamma_{\mathrm{LV}}}} - 1 \qquad (13.2.10)$$

holds. This relation follows from the Girifalco and Good (1959) approximation, $\gamma_{\mathrm{SL}} = (\sqrt{\gamma_{\mathrm{LV}}} - \sqrt{\gamma_{\mathrm{SV}}})^2$, for nonpolar fluids at low vapor pressure ($\gamma_{\mathrm{SV}} \simeq$ constant). In the range $3 < W < 9$ the dependence of $\cos\theta$ on T (or γ_{LV}) is more complicated. It is interesting that Cahn's hypothesis holds for a mean field fluid, whereas renormalization group theory and experiment (Turkovich, 1995) yield $\beta_1 = 0.6$ compared with the experimental value of 0.34 for β.

Let us end this section by displaying the density profiles predicted for thin films at a vapor–solid interface at the temperature $T = 0.14$, which lies between the wetting temperature T_w $(= 0.123)$ and the surface critical temperature T_{cs} $(= 0.145)$ for a solid with interaction parameter $W = 6.4$. The results are shown in Figure 13.8. At bulk vapor density $n_B = 0.0513$, there are two thin film solutions to eq. (13.2.2). These are the coexisting films. As bulk vapor density increases beyond this value, the thick branch of the coexisting films thickens into a perfect wetting layer of liquid as the binodal envelope is approached.

Although the detailed structure of the adsorbed films is not correctly predicted by gradient free energy theory, we see in the next section that the predicted patterns of wetting transitions are qualitatively correct.

Figure 13.8
Thin-film density profiles and disjoining potential versus film thickness at $T = 0.140$. Adapted from Teletzke et al. (1982).

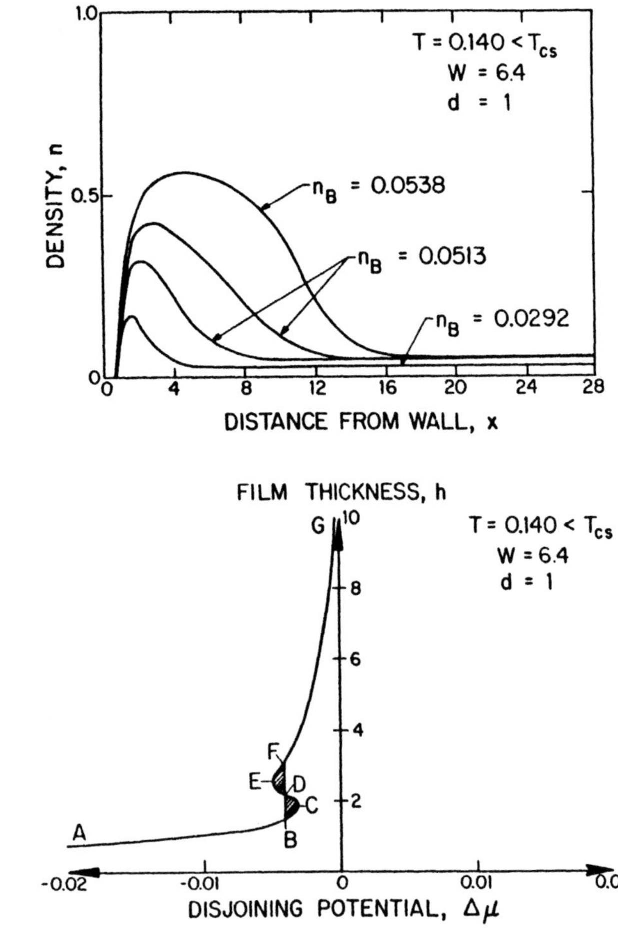

13.3 Nonlocal Density Functional Theory of Wetting Transitions

In this section we examine the wetting behavior of a one-component fluid on a flat solid with a nonlocal density functional model. We explore the Tarazona model for a planar system. The chemical potential for this model is [eq. (12.2.1)]

$$\mu = v_w(x) + kT \ln n(x) + \int \frac{\delta \bar{n}^\sigma(x')}{\delta n(x)} \mathcal{F}(\bar{n}^\tau(x'))dx' \\ \int \frac{\delta \bar{n}^\tau(x')}{\delta n(x)} \bar{n}^\sigma(x')\mathcal{F}'(\bar{n}^\tau(x'))dx' + \int n(x)\bar{u}^A(x'-x)dx'. \tag{13.3.1}$$

The quantities $\bar{n}^\tau$ and $\bar{n}^\sigma$ are given by eqs. (12.2.2) and (12.2.5)–(12.2.8). We use the Carnahan–Starling approximation for $\mathcal{F}(n)$,

$$\mathcal{F}(n) = kT\frac{y(4-3y)}{(1-y)^2}, \quad y = \frac{\pi n d^3}{6}. \tag{13.3.2}$$

We represent the attractive part of the particle–particle pair interaction potential by

$$u_A(r) = 0, \quad r < d$$
$$= 4\varepsilon\left[\left(\frac{d}{r}\right)^{12} - \left(\frac{d}{r}\right)^{6}\right], \quad r \geq d, \tag{13.3.3}$$

and so

$$\bar{u}_A(x) \equiv \int_{-\infty}^{\infty}\int_{-\infty}^{\infty} u_A(r)dydz = -\frac{6}{5}\pi\varepsilon d^2, \ |x| < d$$
$$= 2\pi\varepsilon d^2\left[\frac{2}{5}\left(\frac{d}{x}\right)^{10} - \left(\frac{d}{x}\right)^{4}\right], \quad |x| \geq d. \tag{13.3.4}$$

For the wall–fluid particle potential we take the 9–3 model

$$v_w(x) = \infty, \quad x < \sigma_w$$
$$= W\left[\frac{1}{45}\left(\frac{\sigma_w}{x}\right)^{9} - \frac{1}{6}\left(\frac{\sigma_w}{x}\right)^{3}\right] \tag{13.3.5}$$

In the following text, we present results for this model computed by Dhawan et al. (1991). In the computations, eq. (13.3.1) was discretized using the trapezoidal rule for integration. The resulting set of nonlinear algebraic equations were solved by the Newton–Raphson technique.

Density profiles and the disjoining potential isotherms are presented in Figures 13.9 and 13.10 for the fluid–solid interaction parameters $W = 20\varepsilon$ and $\sigma_w = d$.

In Figure 13.9 the temperatures of ε/k, $0.75\varepsilon/k$, and $0.67\varepsilon/k$ correspond to a temperature above T_{cs}, between T_w and T_{cs}, and below T_w, respectively. A first-order film-film transition occurs at the temperature $T = 0.75\varepsilon/k$. As we expect, the density profiles predicted by nonlocal density functional theory are much more highly structured than those predicted by a local theory such as the gradient density theory. However, as is evident in Figure 13.10, the disjoining pressure isotherms, which determine wetting behavior, are qualitatively the same for local and nonlocal theories.

The parameters of the wall–particle and fluid–fluid iteration potentials used to compute the disjoining isotherms in Figure 13.11 were chosen to approximate argon adsorbed on a flat CO_2 surface. The parameters used were: $\varepsilon/k = 119.8$ K, $d = 3.4$Å, $\sigma_w = 3.727$Å, $\varepsilon_w/k = 153$ K, $n_s\sigma_w^3 = 0.988$, $w = 4\pi\varepsilon_w n_s \sigma_w^3$. In this case the wetting transition temperature is predicted by the model to be $T_w = 0.84\varepsilon/k$ and the surface critical temperature is $T_{cs} = 0.895\varepsilon/k$. The predicted prewetting line (i.e., the film-film coexistence curve) for the Ar $-$ CO_2 system is shown in Figure 13.12. It intersects the vapor branch of the liquid–vapor binodal at T_w (actually the prewetting line is tangent to the binodal at T_w). The prewetting line proved difficult to observe experimentally. A reason why is seen in Figure 13.12: the prewetting line is very

Figure 13.9
Density profiles obtained from the generalized hard-rods model (9–3 Lennard–Jones wall with cutoff). $T^* \equiv KT/\varepsilon$. Top: $T > T_{cs}$; bulk fluid densities corresponding to increasing film thicknesses are 0.09, 0.135, and $0.185d^{-3}$. Center: $T_w < T < T_{cs}$; bulk densities are 0.026263, 0.0262263, and $0.026518d^{-3}$. Films with thicknesses of $h = 0.824$ and $5.7d$ are in coexistence. Bottom: $T > T_w$; bulk fluid densities are 0.008, 0.0126, and $0.0124d^{-3}$. Adapted from Dhawan, et al. (1993).

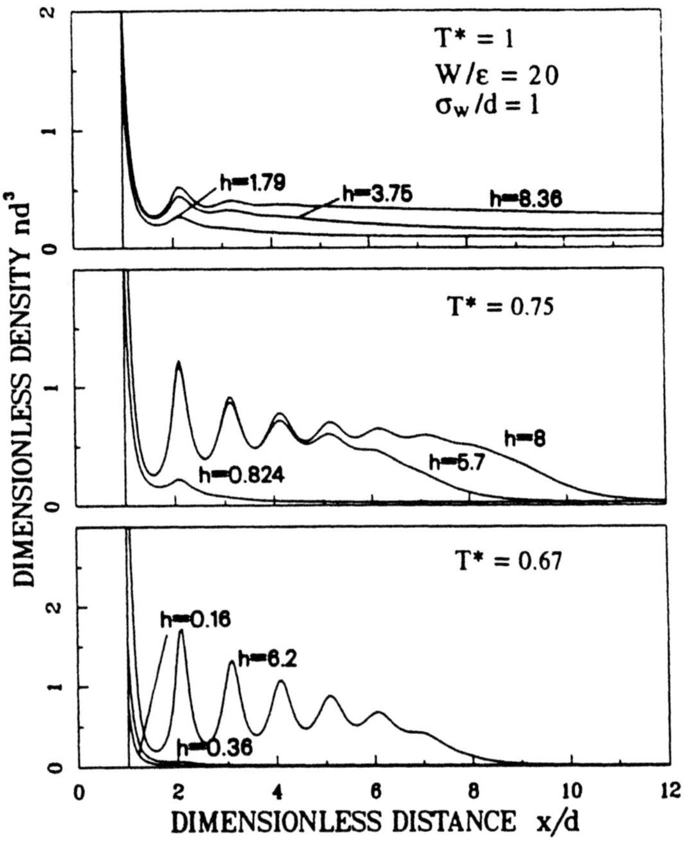

Figure 13.10
Disjoining potential isotherms obtained from the Tarazona model depicting first-order wetting transition. $T^* \equiv kT/\varepsilon$. Adapted from Dhawan, et al.(1991).

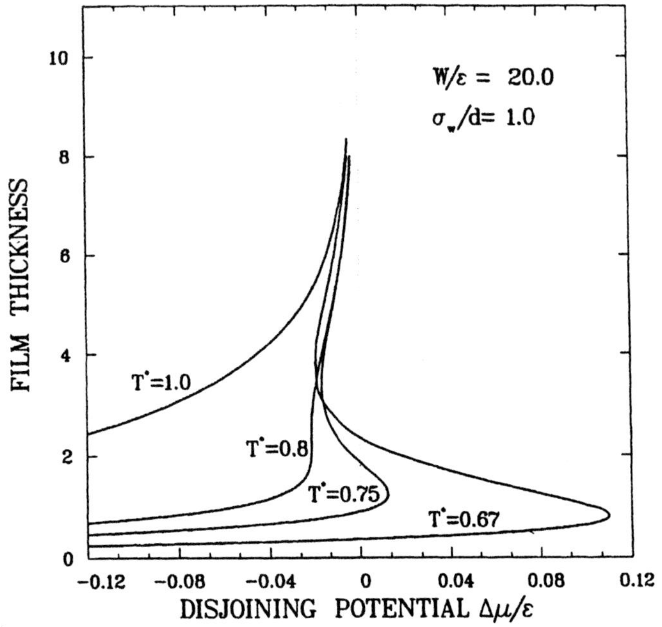

Figure 13.11
Disjoining potential isotherms for the model Ar – CO_2 obtained from the Tarazona Model (9–3 Lennard–Jones wall with cut-off) $T^* \equiv kT/\varepsilon$. The van der Waals loops for the temperature range $T_w^* < T^* < T_{cs}^*$ admit the Maxwell's construction indicating coexisting films. Adapted from Dhawan, et al. (1991).

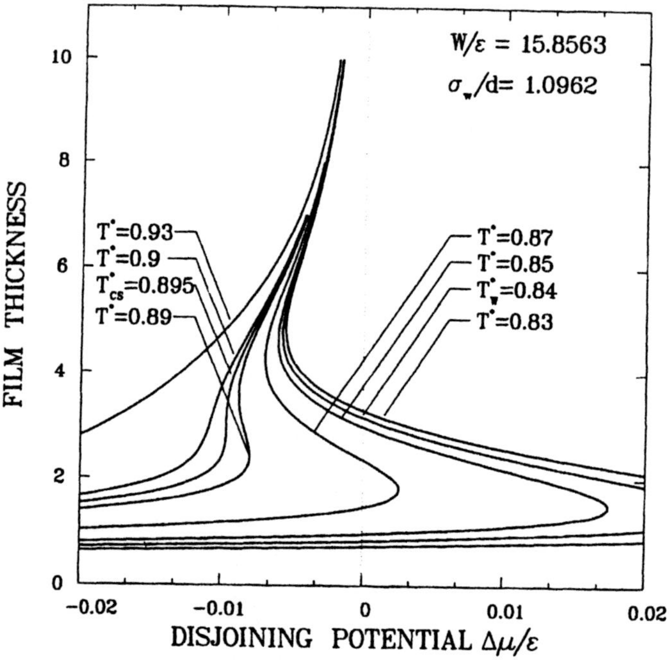

close to the binodal. Observations of the prewetting line will be discussed in Section 13.5.

Finn and Monson (1989) performed Monte Carlo simulations on a 6–12 Lennard–Jones fluid at a 9–3 Lennard–Jones wall [eq. (13.3.5) without the hard sphere cutoff]. They chose the potential

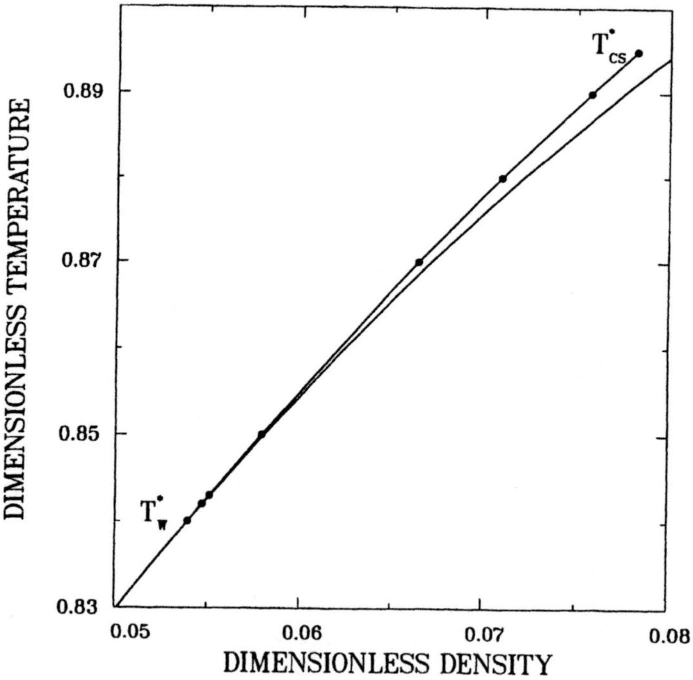

Figure 13.12
The prewetting line for the Ar – CO_2 system from the Tarazona Model (9–3 Lennard–Jones wall with cut-off). $T^* \equiv kT/\varepsilon$. The wetting transition temperature T_w lines at the intersection of the prewetting line and the binodal. Adapted from Dhawan, et al. (1991).

parameters to correspond to fluid argon and solid carbon dioxide. Some of their results for the density profiles are shown in Figure 13.13. They find a first-order film-film transition as predicted by gradient theory and nonlocal density functional theory.

For the 9–3 wall potential without a hard sphere cutoff, the Tarazona model predicts a much lower wetting temperature than it does for a 9–3 wall potential with a cutoff. T_w in this case is predicted by the theory to be less than $0.67\varepsilon/k$. We show in Figure 13.14 the density profiles of Ar at a solid CO_2 surface predicted by the Tarazona model for the 9–3 wall potential without a cutoff. The temperature of the system is $0.67\varepsilon/k$. There is a prewetting transition similar to that found in the Monte Carlo simulations of Finn and Monson. Thus, the Tarazona model is in qualitative but not quantitative agreement with simulations (the predicted T_c is too low). This is not surprising because in the bulk phase limit the Tarazona model is essentially a van der Waals approximation known to give poor predictions of the equation of state of a Lennard–Jones fluid (the predicted critical point is too low). Quantitative improvement would undoubtedly result if the Teletzke–Scriven–Davis model [eq. (12.1.27)] were used instead of the Tarazona one. However, we feel the basic physics is captured by the Tarazona model and so a more complicated calculation is not so interesting.

In fact, although the nonlocal gradient density functional theory misses the detailed structure of the density profile of a fluid near a solid wall, it nevertheless predicts the qualitatively correct wetting transition trends. Thus, in the next section we use

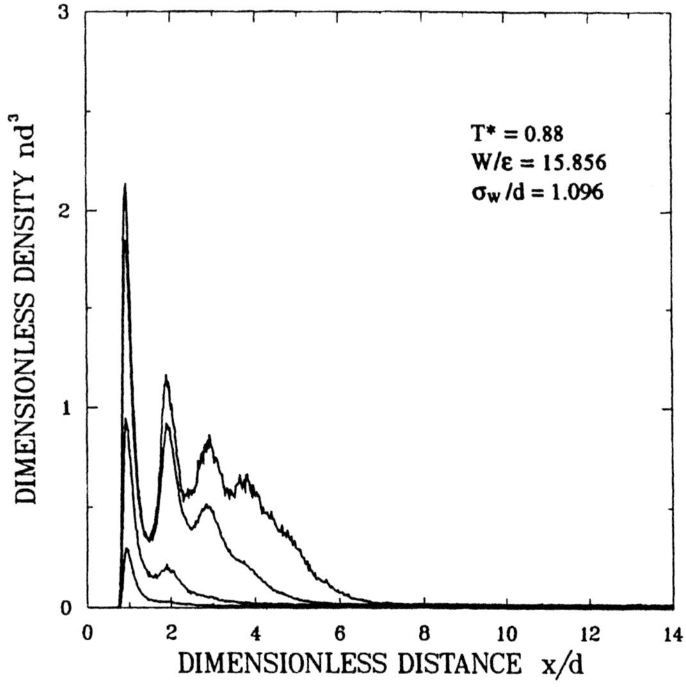

Figure 13.13
Density profiles from the isobaric-isothermal ensemble Monte Carlo method (Finn and Monson, 1989) at a dimensionless temperature $T^*(\equiv kT/\varepsilon)$ of 0.88. Films shown have excess molecular densities of approximately 0.104, 0.460, 1.710 and $2.194d^{-2}$. Films with interfacial excess densities of 0.460 and $1.71d^{-2}$ are approximately in coexistence. Their bulk densities are 0.0184 and $0.0188d^{-3}$, respectively (Private communication, J. E. Finn and P. A. Monson).

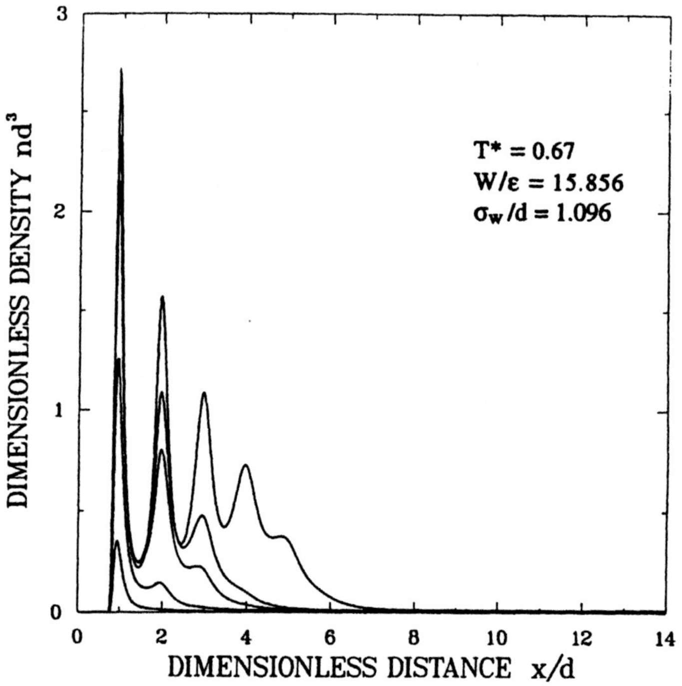

Figure 13.14
Density profiles from the Tarazona model at a dimensionless temperature $T^*(\equiv kT/\varepsilon)$ that gives coexisting films at a bulk density of $0.011513d^{-3}$. Profiles are plotted for interfacial excess densities of 0.1048, 0.4601, 1.2123, 1.6204 and $1.7099d^{-2}$. Films with interfacial excess densities of 1.2123 and $1.6204d^{-2}$ are in coexistence. The equilibrium thicker film is of approximately the same interfacial excess density as that of the equilibrium thick film from Monte Carlo simulations shown in Figure 13.13. Adapted from Dhawan et al. (1991).

a local density functional theory to explore the effect of fluid–wall interactions wetting behavior. Given the sensitivity of the results in the section on the cutoff of the 9–3 fluid–wall potential we anticipate, that the nature of the wall potential may be quite important.

13.4 Local Density Functional Theory of Wetting Transitions

Because of the agreement between the wetting trends predicted by gradient theory and nonlocal density functional theory, we believe local density functional theories suffice for exploring the wetting behavior further in this chapter. Again we consider a one-component fluid at a flat solid surface. We choose the van der Waals approximation,

$$\mu = \mu_h(n(x)) + \int \bar{u}_A(x - x')n(x')dx' + v(x), \qquad (13.4.1)$$

where $\mu_h(n)$ is the chemical potential of a homogeneous hard sphere fluid of density n. We use the usual van der Waals formula for μ_h,

$$\mu_h(n) = \frac{kT}{1 - nb} + kT \ln\left(\frac{n}{1 - nb}\right), \qquad (13.4.2)$$

and study the effects of the forms of the fluid–fluid intermolecular potential and the fluid–wall potential on the nature of wetting transitions.

An especially simple model, which led to a surprising result, was introduced by Sullivan (1981). He assumed that the attractive part of the fluid–fluid pair potential is of the form $u_A(r) = -(a/2\pi r d^2)\exp(-r/d)$, and so

$$\bar{u}_A(x) = -(ad)e^{-|x|/d}. \qquad (13.4.3)$$

He assumed that the fluid–wall potential is of the form

$$\begin{aligned} v(x) &= \infty, \ x < 0 \\ &= -\varepsilon_w e^{-xd}, \quad x > 0. \end{aligned} \qquad (13.4.4)$$

Sullivan made the further assumption that $d = d_w$. We refer to the model defined by eqs. (13.4.3) and (13.4.4) as the exponential model and to Sullivan's special case as Sullivan's exponential model.

The consequence of the equality $d = d_w$ is that eq. (13.4.1) can be converted by differentiation to the equation

$$\sigma^2 \frac{d^2\mu_h}{dx^2} = \mu_h - \mu - 2an. \qquad (13.4.5)$$

Because μ_h is a monotonic function of n, n is a monotonic function of μ_h. Thus, eq. (13.4.5) can be solved by quadrature, which is why Sullivan's exponential model turns out to be so simple.

The difference between eq. (13.4.1) and its first derivative evaluated at $x = 0$ yields the boundary condition

$$\sigma \frac{d\mu_h}{dx}\bigg|_{x=0} = \mu_h(x=0) - \mu - 2\varepsilon_w \equiv Y(\mu_h(0)). \qquad (13.4.6)$$

The other boundary condition is that

$$\mu_h \to \mu_h(n_B) \text{ as } x \to \infty. \qquad (13.4.7)$$

Multiplying eq. (13.4.5) by $d\mu_h/dx$ and integrating the result, we obtain

$$\sigma^2 \left(\frac{d\mu_h}{dx}\right)^2 = \psi(\mu_h) \text{ or } dx = \frac{\sigma\, d\mu_h}{\pm\sqrt{\psi(\mu_h)}}, \qquad (13.4.8)$$

where

$$\psi(\mu_h) = (\mu_h - \mu)^2 - 4a(P_h - P) \qquad (13.4.9)$$

and

$$P = P_o(n_B) = P_h(n_B) - (n_B)^2 a. \qquad (13.4.10)$$

The Gibbs–Duhem equation, $dP_h/d\mu_h = n$, has been used and the properties $\mu = \mu_h(n_B) - 2n_B a$ and $P_h(n_B) - P = -(n_B)^2 a$ guarantee that $\psi(\mu_h) \to 0$ as $x \to 0$, that in turn guarantees that the boundary condition at eq. (13.4.6) is satisfied. The quantity μ_h $(x = 0)$ is determined from the boundary condition

$$Y(\mu_h(0)) = -\pm\sqrt{\psi(\mu_h(0))}, \tag{13.4.11}$$

where the branch $+\sqrt{\psi}$ or $-\sqrt{\psi}$ used in solving for μ_h versus x (from which n versus x can then be computed) is determined by the requirement that x increase from 0 to ∞ as μ_h varies from $\mu_h(0)$ to $\mu_h(n_B)$.

For $\varepsilon_w > kT_c/2$, Sullivan found the disjoining potential isotherms illustrated in Figure 13.15a. Like the models studied in Sections 13.2 and 13.3, Sullivan's model predicts perfect wetting transition at a temperature T_w lying below the critical point of the fluid. However, the model does not predict a prewetting line, that is, according to the models there is no film–film coexistence curve and the *wetting transition is second order* (the derivative $\partial\gamma/\partial\mu$ is continuous at T_w). The situation is illustrated in Figure 13.15b. Consider what happens when the bulk density of vapor is increased isothermally toward the coexistence curve. Below T_w, the thin adsorbed film on the solid changes relatively little until the coexistence curve is reached. Then, further isothermal compression results in formation of a droplet of liquid whose meniscus has a contact angle with the solid lying between 0° and 180°. Above T_w, isothermal compression from the low vapor density state results in the continuous thickening of a uniform film until the coexistence density is reached and a perfect wetting layer of liquid forms on the solid. Unlike the first-order wetting transition, the temperature derivative of the contact angle of a droplet of liquid on the solid is a continuous function of temperature as T_w is approached along the coexistence curve.

In Figure 13.16, the wetting transition temperature for Sullivan's model is shown as a function of $2\varepsilon_w/kT_c$. For $2\varepsilon_w/kT_c < 1.5$,

Figure 13.15
(a) Schematic diagram of disjoining potential isotherms for Sullivan's exponential model. (b) Phase diagram illustrating second-order wetting transition. Adapted from Davis et al. (1986).

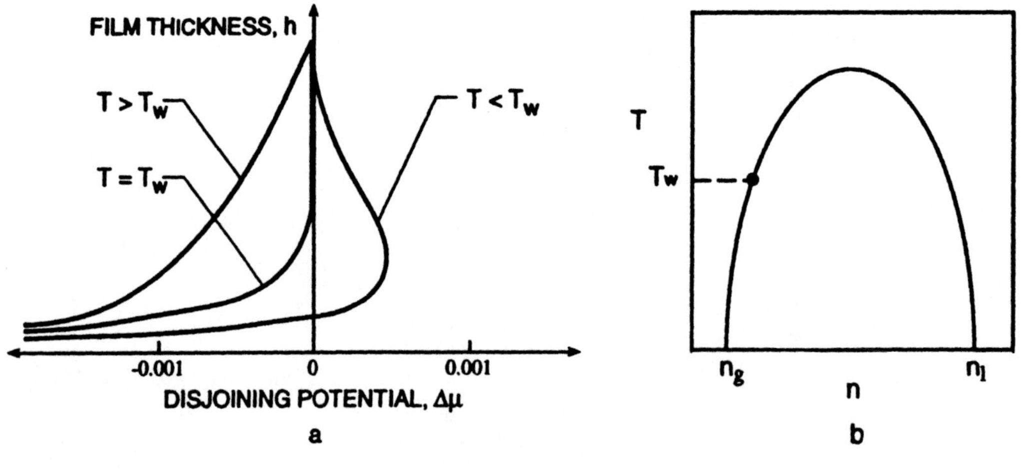

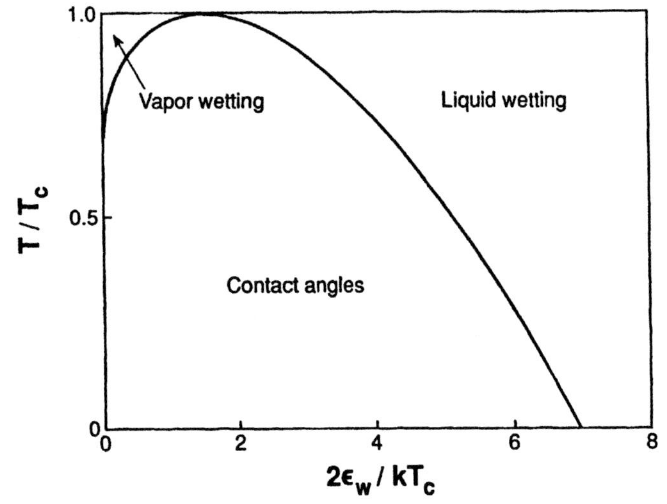

Figure 13.16
Wetting transition temperature as a function of wall–fluid interaction according to Sullivan's theory. Adapted from Davis et al. (1986).

the vapor is the wetting phase (sometimes this is called a drying transition), whereas for $2\varepsilon_w/kT_c > 1.5$, liquid is the wetting phase.

According to Sullivan's model the transition from contact angle forming to perfect wetting by the liquid phase occurs at $an_l(T_w) = 2\varepsilon_w$ and from contact angle forming to perfect wetting by the vapor phase occurs at $an_v(T_w) = 2\varepsilon_w$. Higher values of wall attraction ε_w favor liquid wetting over vapor wetting. In the ε_w range where vapor is the wetting phase, T_w increases with increasing ε_w; in the ε_w range where liquid is the wetting phase, increasing T_w decreases with increasing ε_w. The rather obvious conclusion is that increasing wall attraction (usually thought of as higher energy solids) favors liquid wetting. The behavior seen in Figure 13.16 is qualitatively similar to that predicted by Lennard–Jones interaction models.

That Sullivan's exponential interaction model predicted a second-order wetting transition was surprising when it was presented. The model is very special and unrealistic in ways. First of all, attractive interaction potentials are better represented by inverse power laws than by exponentials. Second, the ranges of the wall–fluid interactions and the fluid–fluid interactions are unlikely to be the same. This second issue can be easily investigated while maintaining much of the simplicity of Sullivan's model (see Teletzke et al., 1983). For the exponential model defined by eqs. (13.4.1)–(13.4.4) differentiation of eq. (13.4.1) twice again generates a differential equation:

$$\sigma^2 \frac{d^2\Xi}{dx^2} = \Xi + 2an \qquad (13.4.12)$$

where

$$\Xi = \mu - \mu_h - v. \qquad (13.4.13)$$

The boundary conditions are

$$\frac{d\Xi}{dx}\Big|_{x=0} = \Xi\Big|_{x=0}; \quad \Xi \to 2an_B \quad \text{as} \quad x \to \infty. \tag{13.4.14}$$

The interfacial tension for the exponential model is

$$\gamma = -\int_{x_0}^{\infty} \left[P_h(n(x)) - P_o(n^B) + \frac{1}{2} n(x)\Xi(x) \right] dx, \tag{13.4.15}$$

where $x_o = 0$ for a solid–fluid interface and $x_0 = -\infty$ for a liquid–vapor interface. For a liquid–vapor interface $v = 0$ and the boundary conditions become $\Xi \to 2an_l^{\text{sat}}$ as $x \to -\infty$ and $\Xi \to 2an_g^{\text{sat}}$ as $x \to \infty$.

The wetting behavior found by Teletzke et al. for the exponential model is shown in Figure 13.17. Several interesting features emerge. The most important is that when the range of the fluid–wall potential is greater than that of the fluid–fluid potential ($\sigma_w/\sigma > 1$), the wetting transition can be either first or second order. Larger σ_w/σ or ε_w favors a first-order transition. There is a region in the upper right-hand corner of Figure 13.17 where liquid wets the solid at all temperatures. In this region the energy benefit of high wall attraction and/or long range of the wall attraction favors perfect wetting at any temperature. Along the curve separating the liquid wetting and the vapor wetting regions the wetting temperature T_w equals the liquid–vapor critical temperature T_c.

The disjoining potential isotherms corresponding to the first-order wetting transition region are shown in Figure 13.18. As expected for a first-order wetting transition, there is prewetting line or film–film coexistence curve for this system. The "van der Waals loop" in the isotherm results from subtle, competing effects. To gain a heuristic appreciation of some of these, let us examine the behavior of a thick thin film.

A simple approximation to a thick liquid like film in the presence of a bulk vapor phase at density n_B is

$$n(x) = n_f, \quad 0 < x < h$$
$$= n_B, \quad h < x < L. \tag{13.4.16}$$

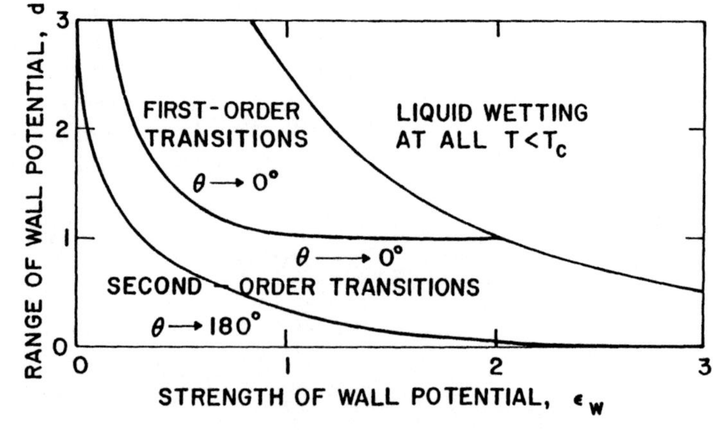

Figure 13.17
Wetting behavior predicted by the exponential model as a function of range and strength of the fluid–solid interaction potential. $T_w = T_c$ along the curve separating the region of liquid setting ($\theta = 0°$) from the region of vapor wetting ($\theta = 180°$). Adapted from Davis et al. (1986).

Figure 13.18
Schematic plot of disjoining isotherms of the modified Sullivan model in the first-order wetting transition parameter region. Adapted from Davis et al. (1986).

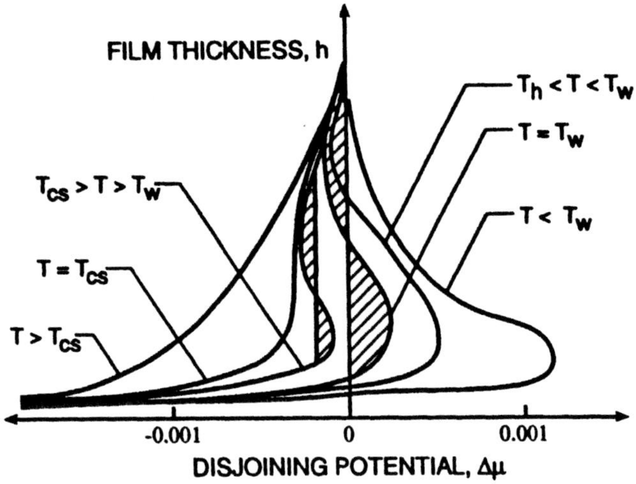

The film density n_f will be approximately equal to the saturated liquid density $n_l(T)$. The van der Waals free energy functional for a planar system is

$$F = A \int_0^L [f_h(n) + nv]dx + \frac{A}{2} \int_0^L \int_0^L n(x)n(x')\bar{u}_A(x, x')dxdx', \qquad (13.4.17)$$

where $f_h(n)$ is the free energy density of a homogeneous fluid of hard spheres.

With the density profile given at eq. (13.4.16) and the disjoining potential calculated from

$$\Delta\mu = \frac{1}{A(n_f - n_B)} \left[\frac{\partial F}{\partial h} - \frac{\partial F(h = \infty)}{\partial h} \right], \qquad (13.4.18)$$

we obtain the thick thin film results given in Table 13.2 for several models of wall–fluid and fluid–fluid attractive potentials.

The thick-film disjoining potential for Sullivan's model is of the form

$$\Delta\mu_{\text{TF}} = (-\varepsilon_w + an_f)e^{-h/d}. \qquad (13.4.19)$$

If temperature is low $an_f > \varepsilon_w$ and so the asymptotic tail of $\Delta\mu$ goes to zero monotonically with positive values as $h \to \infty$. At high enough temperature $an_f < \varepsilon_w$ and so $\Delta\mu$ goes to zero monotonically with negative values. This asymptotic behavior is consistent with the second-order wetting transition and the disjoining potential displayed in Figure 13.15. On the other hand, the thick-film disjoining potential of the exponential model is

$$\Delta\mu_{TF} = -\varepsilon_w e^{-h/d_w} + an_f e^{-h/d}. \qquad (13.4.20)$$

If $d_w \neq d$, $\Delta\mu_{\text{TF}}$ does not approach zero monotonically. This is a necessary condition for an isotherm of $\Delta\mu$ to have a van der Waals loop and thereby to admit film-film coexistence and a first-order wetting transition. The thick-film potential for $d_w \neq d$ suggests the possibility of a first-order wetting transition, whereas for $d_w = d$ it suggests a second-order transition. As indicated in Figure 13.17, whether a first- or second-order transition is predicted for the exponential model is a delicate balance between the relative ranges d_w/d and strengths $\varepsilon_w/\varepsilon$ of the wall–fluid and fluid–fluid interactions.

The wetting behavior of several combinations of wall–fluid and fluid–fluid attractive potentials have been studied with van der Waals density functional theory. The results are indicated in Table 13.2. We have discussed in detail the exponential model. For an inverse power wall–fluid and exponential fluid–fluid interactions (model C) the transition is first-order. For the 6–12 Lennard–Jones system (model E with $\mu = \nu - 1 = 3$), the transition is predicted to be first order. We also saw in the previous section that both computer simulations and a nonlocal density functional theory predict a first-order wetting transition for the 6–12 Lennard–Jones system. In this case the thick-film disjoining potential has the form

$$\Delta\mu_{TF} = \frac{a_H d^3}{6\pi h^3}, \tag{13.4.21}$$

where the Hamaker constant a_{H} equals $6\pi(an_f - \varepsilon_w)$ according to Table 13.2.

Because an inverse power law is a better model than the exponential for fluid–fluid interactions, one is tempted to conclude from Table 13.2 that the wetting transitions in nature will likely be first-order. However, such an inclination has to be tempered by the results of a study carried out by Dietrich and Schick (1985) on the lattice gas model. By assuming the interaction potentials consist of a combination of inverse power laws [e.g., $v(x) = c_1/x^4 + c_2/x^5$

TABLE 13.2 Disjoining Chemical Potential of Thick Film Approximated by Step Profile

Model	*Asymptotic Interaction Potential* *Wall–Fluid* $[u_w(x)]$	*Fluid–Fluid* $[\bar{u}_A(x, x')]$	*Thick Film Disjoining Potential* $(\Delta\mu)$	*Order of Wetting Transition*[a]
A	$-\varepsilon_w e^{-x/d}$	$-ae^{-\lvert x-x'\rvert/d}$	$\left[-\varepsilon_w + an_f\right] e^{-h/d}$	2nd
B	$-\varepsilon_w e^{-x/d_w}$	$-ae^{-\lvert x-x\rvert/d}$	$-\varepsilon_w e^{-h/d_w} + an_f e^{-h/d}$	2nd, $d \leq 1$; $\varepsilon_w > \varepsilon_w^*(d)$
C	$-\dfrac{\varepsilon_w d^\mu}{x^\mu}$	$-ae^{-\lvert x-x\rvert/d}$	$-\dfrac{\varepsilon_w d^\mu}{h^\mu} + an_f e^{-h/d}$	1st
D	$-\varepsilon_w e^{-x/d}$	$-\dfrac{a(\nu-1)d^{\nu-1}}{\lvert x-x'\rvert^\nu}$	$-\varepsilon_w e^{-h/d} + \dfrac{an_f d^{\nu-1}}{h^{\nu-1}}$	Unknown
E	$-\dfrac{\varepsilon_w d^\mu}{x^\mu}$	$-\dfrac{a(\nu-1)d^{\nu-1}}{\lvert x-x'\rvert^\nu}$	$-\dfrac{\varepsilon_w d^\mu}{h^\mu} + \dfrac{an_f d^{\nu-1}}{h^{\nu-1}}$	1st, $\mu = \nu - 1 = 3$; unknown otherwise

d is the fluid molecule diameter and d_w is the wall–fluid molecule interaction length scale.

and $\bar{u}_A(x) = c_3/x^3 + c_4/x^4$], they found that the wetting transition would be first or second order, depending on the values of the parameters (e.g., of $c_1, \ldots, c_4$).

13.5 Experimental Studies of Wetting Transitions

Wetting transitions are common and fairly easy to observe. A small drop of liquid placed at an interface will either completely wet or take on a lenticular (fluid–liquid interface) or pancake (fluid–solid interface) shape with equilibrium contact angles or contact angle. According to theory, the complete wetting transition will occur as the critical point is approached. An example of the wetting transition in a gas–liquid–liquid system is shown in Figure 13.19.

In the absence of water, the methanol-rich phase perfectly wets the air–cyclohexane-rich interface at all compositions and temperatures in the liquid–liquid coexistence region. However, upon addition of a small amount of water to the solution, the liquid–liquid critical point rises rapidly and a wetting transition emerges as shown in the right-hand panel of Figure 13.19. For example, at 1.25 wt % water, the perfect wetting transition temperature T_w is 56°C and the solution critical point is 80°C. Below 56°C a drop of the methanol-rich phase placed at the interface of air and the cyclohexane forms a lens with contact angles. Above 56°C the drop perfectly wets the interface. As water content increases the solution critical point and the perfect wetting temperature increase.

According to theory the wetting transition occurs as the critical point is approached from any direction. For example, if a system has a lower solution critical point the wetting transition will occur as the temperature is lowered. This case is illustrated by 2,6-lutidine and water on glass. The phase diagram and the wetting transition

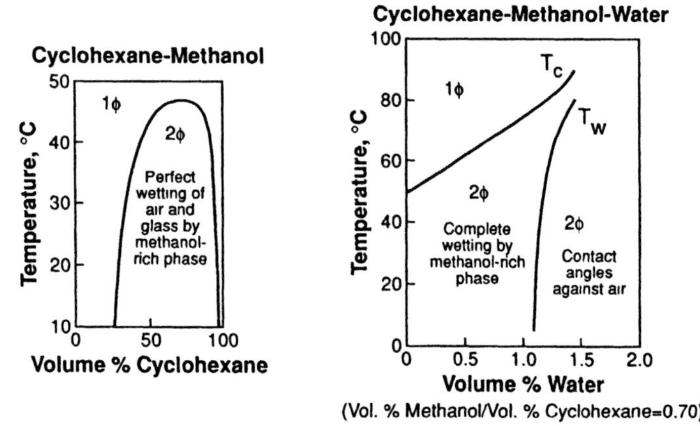

Figure 13.19
Liquid–liquid binodal of the cyclohexane–methanol solution. The right panel shows the effect of adding water on the wetting of the air–cyclohexane-rich phase by the methanol-rich phase. Adapted from Moldover and Cahn (1980) and Davis et al. (1986).

Figure 13.20
Liquid–liquid binodal of the lutidine–water mixture. As the lower critical point is approached from above, a temperature is reached at which the lutidine-rich phase completely wets the interface of the glass and the water-rich phase. (Wetting temperature was determined by Pohl and Goldburg, 1982.)

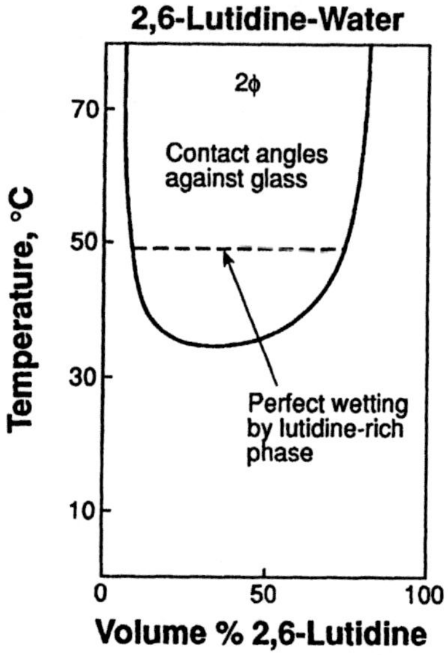

temperature are shown in Figure 13.20. At the lower solution critical point $T_c \simeq 35°C$. The complete wetting transition occurs at $T_w \approx 45°C$.

Any field variable that tends toward a critical point can drive the wetting transition. In fact, in Figure 13.19 water activity is a field that can be used to cause the wetting transition at fixed temperature. For example, at about 80°C, the methanol-rich phase is nonwetting above a water content of 1.4 wt % and is completely wetting below a water content of 1.4 wt %. Thus, the wetting transition water activity corresponds to 1.4 wt % at 80°C. Even "generalized field variables," such as alkane carbon number (which is the same in all coexisting phases), can be varied to accomplish the wetting transition. We saw such an example in Chapter 7, where we described Zisman's work on contact angles of normal alkanes on a Teflon surface in the presence of air. In the case of octane and larger molecular weight alkanes a drop on teflon forms a contact angle. Hexane and smaller molecular weight alkanes completely wet teflon. Thus, the alkane number of the wetting transition is about 7 at room temperature. As alkane number decreases critical temperature of the alkane decreases. Thus, the alkane number acts as a field variable that approaches the critical point as it decreases.

Although the wetting point is relatively easy to observe, establishing that the transition is first-order is a harder matter. Schmidt and Moldover (1983) investigated the film thickness (with ellipsometry) and contact angle behavior in an air, perfluoroheptene (C_7F_{14}), and isopropyl alcohol ($i{-}C_3H_7OH$) system. They observed a wetting transition at $T_w = 311$ K (at the critical point $T_c = 363$ K). On the basis of the temperature dependence of the film thickness and a contact angle, they concluded that the wetting transition is first-order. However, they did not observe the prewetting line.

In a recent ellipsometric study of methanol and cyclohexane Kellay et al. (1993) determined the prewetting line for adsorption at the air–liquid interface. Their prewetting line is shown in Figure 13.21. Note that in agreement with theory in the previous section, the prewetting line is close to the binodal. They observed a first-order wetting transition at $T_w \approx 22°C$ and a surface critical point at $T_{cs} \approx 39°$ C. Kellay et al. do not report the water content in their work, but in the light of the curves shown in Figure 13.19, it appears that their system contained about 1.1 wt % water.

A first order wetting transition has been reported recently for liquid helium at a helium vapor–cesium interface. Using a quartz microbalance technique that detects tiny masses of adsorbed fluids on solids, Taborek and Rutledge (1992) measured the helium adsorption isotherm on a cesium surface at 2.7 K. At a vapor pressure of 131 Torr, there is a sudden increase in adsorbed mass characteristic of a prewetting transition. Thus, they found direct evidence of film–film coexistence and verified that the wetting transition is first-order for the system. Their result is shown in Figure 13.22. Frequency shift is to be interpreted as mass or film thickness. The film thickens continuously into a layer of liquid phase as the vapor pressure approaches its liquid–vapor saturation value. No one has reported experimental determination of a system undergoing a second-order or continuous wetting transition. The situation for first-order transitions is summarized in Figure 13.23. The implications of the pattern is that wettability can be controlled by approaching (for complete wetting) or retreating (for contact angle) from a critical point. Although the pattern depicted in Figure 13.23 evolved largely because of interest stimulated by Cahn's theory of critical point wetting, it is noteworthy that the wetting transition usually occurs relatively far from the critical point.

Figure 13.21
Prewetting line (dashed curve) for adsorption at the air–cyclohexane-rich phase interface. The solid curve is the cyclohexane-rich part of the methanol–cyclohexane binodal. The figure was provided by D. Bonn.

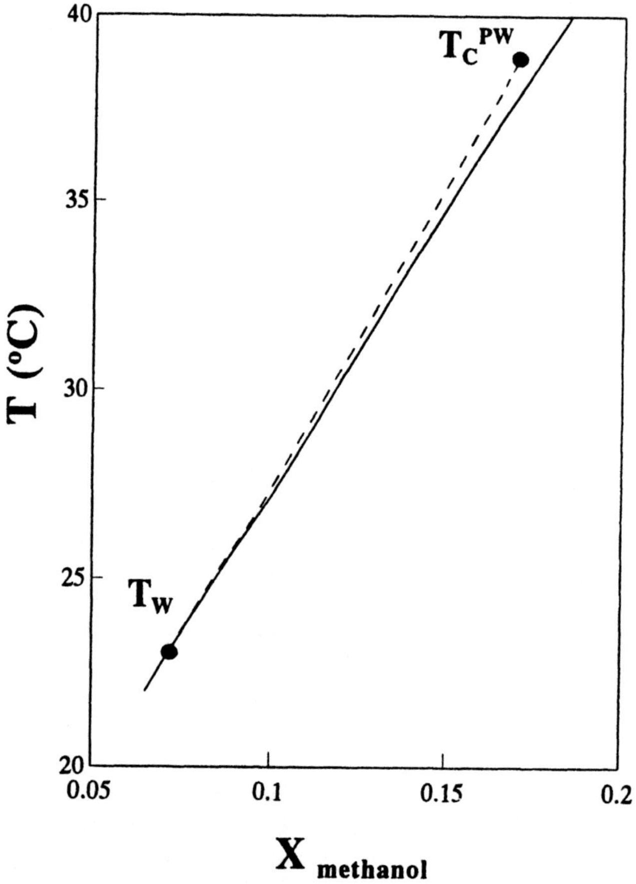

Figure 13.22
Frequency shift (and hence mass adsorbed) versus vapor pressure for helium adsorbed onto a cesium surface at 2.7 K. Adapted from Taborek and Rutledge (1992).

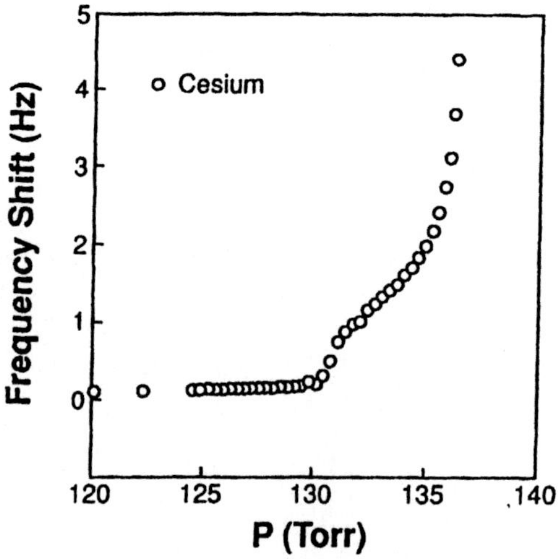

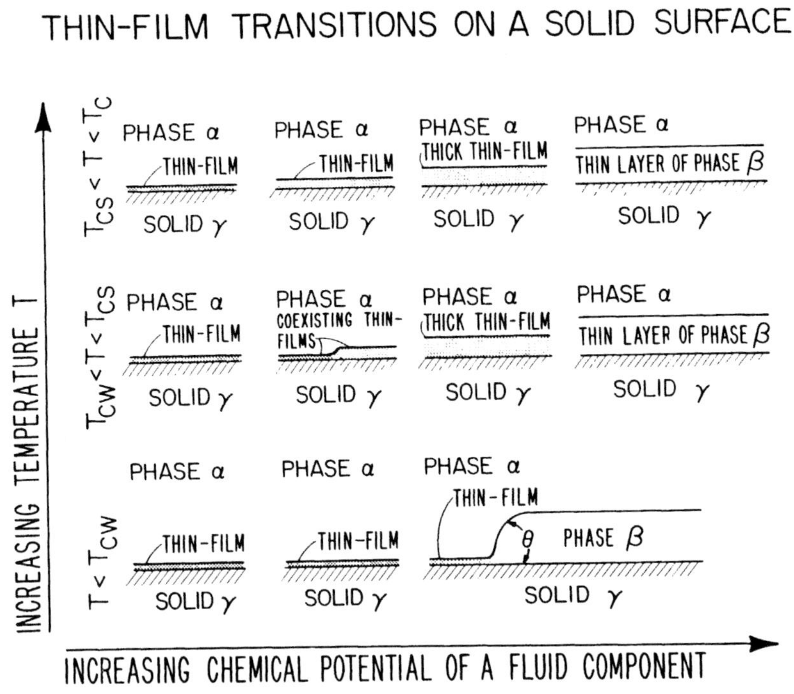

Figure 13.23
Patterns of wetting as a field variable approaches critical and as the activity of a chemical component approaches a third phase. The example is illustrative of an upper critical point. For a lower critical point temperature would decrease toward T_c and the inequalities would become $T_c < T_{cs} < T_w$. Adapted from Davis et al. (1986).

Supplementary Reading

Henderson, D., Ed. 1992. *Fundamentals of Inhomogeneous Fluids*, Dekker, New York.

Safran, S. A. 1994. *Statistical Thermodynamics of Surfaces, Interfaces and Membranes*, Addison-Wesley, Reading, MA.

Teletzke, G. F., Scriven, L. E., and Davis, H. T. 1982. *J. Colloid and Interface Sci.* **87**, 550.

Teletzke, G. F., Davis, H. T., and Scriven, L. E.1988. *Rev. Phys. Appl.*, **23**, 989.

Exercises

1 Consider Sullivan's exponential model. Calculate and plot density profiles and disjoining pressure isotherms for various values of $2\varepsilon_w/kT_c$ in the regime where vapor is the wetting phase ($2\varepsilon_w/kT_c < 1.5$).

2 Same as Problem 13.1 but use density gradient theory.

3 Plot versus film thickness the thick film disjoining potentials, $\Delta\mu$, in Table 13.2 for the parameters $n_f a = 2kT$ and various values of ε_w/kT and d_w/d. Plot $\Delta\mu$ and h in the units kT and d, respectively.

References

Davis, H. T., Benner, R. E., Scriven, L. E., and Teletzke, G. F., K. L. Mittal and P. Bothorel, Eds. 1986. *Surfactants in Solution,* Vol. 6, Plenum, New York.

Dhawan, S., Reimel, M. E., Scriven, L. E., and Davis, H. T. 1991. *J. Chem. Phys.*, **94**, 4479.

Dietrich, S., and Schick, J. 1985. *Phys. Rev. B*, **31**, 4718.

Finn, J. E., and Monson, P. A. 1989. *Phys. Rev. A*, **39**, 6402.

Girifalco, L. A., and Good, R. J. 1959. *J. Phys. Chem.*, **61**, 904.

Kellay, H., Bonn, D., and Meunier, J. 1993. *Phys. Rev. Lett.*, **71**, 2607.

Moldover, M. R., and Cahn, J. W. 1980. *Science*, **207**, 1073.

Pohl, D. W., and Goldburg, W. I. 1982. *Phys. Rev. Lett.*, **48**, 1111.

Schmidt, J. W., and Moldover, M. R. 1983. *J. Chem. Phys.*, **79**, 379.

Sullivan, D. E. 1981. *J. Chem. Phys.*, **74**, 2604.

Taborek, P., and Rutledge, J. E. 1992. *Phys. Rev. Lett.*, **68**, 2184.

Teletzke, G. F., Scriven, L. E., and Davis, H. T. 1982. *J. Colloid Interface Sci.*, **87**, 550; 1983. *J. Chem. Phys.*, **78**, 1431.

Turkovich, L. Private communication, 1995.

14

TERNARY AMPHIPHILIC SYSTEMS

Michael Schick

14.1 Introduction: The Systems and Their Behaviors

We are concerned in this chapter with liquid mixtures of three components: oil, water, and a surfactant. The latter is a molecule that, by design, has a head group that is strongly attracted to water. The tail of the molecule, often composed of one or two hydrocarbon chains, is more attracted to the oil. Because such a molecule is attracted to both oil and water, it is often referred to as an amphiphile, from the Greek "loving both." As a result of its design, amphiphiles tend to self-assemble. To see this, consider a system of pure water to which is added a small amount of amphiphile. The amphiphiles will arrange themselves in a small, spherical assembly in which the hydrophilic head groups are pointing outward to the water, while the hydrophobic tails are inside, shielded from the energetically unfavorable contact with water. This assembly is called a micelle. If a small amount of oil were added to this system, it would go to the center of the micelles. Again because of their construction, it is energetically favorable for amphiphiles to be located at interfaces. If the system of pure water is in a beaker in equilibrium with air, the first amphiphiles to be added to the system will first go to the air–water interface at which the head group will stick into the water, and the tail groups will be in the air, as was illustrated in Figure 7.16. As more amphiphiles are added, most of them will go to the interface, while a smaller amount will remain in solution. Eventually, the interface will be saturated, and the addition of further amphiphiles will cause a large increase in the number in solution, where they will form micelles. The concentration at which this occurs is called the critical micelle concentration or cmc. If instead of a water-air system we consider a water–oil system that, because oil and water do not mix, will phase separately, similar effects will occur. Amphiphile added to this system will first go to the oil–water interface. However, something new can happen if more amphiphile is added to this system. Once the first interface is saturated, additional internal interfaces can be *created* by the amphiphiles causing the water and oil to mix macroscopically, while being separated microscopically by sheets of the amphiphile. This disordered mixture of *coherent* regions of water and *coherent* regions of oil separated by sheets of amphiphile is called a microemulsion. It is an isotropic, fluid phase, but one characterized by a complex structure. In it, the oil and water have been solubilized, and a measure of the strength of an amphiphile is the concentration of it in the microemulsion: only a small concentration of a good, or strong, amphiphile is required to form the microemulsion, but a larger concentration of a weaker one is needed.

If one continues to add amphiphile to the microemulsion, other phases often appear. The sheets may order themselves in a lamellar phase, for example. A correlation found experimentally is that lamellar phases occur in systems containing a strong amphiphile, and do not appear in systems with weaker ones. Other phases that can appear are cylindrical and cubic phases. In the former, cylindrical micelles, with amphiphile heads out and tails in, are arranged in a triangular close packing; in the latter, spherical micelles are packed in a cubic structure. Both of these phases have counterparts in which the concentration of oil is greater than that of water, so that all micelles are "inverted" with tails out and heads in.

Let us flesh out the above introduction with a discussion of some specific amphiphiles, the phase behavior of the ternary system, their scattering behavior, and their interfacial properties.

Although there are many amphiphiles employed commercially, one particular homologous series has been found particularly useful in experimental investigations because the strength of the amphiphile can be systematically changed. This series is that of the *n*-alkyl polyglycol ethers, $H(CH_2)_i(OCH_2CH_2)_jOH$, usually denoted C_iE_j. Dodecylpentaethoxy alcohol, mentioned previously in Section 7.4, is then C_2E_5. The oxyethylene group is hydrophilic because of the presence of the easily polarized oxygen atom; the hydrocarbon tail is hydrophobic. The number of each group can be varied independently to produce a variety of amphiphiles, weak and strong, preferring oil, water, or both equally. C_6E_3 is a weak amphiphile. In a mixture of equal weight fractions of decane and water, one requires a mole fraction of 0.11 to solubilize oil and water. $C_{12}E_6$ is much stronger, requiring only 0.008 mol fraction to solubilize the oil and decane.

The affinity of these amphiphiles for oil and water is temperature dependent. At lower temperatures, the amphiphile is more soluble in water than oil. As indicated in Figure 14.1, two phases are observed to coexist; one is almost a pure oil phase, the other is a water-rich phase that contains most of the amphiphile. As the temperature is increased, a third phase appears. It contains more

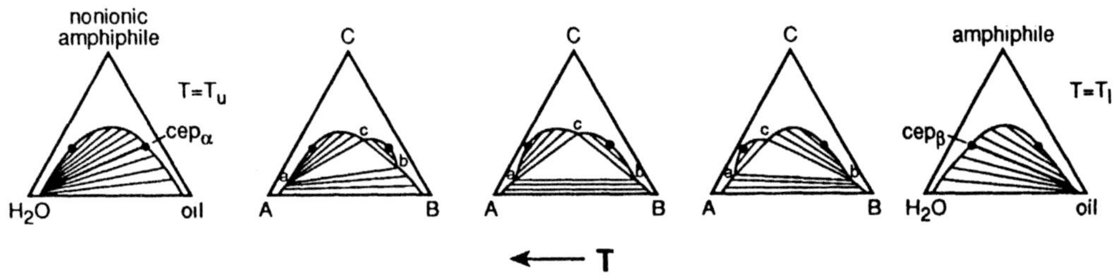

Figure 14.1
Phase behavior of the ternary system as a function of temperature. The critical end points occur at the temperatures T_l and T_u.

amphiphile than the other two, and amounts of water and oil intermediate between the other two phases. Because of its intermediate density, it will be found between the other two phases. Hence, it is often called a "middle phase." If the system produces a microemulsion, it is from the middle phase that it will emerge. As the temperature is increased further, the concentration of oil in the middle phase increases and that of water decreases. At the point where their concentrations are equal, the system is referred to as balanced. Increasing the temperature still further, we increase the concentration of oil in the middle phase, and the middle phase eventually disappears. We are left with two phases again, but now they are a nearly pure water phase that coexists with an oil-rich phase containing most of the amphiphile. The point at which the middle phase and upper oil-rich phase come together is a critical point, of the kind examined in Section 5.4. Because it takes place in the presence of another fluid phase, which acts only as a spectator, it is given the special name "critical end point." Similarly, on reducing the temperature, we come back to the point at which the middle phase and the lower water-rich phase become critical in the presence of the spectator oil-rich phase. This is also a critical end point. For a water–octane-$C_{12}E_6$ system, the two critical end points occur at about 43 and 47°C. There is three-phase coexistence between these temperatures. As the amphiphile is made weaker, the difference between these two end point temperatures first increases; they are at about 36 and 52°C if the amphiphile is C_6E_3. But if the amphiphile is made still weaker, the temperature interval over which three-phase coexistence is observed decreases and eventually disappears at a point at which all three phases become identical. Hence this point is denoted a tricritical point. This phase behavior, in which one has a line of three-phase coexistence bounded by two critical end points, is common in systems of three components. It is not particular to amphiphilic systems.

Much information about the nature of the middle phases comes from X-ray or small angle neutron scattering experiments, of the kind discussed in Chapter 10. Figure 14.2 shows the results of small angle neutron scattering on the middle phase in the system water, toluene, and sodium dodecylsulfate (SDS). Recall from eq. (10.2.4) that the scattering intensity is directly related to the structure factor $S(K)$. The figure shows that the middle phase has a characteristic length l where $l = 2\pi/K_{max}$ and K_{max} is the

Figure 14.2
Small angle neutron scattering intensity versus wavector K. Curves are labeled by the concentration of amphiphile in grams per milliliter. Adapted from Auvray et al. (1984).

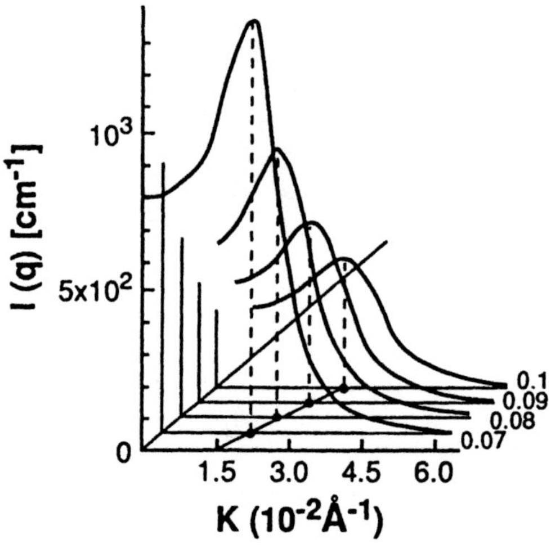

wave vector at which the peak in $S(K)$ occurs. The four curves are labeled by the concentration of amphiphile. We note that as the amphiphile concentration increases, the location of the peak moves to larger K_{max}. This is evidence for the idea that the amphiphiles exist as sheets. The reasoning is as follows. We assume that these sheets form with a definite surface density Γ, much as a two-dimensional liquid would have a definite number of molecules per unit area. The number of amphiphile molecules per unit volume, n, is not fixed, however. From this number per unit volume, n, and number per unit area, Γ, we can form the natural length $l \equiv \Gamma/n$ and natural wave vector $K_{max} = 2\pi/l = 2\pi n/\Gamma$ that increases linearly with the density of amphiphile, and thus with its concentration. This is in agreement with Figure 14.2. From that figure, we also see that $K_{max} \approx 0.03$ Å^{-1} so that the characteristic size of the coherent regions of water and of oil is on the order of 200 Å. This is much larger than the molecular sizes of any of the components. One of the tasks of any theory is to produce a large length such as this.

More direct visualization of the microemulsion can be obtained from the process of freeze fracture electron microscopy in which the liquid sample is quickly frozen and, after several intermediate steps of preparation, is scanned. One such result taken in the system octane, water–$C_{12}E_5$, is shown in Figure 14.3. The stippled regions are octane, and the more uniform gray regions are water. One sees just how complex this fluid is. The bar in the figure is 1000 Å (see also Jahn and Strey, 1988).

Perhaps the most dramatic effect of a good amphiphile is the extreme reduction in the oil–water tension that it can bring about. We know from Section 7.4 that the addition of any surfactant will decrease the oil–water tension, and some such reductions were given in Table 7.3. The reductions shown there are modest, on the order of a factor of 2–3. In contrast, the decrease in tension that a good amphiphile can bring about is of the order of 2–3 *orders of magnitude*. For example, when water and octane coexist with

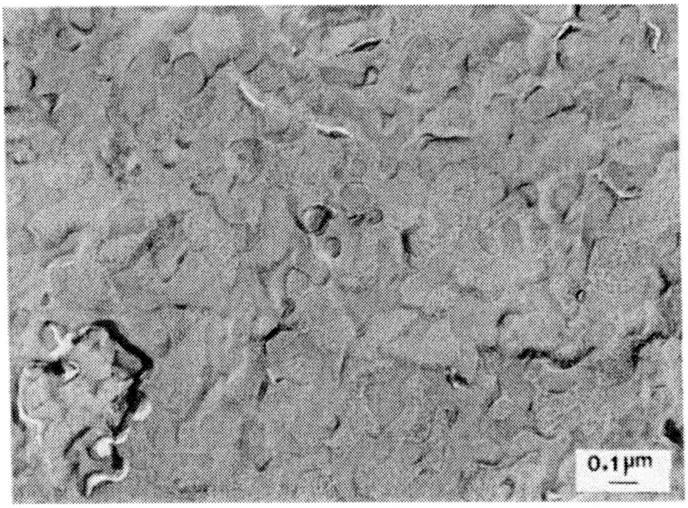

Figure 14.3
Freeze fracture image of the microemulsion in the system water–octane–$C_{12}E_5$. Forty percent of the total weight of oil and water is due to the oil (Courtesy of Reinhard Strey.)

the microemulsion formed by C_8E_4, the oil water tension is only 0.04 dyn/cm! To understand this incredible reduction is another of our tasks. The behavior of all three interfacial tensions is shown schematically in Figure 14.4. Note that the interfacial tension vanishes between two phases that become identical at a critical end point. This must be so as the distinction between the two phases vanishes there. But the oil–water tension does not vanish there. In fact it is usually lowest somewhere between the two end points. Its magnitude is limited from above, of course, by the inequality [eq. (7.5.9)]

$$\gamma_{ab} \leq \gamma_{ac} + \gamma_{bc}, \tag{14.1.1}$$

where a, b, and c stand for water-rich, oil-rich, and middle phases. If the equality holds, the middle phase wets the oil–water interface. The only lower limit to its magnitude is zero.

There is also an intriguing correlation between the wetting behavior of the middle phase and the strength of the amphiphile in

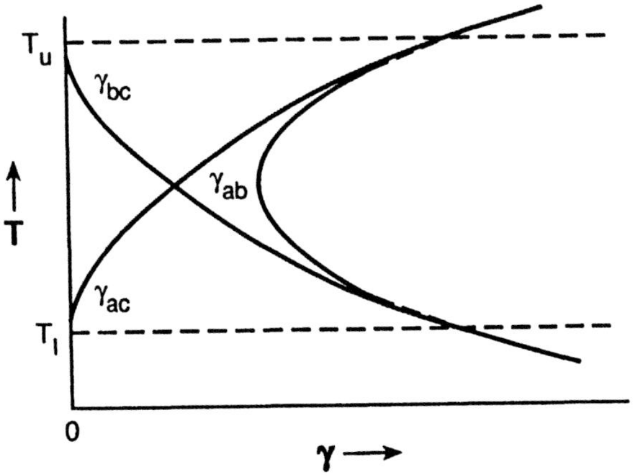

Figure 14.4
Schematic representation of the temperature dependence of the three interfacial tensions.

the system: strong amphiphiles produce middle phases that do not wet the water–oil interface; weak ones produce middle phases that do. This is another behavior that we shall attempt to reproduce. To proceed, we attempt to write down a partition function for a three-component system in which one of the components is an amphiphile.

14.2 Lattice Gases and Lattice Fluids

It is the purpose of this section to show that the partition function of a liquid-gas system, or of a binary fluid system, can be written in an approximate form identical to that of the Ising model of magnetism discussed in Section 3.7 and 3.8. The advantage of this result is that the Ising model is a lattice model, as opposed to a continuum model, which simplifies its treatment enormously. Furthermore, there is a wealth of knowlege about such models that can be of great assistance to an understanding of the original problem. We begin with the grand partition function of a simple liquid.

In Section 2.6, we saw that the grand partition function of a classical system containing N identical molecules in a volume V could be written

$$\Xi(T, \mu, V) = \sum_N Q_N(T, V) e^{\beta\mu N}, \tag{14.2.1}$$

where $\beta \equiv 1/kT$. The canonical partition function, $Q_N(T, V)$ is, from eq. (2.9.6),

$$Q_N(T, V) = \frac{1}{N!h^{3N}} \int \cdots \int e^{-\beta H_{tot}} d^3r_1 \cdots d^3p_N, \tag{14.2.2}$$

where H_{tot} is the total Hamiltonian of the system, a function of all the particle coordinates and momenta. If, as usual, the momenta enter only in the kinetic energy

$$H_{tot} = \sum_j \frac{p_j^2}{2m} + H(\mathbf{r}_1, \ldots, \mathbf{r}_N), \tag{14.2.3}$$

then the integrals over the momenta can be carried out immediately yielding

$$Q_N(T, V) = \frac{Z_N(T, V)}{\Lambda^{3N} N!}, \tag{14.2.4}$$

where Λ is the thermal de Broglie wavelength

$$\Lambda \equiv h/(2\pi mkT)^{1/2}, \tag{14.2.5}$$

and Z_N the "configurational partition function"

$$Z_N(T, V) \equiv \int \cdots \int e^{-\beta H} d^3r_1 \cdots d^3r_N. \tag{14.2.6}$$

If the potential energy is just the sum of pair potentials, then

$$H = \sum_{\langle \mathbf{r}_l, \mathbf{r}_m \rangle} V(\mathbf{r}_l - \mathbf{r}_m), \tag{14.2.7}$$

where $\sum_{\langle \mathbf{r}_l, \mathbf{r}_m \rangle}$ means a sum over all distinct pairs of coordinates.

To formulate the lattice gas model, we divide the total volume V into $\mathcal{N}$ cells of volume v_0 in which one, and only one, particle can be found. We label the cells by discrete indices that we write as i, j, etc. with i standing for a triplet of indices to define the location of the cell in three-dimensional space. Integrals over continuous coordinates are replaced by sums over the lattice

$$\int d\mathbf{r} \rightarrow v_0 \sum_i . \tag{14.2.8}$$

We introduce the site occupation variable

$$\begin{aligned} n_i &= 1 \quad \text{if site } i \text{ is occupied,} \\ &= 0 \quad \text{if site } i \text{ is unoccupied.} \end{aligned} \tag{14.2.9}$$

Then

$$N = \sum_i n_i, \tag{14.2.10}$$

and

$$H(\mathbf{r}_1 \cdots \mathbf{r}_N) = \sum_{\langle \mathbf{r}_l, \mathbf{r}_m \rangle} V(\mathbf{r}_l - \mathbf{r}_m) \rightarrow H(n_1 \cdots n_N) = \sum_{\langle ij \rangle} V_{ij} n_i n_j, \tag{14.2.11}$$

where V_{ij} is the potential between two molecules when one is in cell i and the other in j. No more than single occupancy of any cell is guaranteed by making V_{ii} infinitely large. We make these replacements in the grand partition function, and use one more important observation; that the set of integers $n_1 \cdots n_N$ completely specifies one configuration of the lattice, but that this configuration of the lattice represents $N!$ configuration of the particles, because they can all be interchanged. The grand partition function becomes

$$\Xi(T, \mu, V) = \sum_{n_i = 0,1} \exp \left\{ \beta[\mu - kT \ln(\Lambda^3 / v_0)] \sum_i n_i - \sum_{\langle ij \rangle} V_{ij} n_i n_j \right\}. \tag{14.2.12}$$

This is almost in the form of the partition function of an Ising model in a magnetic field. To bring it to this form explicitly, we introduce the spin variable $s_i = \pm 1$ via

$$n_i = \frac{1 + s_i}{2}, \tag{14.2.13}$$

so that a spin "up" at site i means that there is a particle there, while spin "down" means that there is not. Making this replacement in eq. (14.2.12) we obtain

$$\Xi(T, \mu, V) = \sum_{s_i=\pm 1} \exp\left[\beta(c + h\sum_i s_i + \sum_{\langle ij \rangle} J_{ij} s_i s_j)\right], \tag{14.2.14}$$

where

$$J_{ij} = -V_{ij}/4, \tag{14.2.15}$$

$$\begin{aligned} h &= (\mu - \mu_0)/2, \\ &= \left[\mu - kT \ln(\Lambda^3/v_0) - (1/2)\sum_{j \neq i} V_{ij}\right]/2, \end{aligned} \tag{14.2.16}$$

and

$$c = \left[\mu - kT \ln(\Lambda^3/v_0) - (1/4)\sum_{j \neq i} V_{ij}\right]\mathcal{N}/2. \tag{14.2.17}$$

The right-hand side of eq. (14.2.14) is just $e^{\beta c} Z_{\mathcal{N}}$, where $Z_{\mathcal{N}}(T, h)$ is the partition function of an Ising model of $\mathcal{N} \equiv V/v_0$ spins in a magnetic field h, a system with energy $H - h\sum_i s_i$ where the Hamiltonian is

$$H = -\sum_{\langle ij \rangle} J_{ij} s_i s_j, \tag{14.2.18}$$

a simple generalization of that encountered previously in Section (4.7). Thus, we have mapped, approximately, the liquid–gas system onto a magnetic system. If we solve one, we have solved the other.

A similar mapping applies to a binary liquid mixture. We again divide space into cells, and require that there be either an A or B molecule in each cell. The occupation of each cell is given by the occupation variables

$$\begin{aligned} n_i^A &= 1 \quad \text{if site } i \text{ is occupied by an } A \text{ molecule}, \\ &= 0 \quad \text{if site } i \text{ is unoccupied by an } A \text{ molecule}, \end{aligned} \tag{14.2.19}$$

and similarly for n_i^B. The Hamiltonian of the system is

$$H = \sum_{\langle ij \rangle} V_{ij}^{AA} n_i^A n_j^A + V_{ij}^{AB}(n_i^A n_j^B + n_i^B n_j^A) + V_{ij}^{BB} n_i^B n_j^B. \tag{14.2.20}$$

The two occupation variables can be written in terms of the single spin variable $s_i = \pm 1$ according to

$$\begin{aligned} n_i^A &= \frac{(1 + s_i)}{2}, \\ n_i^B &= \frac{(1 - s_i)}{2}, \end{aligned} \tag{14.2.21}$$

that guarantees that there is always one, and only one molecule in each cell. Repeating the analogous steps of the above derivation, we find that the grand partition function of the binary mixture has the same form as eq. (14.2.14) but that now

$$J_{ij} = -(V_{ij}^{AA} + V_{ij}^{BB} - 2V_{ij}^{AB})/4, \tag{14.2.22}$$

$$h = \left[\mu^A - \mu^B - kT\ln(\Lambda_A^3/\Lambda_B^3) - (1/2)\sum_{j\neq i}(V_{ij}^{AA} - V_{ij}^{BB})\right]/2, \tag{14.2.23}$$

and

$$c = \left[\mu^A + \mu^B - kT\ln(\Lambda_A^3/v_0)(\Lambda_B^3/v_0) - (1/4)\sum_{j\neq i}(V_{ij}^{AA} + V_{ij}^{BB} + 2V_{ij}^{AB})\right]\mathcal{N}/2. \tag{14.2.24}$$

We can perform a similar mapping in the case of a ternary mixture, a mapping we will need in the oil–water–amphiphile system. We label the three different molecules as A, B, and C. In describing ordinary ternary systems, it is often assumed that the Hamiltonian governing it contains only forces between pairs of molecules, and is therefore a generalization of eq. (14.2.20),

$$H = \sum_{\alpha,\beta}\sum_{ij} V_{ij}^{\alpha\beta} n_i^{\alpha} n_j^{\beta}, \tag{14.2.25}$$

where the indices α and β take the values A, B, and C. This is easily mapped onto a problem of interacting spins S_i in which these spins take the three values $\pm 1, 0$. We would like $S_i = 1$ to denote that the site is occupied by a molecule of type A, $S_i = -1$ that it is occupied by type B, and $S_i = 0$, by type C. This is accomplished by the relations

$$\begin{aligned} n_i^A &= S_i(1 + S_i)/2, \\ n_i^B &= -S_i(1 - S_i)/2, \\ n_i^C &= 1 - S_i^2. \end{aligned} \tag{14.2.26}$$

If we now repeat our derivation, we find that the partition function of the ternary system is, to within a multiplicative factor $e^{\beta c}$, the same as that of a spin 1 model coupled to two fields h and Δ. The energy of a given configuration of spins is $H - \sum_i (hS_i - \Delta S_i^2)$ where the Hamiltonian is

$$H = -\sum_{\langle ij\rangle}[J_{ij}S_iS_j + K_{ij}S_i^2S_j^2 + C_{ij}(S_iS_j^2 + S_i^2S_j)], \tag{14.2.27}$$

and

$$\begin{aligned} 4J_{ij} &= -(V_{ij}^{AA} + V_{ij}^{BB} - 2V_{ij}^{AB}), \\ J_{ij} + K_{ij} + 2C_{ij} &= -(V_{ij}^{AA} + V_{ij}^{CC} - 2V_{ij}^{AC}), \\ J_{ij} + K_{ij} - 2C_{ij} &= -(V_{ij}^{BB} + V_{ij}^{CC} - 2V_{ij}^{BC}). \end{aligned} \tag{14.2.28}$$

The magnetic field h is related to the difference in chemical potential between A and B species

$$h = \left[\mu^A - \mu^B - kT \ln(\Lambda_A^3/\Lambda_B^3) - \sum_{j \neq i} (V_{ij}^{AC} - V_{ij}^{BC}) \right] /2, \qquad (14.2.29)$$

while Δ is related to the chemical potential of species C

$$\Delta = \left[2\mu^C - \mu^A - \mu^B - kT \ln(\Lambda_C^6/\Lambda_A^3 \Lambda_B^3) + \sum_{j \neq i} (V_{ij}^{AC} + V_{ij}^{BC} - 2V_{ij}^{CC}) \right] /2. \qquad (14.2.30)$$

As is commonly done, we restrict pair interactions to be between nearest neighbors only so that

$$\begin{aligned} J_{ij} &= J \quad \text{if } i, j \text{ are nearest neighbor sites,} \\ &= 0 \quad \text{otherwise,} \\ K_{ij} &= K \quad \text{if } i, j \text{ are nearest neighbor sites,} \\ &= 0 \quad \text{otherwise,} \\ C_{ij} &= C \quad \text{if } i, j \text{ are nearest neighbor sites,} \\ &= 0 \quad \text{otherwise.} \end{aligned} \qquad (14.2.31)$$

Although quite acceptable for ordinary ternary mixtures, it is not at all obvious that this Hamiltonian will suffice for amphiphilic systems. It is unclear that it distinguishes the amphiphile (species C) from water (A) and oil (B), and that it correctly describes the physical interaction by which the amphiphile gains energy if it sits between water and oil. To put this mechanism in explicitly, we will add to the Hamiltonian eq. (14.2.25) a term whose effect is that the energy is reduced by an amount $|L|$ if the amphiphile sits between an oil and a water, and is increased by the same amount if it sits between two waters or two oils. We hope that as the magnitude of L increases, the model amphiphile will become a more efficient solubilizer of oil and water. This is something that we will have to check later. The term we add is

$$H_{\text{AMP}} = L \sum_{\langle ijk \rangle} (n_i^A n_j^C n_k^B + n_i^B n_j^C n_k^A - n_i^A n_j^C n_k^A - n_i^B n_j^C n_k^B), \qquad (14.2.32)$$

where the sum is over all triplets of adjoining sites that are in a line, and $L < 0$. When expressed in terms of the spin variables via eq. (14.2.26), the term becomes

$$H_{\text{AMP}} = -L \sum_{\langle ijk \rangle} S_i (1 - S_j^2) S_k. \qquad (14.2.33)$$

The final form of the Hamiltonian that we consider for the system of water, oil, and amphiphile is given by eq. (14.2.27) with the nearest neighbor interactions of eq. (14.2.31) and the additional term of eq. (14.2.33). We write it again for convenience,

$$H = -\sum_{\langle ij \rangle} [J S_i S_j + K S_i^2 S_j^2 + C(S_i S_j^2 + S_i^2 S_j)] - L \sum_{\langle ijk \rangle} S_i (1 - S_j^2) S_k, \qquad (14.2.34)$$

where the first sum is now over distinct nearest neighbor pairs only. From this Hamiltonian we can, in principle, calculate the partition function and all the thermodynamic properties of the system. In practice, we can only do this in some manner of approximation. In the next section, we introduce the mean field approximation, apply it first to the Ising model, and then to our model of ternary mixtures.

14.3 Mean Field Theory

Let us begin with a few exact statements from Section 2.5 concerning our system in a canonical ensemble. A particular configuration of the system is specified by the values of all of the $\mathcal{N}$ spins in the system. The energy of the configuration is given by the Hamiltonian [eqs. (14.2.18) or (14.2.34) for example]. The probability of observing a particular state q is given by

$$p_q = \frac{e^{-\beta H_q}}{Q_{\mathcal{N}}}, \tag{14.3.1}$$

where

$$Q_{\mathcal{N}} = \sum_q e^{-\beta H_q}, \tag{14.3.2}$$

and H_q denotes the energy of the qth state. The partition function $Q_{\mathcal{N}}$ gives the Helmholtz free energy F according to

$$F = -kT \ln Q_{\mathcal{N}}. \tag{14.3.3}$$

From thermodynamics, the Helmholtz free energy is related to the energy U and entropy S by

$$F = U - TS. \tag{14.3.4}$$

The energy U is the average value of the energies of all configurations

$$U = \sum_q p_q H_q, \tag{14.3.5}$$

and from eq. (5.5.56) it follows that the entropy is the average value of $-k \ln p_q$

$$S = -k \sum_q p_q \ln p_q. \tag{14.3.6}$$

Therefore, the Helmholtz free energy can be written as an average value

$$F = \sum_q p_q (H_q + kT \ln p_q). \tag{14.3.7}$$

For the Ising model in which each spin takes two possible values, the number of configurations of the $\mathcal{N}$ spins is $2^{\mathcal{N}}$, and p_q can be thought of as the qth element of a diagonal $2^{\mathcal{N}} \times 2^{\mathcal{N}}$ matrix p. Each element of this matrix is the probability of observing a particular configuration, a probability that depends on the energy H_q of that configuration. H_q can also be thought of as the qth element of a diagonal $2^{\mathcal{N}} \times 2^{\mathcal{N}}$ matrix, in which case eq. (14.3.7) can be thought of as a matrix equation, and the sum over configurations q as a trace of the matrix:

$$F = \mathrm{Tr}\ p(H + kT \ln p). \tag{14.3.8}$$

Average values of a particular spin are difficult to compute because the behavior of any given spin is linked to that of its neighbors via the Hamiltonian. In mean field theory, all spins are taken to act independently of the others, an assumption that facilitates calculation. The Helmholtz free energy within mean field approximation is generated by replacing the exact matrix p in (14.3.8) by an approximate one, p_{mft}, in which all spins are independent. The $2^{\mathcal{N}} \times 2^{\mathcal{N}}$ matrix p_{mft} is written as the direct matrix product of $\mathcal{N}$ 2×2 matrices, each of which depends only on one particular spin

$$p_{\mathrm{mft}} = \prod_i p_i. \tag{14.3.9}$$

In the basis $s_i = 1, -1$,

$$p_i = \begin{pmatrix} \frac{(1+m_i)}{2} & 0 \\ 0 & \frac{(1-m_i)}{2} \end{pmatrix}, \tag{14.3.10}$$

so that

$$\mathrm{Tr}\ p_{\mathrm{mft}} = 1, \tag{14.3.11}$$

as it must. The matrix s_i giving the configuration of the ith spin is a direct product of 2 by 2 unit matrices for all sites except the ith site for which the 2×2 matrix has diagonal elements 1 and -1. Thus the average value of s_i is

$$\mathrm{Tr}\ p_{\mathrm{mft}} s_i = m_i, \tag{14.3.12}$$

which identifies m_i as the average magnetization at site i. It is now simple to calculate the mean field approximation to the Helmholtz free energy

$$F_{\mathrm{mft}} = \mathrm{Tr}\ p_{\mathrm{mft}}(H + kT \ln p_{\mathrm{mft}}). \tag{14.3.13}$$

If we consider the Ising model with nearest neighbor interactions only whose Hamiltonian is

$$H = -J \sum_{\langle ij \rangle} s_i s_j, \tag{14.3.14}$$

we obtain

$$F_{\rm mft} = -J\sum_{\langle ij\rangle} m_i m_j + kT\sum_i \left[\frac{(1+m_i)}{2}\ln\frac{(1+m_i)}{2} + \frac{(1-m_i)}{2}\ln\frac{(1-m_i)}{2}\right]. \qquad (14.3.15)$$

In a single homogeneous phase, the average magnetization at each site is the same, $m_i = m$ for all i, so that the free energy per spin $f = F/\mathcal{N}$ of such a phase is

$$f_{\rm mft}(T, m) = -\frac{Jz}{2}m^2 + kT\left[\frac{(1+m)}{2}\ln\frac{(1+m)}{2} + \frac{(1-m)}{2}\ln\frac{(1-m)}{2}\right], \qquad (14.3.16)$$

where z is the number of nearest neighbors any spin has in the lattice. From this free energy, other thermodynamic information can be extracted. For example, the magnetic field $h(T, m)$ can be obtained from the thermodynamic relation

$$h = \frac{\partial f(T, m)}{\partial m}, \qquad (14.3.17)$$

from which

$$h = -zJm + (kT/2)\ln[(1+m)/(1-m)]. \qquad (14.3.18)$$

This relation can be inverted to obtain $m(T, h)$

$$m = \tanh[(h + zJm)/kT], \qquad (14.3.19)$$

that must be solved for m. There can be more than one solution to this equation, in which case, the solution that gives the lowest value of the Helmholtz free energy, eq. (14.3.16) is the stable one. When the magnetic field is nonzero, the stable solution is unique, but in zero field, there can be three solutions. One of these is $m = 0$, as is readily seen from eq. (14.3.19) when the field is zero. This solution corresponds to the disordered state. To see that there can be additional solutions, we expand $\tanh(zJm/kT)$ for small m so that eq. (14.3.19) becomes, at zero field,

$$m = (zJm/kT) - (zJm)^3/3(kT)^3 + O(m^5). \qquad (14.3.20)$$

In addition to the solution $m = 0$, there are two additional solutions $m = \pm m_0$

$$m_0 = 3^{1/2}\left(\frac{kT}{zJ}\right)^{3/2}\left(\frac{zJ}{kT} - 1\right)^{1/2} \qquad (14.3.21)$$

provided that $T < T_c$ where

$$T_c \equiv zJ/k. \qquad (14.3.22)$$

To determine which solution gives the lower free energy, we expand the Helmholtz free energy, eq. (14.3.16), for small m

$$f_{\rm mft}(T, m) = -kT\ln 2 + \frac{m^2}{2}(kT - zJ) + \frac{kT}{12}m^4 + O(m^6). \qquad (14.3.23)$$

This clearly shows that the solution that minimizes the free energy is $m = 0$ for $T > zJ/k = T_c$, and is $m = \pm m_0$ when $T < T_c$. The latter solutions correspond to ordered states. The value of the transition temperature given by mean field theory for the two-dimensional square lattice, for which $z = 4$ is, from eq. (14.3.22), $T_c = 4J/k$, whereas the exact value, given by eq. (3.8.10) is $2.269J/k$. As seen from (14.3.21). the magnetization of the ordered states m_0, vanishes continuously at the transition temperature T_c, so that this transition is called a continuous one. In contrast, we can prepare the system at $T < T_c$ in a positive magnetic field, in which case the magnetization m will be nonzero. If we reduce the field to zero, m will approach the value $m_0(T)$. If we now make the field slightly negative, the magnetization will jump discontinuously to $m = -m_0$. Such a transition, in which there is a discontinuous jump in the magnetization, is called, for historical reasons, a first-order transition.

Below the transition temperature and in zero field, the two ordered phases, one with magnetization m_0 the other with $-m_0$, can coexist with one another. We can calculate the interfacial tension between them within mean field theory. Let us imagine that the interface between them is in a (100) plane, that is, perpendicular to one of the principal axes of the lattice, say, the x axis. Assume that the $-m_0$ state exists in the negative x direction, and the m_0 state exists in the positive direction. We expect, then, that the magnetization profile will vary smoothly from $-m_0$ to m_0 as we traverse the lattice in the x direction. Let m_i now refer to the magnetization per spin of any spin in the ith plane, because we expect all spins in the same plane perpendicular to the x axis to have the same magnetization. We can obtain an equation for m_i just as we did previously when the magnetization was uniform.

From thermodynamics, the magnetic field at the plane i is given, in analogy to eq. (14.3.17), by

$$h_i = \frac{\partial F}{\partial m_i}. \tag{14.3.24}$$

We use the mean field free energy eq. (14.3.15) in this expression. We must also be more specific about the type of lattice. Let us assume that the lattice is a simple cubic one in three dimensions, so that each spin has $z = 6$ nearest neighbors, of which four are in the same (100) plane, and one each in the plane in front and behind it. Thus, the terms in the first sum of eq. (14.3.15) that involve a particular m_i are $-J(4m_i^2 + m_i m_{i+1} + m_i m_{i-1})$. In the absence of any local magnetic field h_i, eq. (14.3.24) yields

$$m_i = \tanh[(4m_i + m_{i+1} + m_{i-1})/kT], \tag{14.3.25}$$

that clearly reduces to the zero field uniform solution in eq. (14.3.19) if all m_i are equal. Note that eq. (14.3.25) is an infinite set of coupled equations for the infinite number of m_i. They are to be solved subject to the boundary conditions that m_i approach one of the bulk solutions m_0 as the index i goes to plus infinity, and that it approach the other bulk solution $-m_0$ as i approaches minus infinity. In practice, the equations can be solved numerically for a large but finite number of planes, on the order of one hundred. When the solution for $m_i(T)$ is obtained, the interfacial tension can be calculated. This is the difference, per unit area, between the Helmholtz free energy of the system with the spatially varying profile, $m_i(T)$, and the uniform profile $m_0(T)$ or $-m_0(T)$:

$$\gamma_{+-}(T) = \frac{1}{A}[F(T, m_i) - F(T, m_0)], \tag{14.3.26}$$

where A is the area of the (100) plane. It should be noted that the value of the interfacial tension will depend on the orientation of the interfacial planes. This is a real effect in solids. However, because we intend to describe liquids, it is simply an artifact of the lattice model. Fortunately, this orientational dependence is not too large except at low temperatures.

Let us now apply this same mean field theory formalism to the model for ternary mixtures governed by the Hamiltonian, eq. (14.2.34). Because there are three possible configurations of the spin variable S_i, representing occupation by water, amphiphile, or oil, the probability p_q to observe a given configuration q of all $\mathcal{N}$ spins can be thought of as the qth diagonal element of a $3^{\mathcal{N}} \times 3^{\mathcal{N}}$ matrix p. In mean field theory, this matrix is approximated by the direct product of $\mathcal{N}$ 3×3 matrices p_i

$$p_{\text{mft}} = \prod p_i, \tag{14.3.27}$$

where, in the basis $S_i = 1, 0, -1$,

$$p_i = \begin{pmatrix} \frac{Q_i + M_i}{2} & 0 & 0 \\ 0 & 1 - Q_i & 0 \\ 0 & 0 & \frac{Q_i - M_i}{2} \end{pmatrix}. \tag{14.3.28}$$

Note that

$$\text{Tr } p_{\text{mft}} = 1. \tag{14.3.29}$$

The spin S_i is represented by the direct product of 3×3 unit matrices for all spin sites except the ith for which the 3×3 matrix has diagonal elements $1, 0, -1$. Similarly S_i^2 has an analogous representation except that the ith matrix has diagonal elements 1, 0, 1. Therfore we find

$$\text{Tr } p_{\text{mft}} S_i = M_i, \tag{14.3.30}$$

and

$$\text{Tr } p_{\text{mft}} S_i^2 = Q_i, \tag{14.3.31}$$

identifying M_i as the average value of S_i, that, from eq. (14.2.26) is the difference in water and oil concentrations at site i. Similarly,

Q_i is identified as the average value of S_i^2 at site i that from the same equations is one minus the concentration of surfactant at that site. The Helmholtz free energy in the mean field approximation is now easily obtained. It is

$$
\begin{aligned}
F_{\text{mft}} = \\
&- \sum_{\langle ij \rangle} [J M_i M_j + K Q_i Q_j + C(M_i Q_j + Q_i M_j)] \\
&- L \sum_{\langle ijk \rangle} M_i (1 - Q_j) M_k \qquad (14.3.32) \\
&+ kT \sum_i \left[\frac{(Q_i + M_i)}{2} \log \frac{(Q_i + M_i)}{2} + \frac{(Q_i - M_i)}{2} \log \frac{(Q_i - M_i)}{2} \right. \\
&\left. + (1 - Q_i) \log(1 - Q_i) \right].
\end{aligned}
$$

From this free energy, other thermodynamic information can be extracted. For example, the field h_i at site i is obtained from

$$h_i(T, M, Q) = \frac{\partial F(T, M, Q)}{\partial M_i}, \qquad (14.3.33)$$

and the field Δ_i from

$$\Delta_i(T, M, Q) = -\frac{\partial F(T, M, Q)}{\partial Q_i}. \qquad (14.3.34)$$

The difference in sign in these two equations reflects the difference present in the (arbitrary) definition of the field in the energy of a configuration, $H - \sum_i (h_i S_i - \Delta_i S_i^2)$. We can carry out the partial differentiations, and then invert the equations to obtain the M_i and Q_i in terms of the fields. Then we will set the fields to be uniform $h_i = h$, $\Delta_i = \Delta$ for all i. In this way we obtain

$$Q_i = \frac{2\cosh(a_i)}{2\cosh(a_i) + \exp(-b_i)}, \qquad (14.3.35)$$

and

$$M_i = \frac{2\sinh(a_i)}{2\cosh(a_i) + \exp(-b_i)}, \qquad (14.3.36)$$

where

$$a_i = \beta \left[\sum_j J M_j + C Q_j + L(1 - Q_j) M_k + h \right], \qquad (14.3.37)$$

and

$$b_i = \beta \left[\sum_j K Q_j + C M_j - \frac{1}{2} L M_h M_j - \Delta \right]. \qquad (14.3.38)$$

The advantage of proceeding in this way is that we obtain equations that are not only applicable in the uniform case, but also in nonuniform situations, as occurs in the description of interfaces.

From the free energy, the phase diagram can be constructed. Continuous transitions can be found in a manner similar to that employed for the Ising model in the preceding section. Coexistence between phases can be sought systematically as follows. Assume that two homogeneous phases coexist, one characterized by M_1 and Q_1, and the other by M_2, Q_2. The Helmholtz free energy per site $f = F/\mathcal{N}$ of uniform phases is, from eq. (14.3.32)

$$\begin{aligned} f_{\text{mft}}(T, M, Q) = & -\frac{z}{2}[JM^2 + KQ^2 + 2CMQ + LM^2(1 - Q)] \\ & + kT\left[\frac{(Q+M)}{2}\ln\frac{(Q+M)}{2} + \frac{(Q-M)}{2}\ln\frac{(Q-M)}{2}\right. \\ & \left. + (1-Q)\ln(1-Q)\right]. \end{aligned} \quad (14.3.39)$$

The phases must coexist not only at the same temperature T, but also in the presence of the same fields

$$h(T, M, Q) = \frac{\partial f(T, M, Q)}{\partial M}, \quad (14.3.40)$$

and

$$\Delta(T, M, Q) = -\frac{\partial f(T, M, Q)}{\partial Q}. \quad (14.3.41)$$

These conditions yield two equations

$$h(T, M_1, Q_1) = h(T, M_2, Q_2), \quad (14.3.42)$$

$$\Delta(T, M_1, Q_1) = \Delta(T, M_2, Q_2). \quad (14.3.43)$$

The phases must also have the same Gibbs free energy, which provides a third equation

$$F(T, M_1, Q_1) - hM_1 + \Delta Q_1 = F(T, M_2, Q_2) - hM_2 + \Delta Q_2, \quad (14.3.44)$$

in the four unknowns M_1, Q_1, M_2, and Q_2. Similarly, we may look for three-phase coexistence in which case we obtain six equations (two each from equating the fields and Gibbs free energy in two different pairs of phases) for the six unknown compositions. We show in Figure 14.5 the phase diagram for a symmetrical system in which the water interacts with itself as strongly as the oil interacts with itself, $V^{AA} = V^{BB}$, and the amphiphile interacts equally with the water and the oil, $V^{AC} = V^{BC}$. Hence, from eqs. (14.2.28), $C = 0$. The phase diagram is shown in the plane $h = 0$ that implies equal concentrations of oil and water, and at a fixed temperature $kT/J = 4.2$, and $K/J = 0.5$. The diagram is plotted as a function of Δ, which is essentially the amphiphile chemical potential, and the amphiphile strength L. Strong amphiphiles should correspond to large negative L.

At small Δ, corresponding to a low amphiphile concentration, the system phase separates into oil-rich (O) and water-rich (W) phases. This is completely analogous to the transition in the Ising model to magnetic up and down phases. As Δ, and therefore

Figure 14.5
Phase diagram of the model ternary mixture at fixed temperature and equal water and oil concentrations as a function of amphiphile chemical potential and amphiphile strength. Continuous transitions are shown by the dashed line, first order transitions by solid lines. Adapted from Lerczak et al. (1992).

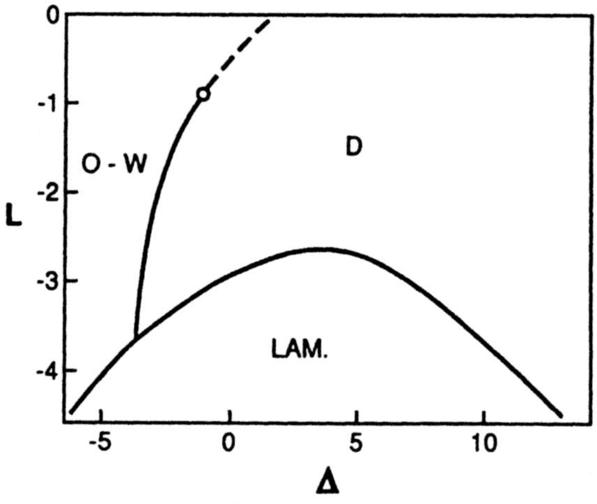

the concentration of amphiphile, increases a transition to a disordered (D) middle phase occurs provided that $|L|$ is not too large. This transition is continuous for weak amphiphiles, and first order for stronger ones, so that there is a coexistence of three-phases. The point that separates these two behaviors is called a tricritical point because here the three-phases undergo a continuous transition at the same point. A lamellar phase (LAM) occurs for large values of $|L|$. In this phase, the various concentrations are not uniform, but vary periodically from one lattice plane to another. They satisfy eqs. (14.3.35) and (14.3.36). In the lamellar phase, the Gibbs free energy evaluated with these nonuniform concentrations is lower than that of any of the other uniform phases. The boundaries of the lamellar phase correspond to the points at which the Gibbs free energy of this phase is equal to that of a neighboring phase. Transitions to the lamellar phase are found to be first order. For sufficiently strong amphiphiles, a point of four-phase coexistence is encountered, below which, the middle phase no longer occurs between the oil and water and lamellar phases. This is precisely the behavior encountered in experiment.

Much of the same phase diagram is shown in Figure 14.6 but is plotted as a function of amphiphile concentration rather than amphiphile chemical potential as in Figure 14.5.

The largest negative value of L/J corresponds to the four-phase point. Note from Figure 14.6 that the concentration of amphiphile in the coexisting middle phase decreases with increasing $|L|$. Thus the strength of this microscopic amphiphilic interaction is, in fact, correlated with the ability of the amphiphile to solubilize oil and water. Furthermore, the oil- and water-rich phases that coexist with the middle phase become increasingly pure as the strength of amphiphile is increased. This is seen in Figure 14.7.

Finally, as the amphiphilic strength increases, a lamellar phase appears at higher concentration. This behavior is again in accord with experiment.

Figure 14.6
The same phase diagram as in Figure 14.5 but shown as a function of amphiphile concentration instead of amphiphile chemical potential.

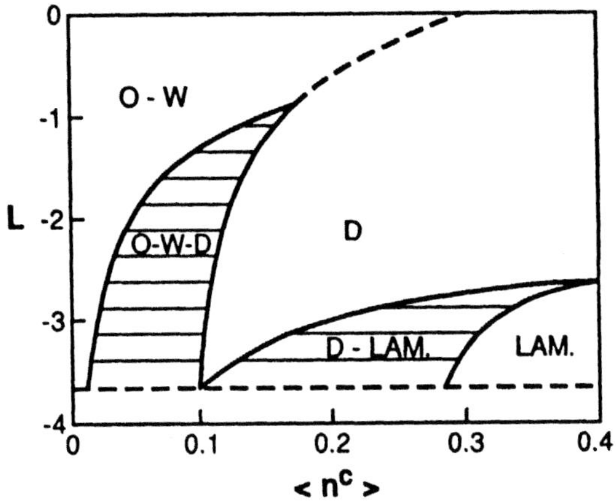

Interfacial profiles can be calculated in this system just as outlined for the Ising model in the previous section. An example is shown in Figure 14.8. There the variation in $\phi_i \equiv \langle n_i^B - n_i^A \rangle$, the local concentration difference of oil and water is shown on passing, lattice plane by lattice plane, from the oil-rich phase to the middle phase. The system is the same as in Figure 14.5. The value of $L/J = -3.5$, which, from Figure 14.5, represents a system quite close to its four-phase point. Note the oscillations in the concentration in the middle phase; in particular there is a water-rich region near the interface with the oil-rich phase. This is due to the presence of amphiphile at the interface. Such excesses have been observed experimentally.

The interfacial profile between oil- and water-rich phases is shown in Figure 14.9. It is clear that the middle phase, for which the local concentration difference is zero, does not wet the oil-water interface. This can be confirmed by calculating the interfacial tensions between the oil and water phases and between the middle and oil phase and middle and water phase. We indeed verify that $\gamma_{ab} < \gamma_{ac} + \gamma_{bc}$. The failure of the middle phase to wet the

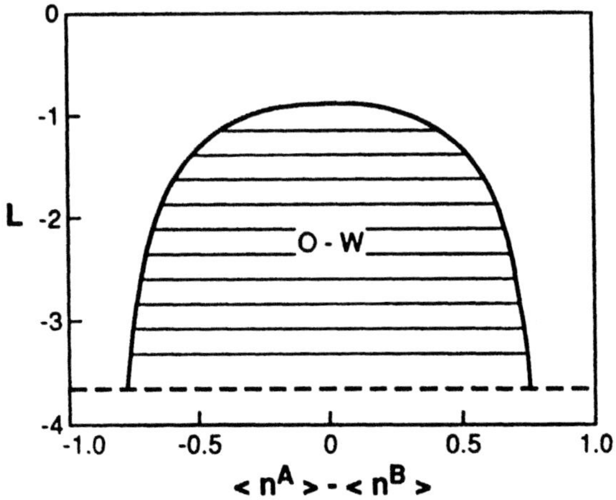

Figure 14.7
Difference in oil and water concentration in the water-rich phase coexisting with the middle phase as a function of amphiphile strength.

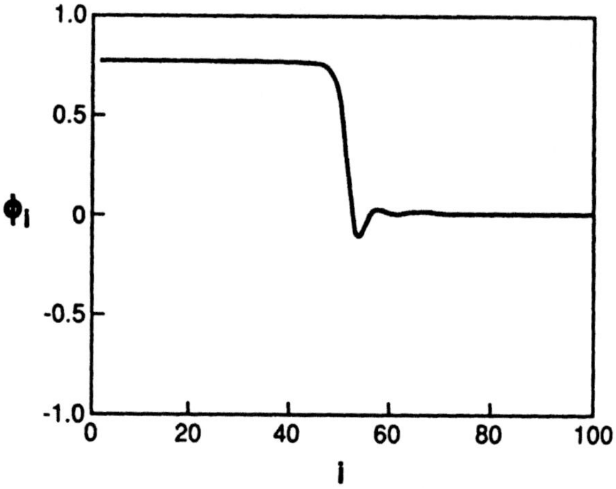

Figure 14.8
Interfacial profile between oil-rich phase and middle phase in a system with a strong amphiphile.

oil–water interface when the middle phase is made from a strong amphiphile is observed experimentally, as noted earlier.

The effect of weakening the amphiphile strength on the oil middle phase profile is shown in Figure 14.10. The value of L has been reduced to $L/J = -1.45$, with all else as in Figure 14.5. There are now no oscillations in the profile, and the middle phase wets the oil–water interface as can be seen in Figure 14.11. A wetting transition occurs in a system with an amphiphile of some strength that is intermediate between these values.

In Figure 14.12, the interfacial tension between oil and water is shown for the system whose phase diagram is shown in Figure 14.5. The tension is shown along the line of three-phase coexistence between oil, water, and middle phases from the tricritical to the four-phase point.

The tension is zero at the tricritical point, rises to a maximum and then falls abruptly to 0.0606 as the four-phase point is approached. The units of the tension are $J/\sqrt{3}d^2$ where d is the lattice spacing. The interface is in the (111) direction, which brings

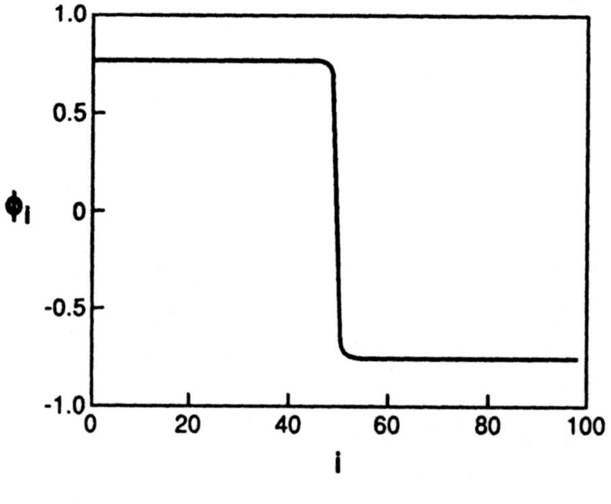

Figure 14.9
Interfacial profile between oil-rich and water-rich phase in the same system as Figure 14.8.

Figure 14.10
Interfacial profile between oil-rich and middle phases in a system with a weak amphiphile.

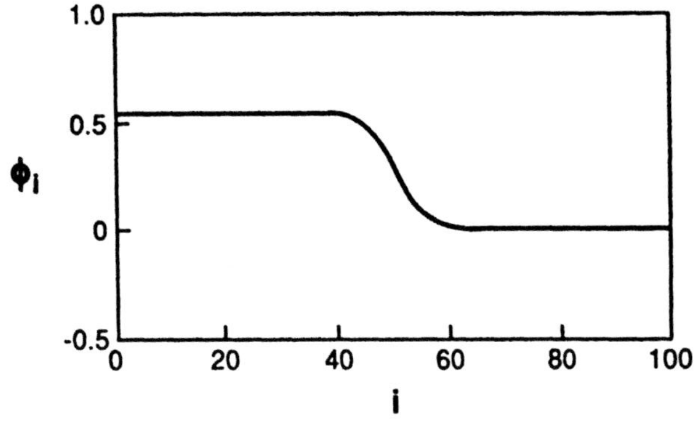

about the factor of $\sqrt{3}$. The value of the tension at the same temperature in the complete absence of amphiphile is 1.71 in these units, so that the reduction we obtain here is about a factor of 30.

The reason for the low oil–water tension along the three-phase coexistence with the middle phase made from a good amphiphile can be understood as follows. The middle phase microemulsion should be thought of as an ensemble of amphiphilic sheets that separate coherent regions of oil and water. The free energy of the system is just the sum of the free energies of the coherent oil regions, the coherent water regions, of the amphiphilic sheets, and of the binding energy of these sheets. Because each sheet separates a large coherent region of water from a similar region of oil, the free energy of each sheet per unit area is simply γ_{ab}, the oil–water interfacial free energy. If the average distance between sheets is L, the free energy of a region of linear extent $2L$ is $L[f_{\text{water}} + f_{\text{oil}}] + 2(\gamma_{ab} - B)$, where B is the binding energy per unit area of the sheets. In three-phase coexistence with the water and oil phases, the free energy of such a region must be the same as a region of the same extent in either of these phase that would be

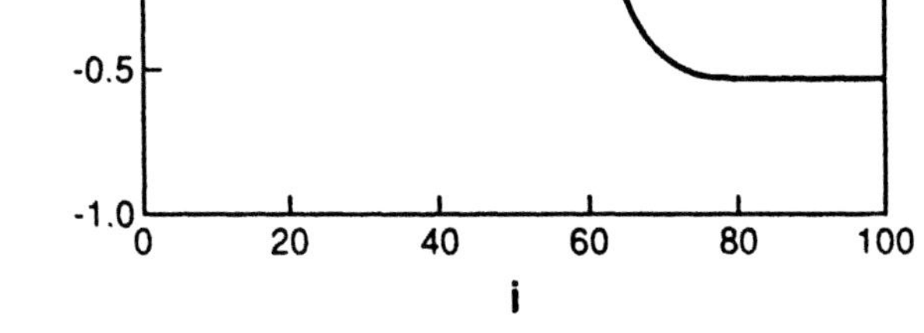

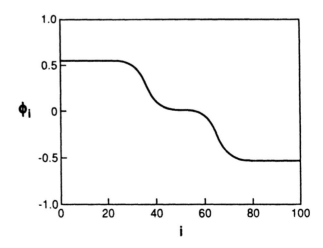

Figure 14.11
Interfacial profile between water- and oil-rich phases in a system with a weak amphiphile.

Figure 14.12
Oil–water interfacial tension along three-phase coexistence with the disordered fluid in the system of Figure 14.5.

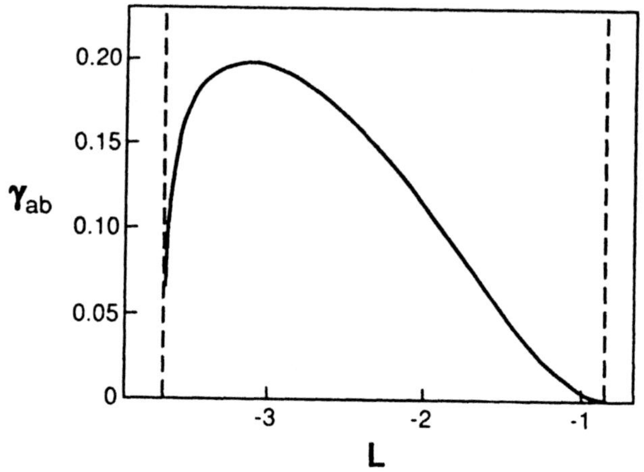

$2Lf_{\text{oil}} = 2Lf_{\text{water}}$ which is equal to $L[f_{\text{water}} + f_{\text{oil}}]$. Equating this free energy to that of the middle phase we obtain

$$\gamma_{ab} = B. \tag{14.3.45}$$

In a good middle phase microemulsion, one in which the amphiphile is a good solubilizer, the coherent regions of oil and water will be large, so that the sheets of amphiphile are far apart. This means that their binding energy must be small and, from (14.3.45), the oil–water tension must be low. It is this effect that we are seeing in Figure 14.12.

Let us summarize the results we have obtained thus far. We have calculated from mean field theory a phase diagram for a symmetric system that resembles that of experimental systems. As we increase the strength of the model amphiphile, its ability to solubilize oil and water increases. For strong amphiphiles, a lamellar phase appears; it eventually preempts the appearance of the disordered middle phase completely. We found that the middle phase made of a weak amphiphile wets the oil–water interface, but that made from a strong amphiphile does not. Finally, after an initial increase, the oil–water tension decreases rapidly as the strength of the amphiphile is increased. All these results are in agreement with experiment. We have interpreted our results using a picture in which the middle phase consists of amphiphilic sheets. It is time to show that this picture of the middle phase as a structured fluid can be supported by calculation of just those quantities that an experimentalist would measure to determine whether the fluid was structured: the structure function of the middle phase itself.

14.4 Structure and Correlation Functions of Middle Phase

In Chapter 15 it is shown that the intensity of scattering at some given angle is directly related to the structure function, $S(\mathbf{K})$ [see eqs. (15.2.2–15.2.4)]. The structure function is, from eq. (15.2.3), just the Fourier transform of the pair correlation function. In other words, it is the average value $\langle n_{\mathbf{K}} n_{-\mathbf{K}} \rangle$, where $n_{\mathbf{K}}$ is the Fourier component of the density. In homogeneous phases, the average values of all $n_{\mathbf{K}}$ vanish except for the $\mathbf{K} = 0$ component. Thus the $n_{\mathbf{K}}$ represent deviations, or fluctuations, about the average value, and $S(\mathbf{K})$ is the average value of the correlations of the fluctuations. The water–water structure function $S_{ww}(\mathbf{K}) = \langle n_{\mathbf{K}}^A n_{-\mathbf{K}}^A \rangle$ is measured in small angle neutron scattering experiments, and it is the quantity we wish to calculate. With water the A component of the ternary mixture, eq. (14.2.26) tells us that

$$\langle n_i^A n_j^A \rangle = \frac{1}{4}(\langle S_i S_j \rangle + \langle S_i^2 S_j^2 \rangle + \langle S_i S_j^2 \rangle + \langle S_i^2 S_j \rangle). \qquad (14.4.1)$$

For the symmetric system with which we have been dealing, the last two terms vanish

$$\langle n_i^A n_j^A \rangle = \frac{1}{4}(\langle S_i S_j \rangle + \langle S_i^2 S_j^2 \rangle). \qquad (14.4.2)$$

In the canonical ensemble in which we have been working, the densities are fixed quantities so that $\langle S_i S_j \rangle = M_i M_j$, $\langle S_i^2 S_j^2 \rangle = Q_i Q_j$ and do not fluctuate. In order to calculate fluctuations about the average values, we must go to the grand canonical ensemble in which the fields that couple to the densities are specified, and the densities fluctuate. We can introduce the Fourier decomposition of these fluctuations so that

$$S_{ww}(\mathbf{K}) = \langle n_{\mathbf{K}}^A n_{-\mathbf{K}}^A \rangle = \frac{1}{4}(\langle M_{\mathbf{K}} M_{-\mathbf{K}} \rangle + \langle Q_{\mathbf{K}} Q_{-\mathbf{K}} \rangle), \qquad (14.4.3)$$

To obtain these averages, we expand M_i and Q_i about the values they attain in the uniform middle phase. For the symmetric system, the average value of M_i is zero, and the average value of Q_i is $1 - \phi_s$, where ϕ_s is the density of surfactant. Thus we write

$$M_i = \sum_q M_q e^{iq \cdot r_i}, \qquad (14.4.4)$$

$$Q_i = (1 - \phi_s) + \sum_q Q_q e^{iq \cdot r_i}, \qquad (14.4.5)$$

where the $q = 0$ term is not included in the sum, and r_i is the position vector of the i'th site. These expressions are substituted into eq. (14.3.32) that is then expanded to second order in the M_q and Q_q with the result

$$F_{\text{mft}} \approx \mathcal{N} f_{\text{mft}}(T, M, Q) + \mathcal{N} \sum_q [\alpha_q M_q M_{-q} + \beta_q Q_q Q_{-q}], \tag{14.4.6}$$

where

$$\alpha_q = \frac{kT}{2(1-\phi_s)} - JD_1(\mathbf{q}) - L(1-Q)D_2(\mathbf{q}), \tag{14.4.7}$$

$$\beta_q = \frac{kT}{2\phi_s(1-\phi_s)} - KD_1(\mathbf{q}), \tag{14.4.8}$$

and

$$D_n(\mathbf{q}) = \cos(nq_x d) + \cos(nq_y d) + \cos(nq_z d), \tag{14.4.9}$$

with d the lattice parameter of the model. In the grand canonical ensemble, the probability distribution is proportional to $\exp\{-\beta(F - \sum_q (h_q M_q - \Delta_q Q_q)]\}$. As there are no physical spatially varying fields h_q, Δ_q at our disposal, we set them to zero. Then we use eq. (14.4.6) to obtain, within mean field theory, the average values we need according to

$$\langle M_{\mathbf{K}} M_{-\mathbf{K}} \rangle = \frac{\prod_{\mathbf{q}} \int M_{\mathbf{K}} M_{-\mathbf{K}} \exp(-\beta F_{\text{mft}}) dM_{\mathbf{q}} dM_{-\mathbf{q}} dQ_{\mathbf{q}} dQ_{-\mathbf{q}}}{\prod_{\mathbf{q}} \int \exp(-\beta F_{\text{mft}}) dM_{\mathbf{q}} dM_{-\mathbf{q}} dQ_{\mathbf{q}} dQ_{-\mathbf{q}}}, \tag{14.4.10}$$

and

$$\langle Q_{\mathbf{K}} Q_{-\mathbf{K}} \rangle = \frac{\prod_{\mathbf{q}} \int Q_{\mathbf{K}} Q_{-\mathbf{K}} \exp(-\beta F_{\text{mft}}) dM_{\mathbf{q}} dM_{-\mathbf{q}} dQ_{\mathbf{q}} dQ_{-\mathbf{q}}}{\prod_{\mathbf{q}} \int \exp(-\beta F_{\text{mft}}) dM_{\mathbf{q}} dM_{-\mathbf{q}} dQ_{\mathbf{q}} dQ_{-\mathbf{q}}}, \tag{14.4.11}$$

The integrals are simply Gaussian, and we obtain

$$S^M(\mathbf{K}) \equiv \langle M_{\mathbf{K}} M_{-\mathbf{K}} \rangle = \frac{kT}{2\alpha_{\mathbf{K}}}, \tag{14.4.12}$$

$$S^Q(\mathbf{K}) \equiv \langle Q_{\mathbf{K}} Q_{-\mathbf{K}} \rangle = \frac{kT}{2\beta_{\mathbf{K}}}. \tag{14.4.13}$$

The structure factors depend on the direction of $\mathbf{K}$ as well as its magnitude reflecting the presence of the lattice, but their qualitative form is independent of direction. We will evaluate them in the (111) direction. Consider first $S^M(\mathbf{K})$. We will show later that this dominates the desired structure function of eq. (14.4.3). From the explicit expression for α_q in eq. (14.4.7), we find that $S^M(\mathbf{K})$ in this direction is given by

$$S^M(1,1,1) = \frac{(kT/J)}{(kT/J)(1-\phi_s)^{-1} - 6\cos(|\mathbf{K}|d/\sqrt{3}) + 6(|L|/J)\phi_s \cos(2|\mathbf{K}|d/\sqrt{3})}. \tag{14.4.14}$$

For a small wave vector, this can be expanded as

$$S^M(1,1,1) = \frac{(kT/J)}{a_2 + c_1|\mathbf{K}|^2 + c_2|\mathbf{K}|^4 + O(|\mathbf{K}|^6)}, \tag{14.4.15}$$

with

$$a_2 = (kT/J)(1 - \phi_s)^{-1} - 6[1 - (\phi_s/4\phi_s^c)], \quad (14.4.16)$$
$$c_1 = [1 - (\phi_s/\phi_s^c)], \quad (14.4.17)$$
$$c_2 = [4(\phi_s/\phi_s^c) - 1]/36, \quad (14.4.18)$$
$$\phi_s^c \equiv J/(4|L|). \quad (14.4.19)$$

This structure function has an extremum when

$$[(\phi_s^c/\phi_s) - \cos(|\mathbf{K}|d/\sqrt{3})]\sin(|\mathbf{K}|d/\sqrt{3}) = 0. \quad (14.4.20)$$

As the amphiphile concentration and the cosine are bounded by unity, the factor in brackets can only vanish if $\phi_s^c \equiv J/(4|L|) \leq 1$. The sine term, of course, always produces an extremum at $K = 0$. Whether this extremum is a maximum or minimum is determined from the second derivative. We find that for weak amphiphilic interactions, $|L| < J/4$, $S^M(\mathbf{K})$ has a maximum at $K = 0$ for all ϕ_s. For a strong amphiphile, $|L| > J/4$, the structure function has its maximum at zero wave vector only for amphiphile concentrations less than the value ϕ_s^c. As ϕ_s increases beyond this value, the position of the maximum moves off of zero wave vector. The position of this maximum increases with amphiphile concentration, as is observed for the structure function $S_{ww}(\mathbf{K})$ in experiment.

The condition that $S_{ww}(\mathbf{K})$ have a peak at a nonzero value of $\mathbf{K}$ provides a convenient defining property of a microemulsion. The line where this first occurs is denoted the Lifshitz line. We emphasize that the free energy per unit volume is analytic on this line so that there is no phase transition when this line is crossed. Thus there is no clear *thermodynamic* distinction between the microemulsion on one side of this line and the ordinary fluid on the other. All correlation functions are also analytic on the Lifshitz line. The experimental distinction is in the dependence of the scattering function on the wave number.

An alternate, and perhaps preferable, definition of the microemulsion is by means of the behavior at large distances of the correlation functions, the Fourier transform of the structure functions. This behavior is determined by the limiting form for small wave vectors of $S^M(\mathbf{K})$, which is quite generally the same as in eq. (14.4.15). From this form for small wave vectors, it follows that the asymptotic form of all correlation functions at large distances is

$$g^{(2)}(r) - 1 \sim r^{-1}\exp(-r/\xi)\sin(2\pi r/\lambda), \quad (14.4.21)$$

where

$$2/\xi^2 = (a_2/c_2)^{\frac{1}{2}} + (c_1/2c_2),$$
$$2(2\pi/\lambda)^2 = (a_2/c_2)^{\frac{1}{2}} - (c_1/2c_2). \quad (14.4.22)$$

The locus of points at which $\lambda^{-1} \to 0$ defines the disorder line. The microemulsion is then defined as that region of the fluid phase in which λ^{-1} is nonzero; the ordinary fluid phase is that region in which λ^{-1} is identically zero. There are no nonanalyticities in the free energy per unit volume of the fluid on this line. Hence there is no phase transition between the microemulsion and the ordinary disordered fluid.

We now consider the other structure function $S^Q(\mathbf{K})$ of eq. (14.4.13). From the explicit expression for $\beta_{\mathbf{K}}$, eq. (14.4.8), we obtain in the (111) direction the expression

$$S^Q(1, 1, 1) = \frac{(kT/J)}{(kT/J)(1-\phi_s)^{-1}\phi_s^{-1} - 6(K/J)\cos(|\mathbf{K}|d/\sqrt{3})}, \qquad (14.4.23)$$

that for small wave vectors, can be expanded as

$$S^Q(1, 1, 1) = \frac{(kT/J)}{\tilde{a}_2 + \tilde{c}_1|\mathbf{K}|^2 + O(|\mathbf{K}|^4)}, \qquad (14.4.24)$$

with

$$\begin{aligned} \tilde{a}_2 &= \frac{(kT/J)}{(1-\phi_s)\phi_s} - 6(K/J) \\ \tilde{c}_1 &= (K/J)a^2. \end{aligned} \qquad (14.4.25)$$

As $\tilde{c}_1$ is positive, $S^Q(\mathbf{K})$ has a maximum at $\mathbf{K} = 0$ and decays monotonically with $|\mathbf{K}|$. We note at this point that the magnitude of S^Q is much smaller than that of S^M due to the presence of the large term kT/ϕ_s in the denominator of the former. Thus, as stated earlier, the behavior of S^M dominates that of the experimentally observable S_{ww} of eq. (14.4.3). The surfactant-surfactant structure function, $S_{ss}(\mathbf{K}) = \langle n_{\mathbf{K}}^C n_{-\mathbf{K}}^C \rangle$ can be measured experimentally, and is easily calculated in mean field theory as it is simply

$$S_{ss}(\mathbf{K}) = \langle Q_{\mathbf{K}} Q_{-\mathbf{K}} \rangle = S^Q(\mathbf{K}). \qquad (14.4.26)$$

In Figure 14.13, we show the water–water structure function for the symmetric system characterized by parameters $kT/J = 4.45$, $K/J = 0.5$, and $L/J = -3.5$. It is shown for five values of the amphiphile concentration, values that span the region of existence of the disordered phase at this temperature, from the triple line to coexistence with the lamellar phase. The wave vector, $\mathbf{K}$ is in the (111) direction, and its magnitude is in units of $\sqrt{3}/d$, where d is the cell size. The edge of the Brillouin zone is $K = \pi$. The magnitude of this peak decreases with increasing amphiphile concentration, while the position of the peak increases. Because our lattice constant must be of a molecular size, it is possible to estimate the absolute values of the peak positions. In particular, if we take the cell size to be $d = 30$ Å, then the peak for $\phi_s = 0.085$ would occur at a physical $K = 3.5 \times 10^{-2}$ Å^{-1}. This is a reasonable value as can be seen by comparison with the experimental results of Figure 14.2. The corresponding wavelength, $2\pi/K$ is 181 Å. Thus we see that the model, while only containing a single length scale of

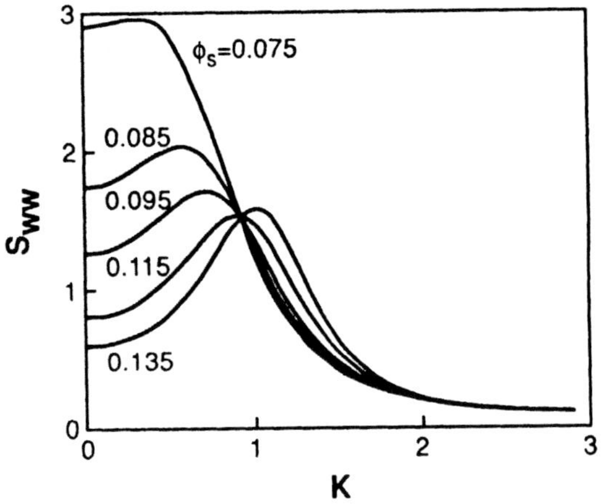

Figure 14.13
Water–water structure function for a balanced system at $kT/J = 4.45$, $K/J = 0.5$, and $L/J = -3.5$ for several different surfactant concentrations. Adapted from Gompper and Schick (1990).

molecular size, produces a water–water structure function that has its peak at a wave vector corresponding to a length much greater than that molecular size.

The surfactant–surfactant structure function is shown in Figure 14.14, and is in qualitative agreement with experiment.

Now that we have been able to calculate the structure functions measured in the scattering experiments, we turn briefly to the calculation of structure in real space, of the form that is observed in freeze fracture images such as those of Figure 14.3.

14.5 Other Structures and Models

It is not difficult to look for solutions of the mean field equations eqs. (14.3.35-14.3.38), that represent cylindrical or cubic phases. Such phases are expected to occur for large amphiphile concentrations. But mean field theory is not capable of giving information on the real space structure of the middle phase, or on the existence of

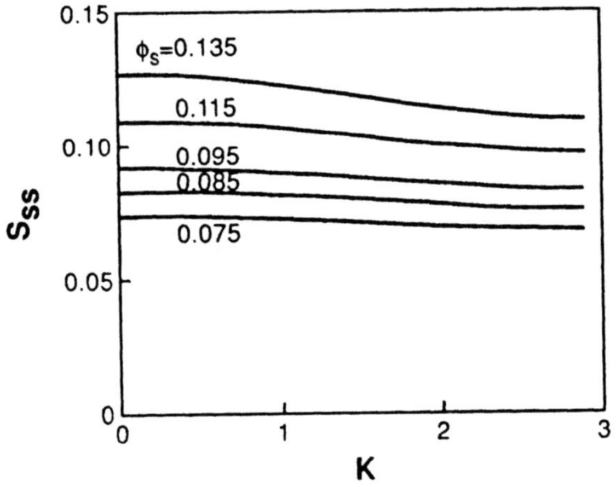

Figure 14.14
Surfactant–surfactant structure function for the same system as in Figure 14.13.

micelles. To extract such information from the model, there is little recourse other than to simulation. A result of a simulation of the spin one model in two dimensions is shown in Figure 14.15. The middle phase is shown there with plus signs representing water, diamonds representing surfactant, and empty spaces representing oil. We see that the fluid is indeed structured, and that almost all amphiphiles are sitting between oil and water. One can observe, however, that not all oil–water contacts are shielded.

The model that we have considered thus far ignores the internal structure of the amphiphile, that it has a head and a tail. Several other models include this structure (e.g., Matsen and Sullivan, 1990). Perhaps one of the most successful is that studied extensively by Larson (1992). In this model, the amphiphile occupies several adjacent cells. There are i head units that behave exactly as water, and j tail units that behave exactly as oil. Extensive simulations have been carried out at various concentrations of amphiphile, oil to water ratios, and for different amphiphiles. A slice through a three-dimensional configuration of the middle phase is shown in Figure 14.16 for a system with 20% amphiphile $i = 4$, $j = 4$, and equal oil and water fractions. Tail and oil units are shown, head and water units are not. Transitions between this phase and the lamellar phase can be monitored in detail, and the integrity of the amphiphilic sheets examined. Small structures such as micelles can be detected, their size distribution measured, and the cmc located.

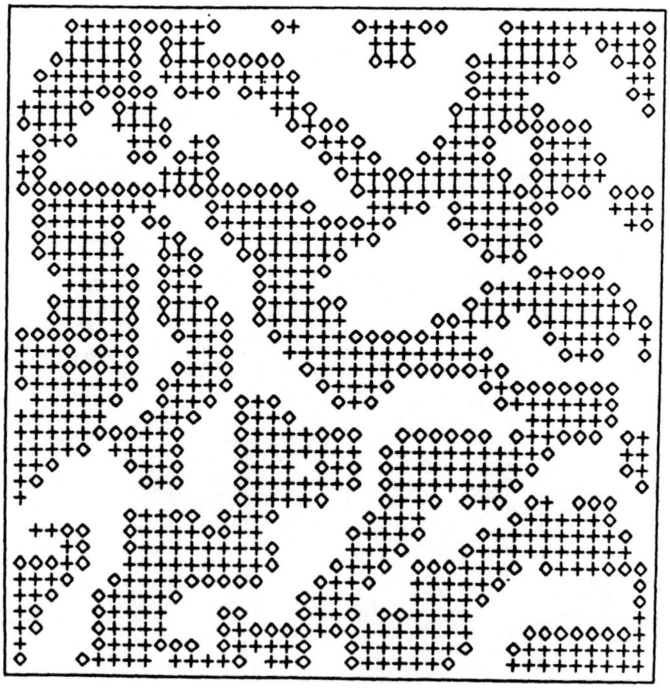

Figure 14.15
Typical configuration in the middle phase as calculated in a two-dimensional spin one model of an amphiphilic system. Adapted from Gompper and Schick (1990).

Figure 14.16
Typical configuration of the middle phase in the model of Larson. Shown is a two-dimensional cut through the three dimensional model. Adapted from Larson (1992).

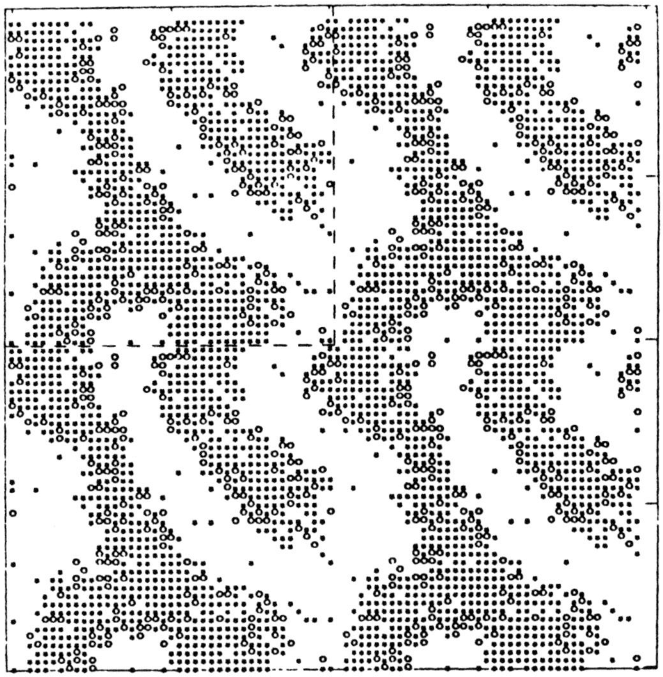

In sum, standard statistical mechanics methods can be applied to models of ternary mixtures containing amphiphiles. Results for the phase behavior, structure functions, interfacial behavior, and local structure are in qualitative agreement with experiment. One expects that additional theoretical work will produce quantitative results for these complex systems comparable to those available now for simple fluids.

Supplementary Reading

Bourrel, M. and Schechter, R. S. 1988. *Microemulsions and Related Systems: Formulations, Solvency and Solvency, and Physical Properties*, Dekker, New York.

El-Nokaly, M. and Cornell, D. 1991. *Microemulsions and Emulsions in Food*, Americal Chemical Society, Washington, D.C.

Gompper, G. and Schick, M. 1994. C. Domb and J. Lebowitz, Eds. *Phase Transitions and Critical Phenomena*, Vol. 16, Academic Press, New York.

Prince, L. M. 1977. *Microemulsions: Theory and Practice*, Academic Press, New York.

Robb, I. D., 19982. *Microemulsions*, Plenum Press, New York.

Safran, S. A. 1994. *Statistical Thermodynamics of Surfaces, Interfaces and Membranes*, Addison–Wesley, Reading, MA.

Exercises

1 A simple calculation of an interfacial energy: Consider the interface that occurs at $T = 0$ and $h = 0$ between the two ground states of the Ising model whose Hamiltonian is given by eq. (14.2.18). Show that the interfacial energy per unit area of a (100) interface is equal to $\gamma = 2J/a^2$, where a is the lattice parameter.

2 The effect of an amphiphile on the interfacial energy: Consider the ternary system governed by the Hamiltonian of eq. (14.2.34). For simplicity, take the parameter $C = 0$.

a) Show that at $T = 0$, the energies per unit volume of the three ground states, corresponding to A-rich, B-rich, and C-rich phases, are

$$E_A(h, \Delta)/V = [-3(J + K) - h + \Delta]/a^3$$
$$E_B(h, \Delta)/V = [-3(J + K) + h + \Delta]/a^3$$
$$E_C(h, \Delta)/V = 0$$

Show that the A and B phases coexist for $h = 0$ and $\Delta \leq 3(J + K)$, and that all three-phases coexist at the triple point $h = 0$, $\Delta = 3(J + K)$.

b) Let $\gamma_{AB}^{(n)}(\Delta)$ denote the energy per unit area of a (100) interface between A and B phases that has n layers of C atoms, the amphiphiles, at the interface between the A and B phases. Show that

$$a^2\gamma_{AB}^{(n)}(\Delta) = J + K + n[3(J + K) - \Delta], \quad n > 1,$$
$$a^2\gamma_{AB}^{(1)}(\Delta) = J + K + [3(J + K) - \Delta] + L,$$
$$a^2\gamma_{AB}^{(0)}(\Delta) = 2(J + K).$$

c) The true interfacial energy per unit area, $\gamma_{AB}(\Delta)$, is the minimum value of $\gamma_{AB}^{(n)}(\Delta)$. Show that for $L < 0$,

$$\begin{aligned} a^2\gamma_{AB}(\Delta) &= 2(J + K), \quad \Delta \leq 2(J + K) - |L| \\ &= 4(J + K) - |L| - \Delta, \quad 2(J + K) - |L| \leq \Delta \leq 3(J + K). \end{aligned}$$

Note that the effect of the amphiphile is to reduce the interfacial tension as the triple point is approached, and that it attains the value zero when $|L| = J + K$. This indicates an instability because the system can introduce interfaces between A and B regions without cost. In fact, for $|L| > J + K$, a lamellar phase appears in which there is an extensive number of internal interfaces between A- and B-rich regions with a layer of amphiphile at each interface. A triple point between A, B, and lamellar phases is now encountered. The appearance of a lamellar phase with increasing amphiphile strength is precisely what is observed in experiment.

3 Wetting: In the model of Problem 2, show that the interfacial energy between the A and C phases is, at the triple point, $\gamma_{AC} = (J + K)/2$ and that $\gamma_{BC} = \gamma_{AC}$ there. From the results of Problem 2, show that if $L \geq 0$, then at the triple point

$$\gamma_{AB} = \gamma_{AC} + \gamma_{BC}.$$

Therefore, the AB interface is wetted by the C phase. In contrast, if $L < 0$, show that

$$\gamma_{AB} < \gamma_{AC} + \gamma_{BC},$$

so that the interface is not wetted. Such a change in wetting behavior as the strength of amphiphile is increased is observed in experiment.

4 The consolute line: For the model of ternary mixtures of eq. (14.1.34) and with $C = 0$, show that there is a line of continuous transitions given by

$$\frac{kT_c(Q)}{zJ} = \left[1 + \frac{L(1-Q)}{J}\right] Q.$$

At a fixed temperature, this can be interpreted as giving a critical value of the interaction L as

$$\frac{L_c(Q)}{J} = \left[\frac{kT}{zJQ} - 1\right] \frac{1}{(1-Q)}.$$

Verify that several points on the line of continuous transitions shown in Figure 14.6 are given by this equation.

5 A convenient measure of amphiphile strength: Consider the form of the structure function $S^M(K)$ given by eq. (14.4.15)

$$S^M(K) = \frac{(kT/J)}{a_2 + c_1 K^2 + c_2 K^4},$$

where terms of order K^6 in the denominator have been ignored.

a) Show that the peak of this function is at $K = 0$ for $c_1 \geq 0$, and at $K_{max} \equiv -c_1/2c_2$ for $c_1 \leq 0$.

b) Define the dimensionless quantity

$$f \equiv \frac{c_1}{\sqrt{4a_2 c_2}},$$

and show that

$$\frac{2\pi\xi}{\lambda} = \sqrt{\frac{1-f}{1+f}}.$$

Thus f is a measure of the ratio of the two lengths ξ and λ. Note that at the disorder line, at which $\lambda^{-1} \to 0$, $f = 1$, at the Lifshitz line at which the peak in the structure function moves off of $K = 0$, $f = 0$, that $f < 0$ in the region in which the peak is at nonzero K, and that $f = -1$ means that ξ diverges. This implies that the system is undergoing a continuous phase transition to a lamellar phase characterized by λ. Thus the strength of an amphiphile can conveniently be characterized by the value of f, a value that lies between 1 (weak) and -1 (strong) if the amphiphile produces a microemulsion.

c) Show that the value of f can be obtained from a ratio of measured quantities via

$$f = -\sqrt{1 - \frac{S^M(0)}{S^M(K_{max})}},$$

if the peak in the structure function occurs at a nonzero value of wave vector, K_{max}. For weaker amphiphiles, it can be obtained by fitting the structure function to the form of $S^M(K)$ above, thereby obtaining the ratios c_1/a_2 and c_2/a_2 from which the value of f follows.

6 Simulate on a two-dimensional square lattice a model of ternary mixtures whose Hamiltonian is given by

$$H = H_1 - K_2 \sum_{\langle\langle ij \rangle\rangle} S_i^2 S_j^2,$$

where the sum is over all distinct second neighbor pairs, and H_1 is the Hamiltonian of eq. (14.1.34) with C and h equal to zero. The new term is added simply to break a degeneracy in the two-dimensional model. Take $K/J = 2$, $K_2/J = -0.2$, and $L/J = -3$. According to Gompper and Schick (1990), the system at temperature $T/J = 1.1$ and chemical potential $\Delta = 5.2$ is in a microemulsion phase. Compare a typical configuration of yours with those shown in the paper.

References

Auvray, L., Cotton, J. P., Ober, R., and Taupin, C. 1984. *J. Phys. Chem.*, **88**, 4586.

Gompper, G. and Schick, M. 1990. *Phys. Rev. B*, **41**, 9148.

Jahn, W. and Strey, R. 1988. *J. Chem. Phys.*, **92** 2244.

Larson, R. G. 1992. *J. Chem. Phys.*, **96**, 7904.

Lerczak, J., Schick, M., and Gompper, G. 1992. *Phys. Rev*, **46**, 985.

Matsen, M. W. and Sullivan, D. E. 1990. *Phys. Rev.*, **A41**, 2021.

15

DETERMINATION OF MICROSTRUCTURE BY SCATTERING METHODS

15.1 Theory of Scattering of Waves by Matter

The intermolecular structure of fluids and solids is of interest for a number of reasons. First, the patterns of X-ray, electron, and neutron scattering arising from intermolecular structure provide fingerprints that allow easy identification of the state and nature of matter one is dealing with. Second, it is of importance in its connection to the development of molecular theory of matter. And, third, the distribution functions representing the intermolecular structure can be used to compute thermodynamics quantities. Earlier we described the theory of distribution functions in some detail and so in this present chapter we explain how one experimentally determines intermolecular structure of fluids and solids by wave diffraction. X-rays and electron waves are scattered primarily by the electrons in the system, whereas neutrons are scattered by the atomic nuclei.

We deal with the scattering of waves under the following two assumptions:

1 The waves arriving at the detector have been scattered once (single scattering assumption).

2 The time it takes the wave to traverse a distance of several interatomic spacings is small compared to the characteristic times of atomic vibrations and the time for an atom to translate a fraction of its diameter in distance. This is the static scatterer approximation.

We also restrict ourselves to elastically scattered waves, that is, to scattered waves that have changed direction through scattering but have not lost energy. Our final results will be valid in particular for X-ray and high energy electron beam determination of fluid and solid structure.

In the scattering event depicted in Figure 15.1, a beam of waves of frequency ν and wavelength λ (or energy $h\nu$ and momentum h/λ, where h is Plancks constant) propagating in the z direction is scattered, by a scatterer at O, through an angle 2θ with respect to the z direction.

The amplitude of the impinging beam is of the form

$$u_o = A_o e^{2\pi i(\nu t - z/\lambda)}, \tag{15.1.1}$$

and the energy flux associated with the impinging beam in energy per unit time per unit area is

$$I_o = C|u_o|^2 = C|A_o|^2, \tag{15.1.2}$$

where C is a constant. The radiation is absorbed at O and reemitted in all directions. The amplitude of the radiation scattered elastically through the angle 2θ and passing through the point P is

$$u_s = A_s e^{2\pi i(\nu t - R/\lambda) - i\alpha_s} \tag{15.1.3}$$

where α_s represents a phase shift resulting from the absorption and reemission process.

The energy per unit times passing through the solid angle $\sin\theta d\theta d\phi$ associated with the scattering angle θ and point P (P locates the ϕ angle of rotation about the z axis) is

$$\dot{E}_s = C|u_s|^2 R^2 \sin\theta d\theta d\phi = C|A_s|^2 R^2 \sin\theta d\theta d\phi. \tag{15.1.4}$$

According to the law of conservation of energy this quantity must be independent of R

$$A_s \propto \frac{1}{R}. \tag{15.1.5}$$

$\dot{E}_s$ will also be proportional to I_o so that

$$A_s \propto \frac{A_o}{R}. \tag{15.1.6}$$

In general the intensity of scattered radiation is greater in some directions than in others so that A_s is a function of θ. Thus, we define a "scattering amplitude" $f_{2\theta}$ by the relation

$$A_s \equiv f_{2\theta}\frac{A_o}{R}, \tag{15.1.7}$$

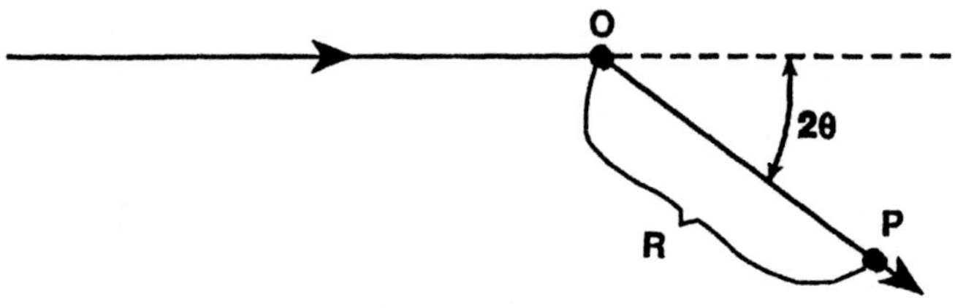

Figure 15.1
Scattering of a beam by a scattering center fixed at O.

and, consequently

$$u_s = f_{2\theta} \frac{A_o}{R} e^{2\pi i(\nu t - R/\lambda) - i\alpha_s}. \quad (15.1.8)$$

The quantity $f_{2\theta}$ depends on the nature of the scattering center and on λ. For X-rays and high energy electrons, theoretical values of $f_{2\theta}$ are available for many atoms. Otherwise, scattering experiments must be done to determine this quantity.

The energy transmitted per unit time per unit solid angle, denoted by $I_{2\theta}$, is by definition

$$I_{2\theta} = \frac{C|u_s|^2 R^2 \sin\theta d\theta d\phi}{\sin\theta d\theta d\phi} \quad (15.1.9)$$

and, for the case considered here, obeys the relation

$$\frac{I_{2\theta}}{I_o} = |f_{2\theta}|^2. \quad (15.1.10)$$

$|f_{2\theta}|^2$ has units of area and is known as the differential elastic scattering cross section. The quantity $\left[\int_o^\pi \int_o^{2\pi} |f_{2\theta}|^2 \sin\theta d\theta d\phi\right] I_o$ represents the total power scattered elastically by the scatterer at O. Thus, because I_o is a flux (power/area), it is clear why $|f_{2\theta}|^2$ is called a cross section.

Consider next scattering from an identical pair of points fixed at O_1 and O_2 and separated by the vector $\mathbf{r}_{12}$ (Fig. 15.2).

From O_1 and O_2 the scattered wave has the forms

$$u_s(O_1) = f_{2\theta} \frac{A_o}{R} e^{2\pi i(\nu t - R/\lambda) - i\alpha_s} \quad (15.1.11)$$

and

$$u_s(O_2) = f_{2\theta} \frac{A_o}{|R + \overrightarrow{O_2D}|} e^{2\pi i(\nu t - \frac{R}{\lambda} - \frac{\overrightarrow{CO_2}}{\lambda} - \frac{\overrightarrow{O_2D}}{\lambda}) - i\alpha_s} \quad (15.1.12)$$

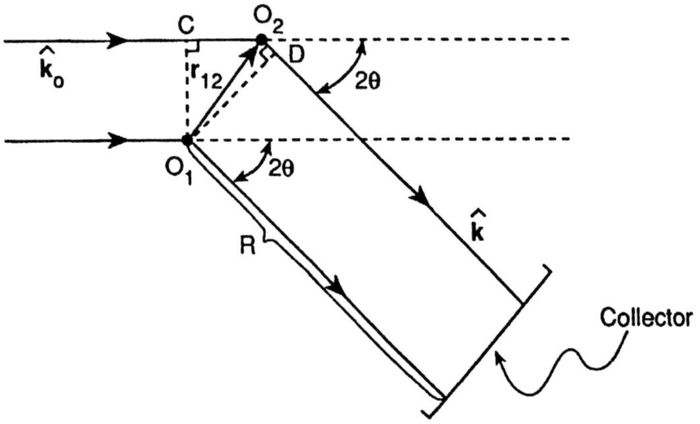

Figure 15.2
Scattering of a beam by scattering centers fixed at O_1 and O_2.

respectively. $\vec{CO_2} + \vec{O_2 D}$ is the difference between the distance a wave striking O_2 has to travel and the distance a wave striking O has to travel. $\hat{\mathbf{k}}_o$ and $\hat{\mathbf{k}}$ are unit vectors lying along the directions of the impinging and scattered rays, respectively.

From the geometry of the situation, we see that

$$\vec{CO_2} = \mathbf{r}_{12} \cdot \hat{\mathbf{k}}_o \tag{15.1.13}$$

and

$$\vec{O_2 D} = -\mathbf{r}_{12} \cdot \hat{\mathbf{k}}. \tag{15.1.14}$$

so that

$$\begin{aligned} -\frac{2\pi}{\lambda}\left[\vec{CO_2} + \vec{O_2 D}\right] &= \frac{2\pi}{\lambda}\mathbf{r}_{12} \cdot (\hat{\mathbf{k}} - \hat{\mathbf{k}}_o) \\ &= 2\pi \mathbf{r}_{12} \cdot \mathbf{s}, \end{aligned} \tag{15.1.15}$$

where

$$\mathbf{s} = \frac{\hat{\mathbf{k}} - \hat{\mathbf{k}}_o}{\lambda}. \tag{15.1.16}$$

Also, it follows that

$$\begin{aligned} |\mathbf{s}|^2 &= \frac{1}{\lambda^2} + \frac{1}{\lambda^2} - \frac{2\hat{\mathbf{k}} \cdot \hat{\mathbf{k}}_0}{\lambda^2} = \frac{2}{\lambda^2}(1 - \cos 2\theta) \\ &= \frac{4}{\lambda^2}\sin^2\theta, \end{aligned}$$

or

$$s \equiv |\mathbf{s}| = \frac{2}{\lambda}|\sin\theta|. \tag{15.1.17}$$

Using the above results and assuming that the collector is far away from O_1 and O_2 compared to the distance $|\mathbf{r}_{12}|$, we can neglect $\vec{O_2 D}$ in $1/|R + \vec{O_2 D}|$ and write

$$u_s(O_2) = f_{2\theta}\frac{A_o}{R}e^{2\pi i(\nu t - R/\lambda) - i\alpha}e^{2\pi i \mathbf{r}_{12}\cdot\mathbf{s}}. \tag{15.1.18}$$

Thus the total scattering amplitude from O_1 and O_2 is

$$u_s = u_s(O_1) + u_s(O_2) = f_{2\theta}\frac{A_o}{R}e^{2\pi i(\nu t - R/\lambda) - i\alpha_s}\left\{1 + e^{2\pi i \mathbf{r}_{12}\cdot\mathbf{s}}\right\}. \tag{15.1.19}$$

We can consider N identical scattering centers in a similar fashion and obtain

$$\begin{aligned} u_s &= \sum_{j=1}^{N} u_s(O_j) = f_{2\theta}\frac{A_o}{R}e^{2\pi i(\nu t - R/\lambda) - i\alpha_s}\sum_{j=1}^{N} e^{2\pi i \mathbf{r}_{1j}\cdot\mathbf{s}} \\ &= f_{2\theta}\frac{A_o}{R}e^{2\pi i(\nu t - R/\lambda) - i\alpha_s + 2\pi i \mathbf{r}_1\cdot\mathbf{s}}\sum_{j=1}^{N} e^{-2\pi i \mathbf{r}_j\cdot\mathbf{s}}. \end{aligned} \tag{15.1.20}$$

If not all the scatterers are identical, then

$$u_s = \frac{A_o}{R} e^{2\pi i(\nu t - R/\lambda) + 2\pi i \mathbf{r}_1 \cdot \mathbf{s}} \sum_{j=1}^{N} F_{2\theta}^{(j)} e^{-2\pi i \mathbf{r}_j \cdot \mathbf{s}} \tag{15.1.21}$$

where $\mathbf{r}_j$ is the position of the jth scatterer and

$$F_{2\theta}^{(j)} \equiv f_{2\theta}^{(j)} e^{-i\alpha_s^{(j)}}, \tag{15.1.22}$$

$f_{2\theta}^{(j)}$ and $\alpha_s^{(j)}$ being the scattering amplitude and phase shift of the jth scatterer.

For a given configuration $\mathbf{r}_1, \ldots, \mathbf{r}_N$ of scatterers, the scattered intensity $I_{2\theta}$ is

$$\frac{I_{2\theta}}{I_o}(r_1, \ldots, \mathbf{r}_N) \equiv \frac{C|u_s|^2 R^2 \sin\theta d\theta d\phi}{I_o \sin\theta d\theta d\phi} = \sum_{j=1}^{N} \sum_{k=1}^{N} F_{2\theta}^{(j)} \overline{F}_{2\theta}^{(k)} e^{-2\pi i \mathbf{r}_{jk} \cdot \mathbf{s}}, \tag{15.1.23}$$

where $\overline{F}_{2\theta}^{k}$ is the complex conjugate of $F_{2\theta}^{(k)}$. If the scatterers are identical, then

$$\frac{I_{2\theta}}{I_o}(\mathbf{r}_1, \ldots, \mathbf{r}_N) = |f_{2\theta}|^2 \sum_{j=1}^{N} \sum_{k=1}^{N} e^{-2\pi \mathbf{r}_{jk} \cdot \mathbf{s}}. \tag{15.1.24}$$

The observed intensity $I_{2\theta}$ is obtained by averaging $I_{2\theta}(\mathbf{r}_1, \ldots, \mathbf{r}_N)$ over an ensemble characteristic of the thermodynamic conditions of the experiment. Thus,

$$\frac{I_{2\theta}}{I_o} = \sum_{j=1}^{N} \sum_{k=1}^{N} F_{2\theta}^{(j)} \overline{F}_{2\theta}^{(k)} \langle e^{-2\pi i \mathbf{r}_{jk} \cdot \mathbf{s}} \rangle. \tag{15.1.25}$$

Equation (15.1.25) is the starting point of our analysis of diffraction studies of ideal crystals and liquids. Note that $I_{2\theta}$ is composed of a product of scattering amplitudes $F_{2\theta}^{(j)} \overline{F}_{2\theta}^{(k)}$, that depends on the nature of the wave-scatterer interaction, and a factor $< e^{-2\pi i \mathbf{r}_{jk} \cdot \mathbf{s}} >$ that depends on the interscatterer structure of the medium.

The quantities $f_{2\theta}$ and α_s must be determined from the experiment or from quantum scattering theory. For atoms the X-ray scattering factors have been determined accurately and are available in tables (e.g., *International Tables of X-Ray Crystallography*, (Ibers and Hamilton, 1974). The X-ray phase shift α_s is always equal to π for scattering from isolated electrons and is usually equal to π for scattering from atoms as long as the frequency of scattered radiation is large compared to the characteristic frequencies at which X-rays are strongly absorbed by electronic excitations. For frequencies near the characteristic frequencies α_s can deviate considerably from π.

Figure 15.3
Scattering amplitude for the carbon atom (MacWeeny, 1951, 1952, 1953, 1954).

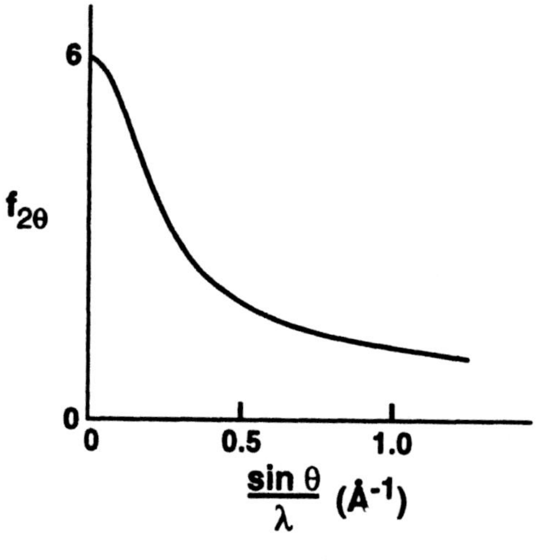

One can estimate electron–atom scattering amplitudes $f_{2\theta}^{\text{el}}$ from the more readily available x-ray scattering amplitudes $f_{2\theta}^{x}$ from the Mott equation (Mott, 1930).

$$f_{2\theta}^{\text{el}}(\mathbf{s}) = \frac{8\pi^2 me^2}{\hbar^2}\left[\frac{n_a - f_{2\theta}^{x}(\mathbf{s})}{s^2}\right], \tag{15.1.26}$$

where n_a is the atomic number of the scattering atom.

Both $f_{2\theta}^{x}$ and $f_{2\theta}^{\text{el}}$ increase roughly linearly with atomic number of a scattering atom. Thus, the relative scattering intensities ($\sim f_{2\theta}^2$) of different atoms go roughly as the ratios of the square of the atomic numbers of the atoms. Neutrons scatter off the atomic nuclei. Their scattering amplitudes do not follow the atomic number in any simple way.

The behavior of $f_{2\theta}$ is illustrated in Figure 15.3 for X-ray scattering from a carbon atom.

15.2 Monatomic Fluids

Consider a one-component fluid of monatomic particles interacting via centrally symmetric forces. In this case, the scattering intensity may be expressed in the form

$$\begin{aligned}\frac{I_{2\theta}}{I_o} &= |f_{2\theta}|^2\left\{N + \sum_{j\neq k}^{N}\sum^{N}\frac{1}{V^2}\int e^{-i\mathbf{K}\cdot\mathbf{r}_{jk}}g^{(2)}(r_{jk})d^3r_j d^3r_k\right\} \\ &= N|f_{2\theta}|^2\left\{1 + \frac{N-1}{V}\int e^{-i\mathbf{K}\cdot\mathbf{r}}[g^{(2)}(r) - 1]d^3r + \frac{N-1}{V}\int e^{-i\mathbf{K}\cdot\mathbf{r}}d^3r\right\},\end{aligned} \tag{15.2.1}$$

where the definition

$$\mathbf{K} = 2\pi\mathbf{s}; \quad K = \frac{4\pi}{\lambda}|\sin\theta| \tag{15.2.2}$$

has been introduced. We define the *structure factor* $S(\mathbf{K})$ by the expression

$$S(\mathbf{K}) \equiv 1 + n \int e^{-i\mathbf{K}\cdot\mathbf{r}}[g^{(2)}(r) - 1]d^3r = \left[1 - n \int e^{-i\mathbf{K}\cdot\mathbf{r}} C^{(2)}(r) d^3r\right]^{-1}. \quad (15.2.3)$$

The second expression for $S(\mathbf{K})$ results from the Ornstein–Zernike equation [eq. (9.3.16)]. With the property $\frac{1}{(2\pi)^3}\int e^{-i\mathbf{K}\cdot\mathbf{r}}d^3r = \delta(\mathbf{K})$, where $\delta(\mathbf{K})$ is the Dirac delta function, we can write eq. (15.2.1) in the form

$$\frac{I_{2\theta}}{N|f_{2\theta}|^2 I_o} = S(\mathbf{K}) + n(2\pi)^3\delta(\mathbf{K}). \quad (15.2.4)$$

The contribution of $\delta(\mathbf{K})$ to eq. (15.2.4) is of no consequence because it is identically zero except when $\mathbf{K} = 0$, that is, except for radiation not being scattered by the atoms of the fluid.

A measurement of the structure factor over a sufficient range of values of K (i.e., a measurement of $I_{2\theta}/N|f_{2\theta}|^2 I_o$ over a sufficient range of θ or λ) will enable one to determine the pair correlation function g_2 of the fluid. This is because of the relationship

$$\frac{1}{(2\pi)^3}\int e^{i\mathbf{K}\cdot\mathbf{r}}\left[S(\mathbf{K}) - 1\right]d^3K = n\left[g^{(2)}(r) - 1\right], \quad (15.2.5)$$

that can be established with the aid of the theory of Fourier transforms.

Because $g^{(2)}$ depends only on the magnitude of r, $S(\mathbf{K})$ depends only on the magnitude of $\mathbf{K}$. To see this, we can express $\mathbf{r}$ in a spherical coordinate system in which $\mathbf{K}$ lies along the z axis so that $\mathbf{K}\cdot\mathbf{r} = Kr\cos\theta$, $d^3r = \sin\theta d\theta d\phi r^2 dr$, and eq. (15.2.3) becomes

$$\begin{aligned} S(\mathbf{K}) &= 1 + n \int_0^{\pi}\int_0^{2\pi}\int_0^{\infty} e^{-iKr\cos\theta}\left[g^{(2)}(r) - 1\right]\sin\theta d\theta d\phi r^2 dr \\ &= 1 + 4\pi n \int_0^{\infty} \frac{\sin Kr}{Kr}[g^{(2)}(r) - 1]r^2 dr, \end{aligned} \quad (15.2.6)$$

a function only of K. A similar integration over the angles of d^3K in eq. (15.2.5) allows us to write

$$g^{(2)}(r) - 1 = \frac{4\pi}{(2\pi)^3 n}\int_0^{\infty}\frac{\sin Kr}{Kr}\left[S(k) - 1\right]K^2 dK. \quad (15.2.7)$$

We saw in Chapter 9 that the isothermal compressibility, κ_T, of a fluid is related to the pair correlation function by the expression

$$nkT\kappa_T = 1 + 4\pi n \int_0^{\infty}[g^{(2)}(r) - 1]r^2 dr. \quad (15.2.8)$$

Thus, in the limit $K \to 0$, the structure factor obeys the equation

$$S(0) = nkT\kappa_T. \qquad (15.2.9)$$

This result represents a convenient check on scattering data because $nkT\kappa_T$ can be obtained from bulk thermodynamic measurements.

For dense fluids not near the critical point a typical pair correlation function is shown in Figure 15.4. The peaks represent the nearest, next nearest, and next-next nearest neighbor structures distributed about an atom centered at $r = 0$. In Figure 15.5 is shown the structure factor corresponding to a fluid similar to that for which the pair correlation function is shown in Figure 15.4.

The fluid illustrated in Figures 15.4 and 15.5 is a liquid near the melting point and is, therefore, relatively incompressible so that $S(0) = nkT\kappa_T = \mathcal{O}(10^{-2})$. In the dilute gas limit $S(0) = 1$, whereas as $T \to T_c$ for $n = n_c$, $S(0) \to +\infty$ because $\kappa_T \to +\infty$ at the critical point.

For a fluid with the structure factor given in Figure 15.5, the radiation scattering experiment is illustrated in Figure 15.6.

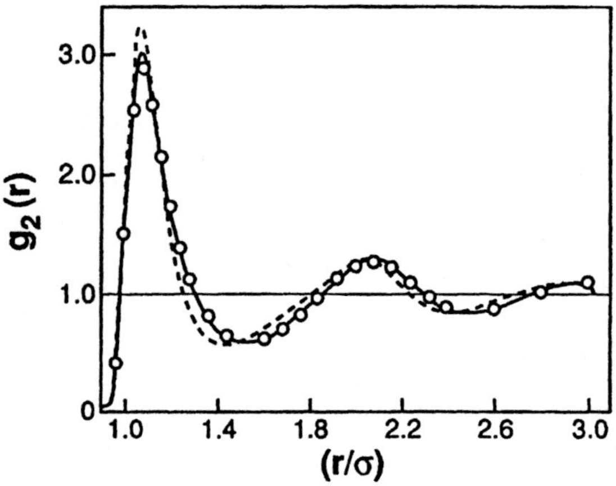

Figure 15.4
Plot of the pair correlation function for a 6–12 Lennard-Jones fluid at $kT/\varepsilon = 0.88$ and $n\sigma^3 = 0.85$. The circles were determined by computer simulation (molecular dynamics) by Verlet (1968). The dashed curve is from the PY theory was discussed in Chapter 9. The solid curve is from the Weeks et al. (1971).

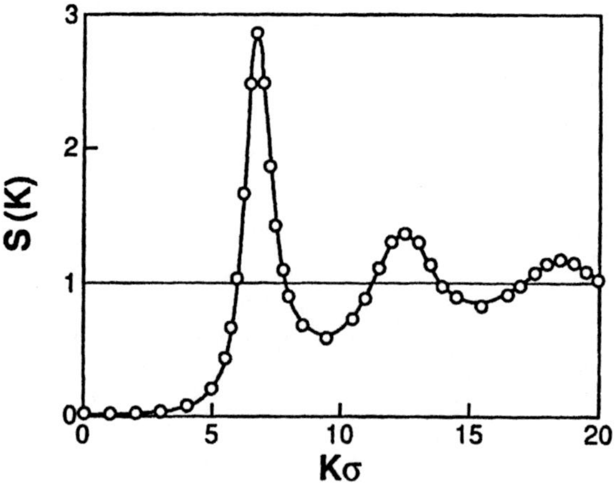

Figure 15.5
Plot of the structure function for a 6–12 Lennard-Jones fluid at $kT/\varepsilon = 0.723$ and $nd^3 = 0.844$. The circles are from the molecular dynamics calculations referenced in the caption of Figure 15.4. The solid curve is from the theory of Weeks et al. (1971).

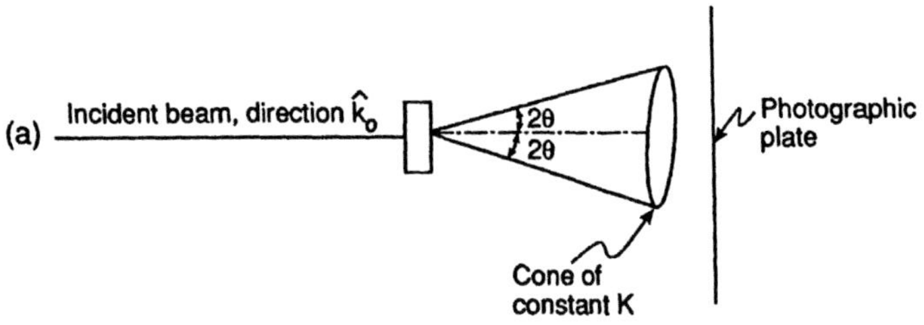

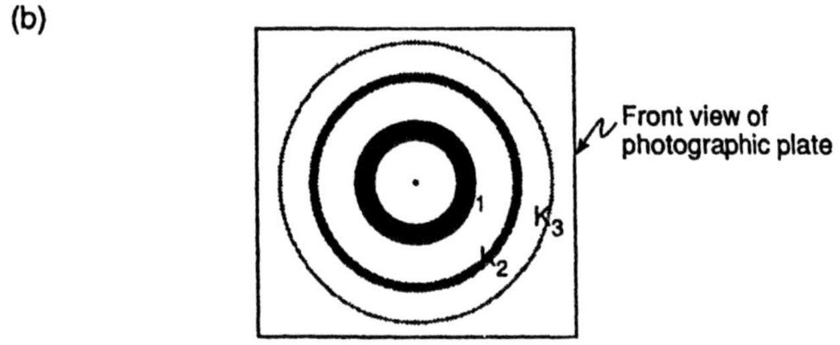

Figure 15.6
Example of X-ray scattering pattern for isotropic fluid. Intensity is constant on a cone of rotation about $\mathbf{k}_o$ at constant angle θ. The K_i in (b) is to denote the successive maxima of $S(K)$. In the center of (b) is the intense spot of undeflected radiation associated with $\delta(\mathbf{K})$.

In the dense fluid region, away from the critical point, the important structure of $S(K)$ occurs in the range for which $K = \mathcal{O}$ (10 Å^{-1}). Thus, one would use radiation with wavelength of the order of a fraction of an angstrom, for example, X-rays or high energy electrons (say 100 eV electrons with wavelength 0.037 Å), to investigate $S(K)$. Near the critical point, however, the pair correlation function becomes a long-ranged function of r because of the long range over which appreciable density fluctuations occur in the fluid. In particular, as described in Chapter 9, the pair correlation function is of the form

$$g^{(2)}(r) = g_{\text{SR}}^{(2)}, \quad r < l$$
$$\simeq 1 + c_1 \frac{e^{-r/\xi}}{r}, \quad r > l \tag{15.2.10}$$

where $g_{\text{SR}}^{(2)}$ behaves as the usual short-range correlation function shown in Figure 15.4, l is a length of the order of a few molecular diameters, c_1 is a constant, and ξ is the characteristic correlation length that depends on temperature as

$$\xi = \xi_o \left| 1 - \frac{T}{T_c} \right|^{-\gamma/2}, \tag{15.2.11}$$

where γ is the critical exponent of isothermal compressibility, $\gamma/2 \simeq 0.6$, and ξ_o is a scale factor. As $T \to T_c$, $\xi \to +\infty$. For the pair correlation function given by eq. (15.2.10) and in the limit that $l/\xi \ll 1$, the structure factor becomes

$$S(K) = S^{\text{SR}}(K) + 2\pi n c_1 \int_l^\infty \frac{e^{-r/\xi}}{r} \frac{\sin Kr}{Kr} r^2 dr$$
$$= S^{\text{SR}}(K) + 4\pi n c_1 \int_0^l e^{-r/\xi} \frac{\sin Kr}{K} dr + \frac{4\pi n c_1}{\xi^{-2} + K^2}. \tag{15.2.12}$$

Sufficiently near the critical point the last term on the right-hand side (rhs) of eq. (15.2.12) is the dominant contribution. However, where this term becomes important, light scattering is more appropriate than X-ray scattering for measuring this effect. To understand this, let us analyze the quantity $4\pi c_1 n/(\xi^{-2} + K^2)$. From the theory presented in Section 9.3 of Chapter 9, it follows that ξ_o and c_1 are of the order of σ, the diameter of the molecular scatterers. Thus, near the critical point, the contribution to $S(K)$ from near critical fluctuations

$$S_{\text{crit}}(K) \equiv \frac{4\pi n c_1}{\xi^{-2} + K^2} \simeq \frac{4\pi n \sigma^3}{|1 - T/T_c|^\gamma + (K\sigma)^2} \simeq \frac{5}{|1 - T/T_c|^{1.2} + (K\sigma)^2}. \tag{15.2.13}$$

Typically in wide angle X-ray scattering, $K \geq 0.3$ Å^{-1} and so, if $\sigma \simeq 4$ Å, the contribution of $S_{\text{crit}}(K)$ to $S(K)$ would be less than about 5 even at $T = T_c$. In the case of light, with a wavelength of say 4000 Å, K is of the order of 10^{-3} Å^{-1}, and so at the critical point $S_{\text{crit}}(K)$ would be of the order of 500 if $|1 - T/T_c|^{1.2} \simeq 10^{-2}$ because $(K\sigma)^2 \simeq 2 \times 10^{-5}$ when $K \simeq 10^{-3}$ Å^{-1} for $\sigma = 4$Å. Light scattering is, in fact, the method of choice for probing near critical point fluid structure. The behavior of $S(K)$ near the critical point is illustrated in Figure 15.7.

In a c-component fluid of monatomic particles interacting via centrally symmetric forces, the scattering intensity may be expressed in the form

$$\frac{I_{2\theta}}{N I_o} = \sum_{\alpha=1}^{c} \sum_{\beta=1}^{c} F_{2\theta}^{(\alpha)} \overline{F}_{2\theta}^{(\beta)} \{S_{\alpha\beta}(K) + X_\alpha X_\beta n (2\pi)^3 \delta(\mathbf{K})\} \tag{15.2.14}$$

where the "partial" structure factor $S_{\alpha\beta}$ is defined by the equation

$$S_{\alpha\beta}(K) = X_\alpha \delta_{\alpha,\beta} + X_\alpha X_\beta n \int e^{-i\mathbf{K}\cdot\mathbf{r}} [g_{\alpha\beta}^{(2)}(r) - 1] d^3 r. \tag{15.2.15}$$

The quantity $\delta_{\alpha\beta}$ is the Kronecker delta, X_α the mole fraction of species α, $n(\equiv N/V)$ the total number density, and $g_{\alpha\beta}^{(2)}$ the pair correlation function for a pair of particles of species α and β, respectively.

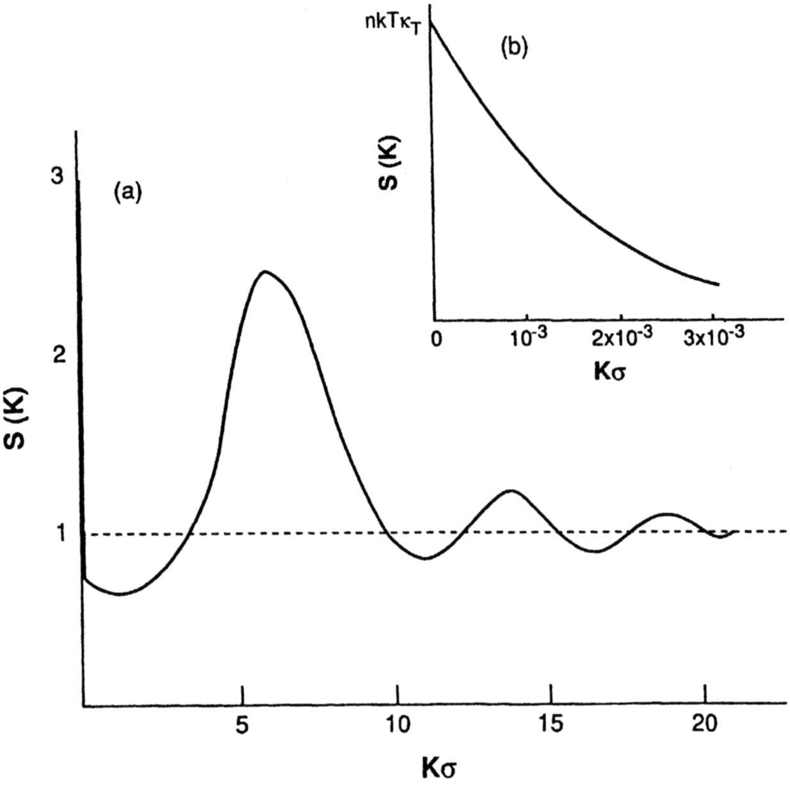

Figure 15.7
Qualitative plot of $S(K)$ versus K for a fluid near its critical point. It is assumed that the fluid is composed of molecules of diameter $\sigma = 4$, $nkT\kappa \gg 1$, and $\xi/\sigma = 1000$. On the scale of (b) for $K\sigma$ characteristic of X-rays, the region shown in (a) would be of almost infinitesimal width.

We conclude from eq. (15.2.14) that an X-ray or electron diffraction experiment does not determine the separate pair correlation functions of the components of a mixture, but, rather yields only a linear combination of these functions. Thus, in a c-component mixture, a diffraction experiment yields only one equation of the $c(c+1)/2$ equations necessary to uniquely determine the pair correlation functions. In principle, but certainly not easy practice, this limitation can be overcome by performing a number of diffraction experiments in which the values of the $F_{2\theta}^{(\alpha)}$ may be varied independently. One way to vary the $F_{2\theta}^{(\alpha)}$ is to vary the energy of the incident X-ray or electron beams. Or one can use neutron beams for which the scattering amplitudes are determined not by the electronic structure (which determines $F_{2\theta}^{(\alpha)}$ for X-rays and electrons) but by the nuclear structure of the scattering atoms. By collecting neutron diffraction data for chemically identical mixtures of different isotopes of the various species, still more independent combinations of the partial structure factors may be generated.

15.3 Polyatomic Fluids

Consider a one-component fluid of N molecules composed of ν atoms. If $F_{2\theta}^{(i)}$ is the scattering amplitude of the ith atom, we can express the scattering intensity in the form

$$\frac{I_{2\theta}}{I_o} = \sum_{j=1}^{N} \sum_{k=1}^{N} \sum_{i=1}^{\nu} \sum_{l=1}^{\nu} F_{2\theta}^{(i)} \overline{F}_{2\theta}^{(l)} e^{-i\mathbf{K}\cdot(\mathcal{R}_i^{(j)} - \mathcal{R}_l^{(k)})} e^{-i\mathbf{K}\cdot\mathbf{R}_{jk}}, \quad (15.3.1)$$

where the position $\mathbf{r}_i^{(j)}$ of the ith atom of the jth molecule has been expressed in the form

$$\mathbf{r}_i^{(j)} = \mathbf{R}_j + \mathcal{R}_i^{(j)}. \quad (15.3.2)$$

$\mathbf{R}_j$ is the position of the center of mass of molecule j and $\mathcal{R}_i^{(j)}$ is the position of atom i relative to the center of mass of molecule j. Introducing the intramolecular scattering factor,

$$\mathcal{F}^{(j)} \equiv \sum_{i=1}^{\nu} F_{2\theta}^{(i)} e^{-i\mathbf{K}\cdot\mathcal{R}_i^{(j)}}, \quad (15.3.3)$$

we can express eq. (15.3.1) in the more abbreviated form

$$\begin{aligned} \frac{I_{2\theta}}{I_o} &= \sum_{j=1}^{N} \sum_{k=1}^{N} \langle \mathcal{F}^{(j)} \overline{\mathcal{F}}^{(k)} e^{-i\mathbf{K}\cdot\mathbf{R}_{jk}} \rangle \\ &= N\left[\langle |\mathcal{F}^{(1)}|^2 \rangle + (N-1) \langle \mathcal{F}^{(1)} \overline{\mathcal{F}}^{(2)} e^{-i\mathbf{K}\cdot\mathbf{R}_{12}} \rangle \right]. \end{aligned} \quad (15.3.4)$$

The ensemble average indicated in eq. (15.3.4) must of course be taken over the vibrational and rotational states (and electronic if temperatures are high enough for electronic transitions) of the molecules of the fluid.

Three factors determine the scattering intensity of a polyatomic fluid. First, there is the "self"-structure factor S_s associated with the single atom sum $\sum_{i=1}^{\nu} |F_{2\theta}^{(i)}|^2$. We define the self-structure factor by the expression

$$\begin{aligned} S_s &= \sum_{i=1}^{\nu} |F_{2\theta}^{(i)}|^2 \Big/ \left| \sum_{i=1}^{\nu} F_{2\theta}^{(i)} \right|^2 \\ &= M \sum_{i=1}^{\nu} |F_{2\theta}^{(i)}|^2, \end{aligned} \quad (15.3.5)$$

where the notation $M = 1/|\sum_{i=1}^{\nu} F_{2\theta}^{(i)}|^2$ is used. We define an intramolecular structure factor by

$$\begin{aligned} S_m &= M\left[\langle |\mathcal{F}^{(1)}|^2 \rangle - \sum_{i=1}^{\nu} |F_{2\theta}^{(i)}|^2 \right] \\ &= M \sum_{j \neq k}^{\nu} F_{2\theta}^{(j)} \overline{F}_{2\theta}^{(k)} \langle e^{-i\mathbf{K}\cdot(\mathcal{R}_j^{(1)} - \mathcal{R}_k^{(1)})} \rangle. \end{aligned} \quad (15.3.6)$$

A third contribution is the intermolecular structure S_d,

$$S_d = 1 + M(N-1) \langle \mathcal{F}^{(1)} \overline{\mathcal{F}}^{(2)} e^{-i\mathbf{K}\cdot\mathbf{R}_{12}} \rangle. \quad (15.3.7)$$

In terms of these quantities, eq. (15.3.4) can be rewritten as

$$\frac{I_{2\theta}}{NI_o|\sum_{i=1}^{\nu} F_{2\theta}^{(i)}|^2} = S_s + S_m + S_d - 1. \tag{15.3.8}$$

The dependence of S_s on **K** is determined by the electronic structure (or nuclear structure in the case of neutron scattering) of the atoms composing molecules of the fluid. That of S_m comes from the relative orientations and vibrations of the atoms of a molecule of the fluid. The behavior of S_d is governed by the intermolecular correlations, including orientational as well as spatial correlations of pairs of molecules.

Experimentally one measures the single quantity $I_{2\theta}/NI_o$ so that experimental separation of the contributions to S_m and $S_d - 1$ is not possible without a model for either S_m or $S_d - 1$. In practice, a formula for S_m is used that can be justified theoretically for certain models and that appears to be useful as an empirical formula for deducing S_d from scattering data. The formula is

$$S_m = M \sum_{j \neq k}^{\mu} F_{2\theta}^{(j)} \overline{F}_{2\theta}^{(k)} e^{-\frac{1}{2}\Delta_{jk}^2 K^2} \frac{\sin K\mathcal{R}_{jk}^{(o)}}{K\mathcal{R}_{jk}^{(o)}}, \tag{15.3.9}$$

where $\mathcal{R}_{jk}^{(o)}$ and Δ_{jk} are interpreted as the mean and root-mean-square deviation from the mean of the separation of atoms j and k of a molecule. As a practical matter, S_d often approaches its limiting value faster than does S_m with increasing K and so Δ_{jk} and $\mathcal{R}_{jk}^{(o)}$ can be determined empirically from the high K part of the scattering data.

To somewhat motivate eq. (15.3.9) let us consider a homonuclear, diatomic molecule behaving as a classical vibrator whose rotational and vibrational motions are independent. For this case

$$\mathcal{F}^{(1)} = F_{2\theta}[e^{-\frac{1}{2}i\mathbf{K}\cdot\mathcal{R}} + e^{\frac{1}{2}i\mathbf{K}\cdot\mathcal{R}}] \tag{15.3.10}$$

and, therefore,

$$\langle|\mathcal{F}^{(1)}|^2\rangle = |F_{2\theta}|^2\langle 2 + e^{-i\mathbf{K}\cdot\mathcal{R}} + e^{i\mathbf{K}\cdot\mathcal{R}}\rangle. \tag{15.3.11}$$

But for the model considered

$$\langle e^{\pm i\mathbf{K}\cdot\mathcal{R}}\rangle = \int_0^{\pi}\int_0^{2\pi}\int_0^{\infty} e^{\pm iK\mathcal{R}\cos\theta} p(\theta,\phi)\sin\theta d\theta d\phi p(\mathcal{R})d\mathcal{R}, \tag{15.3.12}$$

where the orientational and vibrational probability densities are given by

$$p(\theta,\phi) = \frac{1}{4\pi} \tag{15.3.13}$$

and

$$p(\mathcal{R}) = \left(\frac{\lambda}{2\pi kT}\right)^{1/2} e^{-\lambda(\mathcal{R}-\mathcal{R}_0)^2/2kT}, \tag{15.3.14}$$

respectively. λ is the force constant of the molecules. Integrating over θ and ϕ and defining $y = \mathcal{R} - \mathcal{R}_0$, we can express eq. (15.3.11) in the form

$$\langle|\mathcal{F}^{(1)}|^2\rangle = 2|F_{2\theta}|^2 \left\{ 1 + \frac{1}{2iK}\left(\frac{\lambda}{2\pi kT}\right)^{1/2} \int_{-r_0}^{\infty} \frac{e^{\frac{-\lambda y^2}{2kT}}}{y + \mathcal{R}_0} [e^{iK(y+\mathcal{R}_0)} - e^{-iK(y+\mathcal{R}_0)}]dy. \right\} \tag{15.3.15}$$

Because for a stable molecule $\lambda\mathcal{R}_0^2/2kT \gg 0$, we can use with negligible error the approximation

$$\left(\frac{\lambda}{2\pi kT}\right)^{1/2} \int_{-\mathcal{R}_0}^{\infty} \frac{e^{\frac{-\lambda y^2}{2kT} \pm iKy}}{y + \mathcal{R}_0} dy \simeq \left(\frac{\lambda}{2\pi kT}\right)^{1/2} \int_{-\infty}^{\infty} \frac{e^{\frac{-\lambda y^2}{2kT} \pm iKy}}{\mathcal{R}_0} dy = \frac{1}{\mathcal{R}_0} e^{\frac{-kT}{2\lambda}K^2}. \tag{15.3.16}$$

With this result, and $M = 4|F_{2\theta}|^2$, we obtain

$$S_m = e^{\frac{-kT}{2\lambda}K^2} \sin K\mathcal{R}_0/2K\mathcal{R}_0. \tag{15.3.17}$$

For the model considered the mean-square deviation of $\mathcal{R}$ is given by

$$\Delta^2 = \langle(\mathcal{R} - \mathcal{R}_0)^2\rangle = \frac{kT}{\lambda}. \tag{15.3.18}$$

Thus, eq. (15.3.17) has the form proposed at eq. (15.3.9). The quantity $e^{\frac{-kT}{2\lambda}K^2}$ is known as the Debye–Waller factor, and, from its derivation, we see that it accounts for an attenuation in the molecular scattering intensity caused by the vibrational motion of the scattering atoms.

The values of Δ and $\mathcal{R}_0$ have been fitted [using eq. (15.2.9)] to X-ray data on saturated liquid bromine (Norton et al., 1978). The results are $\Delta = 0.0045$ Å and $\mathcal{R}_0 = 2.291$ Å. Their reconstructions of S_m and S_d from X-ray data are shown in Figure 15.8. Spectroscopically, for dilute vapor at 300 K, the values of $\sqrt{\langle(\mathcal{R} - \mathcal{R}_0)^2\rangle}$ and $\mathcal{R}_0$ have been found to be 0.045 and 2.284 Å (Herzberg, (1951). This quite good agreement indicates that the average internal motions of Br_2 are not appreciably different in the vapor and liquid phases. From electron diffraction on dilute oxygen at 300 K, Δ and $\mathcal{R}_0$ have been found to be 0.038 and 1.208 Å (Karle, 1955), whereas from spectroscopy $\sqrt{\langle(\mathcal{R} - \mathcal{R}_0)^2\rangle}$ and $\mathcal{R}_0$ are observed to be 0.037 and 1.207 Å.

For bromine and oxygen, having characteristic vibrational temperatures $\theta_v = 470$ and 2230 K, respectively, the classical mechanical result, $\Delta = \sqrt{kT/\lambda}$ yields 0.040 and 0.019 Å, respectively. We see that the classical value is only 10% low for Br_2 but is a factor of two too low for O_2. The obvious origin of the failure of the classical approximation is the break-down of classical mechanics ($\theta_V/T = 7.43$ for O_2 at 300 K). The quantum mechanical zero-point vibrational motion yields a finite value of $\langle(\mathcal{R} - \mathcal{R}_0)^2\rangle$

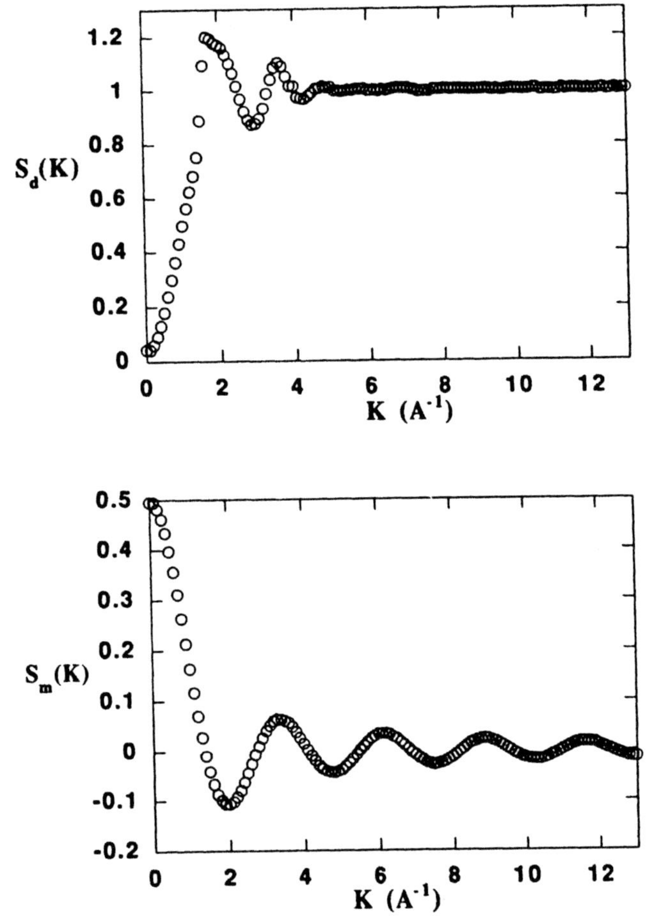

Figure 15.8
Values of S_m and S_d versus scattering vector K for liquid bromine at 20°C. From values deduced by Norton et al. (1978).

even at $T = 0$, so that the observed value at finite T is larger than would be expected classically.

To appreciate the quantum mechanical effect at low temperatures, let us examine the case for which T/θ_V is so low that all the molecules are in their ground vibrational state. Then

$$
\begin{aligned}
\langle e^{\pm i\mathbf{K}\,\mathcal{R}} \rangle &= \frac{1}{iK} \int_0^{\infty} \psi_0^V(\mathcal{R}) \frac{e^{iK\mathcal{R}} - e^{-iK\mathcal{R}}}{\mathcal{R}} \psi_0^V(\mathcal{R}) \mathcal{R}^2 d\mathcal{R} \\
&= \frac{1}{iK} A_0^2 \int_0^{\infty} e^{-(\mathcal{R}-\mathcal{R}_0)^2/x_0^2} \frac{e^{iK\mathcal{R}} - e^{-iK\mathcal{R}}}{\mathcal{R}} d\mathcal{R},
\end{aligned}
\tag{15.3.19}
$$

where the formula given by eq. (4.1.23) has been used for ψ_0^V. We recall that

$$
x_0 = \left(\frac{2\hbar^2}{m\lambda} \right)^{1/4}
$$

and

$$A_0^2 = 1 / \int_0^{\infty} e^{-(\mathcal{R}-\mathcal{R}_0)^2/x_0^2} d\mathcal{R}. \qquad (15.3.20)$$

For stable molecules $x_0 \ll \mathcal{R}_0$, so that $A_0^2 \simeq 1/(x_0\pi^{1/2})$ and

$$\langle e^{\pm i\mathbf{K}\cdot\mathcal{R}} \rangle = e^{-x_0^2K^2/4} \frac{\sin K\mathcal{R}_0}{K\mathcal{R}_0}, \qquad (15.3.21)$$

a result yielding

$$S_m = e^{-x_0^2K^2/4} \frac{\sin K\mathcal{R}_0}{2K\mathcal{R}_0}. \qquad (15.3.22)$$

This expression is formally identical to the classical result. However, here the root-mean-square deviation of $\mathcal{R}$ from $\mathcal{R}_0$ is the temperature independent quantity, $\Delta = x_0/2$, compared to the temperature dependent classical result, $\Delta = \sqrt{kT/\lambda}$.

For Br_2 and O_2, $x_0/2 = 0.036$ and 0.037 Å, respectively. As expected (because $\theta_V/T \gg 1$), the value of Δ observed for O_2 by electron diffraction agrees quite will with quantum mechanical prediction, assuming all the oxygen molecules lie in the ground vibration state. Because for bromine θ_V/kT is of order 1 at 20°C, neither the classical limit nor the $T = 0$ K quantum limit considered here is valid. Thus, to determine Δ theoretically one must calculate the average $\langle(\mathcal{R} - \mathcal{R}_0)^2\rangle$ from the finite temperature quantum mechanical representation (i.e.) from $\langle(\mathcal{R} - \mathcal{R}_0)^2\rangle = \sum_{\nu=0}^{\infty} e^{-\varepsilon_\nu/kT} \int \psi_\nu(\mathcal{R})^2(\mathcal{R} - \mathcal{R}_0)^2\mathcal{R}^2 d\mathcal{R}/q^V$.

In practice one determines Δ and $\mathcal{R}_0$ empirically from the scattering data. In fact, it is the exception and not the rule that dilute gas values of Δ agree with condensed phase values. For example, for compact molecules (say, the diatomics and CH_4) condensed phase values of bonds and bond deviations can agree well with dilute gas values, but already the molecule CS_2 shows substantial condensed phase effects for Δ. This is shown in Table 15.1. The magnitudes of Δ_{jk}, j, k = carbon or sulfur, are larger in the liquid phase than in the dilute gas phase.

TABLE 15.1 X-ray Determined Bond Lengths and Root-Mean-Square Deviations for CS_2 in Liquid and Dilute Gas Phases at 20°C

j	k	Δ_{jk} (Å)	$\mathcal{R}_{jk}^{(0)}$ (Å)	*Phase*
C	S	0.09	1.57	Liquid
		0.036	1.559	Gas
S	S	0.082	3.106	Liquid
		0.045	3.113	Gas

Data from Sandler and Norten (1976).

Let us turn now to the intermolecular structure function S_d. In general it is difficult to unravel the orientation and spatial factors in eq. (15.3.7). The theory of the structure of polyatomic fluids is an important current area of research in statistical physics. Rather than go into any of the current work at this point, let us examine the qualitative behavior of S_d under the simplifying assumption that the orientational correlations between pairs of molecules are independent of the radial separation between molecular centers. Then eq. (15.3.7) becomes

$$S_d = 1 + M|\langle \mathcal{F}^{(1)} \rangle|^2 [S_{\text{cm}} - 1 + n(2\pi)^3 \delta(\mathbf{K})], \qquad (15.3.23)$$

where

$$S_{\text{cm}} = 1 + n \int e^{-i\mathbf{K}\cdot\mathbf{r}} [g_{\text{cm}}^{(2)}(r) - 1] d^3 r. \qquad (15.3.24)$$

$g_{\text{cm}}^{(2)}$ is the center of mass correlation function of a pair of molecules in the fluid.

Because of the factor $M|\langle \mathcal{F}^{(1)} \rangle|^2$, S_d depends not only on the intermolecular structure but also on molecular scattering factors.

If we further specialized to the homonuclear diatomic molecule for which $\mathcal{F}^{(1)}$ is given by eq. (15.3.10), then

$$S_d(K) = 1 + e^{-\Delta^2 K^2/4} \left[\frac{\sin(K\mathcal{R}_o/2)}{K\mathcal{R}_o/2} \right]^2 [S_{\text{cm}}(K) - 1] \quad (15.3.25)$$

for $K \neq 0$. For small K, $S_d(K) \simeq S_{\text{cm}}(K)$ but for K's such that $K\mathcal{R}_0/2$ is of order 1, the molecular factor $\sin(K\mathcal{R}_0/2)/(K\mathcal{R}_0/2)$ affects $S_d(K)$ and at the even higher K's where $\Delta K/2$ is of order 1 $S_d(K)$ is attenuated because of molecular vibrations. Because $MS_s = 1/2$ for the model considered here,

$$\frac{I_{2\theta}}{4|F_{2\theta}|^2 N I_0} = \frac{1}{2} + \frac{1}{2} e^{-\Delta^2 K^2/2} \frac{\sin K\mathcal{R}_0}{K\mathcal{R}_0} + e^{-\Delta^2 K^2/4} \left[\frac{\sin(K\mathcal{R}_0/2)}{K\mathcal{R}_0/2} \right]^2 [S_{\text{cm}}(K) - 1] \quad (15.3.26)$$

for $K = 0$. As $K \to 0$ the rhs of eq. (15.3.26) becomes $S_{\text{cm}}(K)$.

Because the zeroes of $\sin(K\mathcal{R}_0/2)$ [which would occur at integral multiples of $K = 2\pi/\mathcal{R}_0$ if the assumption of orientational independence leading to eq. (15.3.25) actually held] are close to those expected for $S_{\text{cm}} - 1$ for Br_2 it is difficult to judge from Figure 15.8 whether S_d has the shape predicted by eq. (15.3.23). To see at least theoretically how the molecular factors affect S_d, let us consider liquid nitrogen. Suppose the 6–12 Lennard-Jones potential adequately approximates the intermolecular potential of N_2 with parameters $\varepsilon/k = 9.15$ K and $\sigma = 3.681$ Å. Choose T and n such that $n\sigma^3 = 0.844$ and $kT/\varepsilon = 0.723$. Then the values for $S_{\text{cm}}(K)$ are given by the molecular dynamics results in Fig. 15.5. Because $\theta_V = 3340$ K, Δ is accurately predicted by $x_0/2$. Thus we estimate $\Delta = 0.0646$ Å from θ_V. The bond length of N_2 is known to be $\mathcal{R}_0 = 1.1$ Å. The K dependence of S_d predicted by eq. (15.3.25) is shown in Figure 15.9. The molecular factor $e^{-\Delta^2 K^2/4}$ contributes negligibly to S_d over the range of K's used in Figure (15.9). The factor $[\sin(K\mathcal{R}_0/2)/K\mathcal{R}_0/2]$ greatly attentuates S_d however. The first peak of S_d is substantially smaller than that of S_{cm}, and, beyond $K\sigma = 15$ the structure factor S_d has effectively reached its asymptotic value of 1. In agreement with the predictions of eq. (15.2.25), the measured value of the first peak of S_d is substantially smaller than one would expect for a monatomic fluid at the density of liquid Br_2 at 20°C. However, the remaining peaks of S_d in Br_2 do not decay as fast as is predicted by the model equation.

We conclude then that the intermolecular orientational correlations, neglected in obtaining eq. (15.3.25), cannot be neglected even for a diatomic molecule in the condensed phase. The implication of this is that determination of the detailed molecular structure of polyatomic fluids by diffraction can only be accomplished by combining experiment with good theoretical models of the molecular structure of these fluids.

We close this section with a discussion of the correlation function for homonuclear polyatomic fluids (e.g., Br_2, H_2, N_2, O_2, C_n) which *can* be determined directly from diffraction. This is the atom pair correlation function $g_{AA}^{(2)}(r)$, which is defined as the pair correlation function between an atom of one molecule and an atom of another molecule. For a homonuclear molecule with real atomic scattering amplitudes, that is $F_{2\theta} = f_{2\theta}$, eq. (15.3.7) can by rewritten in the form

$$S_d = 1 + \frac{(N-1)}{\nu^2} \sum_j^{\nu}{}' \sum_k^{\nu}{}' \langle e^{-i\mathbf{K}\cdot\mathbf{r}_{jk}} \rangle \qquad (15.3.27)$$

where the primes on the summation symbols mean that j and k correspond to atoms on different molecules. Because the probability density that a pair of atoms be found at separation $\mathbf{r}_{jk}$

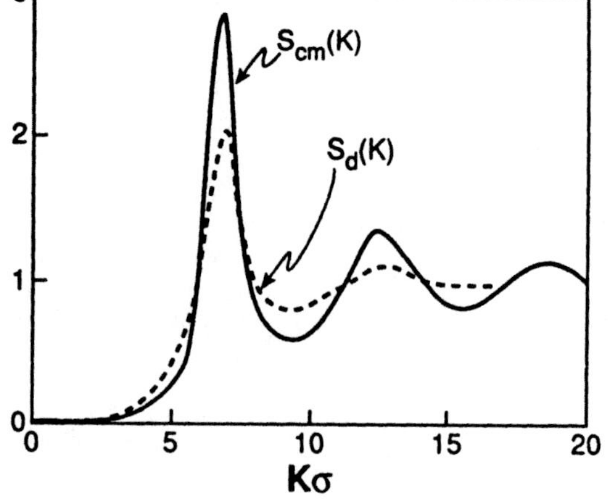

Figure 15.9
Comparison of diatomic structure factor S_d predicted by eq. (15.3.25) with S_{cm} determined by molecular dynamics for 6–12 Lennard-Jones model. $kT/\varepsilon = 0.723$, $n\sigma^3 = 0.884$, $r_0 = 1.1$ Å, and $\Delta = 0.0646$ Å.

is $g_{AA}(r_{jk})V^{-2}$, eq. (15.3.27) becomes with the usual manipulations

$$\begin{aligned} S_d &= 1 + \frac{(N-2)}{\nu^2} \sum_j^{\nu}{}' \sum_k^{\nu}{}' \frac{1}{V^2} \int g_{AA}(r_{jk}) e^{-i\mathbf{K}\cdot\mathbf{r}_{jk}} d^3r_j d^3r_k \\ &= 1 + \frac{(N-1)}{V} \int g_{AA}(r) e^{-i\mathbf{K}\cdot\mathbf{r}} d^3r \\ &= 1 + n \int [g_{AA}(r) - 1] e^{-i\mathbf{K}\cdot\mathbf{r}} d^3r + n \int e^{-i\mathbf{K}\cdot\mathbf{r}} d^3r. \end{aligned} \tag{15.3.28}$$

By Fourier transform inversion

$$\begin{aligned} g_{AA}(r) - 1 &= \frac{1}{(2\pi)^3 n} \int [S_d(K) - 1] e^{i\mathbf{K}\cdot\mathbf{r}} d^3K \\ &= \frac{1}{2\pi^2 n} \int_0^\infty [S_d(K) - 1] \frac{\sin Kr}{r} K dK. \end{aligned} \tag{15.3.29}$$

Thus, $g_{AA}(r)$ can be determined directly from diffraction data after the molecular factor S_m has been subtracted from the data.

As an example, the experimentally determined atom pair correlation function of liquid Br_2 is shown in Figure 15.10. The oscillations in g_{AA} at r's less than 3.5 Å are artifacts of the numerical evaluation of eq. (15.3.29). The irregularities beyond 7 Å are probably numerical artifacts too. Shown also in the figure is $n_{AA}(r)$, the average number of Br atoms around the atom of a reference molecule. For a ν-atom molecule, $n_{AA}(r)$ is defined by the relation

$$n_{AA}(r) = 4\pi\nu \int_0^r \xi^2 g_{AA}(\xi) d\xi. \tag{15.3.30}$$

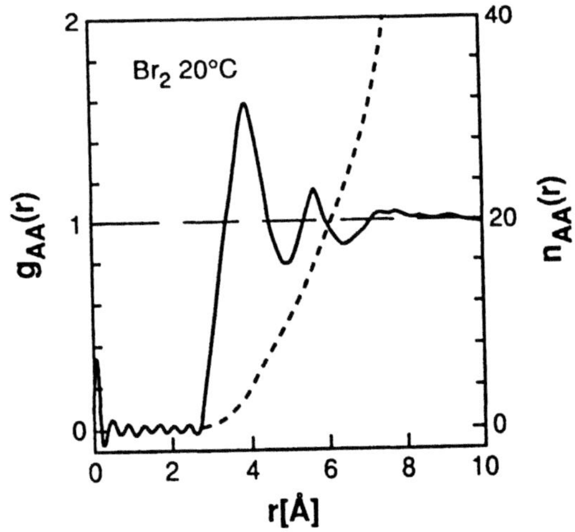

Figure 15.10
The intermolecular atom pair correlation function (g_{AA}) and the number (n_{AA}) of surrounding neighbors in liquid bromine. Adapted from Norten et al. (1978).

From the figure, we see that each Br atom has about 10 atoms on neighboring molecules in the first nearest neighbor range 2.7 Å< r < 4.9 Å.

15.4 Crystalline Solids

Diffraction determination of crystalline structures is generally more precise than that of fluids because of the distinct geometry of crystalline diffraction patterns. Bragg's law, which will be derived in what follows, provides an exceedingly simple description of crystalline diffraction patterns.

First, let us examine ideal crystals, that are defined by the properties that the atoms are fixed in space and that the atoms in the crystal can be located by translational replication of a basic unit cell. Consider the parallelepiped shown in Figure 15.11 with vectors $\mathbf{a}_1$, $\mathbf{a}_2$, and $\mathbf{a}_3$ drawn from a common origin 0 and lying along the three edges of the parallelepiped that intersects at 0.

Relative to the origin 0, the positions of atoms in such a parallelepiped fixed in a crystal may be denoted by the vector $\mathbf{R}^{(i)} = c_1^{(i)}\mathbf{a}_1 + c_2^{(i)}\mathbf{a}_2 + c_3^{(i)}\mathbf{a}_3$, where $c_1^{(i)}$, $c_2^{(i)}$, $c_3^{(i)}$ are positive fractions locating atom i in the parallelepiped. If the parallelepiped is a unit cell for an ideal crystal, the crystal may be constructed by replicating the unit cell such that the ith atom in the reference cell will have translational equivalents at the positions

$$\mathbf{R}_{\mathbf{n}}^{(i)} = \mathbf{R}^{(i)} + \mathbf{R}_n, \qquad (15.4.1)$$

where

$$\mathbf{R}_n = n_1\mathbf{a}_1 + n_2\mathbf{a}_2 + n_3\mathbf{a}_3, \quad n_i = 0, \pm 1, \pm 2, \ldots. \qquad (15.4.2)$$

Another way of stating the unit cell property is that any point $\mathbf{R}$ in a reference unit cell translated by $\mathbf{R}_{\mathbf{n}}$ will arrive at the equivalent point in another cell, that is, the atomic arrangement seen by an observer at $\mathbf{R}$ will be identical to that seen by an observer at $\mathbf{R} + \mathbf{R}_n$.

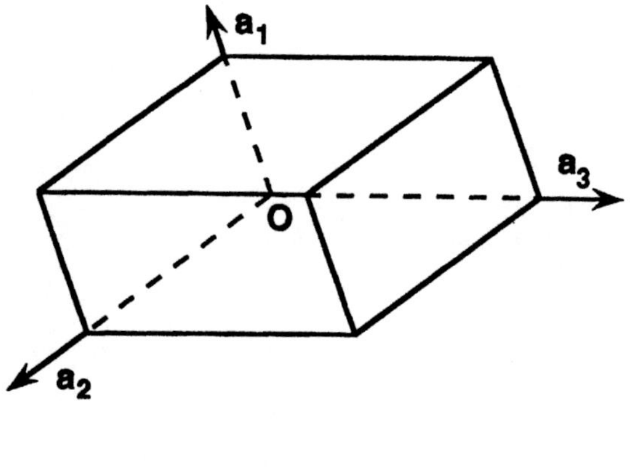

Figure 15.11
Parallepiped. The vectors $\mathbf{a}_1$, $\mathbf{a}_2$, $\mathbf{a}_3$ are neither collinear nor orthogonal in general. The volume of the parellepiped is equal to $\mathbf{a}_1 \cdot [\mathbf{a}_2 \times \mathbf{a}_3]$.

Examples of lattices with a cubic unit cell are shown in Figure 15.12. Figure 15.12a is a simple cubic lattice, $\mathbf{a}_1$, $\mathbf{a}_2$, and $\mathbf{a}_3$ being orthogonal and of the same length a and atoms lying at the corners of the cube shown. Figure 15.12b is a body-centered cubic (bcc) lattice that is the same as Figure 15.12a except there is also an atom in the center of the unit cell.

Consider now an ideal crystal containing c atoms in a unit cell. All or some of the atoms may be different chemical species. Because the positions of the atoms in the crystal can be located by the position vectors, eq. (15.4.1), the summation over particles in eq. (15.1.25) can be expressed in the form

$$\begin{aligned}\frac{I_{2\theta}}{I_0} &= \sum_{n_1,n_2,n_3} \sum_{m_1,m_2,m_3} \sum_{i=1}^{c} \sum_{l=1}^{c} F_{2\theta}^{(i)} \overline{F}_{2\theta}^{(l)} e^{-i\mathbf{K}\cdot(\mathbf{R}^{(i)}-\mathbf{R}^{(l)})-i\mathbf{K}\cdot(\mathbf{R}_n-\mathbf{R}_m)} \\ &= N|\mathcal{F}_c|^2 S_L(\mathbf{K}),\end{aligned} \tag{15.4.3}$$

where $\mathcal{F}_c$ is the *unit cell structure factor*

$$\mathcal{F}_c \equiv \sum_{i=1}^{c} F_{2\theta}^{(i)} e^{-i\mathbf{K}\cdot\mathbf{R}^{(i)}}, \tag{15.4.4}$$

$\mathbf{R}^{(i)}$ being the position of the ith atom in the reference unit cell, and S_L is the *lattice structure factor*

$$S_L(\mathbf{K}) \equiv \frac{1}{N} \sum_{n_1,n_2,n_3} \sum_{m_1,m_2,m_3} e^{-i\mathbf{K}\cdot(\mathbf{R}_n-\mathbf{R}_m)}. \tag{15.4.5}$$

We are free to choose the reference unit cell in the center of the crystal so that the ranges of the summations over n_μ and m_μ are (For large crystals, shape does not affect diffraction properties. Thus, we assume here that the crystal is a parallelepiped of the geometrical shape of the unit cell. This assumption will be dropped later in this section when small or thin crystals are considered.)

$$-\frac{M_\mu}{2} \le n_\mu, \quad m_\mu \le \frac{M_\mu}{2}, \quad \mu = 1, 2, 3, \tag{15.4.6}$$

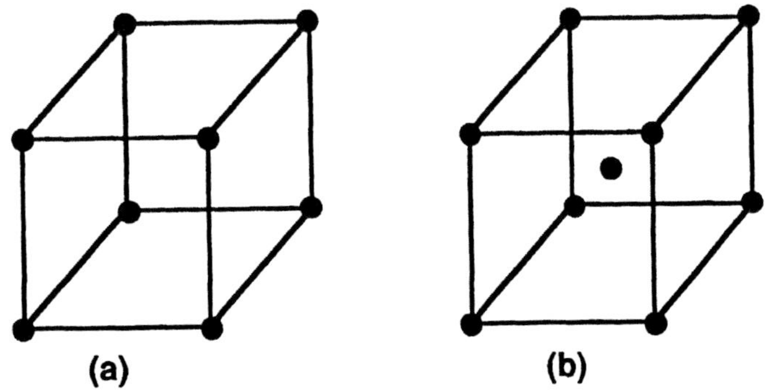

Figure 15.12
Cubic unit cells with (a) atoms at corners of cube and (b) atoms at corners and center of cube. (a) represents the unit cell of a simple cubic lattice and (b) that of a body-centered cubic lattice.

where M_μ is the number of repeated unit cells along the $\mathbf{a}_\mu$ axis of the reference unit cell. For convenience, let us also introduce the reciprocal lattice vectors $\mathbf{a}^1, \mathbf{a}^2, \mathbf{a}^3$, defined by the property

$$\mathbf{a}^\nu \cdot \mathbf{a}_\mu = \delta_{\nu\mu}, \quad \nu, \mu = 1, 2, 3 \tag{15.4.7}$$

Equation (15.4.7) fixes the form of the $\mathbf{a}^\nu$, which can be shown to be

$$\begin{aligned} \mathbf{a}^1 &= \frac{\mathbf{a}_2 \times \mathbf{a}_3}{V_0} \\ \mathbf{a}^2 &= \frac{\mathbf{a}_3 \times \mathbf{a}_1}{V_0} \\ \mathbf{a}^3 &= \frac{\mathbf{a}_1 \times \mathbf{a}_2}{V_0}, \end{aligned} \tag{15.4.8}$$

where

$$V_0 = \mathbf{a}_1 \cdot (\mathbf{a}_2 \times \mathbf{a}_3). \tag{15.4.9}$$

Example
If $|\mathbf{a}_1| = |\mathbf{a}_2| = |\mathbf{a}_3| = a$, and $\mathbf{a}_\nu \cdot \mathbf{a}_\mu = a^2\delta_{\nu\mu}$, find $\mathbf{a}^\nu$, $\nu = 1, 2, 3$. This is a cubic unit cell. $V_0 = a^3$ and, from eq. (15.4.8),

$$\mathbf{a}^1 = \frac{1}{a}\hat{a}_1, \quad \mathbf{a}^2 = \frac{1}{a}\hat{a}_2, \quad \mathbf{a}^3 = \frac{1}{a}\hat{a}_3.$$

Thus, for a cubic unit cell, the reciprocal lattice vectors are collinear with the lattice vectors of the unit cell.

Using the unit vectors $\hat{a}^\nu$, $\nu = 1, 2, 3$, of the reciprocal lattice vectors as a basis for $\mathbf{K}$,

$$\mathbf{K} = K_1\hat{a}^1 + K_2\hat{a}^2 + K_3\hat{a}^3, \tag{15.4.10}$$

and using the condition $\hat{\mathbf{a}}^\nu \cdot \mathbf{a}_\nu = a_\nu$, we find

$$\mathbf{K} \cdot \mathbf{R}_n = n_1K_1a_1 + n_2K_2a_2 + n_3K_3a_3. \tag{15.4.11}$$

Thus, eq. (15.4.5) can be written in the form

$$S_L(\mathbf{K}) = \frac{1}{N}\prod_{\nu=1}^{3}\sum_{m_\nu=-M_\nu/2}^{M_\nu/2}\sum_{n_\nu=-M_\nu/2}^{M_\nu/2} e^{-iK_\nu a_\nu(n_\nu-m_\nu)}. \tag{15.4.12}$$

From the properties of geometric series we can show that if $K_\nu a_\nu$ is not equal to an even multiple of π

$$\sum_{n_\nu=-M_\nu/2}^{M_\nu/2} e^{-iK_\nu a_\nu n_\nu} = \frac{1-e^{-i(M_\nu/2+1)K_\nu a_\nu}}{1-e^{-iK_\nu a_\nu}} + \frac{1-e^{i(M_\nu/2+1)K_\nu a_\nu}}{1-e^{iK_\nu a_\nu}} - 1, \tag{15.4.13}$$

and, if $K_\nu a_\nu$ = even multiple of π

$$\sum_{n_\nu=-M_\nu/2}^{M_\nu/2} e^{-iK_\nu a_\nu n_\nu} = M_\nu + 1. \tag{15.4.14}$$

The rhs of eq. (15.4.13) is of order 1, the rhs of eq. (15.4.14) is of order $N^{1/3}$. [Note that $c(M_1+1)(M_2+1)(M_3+1) = N$, where c is the number of atoms in a unit cell]. From these facts, we conclude that eqs. (15.4.13) and (15.4.14) imply the well known *Laue conditions* for ideal crystals, namely,

$$\begin{aligned} S_L &= \frac{N}{c^2} \quad \text{if } K_1a_1 = 2h\pi,\ K_2a_2 = 2k\pi,\ K_3a_3 = 2l\pi, \text{ and } h, k, l = \text{integers} \\ &= \mathcal{O}(N^{2/3}) \quad \text{if only one of } K_\nu a_\nu/2\pi \neq \text{integer} \\ &= \mathcal{O}(N^{1/3}) \quad \text{if only one of } K_\nu a_\nu/2\pi = \text{integer} \\ &= \mathcal{O}(1) \quad \text{if none of } K_\nu a_\nu/2\pi = \text{integer}. \end{aligned}$$

Thus, if $|\mathcal{F}_c|^2 \neq 0$, diffraction (i.e., a high intensity of the scattered beam) will occur in the direction corresponding to $K_1a_1 = 2\pi h$, $K_2a_2 = 2\pi k$, $K_3a_3 = 2\pi l$, that is,

$$\frac{I_{2\theta}}{N|\mathcal{F}_c|^2 I_0} = \frac{N}{c^2} \quad \text{if} \quad \begin{Bmatrix} K_1 \\ K_2 \\ K_3 \end{Bmatrix} = 2\pi \begin{Bmatrix} h/a_1 \\ k/a_1 \\ l/a_3 \end{Bmatrix}, \tag{15.4.15}$$

where the integers h, k, l are known as *Miller indices*. For any other values of K_1, K_2, and K_3 than those indicated in eq. (15.4.15), the scattered intensity $I_{2\theta}$ will be a factor of $N^{-1/3}$, or less, of the intensity of the diffraction spots.

From eqs. (15.4.13) and (15.4.14), the general form of $S_L(\mathbf{K})$ may be deduced to be

$$S_L(\mathbf{K}) = \frac{1}{N} \prod_{\nu=1}^{3} \frac{\sin^2[(M_\nu+1)K_\nu a_\nu/2]}{\sin^2(K_\nu a_\nu/2)}. \tag{15.4.16}$$

With the aid of the condition

$$\frac{\sin[(M_\nu+1)K_\nu a_\nu/2]}{\sin(K_\nu a_\nu/2)} = M_\nu + 1, \quad K_\nu a_\nu/2\pi = \text{integer},$$

the properties above attributed to $S_L(\mathbf{K})$ follow from eq. (15.4.16).

To gain a geometric appreciation of when diffraction occurs, let us introduce the vector $\mathbf{H}_{hkl}$:

$$\mathbf{H}_{hkl} = \frac{h}{a_1}\hat{a}^1 + \frac{k}{a_2}\hat{a}^2 + \frac{l}{a_3}\hat{a}^3. \qquad (15.4.17)$$

This vector is normal to the plane that intersects the unit cell at a_1/h, a_2/k, and a_3/l along $\mathbf{a}_1$, $\mathbf{a}_2$, and $\mathbf{a}_3$. The vector and plane are shown in Figure 15.13. Although for convenience of drawing, $\mathbf{a}_1$, $\mathbf{a}_2$, and $\mathbf{a}_3$ are shown as orthogonal axes, this has not been assumed in the theoretical development.

To prove that the vector $\mathbf{H}_{hkl}$ is normal to the plane shown in Figure 15.13, consider the unite vector $\hat{n}_{hkl}$ normal to the plane and originating from the origin 0. The distance d_{nkl} from the origin to the plane along $\hat{n}_{hkl}$ is equal to either $\mathbf{a}_1/h$, $\mathbf{a}_2/k$, or $\mathbf{a}_3/l$ onto $\hat{n}_{nkl}$, that is,

$$d_{nkl} = \frac{\mathbf{a}_1}{h}\cdot\hat{n}_{hkl} = \frac{\mathbf{a}_2}{k}\cdot\hat{n}_{hkl} = \frac{\mathbf{a}_3}{l}\cdot\hat{n}_{hkl} \qquad (15.4.18)$$

But if $\hat{n}_{hkl}$ is expressed in the reciprocal vector basis,

$$\hat{n}_{nkl} = \alpha_1\hat{\mathbf{a}}^1 + \alpha_2\hat{\mathbf{a}}^2 + \alpha_3\hat{\mathbf{a}}^3,$$

then the conditions $\hat{\mathbf{a}}_i \cdot \hat{\mathbf{a}}^j = \delta_{ij}$ and eq. (15.4.18) imply $\alpha_1 = d_{hkl}\frac{h}{a_1}$, $\alpha_2 = d_{hkl}\frac{k}{a_2}$, and $\alpha_3 = d_{hkl}\frac{l}{a_3}$, or

$$\hat{n}_{hkl} = d_{hkl}\left[\frac{h}{a_1}\hat{a}^1 + \frac{k}{a_2}\hat{a}^2 + \frac{l}{a_3}\hat{a}^3\right] \equiv d_{hkl}\mathbf{H}_{hkl} \qquad (15.4.19)$$

verifying the claim the the vector $\mathbf{H}_{hkl}$ is normal to the plane. Moreover, from this result we find

$$\mathbf{H}_{hkl} = \frac{1}{d_{hkl}}\hat{n}_{hkl}, \qquad (15.4.20)$$

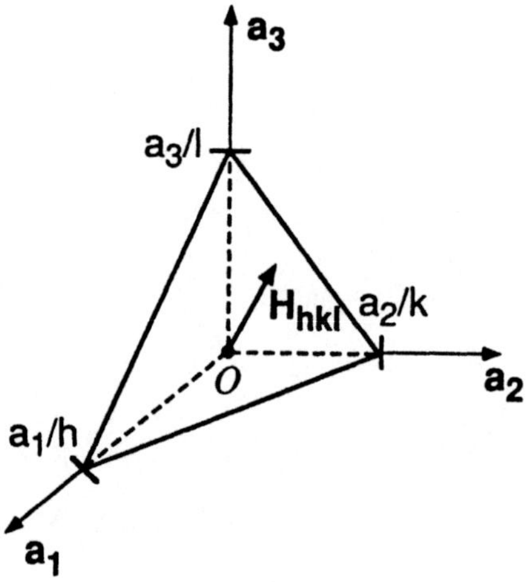

Figure 15.13
Plane corresponding to the Miller indices (h, k, l) and the vector $\mathbf{H}_{hkl}$ normal to the plane.

implying that the scalar value (length) of $\mathbf{H}_{hkl}$ is equal to the inverse of distance between the plane shown in Figure 15.13 and parallel plane passing through the origin 0.

All planes constructed as the one shown in Figure 15.13 and corresponding to the Miller indices $(\frac{h}{n}, \frac{k}{n}, \frac{l}{n})$, $n = \pm 1, \pm 2, \ldots$ are parallel to the plane (h, k, l) and d_{hkl} is the distance between any adjacent pair of these planes.

Physically, then, we obtain diffraction when the quantity $\mathbf{K}/2\pi$ is normal to the plane described by an integer set of Miller indices (h, k, l) and is equal to magnitude to d_{hkl}. This follows from eq. (15.4.16) in which the Laue conditions are stated in the form.

$$\frac{\mathbf{K}}{2\pi} = \frac{h}{a_1}\hat{a}^1 + \frac{k}{a_2}\hat{a}^2 + \frac{l}{a_3}\hat{a}^3 = \mathbf{H}_{hkl}. \tag{15.4.21}$$

Because the relation between the scattering angle θ and $\mathbf{K}$ is $K = \frac{4\pi}{\lambda}\sin\theta$ and $K/2\pi = d_{hkl}$, we find the familiar statement of the diffraction conditions of *Bragg's law*,

$$2d_{hkl}\sin\theta = \lambda, \tag{15.4.22}$$

relating the diffraction angle to the wavelength of the incident wave and to the spacing between the parallel plane from which the scattering is coherent.

An orthorhombic crystal is defined as one whose unit cell lattice vectors are orthogonal, $\mathbf{a}_\nu \cdot \mathbf{a}_\mu = a_\nu a_\mu \delta_{\nu\mu}$, and, therefore, whose reciprocal lattice vectors obey the relations $\hat{a}^\nu = \hat{a}_\nu$ and $\hat{a}^\nu \cdot \hat{a}^\mu = \delta_{\nu\mu}$. In this case,

$$d_{hkl}^2 = \frac{h^2}{a_1^2} + \frac{k^2}{a_2^2} + \frac{l^2}{a_3^2} \tag{15.4.23}$$

and Bragg's law may be written in the form

$$\frac{h^2}{a_1^2} + \frac{k^2}{a_2^2} + \frac{l^2}{a_3^2} = \frac{4}{\lambda^2}\sin^2\theta. \tag{15.4.24}$$

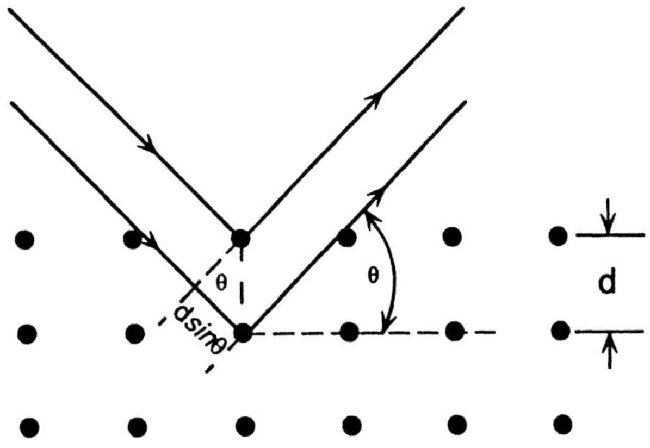

Figure 15.14
Bragg diffraction from a tetragonal crystal. The waves scattered from successive parallel planes will be in phase if the extra distance $2d\sin\theta$ traveled by the wave reflected from the second plane over that traveled by the wave reflected from the first plane is equal to an integral multiple of the wavelength, that is, if $2d\sin\theta = n\lambda$, $n = 1, 2, \ldots$, the scattered beam will be intense. This is equivalent to eq. (15.4.22) for Bragg diffraction.

An example of an ideal crystal diffraction pattern is given in Figure 15.15. Each of the spots on the film corresponds to diffraction from a particular set of parallel planes (h, k, l). For a given set of Miller indices, (h, k, l), the direction of the diffracted beam relative to the incident beam will be $\hat{\mathbf{k}} - \hat{\mathbf{k}}_0 = \lambda \mathbf{H}_{hkl}$ and the angle of deflection may be determined by eq. (15.4.22). The number and arrangement of spots seen in an experiment such as that shown in Figure 15.15 will depend on the origination of $\mathbf{a}_1$, $\mathbf{a}_2$, and $\mathbf{a}_3$ relative to $\hat{\mathbf{k}}_0$, and, as we shall discuss shortly, the values of $|\mathcal{F}_c|^2$ for the Bragg values of $\mathbf{K}$.

If instead of a single crystal, one studies powders of crystals or a polycrystalline sample, then the spots that would show up in single crystal pattern will become rings. Imagine a small crystal that gives a Bragg spot in the direction $\hat{\mathbf{k}} = \hat{\mathbf{k}}_0 + \lambda \mathbf{H}_{hkl}$. The plane of this crystal makes an angle θ with respect to $\hat{k}_0$. If the crystal is rotated keeping the angle between the (h, k, l) plane and $\hat{\mathbf{k}}_0$ fixed at θ, then the diffraction spot will rotate about the $\hat{\mathbf{k}}_0$ axis and will trace out on the film in Figure 15.15 a circle corresponding to a cone whose side makes an angle 2θ with its axis. If the powder of crystals of the polycrystalline sample has enough randomly oriented single crystal regions, all the diffraction spots seen in a single crystal experiment of the type illustrated in Figure 15.15 will become rings. In this sense a polycrystalline X-ray scattering pattern is similar to that of a liquid (Fig. 15.6). However, the diffraction rings of a crystalline sample are much more intense than those of a liquid. Also, more rings are resolved for a crystalline sample.

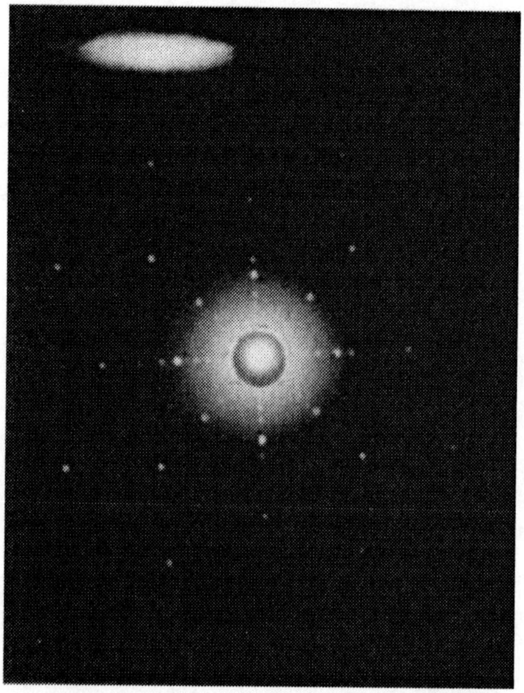

Figure 15.15
Laue pattern for 001 plane in crystalline silicon. Courtesy of M. D. Ward.

Thus far we have concentrated on the diffraction pattern allowed by the lattice structure factor S_L. However, if the scattering planes corresponding to a given Miller indices (h, k, l) have no scattering atoms or, more likely, have bisecting planes with an identical atomic occupation, diffraction by these planes will be absent. The absence of diffraction by a given crystal plane in our analysis is equivalent to the cell structure factor $|\mathcal{F}_c|^2$ being zero for that plane. The easiest way to illustrate the role of the cell structure factor in diffraction is to consider a cubic lattice ($a_1 = a_2 = a_3 = a$, $\mathbf{a}_\nu \cdot \mathbf{a}_\mu = a^2\delta_{\nu\mu}$) of identical atoms for the simple cubic (sc), the face-centered cubic (fcc), and the bcc structures. In the sc crystal, there is one atom at the origin of the unit cell axes ($\mathbf{R}^{(1)} = 0$), the other atoms being reached by the translations $\mathbf{R}_n$; in the fcc crystal, there are atoms at $\mathbf{R}^{(1)} = 0$, $\mathbf{R}^{(2)} = (a/2)(\hat{a}_1 + \hat{a}_2)$, $\mathbf{R}^{(3)} = (a/2)(\hat{a}_1 + \hat{a}_3)$, and $\mathbf{R}^{(4)} = (a/2)(\hat{a}_2 + \hat{a}_3)$, the other atoms being reached by the translations $\mathbf{R}_n$; in the bcc crystal, there are atoms at $\mathbf{R}^{(1)} = 0$ and $\mathbf{R}^{(2)} = (a/2)(\hat{a}_1 + \hat{a}_2 + \hat{a}_3)$, the other atoms being reached by the translation $\mathbf{R}_n$. Thus, from the definition of $\mathcal{F}_c$, eq. (15.4.4), we find for diffraction values of $\mathbf{K}$ for the cubic lattice

$$\text{sc:}\quad \mathcal{F}_c = F_{2\theta} \tag{15.4.25}$$

$$\text{fcc:}\quad \mathcal{F}_c = F_{2\theta}[1 + e^{-i\pi(h+k)} + e^{-i\pi(h+l)} + e^{-i\pi(k+l)}] \tag{15.4.26}$$

$$\text{bcc:}\quad \mathcal{F}_c = F_{2\theta}[1 + e^{-i\pi(h+k+l)}]. \tag{15.4.27}$$

In Table 15.2, we list values of $|\mathcal{F}_c|^2/|f_{2\theta}|^2$ for the three structures for several planes of the cubic lattice.

From a powder pattern strip of the type shown in Figure 15.16, it would be an easy matter to distinguish the three crystals of Table 15.2.

TABLE 15.2 Unit Cell Structure Factors, $|\mathcal{F}_c|^2$ for sc, fcc, and bcc One-Component Crystals

Miller Indices $(h, k, l)^a$	$\lvert\mathcal{F}_c\rvert/\lvert f_{2\theta}\rvert$ *(sc)*	$\lvert\mathcal{F}_c\rvert^2/\lvert f_{2\theta}\rvert^2$ *(fcc)*	$\lvert\mathcal{F}_c\rvert^2/\lvert f_{2\theta}\rvert^2$ *(bcc)*	$h^2 + k^2 + l^2$
(1, 0, 0)	1	0	0	1
(1, 1, 0)	1	0	4	2
(1, 1, 1)	1	16	0	3
(2, 0, 0)	1	16	4	4
(2, 1, 0)	1	0	0	5
(2, 1, 1)	1	0	4	6
(2, 2, 0)	1	16	4	8
(3, 0, 0)	1	0	0	9
(2, 2, 1)	1	0	0	9
(3, 1, 0)	1	0	4	10
(3, 1, 1)	1	16	0	11

Also given are values of $h^2 + k^2 + l^2$ $(= \frac{4a^2}{\lambda^2}\sin^2\theta)$.

[a] Only one example of equivalent indices, such as (1, 0, 0), (0, 1, 0) and (0, 0, 1), is given.

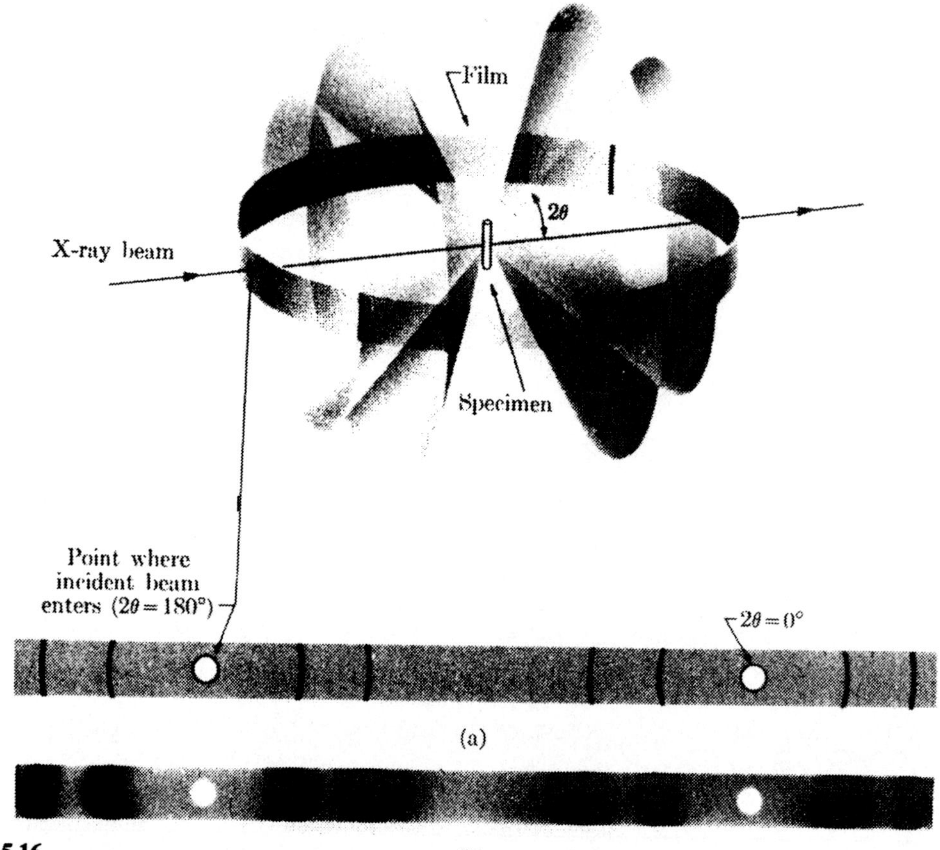

Figure 15.16
Powder pattern. Each diffraction line arises from a specific interplanar spacing. Through the use of monochromatic x-rays, the value of d_{hkl} may be calculated eq. (7.1.84). (Adapted from Cullity, 1956).

For the sc crystal the ratios of observed values of $\sin^2\theta$ from the successive bands (measured from the direction $\hat{\mathbf{k}}_0$) will be 1:2:3:4. . ., for the fcc crystal the ratios will be 3:4:8:11. . . , and for the bcc crystal the ratios will be 2:4:6:8:10:12:14:. . . . Note that the first band to distinguish between sc and bcc is the seventh one, because bcc has a band for $h^2 + k^2 + l^2 = 14$ but sc does not have one at $h^2 + k^2 + l^2 = 7$.

Next let us investigate the effect of atomic vibrations on the diffraction pattern of a crystal. For simplicity, let us assume that the vibrations are described by the classical Einstein model, that is, each atom behaves as an independent harmonic oscillation. Thus, the probability density that the ith atom of the reference unit cell is at $\mathbf{r}_i$ is given by

$$p(\mathbf{r}_i) = \frac{1}{(2\pi\Delta^2)^{3/2}} e^{-(\mathbf{r}_i - \mathbf{R}^{(i)})^2/2\Delta_i^2}, \tag{15.4.28}$$

where

$$\Delta_i^2 = kT/\lambda_i \tag{15.4.29}$$

λ_i is the lattice force constant of the ith atom. The lattice force constant is related to the characteristic Einstein temperature θ_E by

$\theta_E = (\hbar/k)\sqrt{\lambda/m}$ for a monatomic crystal. For this model, we obtain

$$\begin{aligned}\langle|\mathcal{F}_c|^2\rangle &= \sum_{i=1}^{c}\sum_{j=1}^{c} F_{2\theta}^{(i)}\overline{F}_{2\theta}^{(j)}\langle e^{-i\mathbf{r}_{ij}\mathbf{K}}\rangle \\ &= \sum_{i=1}^{c}|F_{2\theta}^{(i)}|^2 + \sum_{i\neq j} F_{2\theta}^{(i)}\overline{F}_{2\theta}^{(j)} \int e^{-i\mathbf{r}_i\mathbf{K}+i\mathbf{r}_j\cdot\mathbf{K}} p(\mathbf{r}_i)p(\mathbf{r}_j)d^3r_i d^3r_j \quad (15.4.30)\\ & \sum_{i=1}^{c}|F_{2\theta}^{(i)}|^2 + \sum_{i\neq j} F_{2\theta}^{(i)}\overline{F}_{2\theta}^{(j)} e^{-i\mathbf{K}\cdot(\mathbf{R}^{(i)}-\mathbf{R}^{(j)})} e^{-(K^2/2)(\Delta_i^2+\Delta_j^2)},\end{aligned}$$

where we have used the property

$$\frac{1}{(2\pi\Delta^2)^{3/2}}\int e^{-\xi^2/2\Delta^2} e^{-i\mathbf{K}\cdot\boldsymbol{\xi}} d^3\xi = e^{-K^2\Delta^2/2}. \qquad (15.4.31)$$

Equation (15.4.30) can be used to express the scattering intensity of an Einstein crystal in the form

$$\frac{I_{2\theta}}{NI_0} = \left[\sum_{i=1}^{c} F_{2\theta}^{(i)2} + \sum_{i\neq j}^{c} F_{2\theta}^{(i)}\overline{F}_{2\theta}^{(j)} e^{-K^2(\Delta_i^2+\Delta_j^2)/2} e^{-i\mathbf{K}\cdot(\mathbf{R}^{(i)}-\mathbf{R}^{(j)})}\right] S_L^{Id}(\mathbf{K}). \quad (15.4.32)$$

Thus, for an Einstein crystal the lattice structure factor is the same as that of the ideal crystal of fixed particles discussed in the previous paragraphs. However, the cell structure factor is affected by the vibrational motions of the atom; the larger the values of **K** (and, therefore, of $h^2 + k^2 + l^2$ or θ) the more the thermal vibrations affect $\langle|\mathcal{F}_c|^2\rangle$. The effect also increases with T because $\Delta^2 \propto T$. As mentioned earlier the quantities $e^{-K^2\Delta_i^2/2}$ are known as *Debye–Waller factors*.

Example

Consider a monatomic cubic crystal with the following characteristics:

$$a = 3 \times 10^{-8}\ cm$$
$$\theta_E = 400K$$
$$\text{molecular weight} = 25.$$

Find the Miller indices for which $e^{-K^2\Delta^2} \leq 0.9$ or $K^2\Delta^2 \geq 0.105$ at $T = 300$ K and $T = 1000$ K for the Laue values of K.

For a cubic crystal Laue diffraction occurs for

$$K = 4\pi^2(h^2 + k^2 + l^2)/a^2.$$

At $T = 300$ K,

$$\frac{4\pi^2}{a^2}\frac{T\hbar^2}{km\theta_E^2} = 1.6087 \times 10^{-2},$$

TABLE 15.3 Electron Diffraction of a Polyethylene Single Crystal

Reflection	d_{hkl} (Å)	$T = 0$ K $I_{2\theta}/[(N/c^2)I_0]$ (Å^2)[a]	$T = 300$ K $I_{2\theta}/[(N/c^2)I_0]$ Å^2
100	7.40	–	–
010	4.93	–	–
110	4.10	119	105.5
200	3.70	84	72
020	2.46	45	33
210	2.95	6	4.9
120	2.34	4	2.75
310	2.20	17	11.1
130	1.60	33	15
220	2.05	9	5.6

Courtesy of Thomas and Sass (1973).

[a] Based on 4C and 8H per unit cell with 200 keV electrons, correcting for relativistic effects and using a C–H distance of 1.12 Å.

so that for

$$h^2 + k^2 + l^2 \geq 6.5$$

$e^{-K^2\Delta^2} \leq 0.9$ at $T = 300$ K. Similarly we find that for

$$h^2 + k^2 + l^2 \geq 2$$

$e^{-K^2\Delta^2} \leq 0.9$ at $T = 1000$ K.

From the example given here, we see that thermal vibrations can cause severe attentuation of the diffraction intensities for sufficiently large angles θ (i.e., sufficiently large $h^2 + k^2 + l^2$). An example of such attentuation is given in Table 15.3, where the quantity $I_{2\theta}/[(N/c^2)I_0]$ is computed for polyethylene at $T = 0$ K ($\Delta = 0$) and at $T = 300$ K ($\Delta^2 = 0.16$ Å^2 for C–C vibrations. Hydrogen contributions are ignored in calculation of Δ.).

It is interesting to note that because the vibrational attentuation of the diffraction intensity goes as $e^{-K^2\Delta^2}$, experimental measurement of this attenuation provides an estimate of the mean square vibrational displacement Δ^2 of the atoms of the lattice.

15.5 Small Crystallites

In this section we examine the diffraction behavior of ideal crystals that are thin in at least one dimension. As seen in the following the effect of thinness is to cause the diffraction spots to have resolvable satellite structures.

Consider the behavior of $S_L(\mathbf{K})$ near a diffraction spot. Define s_1, s_2, s_3 to be small deviations of K_1, K_2, and K_3 from $2\pi h$, $2\pi k$, and $2\pi l$. Thus,

$$e^{-iK_\nu a_\nu} = e^{-is_\nu a_\nu}, \qquad (15.5.1)$$

and for a parallelepiped crystal geometrically similar to the unit cell, eq. (15.4.16) enables us to express S_L in the form

$$S_L = \frac{N}{c^2} \prod_{\nu=1}^{3} \mathcal{I}(s_\nu), \qquad (15.5.2)$$

where

$$\mathcal{I}(s_\nu) \equiv \frac{\sin^2[(M_\nu + 1)s_\nu a_\nu/2]}{(M_\nu + 1)^2 \sin^2(s_\nu a_\nu/2)}. \qquad (15.5.3)$$

Beecause s_ν denotes a small deviation of K_ν from its Bragg value, then $s_\nu a_\nu$ will be small compared to unity so that

$$\sin(s_\nu a_\nu/2) \simeq s_\nu a_\nu/2, \qquad (15.5.4)$$

a result that may be used to simplify $\mathcal{I}(s_\nu)$ to the form

$$\mathcal{I}(s_\nu) = \frac{\sin^2[(M_\nu + 1)s_\nu a_\nu/2]}{[(M_\nu + 1)s_\nu a_\nu/2]^2}. \qquad (15.5.5)$$

The quantity $\mathcal{I}(s_\nu)$ is plotted in Figure 15.17. As is illustrated in Figure 15.17, $\mathcal{I}(s_\nu)$ is of appreciable magnitude only for $s_\nu a$ equal to a few multiples of $1/(M_\nu + 1)$.

A few special cases will serve to emphasize the effect of crystal size on the diffraction pattern.

Case 1. Large crystal. $M_\nu = O(N^{1/3})$, $\nu = 1, 2, 3$. In this case the factor $\mathcal{I}(s_\nu)$ is of negligible magnitude for $s_\nu a_\nu$ greater than a few multiples of $N^{-1/3}$. Thus, the satellite structure on a Bragg spot would not be resolved for a crystal thick in every dimension.

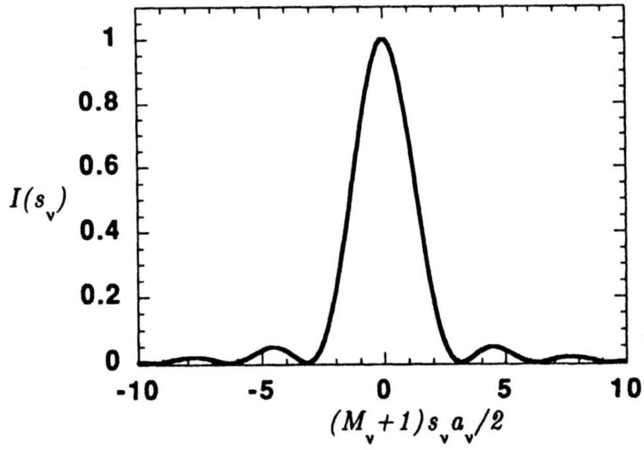

Figure 15.17
Satellite structure of diffraction spot at $s_\nu = 0$. $\mathcal{I}(s_\nu)$ is defined by eq. (15.5.5).

Case 2. Crystal thin in one dimension, say along $\mathbf{a}_1$, $M_1 \ll M_2$, M_3, *and so* $M_2, M_3 = \mathcal{O}(N^{1/2})$. In this case, $\mathcal{I}(\mathfrak{s}_\nu)$, $\nu = 2, 3$, will be negligible for $\mathfrak{s}_\nu a_\nu$ larger than a few multiples of $N^{-1/2}$, so that the diffraction spot is very sharp with respect to variations of $\mathbf{K}$ in the directions $\hat{\mathbf{a}}^2$ and $\hat{\mathbf{a}}^3$. However, if M_1 is sufficiently small, then $\mathcal{I}(\mathfrak{s}_1)$ may be appreciable over a broad detectable range of $\mathfrak{s}_1$ [because $\mathcal{I}(\mathfrak{s}_1)$ is of order 1 for $0 < \mathfrak{s}_1 a_1 < a$ few multiples of $(M_1 + 1)^{-1}$]. The diffraction patterns typical of a thin crystal are illustrated in Figure 15.18. If the beam is in the direction $\mathbf{a}_1$, then a planar film whose normal is $\hat{\mathbf{a}}_1$ will pick up the spots corresponding to $\mathbf{K} = (2\pi k/a_2)\hat{a}^2 + (2\pi l/a_3)\hat{a}^3$, that are very sharp because $\mathbf{a}_2$ and $\mathbf{a}_3$ are the thick directions of the crystal. If the beam is in the direction of $\mathbf{a}_2$, then a planar film normal to the beam will pick up the spots corresponding to $\mathbf{K} = (2\pi h/a_1)\hat{\mathbf{a}}^1 + (2\pi l/a_3)\hat{\mathbf{a}}^3$. However, instead of sharp circular spots, there will be rodlike spots or streaks, with the axis of the rod running in the direction $\hat{\mathbf{a}}^1$ corresponding to the finite range $\mathfrak{s}_1$ over which $\mathcal{I}(\mathfrak{s}_1)$ is appreciable. These two situations are illustrated in (2-A) and (2-B) of Figure 15.18.

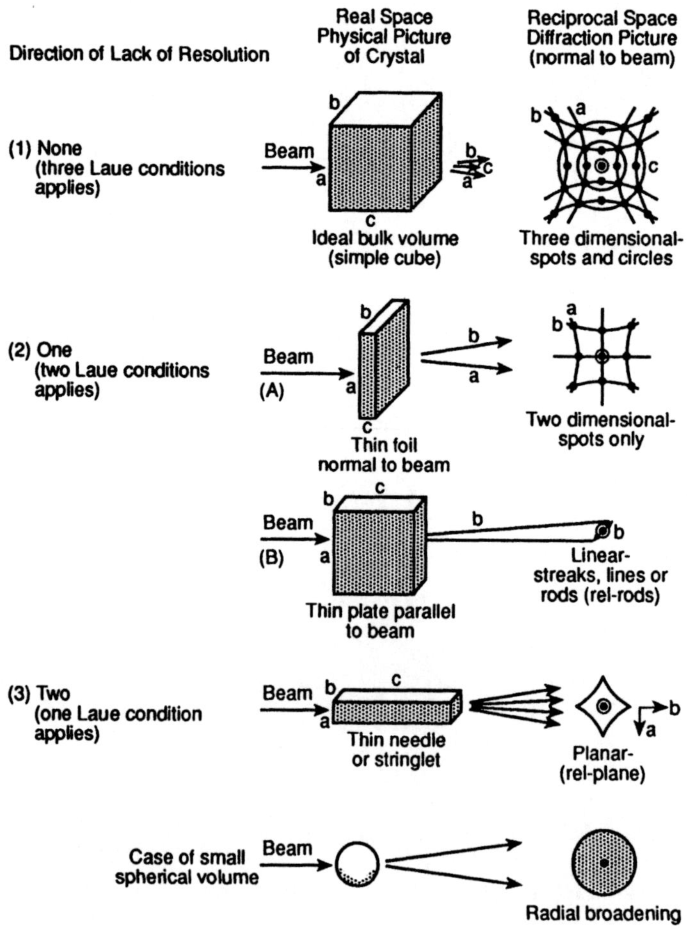

Figure 15.18
Summary of diffraction effects from crystals.

Case 3. Crystal thin in two dimensions, say $\mathbf{a}_1$ *and* $\mathbf{a}_2$, M_1, $M_2 \ll M_3$ *and so* $M_3 = \mathcal{O}(N)$. In this case, $\mathcal{I}(\mathfrak{s}_3)$ will be negligible for $\mathfrak{s}_3 a_3$ larger than a few multiples of N^{-1}, whereas $\mathcal{I}(\mathfrak{s}_1)$ and $\mathcal{I}(\mathfrak{s}_2)$ will be detectable for $\mathfrak{s}_1 a_1$ and $\mathfrak{s}_2 a_2$ ranging up to a few multiples of $(M_1 + 1)^{-1}$ and $(M_2 + 1)^{-1}$. For a beam in the direction of $\mathbf{a}_3$, the spots corresponding to $\mathbf{K} = (2\pi h/a_1)\hat{a}^1 + (2\pi l/a_2)\hat{a}^2$ will be spread out as illustrated in (3) of Figure 15.18, the spreading arising from the range of values of $\mathfrak{s}_1$ and $\mathfrak{s}_2$ over which the product $\mathcal{I}(\mathfrak{s}_1)\mathcal{I}(\mathfrak{s}_2)$ is large enough to give a detectable signal.

Let us next consider crystals of arbitrary shape and having at least one thin dimension. For $\mathbf{K}$ values in the vicinity of a diffraction spot, eq. (15.4.12) can be written as

$$S_L = \frac{N}{c^2}|\mathcal{T}_{\mathfrak{s}}|^2, \qquad (15.5.6)$$

where

$$\mathcal{T}_{\mathfrak{s}} = \sum_{n_1,n_2,n_3} e^{-i(n_1 a_2 \mathfrak{s}_1 + n_2 a_2 \mathfrak{s}_2 + n_3 a_3 \mathfrak{s}_3)}. \qquad (15.5.7)$$

To reduce eq. (15.5.7) further for an arbitrarily shaped crystal of arbitrary size is rather difficult. However, if we assume that the crystal is sufficiently large that the $a_\nu \mathfrak{s}_\nu$ are quite small compared to unity over the range of $\mathfrak{s}_\nu$'s for which the $\mathcal{I}(\mathfrak{s}_\nu)$'s are appreciable, then the sums in eq. (15.5.7) may be approximated by integrals. Thus, if we define

$$\begin{aligned} x_\nu &\equiv n_\nu a_\nu \\ dx_\nu &\equiv a_\nu \Delta n_\nu, \quad \Delta n_\nu \equiv 1, \end{aligned} \qquad (15.5.8)$$

then

$$\sum_{n_\nu} e^{-i n_\nu a_\nu \mathfrak{s}_\nu} \simeq \frac{1}{a_\nu}\int e^{-i x_\nu \mathfrak{s}_\nu} dx_\nu \qquad (15.5.9)$$

and

$$\mathcal{T}_{\mathfrak{s}} = \frac{c}{a_1 a_2 a_3 N}\int\int\int e^{i(x_1\mathfrak{s}_1 + x_2\mathfrak{s}_2 + x_3\mathfrak{s}_3)} dx_1 dx_2 dx_3. \qquad (15.5.10)$$

The quantity $d^3r \equiv \hat{\mathbf{a}}_1 dx_1 \cdot [(\hat{\mathbf{a}}_2 dx_2) \times (\hat{\mathbf{a}}_3 dx_3)] = \hat{\mathbf{a}}_1 \cdot [\hat{\mathbf{a}}_2 \times \hat{\mathbf{a}}_3]\, dx_1 dx_2 dx_3$ is the volume of an infinitesimal parallelepiped with side along $\hat{\mathbf{a}}_1$, $\hat{\mathbf{a}}_2$, and $\hat{\mathbf{a}}_3$ and of lengths dx_1, dx_2, and dx_3, respectively. Thus, we may write eq. (15.5.10) in the form

$$\mathcal{T}_{\mathfrak{s}} = \frac{1}{V}\int_V e^{i\mathbf{r}\cdot\mathfrak{s}} d^3r, \qquad (15.5.11)$$

where we have used the relation $V = NV_0/c$, $V_0 = \mathbf{a}_1 \cdot (\mathbf{a}_2 \times \mathbf{a}_3)$ being the volume of the unit cell and $\mathbf{r}$ the position vector locating d^3r. The integration is over the volume V of the crystal. It is interesting that eq. (15.5.11) depends only on the shape V of the crystal, implying that the satellite structure of the Bragg spots for small or thin crystals depend only on the shape of the crystal and not on the geometry of the unit cells.

If the crystal is a sphere of radius R, it is easy to compute $\mathcal{T}_s$. Using a spherical coordinate system centered in the sphere, we find

$$\mathcal{T}_s = \frac{1}{V}\int_0^{\pi}\int_0^{2\pi}\int_0^{R} e^{irs\cos\theta}\sin\theta d\theta d\phi r^2 dr = \frac{3(\sin Rs - Rs\cos Rs)}{(Rs)^3}$$

and so

$$|\mathcal{T}_s|^2 = \left[3\left(\frac{\sin Rs - Rs\cos Rs}{(Rs)^3}\right)\right]^2 \tag{15.5.12}$$

for a small spherical system. At $s = 0, \mathcal{T}_s = 1$. For sufficiently large R, $\mathcal{T}_s$ will be negligibly small for s greater than a few multiples of R^{-1} so that the Bragg spots will be sharp. For sufficiently small spheres, however, all the Bragg spots will be diffuse, the satellite structure depending only on the magnitude of s and not on its direction, that is, the structure is spherically symmetric in s. A small sphere is included in Figure 15.18.

Exercise

Consider a crystal with a cubic unit cell. Derive from eq. (15.5.11) an expression for $|\mathcal{T}_s|^2$ for a rectangular parallelepiped thin in the x direction, that is, sides $L_x \ll L_y$, L_z with L_y and L_z very large compared to the unit cell side a. If the x direction lies along the direction of the body diagonals of the unit cells of the crystal, describe the diffraction pattern for the crystal.

When the small particles are distributed randomly (e.g., in a powder pack) the observed value of $|\mathcal{T}_s|^2$ will be averaged over all orientations. The formulas for $\langle|\mathcal{T}_s|^2\rangle$ for several simple volumes averaged over a uniform distribution or orientations ard given below

Very thin cylinder of length $2L$

$$\langle|\mathcal{T}_s|^2\rangle = \left[\frac{Si(2sL)}{sL} - \frac{\sin^2 sL}{(sL)^2}\right], \tag{15.5.13}$$

where

$$Si(x) = \int_0^x \frac{\sin t}{t} dt. \tag{15.5.14}$$

Flat disks of radius R

$$\langle|\mathcal{T}_{\mathfrak{s}}|^2\rangle = \frac{2}{(\mathfrak{s}R)^2}\left[1 - \frac{1}{\mathfrak{s}R}J_1(2\mathfrak{s}R)\right], \tag{15.5.15}$$

where J_1 is the Bessel function of first order.

Ellipsoids of revolution of axes 2a, 2a, and 2b.

$$\langle|\mathcal{T}_{\mathfrak{s}}|^2\rangle = \int_0^{\pi/2}\left[\Phi\left(\mathfrak{s}a\sqrt{\cos^2\theta + (b/a)^2\sin^2\theta}\right)\right]^2 \cos\theta d\theta \tag{15.5.16}$$

where

$$\Phi(x) = 3\frac{\sin x - x\cos x}{x^3}. \tag{15.5.17}$$

15.6 Colloidal Dispersions and Small Angle Scattering

Colloids are aggregates of molecules dispersed in a liquid phase. Examples of colloids are surfactant micelles in water (soaplike molecules that spontaneously aggregate into spherical, cylindrical, or discoid objects with polar beads pointing outward and oily tails pointing inward), proteins in solution, latex spheres suspended in a solvent, gold sols, dispersed crystallites of silicas or clays, and the like. The colloidal particles range in size from tens of angstroms to microns.

The X-ray or electron scattering amplitude $F_{2\theta}^{(j)}$ of an atom is actually the integrated amplitudes of all the electrons of the atom. If the density of electrons were $\rho^e(\mathbf{r})$ and the scattering amplitude per electron were $f_{2\theta}$, then the scattering amplitude of the electrons in the volume element d^3r would be $f_{2\theta}\rho^e(\mathbf{r})d^3r$. In general we can express the local scattering amplitude as $\gamma(\mathbf{r})d^3r$, where $\gamma(\mathbf{r})$ is the product of the local electron scattering amplitude and the local electron density. In terms of $\gamma(\mathbf{r})$, the scattering amplitude of a molecule or a colloidal particle whose center of mass is at $\mathbf{R}_j$ is given by

$$\mathcal{F}^{(j)} = \int_{V_c} \gamma(\mathbf{r} - \mathbf{R}_j)e^{-i\mathbf{K}\cdot(\mathbf{r}-\mathbf{R}_j)}d^3r \tag{15.6.1}$$

where V_c is the volume of the molecule or particle.

Scattering techniques are often used to estimate the size and interparticle structure of colloids. From the analysis presented in earlier sections for multicomponent polyatomic systems it is clear that the theory of scattering by a colloidal dispersion is extremely complicated. However, if the colloidal particles are large compared to the solvent molecules, then the scattering vector **K** at which colloidal particle scattering is appreciable is so small that the intermolecular structure factor of the solvent is approximately constant and so contributes primarily to nearly constant background scattering that can be subtracted from the intensity data by running a solvent blank. Because $K = (4\pi/\lambda)|\sin\theta|$ one refers to small K scattering as *small angle scattering*.

The scattering intensity of a colloidal dispersion minus that of the pure solvent blank can be expressed as

$$\frac{\Delta I_{2\theta}}{I_o} = \sum_{j,k} \langle \mathcal{F}^{(j)} \overline{\mathcal{F}}^{(k)} e^{-i\mathbf{K}\cdot\mathbf{R}_{jk}} \rangle, \tag{15.6.2}$$

where

$$\mathcal{F}^{(j)} = \int_{V_c} [\gamma_c^{(j)}(\mathcal{R}) - \gamma_s(\mathcal{R})] e^{-i\mathbf{K}\cdot\mathcal{R}} d^3\mathcal{R}. \tag{15.6.3}$$

$\gamma_c^{(j)}(\mathcal{R})$ denotes the scattering amplitude density of the jth colloid and $\gamma_s(\mathcal{R})$ denotes that of the solvent molecules.

In general, the amplitudes $\mathcal{F}^{(j)}$ and $\overline{\mathcal{F}}^{(k)}$ depend on the orientations of j and k *and* on the separations $\mathbf{R}_{jk}$ of their centers of mass. Thus, except in the simplest case of identical spheres, a theoretical model is needed to derive particle size and structure information from eq. (15.6.2).

For identical colloidal particles at low concentration eq. (15.6.2) reduces to

$$\frac{\Delta I_{2\theta}}{N I_o} = \langle |\mathcal{F}^{(1)}|^2 \rangle. \tag{15.6.4}$$

An approximation frequently used is that γ_c and γ_s are constant. This approximation is justified only when the colloid particles are sufficiently large that $V_c^{1/3} \gg a$ and $Kl \ll 1$, where l is the mean separation of the atoms composing the colloidal particle and K is the magnitude of the wave vector at which colloidal particle scattering is appreciable. In this case

$$\frac{I_{2\theta}}{N I_o} = (V_c \Delta\gamma)^2 P(K), \tag{15.6.5}$$

where $\Delta\gamma \equiv \gamma_c - \gamma_s$ and the *form factor*, $P(K)$, is defined by

$$P(K) = \langle |\mathcal{T}_K|^2 \rangle, \tag{15.6.6}$$

with

$$\mathcal{T}_{\mathbf{K}} = \frac{1}{V_c} \int_{V_c} e^{-i\mathbf{K}\cdot\mathcal{R}} d^3\mathcal{R}. \tag{15.6.7}$$

$\mathcal{T}_{\mathbf{K}}$ has the same form as $\mathcal{T}_{\mathbf{s}}$, the form factor of the diffraction spots of a crystal [see eq. (15.5.11)]. Thus from results given in the previous section it follows that the form factors of a few simple colloids are given in the following.

Sphere of radius a

$$P(K) = \left[\Phi(Ka)\right]^2, \tag{15.6.8}$$

where

$$\Phi(x) = 3\frac{\sin x - x\cos x}{x^3} \tag{15.6.9}$$

Ellipsoid of revolution of axes 2a, 2a, and 2b

$$P(K) = \int_0^{\pi/2} \left[\Phi(Ka\sqrt{\cos^2\theta + (b/a)^2 \sin^2\theta})\right]^2 \cos\theta d\theta. \tag{15.6.10}$$

Very thin cylinder of length 2L

$$P(K) = \frac{Si(2KL)}{KL} - \frac{\sin^2 KL}{(KL)^2}, \tag{15.6.11}$$

where

$$Si(x) = \int_0^x \frac{\sin t}{t} dt. \tag{15.6.12}$$

Flat disk of radius a

$$P(K) = \frac{2}{(Ka)^2}\left[1 - \frac{1}{Ka}J_1(2Ka)\right], \tag{15.6.13}$$

where J_1 is the Bessel function of first order.

Spherical shell of outer radius a_o and inner radius a_i

$$P(K) = \left\{\frac{3}{K^3(a_o^3 - a_i^3)}\left[\sin Ka_o - \sin Ka_i - Ka_o\cos Ka_o - Ka_i\cos Ka_i\right]\right\}^2. \tag{15.6.14}$$

The form factors for these shapes are shown in Figure 15.19. Comparison of low angle scattering data for dilute colloidal dispersions with theoretical form factors sometimes enables one to estimate particle size and shape. A complicating factor is that polydispersity (i.e., a range of colloid sizes and shapes) blurs the distinct features seen in Figure 15.19. An example of polydispersity will be shown later in this section.

A comparison between the predicted and measured form factor of a disperson of polymethylmethacrylate (PMMA) latex spheres in water is given in Figure 15.20. For this dilute, monodisperse system the theory agrees well with the experiment. Scattering from a spherical shell can be accomplished by neutron scattering from deuterated latex particles in a protonated and deuterated water mixture. In Figure 15.21, the results of small angle neutron scattering from PMMA latex particles (radius $a = 495$ Å) coated with a deuterated PMMA latex shell

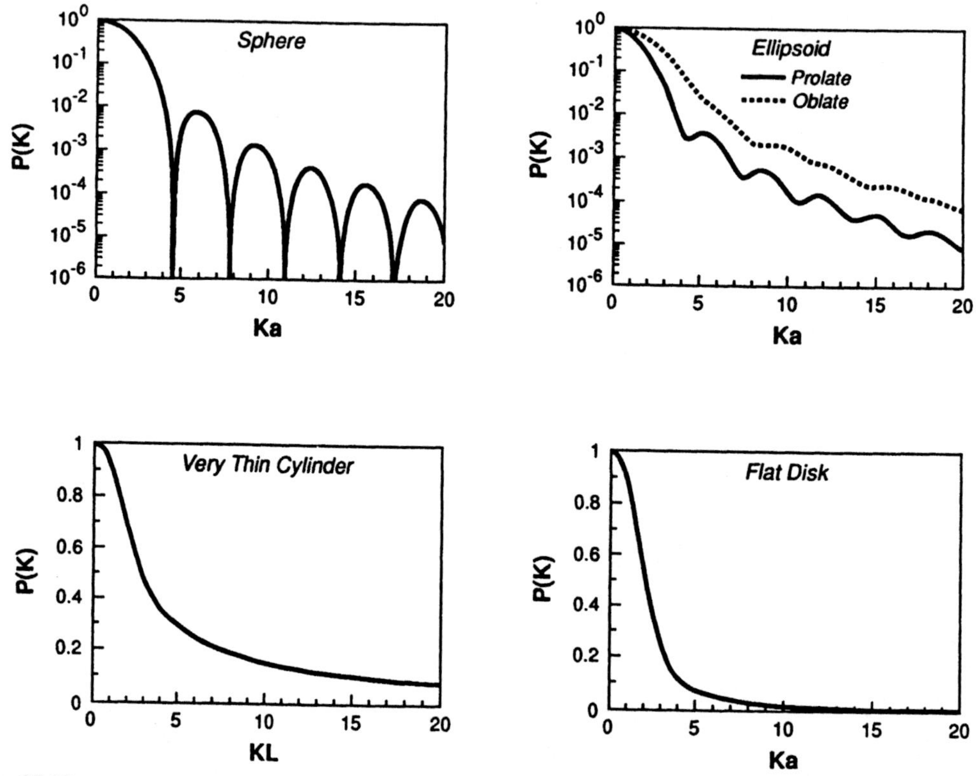

Figure 15.19
The form factor for the low angle scattering of a few simple colloidal dispersions. The symbol a denotes the radii and major axis diameter of the sphere, disk, and ellipsoids. L denotes the length of the thin cylinder.

(thickness $a_o - a_i = 30$ Å) are compared with theory. In the experiment the ratio of H_2O to D_2O was adjusted so that the latex core scattering length density was the same as the aqueous phase, and so everything except the deuterated PMMA spherical shell was transparent to the coherently scattered neutrons. The agreement between theory and experiment in Figure 15.22 is excellent.

Analysis of eq. (15.6.2) in the nondilute concentration regime is quite complicated. However, to illustrate the qualitative features of scattering by a colloidal dispersion of appreciable concentration, we consider a system of identical spherical colloids and assume again that γ_c and γ_s are constant. Equation (15.6.2) then becomes

$$\frac{I_{2\theta}}{N(V_c \Delta\gamma)^2 I_0} = P(K)[S^{cm}(K) + n(2\pi)^3\delta(\mathbf{K})], \qquad (15.6.15)$$

where the quantity $S^{cm}(K)$ is the intercolloid structure factor:

$$S^{cm}(K) = 1 + 4\pi n \int_0^\infty \frac{\sin Kr}{Kr}[g^{(2)}(r) - 1]r^2 dr. \qquad (15.6.16)$$

According to eq. (15.6.16), the particle form factor and the intercolloid structure factor contribute jointly to the scattering intensity. To illustrate this joint effect, suppose the intercolloidal interactions are

Figure 15.20
Comparison of theory and small angle neutron scattering from an aqueous dispersion of PMMA latex spheres. $a = 495$ Å. Adapted from Fisher et al. (1988).

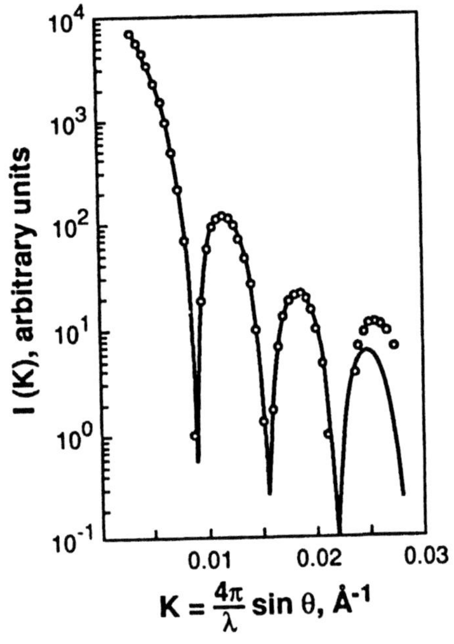

well approximated by hard sphere interactions and use the Percus–Yevick (PY) approximation to compute the $S^{\text{cm}}(K)$. The PY result is

$$S^{\text{cm}} = \frac{1}{1 - 4\pi n \int_0^\infty C^{(2)}(r) \frac{\sin Kr}{Kr} r^2 dr}, \tag{15.6.17}$$

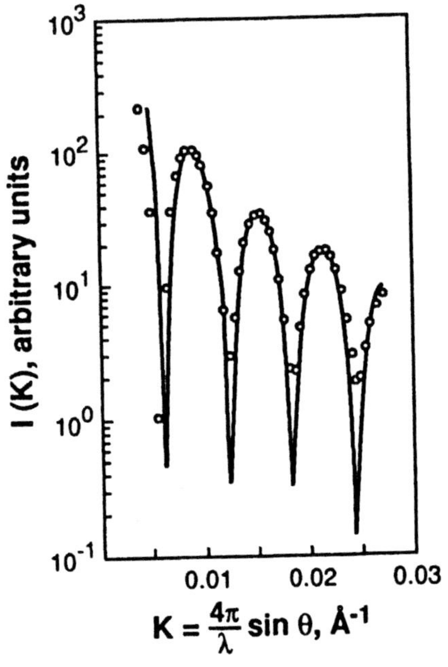

Figure 15.21
Comparison of theory and small angle neutron scattering form an aqueous disperson of hollow spheres. The inner radius of the spheres is $a_1 = 495$ Åand the shell thickness is 30 Å. Adapted from Fisher et al. (1988).

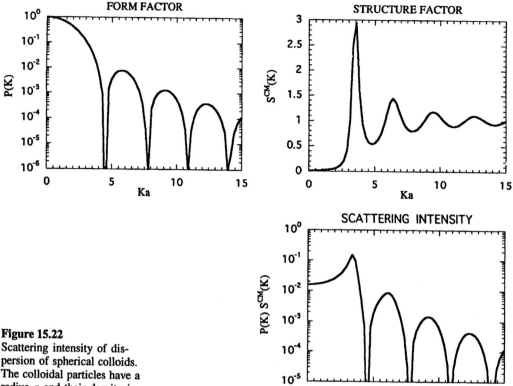

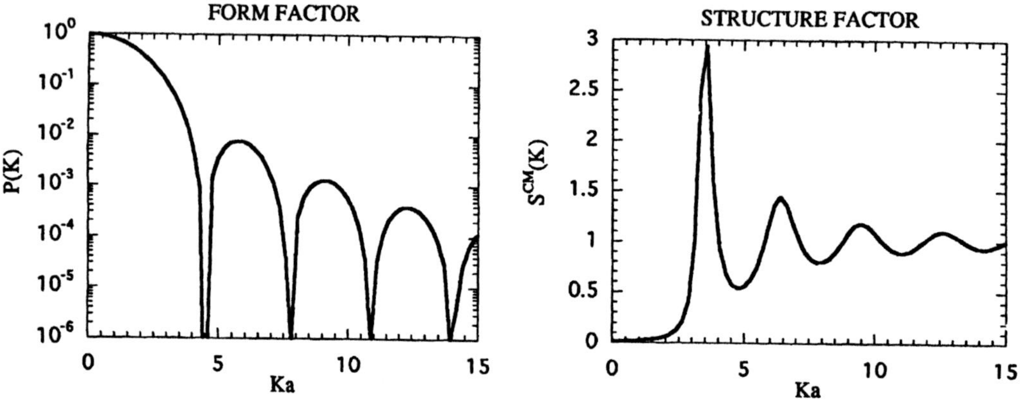

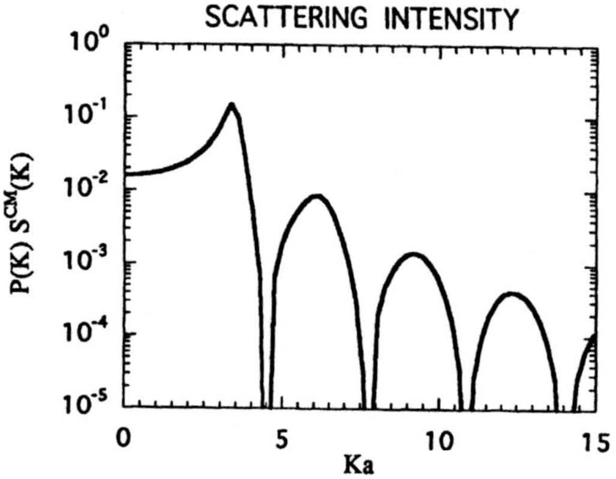

Figure 15.22
Scattering intensity of dispersion of spherical colloids. The colloidal particles have a radius a and their density is $n = 0.5a^{-3}$.

where [see eq. (9.4.19)]

$$C^{(2)}(r) = -(1-\eta)^{-4}\left[(1+2\eta)^2 - 6\eta\left(1+\frac{1}{2}\eta\right)(r/2a) + \frac{\eta}{2}(1+2\eta)^2(r/2a)^3\right], \quad r < 2a$$
$$= 0, \quad r > 2a, \tag{15.6.18}$$

with $\eta = (4/3)\pi n a^3$. The results in Figure 15.22 show that for colloid sized particles, interpretation of the scattering data is complicated by the competition between the form factor and the intercolloidal structure factor.

If the spherical colloids are polydisperse, that is, if the spheres are distributed over a range of sizes, then the scattering data are even more difficult to unravel. In this case,

$$\frac{\Delta I_{2\theta}}{N(\Delta\gamma)^2 I_0} = \int \left(\frac{4}{3}\pi a^3\right)^2 \alpha(a)[\Phi(Ka)]^2 da$$
$$+ \iint \left(\frac{4}{3}\pi a_1^3\right)\left(\frac{4}{3}\pi a_2^3\right)\alpha(a_1)\alpha(a_2)\Phi(Ka_1)\Phi(Ka_2)[S^{\rm cm}_{a_1a_2}(K) - 1]da_1 da_2, \tag{15.6.19}$$

for $K \neq 0$ where $\Phi(x)$ is given by eq. (15.6.9) and $N\alpha(a)da$ is the probable number of colloidal spheres having a radius between a and $a + da$. Thus, we see that particle size distribution information can be gotten from scattering data only with the aid of models for

$\alpha(a)$ and $S^{cm}_{a_1 a_2}(K)$. Ways to do this are discussed and illustrated in papers by Hayter and Penfold (1981) and Kaler (1988).

To illustrate the effect of polydispersity, consider a binary dispersion sufficiently dilute that $S^{cm}_{a_1 a_2} = 1$ and in which the particle distribution is

$$\alpha(a) = x_1 \delta(a - a_1) + x_2 \delta(a - a_2), \tag{15.6.20}$$

that is, a fraction x_1 of particles have a radius a_1 and a fraction $x_2 = (1 - x_1)$ have a radius a_2. Equation (15.6.19) then reduces to

$$\frac{\Delta I_{2\theta}}{N(\Delta\gamma \overline{V}_c)^2 I_0} = \overline{P}(K) = \frac{x_1 a_1^6 P(Ka_1) + x_2 a_2^6 P(Ka_2)}{x_1 a_1^6 + x_2 a_2^6}, \tag{15.6.21}$$

where $\overline{V}_c^2 \equiv x_1(4\pi a_1^3/3)^2 + x_2(4\pi a_2^3/3)^2$. The average form factor $\overline{P}(K)$ for a dispersion in which $a_2 = 1.2a_1$ and $x_1 = 1/4$ is given by the equation

$$\overline{P}(K) = 0.1P(Ka_1) + 0.9P(Ka_2). \tag{15.6.22}$$

The binary dispersion form factor is compared in Figure 15.23 with the form factors of the corresponding monodispersions. Because of the sixth power dependence of the scattering on the particle radius, the binary form factor is heavily weighted toward that of colloidal particles 2. Nevertheless, there is aleady considerable smoothing of the form factor in such a simple binary dispersion. In fact, the form factor of the dispersion of spheres is similar to that of prolate ellipsoids (Figure 15.19).

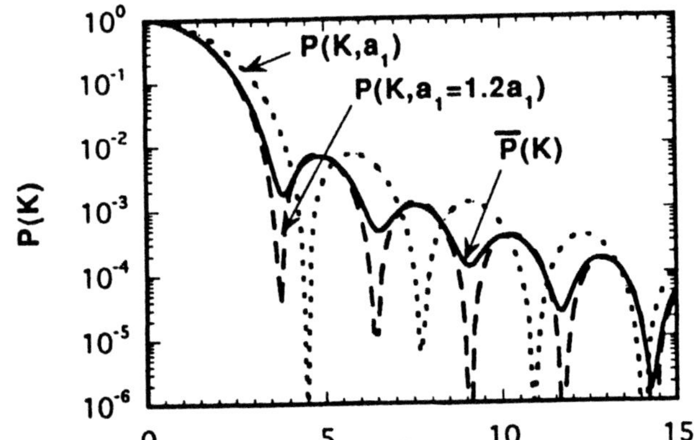

Figure 15.23
The form factors for monodispersed spheres of radius a_1 and a_2, respectively, and a mixture of spheres of radii a_1 and a_2.

Supplementary Reading

Glatter, O. and Kratky, O. 1982. *Small Angle X-ray Scattering,*, Academic Press, New York.

Guinier, A. and Fournet, G. 1955. *Small-Angle Scattering of X-rays*, Wiley, New York.

Exercises

1 From eq. (9.3.16) it follows that the structure factor of a homogeneous monatomic fluid obeys the equation

$$S(K) = \frac{1}{1 - n(2\pi)^{3/2}\tilde{C}^{(2)}(K)}$$

where

$$\tilde{C}^{(2)}(K) \equiv \frac{1}{(2\pi)^{3/2}} \int e^{-i\mathbf{K}\cdot\mathbf{r}} C^{(2)}(r) d^3r.$$

$C^{(2)}(r)$ is the direct correlation function. For the PY/MSA model of hard spheres the direct correlation function is given by eq. (9.4.19). Using this equation compute and plot the structure factor for a hard sphere fluid at the densities $n = o.1/d^3$, $0.5/d^3$, $0.75/d^3$, and $1.5/d^3$.

2 For the Debye–Hückel limiting law for electrolytes, the correlation function $h_{\alpha\beta}(R) \equiv g^{(2)}(R) - 1$ is given by eq. (9.4.7). Use this to calculate and plot the partial structure factors $S_{\alpha\beta}(K)$ for a 1–1 electrolyte at room temperature and at the concentrations 1, 5, 10, and 25 mol/L. Assume that the dielectric constant is $\varepsilon = 78$ and is independent of electrolyte concentration (certainly a bad approximation at 25 M and probably a poor one at 10 M concentrations).

3 Treating water as a hard sphere fluid and ions as point charges, combine the results of Problems 15.1 and 15.2 to approximate the structure factor of an electrolyte solution. Set the dielectric constant equal to 78, the diameter of water to 2.6 Å, the temperature to 298 K and the fluid density to that of water. Assume that $F_{2\theta}^{(\alpha)}$ is real and is the same for water and the ions. Then the structure factor is $S(K) = \sum_{\alpha,\beta} S_{\alpha\beta}(K)$. Calculate and plot $S(K)$ for a 0 M and a 5 M solution of a 1-1 electrolyte.

4 Consider a dilute ($S^{\text{cm}}_{a_1 a_2} = 1$) polydisperse system of spherical colloids. Assume the dispersion obeys the particle distribution

$$\alpha(a) = \sqrt{\frac{2}{\pi\sigma^2}} e^{-a^2/2\sigma^2}, \quad 0 < a < \infty,$$

where $\sigma = 50$ Å. Define the average form factor by

$$\bar{P}(K) \equiv \frac{1}{\bar{V}_c^2} \int_0^\infty \left(\frac{4}{3}\pi a^3\right)^2 \alpha(a) P(Ka) da,$$

where

$$\bar{V}_c^2 \equiv \int_0^\infty \left(\frac{4}{3}\pi a^3\right)^2 \alpha(a)da.$$

Compute and plot $\bar{P}(K)$ versus K for this dispersion. Compare this average form factor with that of a monodisperse system in which all the colloids are of radius

$$\bar{a} = \int_0^\infty a\alpha(a)da.$$

5 Calculate and plot the form factor for a dilute dispersion of disks having a radius R and length $L = 2R$. Plot $P(K)$ versus KR. Explain the differences between this form factor and that of long cylinders.

6 Consider a binary solution of hard spheres each with scattering length $F_{2\theta}^{(\alpha)}$. Use the PY theory of hard sphere solutions to predict the structure factor $S(K)$ for various concentrations and densities. The solution to PY equations of mixtures of hard spheres is given in Lebowitz (1964).

7 Using the Waisman–Lebowitz (1972) solution of the mean spherical approximation equations calculate the partial structure factors $S_{\alpha\beta}$ for a 1-1 primitive electrolyte (charged hard spheres in a dielectric continuum of dielectric constant 78). Compare structure factors at several densities and concentrations.

8 Using the Weeks–Chandler–Anderson (WCA) approximation with the PY theory of the pair correlation function [eq. (9.4.69)] predict the pair correlation function of a 6–12 Lennard–Jones fluid at densities $nd^3 = -0.75$ and $nd^3 = 1$. Use argon parameters and a temperature of 175 K.

9 Calculate the structure factor for argon using the WCA approximation and for the conditions given in Problem 15.8. *Hint:* Calculate the hard sphere diameter d from eq. (9.4.69) and use this in the PY hard sphere formula for $C^{(2)}(R)$.

References

Cullity, B. D. 1956. *Elements of X-ray Diffraction,* Addison–Wesley, Reading, MA.

Fisher, L. W., Meldopolder, S. M., O'Reilly, J. M., Ramkrishnan, V., and Wignall, G. D. 1988. *J. Collois and Interface Sci.,* **123,** 24.

Hayter, J. B. and Penfold, J. 1981. *Molecular Physics,* **42,** 109.

Herzberg, G. 1951. *Spectra of Diatomic Molecules,* Van Nostrand, New York.

International Tables of X-Ray Crystallography. 1974. Kynoch Press, Birmingham. Editors: J. A. Ibers and W. C. Hamilton.

Kaler, E. W. 1988. *J. Appl. Cryst.,* **21,** 729.

Karle, J. T. 1955. *J. Chem. Phys.,* **23,** 1739.

Lebowitz, J. L. 1964. *Phys. Rev.,* **133,** A895.

Norton, A. H., Agrawal, R., and Sandler, S. I. 1978. *Molecular Physics,* **35,** 1077.

MacWeeny, R. 1951. *Acta. Cryst.,* **4,** 513; 1952. *Acta. Cryst.* **5,** 463; 1953. *Acta. Cryst.* **6,** 631; 1954. *Acta. Cryst.* **7,** 180.

Mott, N. F. 1930. *Proc. Roy. Soc. A,* **127,** 658.

Sandler, S. I. and Norten, A. H. 1976. *Mol. Phys.,* **32,** 1543.

Thomas, E. L. and Sass, S. L. 1973. *Die Makro. Chem.,* **164,** 333.

Berlet, L. 1968. *Phys. Rev.,* **165,** 201.

Waisman, R. and Lebowitz, J. L. 1972. *J. Chem. Phys.,* **56,** 3086.

Weeks, J. D., Chandler, D., and Andersen, H. C. 1971. *J. Chem. Phys.* **54,** 5237; 1972. *J. Chem. Phys.,* **55,** 5422.

INDEX